Hermann Kreutzmann

HUNZA MATTERS

Hermann Kreutzmann

HUNZA MATTERS

Bordering and ordering between ancient and new Silk Roads

2020
Harrassowitz Verlag · Wiesbaden

Dedicated to Sabine Felmy and all our companions and friends in the Hunza Valley and beyond

Bibliografische Information der Deutschen Bibliothek:

Die Deutsche Bibliothek verzeichnet diese Publikation
in der Deutschen Nationalbibliografie;
detaillierte bibliografische Daten sind im Internet über http://dnb.dnb.de abrufbar.
Bibliographic information published by Die Deutsche Bibliothek:
Die Deutsche Bibliothek lists this publication in
the Deutsche Nationalbibliografie; detailed bibliographic data is available in the
Internet at http://dnb.dnb.de

For further information about our publishing program please consult our
website http//:www.harrassowitz.de/verlag

Cover illustration: Hermann Kreutzmann 29 September 1985
Back cover: Hermann Kreutzmann 4 April 1985

Editors: Anne Beck and Sabine Felmy
Photography: Hermann Kreutzmann except where dedicated otherwise
Design and typography: Hermann Kreutzmann and Harald Weller
Desktop publishing and image processing: Harald Weller
Cartography: Bernd Hilberer and Markus Hauser
Printing and binding: Memminger MedienCentrum AG, Memmingen
Printed on permanent/durable paper.

Printed in Germany · ISBN 978-3-447-11369-4

Contents

The 114-metre-long bridge connects Gulmit with Shishket across the Hunza River. In the background the Pasu pinnacles are visible. The image is the more impressive in 1954 as the location has changed its appearance over time. The Shimshal floods of the 1960s and twenty years later the surge of the Balt Bar glacier had led to the accumulation of several metres of sediments when an eight-kilometre long lake was formed behind the barrier. The bridge which was not part of the Hunza trade route fascinated many visitors who took similar photographs; see Aurel Stein (1930), Jean and Franc Shor (1953), and it became the visible icon of Eugen Schumacher's (1955) cinema film 'Im Schatten des Karakorum'. During the construction of the Karakoram Highway further surges occurred (Wang Wenying, Huang Maohuan and Chen Jianming 1984). The valley bottom is lying much higher now, and the 'Chinese bridge' was moved further north. Source: German-Austrian Himalaya-Karakoram expedition 1954; Wolfgang Pillewizer (1960, following page 112). Courtesy of Gerhard Klamert

Avant-propos

Approaching Hunza and Nagar

Microstates in remote mountain valleys have been exposed to all kinds of vagaries and winds of change, especially when they are located at thoroughfares and mountain passes that have become involved in geopolitical skirmishes and power struggles for domination in certain spheres of influence. Hunza and Nagar were two microstates that seemed to have been assigned either roles of subaltern players or functions as arenas for indirect rule when nineteenth century imperial powers were struggling for the control of the Pamirian Crossroads and the Wakhan Quadrangle.[1] Their location in highly contested spaces at the fringes of moving spheres of influence resulted in power games leading to certain political affiliations and boundary policies that have prefabricated and shaped the future of these microstates. Nevertheless, within a span of two and a half centuries there have been several incidents of conflict and confrontation, proactive and reactive participation, and of changing sides that have contributed to what are known as the mountain districts of Hunza and Nagar within Gilgit-Baltistan. Among those repeated interventions leading to a lack of balance, two momentous events – the abolition of the microstates and the advent of a 'modern' road system – happened around the same time and have significantly changed the internal structure and the external relations for the inhabitants of the Hunza Valley.

More than forty years ago, the hereditary rule of the *tham* of Hunza and Nagar ended between 1972 and 1974 and the *thamokuṣ* systems in their respective principalities or – as the people call them – the Hunza and Nagar States were abolished once and for all. The Government of Pakistan retired both *thamo* as state pensioners from their awarded military ranks and finally allowed them to keep their ceremonial titles as the last *mīr* of Hunza and Nagar. Both *mīrdoms* had maintained a semi-autonomous status until their termination, followed by their administrative integration into Pakistan's Northern Areas and transformation into mere sub-divisions of the Gilgit District. More than a century earlier, power and rule had been slightly modified when allegiance to the Maharaja of Kashmir began and was further amended with the establishment of the Gilgit Agency and the conquest of the Hunza Valley in 1891. British colonial intervention had reached the Hunza Valley and became part of a dual strategy; external interference by the Kashmiri Dogra rule and British-Indian frontier administration changed outside relations, military control and internal developments. The lease of the Gilgit Agency by the Maharaja of Kashmir to the British authorities in 1935 severed the Kashmir connection and enhanced British influence for an envisaged period of sixty years. Two years later Hunza State's relations with the north terminated when the colonial authorities pressurised the *tham* to unilaterally cut off his relations with China. The mid-1930s became the period when Hunza and Nagar were unilaterally bound to the British Raj, leading to the final decade of colonial domination based on subsidies and backing the appointed local rulers.

The fight for independence from British colonial rule and the Kashmiri yoke led to the short-lived 'Independent Republic of Gilgit' in November 1947 in which junior officers from Hunza and Nagar played gallant and prominent roles. This ephemeral entity was an intermediate step paving the way for the subsequent unanimous accession as Gilgit Agency to Pakistan. Territorial changes and administrative reforms during the early years of independent Pakistan affected the status of Hunza and Nagar within the Northern Areas or Gilgit-Baltistan, as the administrative entity has been called since 2009. No peace treaty has been concluded and signed between India and Pakistan as the outcome of the Kashmir conflict; consequently the stalemate in this issue of liminality continues.[2] India is treating the region as part of Pakistan-administered/occupied Jammu and Kashmir. The dispute constellation has left the region stigmatised as part of an unsolved South Asian crisis that has led to ambiguity under international law and created a constitutional limbo that limits political leverage and participation in decision-making.

The Pak-China Friendship Highway was officially opened within five years after the abolition of Hunza and Nagar States, and the modern road artery connected the Hunza Valley with down-country Pakistan and Xinjiang in China. This symbolic turning-point marks a new era, and the Karakoram Highway has changed the pace of movement, the exchange of goods and people, accelerated travel and traffic, and has strengthened bonds within Pakistan.[3] Nevertheless, the controversy about Hunza and Nagar's territorial belonging and that of all other administrative units within

Opposite
Mount Rakaposhi (7,788 m) towering above Chulmet (2,000 m) in Nagar impressed every traveller on the Hunza Road by its steep and visible slopes leading from the desert steppe to the glaciated top. Source: Oil on canvas painting by Alexandre Evgenievich Iacovleff (1934; plate 27), reproduced with kind permission from Gallica. Bibliothèque nationale de France

Baltit khan has remained an impressive fortified structure with a compact settlement surrounding the former seat of the Hunza ruler (tham) which is elevated on a moraine and protected at the rear by a steep cliff. Source: Author's own water colour and pencil drawing in 1985

Gilgit-Baltistan has not been solved as the whole debate is still embedded in the continuing Kashmir dispute today.

The debate has reached a new turn with the announcement of the China-Pakistan Economic Corridor in recent years. Its integration into China's Belt and Road Initiative has led to a reinterpretation of the Hunza Valley's role as Pakistan's trade corridor on the way across the Karakoram Range before reaching Kashgar as the main hub of distribution between China, Central and South Asia. The latest turn might prove to be the most significant change as never before have so many claims been put on local resources by capital-rich investors and entrepreneurs. The imbalance between local stakeholders and external actors is reflected in an administrative setup where rules and regulations often fail to provide security as the liminal state of constitutional rights has not been sufficiently advanced. The immediate future will show the direction of change.

Personal passage to the Hunza Valley

When Sabine Felmy and I first arrived in Pakistan as graduate students, more than forty years ago, we were inspired by imperialism theory and dependency debates to explain relations between colonisers and colonised, to analyse asymmetries, deformations and marginalisation in world market relations. Self-reliance, autocentric development and dissociation strategies in the search for viable alternatives to modernisation within metropolitan capitalism were being promoted in academic discussion circles to overcome unequal development and resulting economic deficits. These approaches provided useful insights into dependent relations and structural patterns of international exchange. To go beyond the state-to-state relationships, analytical tools were required allowing deviations from switchback interpretations that changed similar state actions from metropolitan to dependent once international boundaries were crossed. The fine structure of economies and societies needed to be understood in greater depth and historical detail. Braudelian global histories, Chaudhurian perspectives from Asia and Wallersteinian world systems analyses allowed a closer view on the inside story, on exchange relations for single commodities, and on mobilities of goods and people, as Eric Wolf proposed in his 'peoples without history'.[4] Attention shifted from the prime producers of history who were writing their own to the neglected actors and products of historical events and exchange relations. An alleged vectorial entropy could be understood in energetic categories as an accumulative unilinear direction of modernisation processes which built on Hegelian perceptions postulating that 'world history is moving from East to West, and Europe represents the final stage of world history per se'.[5] Western supremacy emerged as an unchallenged and constitutional element of the modernisation process on a path without a return option. The 'demystification of Asia' implied a shift from European admiration of the culturally superior Asian continent to a region that could and/or should be forcefully conquered and economically dominated. The mid-18th century shift of global importance and influence from Asia to Europe had all kinds of repercussions for hegemonic world-views, controlled economic exchange relations and forceful interferences, eventually leading to imperial divisions that created and demarcated spheres of influence. The discourse about the Great Game between Russia and Great Britain has revealed actors and supporters, thinkers and policy-makers, and resulting academic affirmations of allegedly inevitable modernisation processes.[6]

The above-mentioned authors, their and others' dense descriptions and structural analyses helped to alter such world-views. Shifting attention from the centres of decision-making and coastal ports to the interior margins and mountain peripheries allowed, for example, to place regions located in between imperial spaces and/or post-colonial states into focus. Besides looking at whole countries, so-called marginal regional areas of interest within a single country such as Afghanistan, Pakistan, or India could tell a complex story of domination and unequal relations as well. Based on personal travel experiences, our regional interest led to first contact with the works of critical Pakistani thinkers such as Hamza Alavi, Tariq Ali, Mahbub-ul-Haq, and the poetry of Faiz Ahmad Faiz.[8] So far unheard voices offered interpretations and views from ground experience that slightly shifted Eurocentric views towards Asia.

Arthur McCormick contributed this view of the tham of Nagar's palace in Uyum Nagar in June 1892. Source: The sketch is reproduced in William Martin Conway (1894a: 28[illegible])

The Cold War prevented a western student from gaining much knowledge about the neighbouring regions in the People's Republic of China and the Soviet Union. People like Lawrence Krader, Pavel Luknitsky and Jan Myrdal contributed to our scant knowledge base.[9] My personal interest in Hunza and the mountainous interface between Central and South Asia hardly originated from any broader theoretical compendia that were read in human geography and social science circles; these areas neither existed nor shone in such discourses. The basic question that developed over time was informed by surmising that the effects of century-long colonialism and dependency relations would also be felt on all spatial levels and by all social strata; thus an area of personal preference would provide the same. An accessible and comprehensible spatial scale would be a rural setting in a region between the centres of decision-making. Empirical ethnographic studies had already shifted their focus and contributed a vast number of village studies. Why not select a remote, formerly semi-independent mountain principal-

'The 'Map of Central and Western Asia 1873' shows the powers contesting for territorial dominance in the Pamirian Crossroads in bright colours: 'Independent Turkestan' of Yakub Beg in yellow is surrounded by British (red), Persian (green), Russian (brown) and Chinese (green) claimants, while purple represents Afghanistan. Kashmir is included in the British realm; the Hunza Valley is included in Kashmir and borders with Yakub Beg's Tien Shan-centred empire. Courtesy of the Pamir Archive Collection

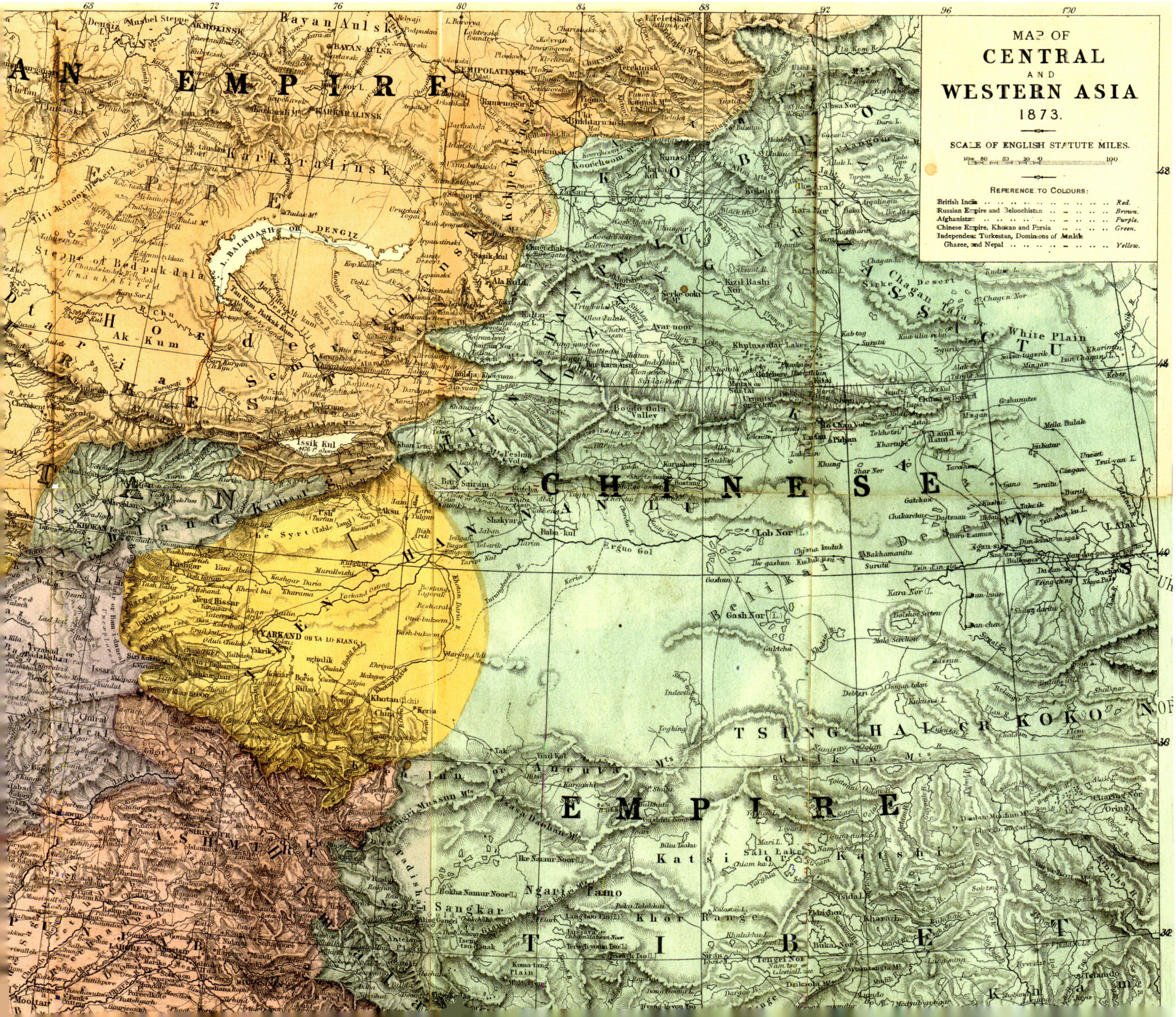
MAP OF
CENTRAL
AND
WESTERN ASIA
1873.
SCALE OF ENGLISH STATUTE MILES.
REFERENCE TO COLOURS:
British India .. Red.
Russian Empire and Beloochistan .. Brown.
Afghanistan .. Purple.
Chinese Empire, Khokan and Persia .. Green.
Independent Turkestan, Dominions of Atalik Ghazee, and Nepal .. Yellow.
EMPIRE
CHINESE
EMPIRE
TIBET
KOKO NOR
TSING HAI
Issik Kul
Lob Nor
YARKAND
Kashgar
Khotan
White Plain

Chitral and Gilgit-Baltistan comprise the former northern frontier of British India and contain a number of microstates and principalities that were incorporated into Pakistan's administration by 1974 at the latest. Presently they are part of two distinct administrative units. Chitral consists of two districts that are part of the 'Provincially Administered Tribal Areas' of Khyber-Pakhtunkhwa which were fully integrated into the province only in May 2018. Gilgit-Baltistan is configured as ten districts under federal administration by the Ministry of Kashmir Affairs; the Government Order of 2018 confirmed Gilgit-Baltistan's undecided status that has failed to end its disputed state and grant it the rights of a full-fledged province. With 10,109 square kilometres, Hunza is one of the three large districts, together with Ghizer and Skardu, while Nagar is the smallest with an area of 4,137 square kilometres. Source: Data according to Government of Gilgit-Baltistan (2018). Design by Hermann Kreutzmann

ity to study what 'under-development' meant in the definition and perception of Andre Gunder Frank and how development efforts initiated change and led to effects on the ground?[10] Later on, this restless and globally informed conceptual thinker encouraged us to understand the 'centrality of Central Asia' better, and he opted to 'ReOrient' our views on non-European societies and their historically transformed exchange relations.[11] Suddenly the interface between China, Central and South Asia became quite a prominent space with historical depth, and the Karakoram Mountains – previously stigmatised as 'remote' and negligible – were suddenly and surprisingly placed in a central location of attention. Their geopolitical significance as contested spaces in boundary-making and providing exchange corridors was obvious even when perceived as sparsely populated valleys of insignificant economic importance. For our empirical undertakings the advantage of a mountain abode was evident: in a less crowded place there would be ample opportunities to get into close contact with actors and stakeholders. The chosen area had never been one of the mainstream examples for testing various theories in development studies and mountain research. Pakistan was not as popular as India and Nepal for empirical studies, but I was convinced that the interlocutors already known to us would welcome outside researchers and would share our curiosity for a better understanding of the internal effects of external interventions. The prevalent debates in development studies and human geography provided the theoretical context for gaining empirical experience during extended visits in northern Pakistan. Beyond any academic discourse we were attracted by the openness and hospitality that welcomed visitors in Pakistan. Travel opportunities provided access to rarely visited parts of the Hindukush, Karakoram and Himalaya that opened up a complex and fascinating new world to us. Available literature was quite limited and consisted of reprinted colonial travelogues and gazetteers. The body of literature was augmented by very few post-independence scholarly works from Pakistan, some anthropological studies from international researchers and a growing number of esoteric and popular accounts. The marketing and popularisation of the Hunza myth were coordinated by the authority of the *mīr* of Hunza and his relatives who enjoyed the help of diplomats, journalists, travellers, writers and film-makers. Providing a monopolistic protective screen to any visiting guest, he succeeded in promoting himself as a benevolent ruler in a happy country of centenarians devoid of crime and disease, thus supporting the expectations of searchers looking for and finding a Shangri-La as a self-fulfilling prophecy.

When Tahir Ali's anthropological dissertation on the Burusho of Hunza was published in 1983 it was the only academic work to date that was not curtailed by the *mīr* of Hunza's influential narratives or restrictively informed through consulting representatives of the elite classes only.[12] Tahir Ali provided a less filtered and less blurred view on the social processes and political formations that were taking place at the time when the *mīri* rule was challenged and finally abolished, and when the Karakoram Highway was built. At the same time my thoughts about a project fermented and suggested an urgent need for a research effort about the impact of the just opened Karakoram Highway on the Hunza Valley. It would allow a comprehensive assemblage of all kinds of transformations by applying a diachronic analysis of the winds of change. In this context the previous rootedness in state-of-the-art development theories was very helpful, as a timeline was envisaged beginning from pre-colonial settings and relationships via colonial domination and internal transformations to the contemporary period of changing infrastructures, competing political factions, efforts of development agents, and growing state interference.

Now, four decades later, when the former graduate student is confronted with the challenges of retirement from his academic chair, we have experienced and have been exposed to several new approaches and turns in academic debates in general and development studies in particular. Structural global debates reached their climax and exposed their limits in the 1980s; they were replaced by theories in which actors, individuals and institutions took the explanatory position of classes, groups and states.[13] Analytical frameworks that had tried to explain structural elements and their interdependencies or that emphasised survival strategies in harsh environments took turns with vulnerability and livelihood concepts, and accessibility, resilience and sustainable development prose.[14] Geographers had been struggling for centuries with their space-oriented thinking and its theoretical framing. Having given up space as their central category, they missed the chance to participate in emerging interdisciplinary state-of-the-art debates. Thus, they were sidelined in social science debates in which the mobility and spatial turn emphasised space-based relations. Recent debates moved from colonial and imperial ascriptions to hegemonic and post-colonial approaches. During forty years of focusing on 'developments' in the Hunza Valley, our research has always been affected and inspired by new spinning thoughts and evolving debates. The 'Culture Area Karakorum' project attempted to rejuvenate a traditional geographical concept in order to contribute to the advancement of area studies. Its multidisciplinary approach offered ample opportunities to pursue individually promoted thematic issues. The subsequent 'Crossroads Asia' consortium put emphasis on Norbert Elias' concept of figurations and attempted to overcome conventional area studies approaches, which had featured prominently in the 'Culture Area Karakorum' project's title.[15] The most recent dissertation that was based on empirical fieldwork in the Hunza Valley applied an assemblage concept for Nagar.[16] Thus, my authoring of more than a hundred research articles, books and chapters in compendia has resulted in changed foci of interest, positional flavours, diverging and shifting viewpoints. The Karakoram Highway was initially seen as an innovative bilateral infrastructure project providing accessibility and mobility by linking the Hunza Valley with down-country Pakistan during its early stages when nation-building, ethnicity and overcoming colonial legacies were at stake in intellectual debates.[17] A second aspect was attributed to the Karakoram Highway's geopolitical impact and strategic perception within the framework of the Kashmir conflict by connecting China and Pakistan through a modern road artery which was suspiciously monitored by India. An international exchange corridor traversing disputed territories without everyone's acknowledgement under international law and containing spaces excluded from full representation within Pakistan's constitution posed a challenge for neighbourly political and economic relations. A third dimension was envisaged by looking at the various activities of national and international

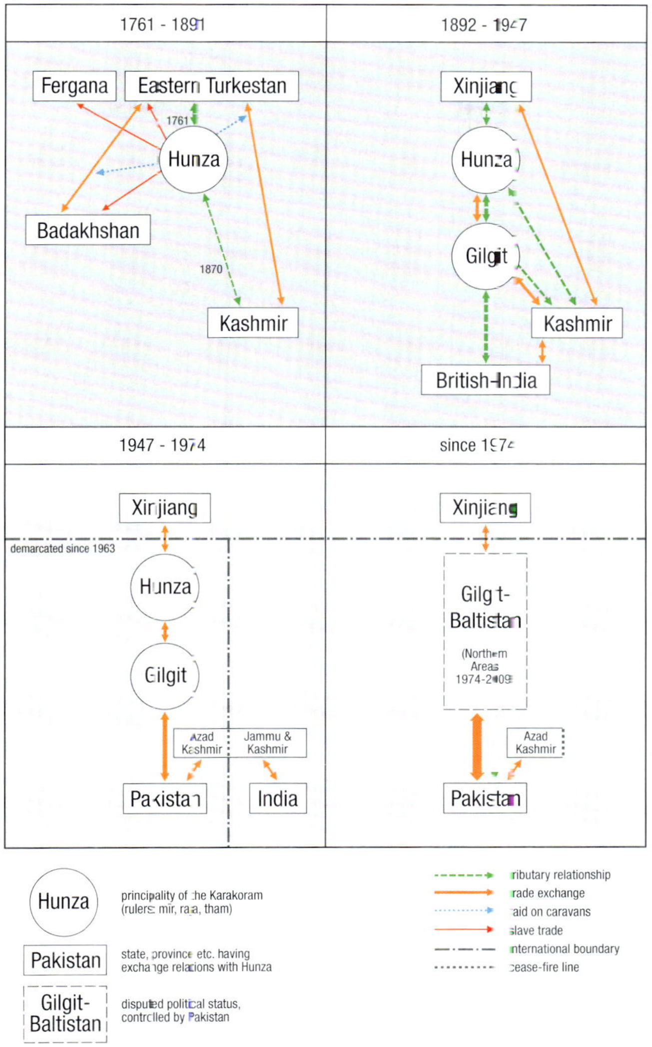

The historical development of Hunza's exchange relations shows a transition from a situation where the principality was embedded in various diplomatic and economic relations until 1891 to a constellation where the dominant exchange axes from north to south are somehow structuring an exchange that is definitely dominated by Hunza's relations with down-country Pakistan. Source: Modified and expanded from Hermann Kreutzmann (1991a: 717).

The fourth edition of the map 'Gilgit Agency and Jammu & Kashmir' in 1995 still repeats the colonial administrative terminology in northern Pakistan although major reforms and renaming have taken place since the 1970s when the Gilgit Agency and Baltistan became the Northern Areas. An exception is the 1983 name change from Baltit to Karimabad. Source: Survey of Pakistan (1995)

development agents that allowed putting significant emphasis on transformations in combined mountain agriculture and fostering market access.[18] Survival strategies in harsh mountain environments sounded somewhat sensational as a heading and description for such endeavours that focused on transformations in the agricultural sector and resource base. In present times, these three thrusts would remain important and purposeful, but the Karakoram Highway would have become an actor itself among others in network relations. The assemblage approach would leave more degrees of freedom to look at certain aspects, while the results of enquiry and scrutiny would probably not differ as much as had been aspired and expected.

The principal challenges in utilising available local resources, in handling changing sets of external interests and in designing coping strategies for making a living by investing in education and mobilising the younger generation would always play a prominent role in assessing the chances and constraints, the limitations and potentials for mountain communities in such harsh environments that increasingly depended on external relations. Accelerated mobility and access to motor transport have become visible signs of economic activities that are no longer restricted to their place in the Karakoram; people have been on the move between their multi-local habitations and sources of income. After the end of the Cold War the road's impact has been interpreted in the framework of a wider world view. In this part of the world the early age of globalisation[19] was succeeded by the new Silk Road or Chinese Belt and Road initiative as a dominant feature of positioning external interferences. Its salient effects can be perceived in the road infrastructure, which in its fourth remake poses a significant bottleneck along the Karakoram Highway as the vital part of the China-Pakistan Economic Corridor functions as the sole connection between the lowlands of China and Pakistan. Now, remote mountain areas again seem to be revalued in terms of accessibility, connectivity and resource exploitation. Nevertheless, a void has remained that needs increased attention.

George Nathaniel Curzon acted as under secretary of state for India in Lord Salisbury's Tory government. After extensive travels in Asia he became the influential protagonist of their 'forward policy' that pushed British India's western and northern borders to new limits. The photograph was taken in front of the British Residency in Gilgit on 13 September 1894 to honour him 'before starting for the Pamirs'. Marquess Curzon is sitting in the middle; on his right is Colonel Neville Francis Fitzgerald Chamberlain, the Military Secretary to the Kashmir Government, on his left Henry Lennard (standing); both accompanied him during his journey. Five years later he would become viceroy of India. Source: George Nathaniel Curzon (2012: 32, 106); Curzon Collection; reproduced with permission © The British Library Board Photo 430/67(32)

The grand debates about colonialism and imperialism have focused on the centres of decision-making in European capitals and its effects on important and exploitable regions overseas. Port cities and seats of power represent the linkages. In many ways inaccessible border and mountain regions that were devoid of significant valuable and exploitable resources have not only been marginalised in terms of focus, but have been neglected and sidelined owing to their economic and physical remoteness. Because of limited available evidence they seem to be difficult to come by as they fail to be adequately represented. Their subordination under political developments and economic processes in colonial centres led to a critical ignorance and neglect. The purpose of this book is to shed some light on socioeconomic transformations in the Hunza Valley that enable some insights into the structuration and transition from colonial interference to postcolonial integration, and that identify long-lasting effects of previous deformations and hint at elements of path-dependency and rootedness. However, it becomes obvious that the main interest of understanding the processes of economic, political and societal transformation still requires the analysis of actors and factors, the interlinkages between local, regional, national and global influences and interventions, and the embedding in a historical setting that builds on the interpretation of scanty written sources. The jargon might have changed and become more elaborate and sophisticated; in essence, the challenges of understanding have remained similar. Consequently, I have taken the liberty to look back on previous work and to include in this book descriptions and illustrations that originated from different visits and publications; they may incorporate a growing maturity and reflectivity of the researcher and his professional approaches and visions.

In military attire, tham M. Nazim Khan of Hunza, honorary officer of the Gilgit Scouts is photographed on Baltit's polo ground sitting on a white horse in the mid-1930s. The British and Kashmir-backed ruler of Hunza is at the peak of his local authority and respect by the colonial administration. Source: Lorimer Collection in the School of Oriental and African Studies, University of London (PP MS 66), Box II.45; reproduced with kind permission

Muhabbattullah Beg, son of M. Raza Beg and father of Qudratullah Beg, is photographed by Aurel Stein in Mintaka on 8 September 1913. Source: Stein Collection in the British Library. The latter is reproduced here with kind permission of the International Dunhuang Project Photo 392/28 (163)

CHANGING KNOWLEDGE PRODUCTION OVER TIME

Ever since I purchased my copy of John Biddulph's 'Tribes of the Hindoo Koosh' from Ghulam M. Beg at his bookstall in Gilgit's Jamatkhana Bazaar in August 1981, I have not stopped collecting publications and reading about Gilgit-Baltistan. Even then the poor state of printed evidence was more than obvious. Mainly reprints of colonial travelogues and reports were available in Pakistan during those times; a situation that has changed slightly in the meantime but not led to anything like a satisfactory coverage. The challenge of interpreting historical material is the limited number of reference points such as colonial records written by the British Political Agents and employed local news-writers and spies at their duty stations. Locally a history of Hunza was initially written as the *Tarikh-i Hunza* by M. Raza Beg in the Persian language, which was augmented by his son Muhabbathullah Beg and expanded by his grandson Qudratullah Beg to become the *Tarikh-i 'ahd-i 'atīq-i riyāsat-i Hunza*. These authors from a *wazīri* family background emphasised the 'history of the Hunza principality' by providing their perspective.[20] With the help of his grandson Jamal Khan, the longest-ever-serving ruler *mīr* M. Nazim Khan published his own autobiography after forty-four years on the *takht* in which he attempted to paint a bright picture of his rule under British paramountcy in 1936.[21] Following his death two years after the completion of the manuscript, his successors distributed copies of his duplicated autobiography on several occasions for decades; especially visitors to Hunza who stayed with the *mīr* in his guesthouse received copies as an exclusive token of appreciation. An exceptional occasion occurred when David Lockhart Robertson and Emily Overend Lorimer spent fifteen months in Aliabad and profited from their close relationship to notables until September 1935. The Lorimers contributed a bulk of material to anthropology and language studies that has survived as valuable reference material in archives and publications. They were well connected with the ruling classes of Hunza, they admired *mīr* M. Nazim Khan who took a close interest in their studies and functioned together with the representatives from the *wazīr* group as their chief informant and advisor. It is safe to state that these publications contain the elite view on Hunza history and society.

Fifty years after the Lorimers' stay in Hunza we got the opportunity to spend a year in Hunza. Again a couple – Sabine Felmy and I – lived in Central Hunza. By then, the *mīri* rule had been abolished, giving us the freedom to reside in the premises of his former servants, meet a wide spectrum of interlocutors and have social intercourse with all kinds of people from different positions and professions. Especially our research on agricultural practices profited as we could accompany farmers and shepherds during their work at all kinds of locations in the valley. A complex, diverse and vast knowledge about economic strategies, conflict constellations, memory and history is embedded in numerous persons. We were glad and felt immensely honoured to find so many open minds and curious interlocutors who shared their specific knowledge, wit and individual viewpoints with us. On the other hand, the prevalent paranoia of state authorities in the Kashmir conflict arena affected basic undertakings in geographical fieldwork. Sketch mapping of infrastructure and settlements raised the attention of several 'services', as did taking residence in people's homes in remote villages for several months in a row. Their local representatives became suspicious about our intentions and regularly had to report about our movements to their superiors. Nevertheless, over four decades no serious problems and restrictions arose from this setup as all our relevant findings had been made accessible and were published in academic journals and books. Rather, we became the laughing-stock of the village because of recording all names for hamlets and places, following the courses of irrigation channels and exploring the routes to high pastures and seasonal settlements. On several occasions it was doubted whether such work could lead to a doctoral degree. There remained a reservation as our friend Hoor Shah publicly phrased it during his son's wedding celebrations: "Our elders know much more about irrigation channels than Dr Hermann, but they never received a PhD for their wisdom." Others acknowledged that the Hunza Valley had become a household name in comparative mountain research.

In an area where good and precise topographical maps are unknown, where cadastral surveys and land-use mapping were never conducted, a wide range of possible contributions occurred by applying cartographic and geographic methods. The rough mapping during colonial times and the post-colonial compilations of rather imprecise and sometimes misleading maps obviously served the purpose of presenting opaque references to topographic knowledge.[22] Only two maps of adequate scale existed for the Hunza Valley. The German-Austrian Himalaya-Karakoram expedition of 1954 had included the geodesist and map-maker Karl Heckler who worked very hard during his stay in Hunza and unfortunately found his death between Gulmit and Sarat just before the end of the expedition.[23] The material he had collected was augmented by further cartographic surveys during the German Karakoram Expedition of 1959; the central Hunza Valley was covered under the leadership of Hans-Jochen Schneider. It took until 1967 before the first map sheet covering Minapin and part of the Rakaposhi Range was published.[24] It would remain the only outcome of the cartographic project until Rüdiger Finsterwalder based his map 'Hunza-Karakorum' on the data which had been recorded forty years earlier.[25] The second high-quality map available for a section of the Hunza Valley is 'The map of Batura Glacier' that was prepared by the Institute of Glaciology, Cryopedology and Desert Research in Lanchow in 1978 as a side effect of the Chinese survey in preparation for the construction of the Karakoram Highway.[26] Both maps primarily cover glacier areas in two small sections of the Hunza Valley. For my own research a lucky strike occurred when during our first meeting in 1983 Hans-Jochen Schneider offered me the use of his unpublished contour-plan which was based on his and Karl Heckler's surveys during the 1954 and 1959 expeditions. As the plan covered most of the densely populated oases of Central Hunza and Nagar it became a valuable source for my own large-scale maps. Sections of this contour plan were enlarged and became the basis for land-use surveys and recordings of settlement patterns, village lands and pasture areas. In recent years we were able to make use of a whole range of satellite images from public domains.

Mountain researchers at work; Hans-Jochen Schneider and his assistant, Wali Khan, in 1954. Source: German-Austrian Himalaya-Karakoram expedition 1954. Courtesy of Gerhard Klamert

Qudratullah Beg photographed by David or Emily Lorimer in 1934. Source: Lorimer Collection in the School of Oriental and African Studies, University of London (PP MS 66), Box 10; reproduced with kind permission

Previous page
The suspension bridge between Gulmit and Shishket became the symbol of the German-Austrian Himalaya-Karakoram expedition of 1954. It was depicted on flyers, book covers and regularly reproduced in articles. Source: Information leaflet for Eugen Schumacher's (1955) cinema film 'Im Schatten des Karakorum'

The available literature has grown over time. Newspaper reporting expanded with the construction of the Karakoram Highway and the activities of national and international development agents. Journal articles and reports covered their activities and public responses. Educated professionals have put their own views and visions into books; retired army engineers shared their reminiscences from roadbuilding adventures.[27] The market of 'grey literature' is ever growing, and recently the coverage of relevant themes for Gilgit-Baltistan in websites and blogposts has become a daily practice and offers an abundance of snippets and opinionated comments.

Archival material remained an important access corridor to historical developments and needed to be interpreted with great prudence. The bulk of archival sources was created in the Political Agent's office in Gilgit; the Gilgit Monthly Political Diaries contain a wealth of detailed information about events and public opinions, the judgement and performance of rulers, their relations with their 'subjects' and the reception of traders and visitors. From the late 19th century until Pakistan's independence, such details were regularly recorded by white male officers from Great Britain with the help of indigenous intermediaries (*pundit*) in their service.[28] Prior to that there are travelogues by visitors on political missions who prepared detailed accounts of their missions for their clients in the intelligence departments who commissioned these espionage-related reconnaissance trips. Many of these are well-known authors of published articles, books, and reminiscences containing declassified descriptions and opinions. Most travellers arrived from the south via Kashmir, but a few – and among them Russian – explorers marched across the northern passes and documented road geometries and the living conditions along their routes. The Polish officer Bronisław Grąbczewski was among the pioneer travellers whose insightful diaries and photographs produced during his service in the Tsarist army were recently made accessible and published.[29] Patrick Vans Agnew and John Biddulph were the first British travellers visiting the Hunza Valley, and their diaries contain a similar set of information.[30] Subsequent and more frequent travellers and their reports augmented the scanty geographic information and contributed to the set of externally-induced knowledge generation that was regularly compiled in various sets of gazetteers, military reports and route handbooks. Most of them contained classified information about controlling the political situation on the spot, using local resources in case of military advances and expeditions, and about potential revenue generation and the cost of administration. These classified reports were kept as confidential sources for the perusal of colonial officers only. In recent decades some academic use could be made of them when they became accessible in archives. Pamirian files from the Russian State Military Historical Archive in Moscow offered some insights about strategic considerations, trans-boundary connections of religious leaders and military officers.[31] Comprehensive surveys were made in the context of military manoeuvrings, administrative reforms, the installation of new rulers, borderland and trade politics. Certain insights can be derived from critical reports within the British-Kashmiri colonial bureaucracy when competition for posts and individual vanity broke cover and revealed shortcomings, failures and grave mistakes. The sometimes erratic critiques and descriptions by Reginald Charles Francis Schomberg are a case in point that caused broad discussions and revealing replies.[32] The Royal Society for Asian Affairs preserves Schomberg's personal records and his unpublished book manuscripts.[33] His sarcastic views and hurtful assessments of rulers, acquaintances and servants, farmers and shepherds, and whole communities need to be carefully analysed and contextualised. Nevertheless, they provide the emotional and unexpurgated views of an eccentric traveller and spy who often promoted politically

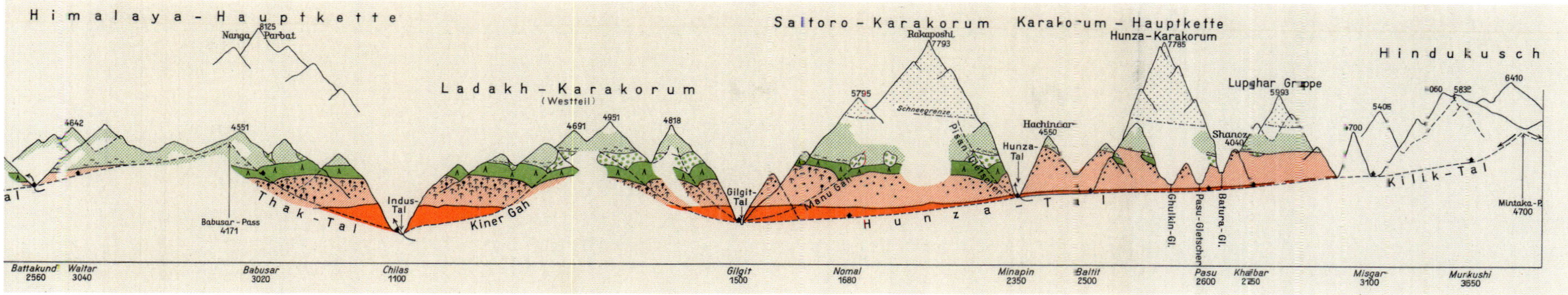

The orographic and vegetation profile was compiled by Karlheinz Paffen as a member of the scientific team within the German-Austrian Himalaya-Karakoram expedition of 1954. Source: Karlheinz Paffen, Wolfgang Pillewizer and Hans-Jochen Schneider (1956, supplement)

non-correct viewpoints. Careful interpretation offers a wide variety of insights: Reginald Schomberg was probably the most experienced colonial officer who had seen more remote locations than anybody else and who was very close to his indigenous intermediaries.

Comparatively few written records exist from local chroniclers as was mentioned earlier on. The longest-ruling *tham* M. Nazim Khan (1892–1938) presented in an 'autobiography' his views and interpretations of Hunza history and society.[34] The mimeographed version of his achievements was distributed to interested visitors even long after his death, as it remained a rare and unique written record from the *thamo* perspective. The mid-1930s are a well-covered period, probably due to the presence of David and Emily Lorimer.[35] For the purpose of gathering ethnographic and linguistic data they had hired Qudratullah Beg who himself used this exposure to become a local chronicler. The 'history of Hunza' that is connected with his name is a project of several generations of *wazīrkuć*. From this lineage a number of Hunza's viziers (*wazīr*) originate, a few of whom have become pioneer recorders of local history and collectors of local traditions. The son of the famous *wazīr* Asadullah Beg and younger brother of *wazīr* Humayun Beg, M. Raza Beg, began in the 1920s to generate a written collection of historical narratives and stories in the Persian language. He fulfilled the function of treasurer (*faráaji*) of the private property in the household of the *tham*; thus he was quite close to *mīr* M. Nazim Khan in the early phase of his rule and had access to his documents.[36] After serious disagreements with the *tham* M. Raza Beg tried to complete his historical account at the imamat seat (*dar-xāna*) of the Aga Khan in Bombay. Being the head of the first Hunza delegation, M. Raza Beg had previously visited their *imām* at his *darbār*. His son Muhabbathullah Beg and grandson Qudratullah Beg seconded the work on the historical account as scribe and continued after his sudden death in exile in 1923. A full manuscript emerged only after they had worked on their collection over a span of forty-two years before it was finally completed in December 1962. A proper publication of the Persian manuscript took another two decades, and in 2006 Qudratullah Beg's son Saadullah Beg presented an English translation as an effort spanning four generations of contributors.[37] Their narrative is the proof of a *wazīr*-centred interpretation of Hunza's history that focuses on the dwellers of the central oasis. British archival sources that have merged accounts and reports from daily practice in the Gilgit Agency and beyond paint a favourable picture of the benevolent actions of a colonial power towards mountain dwellers who will be exposed to the boons of modernisation. The significant attention given to such remote and scarcely populated mountain valleys is mainly due to their part in the 'Great Game' in which international boundaries were staked out and spheres of influence marked. The autobiography of *mīr* M. Nazim Khan has produced their dynasty's (*ayaśó*) version of interpreting its hereditary role in the microstate of Hunza. The accessible archival material provides three differing perspectives from influential and powerful actors in their respective realms. Kashmiri, Chinese and Russian archival documents are a rare commodity; the same applies to views from Nagar's chroniclers and Hunza's non-Burusho and non-Tarakuć communities. Consequently, the biased evidence from accessible sources has to be acknowledged and cautiously interpreted.

For the postcolonial period no monthly records from office-bearers, contemporary gazetteers and route descriptions exist or could be used in the same format as was established within the standardised colonial reporting system. From then onwards, evidence is based on a different set and

The 1959 Karakoram expedition was accompanied by the liaison officer Sahib Shah who was sent along by the Survey of Pakistan for cooperation in map-making. Source: German Karakoram expedition 1959. Courtesy of Hans-Jochen Schneider

The two tham of Hunza and Nagar, M. Nazim Khan and Shah Sikandar Khan, are flanking the Political Agent George Kirkbride during the Silver Jubilee Celebrations in Gilgit on 6 May 1935. Source: Lorimer Collection in the School of Oriental and African Studies, University of London (PP MS 66), Box 10; reproduced with kind permission

Sultan M. Shah, Aga Khan III, increased his influence in Hunza in the 20th century, initially by receiving delegations of Hunzukuć, later by sending messengers and missionaries who implemented reforms and introduced communal institutions. Source: Image taken by an unknown photographer for World Wide Photos on 27 June 1939. Courtesy of the National Portrait Gallery

wide variety of sources. Newspaper cuttings, travelogues, and diaries from mountaineering and scientific expeditions, reminiscences from local leaders and respected personalities have provided documents. In recent years articles, autobiographies and collections from local scholars occur in new media and are sometimes offered in print. Since the appearance of development organisations a countless number of annual and quarterly reports, consultancy write-ups and 'grey literature' have been produced that occasionally contain useful information and sometimes enlightening narratives.

In addition, there is a severe bias in dealing with Hunza or Nagar. Much more has been written by academics, officers, local scholars and international visitors on Hunza; there are many more local authors from Hunza than from Nagar, and the transformations related to accessibility, modernisation, political representations, strategic considerations and development activities have always shifted the pendulum of attention in favour of Hunza. Ever since the British-Kashmiri forces conquered the Hunza Valley in the final decade of the 19th century they put Hunza in the limelight and needed Nagar only as a balancing measure for two rival neighbours.[39] In colonial times the border issue dominated their attention and actions as was expressed by many administrators and visitors. In an exemplary manner the Political Agent George Kirkbride's brief statement suffices here to underline the point: "It is true that Nagar has not the importance of Hunza, chiefly because its territories do not march with those of any foreign state, …"[40] This assessment was made when the British position was strongest in the 'Northern Frontier', in 1935 their Gilgit Agency lease would commence. Reginald Schomberg endorsed the same view using even stronger words and mentioned that Nagar has "… played an insignificant rôle in history; even the States of Punyal and Yasin have been more in evidence. … Nagir is such a weak partner [for the British] that it has rather been ignored."[41] There might be an outsiders' distinction due to location and exchange relations, but there has always been an active process of self-promoting Hunza as well. The members of Hunza's elite have been quite successful in keeping the microstate in the public eye. Nevertheless, mountaineering and scientific expeditions have been equally interested in both regions; the major difference occurs in terms of attention given to anthropology, archaeology, development studies and human geography. This is an explanation rather than a justification. The present book maintains a similar approach as external and internal exchange relations, geopolitical interest and interference, power games and mystifications are under scrutiny. The unequal distribution of archival evidence, background information and time allocation for fieldwork in Hunza and Nagar probably explains the greater emphasis on the former microstate or district. Nevertheless, cordial relations exist with interlocutors and friends in both parts of the valley.

A major motivation for publishing this trilogy – Pamirian Crossroads, Wakhan Quadrangle, and Hunza Matters – in book form has been the desire to make accessible to interested readers some selections of the valuable existing resource material gathered from a variety of scattered sources for a location and space that is often at the margins of academic and disciplinary interest.[42] By putting the peripheral space between major players into the centre of attention and by acknowledging various forms of knowledge production the attempt is made to recognise and substantiate dispersed sources that are accessible in personal diaries and travelogues, newspaper articles and grey literature, public and private archives, libraries and repositories in various locations. My own interpretations are offered on the basis of the collected material that can be traced and verified from the given sources and presented evidence.

The two sketches by Arthur McCormick were prepared in Uyum Nagar and Baltit on 14 and 19 June 1892, respectively. They depict the seats of power in both microstates. The sketches illustrated the narrative in William Martin Conway (1894a: 248, 274)

The photograph titled 'Wazirs of Hunza'; shows three persons sitting in festive robes; the prime position is occupied in the centre by the acting wazīr Humayun Beg who was deposed by Safdar Ali Khan shortly after the visit of the Gilgit Mission. The expedition's surgeon George Michael Giles took the photograph in April 1886 when Humayun Beg had just taken over the influential position from his deceased father Asadullah Beg. Source: William Lockhart and Robert Woodthorpe (1889, preceding page 391); the image is reproduced here with permission from the Royal Geographical Society in London, call number S0025146

Illustrations – mapping and visualisation

Printed versions of artistically drawn and professionally published maps often accompanied books, reports and articles, as well as handwritten and printed matter. One aim of this book is to acknowledge the path cartography has taken over time in presenting highly selective information. The persuasive character of maps cannot be over-estimated; therefore exemplary specimens have been reproduced that have shaped the perception of the Hunza Valley. It is surprising how poor the state of map-making was during colonial times. Remoteness and the perception of the Hunza Valley as mainly a thoroughfare have contributed to the neglect of large-scale cartography. The best maps were produced for the approach to passes and control posts, monitoring glacial activity and leading to peaks of interest to mountaineers.[43] In addition, there exists a wealth of illustrations initially published as drawings and sketches prepared in oil, pastel and water colours. Later on, photography mainly replaced these artistic expressions although the usage of various forms of handmade drawings and modern technology-based pictorial evidence continued to exist along parallel lines. Nearly all pictures that represent views of the Hunza Valley until Pakistan's independence and for some time afterwards were taken by outsiders as visual aids for their reports. The mainly male visitors exclusively controlled and possessed the

During Bronislaw Grombchevsky's stay in Baltit, tham Safdar Ali Khan entertained his guest with local hospitality during a polo match, Dom musicians performed as part of their duties in early September 1888. Source: Bronislaw Grombchevsky (1888: 127–129), the image is reproduced with permission of the Scientific Archive of the Russian Geographical Society, font 45, opis 1, file 4, sheet 23

modern technology of camera and film. There is not much evidence from local artists found in the field of painting and sketching either. Readers should be aware that the illustrations represented in this book were mainly produced by outsiders as well.

A host of illustrations were collected from various archives and repositories. The first ever photographs were taken by Captain George Michael Giles, the surgeon of the Gilgit Mission in 1886, followed by Bronislaw Grombchevsky, the Polish officer on special duty in Russian Turkestan, who approached Hunza from the northern passes two years later.[44] Their reports covered route conditions, fortifications and people of importance. Successive visitors have contributed more photographs that allow documenting a sequence of changes in the landscape, building styles and road conditions, and portraying actors and leading personalities from various generations. Another surgeon, Richard William George Hingston, visited the Hunza Valley and took photographs during the Indo-Russian Triangulation expedition from April to September 1913.[45] The same period of coverage applies to Marc Aurel Stein who visited the Hunza Valley and left us with a substantial set of various photographs.[46] Philips Christiaan Visser and Jenny Visser-Hooft took a number of photographs during their 1922 and 1925 expeditions into the Karakoram; the images document their prime interest in the road conditions and glaciology of Go-

David Lorimer photographed a shamanistic session during an official visit as Political Agent in Baltit in 1924. The female biṭán attracted a huge audience. Source: Lorimer Collection in the School of Oriental and African Studies, University of London (PP MS 66: V 8); reproduced with kind permission

jal.[47] Jenny Visser-Hooft was probably the first woman to take photographs in the Hunza Valley; these have been preserved in the Dutch National Archives.

About the same time the Political Agent David Lorimer had begun his Burushaski studies, made early recordings and took photographs. A decade later he and his wife Emily Overend Lorimer produced a wealth of images during their stay in Hunza between July 1934 and September 1935. They covered all aspects of their endeavour commencing with their journey from Srinagar to Hunza and depicting personalities and everyday life in the central oasis.[48] During their stay they experienced all cultivation seasons and participated in major events in Central Hunza and Gilgit. Emily Overend Lorimer had access to houses and exchange with women, made friends with their neighbours and took more than nine hundred photographs herself. Their contemporary guest and repeated visitor Reginald Schomberg contributed a peculiar set of photographs from the remoter regions such as Shimshal close to the passes leading into China.[49] The Hunza Valley's road infrastructure was thoroughly investigated by a reconnaissance party commissioned by the Royal Air Force Intelligence in May 1937; Flight Lieutenant Francis Claude Sturgiss documented bridges and infrastructure assets during their ground survey.[50] The first aerial photographs originate from the mid-1930s when landing grounds were sought between Gilgit and Kashgar.[51]

Beginning in the 20th century, photography had more or less replaced drawings and sketches although some visitors preferred hand-made illustrations. The mountaineering pioneer William Martin Conway took along the artist Arthur McCormick who prepared three hundred illustrations for their expedition's book on 'climbing and exploration in the Karakoram-Himalayas', an undertaking that was made possible already shortly after the conquest of Hunza and Nagar in 1891.[52] Both were proud of tomb raiding in Nagar; they sent the excavated skulls back home for anthropological studies; Arthur McCormick sketched their actions.[53] Later expeditions trusted in the ability and qualities of profession-

In his painting titled the 'dancers of the Mir of Hunza' Aleksandr Yakovlev captured a rare representation of court dancers performing in front of the local Ḍom musicians. Female dancers (B. buġóṛ, yurġol) have been known in Hunza since the time of mīr Silum Khan who introduced the entertainment gatherings (P. mushrab) at his court (Qudratullah Beg 1962: 351). In the caption of his charcoal sketch (opposite) the artist refers to a male dancer with an effeminate appearance and dress. Georges Le Fèvre (1935: 207), who was present at the occasion, referred to male dancers as well and described them as wearing 'braids of false hair – which had formerly graced the heads of Kashgar beauties'; see as well Peter Fleming (1936: 375); Ella Maillart (1937: 289); N. N. (1938: 593). Source: Tempera on canvas laid on board by Aleksandr Yakovlev, August 1931; reproduced with kind permission © courtesy of Sotheby's. Charcoal sketch reproduced from Alexandre Iacovleff (1934: 12) with kind permission from Gallica. Bibliothèque nationale de France

al artistic painting as well. The leader of the 1931 'Croisière jaune' expedition, Georges-Marie Haardt, had commissioned the reputed Russian artist Aleksandr Yevgeniyevich Yakovlev who delivered more than one hundred pencil drawings and sketches, sanguine, pastel, charcoal drawings on paper and oil paintings that introduced detailed portraits of actors and visions of landscapes in his version of realistic expressionism.[54] Aleksandr Yevgeniyevich Yakovlev became a celebrated artist of his time; nevertheless, his connection with the Hunza Valley is less well-known, as paintings from his prolonged visit there were rarely part of contemporary exhibitions. These paintings have been rediscovered in recent years although the ones related to the Hunza Valley have been mainly appreciated because of their landscape art. Nevertheless, his portraits show some of the contemporary and influential protagonists of Hunza and Nagar.

The 1930s were a period when a host of visitors with different interests and viewpoints became more interested in a region that offered recreation and sports beyond colonial and geopolitical interests. Mountaineering and hunting, esotericism and exploration, map-making and travel attracted more visitors who contributed photographic images as illustrations for their articles, reports and travelogues.

After World War II a growing number of academic and mountaineering expeditions arrived in the Hunza Valley; valuable photographs were contributed by Gerhart Klamert, Wolfgang Pillewizer and Hans-Jochen Schneider, participants of the German-Austrian Himalaya-Karakoram expedition of 1954 and the German Karakoram Expedition 1959, respectively.[55] Irene von Unruh visited Hunza at the same time and contributed her narrative illustrated with photographs of daily practices by men and women.[56]

The mutual interest of service providers and visitors in creating a mythical land of a Shangri La-type drew even more attention to the remote mountain region which was still not easily reached. The Hunza Valley was perceived as the hub of healthy elders in the Karakorams; consequently the pictorial coverage in the 1950s and 1960s has addressed

One of the first and most famous teachers of Hunza and advisor to the mīr, Ustad Sultan Ali aka Samarkhan, is shown in this 1954 photograph in full official dress similar to British colonial representatives. Working as a translator for mīr Jamal Khan, he significantly distinguished himself from the local tradition by his dress code. During mīr M. Nazim Khan's rule only members of the ayáśo were allowed to wear ties. Source: German-Austrian Himalaya-Karakoram expedition 1954. Courtesy of Gerhard Klamert

this issue in the 'land of just enough'.[57] Mountain scenery, irrigated terraces and villages inhabited by strong and sturdy mountain dwellers became the stereotype representations.[58] Travellers of this mindset and journalists for international magazines visited Hunza in growing numbers; since then, less exclusively male-dominated photographs have been published. From the 1980s onwards in terms of illustrations we can rely on our own photographs, some were initially published in previous publications,[59] the vast majority have been especially selected for use in this book.

In terms of illustrations it was possible to collect evidence for nearly every decade since the 1880s. The focus has shifted over time, but two genres were dominant. First, the ruling classes and members of the elite were primarily portrayed; later on photographic samples included guides, porters and service providers. Personalities such as the teacher (*ustād*) Sultan Ali, who was deputed by *mīr* Jamal Khan to act as guide and interpreter for any foreign visitor, are depicted in their photographs and represented a 'typical' Hunzukuċ. Beginning with the Lorimers, scenes of daily routine and their actors received attention and documentation. Second, painters and photographers were overwhelmed by the landscape beauty and extreme geomorphological features of the Karakoram Mountains and the Hunza Valley. Especially scientific photography attempted to record these artefacts for disciplines such as architecture, geology and glaciology. Ethnographers and geographers became interested in documenting a more comprehensive picture of environmental assets and livelihood conditions, of settlement patterns, customs and festivals. The illustrations reproduced here were selected as accompanying evidence for the presented argument. They are divided into four perspectives making up the structure of the book: road and infrastructure, environment and resources, actors and their arena, and myth creation.

More than a decade after the edited volume of 'Karakoram in Transition' was published in 2006 it seems justifiable to reconsider the transitional processes that have affected and altered the physical and infrastructure environment, influenced cultural processes, transformed living and socio-political conditions for the communities and inhabitants of the valley. Since then they have been exposed to significant changes due to the collapse of international tourism in the aftermath of 9/11, the implementation of the Governance Order of 2009, the interruption of the Karakoram Highway caused by the Atabad landslide a year later and the 2013 announcement of the New Silk Road initiative leading to the implementation of the China-Pakistan Economic Corridor through the Hunza Valley. New evidence could be found for further insights into earlier developments in various archives, personal records of eyewitnesses of the 20th century. The search for a more sophisticated understanding of the nexus between colonial intervention, the 'opening-up' of the Karakoram valleys and its impact on local power structures and economic developments can be based now on a broader footing and expanded evidence. It is a necessary precondition for fitting recent developments into a narrative that is primarily based on our own empirical studies during annual visits.

One of the steepest flanks on earth is the pinnacle of Bubulimiting (5,990 m) in the west; Bojohagur Duanasir (7,328 m) in the centre and Ultar Sar (7,350 m) in the east tower above the valley bottom of the Hunza River at 2,100 m. Baltit Fort (2,400 m) is located on the western lateral moraine and illuminated in the early morning hours. The cultivated lands of Karimabad are located on the river terraces and scree slopes and stretch up to the highest irrigation mark at about 2,800 m. The photograph was taken on 29 September 1985 from the upper part of the Tsilkiang nālā.

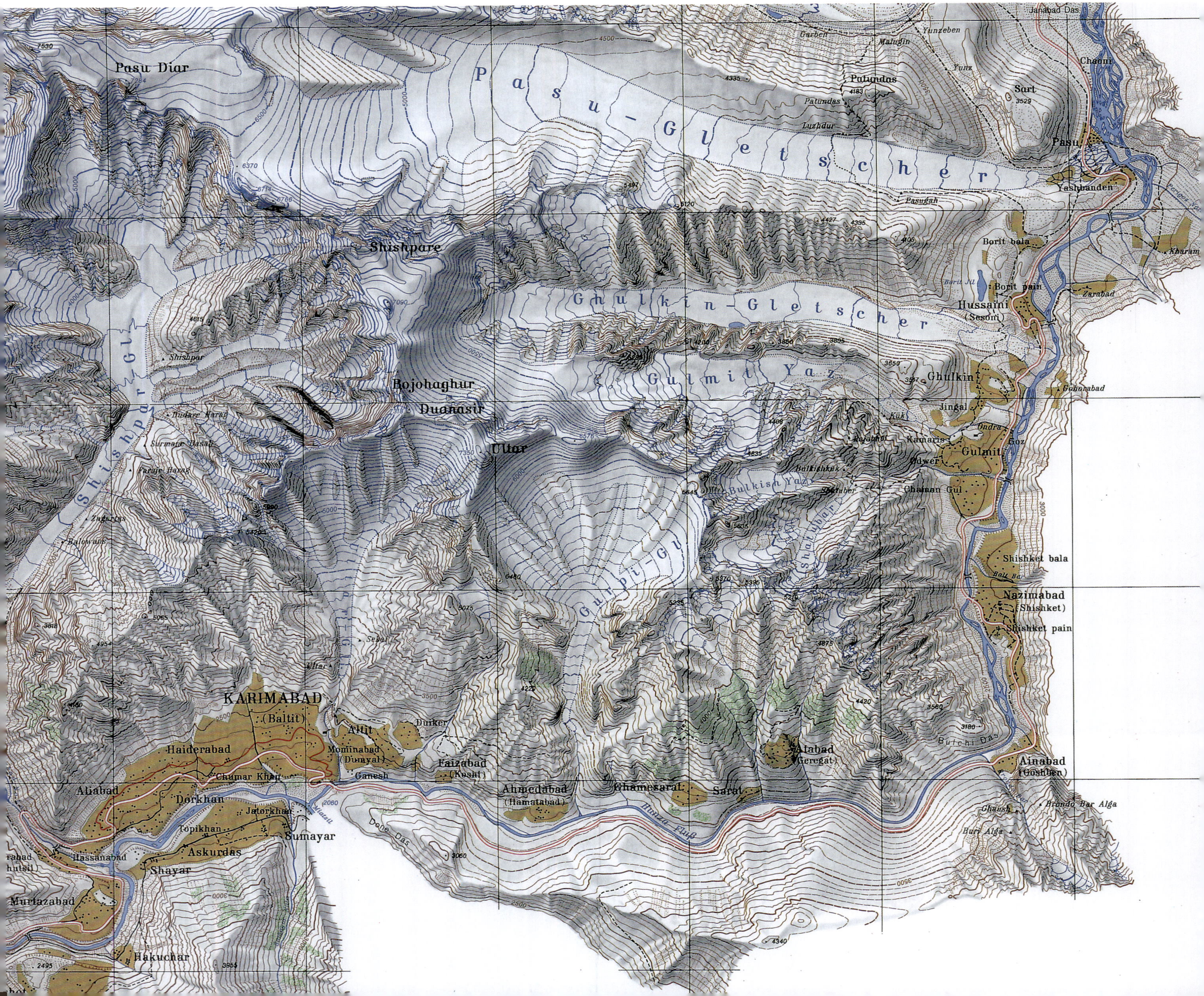
Pasu Diar
Pasu-Gletscher
Patundas
Sart
Pasu
Yashbanden
Shishpare
Borit bala
Borit pain
Hussaini
(Sesoni)
Ghulkin-Gletscher
Ghulkin
Gulmit Yaz
Bojohaghur
Duanasir
Ultar
Gulmit
Kamaris
Odwer
Chaman Gul
Shishpar Gl.
Shishket bala
Nazimabad
(Shishket)
Shishket pain
KARIMABAD
(Baltit)
Haiderabad
Altit
Mominabad
(Dumyal)
Faizabad
(Kushi)
Ahmedabad
(Hamatabad)
Ghamesaral
Sarat
Aliabad
(Geregat)
Ainabad
(Goshben)
Aliabad
Dorkhan
Hassanabad
Askurdas
Shayar
Sumayar
Murtazabad
Hakuchar
Hunza Fluß

Acknowledgments

This book is dedicated first and foremost to our long-time hosts. During our first one-year-long stay in Karimabad from 1984 to 1985, our home base was provided by three brothers – Ibadat Shah, Sadat Ali, and Rahmat Ali – formerly in employ with the *thamo* household and by the time of our first visit in 1981 already successful hospitality entrepreneurs themselves. Our friendship and regular meetings with this extended family have continued and lasted until today; their wives – Kaniza Begum, Sikem Bano, Taj Khanum – and children have survived their husbands and fathers. An extraordinary friend was the late Rai Ghulam-uddin in Gulmit. He and his wife Nikab Sultana provided not only shelter and a secure place during sixteen months in 1990 and 1991; for more than thirty years we have enjoyed our annual visits; they shared with us their vast knowledge of people, Wakhi language and culture, Hunza history and politics. Nikab Sultana introduced us to daily household routines, agricultural practices and women's organisations to name only a few aspects. Their house was our home, meeting-place and arena of serious debates intermingled with humour and laughter. Many other families in Hunza, Nagar and other parts of Gilgit-Baltistan provided a homely atmosphere and shelter for us and shared their experiences, knowledge, narratives and stories with us. In Karimabad and Gulmit several people observed our wanderings and enquiries for a variety of agencies and services; direct and indirect encounters led to mutual understanding, respect and friendship. Ghulam M. Beg in Gilgit, Mohmin Hayat from Aminabad, Nazimullah Beg from Karimabad, and Jalal-ud-din from Gulmit shared their knowledge with us, kept us company while fulfilling their duties among other colleagues who were not personally known to us as members of Hunza society. Others like Dadali Shah helped to open doors and provide access for meetings with important personalities. The headmaster of Karimabad High School, Iftikhar Hussein, requested me to teach physics at his school as the science teacher had taken 'casual leave'; after he introduced me in the classroom as 'Hermann the German' my name was known in every village. These encounters broadened the spectrum of our acquaintances and interlocutors so that basically an entry point was given in every village. In case we didn't know anybody, a name scribbled on a piece of paper (*ćiṭ*) with some friendly recommendation by a respected friend would help. Walking the Hunza Valley was a prime activity in the early years when little traffic was seen on the Karakoram Highway. Minibuses called 'wagons' would ply in the morning from Gulmit or Karimabad to Gilgit and return fully loaded in the afternoon from the bazaars. The drivers became our friends and helped us to return home in late afternoons; to name only two from a much bigger professional group Khano from Gaimush and Gouda Ali from Pasu were especially supportive of our cause and collected exhausted fieldworkers at the end of the day somewhere along the Karakoram Highway. During the early years leaving the valley was more difficult than it is today; especially helpful was the growing number of white vehicles that belonged to development agents. Rahmat Ali from Aga Khan Housing Board took us along when on mission to construct schools in remote villages. Many others were interested in an exchange between development agencies and academic research. The first General Manager of the Aga Khan Rural Support Programme Shoaib Sultan Khan invited me to first, second and third dialogues when eventually Hussain Wali Khan would hand over the cheque for a physical infrastructure project. Shoaib Sultan Khan introduced me to the charismatic Akhtar Hameed Khan who taught us grass-roots development thinking on the ground;

The gabled roof is characteristic for Gulmit's jamáat khana. Next to it is attached the local council office and in front the Ismaili study centre. In the background is Ghulam-ud-din's house; our residence for the next sixteen months was in a separate room between the local council and his house. Potato and wheat field parcels are nearby. The photograph was taken on 29 April 1990

Opposite
Our two homes during fieldwork were located in Karimabad (1984 to 1985) and Gulmit (1990 to 1991).
Source: Map 'Hunza-Karakorum 1:100,000' prepared by Rüdiger Finsterwalder on the basis of expedition surveys in 1954 and 1959; see Hermann Kreutzmann (1996b), published by Deutscher Alpenverein (1995).

Shah Khan presenting his recently published account of the Gilgit Scouts and his participation in the struggle for liberation in 1947. The photograph was taken on 25 September 2000

something we had never learnt at university. His modest and unpretentious appearance and open ears for any kind of human challenges led to a complex analysis and solution-based advice always incorporating communal and cooperative efforts. In the course of time we met and befriended all general managers. Izhar Ali Hunzai started his career in Gilgit about the same time as our research project began; so we have a common set of experiences and much more in common beyond that. Khalida Nasir was the first woman organiser in the project; with her, Yasmin Karim and Ameneh Azam Ali we link valuable experiences and exposures that would otherwise have escaped our attention. For more than three decades we could trust in logistical support by branches of the Aga Khan Development Network, and we are grateful to Stephen Rasmussen, Abdul Malik and Khadija Khan to name only a few. Logistical support was provided by many other friends; especially in times when few vehicles were available Naunihal Shah and Amjad Ayub solved any problem. Naunihal Shah and Bibi Shagufta extended great hospitality to us and shared their knowledge about the *guśpúr* microcosm. The family of *mīrzāda* Shah Khan belongs to our longest acquaintances; from the times of the Village Hotel in Gulmit to becoming neighbours there, we enjoyed ample gatherings and occasions of great hospitality; Group Captain Shah Khan kindly shared the first version of his autobiographic manuscript with us. From the civil society in Karimabad Iqbal Muhammad has become a trusted friend among activists in schemes where social responsibility is required and found in civil servants and private entrepreneurs alike.

In places where no vehicle can go we always had confidence in good friends such as Sanjar Beg of Pasu who was our first companion to visit Shimshal in 1985. Later Inayat Ali introduced us to Didar Karim; since then we have had the opportunity and joy of good company to visit Shimshal Pass several times with knowledgeable experts and relatives in place. His mother, Safida Begum, kindly hosted us several times in her Shuwart house. Aslam Kotoshi from Faizabad led us to Snow Lake from Biafo to Hispar. All have significantly contributed to our safe and sound return and deserve great gratitude. Didar Ali from Haiderabad and now Gilgit has constantly supported us for three decades; he and his wife Parizadi have provided a shelter at numerous occasions, during and in the aftermath of 9/11.

The last *tham* of Nagar, Shaukat Ali Khan, and his sons Barkat and Heibat represented tradition and modernity of Nagar State; their opinions about it and resulting strategies of adaptation to changed frame conditions were shared with us. In the context of local educational and infrastructure development efforts, the exchanges and meetings with the leading activist Syed Yahya Khan from Bar were enlightening; his vision and energy have been instrumental for early efforts of land development and nature protection. When meeting the social organiser Ali Mohammad from Askurdas we learnt that he had studied our society in Berlin before deciding to put his energies in developing Nagar. Syed Israr Hussein from Minapin has remained a long-term friend, advisor and supporter for all activities in Shinber. Two *guśpúr* members from Nagar residing in Gilgit supported our work: Raja Lateef from Chalt and the then Superintendent of Police, Ali Ahmad Jan, who among other acts of support wrote for us a *ćiṭ* that allowed us to visit Khunjerab Pass in 1985 before it was opened for foreigners. Thus he actively supported my gathering of the required data for my very first academic publication on 'Yak-keeping in Hunza'.[60] Our friendship with Nagarkuć has continued and deepened as three development experts were selected for training within the framework of the mountain development programme conducted by Capacity Building International in Feldafing; Essa Muhammad from Shayar has remained a constant supporter and knowledgeable interlocutor to date.

In Gilgit it was quite difficult during Cold War days of omnipresent paranoia to get hold of statistical data and background information. A hub of knowledge was the Municipal Library, located in the nowadays renamed Biddulph House, in the person of the librarian Sherbas Barcha. He was instrumental in awareness creation for local historiography, language preservation and Shina studies, promoter of the Karakoram Writers' Forum and intellectual advisor. Within Gilgit's administration and planning organisations, Sabit Rahim from Ishkoman and Yasir Hussain from Kashrote gave valuable advice and statistical background to the social and economic performance of the region.

The majority of persons mentioned so far are either retired or no longer among us; these acquaintances from the 1980s have found their successors in their sons and daughters; other educated persons have emerged in important positions. The Aga Khan Rural Support Programme was managed by Abdul Malik from Aliabad, and recently the first Wakhi in the person of Muzaffar-ud-din has taken charge. Their support has been a great boon to us as well as the intellectual support by Amin Beg and Aziz Ali Dad. From government circles we have always gained from our exchanges with the educational advisor Majeed Khan. Zafar Iqbal has recently shared his vast knowledge in two books that are less related to his position as head of the Northern Areas Transport Corporation or the Sost Dry Port; his writer's attention was devoted to Hunza's history and the position of Wakhi and members of the Pakistan People's Party in the affairs that led to the abolition of Hunza State. New forms of knowledge gathering and preservation have emerged in electronic media. Fazal Amin Beg is among the pioneers of recording poetry, songs and interviews that he has made available for the general public; his decade-long support is duly acknowledged.

My work would have been impossible without the help, friendship and support of numerous further acquaintances; the name of my late friend Adam Nayyar stands for so many others. He was the witty colleague, challenging my work and viewpoints, and the critical companion and friend during two decades in which his Islamabad home was my point of arrival and departure. During night-long discussions we debated the essence of fieldwork, the performance of more or less philanthropic actors including our own positions, as well as the generation of knowledge, and we shared a common sympathy for our friends from Hunza and Nagar. His house was the base to meet delegations and visitors in Islamabad and to prepare the meetings with academics, administrators and bureaucrats in a true Pakistani fashion. Even during martial law times we enjoyed lively and inspired debates, cultural performances and thoughtful exchanges of differing views. Since his demise Islamabad has become a different place.

The research for this book was made possible by the universities I have been affiliated to and employed by, namely Freie Universität Berlin (1984–1989, 2005–2020) and Friedrich-Alexander University Erlangen-Nürnberg (1996–2005). Substantial research funds for archival research, fieldwork, expeditions and travels were made available by various grants from the Deutsche Forschungsgemeinschaft. Several individual research projects could be implemented as several grants were awarded between 1984 and 2017, beginning with my doctoral studies in the team of Fred Scholz; logistical support was provided by Karl Jettmar from the South Asia Institute in Heidelberg. Follow-up projects included my participation in the 'Culture Area Karakoram' priority programme followed by a fellowship for professorial candidates (1993–1994) and a fellowship in the Deutsche Forschungsgemeinschaft's generous Heisenberg programme (1996). The latter endeavours, pointing towards my future professional career path, were made possible thanks to Eckart Ehlers who always backed my efforts by his genuine sympathy and full trust. From 2010 to 2016 the Federal Ministry of Education and Research (BMBF) funded the competence network 'Crossroads Asia' which paid special attention to cross-border linkages and networks in Central and South Asia. The human capacity wing of the German Society for International Development (formerly InWEnt, now GIZ) offered valuable insights into effects of development-related activities during my participation as the academic advisor to their mountain development programme (2006–2012) and afterwards as backstopper and trainer.

Helpful and professional team members from our Centre for Development Studies at Freie Universität Berlin supported the book project in various capacities from secretarial and office work to copyright and reproduction handling, search for data and publications, scanning and processing of slides; my gratitude goes to Emilia Adam, Maksat Anarbaev, Lilly Aufdembrinke, Sarah van Bentum, Sandra Cacciarru, Camille Dammann, Andrei Dörre, Marek Felten, Mari-na-Elena Heyink, Dennis Scholl, Stefan Schütte, Mayinur Shanatibieke, and Michael Spies.

Special thanks go to Anne Beck who often advised and helped to improve my use of the English language in its written form. She and Sabine Felmy helped editing the whole manuscript; any mistakes and shortcomings solely remain my responsibility. More than linguistic support was kindly provided by the eminent language scholars of the area. Since the early days when Wilhelm von Humboldt detected and realised the importance of studying foreign and minority languages in order to learn more about his own, but also about the peculiarities of other societies, the Hindukush and Karakoram have been a target for most eminent language researchers, just to name the nestor Georg Morgenstierne from Norway.[61] It was my great fortune to have been able to solicit the support of the leading international language researchers during my fieldwork. The late Alexandr Leonovich Gryunberg and Ivan Mikhailovich Steblin-Kamensky in St. Petersburg, who provided me with access to their personal records and to archives in St. Petersburg. Repeatedly and gratefully I acknowledge the valuable intellectual

Syed Yahya Shah has been an activist in various innovative undertakings in Nagar for decades. He was one of the early promotors of female education, nature protection and an influential counterpart for development organisations. The photograph was taken in Bar on 13 May 2015

Altit and Baltit Fort tower high above the Hunza River. The valley (2,100 m) is incised up to one hundred metres deep and marks the border between Hunza and Nagar. Both central oases are located on the river terraces and scree slopes that have been converted into fertile agricultural lands by gravity-fed irrigation channels. Nagar is towered over by the massif of Diran and Rakaposhi (7,788 m), separated by the deeply incised Sumayar Valley. Baltit Fort (2,480 m) in the foreground is elevated on a moraine that has a steep flank on the left to Harchi Har, a valley that separates Altit and Karimabad. On the right, houses of Diramishal and Qhurukushal form the compact settlement of the former Baltit khan. The photograph was taken on 26 March 2019

advice and kind support by Georg Buddruss from Mainz whose command of so many languages enhanced the quality of my glossary over a period of thirty years. The late Hermann Berger kindly allowed me to use his word cards long prior to the publication of his Burushaski dictionary and grammar. The cooperation between him and the eminent local scholar Allama Nasir-ud-din Hunzai could be established as long ago as 1985, when his son Izhar Ali Hunzai and the author photocopied all his collected word lists in Jamaatkhana Bazaar Gilgit when electricity supply, toner for copying and printer paper were rare and often unavailable commodities. Thus, two bodies of collections came together at Hermann Berger's desk: he not only acknowledged the substantial addition to his own collection of terms, but made Nasir-ud-din Hunzai his cooperation partner in publishing the three volumes of the Burushaski to German dictionary in 1998.[62] The pioneers of language research have initiated an enhanced awareness for the recordings of linguistic specimens, for applying scripts to non-written languages in order to prepare radio plays and write down poetry and narratives. A new generation of authors and poets, journalists and reporters has emerged who have contributed in their own capacities to this book. Articles from the Pamir Times and other newspapers have been instrumental for keeping up to date about events and debates; special exemplary thanks go to Noor Mohammad and Zulfiqar Ali Khan, Farman Ali and Aziz Ali Dad who keep the flow of news and background information going.

The quality of illustrations was significantly enhanced by the expertise of Yasmin Khan, Head of Paper Conservation Section at Library of Congress in Washington DC who again helped in searching and securing valuable material and advised on reproductions. My work was most significantly and constantly supported in manifold ways by Markus Hauser and his valuable Pamir Archive Collection. From his vast assembly of rare historical documents and valuable maps, a number of those relevant to my topic have been reproduced in sections or in full size in the book. His expertise in cartography and map-making is reflected in the book's reproductions. The professional cartographic skills of our late colleagues, Detlef Engel, Joachim Kruhöffer and Rudolph Rössler, have provided the early versions of sketches that have been processed and expertly redrawn by Bernd Hilberer who supplied the majority of original illustrations for this book; all professional cartographers have succeeded in converting my often rough sketches into meticulous illustrations. Harald Weller contributed his professional expertise in improving the layout, typography and reproduction of all illustrations. From the publisher Harrassowitz I would like to thank Barbara Krauß and Jens Fetkenheuer for their interest in the topic and excellent handling of the trilogy project.

Routes explored in the Hindukush, Karakoram, Pamir, and Himalaya by Hermann Kreutzmann between 1977 and 2019. Design by Hermann Kreutzmann

Copyright

Endnotes of pages 9–35

1 The regional denomination refers to the two predecessor volumes for the present and final one of my trilogy; see Hermann Kreutzmann (2015a; 2017a).
2 Several contributions have dealt with the issues of the Kashmir dispute and stalemate; see Antía Mato Bouzas (2012; 2019); Hermann Kreutzmann (2008a; 2015b; 2016).
3 Hermann Kreutzmann (1987; 1989a; 1991a).
4 Fernand Braudel (1960); Kirti Narayan Chaudhuri (1981; 1990); Immanuel Wallerstein (1974; 1988); Eric Wolf (1982).
5 Translation of the original phrase 'Die Weltgeschichte geht von Osten nach Westen, denn Europa ist schlechthin das Ende der Weltgeschichte', quoted after Ernst Schulin (1958: 52–53).
6 See for an extensive discussion and evidence Hermann Kreutzmann (2013a; 2015a: 69–70; 2017a: 35–40).
7 See Matthew Edney (2009); Ainslee Embree (1977); Hassan Nawaz Gardezi (1991); Daniel Headrick (1981); Brian Hudson (1977); Wolfgang Mommsen (1981); Hermann Kreutzmann (1997); David Spurr (1993).
8 See Hamza Alavi (1971; 1989); Tariq Ali (1983).
9 Lawrence Krader (1959; 1963); Pavel Luknitzky (1954); Jan Myrdal (1981).
10 Andre Gunder Frank (1978; 1980).
11 Andre Gunder Frank (1992; 1998).
12 Tahir Ali (1983; 1984); see in contrast the works beginning with David Lorimer (1979-1980) and Emily Lorimer (1936a–g; 1939) up to Irmtraud Müller-Stellrecht (1979–1980; 1983); a constellation which is critically reflected upon by Irmtraud Stellrecht (2006: 211, footnote 4; 2007).
13 Robert Chambers (1983; 2006).
14 Nigel Allan (1986); Robert Chambers and Gordon Conway (1992).
15 Katja Mielke and Anna-Katharina Hornidge (2017); Irmtraud Stellrecht (1998).
16 Michael Spies (2019).
17 Hermann Kreutzmann (1991a).
18 Eckart Ehlers and Hermann Kreutzmann (2000); Hermann Kreutzmann, Ghulam Amin Beg, Lu Zhaohui and Jürgen Richter (2009).
19 Arjun Appadurai (1996); Neil Brenner (1999).
20 See the remarks of Wolfgang Holzwarth (2006) and Siegfried Weber (2007: 61–64) on the two versions of the 'History of Hunza'. In the following quotes are taken from the English version by Saadullah Beg, son of Qudratullah Beg, which was published in the former's name in 2006.
21 Mohomed Nazim Khan (1936).
22 This especially holds true for the widely distributed set of compiled maps titled 'India and Pakistan 1 : 250,000. Series U 502' published between 1959 and 1963.
23 The circumstances of the fatal accident of Karl Heckler were described by the expedition leaders Matthias Rebitsch and Wolfgang Pillewizer (1955); excerpts from Karl Heckler's diary were published in the yearbook of the Austrian Alpine Club (Karl Heckler 1954).
24 Hans-Jochen Schneider and Hans Baumert prepared the map 'Minapin (Rakaposhi Range) NW-Karakorum 1:50,000' in 1968.
25 The 'Hunza-Karakorum' map was published at a scale of 1: 100,000 by the German Alpine Club in 1995; see the special issue of the journal 'Erdkunde' (50 (3) 1996) with explanatory articles and the map attached.
26 The map was commissioned by the Academia Sinica and published at a scale of 1 : 60,000 in 1978.
27 See Muhammad Mumtaz Khalid (2006; 2009); Zafar Iqbal (2018a, b).
28 See Felix Driver (2001; 2015); Hermann Kreutzmann (2017a).
29 Bronisław L. Grąbczewski [Grombchevsky] (1888/2015); Wawrzyniec Popiel-Machnicki, Adam Pleskaczyński and Konstancja Pleskaczyńska (2017).
30 John Biddulph (1876; 1878; 1881); Patrick Alexander Vans Agnew (1915).
31 The Pamirian files are located in the Российский Государственний Военно-Исторический Архив (РГВИА), RGVIA, font 1396, opis 2, file 2168.
32 See for example 'Criticism by Colonel Schomberg of the administration of the Gilgit Agency' (IOL/P&S/12/3293).
33 The unpublished book manuscripts and a great number of photographs are kept in the archives (RSAA/SC/SCH/2/47).
34 Mohomed Nazim Khan (1936). A badly edited version full of typographical mistakes and terminological changes by a relative was published in book form in 2014.
35 Their archival records are accessible in the British Library (India Office Records and Private Papers Mss Eur D1168) and in the School of Oriental and African Studies (SOAS) in the University of London as 'Papers of Lieutenant-Colonel David Lockhart Robertson' (1906–1948). A substantial body of his manuscripts related to Hunza and Nagar, Gilgit Chi-

tral and Yasin has been edited and commented by Irmtraud Müller-Stellrecht (1979–1980).

36 See Stuart Hill Godfrey (1898: 75, 82).

37 Haji Qudratullah Beg (1962); see for a comparison of Hunza and Nagar chroniclers Jürgen Frembgen (1996); Siegfried Weber (2007: 61–65).

38 See for example the documentation and diaries from the German-Austrian Himalaya-Karakoram expedition (1954) and the German Karakoram Expedition (1959). Newspaper cuttings from Pakistani media have been listed in the bibliography; Aziz Ali Dad has become a regular commentator on issues concerning Gilgit-Baltistan. A detailed account on the construction of the Karakoram Highway was published by Muhammad Mumtaz Khalid (2006; 2009); the personal view of Shah Khan (1985) and a history providing new documents and special attention to Gojal by Zafar Iqbal (2018a, b) reflect the wide range of contributions. Other scholars offer their interviews, reports and writings in the internet such as the 'Hunza Development Forum' (https://hisamullahbeg.blogspot.com/) and Fazal Amin Beg (www.fazalamin.com).

39 Stuart Colvin Bayley (1895: 2); Algernon Durand (1899: 139).

40 Demi-official letter from Major George Kirkbride, Political Agent Gilgit, to Lieutenant Colonel Lionel Edward Lang, Resident in Kashmir, dated 15 July 1935 (FO 371/19335: 226).

41 Reginald Schomberg (1935a: 158).

42 Wilhelm van Schendel (2002) has enriched the debate and drawn attention to neglected borderland areas.

43 Detailed maps from colonial times have been Survey of India maps prepared by Reginald Schomberg (Shingshāl Mustāgh Area) and the Baltit map (No. 42 L) during the 1930s. The above-mentioned Chinese map of the Batura glacier (1975) and the Hunza-Karakorum map by the German expeditions during the 1950s are among the rare specimens of high-quality large-scale maps.

44 See for the reproduction William Lockhart and Robert Woodthorpe (1889); the original photographs of the Gilgit Mission are archived in the Royal Geographical Society. The photographs of Bronisław L. Grąbczewski [Grombchevsky] are kept in the Scientific Archive of the Russian Geographical Society (НАРГО – Научный Архив Русского Географического Общества) in St. Petersburg; some selected photographs were kindly provided for reproduction in this book.

45 The photographs of Richard William George Hingston are kept in the Trinity College Library in Dublin (MS 10484/7).

46 The photographs of Marc Aurel Stein are accessible through the webpages of the International Dunhuang Project and the British Library's Stein Collection (India Office Records and Private Papers Photo 392 series).

47 See Jenny Visser-Hooft (1926) and Philips Christiaan Visser (1928) who illustrated their books with more than fifty photographs from the Hunza Valley. The photo albums of the two Karakoram expeditions (1922 and 1925) contain more images that are kept with the National Archives of the Netherlands in Den Haag, Collection 558.

48 The collection of approximately 400 photographs taken by David Lorimer in Hunza during the 1920s and 1930s in Hunza is kept in the School of Oriental and African Studies archives in London along with 22 reels of cinefilm (MS 66). In their report they stated that Emily Lorimer took 918 and David Lorimer 809 photographs (RAS/LOR/1/2/2: 21) during their stay in 1934–1935. Several photographs have been donated only recently to the archives and have remained undocumented to date.

49 Reginald Charles Francis Schomberg's photographs are preserved in the Royal Society for Asian Affairs (RSAA/SC/SCH/1).

50 The photographs taken by Francis Claude Sturgiss are included in his confidential report about the 'proposed establishment of further R.A.F. landing grounds in the Gilgit Agency in general and at Pasu and Sost in particular' from 1937 (NAI/PR/40002107. File No. 82(2)-X/39).

51 Report on 'Aircraft & aviation. Flight of R. A. F. aeroplanes to Gilgit' 1934 (IOL/P&S/12/1993); Report on 'Proposed establishment of further R.A.F. landing grounds in the Gilgit Agency in general and at Pasu and Sost in particular' 1937 (NAI/PR/40002107).

52 Arthur David McCormick (1895) titled the book in which he presented his drawings 'An artist in the Himalayas'; others are contained in William Martin Conway (1894a).

53 His sketch is reproduced in William Martin Conway (1894a: 244).

54 The sketches and paintings of Aleksandr Yevgeniyevich Yakovlev are contained in a number of publications and exhibition catalogues; see Alexandre Evgenievich Iacovleff (1934; 2004); Alexandre Iacovleff, Caroline Haardt de La Baume and Jean-François Ruchaud (2012).

55 See for example John Edwards (1961); Wolfgang Pillewizer (1957; 1960); Hans-Jochen Schneider (1956; 1961). Gerhart Klamert kindly offered some of his invaluable photographs for reproduction here.

56 See Irene von Unruh (1955).

57 Jean Shor and Frank Shor (1953); Renee Taylor (1964).

58 See for example popular accounts by Allen E. Banik and Renée Taylor (1960); Daud Beg (1964); Jewel Hatcher Henrickson (1960); Jay Milton Hoffman (1968); Sabrina and Roland Michaud (1975); Jean Bowie Shor (1955); John Tobe (1960). Some visitors of that period made somewhat more critical and differentiated observations; a few evidence-based contributions stem from this period such as John Clark (1963); Naohiko Harada and Atsuko Miyoshi (1963); Barbara Mons (1958); Laurence Frederic Rushbrook Williams (1964).

59 Sabine Felmy (1986; 1993; 1997); Hermann Kreutzmann (1989a; 1993a, b; 1996a, b; 2005b; 2006; 2015a; 2017a).

60 Hermann Kreutzmann (1986).

61 Wilhelm von Humboldt (1838); Georg Morgenstierne (1932).

62 Hermann Berger (1998) who had published articles on certain aspects of Burushaski from 1960 onwards; see for the development of linguistic studies Georg Buddruss (1985; 1993; 2006).

Baltit Fort was renovated and restored by the Aga Khan Trust for Culture and has been converted into a museum. The photograph was taken on 3 August 2017

Introduction

On 14 July 2017 – a few days after Aga Khan IV Karim-al-Hussaini had celebrated his Diamond Jubilee as *imām* of the Shia Imami Ismaili community – a gathering took place which village elders and respected persons from all over Hunza were invited to attend in Karimabad's palace garden. The venue was well selected and of significance as the so-called new palace and the surrounding property would be on centre stage during the discussions. Two former seats of power are symbols of hereditary rule in Hunza. From pre-colonial times the autonomy and independence of the autocratic *tham* of Hunza had been vested in their fortified abodes and seats of power. Altit and Baltit forts are the landmark buildings that have been preserved and renovated under the professional leadership of the Aga Khan Trust for Culture. Nowadays they function as museums and attract visitors to Hunza. Baltit Fort was perceived as an impressive *khan* that had never been conquered by outsiders up to the rule of *tham* Safdar Ali Khan. During the Hunza-Nagar Campaign combined British-Kashmiri forces conquered the principalities in the Hunza Valley in December 1891; they looted Baltit fort and initiated their 'paramountcy' with sumptuous Christmas celebrations there.[1] Safdar Ali Khan fled to China and was succeeded by his half-brother M. Nazim Khan who functioned as an 'indirect' ruler, appointed and backed by the colonial authorities.

Both forts are landmark remnants of Hunza's earlier history as a fortified bastion. The new palace originates from the mid-1940s and symbolises the transition from colonial domination to semi-autonomy after Pakistan's independence.[2] Earlier on M. Nazim Khan had established a small guesthouse on a big boulder which he named 'Tash Mahal' in the Karimabad *muḥalla* of Baltit in 1925. The British-style bungalow building dominated his orchard which was cleared step-by-step to make space for more functional buildings such as a *darbār* hall for receptions and courtiers' quarters. The growing numbers of visitors – colonial officers and administrators, emissaries of the Aga Khan, travellers and traders between Kashmir and Kashgar, pilgrims and spies, hunters and explorers – were accommodated in this comfortable and exposed spot where the *tham* could receive and entertain them, organise and provide requested and required services, and also personally control and monopolise the hospitality business. The built-up area in Karimabad was expanded and the daily receptions (*darbār*) and court sessions were shifted from Baltit Fort to Karimabad orchard. His grandson, *mīr* Jamal Khan, built the 'new palace' towards the end of colonial rule, and Aga Khan III Sultan M. Shah supported the setup here. Ever since then, the last *tham* of Hunza, M. Jamal Khan, resided here until his death in 1976. The public graveyard in Suryas Das is located just above the garden and contains the tombs of the two rulers who were most fond of this green oasis. M. Nazim Khan had initiated the shift from Baltit Fort which was then abandoned and remained more or less empty during *mīr* Jamal Khan's rule. After the end of Hunza State in 1974 and his death two years later, his son Ghazanfar Ali Khan was urged to leave Pakistan's Foreign Service and henceforth managed family affairs and commercial dealings from here. In honour of the regular support that the Aga Khan had provided, the village of Baltit was renamed Karimabad in 1983. Baltit Fort was handed over to a heritage trust in which office-bearers in the Aga Khan Development Network, Ghazanfar Ali Khan's family, and some selected personalities are represented. With the growing tourism boom after the opening of the Karakoram Highway, the Karimabad orchard shrunk further, providing space for quarters used by the two brothers of Ghazanfar Ali Khan and other family members. The lower part was separated for the building of a hotel, a massive complex much larger than the modest palace. The venue of the meeting has been an arena for discussion, entertainment and hospitality for nearly one century now.

In the summer of 2017 Ghazanfar Ali Khan, then Governor of Gilgit-Baltistan, had invited the village elders on to his lawn, which was situated between the hotel and the palace.[3] More than two hundred and fifty persons were present and they expected a celebration in the context of the Aga Khan's Diamond Jubilee, which had taken place three days earlier and inaugurated a full year of festive events. Hunza occupies a significant position in the Aga Khan's cosmos, being one of the few areas with an Ismaili majority population. After the demise of their *imām* Sultan M. Shah, the date of 11 July marks the handing-over and transition of the leadership of the Ismaili *jamáat* to his grandson Karim al-Hussaini in 1957. Since then Aga Khan IV has significantly influenced and shaped the welfare conditions and infrastructure development in Hunza as well as the fortunes of the ruling dynasty. But the reason for gathering so many respected repre-

The last flag of the Hunza State was the result of several amendments and additions. Qudratullah Beg (1962: 304–305) described it: "Since ancient times and time immemorial the tiny kingdom of Hunza and its ruler had a flag which contained the signs of a bow and arrow." Supposedly mīr Muhammad Jamal Khan 'added a sign of a male lion bearing a flag in his mouth'; the Pakistan crescent and star were probably added after 1947. Source: Reproduced from the imprint on the front cover of Wolfgang Pillewizer (1960)

The first guesthouse for visitors was built in Karimabad garden in 1925 by tham M. Nazim Khan to accommodate a growing number of visitors. Tash Mahal was constructed on a rock and is visible on the left in the background. The 'new' palace was built in the garden twenty years later by mīr Jamal Khan with support from Aga Khan III and subsequently extended and modified. Since then it has been his residence after vacating Baltit Fort. The photograph was taken on 22 March 2019

sentatives of the Hunzukuċ was related to another occasion and set of events. In his speech Ghazanfar Ali Khan asked for genuine help and support from the Hunza people. This constellation was quite a new one although the episode did not occur completely unexpectedly. First, it had been long known to the public that different factions within his family were struggling with each other to get hold of the remaining assets that were left from hereditary rule (*thámkuṣ*). The wealth accumulated during former power constellations had shrunk significantly, but the lifestyle of the family members still required financial means and substantial funds. Second, the effects of the China-Pakistan Economic Corridor as part of China's New Silk Road Initiative were strongly felt in the Hunza Valley. The road infrastructure had been considerably improved; accessibility and communication had eased. For the past two years a new domestic tourism boom had multiplied the numbers of visitors and offered new income sources for all kinds of service providers. Whenever an economic opportunity arose in the Karakoram Mountains, interested parties and potential investors from outside found their way into the Hunza Valley. Economic enterprises in commerce and trade, prospecting for valuable mineral resources and gemstones, and a variety of tourism-related opportunities had attracted persons who, in Hunza's conventional system of customary property access and ownership, would have not been eligible to act at all.[4] A close-knit set of rules was observed that had banned outsiders from intruding and acquiring property. In the 1980s the Hunza Valley had experienced the expulsion of an independence war hero who arrived as a tourism investor from outside. The investor had managed to acquire land for a hotel complex, but could never build it as the community opposed any outsider's purchase and claim to land property to date.[5] The only person formerly exempted was the Aga Khan who as *imām* and spiritual leader of the Ismailiya has occupied an elevated and exceptional position among the Hunzukuċ. In earlier times all the religious community centres (*jamáat khana*) had been the personal property of the Aga Khan, and his estate agent controlled and managed his assets. But now a new phase in Hunza's history has commenced, and new forces entered the valley hitherto unknown in their potential for social change. With the advent of new external players in the China-Pakistan Economic Corridor all frame conditions had been altered and inherited rules in property

Opposite
The central oases of Hunza and Nagar are densely populated and terraced in the middle elevation located a hundred metres above the valley bottom, where comparatively flat land provided the space for settlement. The meltwaters from the glaciated zone – here visible in the higher elevations with Rakaposhi (7,788 m, towering above – provide the means for irrigation. Emily Overend Lorimer (1936e: 159–160) described the scene: "… the whole landscape is a study in sepia and khaki, and the only colour is in the brilliant blue sky and the dazzling snow on the higher peaks. This is what makes you appreciate the green of the oases. … there is no flat country, and consequently there are no horizons … only the jagged outlines of peaks and mountain ridges against the sky." The photograph was taken on 28 March 2008

On festive occasions inherited leadership is required for leading the dances as a sign of persisting respect. Ghazanfar Ali Khan and Hisamullah Beg are representing former thámkuṣ and wazīrkuč influence that has vanished in many instances, but significantly survived in ceremonial and political affairs. The photograph was taken on 9 August 2014

dealings were to be modified by government institutions and potent investors.

The occasion of Ghazanfar Ali Khan's invitation in 2017 was his fear that certain properties such as the 'new palace' would prematurely get out of his hands as his eldest son was pursuing an untimely early inheritance of this property. The first-born son tried to play the heredity game of the *thamo* family although more than forty years had passed since *thámkuṣ* was abolished. His father had never been *mīr,* nor was he himself heir to any position (*mīrzāda*) coming close to rulership. Ghazanfar Ali Khan had been a life-long politician embedded in an increasingly powerless and grossly impotent economic setting and his position was characterised by dwindling assets. These losses of authority and decision-making power were paired with the main actor's simplistic mindset trying to regain bygone influence and wealth. Herewith Ghazanfar Ali Khan has been displaying an odd and unrealistic world-view. Confrontation and conflict seem to be unavoidable in a household setting where all members seem to display unjustified arrogance and adhere to misconceptions of influence and power.

Ghazanfar Ali Khan confided to the gathered audience that all his three sons had been disobedient (*nā-farmān*) and thus failed to convince him that they would live in and represent Hunza. Ironically the same statement had been made about him in 1968 before he returned home.[6] To the audience it seemed an obvious allegation given his three sons' dual relations with Hunza and Punjab – where their mother came from – and their behaviour and performance in Hunza. Consequently, they would not be eligible to inherit the palace if this frame of mind prevailed and should definitely be prevented from selling out to an outsider. In Karimabad rumours were spreading that Islamabad's notorious property tycoon and other potent investors might be interested in acquiring the palace as the actual tourism boom would dearly need further unique landmarks as attractions and selling-points. Ghazanfar Ali Khan told the gathering of village elders that the 'palace is the property of the Hunza people' and added: "Not a hand full of Hunza soil should ever be sold to an outsider." That statement was repeated by several attendants of the gathering and complied well with the old local rule in property dealings where land should be offered to relatives and community members on a priority basis if a sale was unavoidable. He proposed that he himself wanted to live in the palace for the rest of his life and then would hand it over to the people or to His Highness the Aga Khan as he had done before with Baltit Fort. The palace could become a museum as had happened to Altit and Baltit forts after their successful renovation. His reasoning included quite a new perception of dealing with property in Hunza. In former times the *tham* had always claimed that all land was his property and that final decisions were vested in his ubiquitous authority. In post-independence days villagers had successfully struggled to gain control of the fields and pastures they had tilled and used for generations. These confrontations and negotiations lasted well into the 1980s and had contributed to hardening attitudes among political parties of pro-*mīr* and anti-state factions.

Now on this occasion one witty attendant mentioned that he would only believe in the handover of the palace to the people if the transfer were written and documented in the land record (*paṭwārī*) books. Such a move has not taken place since. As witnesses, all participants of the gathering signed a document in which Ghazanfar Ali Khan's statements and wishes were summarised. His formal announcement was publicly repeated two days later in a gathering on Karimabad's polo ground (*šabáran*) where a substantial crowd of spectators had gathered. It became quite clear that a handover to the 'people' would not be feasible as inner-community conflicts would immediately emerge. The enmity and competition between different villages, lineage groups, social strata and elite groups have been reflected in the internal politics for many generations. Hunza society appears homogeneous only from an external and distant perspective. With the abolition of *thámkuṣ* at the latest, its community was split into anti- and pro-Hunza State factions.

The only authority that would find unanimous acceptance and respect lies in the Imamat of the Ismailiya. The conflicting interests have not been reconciled, the inheritance case is still pending – in the meantime Hunza House, the only remaining Islamabad property, has been sold. A new phase of Hunza's internal and external relations has commenced where existing rules can be superseded and formerly unaccepted actions have materialised.

The episode described above shows a shift in political relations and reveals inner-societal dynamics linked to real estate and symbolic property, economic and political functions of sidelined personalities. They claim to act as representatives of Hunza although they fail to enjoy inherited esteem; on the political stage they pretend to be Hunza's stakeholders although they only pursue personal monetary interests. The conflict constellation incorporates a number of traits that are rooted in Hunza's history while at the same time new actors and elements have emerged that significantly and forcefully affect the path of developments. The colonial legacy combined with reinforcement of hereditary rule has translated into a societal setup deeply rooted in earlier settings.

The Hunza Valley – an area between bordering and ordering

My focus on developments in the Hunza Valley might seem to suggest that Hunza and Nagar – the two (semi-) autonomous microstates in the valley – were the main object of colonial desires. During the rivals' struggle to expand their respective spheres of influence the Hunza Valley appeared to be an important corridor and an 'area between' the real valuable entities. Early explorers such as Patrick Vans Agnew, John Biddulph and Bronislaw Grombchevski directed their travels into the valley; the latter two managed to visit Baltit as their focal point from where they would return to the territories controlled by British and Russian imperial troops, respectively.[7] The mountainous interface located in the quadrangle between Afghanistan, Russia, China and British India may have seemed to be the central arena in which the 'Great Game' would find its decisive climax. At the same time Hunza and Nagar were bordered by a triangle composed of Badakhshan, Kashgaria, and Kashmir. Later on, the principality of Badakhshan would be split into an Afghan and a Russian realm. However, entities such as Wakhan, Hunza and Nagar merely filled a void in an arena of power struggles; some actors perceived the region as a buffer zone, others as a negligible spatial quantity, and several as both, owing to the area's lack of wealth and resources. The attention given to these microstates originates from challenging and difficult access routes. The short and – from a British-Kashmiri viewpoint – successful Hunza-Nagar Campaign of 1891 terminated the initial struggle for a passage to Hunza by 'opening-up' and 'pacifying' the valley. This wording typically represents imperial policies that translated well into the subsequent modernisation project. Infrastructure development has played a major role in debates focusing on development and modernisation. From time to

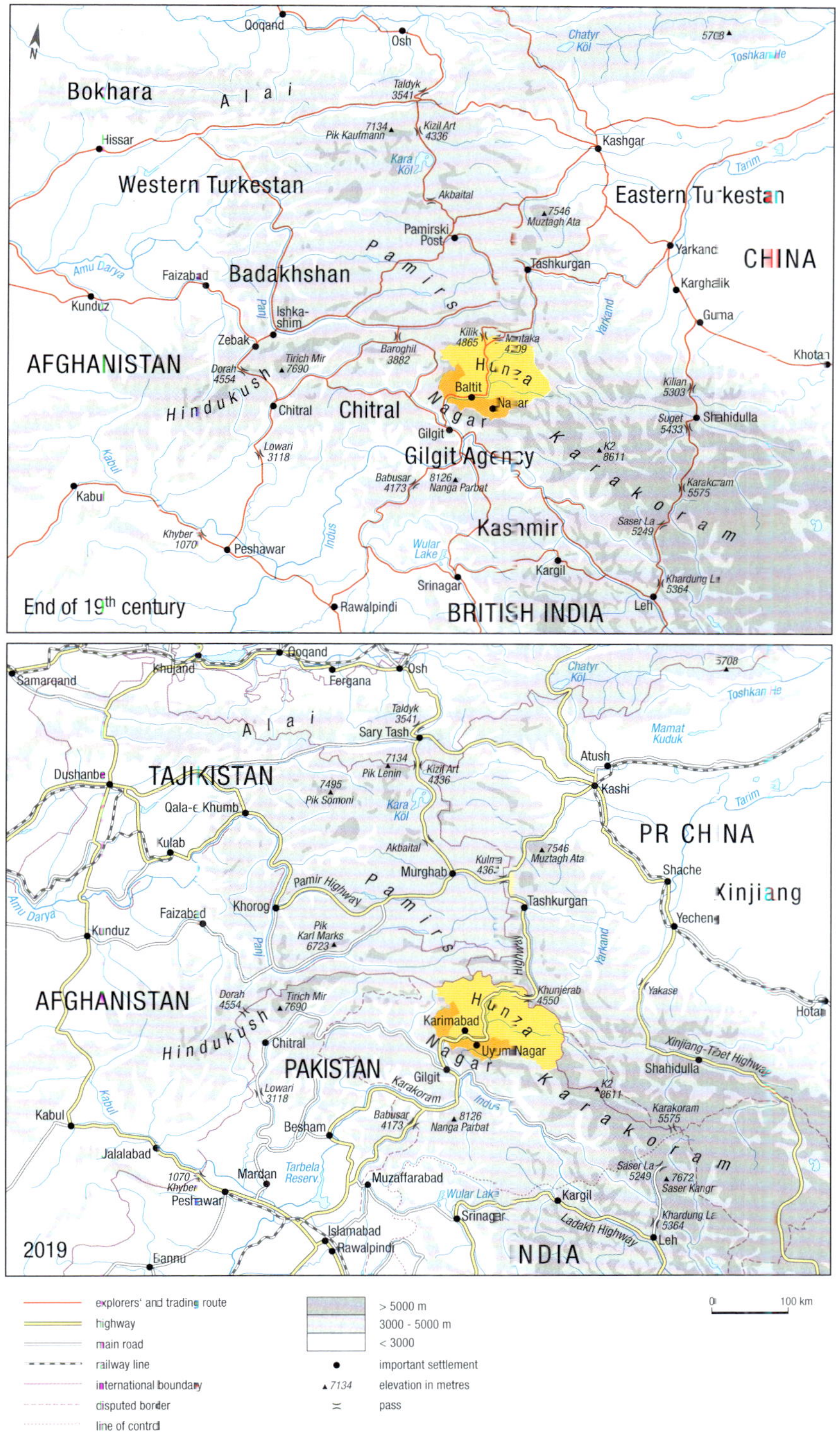

Hunza and Nagar had been conquered by the end of the 19th century, being positioned between Afghanistan, Bokhara and Russian or Western Turkestan, Xinjiang or Eastern Turkestan under China's domination, and Kashmir and British India. The indicated routes were used by trade caravans, military troops, explorers and pilgrims. Only one route followed the Hunza Valley. Presently the northern and southern rail and road network has been significantly expanded; the Hunza Valley connects with the lowlands via the Karakoram Highway. Design by Hermann Kreutzmann

At the crossing of the Kilik Pass (4,765 m) two contrasting landscapes became visible that separate the Central and South Asian highlands and their general features as plateaux in the north of the subcontinental watershed and as deeply incised valleys draining to the south. From the south the steep ascent is reflected in the rugged Karakoram Mountains where Burusho and Wakhi mountain farmers have founded their dwellings, while the view north towards the Kirghiz encampment of Paik resembles the high Pamir plateau with undulating surfaces providing the grazing ground for Kirghiz pastoralists. Source: Tempera on canvas laid down on board by Aleksandr Yakovlev, August 1931; reproduced with kind permission of Sotheby's © collection particulière française

Opposite
The map 'Asia' was published by John Arrowsmith in 1842; the section represented here shows the Pamirian Crossroads and the limits of respective spheres of influence. Four realms – Tartary including Bokhara and Badakhshan, the Chinese Empire stretching east to the Tien Shan and Little Tibet, the British Empire having expanded its control to the Punjab, and Afghanistan in which Kabul was a northeastern stronghold – met in the 'Wakhan Quadrangle' where mountains such as the Bolor, Hindukush, Pamir and the Sulaiman Mountains seem to be structuring features for bordering and ordering. Source: The London atlas of universal geography, exhibiting the physical & political divisions of the various countries of the World, constructed from original materials. London 1842, plate 27. Courtesy of the Pamir Archive Collection

Desert of Kizil Koom
Desert of Kharesm or Khiva
Balkash Nor or L. Tengiz
Tarbagatai
Turkestan
Tashkand
Khojend
KOKAN
Margilan
Andejan
Alatau
Mussur-tagh
Samarkand
BUKHARA
Kara ata
Dizzikh
Uratuppa
Merve
Charjoo
Karshi
Hissar
Kilif
Andkhoo
Balkh
Kulum
Badakshan
Bolor Ms.
Bolor
Chitral
Hindoo Coosh
Kashgar
Yarkand
Yengishar
Aksou
Ouchi
Karatal
THIAN CHAN
Khoten or Ilitsi
Keldia
Tchera
Picaiya
Tak
Desert
Kouenlou
LITTLE BOKHARA
Balti
Ladak
Leh
Nobra
Pang-koong
Cashmere
Islamabad
Attock
Peshawer
CABUL
Bamian
Ghiznee
Candahar
Herat
Furrah
Dooshak
Seistan
Afghanistan
Kohat
Dera Ism! Khan
Kahoree
Soliman Ms
Moultan
Lahore
Umritsar
PUNJAB
Belaspore
Bhutneer
Bickaneer
DELHI
Meerut
KELAT
Jhalawan
Khozdar
Sarawan
BELOOCHISTAN
Mekran
Kohistan
Bumpoor
Gundava
Shikarpur
Meshed
Nishapoore
Toorshect
Kelat
Nissa
Abiverd
Kaboochan
KHIVA
Khizarist
Urghendi
KHARESM
Mangut
Kazabad
Kurkusa
Shaktemir
Old Urghendi
Konrat
Selzira
Caravan
Kultchuktam
Kalpaks
Kizil Klak
Juz Kuduk
Kushkat
Bala Hissar
Kara
Wakend
Vale of Sood
Kermina
Karakul
Andkhoo
Meimoona
Ghoor Ms
Ghoree
Kundez
Talikan
Durwaz
Wachan
Karategin
Tashbalik
Hindoo Koosh
Sub zawar
Ghorian
Ghayn
Beerjoon
Tubbus
Toon
Ispuk
Neh
Khubbes
Teheroot
Bumm
Regan
Surhud
Booloo
Desert
Moostoong
Shawl
Kwettah
Sebee
Lheree
Dadur
Bagh
Zuhuree
Punjgoor
Kohuk
Kussurkund
Sirroo
Keeh
Teez
Busman
Gull
Jalk
Kharan
Much Ms.
Soherab
Noshky
Chupper
Tull
Boree
Kund
Sira Killa
Kojeh
Amran Hills
Urghundab
Gooshta
Girishk
Dilaram
Bakwa
Zamin Dawer
Saksir
Deh Zungee
Deh Koondee
Logur
Jellalabad
Muzufferabad
Kishtewar
Jummoo
Chamba
Koolloo
Mundee
Tandee
Rawil Pindee
Attock
Bimbur
Wuzierabad
Dera Deenpunah
Toolumba
Ludhiana
Umbala
Hansee
Futtehabad
Hurdwar
Almora
Kumaon
Sreenuggur
Gurhwal
Jumna
Ganges
Rampur
Bareilly
Aligurh
Chooroo
Rewaree
Narnoul
Madoopr
Beekanpr
Jesulmeer
Desert
Bahawulpur
Ooch
Mittenkote
Thanote
Alamudpr
Surwur
Manas sarowar
L. Rawan Rhud
Gortop or Gurdokh
Mansarowar
Chanthan
Rodok
Gheea
Spiti
Dunkir
Garetop
Himalaya
Dhawalagiri
Mustang
Sabana
Nepal
Tibet
Ladak

time the infrastructure argument has been renewed and excavated from the imperial toolbox: most recently in the controversial debate about the 'case for colonialism' where infrastructure development and cost of administration played a prominent role.[8] Opponents of this view calculate the high price that had to be paid locally for infrastructural assets that served extractive and external purposes and contributed to structural deformations.[9] While both arguments are often made from a bureaucracy-dominated and political stability-related point of view centred in capital cities, the analysis falls short when it comes to present evidence from remote and less productive regions.[10] Evidence has been mainly provided from the centres of economic activities and power. Remote rural regions gained attention only as areas of mineral extraction, forest depletion or exploitative agricultural plantation zones. High Asian mountain regions were mainly excluded from such considerations as they often functioned mainly as political buffer zones which attracted geopolitical interests as corridors to channel transition flows or to ban exchange. Infrastructure development that took place in mountains was an investment in a passage and thoroughfare rather than in a specific space lying in between. Infrastructure assets have been transformed over time; their roots and impact need to be considered in terms of long-lasting effects that have affected internal social structures and external exchange relations.

A significant mismatch exists between the perceived importance of these microstates, their role in the imperial design and degree of interference on the one hand, and the impact of colonialism for internal transformations and path-dependent developments on the other. The plain argument would be that Hunza and Nagar were important bottlenecks en route to gaining spatial control in a wider territory across the Karakoram Mountains. Once they were conquered, the realms became in no time compliant 'subjects' and behaved as submissive principalities within the British sphere of influence. Borders were designed and partly demarcated – although the question of international and outer boundaries deliberately remained vague. Internal administration followed a prescribed and tested role model of indirectly ruled entities. The cost of their domination was kept rather low – as envisaged for any remote mountain valley of limited importance – because subsidies for the ruling dynasties and few elite groups remained quite moderate. In the aftermath of the conquest, British attention immediately

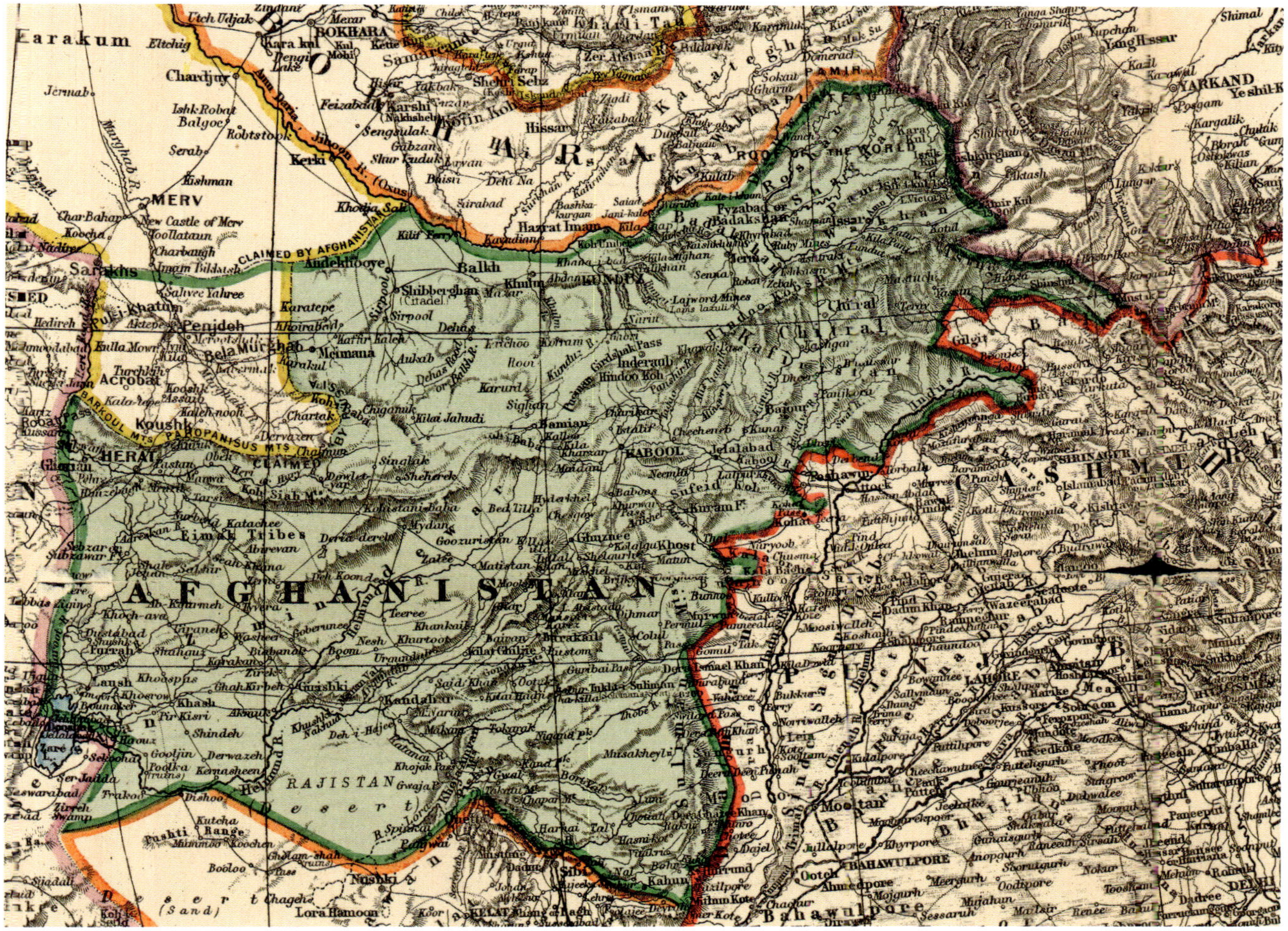

The map 'Central Asia – Afghanistan and her relation to British and Russian territories' was prepared for the Library of Congress in 1885 and shows peculiar boundaries of Afghanistan prior to the delimitation of the Durand Line. In the represented section the northeastern appendix of Afghanistan is extending to territory north of Gilgit, attributed and coloured green as part of the Afghan Kingdom. Thus, Chitral, Ghizer, Hunza and Nagar are included in Afghanistan and seem to be frontier territories to the red line that represents the British sphere of influence. The New York-based company G. W. & C. B. Colton was documenting this worldview that would be completely amended in less than a decade. Courtesy of the Pamir Archive Collection

drifted and shifted to other locations within the Great Game arena; Afghanistan, Chitral and Kashgar remained in focus as contested hotspots for quite some time.

In the case of the Hunza Valley it is important to analyse how its opening-up through road development – from a bridle path for Central Asian trade to the China-Pakistan Economic Corridor – has affected the inner structure of the microstates which have recently become separate and self-contained districts within Gilgit-Baltistan's administrative setup. In the following, I shall attempt to analyse the transformative processes connected with the fickle implementation of more sophisticated passages to Hunza in order to gain further insight into the changing political status from autonomous microstate to administrative district, the impact of external players on internal change processes, and the transforming social setup ranging from inherited elite group dominance to merit- and wealth-based social stratifications. All processes mentioned were not necessarily replacements of previous structures in the course of transition from pre-colonial to post-colonial times. Certain structures enjoyed a higher persistence than others. Consequently, we shall encounter remnants and find relics with a longer than expected shelf life. Two questions are connected with such an observation: Are the unwritten laws of community setup bygones? Has vulnerability reached the valley or is the community linked to communal, national and global networks to such an extent that previous bonds are irrelevant and have been superseded by new linkages? At least two different models coexist in the Hunza Valley: on the one hand, inherited bonds with sets of rules and obligations

Opposite
Access to the Hunza Valley was a challenging affair. The sketch by Richard Caton Woodville titled 'The trouble in Gilgit. On the march' illustrates the conquest of Hunza and Nagar during the campaign of 1891. The difficulties posed by the mountain terrain and the prevailing military hierarchies distinguishing British officers and 'native' soldiers are depicted. Source: The Illustrated London News, 20 August 1892, page 236

In the framework of beginning negotiations about the border alignment with China Pakistan's President M. Ayub Khan realised Hunza's importance and paid a visit to the Gilgit Agency and Hunza in February 1962; he promoted mīr Jamal Khan to Major-General. Source: Private Collection

provide help in times of need and, on the other hand, new degrees of freedom have emerged that offer opportunities in terms of educational development, out-migration and remittances. Multilocality seems to be one answer to previous seclusion and remoteness. To support these considerations it could be argued that isolation never existed, only the degree and quality of interference have changed over time. A sequence beginning with pre-colonial relations to neighbouring powerhouses would prepare the ground for colonial interventions by rule-makers and administrators. From a state of 'apprehensive territoriality'[11] that is linked to undefined boundaries, uncertain power relations, and unsolved administrative responsibility, a troubled path led to post-colonial attempts to integrate Karakoram mountain communities as part of the former Gilgit Agency and present-day Gilgit-Baltistan into the Pakistani state. Spatial relations transgress these confines; inhabitants from Gilgit-Baltistan cross borders into Central Asia for different purposes; households are dependent on proceeds and remittances that are generated in a multitude of locations within Pakistan and beyond. Such relations have changed over time; certain policies and resulting structures have been instrumental in enabling these adaptations and transformations.

Modernisation strategies as instruments for access, control and exchange

Constructing the Hunza Road, the Karakoram Highway and the China-Pakistan Economic Corridor impacted on the valley's communication and exchange relations in quite different ways that appear as structural features in a multidimensional time-space continuum. Technologies of their respective times materialised in visible symbols of uni-directional intervention. All features of modernisation were initiated outside the mountain valleys and followed external interests characterised as 'pacification', 'opening-up', 'development', and 'globalisation'. Planning and implementation followed paramount goals and strategies in which mountain communities filled the gap between major markets and players.

The first telegraph line that connected Misgar – functioning as the northernmost post-office in Hunza – by wire with Gilgit, Srinagar and Delhi in 1918 eased outward communication. A side-effect was that it enabled the ruler to exercise internal surveillance with the establishment of wire stations in all Hunza villages along the line. Henceforth the *tham* could easily communicate with his local agents and office-bearers. Information was controlled and exchanged on a daily basis; hitherto such a speedy news exchange had been unknown. Sixty years later, the Karakoram Highway symbolised China's and Pakistan's mutual strategic interests of spatial control and safeguarding their respective spheres of influence. External interest had triggered the road development; in the final years of President Ayub Khan's reign, *mīr* Jamal Khan had become suspicious of the pace of change and condemned his pro-Chinese policies which were later endorsed by his successors. Allegedly 15,000 Chinese workers were active in the Hunza Valley. He disclosed his reservations at the beginning of the 1970s while flying back from Karachi to Rawalpindi in the company of the American ambassador Farland. The diplomat related that the *mīr* feared that the Chinese would 'eventually take over Hunza'.[12] In his helpless position *mīr* Jamal Khan told him that he would attempt 'to enlist the aid of the Aga Khan in warning the Pakistan Government where their pro-Chinese policies might lead'.[13] The winds of change seem to have become stronger and *mīr* Jamal Khan obviously feared to lose his authority and strong influence in Hunza. It seems to be somewhat coincidental that infrastructural transformations and political reforms would occur around the same time. The processes leading to this point of departure to a different destination demand greater scrutiny in the course of my argument. Persistence and change are accompanying the construction of the Karakoram Highway which initiated a new phase of material and personal exchange relations by triggering mo-

The central oases of Hunza and Nagar before the advent of motor transport. Rakaposhi (7,788 m) towers above Nagar. The village lands are clearly demarcated by the upper irrigation channels and the deeply incised canyon of the Hunza River. Orchards and tree plantations are distinguishable from the cultivated lands for crop-farming and irrigated meadows. Source: German-Austrian Himalaya-Karakoram expedition on [illegible]7 July 1954. Courtesy of Gerhard Klamert

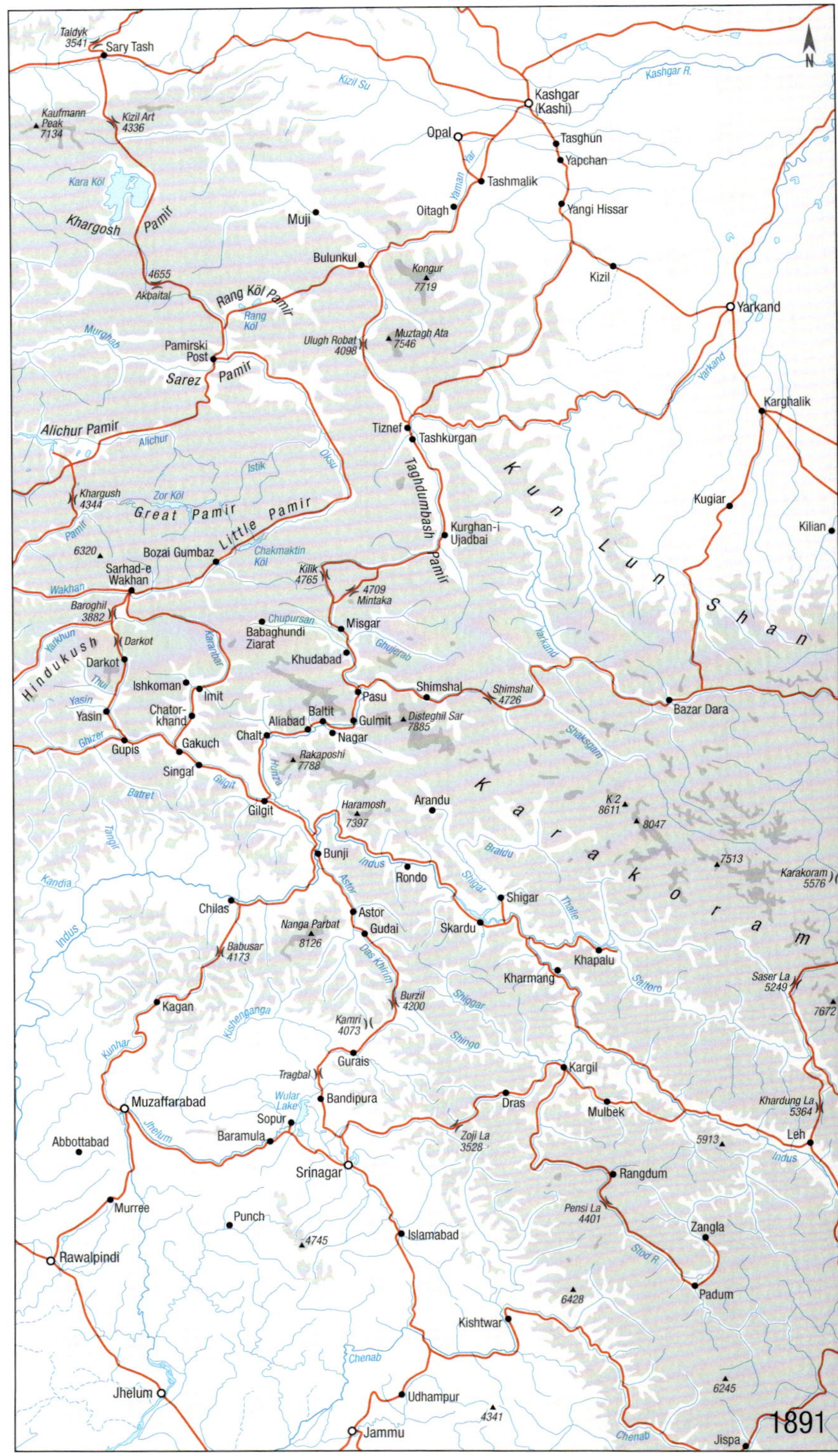

For 1891, the routes taken by traders and explorers are represented. Some tracks allowed only foot travel while most of them permitted pack animals to be taken along.
By 1978, motor transportation had been in use along the Pamir Highway since the 1930s. Motorised transport was possible in the Tarim Basin as well. Places like Gilgit received the first light vehicles in 1949; from the 1950s onwards regular supply via Babusar Pass was possible. Tashkurgan was connected with Kashgar in 1958 and, not much later, the Jammu to Srinagar road was extended into Ladakh.
In 2018, the Chinese road network has made a great leap forward; Xinjiang is now endowed with a web of high-quality highways that connect with the China-Pakistan Economic Corridor as well as with the cross-boundary roads towards Tajikistan, Kyrgyzstan and Kazakhstan that lead further into Eurasian countries in the west. India and Pakistan have improved the quality of their road systems in the plains, whereas little progress is seen in the Himalaya, Karakoram and Hindukush. Source: Based on Hermann Kreutzmann (2015a: 120) and updated from various sources.

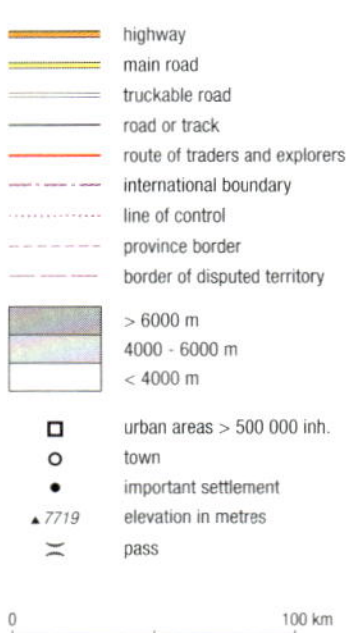

bility of goods, knowledge dissemination and human migrations.

After a century of wire-based exchanges we are now faced with a new phase that Aziz Ali Dad has termed a 'game-changer' for future exclusion or participation in decision-making processes and well-being.[14] A glass-fibre cable was put into the ground along the Karakoram Highway in 2017, thus connecting the Chinese telecommunication grid from Kashgar with down-country Pakistan's network.

Postal runners carrying letters (*ḍāk*) and secret files in a mailbag were instrumental in closing the information gap between Misgar and British India's northernmost consulate in Kashgar from 1918 onwards. After the Chinese Revolution had reached Xinjiang the border to Pakistan had been closed and communication interrupted since 1951.[15] Eighteen years later the first barter trade recommenced, while a regular postal service across Khunjerab Pass was only reopened by motor transport in September 1985. Since then mailbags have been periodically exchanged across borders.[16] Motor traffic on the Karakoram Highway follows the same route; the main artery has been the only functioning cross-boundary communication line for heavy traffic and passenger vehicles between Central and South Asia.

The Pamirian Crossroads were a hotspot of imperial boundary-making during and after the 'Great Game'. The search for natural and/or scientific boundaries hardly disguised the quest for territorial appropriation that was qualified in areas of direct and indirect control. The latter often functioned as territorial buffers separating imperial powers. Thus, armed conflict between the major actors was supposed to be avoided, while the buffer areas were constrained in their external affairs, neutralised in international relations and augmented by subsidies that fulfilled the function of strengthening the ruling elite in order to guarantee peaceful conditions and stability within the buffer zone. The discussion about boundary-making and its

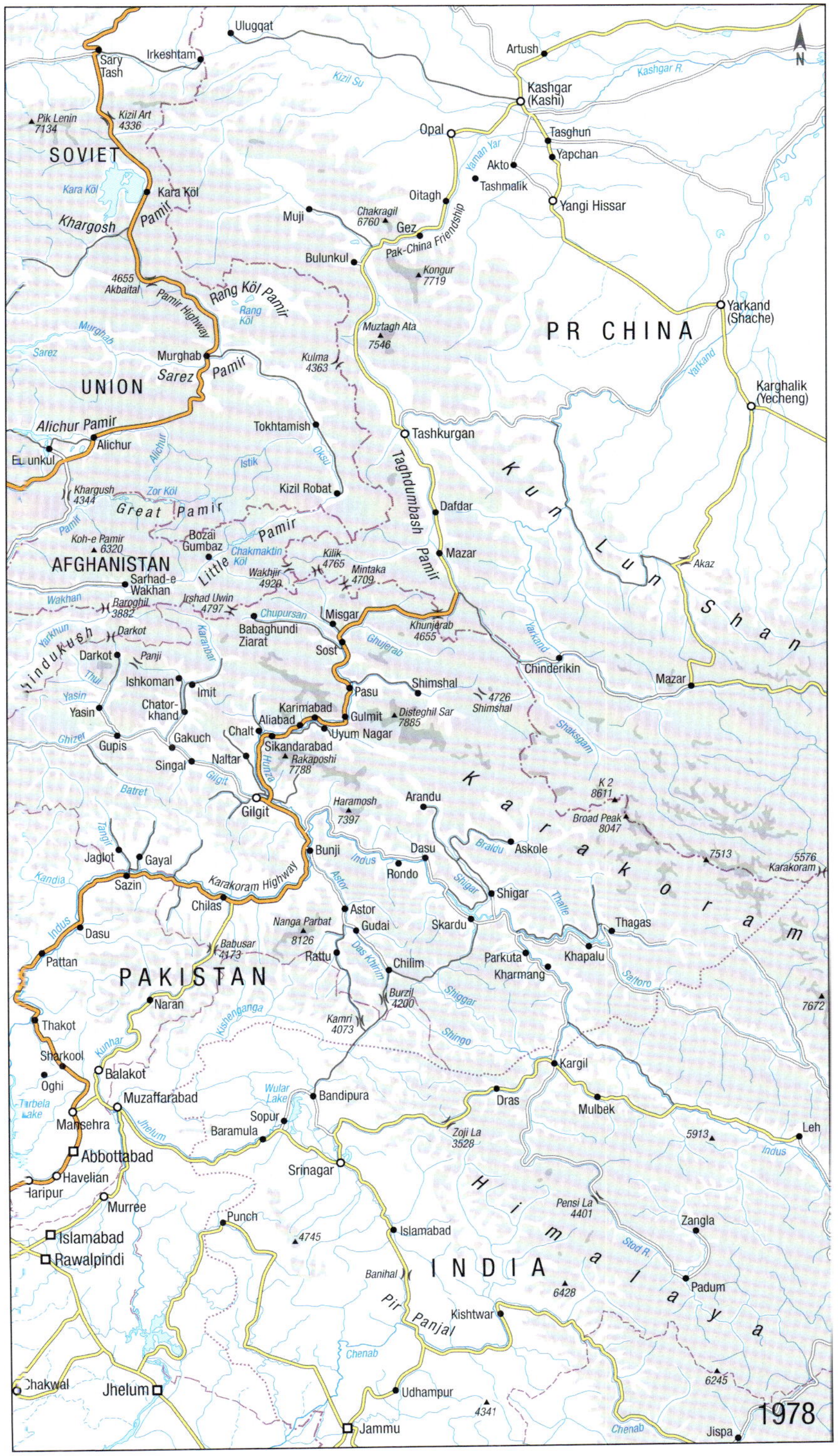
Ulugqat
Artush
Irkeshtam
Sary Tash
Kashgar (Kashi)
Kizil Su
Kashgar R.
N
Pik Lenin 7134
Kizil Art 4336
Opal
Tasghun
SOVIET
Yapchan
Akto
Tashmalik
Kara Köl
Kara Köl
Oitagh
Yangi Hissar
Khargosh
Pamir
Muji
Chakragil 6760
Gez
Pak-China Friendship
Bulunkul
Kongur 7719
4655 Akbaital
Rang Köl Pamir
Pamir Highway
Rang Köl
Muztagh Ata 7546
Yarkand (Shache)
PR CHINA
Murghab
Sarez Pamir
Kulma 4363
Karghalik (Yecheng)
UNION
Alichur Pamir
Alichur
Tokhtamish
Tashkurgan
Bulunkul
Kizil Robat
Kargush 4344
Great Pamir
Taghdumbash Pamir
Dafdar
Kun Lun Shan
Koh-e Pamir 6320
Bozai Gumbaz
Little Pamir
Chakmaktin Kol
Killik 4765
Mintaka 4709
Wakhjir 4923
Mazar
Akaz
AFGHANISTAN
Sarhad-e Wakhan
Wakhan
Baroghil 3882
Irshad Uwin 4797
Chupursan
Misgar
Khunjerab 4655
Babaghundi Ziarat
Darkot
Hindukush
Darkot
Panji
Sost
Chinderikin
Mazar
Ishkoman
Imit
Pasu
Shimshal
4726 Shimshal
Yasin
Chator-khand
Karimabad
Aliabad
Gulmit
Disteghil Sar 7885
Uyum Nagar
Ghizer
Gupis
Gakuch
Chalt
Sikandarabad
Rakaposhi 7788
Singal
Naltar
Gilgit
Batret
Gilgit
Haramosh 7397
Arandu
K 2 8611
Broad Peak 8047
Jaglot
Gayal
Bunji
Dasu
Askole
Braldu
Karakoram
7513
5576 Karakoram
Sazin
Karakoram Highway
Rondo
Indus
Shigar
Chilas
Astor
Kandia
Dasu
Nanga Parbat 8126
Gudai
Skardu
Thagas
Babusar 4173
Rattu
Chilm
Parkuta
Kharmang
Khapalu
Pattan
PAKISTAN
Naran
Burzil 4200
Saltoro
7672
Thakot
Kamri 4073
Shingo
Sharkool
Kunhar
Kargil
Balakot
Ogi
Muzaffarabad
Wular Lake
Bandipura
Dras
Mulbek
Tarbela Lake
Mansehra
Sopur
Baramula
Zoji La 3528
5913
Leh
Jhelum
Abbottabad
Srinagar
Indus
Havelian
Haripur
Murree
Himalaya
Pensi La 4401
Punch
4745
Islamabad
Zangla
Islamabad
Rawalpindi
Banihal
INDIA
Stod R.
Padum
Pir Panjal
6428
Kishtwar
Chenab
6245
Chakwal
Jhelum
Udhampur
4341
Jammu
Chenab
Jispa
1978

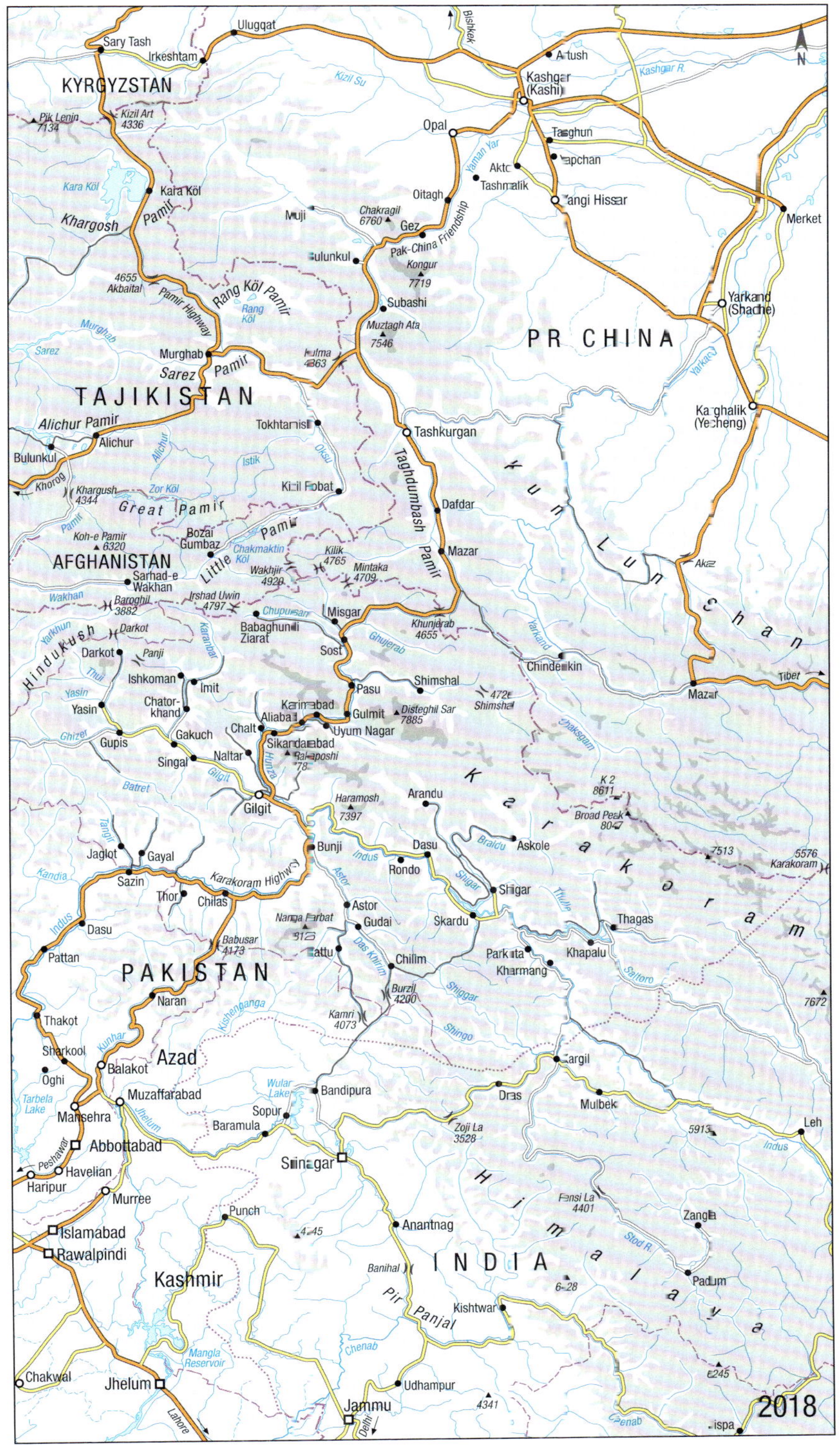
Bishkek
Ulugqat
Sary Tash
Irkeshtam
Artush
KYRGYZSTAN
Kashgar (Kashi)
Kizil Su
Kashgar R.
N
Pik Lenin 7134
Kizil Art 4336
Opal
Tasghun
Yapchan
Akto
Tashmalik
Kara Köl
Kara Köl
Oitagh
Yangi Hissar
Khargosh
Pamir
Muji
Chakragil 6760
Gez
Merket
Pak-China Friendship
Bulunkul
Kongur 7719
4655 Akbaital
Rang Köl Pamir
Pamir Highway
Rang Köl
Subashi
Yarkand (Shache)
Muztagh Ata 7546
Murghab
PR CHINA
Sarez Pamir
Kulma 4363
TAJIKISTAN
Karghalik (Yecheng)
Alichur Pamir
Alichur
Tokhtamish
Tashkurgan
Bulunkul
Khorog
Kizil Robat
Kargush 4344
Great Pamir
Taghdumbash Pamir
Dafdar
Kun Lun Shan
Koh-e Pamir 6320
Bozai Gumbaz
Little Pamir
Chakmaktin Kol
Killik 4765
Mintaka 4709
Mazar
Akaz
AFGHANISTAN
Wakhjir 4923
Sarhad-e Wakhan
Wakhan
Baroghil 3882
Irshad Uwin 4797
Chupursan
Misgar
Khunjerab 4655
Babaghundi Ziarat
Darkot
Hindukush
Darkot
Panji
Sost
Chinderikin
Mazar
Tibet
Ishkoman
Imit
Pasu
Shimshal
4726 Shimshal
Yasin
Chator-khand
Karimabad
Aliabad
Gulmit
Disteghil Sar 7885
Uyum Nagar
Ghizer
Gupis
Gakuch
Chalt
Sikandarabad
Rakaposhi
Singal
Naltar
Gilgit
Batret
Gilgit
Haramosh 7397
Arandu
K 2 8611
Broad Peak 8047
Jaglot
Gayal
Bunji
Dasu
Askole
Braldu
Karakoram
7513
5576 Karakoram
Sazin
Karakoram Highway
Rondo
Indus
Shigar
Thor
Chilas
Astor
Kandia
Dasu
Nanga Parbat
Gudai
Skardu
Thagas
Babusar 4173
Chilm
Parkuta
Kharmang
Khapalu
Pattan
PAKISTAN
Naran
Burzil 4200
Saltoro
7672
Thakot
Kamri 4073
Shingo
Sharkool
Kunhar
Azad
Kargil
Balakot
Ogi
Muzaffarabad
Wular Lake
Bandipura
Dras
Mulbek
Tarbela Lake
Mansehra
Sopur
Baramula
Zoji La 3528
5913
Leh
Jhelum
Abbottabad
Srinagar
Indus
Peshawar
Havelian
Haripur
Murree
Himalaya
Pensi La 4401
Punch
Islamabad
Anantnag
Zangla
Rawalpindi
Banihal
INDIA
Stod R.
Padum
Kashmir
Pir Panjal
Kishtwar
Chenab
Mangla Reservoir
Chakwal
Jhelum
Udhampur
Lahore
Jammu
Delhi
4341
Chenab
Jispa
2018

During the celebrations of the Diamond Jubilee of Aga Khan IV several villages greeted their imām with graffiti on mountain slopes. The photograph was taken in Khaiber on 30 July 2017

effects was explicitly discussed in 'Pamirian Crossroads' and the 'Wakhan Quadrangle' in great detail and with particular reference to Kirghiz and Wakhi. The debate will only be resuscitated in the present regional setting and local contexts when relevant issues need to be exemplified.[17] The general and diversified interventions, negotiations and agreements covering the wider region have significantly impacted on the Hunza Valley. Boundaries and their effects in space and time will indirectly be a central theme in the course of the argument. All aspects of development – from pre-colonial to post-colonial periods – have been strongly linked and interrelated with the constitutional, diplomatic, economic and trade-related setup in the borderlands. The colonial project of infrastructure, commercial and rural development easily translated into post-colonial attempts at modernisation and experiments with globalisation that eventually led to the idea and conceptual framework of a China-Pakistan Economic Corridor across the Karakoram Mountains.

The Kashmir conflict arena has remained a salient spatial configuration that has affected and structured the position of Hunza and Nagar in the context of post-colonial bordering and subsequent ordering. After the abolishment of the Hunza State, British observers were convinced that Sheikh Abdullah, who had opted for Kashmir's accession to India by then, would follow a pragmatic approach to solving the crisis. As the leader from the Kashmir Valley he had realised that the intervention in Azad Kashmir and termination of the Hunza State "… virtually amounted to the integration of these areas into Pakistan. He did not appear agitated on this subject; on the contrary he emphasised the geographical, cultural and linguistic differences which separated the people of Gilgit and Hunza from the rest of Kashmir."[18] What was articulated then is still valid today, but this event contributed neither to any consolidation of territorial claims nor to any reconciliation since; the constitutional stalemate and regional conflict constellation continues.

The communities of Gilgit-Baltistan have been trapped in a constitutional limbo or a state of liminality where the effects of undefined borders and disputed boundaries play a prominent role for a prolonged status quo.[19] The indecisiveness of negotiation partners and stakeholders has prevented mutual agreements based on compromises and kept the region in a kind of cul-de-sac position. The provisional status that awaits clarification has become a permanent and regular feature which sometimes is interrupted by armed conflict and prolonged phases of tension in the Kashmir arena.

Furthermore, the Cold War constellation in the second half of the 20th century has catapulted the Gilgit Agency/Northern Areas into a geopolitical pivotal position that initially was characterised by its frontier location towards the Soviet Union and the People's Republic of China. At a time when Pakistan was seen as a bulwark of the Western world international donors were prepared to deliver aid packages for rural development. Later on, Pakistan's Northern Areas adopted the role of bridging the gap between its economic centres in down country and the emerging Chinese markets. The Karakoram Highway was the proper infrastructural asset to symbolise this connectivity and functioned as an enabling structure. Conflict-laden constellations characterised the neighbourhood of this bilateral corridor, thus directly and indirectly influencing the functioning and perception of this commercial and strategic artery.

The Saur Revolution in Afghanistan, the subsequent invasion of Soviet troops, the counteraction by Mujaheddin groups operating from Pakistan territory, and the never-ending crisis constellation in today's Afghanistan have contributed to the special attention given to the Hindukush and Karakoram that has translated into an hitherto unexperienced allocation of development funds. Gilgit-Baltistan was discovered by international donors and government institutions and made up for a difficult start in comparison with other provinces of Pakistan; since the 1980s major development packages have been made available for education and health infrastructure, professional and rural development.[20] The completion of the Karakoram Highway gave a boost and was a welcome asset and symbol in this developmentalist endeavour that was proactively mediated and prominently represented by the Aga Khan Development Network. The various sub-organisations have significantly impacted on the social and physical infrastructure in northern Pakistan.

The roots of modernisation are much older and go back to colonial times when packages for 'modern' crop-farming and animal husbandry were implemented in addition to administrative reforms and road development that served mainly military and strategic purposes.[21] Development strategies inspired by modernisation theory found application in a much wider spatial and temporal context.[22] The general justification for implementing special development packages in northern Pakistan was the assumption that the per capita income in the Northern Areas was only half of that of an average citizen in Pakistan at the beginning of the 1980s. Later on statistical evidence was presented that painted an even bleaker picture for the 1990s when an average per capita income of less than one third of Pakistan's was recorded in the Northern Areas and Chitral. The Gilgit District seemed to have been responding best to the development packages as its per capita income more than tripled up to 2005, although it still was just sixty percent of Pakistan's average after a quarter century of development activities.[23] The validity of this quantitative dealing with achievements is highly questionable as the data were taken from a sample survey executed by the Aga Khan Rural Support Programme and might have grossly misinterpreted the aspect of multi-local incomes and transfer receipts of mountain households.

The disputed state of affairs in boundary-making and the constitutional limbo of the Western High Asian borderlands are vividly depicted in the representation of administrative affiliations and actual control by differing governments. Under international law there appear to be only a few sections that have been undisputed and acknowledged by all neighbours. Source: Modified and updated on the basis of Hermann Kreutzmann (2015b: 281).

The mask dance on Baltit's polo ground was part of the usual entertainment that was put on for the European visitors in 1954. Later on it vanished and only the shaman's dance and the sword dance survived. Source: German-Austrian Himalaya-Karakoram expedition 1954. Courtesy of Gerhard Klamert

SETTING THE STAGE – FOUR PERSPECTIVES

The exemplary selection of the Hunza Valley for this study is linked to a combination of actors and factors. The special setting that has placed the two former principalities of little economic wealth and power in the centre of a contested geopolitical arena has given unexpected attention to the Hunza Valley, which has played a continuous role in terms of cross-boundary communication and trans-mountain exchange. The Hunza Road became a colonial topos that had emerged when access was forbidden and restricted; it came to function as a vital life-line when the northernmost British Consulate in Kashgar needed accessibility and supplies. Ever since, improvement schemes of road conditions have been discussed and led to the Karakoram Highway, which has been reconfigured and rebuilt four times by now and was integrated into the China-Pakistan Economic Corridor as part of the New Silk Road project.[24] Actors have changed over time; in the approach pursued here, the focus is laid on the agents of change and the effects of intervention that trigger adaptations and reactions. The presentation of an evidence-based sequence of economic, infrastructural and political transformations that have affected locally-embedded livelihood strategies as much as external exchange relations is rooted in the specific setting that provides valuable sources along the timeline. Four perspectives are offered that enable a focus on modernisation and path-dependency in terms of accessibility and infrastructure development, resource exploitation and utilisation strategies, agency in development and power constellations, and finally the creation and persistence of myths about the Hunza Valley.

First, accessibility and road-building represent a classical modernisation strategy that is based on the axiomatic paradigm that development is linked to mobility and exchange. Top-down development strategies are reflected in the planning and implementation of infrastructure projects, educational and health facilities, agricultural extension and professionalisation of trades as well as the promotion of tourism. The lack of urban regional centres to absorb rural migrants contributes to the stimulation of out-migration: in our context to highland-lowland mobility. All these features have been observed in state-led planning and external development programmes initiated by philanthropic and private sector organisations. The road system has been significantly transformed over time; each phase is linked to a different set of actors and power constellations that contribute to changing sets of factors in need of analysis and recognition.

Second, the setting has been transformed through settlement expansion, as well as mobility of people within the valley and from outside. The integration of migrants stimulated internal colonisation processes; the distinction and stigmatisation of social groups heralded the emergence of social stratification that eventually amplified conflict constellations. The structuring of the Hunza Valley through settlement in irrigated oases located in different agro-ecological zones and the inclusion of natural pastures in the cultivation strategies resulted in quite diverse resource configurations.[25] The exploitation and utilisation of local resources have resulted in a process of expansion and attempted optimisation that has been exposed to a wide repertoire of rural development packages and agricultural extension services. At the same time a grave misunderstanding of these economies as being solely based on combined mountain agricultural practices – irrigated crop-farming in combination with animal husbandry – has perpetuated for decades the assumption that mountain farming could provide sufficient sustenance for a growing population. This misconception neglected the importance of non-agrarian local and regional resources that have always augmented the income base of mountain farmers. In addition, the high degree of mobility in mountain communities, the widespread tapping of resources, and the significance of remittances transferred to mountain households have remained underestimated in most development projects that focused on the rural setting, disregarding the multiple localities outside the village settings that contributed to household incomes. With the advent of the China-Pakistan Economic Corridor and the awakening of domestic tourism a growing number of entrepreneurs have become aware of a 'gold rush' situation that motivated outsiders to fathom the potential for quick and substantial monetary returns. Underutilised regional potentials have recently entered the focus of scrutiny. Mineral resources and service industries such as tourism and trade have aroused the interest and desire of newcomers, thus creating a new constellation of external challenge and local response.

Third, the observation that agency in development is strongly linked to power constellations seems to be a permanent and persistent feature in the mountain communities concerned. The *tham* as initiators of irrigation projects and rural colonisation occupy a central part in the collective memory of Hunza and Nagar. They stimulated spatial appropriation,

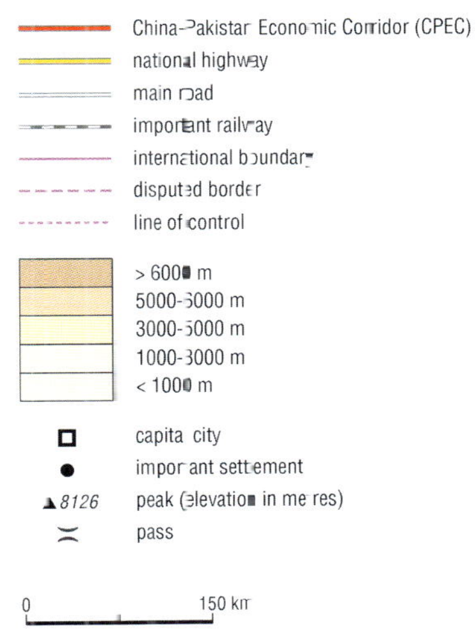

The China-Pakistan Economic Corridor essentially depends on the thoroughfare and quality of the Karakoram Highway. In down-country Pakistan various routings have been discussed. It remains to be seen whether all variants will be implemented in future. The terminus of all is Gwadar Port. Design by Hermann Kreutzmann

Prime Minister Zulfiqar Ali Bhutto and his wife Nusrat visited Hunza in September 1974 and were hosted by mīr Jamal Khan. His son Ghazanfar had joined diplomatic service in the Foreign Office. Source: Private Collection

the settling of migrants and refugees, and induced infrastructure development such as building bridges and roads. After colonial intervention, the British-Kashmiri administration appointed all subsequent *tham* as 'indirect rulers' with limited decision-making powers, backed by tightly structured subsidies and sufficient protective forces. The semi-autonomous *tham* functioned as implementers for externally designed projects and maintainers of the administrative, military and trade infrastructure. The general setup did not significantly change after independence; it was only after the abolition of Hunza and Nagar states and the demise of both *tham* by 1974 that structural changes in power relations were introduced while hereditary influence partly survived. Parallel systems of surviving dependency relations within local communities and regulated bodies of representation and decision-making have existed and shifted power relations in a pendular manner of movement. The constellation has become more complex as religious affiliations have influenced internal power relations and offered opportunities for surmounting traditional social barriers. The obvious contradiction between a strengthening of hereditary rule by investing communal functions in it and promoting modernisation with attractive options for social mobility has remained a characteristic of the Ismaili communal setup in Hunza and a challenge for developmental and societal transformation.

Fourth, several myths were created about the Hunza Valley that require reflection in terms of romanticism and connotations of brave and resilient mountain dwellers enduring their fate in isolated valleys. The standard picture involves impoverished residents following traditional occupations in a stagnant society. Outsiders brought their expectations and perspectives into the valley, resulting in the cynical constellation of promoting a longevity myth in an area of grave food deficiency and starvation. The visitors' search for a Shangri-La in the Karakoram lifted Hunza into a state of happiness, crimelessness, order and contentedness that was exploited by the ruling elite for profits in the hospitality business. Both myths entered the Hunza Valley from outside and uplifted the one group – the Burusho of Central Hunza – at the expense of pejoratively diminishing the others. The discriminated groups were mainly composed of Nagarkuč and Wakhi. The myth creation by outsiders has matured into a myth adoption and instrumentalisation by insiders. Its function needs to be analysed in terms of disguising structural deficits and taking advantage at the expense of others. The longevity and happiness myth is not the only one centred in the Hunza Valley. More recent myths are related to political affiliations and external influence. Adversaries of different backgrounds suggest either that disguised powers clandestinely attempt to establish an Ismaili state[26] in Hunza or that Gilgit-Baltistan has been occupied by Chinese troops that try to incorporate northern Pakistan into the Chinese realm.[27] These allegations will be treated under the myth headline as no evidence whatsoever could be provided. However, the debates shape certain perceptions and reactions that should be contextualised with the state of liminality in which the region's constitutional position has been lingering without seeing significant representational changes for the near future.

These four perspectives – road, resources, actors and factors, and myths – structure the following chapters, which present an evidence-based approach to promote my arguments about the adaptation and persistence of colonial structures and thinking in a setting of path-dependent development that leaves limited leverage for fundamental and significant change in personal dependencies and power relations. The recent conflicts among the offspring of the last *tham* of Hunza – mutually challenging their respective claims and positions, overrunning their inherited influence and exaggerating their supposed authority – shed light on the limbo in which the communities of the Hunza Valley are pending. Tentative rules and regulations, battles for positionality and representation, misconceptions of authority and power are the main characteristics of relations between actors and stakeholders. All four perspectives open viewpoints from different angles on the persistence of long-established clientelistic relations, attempts to overcome such structures, opportunities and niches that have been offered, and the mythical instrumentalisation of perceived and reclaimed superiority. The 'red thread' of event perception is connected to ideas of modernisation that have become perceivable first of all in infrastructural changes, followed by resource exploitation. Consequently, roads have been chosen as visible witnesses of external interventions that have affected all walks of life. Opportunities for additional local income generation as well as the ability to leave the valley and change one's own state of affluence have affected previous social stratifications, the effectiveness of inherited authority, and the emergence of a new group of influential stakeholders. Consequently, a thorough investigation will be provided into actors and change agents such as the previous elite during *thámkuṣ*, rooted in colonial support structures and post-colonial schemes of recognition and political representation within the liminal state institutions. External support projects have been embedded in development packages that have persisted since colonial times and are linked to religious affiliation as much as to political instrumentalisation. The scope for semi-autonomy and independent manoeuvring needs to be assessed in terms of adopting ideas of modernisation.

Endnotes of pages 39–57

1 Edward Frederik Knight (1893: 481–487).
2 Gerhart Klamert (1959: 53).
3 The narrative of the event is based on several interviews with participants and observers in Karimabad in 2017 and 2019.
4 Sultan Abbas and Aziz Ali Dad (2017); Mueezuddin Hakal (2019).
5 The property purchased by Brigadier Aslam in the village of Gircha has remained in the same state since then. Except for a boundary-wall no further construction has taken place.
6 Statement from the report by the British High Commissioner to Pakistan Cyril Pickard on 'Pakistan. Visit to Hunza', dated 11 October 1968 that was labelled 'Confidential – U. K. eyes only' (FCO 21/71: 4).
7 Patrick Vans Agnew (1915); John Biddulph (1880), Bronislaw Grombchevsky (1888/2015).
8 This debate was again stimulated by the provocative article by Bruce Gilley (2017).
9 Shashi Tharoor (2016) managed to summarise central aspects of colonial exploitative schemes and deformation strategies in his well-received book.
10 See Edwin Ardener (2012); Hermann Kreutzmann (2015a: 32–60).
11 Wilhelm van Schendel (2014).
12 Minutes by Ian Sutherland, Head of South Asia Department in the Foreign Commonwealth Office, dated 11 May 1970 (FCO 37/694).
13 FCO 37/694.
14 High Asia Herald, 13 September 2017.
15 IOR/2/1085/295: Establishment of regular postal service between India and Sinkiang. Conditions of service of mail runners on the Misgar-Tashkurgan section (1936–1939); Hermann Kreutzmann (2015a).
16 The Muslim, 11 August 1985.
17 Hermann Kreutzmann (2015a: 51–125; 2017a: 20–56; 2017b).
18 Confidential report of 'visit to Jammu and Kashmir' in October 1974 by Duncan Robin Carmichael Christopher being the first British diplomat visiting Srinagar after India's and Pakistan's independence (FCO 37/1453).
19 Antía Mato Bouzas (2012); Chad Haines (2004; 2012); Cayley Hong (2013): Hermann Kreutzmann (2008a; 2013a; 2015c); Alastair Lamb (1991); Anita Raman (2004).
20 Quantitative data are difficult to come by for this assessment as development funds are split and sometimes hidden in a complex budget allocation and internal financial transfers. The Aga Khan Rural Support Programme received the equivalent of close to 89 million US dollars in international aid between 1982 and 2000; its share was quite extraordinary in comparison with other rural support programmes in Pakistan (World Bank 2002: 96–98; Mahmood Hasan Khan 2009: 71–72, 139–141).
21 See the report on 'Economic development of the reconstituted Gilgit Agency. Grant by Government of India' containing the Letter of P. A. Gilgit to Ass. to Resident in Kashmir No. 13-T of 1936, dated Camp Chilas, the 21st February 1936' (IOL/P&S/12/3294) in which development packages are suggested that do not significantly differ from those fifty years later; reprinted in Hermann Kreutzmann (1989a: 225).
22 See for the implementation of development packages and conceptual approaches Geoffrey Wood, Abdul Malik and Sumaira Sagheer (2006); World Bank (2011).
23 All data referred to were taken from Mahmood Hasan Khan (2009: 138); see as well Geoffrey Wood, Abdul Malik and Sumaira Sagheer (2006: 63). The first promotional movie commissioned by the Aga Khan Foundation and titled 'First harvest. Valleys in transition' exactly conveyed this message (Nicholas Kendall and Jack Silverman 1983).
24 See Aga Khan Rural Support Programme (2017); Aziz Ali Dad in High Asia Herald, 13 September 2017 and The News, 5 May 2018; Jeremy Garlick (2018); Government of Pakistan (2017); Hermann Kreutzmann (2015a).
25 See for a detailed discussion Hermann Kreutzmann (2000b; 2004a; 2005b; 2006); Michael Spies (2016; 2018; 2019).
26 See for the broader debate in the Ismailiya's position Michel Boivin (2003; 2013); Faisal Devji (2014); Wolfgang Holzwarth (1994); Faquir M. Hunzai (2004); Soumen Mukherjee (2016); Jonah Steinberg (2011).
27 See Selig Harrison in the New York Times, 26 August 2010; Hindustan Times, 2 September 2010; Senge Sering (2012), and critical assessments by Izhar Ali Hunzai (2013); Hermann Kreutzmann (2015c: 285–286).

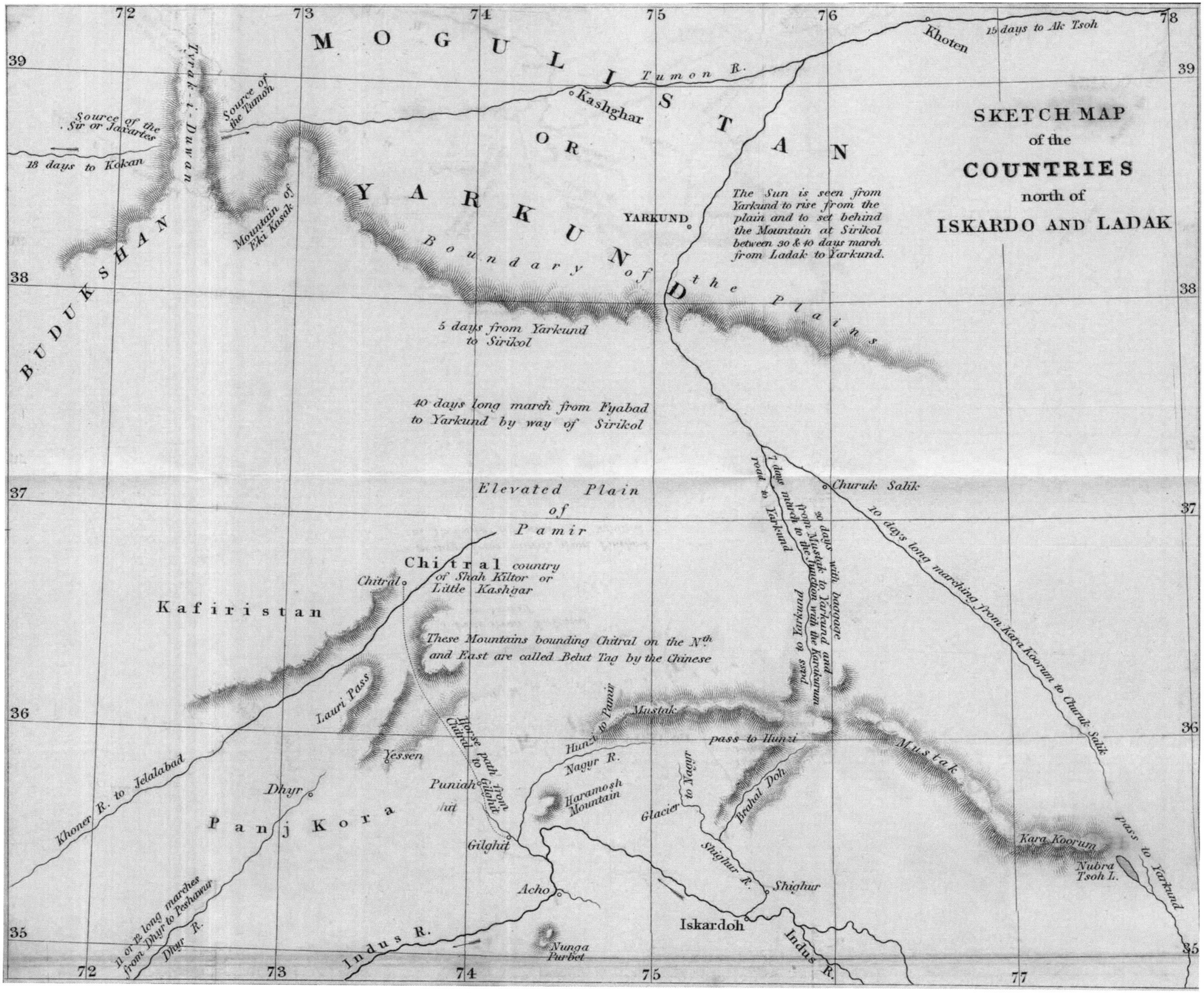
SKETCH MAP of the COUNTRIES north of ISKARDO AND LADAK
MOGULISTAN OR YARKUND
Boundary of the Plains
Tumon R.
Kashghar
Khoten
15 days to Ak Tsoh
YARKUND
The Sun is seen from Yarkund to rise from the plain and to set behind the Mountain at Sirikol between 30 & 40 days march from Ladak to Yarkund.
Source of the Sir or Jaxartes
Source of the Tumoh
Tyrak-i-Duwan
18 days to Kokan
Mountain of Eki Kasak
BUDUKSHAN
5 days from Yarkund to Sirikol
40 days long march from Fyabad to Yarkund by way of Sirikol
Elevated Plain of Pamir
Churuk Salik
10 days long marching from Kara Koorum to Churuk Salik
7 days march to Yarkund
road to Yarkund
20 days with baggage from Mustak to Yarkund and Junction with the Karakorum
pass to Yarkund
Chitral country of Shah Kiltor or Little Kashgar
Chitral
Kafiristan
These Mountains bounding Chitral on the Nth and East are called Belut Tag by the Chinese
Lauri Pass
Yessen
Horse path from Chitral to Gilghit
Mustak
Hunzi to Pamir
pass to Hunzi
Nagyr R.
Glacier to Nagyr
Brahal Doh
Haramosh Mountain
Puniah
Gilghit
Dhyr
Panj Kora
Khoner R. to Jelalabad
11 or 12 long marches from Dhyr to Peshawur
Dhyr R.
Acho
Indus R.
Nunga Purbet
Shighur R.
Shighur
Iskardoh
Kara Koorum
Nubra Tsoh L.
pass to Yarkund
39
38
37
36
35
72
73
74
75
76
77
78

From Hunza Road to China-Pakistan Economic Corridor

Routes to roads – changing perceptions

By the middle of the 19th century, the Hunza Road seemed to represent the hope for an alternative access from Kashmir to Kashgar and an opportunity to control a larger area of the northwest of the continent. British India and Kashmir were the dominant forces in this endeavour to enter an area that had managed to maintain its autonomy by preserving a difficult physical approach. The road had become a symbol of access, although it was barely a dirt track for pack animals and porters. A century later, the Karakoram Highway could be perceived as the materialised colonial dream of accessing and opening up an area of attempted control. Considerations about providing the necessary infrastructure to control and dominate a region of doubtful material wealth but strategic importance were inspired by spatial thrusts from various directions. A region that fulfilled the function of a buffer space rather than being of importance in itself was part of expanding spheres of influence. Their prime function lay in strategic considerations to keep competitors at bay. Infrastructure development in the 19th century meant the provision of rest-houses, fodder and fuel; sometimes pack animals and/or coolie labour were provided for a quick relay from station to station. The Urdu term *paṛā'o* or in Burushaski *sarúu*, meaning a stage of a journey or a halting-place, was the unit of measurement in penetrating inhospitable regions and separating spheres of influence. The more stages between areas of interest, the better a buffer worked as a distance-keeper; the *paṛā'o* became the unit grid of perceiving unknown territories. Infrastructure development in colonial times enabled military troops and expeditions to move into the inhospitable and remote border regions: colonial parlance often distinguished between settled and non-settled, administered and tribal regions, between plains and hills, religious and pagan beliefs, standardised languages and non-written vernaculars. The Karakoram Mountains match most of these characteristics, and they were long considered to be the antithesis of a suitable region for permanent administration and control. In fact, most of the mountainous northern rimlands of the Indian subcontinent – from Afghanistan to Assam – have been perceived as such inhospitable valleys and hostile areas over long periods.

Nevertheless, the first motor road across the Pamirs was built by the Soviet Union and posed a major challenge to the grid-based approach to remote locations. The British colonial *ḍāk* bungalow, where travellers could stay overnight, where horses were changed and runners handed over the mail bag, was regarded as the comfort zone in the *paṛā'o* from which the next destination could be accessed if road

The Karakoram Highway memorial near Thakot was erected by the Road Maintenance Battalion 142 that was involved in the implementation of the bilateral road project from the Pakistan side prior to the opening. The photograph was taken on 17 July 2013

Opposite
The 'Sketch map of the countries north of Iskardo and Ladak' shows the state of route knowledge prior to Patrick Vans Agnew's pioneering visit to the Hunza Valley. The only specified corridors crossing the mighty mountain ranges were the Pamir route from Chitral and the Karakoram route from Ladakh to Yarkand. A potential connection between the Hunza Valley and the southern Silk Road oases is vaguely indicated by traversing the Pamirs. An additional option to reach the oasis towns existed only further west from Badakhshan and Qoqand in the Fergana Valley. Source: Insert map to 'Map of Kashmir with its passes' surveyed by Godfrey Thomas Vigne and published by John Walker 16 May 1842. Courtesy of the Pamir Archive Collection

Road conditions near Nilt in Nagar during 1934. Narrow passages were defence measures and control posts at the same time. Source: Photo collection of David Lorimer during his journey from Srinagar to Aliabad in June 1934; SOAS Library Box II.1

conditions permitted. Some called the Pamir Highway a threat to British domination in the mountainous buffer zone where they were supposedly in control of communication and speedy exchange of news. The 20th century experienced the transition from step-by-step stages in covering major distances to a homogenous perception of distance supported by an expanding railway network that connected with a web of motor roads in the next phase.

Accessibility is an asset that has changed its meaning over time. Nowadays it seems to be a precondition for any kind of development and modernisation: mobility and migration require flexible movement. Roads are a common feature and would need no specific attention if they were part of a wider traffic network. In cases where a modern infrastructure does not exist or where a new road connects major traffic systems for the first time, phrases such as the 'eighth wonder of the world' have sometimes been applied.[1] This happened with the Karakoram Highway when it was inaugurated in 1978, thus connecting the Moghul and subsequent colonial Grand Trunk Road of South Asia with the Chinese traffic network of the historic 'silk roads' in Xinjiang for the first time. The People's Daily correspondent Li Yun-fei reported from the opening ceremony at Thakot on 18 June 1978: "'Silk Road' replaced by new highway".[2] At the ceremony General Zia-ul-Haq likened the new road to the Great Wall that 'was regarded as an extraordinary accomplishment in ancient times, the Karakoram Highway will be considered a wonder of the present age.'[3] The lack of a road would not necessarily mean lack of contact. The exclusive and spectacular advertisement of a new road narrates the beginning of a new and modern era, supporting strategic political partnership between neighbours. In a rather young country such as Pakistan, the opening up of a mountain area was an effort of nation-building, or – as a senior research fellow of the Institute of Strategic Studies, Nazir Kamal, described it – the endeavour to bring 'to fruition a national aspiration for physical and with it socio-economic unity'.[4] Whenever such symbolic events take place, historical references are consulted. In the case of the Karakoram Highway, it has predominantly been the metaphor of 'silk roads' that was referred to in a rather unspecific manner and that is reflected in the Wakhi term *šarā-e rešum* for the Karakoram Highway.[5] Epigraphical evidence was taken as such without giving appropriate attention to its interpretation and chronological value. In the silk roads context little evidence is presented about specific events and developments connected with migrating and mobile people, little or bigger mountain kingdoms and changing power structures. Cultural transfer and economic exchange remained rather opaque and were part of an ambiguous historical narrative. Consequently, the political discourse about historical depth singled out the silk roads metaphor and neglected to pay attention to more recent events related to the transformation of communication between communities along preferred routes and the commercial exchange of commodities, goods and people. The route development from the Hunza Road to the Karakoram Highway contains many more aspects that need to be contextualised with political and technological developments outside the region as well with internal power games and diplomatic relations. Colonial interference and competition between neighbouring dominating powers have more or less continuously affected the upgrading of roads and extension of control measures. Accessibility and infrastructure development seem to be rather abstract categories that should be filled with historical evidence related to external interest, power games, societal transformations and economic developments. The Hunza Road (*Hunzae róoṭ*) and its subsequent appellations are the guiding lifeline of enquiry in order to establish a record of an eventful set of interventions over time.

Areas excluded from proper access face severe challenges. Over time the notion of accessibility and its qualities has varied. Going back in history for more than a millennium, the meagre evidence that can be produced has mainly been interpreted in terms of dissemination of knowledge, religions, languages and technologies. Not only epigraphical inscriptions by pilgrims and traders but also missionaries' reports produced and left for posterity circumstantial evidence. These valuable witnesses of pre-modern exchange add historical depth to the connectivity and relationships between places that lie across mountains.

Theories about power structures must consider accessibility in order to grasp the means and modes of control and expansion. Imperial configurations incorporate power structures on various levels. On the highest configurative level, we find subcontinental entities such as the Chinese or Mongolian Empires. Within these imperial settings, numerous smaller entities are embedded with more or less strong bonds to the central seat of power. Distance and accessibility issues allowed the existence of emirates and *mīr*-doms and the survival of microstates and principalities that were subordinated to the centre by regularly demonstrating their loyalty through dues, gift exchange and taxes. Starting at the lowest level of organisation, this could mean that communities that had the advantage of poor accessibility could preserve a certain degree of autonomy and keep direct interference by their powerful neighbours at bay. In fact, their delivery of gifts was often remunerated with presents of higher value from the representatives of the empire. We shall see in the case of Hunza that advantageous and favourable gift exchange was practised with more than one neighbour. Such a quasi-equilibrium could reduce external dependency to a remote control mechanism. Owing to difficult travelling conditions, bad routes and poor accessibility, Chinese and Kashmiri officials have rarely or never been in Hunza.[6] In most cases, the powerful actors of the centre are perceived as setting the political agenda that others have to accept. Enclosure within a dominant system and its incorporation into long-distance administrative structures are assumed to be the fate of all participants if state action pursues its goal of ultimate control mediated through spanning spatial distances and creating accessibility. Beyond central control are 'regions of refuge'[7] that have somehow managed to follow their own rules and regulations. The creation of escape strategies and the search for regions of refuge may be attempts to reduce risk and to avoid conflicts with powerful actors. Successful communities are often very remote, less powerful and poverty-stricken, so strict control would by no means be cost-efficient. The absence of roads might have contributed to maintaining their refuge status and the enjoyment of certain degrees of freedom over time. Similar ideas could be applied to a number of communities in the northwestern mountain ranges and rimlands of the South Asian subcontinent where Central Asian oases can be reached across the high passes. The combination of 'roads and rivals'[8] illustrates that, beyond accessibility, the ability of actors to maintain their status quo of autonomy and/or their successful strategies to avoid external interference with their internal administration needs to be analysed when the persistence of dominance in a given setting is perceived as a reflection of power and political structures.

During his Pakistan visit in February 1964 the first Premier of the People's Republic of China, Zhou Enlai, had discussions with mīr Jamal Khan who had played his part in securing Hunza territory during the negotiations that had led to the Sino-Pakistan Boundary Agreement the year before. As an outcome of this bilateral treaty the future relationship based on cross-border exchange and infrastructure development would emerge in the coming years. Source: Pamir Times, 24 June 2019

In present discussions, accessibility is often a necessary precondition for participation in market-based economies and globalisation.[9] Especially the 19th and 20th centuries are periods when special emphasis was directed towards routes of access, implementation of traffic infrastructure for administrative and military control, development and research. In this context the observation of Robert Chambers seems to be a self-fulfilling prophecy: "Most learning about rural conditions is mediated by vehicles. This applies not only to rural tourism, but also to research. Starting and ending in urban centres, visits follow networks of roads."[10] Consequently, in the following the development of communication and traffic across the Karakoram passes is analysed and introduced through the lens of routes and roads in space and time.

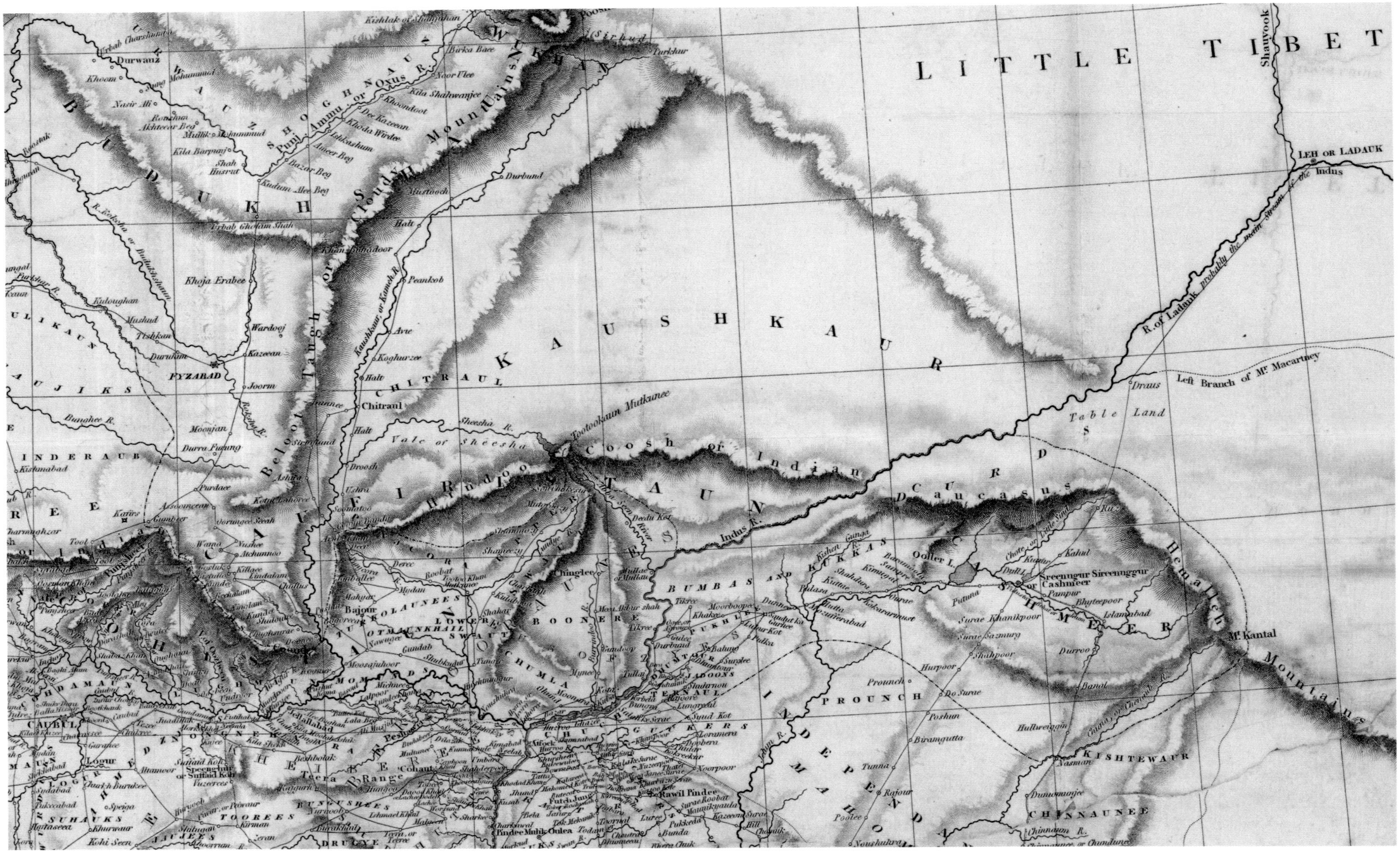

The section from 'A Map of the Kingdom of Caubul and some of the neighbouring countries, altered from a Map constructed in the Year 1809. By Lieut. John Macartney, 5th Regt. Bengal Native Cavalry. A New Edition: Redrawn and engraved with considerable improvements' was published in 1838. The Oxus is bordered in the south by a range called 'Budukhshaun' merging with the 'Beloot Taugh or Cloudy Mountains' that separate Fyzabad from 'Kaushkaur', with the centre in 'Chitraul', and touch 'Caufiristaun'. Further south, 'Cohistaun' leads to the 'Hindoo Coosh' or 'Indian Caucasus'. The major empty spaces north of the crest are 'Kaushkaur' and, further northeast, Little Tibet. Kashkar was a contemporary synonym for Chitral and suggests here that the whole area was controlled by Chitral although no settlements are identified. 'Kafiristaun' is bordered in the south by the Indus River, and 'Sreenugur' in 'Cashmeer' was contained in the Kashmir Valley. No indication of Gilgit and the Hunza Valley is found in the map. Less than a decade later the British would let the Dogra rulers have control over these 'blank' territories which would be codified in the Amritsar Treaty of 1846. Maharaja Gulab Singh is remembered as saying that Gilgit is the gateway for Badakhshan and Turkestan; no other additional value could be attributed to the control of Gilgit (quoted after Siegfried Weber (2007: 39)). Courtesy of the Pamir Archive Collection

Antecedents

Especially in a mountainous setting such as the transition zone between Central and South Asia – represented by a mountain belt – aspects of traverse and cross-mountain passages have been reported by early travellers. During an important period of early communication, the northern route (Sanskrit *Uttarāpatha*) seems to have been the link between Gandhara and Kashmir in the south, and Bactria and Sogdia in the north, appearing as a 'broad collection of flexible arteries on a loosely defined northwest–southeast axis intertwined with multiple feeder routes'.[11] Chinese pilgrims in search of Buddhist wisdom were among the first to publish extensive route reports and descriptions about the challenges they met on their way. These narratives contain a host of valuable information regarding the inhabitants of the valleys and mountains traversed. The sheer height of high passes and glaciated regions were physical obstacles at a time when the equipment was quite basic. The passages without food and water supplies required substantial logistics between stages and the reliability of companions and caravans. Fodder and meadows for pack animals, timber fuel for food preparation in high mountain steppes and deserts were constraining factors in finding suitable routes across the mountains. The Chinese pilgrims used locally well-known routes as possible corridors, but knowledge about them was embedded in mountain farmers and nomads dwelling along those routes and using them to access seasonal pastures. Successors of Faxian, Songyun and Xuanzang could make use of previous experiences.[12] Pilgrims and traders who succeeded in surviving their journeys managed to connect places. At the same time, these mountain regions functioned as peripheral zones of regional principalities, microstates and empires that were interested in keeping at bay potential conquerors and intruders. Their rulers aimed to control all potential passages, blocking routes and refusing travellers' applications for passes. The fear of losing dominance in their fiefdoms to powerful neighbours was quite realistic, and many stories could be told of such events.

Some traverses had become popular, such as the trade route between Kashmir and Yarkand or the routes in the Karakoram that have preserved a wealth of more than 50,000 petroglyphs and 5,000 inscriptions providing ample evidence of activities spanning more than a millennium.[13] A multilingual communication structure and travellers from

The monkey Su Wukong is leading the pilgrim Xuanzang and his companions on their 'Journey to the West' (Xiyouji) which took place between 629 and 645 CE. The popular novel by Wu Cheng'en was published in 1570 and is based on the pilgrimage to Gandhara in present-day Pakistan. Under the protection of the far-reaching Tang Empire Xuanzang's pilgrimage was possible and left detailed descriptions about the routes and the economic, religious and linguistic situation in Inner Asia; see Samuel Beal (1884); René Grousset (1986); Albert von Le Coq (1926); Mishi Saran (2005). The photograph of the monument commemorating his journey was taken in Bezeklik on 25 September 2010

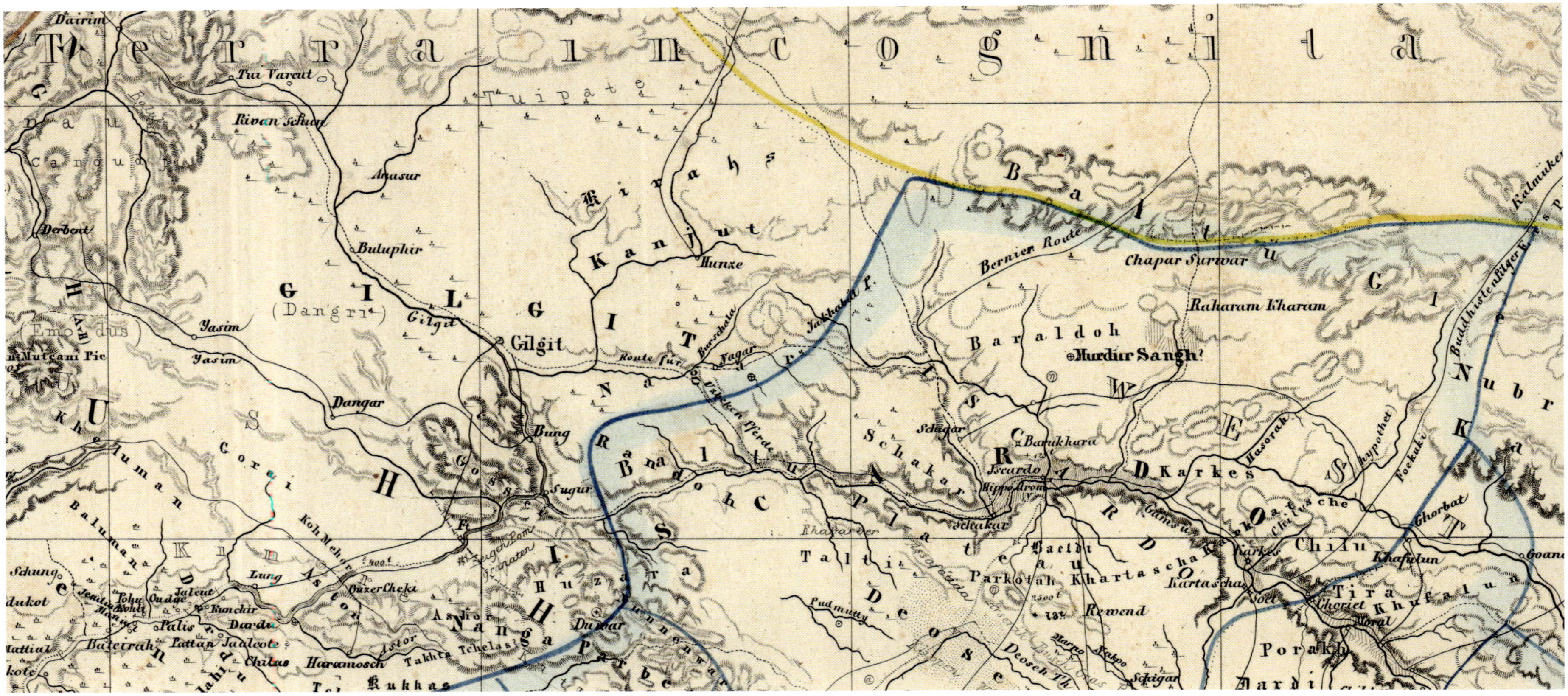

The section from the 'Karte Inner-Asien's zu Carl Ritter's Erdkunde Buch 3' of 1841 was prepared by Carl Zimmermann in Berlin. It provides the cartographic evidence for the compilation work of Carl Ritter who gathered contemporary knowledge and tried to present it in a comprehensive and holistic manner. The 'Zimmermann map' clearly shows the variegation in knowledge for certain Asian areas. Gilgit occupies quite a significant area north of the blue line representing the 'Indian Caucasus' or Hindukush. Hunza is located north of Gilgit with the appellation 'Kanjut' from where a northward continuing valley makes the connection towards 'terra incognita'. Gilgit and Kanjut are contained in the little-known Central Asian realm separating the Kun Lun Shan and the Silk Road oases from the better known Kashmir in the south and Afghanistan in the west. Gilgit and Hunza are part of the 'void' in between. Twenty-five years later Gottlieb Wilhelm Leitner (1889: 73) augmented the list with synonyms such as 'Khajuna, Kanjut, Hunzije, Khajuni' for the Hunza Valley. Source: Karte Inner-Asien's by Carl Zimmermann (1841). Courtesy of the Pamir Archive Collection

various origins are reflected in Sanskrit and Indic languages as well as Middle Indian, Chinese, Tibetan, Hebrew and Syriac inscriptions that are found along the routes.[14] In a welcome new approach Jason Neelis has pointed to the physical and geographical properties of ancient routes and drawn renewed attention to 'capillary networks' that followed a mesh of routes across well-known passes that were used in early Buddhist travels across the Karakoram Mountains.[15] The perception of single dominant routes needs to be replaced by a network of routes that could lead travellers, itinerant traders, pilgrims and missionaries through a variety of valleys and passes. They are proof of a millennium-old connectivity between Central and South Asia. Old Chinese sources document the millennium-old knowledge about the southern rim of their empire, but document as well that traffic across those passes was rather limited and under surveillance.[16] The question remains how these connections continued to develop. Historical evidence is available for a long period from the early reports until the 13th century travels of international visitors that connected Europe with the Mongol courts and contributed to an emergent fascination for these challenging routes. The southern passes leading from the Tarim Basin towards the Karakoram and Himalaya – the Hunza Valley seems to have been bypassed – appeared in the 18th century Qianlong Atlas as a necklace of guard posts fading away at the foot of the northern passes.[17] The depictions suggest quite a thorough control of passages. The main connections followed routes originating either in Kabul or in Kashmir and meeting their northern destination beyond the Karakoram passes in Kashgar and Yarkand.

The 'Apotheosis of War' was drawn by Vasily Vereshchagin, once famous war artist of the Russian Empire in 1871. The theme alludes to the cruel regime of Wali Khan in Kashgar and might have been inspired by the murder of Adoph Schlagintweit there on 26 August 1857; see Hermann Kreutzmann (2015c). The display of skulls relates to the autocrat's practice of executing local enemies and foreign intruders alike. Kashgar – its city wall is supposedly visible in the painting's background – was a forbidden city for British and Russian visitors during Wali Khan's reign. Vasily Vereshchagin became a renowned realistic painter who travelled widely; the 'Apotheosis of War' is part of his second Turkestan series and precedes photographic documentation. Courtesy of the Tretyakov Gallery Moscow

Forbidden routes – explorers as trespassers

In the context of the Hindukush, Karakoram, Pamirs and Himalaya, these routes remained the lines of communication and control well into the 18th and 19th centuries. On Carl Ritter's map published in 1852, the only routes that traverse the *Tsungling* (Onion Mountains) and the *Muztagh* (Karakoram) lead from Chitral to Wakhan and from Gilgit and Nagar to a hub called Kandschat (*Hiundês*) which resembles Kanjut, a toponym frequently used as a synonym for Hunza, from where the route continues to Yarkand.[18] Kashgar had been declared forbidden for foreign traders and visitors when the Qoqandi rulers established their interregnum in Kashgaria in the mid-19th century. The ban on visiting Kashgar motivated adventurous explorers such as Adolph Schlagintweit, employee of the East India Company, and Chokan Valikhanov, disguised as a merchant who was on duty for the Imperial Russian Geographical Society, to attempt a visit to the forbidden city. Whereas Adolph Schlagintweit was murdered in Kashgar in 1857, his rival explorer Chokan Valikhanov succeeded in his daring endeavour and provided St. Petersburg with a detailed report on the conditions in Kashgaria.[19] The two adventurers had approached Kashgar from opposite directions, either from the Kun Lun Shan or from the Tien Shan. The Karakoram passes were out of reach. On a map published in 1842 Gilgit seems to be an isolated place surrounded by white spots while Badakhshan and Baltistan are represented in detail. The map was prepared by the later Superintendent of the Great Trigonometrical Survey of India, Colonel James Thomas Walker, who, a quarter century later, prepared another map titled 'Turkestan, with the adjoining portions of the Russian and the British territories' in which he summarised the existing British and Russian knowledge. In his commentary of 1873, he lauded not only the cordial professional relations of the time, but also the support received from influential circles interested in these maps: "It appeared that our knowledge of the Geography of these regions was almost solely de-

There were only a few foreign visitors to the Hunza Valley prior to the Hunza-Nagar Campaign. Patrick Vans Agnew reached only as far as Chalt in September 1847, followed by John Biddulph who was the first visitor in Baltit Fort in August 1876. William Barrow, William Stephen Lockhart and Robert Gosset Woodthorpe were the prominent members of the 'Gilgit Mission' who wrote about their pioneer travels covering the whole length of the Hunza Valley in April–May 1886. The account of Bronisław Ludwig Grąbczewski's journey from the northern passes down to Baltit during August–September 1888 is among the most detailed route descriptions of their time. Prior to the British military campaign Francis Younghusband was sent to augment the hitherto collected information. Design by Hermann Kreutzmann based on the diaries and reports of the expeditions

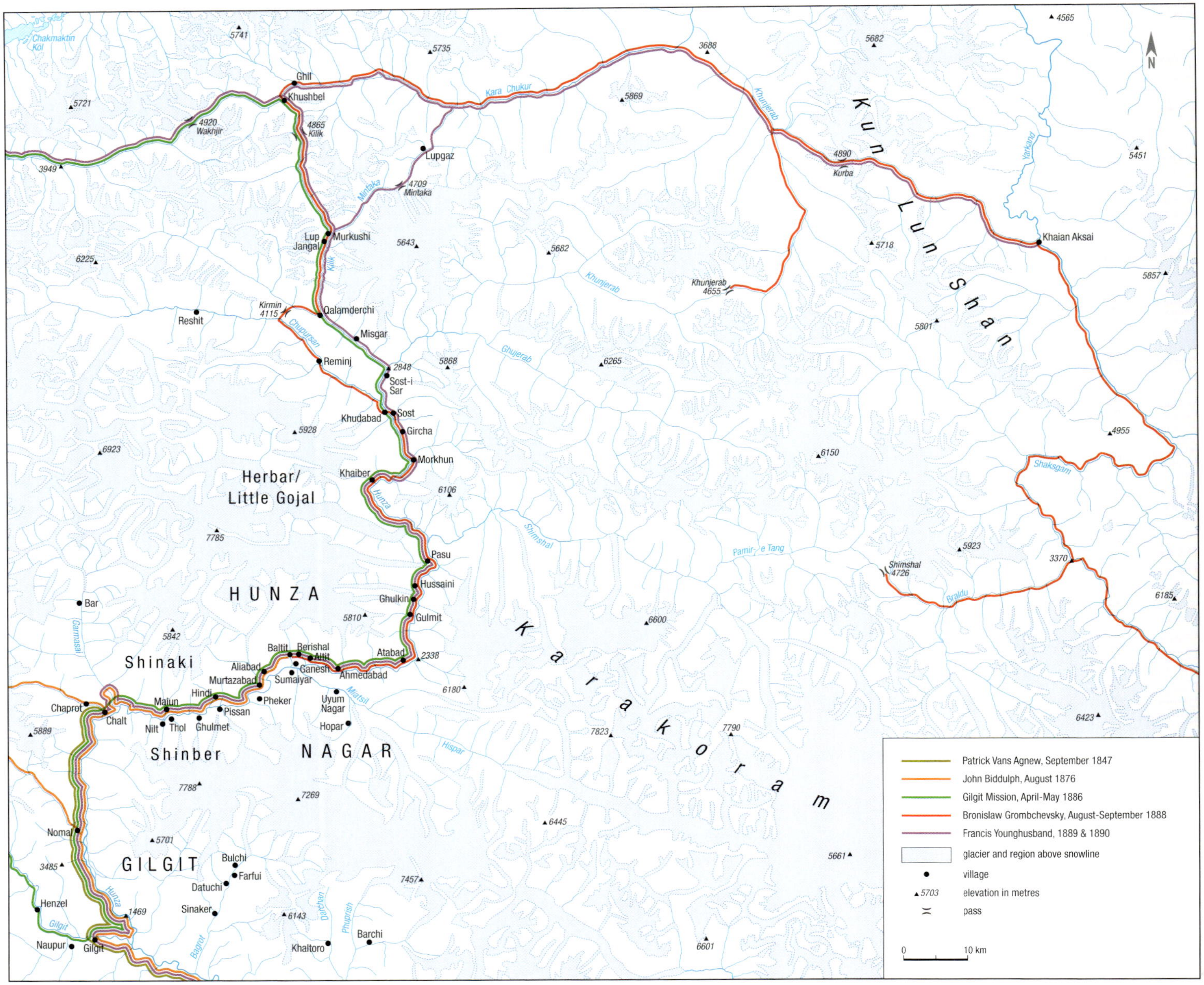

rived from the maps which had been compiled in 1711 by the French Jesuits in the service of the Chinese Emperor ... This state of things being most unsatisfactory, I obtained the permission of the Secretary of State for India to proceed to St. Petersburg, to ascertain whether any more accurate and valuable information might be obtained from Russian sources."[20] Hunza and Gilgit were far away from any supra-regional interest. Subsequent events drew further attention towards the exploration of routes leading from south of the Hindukush and Karakoram passes towards the Tarim Basin.

Pre-colonial route development and the battle for knowledge

The route story changed with the advent of British and Russian military officers on espionage and reconnaissance missions. Promising young officers were selected for the physically challenging mountain trips and delivered extensive reports about their impressions. The content focused on the diplomatic relations with local rulers; the quality of route conditions and their potential feasibility for army movements and relocations also played a major role.

Patrick Alexander Vans Agnew was accompanied by Jasper Edward Winterbottom and Lieutenant Ralph Young on his pioneering journey to Hunza and Nagar in August-September 1847 when he noticed after leaving Nomal: "The valley narrowed much, and beyond the little village of Gooech [Gwachi] becomes a chasm between immense perpendicular cliffs. After with some difficulty getting down a small precipice, where the road is purposely kept difficult by the Gilgitees, we reached an oval basin of cultivation ... A vista in the hills showed the course of a considerable valley in that direction; and a low spur on the other side of the river, intercepting the view, at a distance of about two miles, was pointed out to us as the boundary of Nuggur [Nagar]. At this place the Gilgitees have four or five small villages. These were seized by Shah Ghuzunfur of Hunza on the occasion of Secundur Khan's assassination. Nattu Shah urged him to restore them, but without success. Then, to use his own words, 'I made great friends with him, and married his daughter, and when his vigilance was thoroughly lulled I sent a party and took his six forts' (there is no other word for their walled inclosures) 'in one night, and have kept them ever since.'"[21] The British party came to a halt in Chalt and Chaprot. An earlier invitation to visit Nagar was cancelled, and the party had to retreat after exploring the Chaprot and Garmasai valleys.[22] Patrick Alexander Vans Agnew estimated the population strength of Hunza and Nagar at about seven thousand houses, not such a big danger for a well-equipped British and/or Kashmiri force, although the depopulated Gilgit was supposed to have only one thousand houses left. He reported that 'the only road into Nuggur is across a rope-bridge, where they have a guard.'[23]

Thirty years later, in August 1876, John Biddulph was the first Englishman to visit Hunza.[24] During his week-long stay in Hunza, Tara Beg, son of *wazīr* Asadullah Beg, was held as a hostage in Gilgit.[25] John Biddulph was amazed by the giant mountains and glaciers in the valley. On his way from Shinaki to Central Hunza he experienced the difficult passages of the Hunza Road: "Immediately on leaving Hini the road winds up and down staircases of rock and along the face of steep gravelly hills, and continues so for about three miles. The ascents and descents are innumerable, and the road though good is very narrow, and so steep in places that horses have to be let down by their tails."[26] After a lengthy stay at *mīr* Ghazan Khan's fort and lengthy discussions about the transfer of Chaprote to Hunza, John Biddulph wanted to proceed to the Pamirs. Ghazan Khan tried to dissuade him by stating that "... the Kirghiz would attack me ... The road was quite impassable on account of the water, &c. ... I answered that I was not afraid of the Kirghiz; that I was accustomed to bad roads (Goodness knows nothing could be worse than the road between Hunza and Gilgit!); and finally I did not intend to go to Punja [Qala-e-Panja], but should return to Gilgit by Ishkoman."[27] Every kind of excuse was tried by *mīr* Ghazan Khan to hinder John Biddulph from travelling further up the valley; consequently he returned the same way he had come. He could not report about the political conditions in Gojal nor explore the route conditions to Sarikol and Wakhan. The information John Biddulph collected was from Beehoo, his companion on the road from Gilgit: "He told me that horses and camels traverse them the whole year. The presence of several camels (double humped) which I saw in Hunza partly confirms this. They would hardly be brought here if the roads are closed for any length of time. ... The first two marches to Goohjal (which is a district not a place) are not very good, but beyond that the road is excellent. The hills apparently are not so steep, and approximate to the Pamir type."[28] He continues by giving the distances between villages, and remarking where there are 'marches without population'. The distinction between Central Hunza and Gojal is mentioned; Gojal is the Wakhi section and begins beyond the Gulmit Bridge (*baṣ-talénum maqsoó*). This was later confirmed by Gottlieb Wilhelm Leitner who had heard that 'Chilásis call the people between Hunza and Pamêr on the Yarkand road Gojal'.[29] On his way back John Biddulph was accompanied by the *yarpá* Murad who was sent along by *mīr* Ghazan Khan to make sure that he left Hunza on the agreed-upon route. His report contains valuable information about the environment and the political troubles in the Hunza Valley. The road con-

One of the rare portraits of John Biddulph is on display in the Gilgit Municipal Library which has received the title 'Biddulph House' as it is accommodated in his former residence which was constructed during the first Gilgit Agency. The photograph was taken on 7 August 2017

The 'Sketch map constructed to illustrate explorations made in connections with the operations of the Great Trigonometrical Survey of India by Mullah in 1876' made a significant contribution to hitherto unknown routes and was published in the Geographical Magazine in May 1878. It summarised the topographical knowledge of the 'Northern Frontier' and clearly shows the emptiness of the Gojal section in the Hunza Valley at the time of John Biddulph's reconnaissance. Courtesy of the Pamir Archive Collection

dition was an interesting challenge for John Biddulph rather than anything to worry about. Nevertheless, he was the first person to state that the route was too difficult to provide an approach for a huge hostile army. The information gathered was a valuable hint on the strategic importance of controlling Chalt and Chaprot. When held by Nagar, which was the case between 1875 and 1886, the two fertile oases formed a separating belt between Gilgit and Hunza. At other times – Hunza had occupied the fertile oases between 1852 and 1875, and would repeat its hold between 1886 and 1891 – its advanced position allowed Hunza easy attacks on Nomal and thrusts as far as Gilgit.[30] The complex constellation of changing spatial dominance reflected the persisting importance of road access.

As an expert on the region with first-hand experience, he was wooed away from his military career to join the political department on the basis of permanent employment there. The euphemism 'political' as a disguise for espionage made John Biddulph an 'officer on special duty' for four years during which he established the Gilgit Agency in 1877. His mission was less successful than his supervising officers had expected: he had serious problems with the representatives of the Maharaja of Kashmir in Gilgit, and the recently established Gilgit Agency was closed again in 1881. Besides his political mission, John Biddulph's personal interest was mainly directed towards ethnography and ornithology. His main findings were published in his 'Tribes of the Hindoo Koosh' in 1880, which contained valuable information about the Hunza Valley.[31] He produced the background papers that served Colonel William Lockhart – former Deputy Quartermaster-General in the Intelligence Branch before being appointed envoy by the foreign department on a political reconnaissance trip to Chitral, Wakhan and Hunza – for the preparation of what became known as the 'Gilgit Mission' in 1885-1886.[32] Map-making was commissioned to Lieutenant-Colonel Robert Gosset Woodthorpe, Indian Survey Department, while the surgeon George Michael Giles, Indian Medical Service, was the photographer and botanist of the mission. From the Intelligence Branch of the Indian Army Captain Edmund Barrow joined the group; he was an acquaintance and subsequently a rival of John Biddulph. Both intelligence officers knew each other well from their previous cooperation during the Forsyth Mission to Yarkand; Edmund Barrow was given the task of preparing a gazetteer, as he was a keen observer in the tradition of gazetteer writing and route handbook preparation. He frequently mentions in his diary the long time it takes to convert the gathered information into a readable gazetteer that includes a detailed section on Dardistan.[33] Four Indian military surveyors, trained at Roorkee in the field of 'route-sketching and subsidiary reconnaissance', supported him in gathering 'intelligence'.[34]

In the opening chapter of his diary about the 'Chitral-Kafiristan Mission' Edmund Barrow summarises the mission statement: "In 1885 the relations between the British and Russian Empires were very strained and war seemed imminent. At that time our knowledge of the mountainous region commonly known as the 'Hindu Kush' which is the watershed between the Oxus and the Indus, was somewhat vague, limited and confused. It was essential to the Government of India to be more correctly informed thereon for military and strategical reasons. It was also desirable, if possible, to enter friendly political relations with the Chiefs who ruled over the wild tribes inhabiting that little known area. It was therefore decided to send a small secret Mission through Kashmir and Gilgit to explore and survey that interesting

The section of the 'Sketch map of territory between Cabul and Muztagh Pass' was compiled by Colonel Henry Tanner on the basis of 'various details supplied by Major Biddulph from native information; it is based on the latest surveys'. Little detail is given on the routes north of Baltit; Ahmedabad is located on the opposite left bank, Atabad is put on the same wrong bank and close to Pasu in the location of Hussaini. Hunza is identified as 'Kanjut or Little Guhjal'. John Biddulph attached the map to illustrate locations in his book on the 'Tribes of the Hindoo Koosh' Source: John Biddulph (1880, map in pocket)

and important mountain region with a view of supplying the Government of India with correct up-to-date knowledge of its population and resources and of its bearing on the general military situation."[35]

Edmund Barrow became a well-known expert, lauding himself for "... having traversed every road and pass in the region between the Indus and Kafiristán, the Pamirs and the Hindu Ráj, having spent 15 months among the people, and having given much thought and labour to the subject in compiling the gazetteer of the Hindu Kush."[36] In line with his companions and military comrades he made a good career: all were subsequently promoted to the rank of general. His personal diaries contain some unpublished descriptions of the circumstances and the terrain the mission was traversing. Edmund Barrow was quite impressed by the difficulties posed by the Hunza Road in terms of political, physical and social challenges. The mission members reached Gilgit on 9 December 1885 and began the year 1886 in Gilgit waiting for permission to enter the Hunza Valley. From January to April they passed their time with polo matches and hunting

The Gilgit Mission consisted from left to right of Colonel Robert Gosset Woodthorpe of the Bengal Army and officiating deputy superintendent, Survey of India, Captain Edmund Barrow of the Quarter-Master General's Department, under the command of Colonel Sir William Stephen Alexander Lockhart of the Royal Engineers, and Surgeon Dr. George Michael Giles of the Indian Medical Department. The latter was the photographer of the mission. The photograph was used as the frontispiece in the 'strictly confidential' report of the 'Gilgit Mission' (1889)

for duck and *markhor*. Finally, after four months in Gilgit, the good news arrived on 11 April 1886 that the son of *mīr* Ghazan Khan, M. Nazim Khan, a son of a morganatic marriage, would be sent to Gilgit as a hostage (*badli*) in exchange and that their safe passage to Hunza would be guaranteed. The timing was good as *wazīr* Asadullah Beg, a stern critic of the British, had passed away, giving *mīr* Ghazan Khan more leverage for his own decision-making and issuing an invitation to the visitors.

The party, consisting of four British officers, Bakshi Mulraj[37], the governor of Gilgit, a local go-between and translator, sepoy orderlies, staff, porters and servants, received a salute by the Kashmir troops and set off on 14 April; M. Nazim Khan had reached Gilgit territory just in time and would remain as a hostage in the Gilgit garrison.[38] This safety measure was quite common and obviously liked by some: "He and his followers seem rather pleased to be hostages, as it means 'free dinners' for the time they are in Gilgit, as well as presents of money all round."[39] The first physical obstacle was the crossing of a rope bridge, which consisted of three shaky sagging 'ropes' plaited from willow twigs that were bound together. The middle rope is meant to be walked on, while the other two are hand-ropes that are intermittently tied to the foot-rope by cross-ties; all of them are often in bad repair: "The crossing of the rope bridge was a long tedious process as of course only 3 or 4 men could be on it at one time."[40] The 'ponies were swum across higher up'.[41] On 17 April they continued their journey from Pilche to Nomal: "The road lay along the bed of the river and was on the whole fairly good, but in summer it is impracticable on account of the water and a path along the cliffs has to be taken which is somewhat difficult."[42] The next rope bridge came after they left Nomal on 18 April: "Half a mile from the fort we crossed the Hunza river by a somewhat difficult rope bridge, horses fording. Except the gorge above noted the road was easy for laden animals the whole way to Safed Áb (8 miles) or Shai Wai as the Dards call it."[43] The landscape and scenery impressed the travellers: "The mountains which here hem in the valley end abruptly in sheer lofty precipices, the like of which we have not yet seen anywhere and the water gushing out of the rock in the midst of all this sterile grandeur reminds one of the story of Moses. The summer road on the opposite bank looks and is a most fearsome place, a mere scratch by of clefts and ledges along the face of a most awful precipice, I was thankful we had not got to use it. It is of course quite impractible for animals, and almost so for men."[44] On 19 April the group left for Chalt, again encountering some obstacles. They made the acquaintance of a suspended mountain path (*parri*) supported by scaffolding, which is quite common in High Asia. Edmund Barrow faced the inevitable on the Hunza Road: "Short march of 6 miles, but not an easy one. At ½ mile there is twig bridge, then a mile further on there is a bad 'Pari' most difficult for horses, and yet a mile further a second Pari, only practicable for men on foot, and very difficult even for them. Horses have to be swum round the Pari. In summer this part of the road is quite impracticable, and a much higher level must be used. The rest of the road was fairly easy, bar an avalanche of snow which had to be crossed. ... Chalt is a cheerless looking place, the fort standing at the upper end of a broad sloping treeless plain, irrigated but poorly cultivated. It is the last place in Gilgit territory."[45] On 21 April they left Chalt for Hunza: "A mile or so from Chalt we crossed the Budlas stream which is the boundary between Hunza and Chalt. Close to the mouth of this stream we had to ford

Bakhshi Mulraj, Governor of Gilgit during the 'Gilgit Mission' and after, was 'the first Hindú who has yet visited Hunza; the Kashmir officials sent here are always Mussulmans. I can't speak too highly of Bakshi Mulraj. He has great tact, readiness, and resolution' stated William Lockhart in a letter to Algernon Durand, dated 26 April 1886 (William Lockhart and Robert Woodthorpe (1889: 393). The governor kept a close vigilance on the developments in Hunza and Nagar. He informed Maharaja Pratap Singh immediately after Safdar Ali Khan had murdered his father, tham Ghazan Khan, on 26 August 1886. Source: Section of a photograph reproduced in William Lockhart and Robert Woodthorpe (1889, following page 284)

The section taken from the 'Map of Chitrál, Hunza and parts of Wakhán and Káfiristán, surveyed by R. G. Woodthorpe, R.E., assisted by Sub-Surveyor Bápu Jádu' shows the area in the Hunza Valley that was visited by the Gilgit Mission in 1886 and was published in Dehra Dun in November 1888. The map shows the aggregated knowledge that was based on the mission's surveys and the incorporated knowledge from earlier undertakings including Munshi Abdul Rahim's journey to Wakhan and Badakhshan. Source: IOL&R: W/LPS/21/J9 and Public Record Office FO 925/2661. Courtesy of the Public Record Office

People carrying loads near Chalt Fort in Shinber; on its left the Gilgit Mission's tent camp was set up on 19 April 1886. Source: William Lockhart and Robert Woodthorpe (1889); the image is reproduced here with permission from the Royal Geographical Society in London, call number S0025142

Opposite
The central oases of Hunza and Nagar in April 1886. The snowline is still rather low; the apricot trees are in blossom, but the poplars are still leafless. The village lands are limited by the major irrigation channels that mark the border between crop-farming and pastures for livestock higher up. In the background Rakaposhi (7,788 m) towers above the Nagar oasis. Source: William Lockhart and Robert Woodthorpe (1889, following page 84); the image is reproduced here with permission from the Royal Geographical Society in London, call number S0025137

the Hunza river, and then a mile further on, we recrossed to the right bank by fording. Both fords very difficult for the coolies, owing to the strength of the current, but this double passage can only be avoided by taking a very difficult path along the face of the cliffs on the right bank. The rest of the road was easy, with the exception of two bad 'Paris'. How our horses managed them is a mystery to me. The total march was about 8 miles. Mayún is a fort situated on a small promontory between the Hunza river and the deep ravine of the Mayun stream. This promontory is bounded on both its sides by precipices 300' high, and backed at its base by a precipitous mountain, so for Dard warfare it may be considered impregnable. The fields of Mayun are all on the west side of the ravine. Opposite Mayun are the Nagar villages of Nilt and Tol separated by another ravine. It was at this ravine that a Kashmir Army was defeated some 20 years ago, and almost annihilated."[46] Edmund Barrow had spotted the strategically important place in the valley. Coincidentally, five years after their visit, the 'Hunza-Nagar Campaign' under the leadership of Algernon Durand would decide at exactly this spot the fate of the valley and its inhabitants.

On 22 April they left Maiun for Hindi: "A short march, only 6 ¾ miles, but the first 5 miles were a succession of difficult 'Paris', the path in many places is scarcely a foot wide along the edge of an appalling precipice."[47] Colonel William Lockhart remarked in his diary: "Road awful in places; rocky staircases and narrow paths over long bits of shale that kept sliding down from above as we disturbed it below. The opposite (Nagar) side is much more fertile. The cultivated ground, terraced from the foot of the mountains to the river, is in broader and longer strips than on this bank."[48] The next day they left Shinaki for Central Hunza; as a succession of hindrances, the road that separated the Shina-speakers from the Burusho area of Central Hunza offered its own challenges: "The road after the first ½ mile was a succession of 'Paris' for about five miles. One or two very jumpy places where the narrow path overhung a precipice. Even a mountaineer from his youth like Sekundar [his serv-

The detailed view of the balconies in two wooden mosques in Baltit was taken by George Michael Giles in April 1886. Source: William Lockhart and Robert Woodthorpe (1889, following page 388); the image is reproduced here with permission from the Royal Geographical Society in London, call number S0025136

ant] who was next behind me said: 'Go on Sahib, for God's sake don't stop to look.' At the end of the 'Pari' we came to Motazabád [Murtazabad], a poor miserable looking place with a couple of forts. Here we were met by the 'heir-apparent' Safdar Ali Khan a fat dissipated youth, got up in a really tasteful sky blue silk choga trimmed with gold. A Chinese look about it, probably Yarkandi. From Motazabád there is a steep descent into a deep ravine which separates the Hunza district from that of Hini. On the other side there is a short steep ascent which is fortified at the top and this is said to be the only entrance into Hunza from the south and west. The first glimpse of Hunza as gained from the Motazabád hill is not unpleasing. A great bay of rocky mountains, from whose base the ground slopes away towards the river ending in cliffs several hundred feet high. On the opposite bank is a similar but narrower slope on which are the villages appertaining to Nagar. Generally speaking the villages are lost in a mass of fruit trees which at this season are all in bloom. At the further end of the Húnza valley was a low hill covered with houses and crowned by the 'Tham's' palace. We camped at Aliabád, the second of the Hunza villages, Hasanabád being the first. At Aliabád there is a mud fort."[49] It took the Gilgit Mission ten full days to cover the distance of less than one hundred kilometres between Gilgit and Aliabad. Edmund Barrow was fully convinced about the bad state of the 'road' and its obstacles for bulk transport and troops. William Lockhart envisaged the broader picture already upon arrival in Aliabad: "This route is of no importance, and may be disregarded as in any way a weak point. The case lies in a nutshell. When the Killik (now deep in snow) becomes practicable for the passage of troops, the river is swollen, and the road absolutely barred for any but the most daring cragsmen. When the river-road is fit for traffic in winter and spring, the Killik 'is unfit for anything but the passage of birds,' as Ghazan Khan wrote to me at Gilgit. We have just got here in time, the Killik being reported still deep in snow, whilst the river is beginning to rise. Of course an army could blast a way for itself with time and immense labour, from the north, but the country could not give any supplies whatsoever – even our small party is a difficulty. The Killik is the best pass on the Hunza frontier, so all others may be left out of our consideration. The road we have come by is nearly the worst in our experience in this trip. Narrow paths on the faces of huge cliffs, corkscrew staircases in the rock, or else a track for miles along the steep, shingly mountain side, everything on the slide, one's passage disturbing the mass and bringing down quantities of *débris* from above. Had no accident to man, horse, or baggage, but only good hill-ponies could have done the journey – and even with them some bits looked as if our beasts could never be led across them in safety. Of course *laden baggage-ponies* are altogether out of the question, but a Baltí coolie will go anywhere. The Hunza bank is much wilder and more barren than the opposite, or Nagar one. On the latter, the strips of cultivated land between the mountain-foot and the river are broader, longer, and of more frequent occurrence than on this bank. Nagar seems, indeed, very fertile, every bit of ground not absolute precipice, being tilled; fruit-trees now in blossom everywhere. ... What strikes one in all this country is the very careful husbandry, every scrap of fairly level ground being tilled, and the stone walls and terracing show much labour. There is nothing slovenly or untidy about the fields – they would do credit to any European country."[50] Before they met *tham* Ghazan Khan in his fort several letters were exchanged with pleasantries, offers of hospitality and his expectation that the British would help him regain Chalt and Chaprot in exchange for a free passage to Kilik Pass and Wakhan.

The next day the party had an easy march through the central oasis along a path that is still called *mákučim gan* – the middle road, the main trade route from Gilgit to Yarkand – leading to the polo ground (*šabáran*) just below Baltit Fort: "A short march of 3 ½ miles through terraced fields by stony lanes brought us to the Polo ground under the fort, the road being lined with thousands of spectators, all in sombre grey chogas with grey Dard caps so that the gathering could hardly be called gay or even picturesque. We encamped in an orchard just below the Polo ground the view from which I think is the finest I ever saw. Such a panorama of stupendous snow mountains all within 12 or 20 miles of us and miles of terraced slopes with forts perched on every knoll."[51] After having reached the seat of power (*thámkuṣe diś*) in Baltit, political troubles replaced the physical ones as *mīr* Ghazan Khan stated an ultimatum demanding the return or transfer of Chaprot back to him as the condition for their safe passage through his country.[52] Difficult negotiations ended in leaving a 'ferocious' *tham* Ghazan Khan in his fort who would be murdered and replaced by his first-born son Safdar Ali Khan less than four months later. The mission proceeded towards Gojal where more physical obstacles blocked their way: "From Mauhamadabad [Ahmedabad] there is a very steep descent to the bed of the river. About a mile beyond this our march was arrested by a torrent of mud and melted snow which came tumbling down from the mountains above and formed a liquid rolling mass of mud and water and boulders across our path. It was very difficult and even dangerous to cross and still more so for those who came after us as every minute it increased in volume. Beyond this, there was another steep ascent and descent, and then moving across the river bed we encamped. Though only 7 ½ miles it was a tiring march, all rock staircases or heavy sand."[53] On 28 April, they reached Gulmit in Gojal: "For us horsemen a very easy march as for the greater part of the way it lay over the broad sandy bed of the river between stupendous precipices. The river had to be forded twice. The coolies however, had to take the 'Pari' road along the right bank and this is very difficult, dangerous even for men on foot."[54] The party must have sighed with relief after reaching Gulmit: "This is an important place containing about 150 houses, supplies plentiful, and readily brought in. The people are all Wakhis and much more friendly than the Kanjuti devils we have been dealing with."[55] The onward journey was much easier than crossing the *parri* or *perien*, but in Gojal the river needed to be forded about two dozen times before reaching the Kilik Pass. With rising water levels due to the seasonal increase in glacier melt: "The march altogether was a very fatiguing one for the coolies."[56] The porters had to cross the rivers on foot, while the officers were mounted on horses. The party left Hunza on 7 May 1886; it had taken them more than three weeks to negotiate the difficult Hunza Road. The diary entries about road conditions cover a substantial space in Edmund Barrow's notes, while the leader of the mission, Colonel William Lockhart, mentions difficulties only when men and horses were endangered and that was the case when the Kilik Pass was crossed rather early in the season.[57]

The two British expeditions brought the first Englishmen ever to visit the Hunza Valley in a span of a decade. Comparing the missions of John Biddulph and Edmund Barrow – both spent a rather short time along the Hunza

A group of Hunzukuc resting near Harchi Har. On top of Ultar Glacier's moraine is Baltit Fort, photographed by George Michael Giles in April 1886 from behind. Source. William Lockhart and Robert Woodthorpe (1889, following page 84); the image is reproduced here with permission from the Royal Geographical Society in London, call number S0025138

The portrait of mīr Ghazan Khan was drawn by Robert Gosset Woodthorpe in April 1886, a few months before he was murdered by his son Safdar Ali Khan. Source: William Lockhart and Robert Woodthorpe (1889, following page 390)

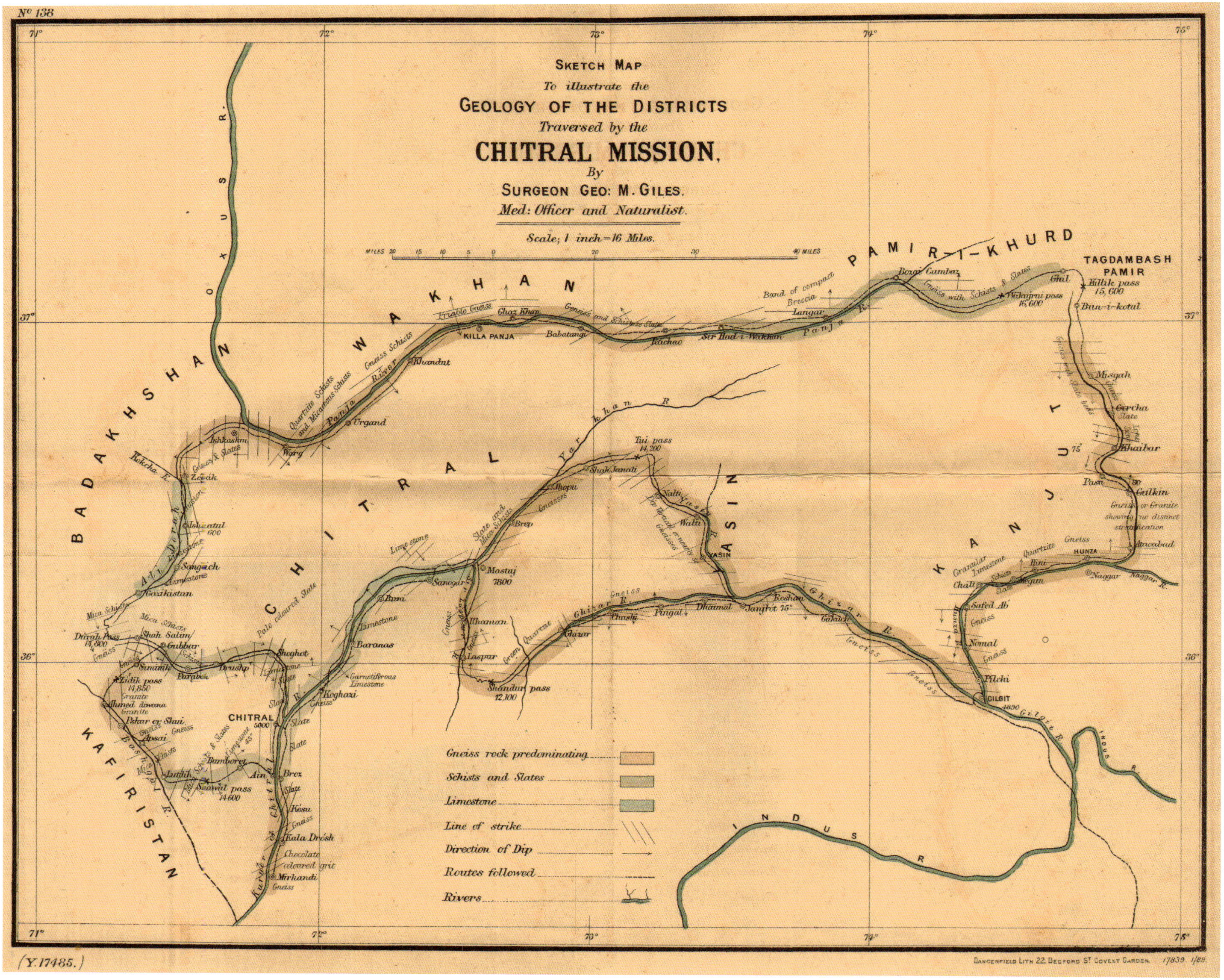

No 138
Sketch Map
To illustrate the
Geology of the Districts
Traversed by the
CHITRAL MISSION.
By
Surgeon Geo: M. Giles.
Med: Officer and Naturalist.
Scale; 1 inch = 16 Miles.
MILES 20 15 10 5 0 20 30 40 MILES
PAMIR-I-KHURD
TAGDAMBASH PAMIR
WAKHAN
BADAKHSHAN
CHITRAL
YASIN
KANJUT
KAFIRISTAN
OXUS R.
Panja R.
Yarkhan R.
Ghizar R.
Gilgit R.
INDUS R.
KILLA PANJA
Ishkashm
Urgand
Khandut
Babatangi
Langar
Bozai Gumbaz
Kilik pass 15,600
Bun-i-kotal
Misgah
Gircha
Khaibar
Pasu
Gulkin
HUNZA
Naggar
Chalt
Nomal
Pilchi
GILBIT
Mastuj 7800
Shandur pass 12,100
Laspar
Tui pass 14,200
Shah Janali
Brep
Sanogar
Barenas
Koghazi
CHITRAL 5000
Drushp
Shoghot
Dorah Pass 14,800
Gazikistan
Sanglich
Zebak
Ishkashm
Zidik pass 14,850
Bamboret
Ain
Brez
Kesu
Kala Drosh
Mirkandi
Chocolate coloured grit
Gneiss rock predominating
Schists and Slates
Limestone
Line of strike
Direction of Dip
Routes followed
Rivers
71°
72°
73°
74°
75°
37°
36°
(Y.17485.)
Dangerfield Lith 22, Bedford St Covent Garden. 17839. 1/89.

The condition of the Hunza Road between Atabad and Gulmit is clearly visible; if the seasonal water level permits, the track follows the valley bottom; during summers the more difficult high road has to be taken along cliffs and across scaffolded passages (parīi) supported by wooden beams. Source: William Lockhart and Robert Woodthorpe (1889, following page 160); the image is reproduced here with permission from the Royal Geographical Society in London, call number S0025118

Road – by the entries in their diaries, it becomes obvious that the two journeys yielded quite different results and interpretations of the local conditions. First of all, while John Biddulph travelling on his own could converse in Persian without the help of interpreters and got along well with the local *mīr* and their representatives, it was not given to Edmund Barrow, who recorded his observations as part of a large mission, to record any pleasant descriptions of them. The ruler of Hunza, *mīr* Ghazan Khan, was unfavourably described: "… the monarch received us being too ill and too lame to go downstairs. He is an obese and intensely ugly old man, with a broad flat nose and a wrinkled parchment face shot with swollen blood vessels."[58] About the Nagarkuč and their *tham* he remarked: "The Rá of Nagar, Uzur Khan by name, came to call accompanied by five picturesque looking 'stink-pots', who obviously did not often wash."[59] Even forty years later he was full of revulsion towards the *tham*, but his appreciation of the road conditions had changed: "I do not think I ever met a greater ruffian than Ghazan Khan. He not only used every means short of force to obstruct our passage, such as preventing supplies reaching us. … We were almost the first Europeans to travel right through his country, and when we reached Hunza – his capital and residence – we received a most inhospitable welcome from him … by bribing his officials and buying direct from the villagers *en route* we successfully managed to make our way across the

Opposite
The expedition's surgeon George Michael Giles contributed the 'sketch map to illustrate the geology of the districts traversed by the Chitral Mission' in which he annotated the rock formations. In the Hunza Valley he observed and distinguished gneiss, granite, quarzite, limestone and slate in a rough attribution for Kanjut. Source: William Stephen Alexander Lockhart and Robert Gosset Woodthorpe. The Gilgit Mission 1885–86. London 1889, map 3. Courtesy of the Pamir Archive Collection

Altit Fort towers above the flat village lands in the centre of Altit rat. Source: William Lockhart and Robert Woodthorpe (1889, following page 394); the image was taken on 26 April 1886 and is reproduced here with permission from the Royal Geographical Society in London, call number S0025141

Hindu Kush by the Kilik Pass, which, though a snow pass over 15,000 feet, is not really difficult if crossed before sunrise while the snow is still fairly hard."[60]

Edmund Barrow mainly gathered information with the help of indigenous intermediaries; in this case the Raja of Astor, Bahadur Khan, was translating for him. Consequently, his narrative is somehow flavoured by the respective relationships with his interpreters. In a confidential memorandum written after completing the work on the 'Gazetteer of the Eastern Hindu Kush' he offered himself for the position of re-opening the agency in Gilgit: "I should welcome the opportunity of doing useful work among these barbarous tribes."[61] His ambition seemed to have been to promote himself as the pioneering and sole authority on the area, as he repeated at the end of 1885.[62] "Many of these places I have been one of the first Europeans to visit, and I am proud to think that it is Englishmen and not Russians to whom that honour is due."[63] He deliberately omitted the names of his predecessors along the routes, thus neglecting the contributions made by John Biddulph, Munshi Abdul Rahim and other 'native explorers'. Their findings were used in compiling the gazetteer; in several instances he failed to acknowledge their contributions and to name the sources he consulted.[64] It is revealing that Edmund Barrow is mainly appreciated as a highly respected Major General who at the peak of his career became Military Secretary to the India Office, while John Biddulph is mainly remembered as the author of an insightful ethnographic book and as an ornithologist.[65]

Even his strong supporters started to understand that John Biddulph might be the wrong ‘officer on special duty’ for the task envisaged. His ethnographic and environmental curiosity, ability to travel in most difficult terrain and his language skills did not obviously qualify him to deal well on a day-to-day basis with Kashmiri officials in Gilgit and with some of the local rulers. The sympathetic Viceroy Lord Ripon summarised the outcome of the first Gilgit Agency in a despatch to the Secretary of State Earl Granville on 15 July 1881: “Major Biddulph was deputed to Gilgit in 1877 with the object first, of obtaining trustworthy information in regard to occurrences beyond the border; and secondly, of gaining influence among the neighbouring tribes. There can be no doubt that Major Biddulph, who is an active and enterprising officer, has during his residence in Gilgit added much to our knowledge of the country round, and of the tribes inhabiting it; the intelligence which he has from time to time obtained of the course of events in Chitral and Badakhshan has not been unimportant, and he was well placed for observing, during our recent occupation of Kabul, the general state and tendency of affairs in the countries bordering on North-east Afghanistan. On the other hand, his position in the remote corner of a foreign State obliged him to deal chiefly with these tribes through the medium, not always trustworthy or favourable, of the local officials. Any attempts to exercise direct influence would have been met by suspicion and jealousy; while he was continually hampered by the double-dealing, intrigues and feuds of petty Chiefs, over whom he had no real control.”[66] His mission was perceived as a failure, and consequently the first Gilgit Agency was closed.

John Biddulph mastered the difficult road easily in terms of travelling conditions; he was eagerly looking for new challenges, spectacular landscapes and fauna, while Edmund Barrow suffered a great deal on the narrow paths, *parri*, and river crossings. Their personal impressions might have influenced their conclusions, which differed significantly. As the ‘Gilgit Mission’ reported the threat of an invasion from the north – only for small groups and in compliance with local elites – their verdict was more inspired by strategic considerations: “… when these passes into Hunza are not closed by snow, the road out of Hunza to Gilgit is closed by reason of the swollen torrents, and a force entering Hunza in the summer would have to wait in this inhospitable country till such time as the road below was open, unless a high level road were engineered round the crags and precipices, where there at present exists a summer route. This, though not impossible, would require time and an absence of opposition which should not be allowed to an invader.”[67] Robert Woodthorpe, though, tended to side with John Biddulph, who had ruled out such a proposition in his report ‘of a journey to Gilgit, Hunza, and the Karoombur valley’ where he attributed an important role to Yasin.[68] Similarly, Francis Younghusband, an able explorer and traveller like John Biddulph himself, would confirm at a later stage that ‘the route cannot be declared as practicable for laden animals’ from the Khunjerab Pass down into the Hunza Valley.[69] In addition, there would be no food supplies available along the route: “Hunza cannot support a single man of the garrison, and Nagar under fifty.”[70] Furthermore, Francis Younghusband doubted that the Russians could ever send more than ‘500 or 600’ soldiers through the Hunza Valley and their seasonal window of activity would be only open for four months to bring troops from Turkestan across the ‘ of India’ down the

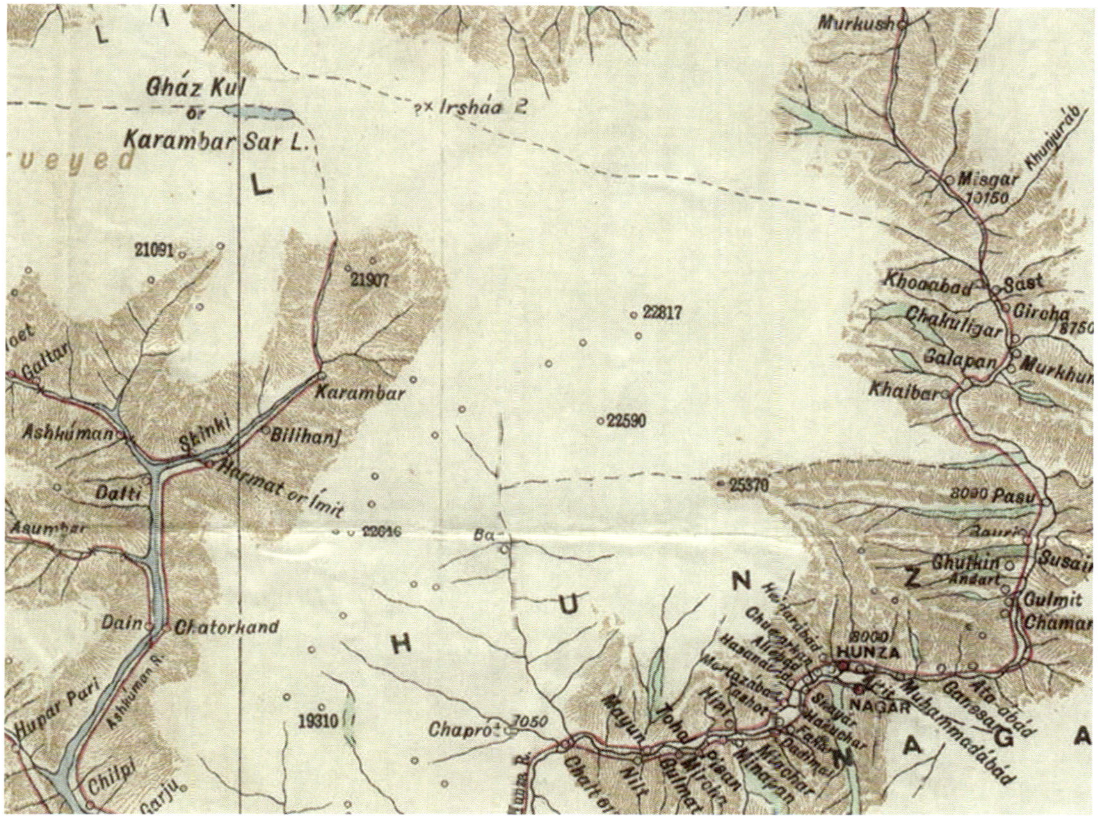

John Biddulph had visited the Yasin and Hunza valleys, but only reached Baltit. The Gilgit Mission succeeded in traversing the Hunza Valley towards Misgar and Markushi from where they left via Kilik Pass to the Taghdumbash Pamir and Wakhan. Source: IOL&R: W/LPS/21/J9 and Public Record Office FO 925/2661. Courtesy of the Public Record Office

The Gilgit Mission left the Hunza watershed via Kilik Pass on 7 May 1886 on their way to the Little Pamir. Source: William Lockhart and Robert Woodthorpe (1889, following page 402); the image is reproduced here with permission from the Royal Geographical Society in London, call number S0025134

passes.[71] Nevertheless, the promotors of a more active involvement and 'forward policy' interpreted the road conditions quite differently. Many authors supported the view of imminent danger, such as Henry Lansdell who endorsed the opinion that the passes to the north could easily be crossed by a 'waggon', a symbol similarly used by Lord Curzon stating about the Kilik: "This is the pass of which Captain Grombchevski, who crossed it in August, 1888, penned the somewhat hyperbolic report that it is 'exceedingly easy, so that a cart with a full team of horses could follow it'."[72] Needless to mention that in Bronisław Grąbczewski's diary there is no reference to such possibilities; contrary to Henry Lansdell's quote he stated that in places even for single horses the path was unpassable and that their losses were high.[73] The judgement of others was more realistic. People like Captain Henry Deasy experienced the Hunza Road above Baltit as "... a track [that] became narrower, and often steep and stony to be impassable for laden ponies, so that it was necessary for a time to transfer all the baggage to coolies."[74] The contrasting views depended on the commentator's position, either being with the forward politicians or opposing them.

There existed only one Englishman among the two – in Edmund Barrow's view – who had gathered valuable insights into local conditions and was qualified for promotion. "Major Biddulph's experiences ought to be sufficient to prevent us ever again sending an 'officer on special duty' to Gilgit, unless we wish once more to suffer humiliation of seeing a British officer taking refuge in a Kashmir fort from violence of the tribesmen, whose best friend he should be. ... I would gladly take up to-morrow the more extended and onerous sphere of duty ... on any pay the Government might consider adequate."[75] Edmund Barrow tried to gain the position to be filled by openly criticising John Biddulph and running down Algernon Durand, his main rival for the post. Edmund Barrow failed to get the seemingly attractive post; one reason might be doubts about the accuracy of his data collection.[76] Subsequently the brother of the eponymous negotiator of the Anglo-Afghan boundary agreement was selected for the Gilgit post; Algernon Durand changed the route perception and assessment of accessibility, earning fame for his braveness and for stabilising the re-opened agency.

A group of porters servicing the Gilgit Mission is resting on Kilik Pass (4,765 m). On 7 May 1886 they left Hunza and proceeded toward Little Pamir and Wakhan.
Source: William Lockhart and Robert Woodthorpe (1889, following page 402); the image is reproduced here with permission from the Royal Geographical Society in London, call number S0025135

The section is part of a map centred on Afghanistan which is devoted to identifying the British and Russian spheres of influence in 1885. The Hunza Valley is shown as an extensive region titled 'Kundjut' bordering with the Little Pamir which is divided between Afghanistan and China. Gilgit, Hunza and Nagar are included in the red coloured area that is part of Kashmir and dominated by British India. Source: Karte von Afghanistan mit den angrenzenden russischen und englischen Gebieten. Leipzig 1885. Courtesy of Pamir Archive Collection

Gilgit Fort is located on the bank of the river; from the garrison the rope bridge that allows access to the Hunza Road can be overseen and controlled. William Lockhart (1889: 281) described the favourable position: "A permanent bridge, suspension or other, could easily be thrown across the river, either opposite the present fort, where the width between the crests of the banks is 400 feet., or at any other suitable place in the immediate neighbourhood. The fort would then form a bridge-head, and under its wall would be erected the bázár." Source: William Lockhart and Robert Woodthorpe (1889, following page 256), the image is reproduced here with permission from the Royal Geographical Society in London, call number S0025144

The second Gilgit Agency and the 'pacification' of Hunza

In reaction to a perceived disobedience and opposition by the rulers of the Hunza Valley and in response to their alleged duplicity and outrageous demands, Algernon Durand was commissioned to re-establish the Gilgit Agency and to safeguard approaches and control of the territory south of the Hindukush and Karakoram passes.[77] The local rulers were to be taught a lesson by a joint Anglo-Kashmiri effort they would not easily forget: "The Chiefs of Hunza and Nagar, especially the former, owing to their isolation, and the natural difficulties of their country, have the most exaggerated idea of their own power and importance. This has led them to assume a tone of insolence in dealing with Kashmir, and, since they have been brought in contact with the officers of the Supreme Government, towards the latter, which has up to the present time received but slight checks. Their independent position, and the numerous defeats which they have inflicted on Kashmiri troops in past days, have given them a fictitious importance in the eyes of the neighbouring tribes. ... Until the Chiefs of Hunza and Nagar have been brought into complete subjection, and have felt or realized our power, their attitude will always remain the same, and they will be a cause of constant anxiety and of possible disturbance on the frontier."[78] After having assessed the general situation, Algernon Durand's initial attention was drawn towards route conditions, which he described graphically: "It is difficult for any one who has not traversed the country to realise what a road in the heart of the Hindu Kush means. When I first visited Gilgit, five years ago, there was not a yard of what we should call a road in the whole region, and only one permanent bridge, that over the Chitral river at Chitral itself. Narrow paths – so narrow that often while the rider's boot on one side brushed the cliff, his outer foot overhung a precipice – followed the course of the streams. Often in the course of one short march the path ascends a thousand feet or more to avoid crossing some precipitous cliff, and the repeated ascents and descents render riding a weariness to the flesh. Frequently the path is carried across the face of a cliff on roughly constructed galleries, upheld by shaky timbers jammed into interstices in the rock. In many of the valleys, when the summer sun melts the accumulations of snow, and the mighty glaciers pour down their flooded torrents, the lower paths become impassable for animals. For months at a time all animal traffic is suspended, and men on foot alone, following giddy tracks skirting gigantic precipices, can with difficulty find their way from valley to valley."[79]

In Algernon Durand's negotiations the control of communication and the right of passage were always of prime importance; therefore he suggested 'the establishment of a suspension bridge at Gilgit under the guns of the fort'.[80] The British government tried to control who entered the Hunza Valley and with whom the *mīr* of Hunza would have exchange relations. The Hunza Road played a major role as

The portrait shows General Bronislaw Grombchevsky at the time when he ventured on his first Central Asian expedition to Southern Kashgaria in 1885. Source: Mikhail Kazbekovich Bashkanov, Alexander Kolesnikov, Maria Matveeva and Alexander Ivanovich Glukhov (2017: 4); Scientific Archive of the Russian Geographical Society, font 45, opis 1, file 4, sheet 2

a thoroughfare and line of communication. The fear of a military intervention from the north had been negated by John Biddulph, but his successors made decisions under the premise that Russian forces had advanced in the Pamirs. The only Russian officer to have visited Hunza, Bronislaw Grombchevsky, – who had stated that he was on a private mission with the permission of his government – had come to a similar conclusion when he was travelling from Kilik Pass to Baltit. His group could not rely on pack animals for his luggage as the route was too difficult and could only be improved by blasting. In addition the supply situation was grave in Upper Gojal where food surplus was practically unknown. Especially summers were difficult, as rivers had to be forded in many deep and difficult places.[81] During Bronislaw Grombchevsky's approach to Baltit he heard rumours that *mīr* Safdar Ali Khan had tried to instigate the Chinese representatives in Kashgar to show presence in Hunza in order to keep both imperial powers at bay. He realised that a power game was going on.[82]

Hunza's relations with China were another cause of disagreement because the British authorities perceived all areas 'south of the Hindukush' as their sphere of influence.[83] They informed *mīr* Safdar Ali Khan: "Your State lies to the south of these [Pamir] mountains, which here form the boundary of the Indian empire, and is within the borders of British India. Considering these facts it is necessary for the protection of Kashmir and your State, that these troops of the Supreme Government should have free access to your country. By this means alone can the requisite arrangements be made for safeguarding here the frontiers of the Empire. It is not the intention of the Supreme Government in any way to interfere with the form of rule in your State, but the arrangements for protecting its frontier necessitate on the part of Government the making of roads which will enable it, should occasion arise, to place troops rapidly in positions guarding the passes leading from the Pamirs."[84] Algernon Durand makes it quite clear who on the one side is 'Supreme Government', 'Empire' and 'India' and on the other hand the 'State' and the 'country'; he continues with his demands and threats: "The Supreme Government has therefore decided to make a road from Gilgit to Chalt, where a fort will be built, and from Chalt to Hunza-Nagar, or so far beyond that as may be necessary. As a feudatory of the British Government you are now called upon to give any aid in your power towards the construction of the road. I am further directed to inform you that, so far as regards the road beyond Chalt, which will pass through your territory, no refusal on your part to permit its construction will be accepted. The road must be made. Unless you instantly comply with the demands of the Supreme Government troops will enter your territory, and the road will be constructed in spite of any opposition you may offer."[85] Abandoning raiding, granting the right of passage, and improved bilateral diplomatic relations were the parameters for judging whether or not an intervention was justified. It was *mīr* Safdar Ali Khan who, as expected, seized the occasion and provided the pretext for the self-appointed arbitrators to move on towards Hunza. Thus, Algernon Durand stated in his memorandum about Hunza: "His agreement with us is to abstain from raiding on the Yarkand road, to exclude foreigners from his State, not to allow Chinese officials to interfere in the internal affairs of Hunza, and to give free passage to British officers. In direct contravention to his engagements he has received Chinese officials in Hunza, has invited Russian officials to enter …, has extorted black mail through his representative in Kashgar under threats of raiding the trade routes, and has calls on the Chinese and Russians to aid him against us."[86] Consequently, punishment seemed to be justified, the control over the external relations of Hunza rightful, and the improvement of road access necessary to speed up military intervention when required and to reduce the costs of supplying administrative and military posts.

Opposite

The competition between Russia and Great Britain about their respective influence in the Pamirs, Wakhan and Hunza in the final quarter of the 19th century was stimulated when the Polish-born explorer in Russian military service Colonel Bronisław Grąbczewski (Grombchevsky) was sent by the St. Petersburg Geographical Society to explore the Hunza Valley. His collected data and observations from three Central Asian expeditions were included in a major map which was produced by the Turkestan Military-Topographical Department and published in 1895. The represented section from this map shows the results of the second expedition to Kanjut as the southernmost extension of his exploratory field in the Pamirian Crossroads. Bronislaw Grombchevsky crossed Kilik Pass on 12 August 1888 and met with Mir Safdar Ali Khan in Baltit Fort as his ultimate southern destination before leaving Hunza via Mintaka Pass on 6 October; see Bronislaw Grombchevsky (1888); Mikhail Kazbekovich Bashkanov, Alexander Kolesnikov and Maria Matveeva (2015). Courtesy of the Staatsbibliothek zu Berlin – Preußischer Kulturbesitz

A Russian threat? – fellow explorers on the Hunza Road

The pretext for intervention that could easily be communicated from India to London was a Russian threat of invasion from the north. Rumours had been around since the 1870s that Russian explorers had expanded their knowledge base further down the Pamirs and were approaching the Afghan and Kashmir spheres of influence. Chitral, Gilgit and Hunza became early warning posts for this perceived threat.

When John Biddulph visited the Hunza Valley, a rumour reached his ears that two years earlier, in 1874, "… a very fair man with reddish beard came from Iskardo to Gilgit, and said he was a Hadji from Yarkund. He lived in Gilgit for some time and passed as a native, and finally went to Cashmere. At Gurais he was met by servants with tents, &c.; and then it appeared that he was a European. Gholam Hyder spoke of this as being well known in Gilgit, and said that the common opinion is that he was a Russian. I asked him if he had ever heard of Russian agents coming to Yassin and Chitral. He said he had not, but he had heard of their coming to Badakhshan."[87] Whether these rumours were unfounded or not, it was one of the military tasks of the Gilgit Mission to assess the potential danger of an invasion from the North. Colonel William Lockhart was a bit uncertain about the probability of a threat when he wrote in his report: "Although it is certain that no army could penetrate this portion of the range until a military road had been engineered for it, it is nevertheless manifest that, after winning over local chiefs, an enemy could throw lightly equipped bands across the passes into Chitral, and even into Hunza, and thence kindle a flame that would touch a considerable portion of our northwestern frontier, while the real attack took place further west."[88] Nevertheless, he recommended in his memorandum: "The acquisition of Gilgit would secure for us the continued loyalty of Chitral, carrying with it our right of way through the Mehtar's dominions, and his active co-operation in time of need. In my opinion it would ensure the safety of the Hindu Kush."[89] It seems that a potential danger was seen rather on the Chitral and the Eastern Hindukush side than on the Hunza Road which was still unknown to them at this point.

Although there is no evidence of a Russian visiting Gilgit that early, it would not take long before the first Russian explorers approached from a northerly direction, doing exactly the same as the British did from the south. Initially

The photograph of the local style of a one-sided cantilever bridge was taken by Bronislaw Grombchevsky close to Qalamderchi on 16 August 1888 on his way from Kilik Pass to Misgar. Source: Bronislaw Grombchevsky (1888: 144–146); Mikhail Kazbekovich Bashkanov, Alexander Kolesnikov, Maria Matveeva and Alexander Ivanovich Glukhov (2017: 55); the image is reproduced with permission of the Scientific Archive of the Russian Geographical Society, font 45 opis 1, file 4, sheet 36

the brothers Grigorii and Mikhail Grum-Grzhimailo had attempted to reach Hunza, but were stopped at Tashkurgan in 1887. When *tham* Safdar Ali Khan heard about it he was convinced that the Russian Tsar had sent them for Hunza.[90] This is the time when pairs of travellers represented the Anglo-Russian rivalry. The ambitious and highly respected Bronisław Grąbczewski became an elected member of the Imperial Russian Geographical Society, which financed the major part of his expedition; a smaller amount was contributed by the War Ministry. With the equivalent of more than three thousand roubles the future 'Dervish of the Hindukush' left St. Petersburg at the end of May 1888.[91] Francis Younghusband met him in the Pamirs, the first Russian to visit Hunza. The former was obviously quite impressed, admitting: "... as to the road, Colonel Grombehevsky has said that even he, an experienced hunter in the Turkistan mountains had to be helped along in parts."[92] Both explorers had similar tasks and felt themselves to be professional colleagues in a challenging terrain providing them with the right kind of adventure. The reconnaissance of routes and passes, the preparation of route maps and additional information about the supply situation regarding food, fodder and fuel as well as the maintaining of diplomatic relations with local rulers fell within their portfolios.[93]

On 12 August 1888 Colonel Bronisław Ludwig Grąbczewski along with twelve Cossacks crossed the Kilik Pass and noted in his diary that 'today we entered the robber's nest'.[94] Coming from the Kushbel Valley in the north they visited *mīr* Safdar Ali Khan in Baltit; the expedition returned via Mintaka Pass on 6 October. He gives a route description with particulars in the same fashion as his British counterparts, but including more detail. Mentioning the environmental challenges on ascending Kilik Pass, the number of river crossings involved, and the terrain conditions for travel by horse, his assessment is that the Kilik Pass is more difficult than Wakhjir Pass, and that this section of the route was the most challenging since leaving the garrison in Margelan.[95] At the same time he acknowledges that with little investment the route could be vastly improved and made suitable for travel. Bronislaw Grombchevsky provided the popular etymology for the Mintaka Pass with the suggested translation of 'one thousand goats', recorded the bridge across the Kilik before the confluence, and noted that the river section between Murkushi and Kalamderchi below the junction of Kilik and Mintaka rivers was called Kanjut.[96] Bronislaw Grombchevsky was equipped with a letter from the Russian Consul in Kashgar and asked for a personal interview at Baltit. In advance he informed *mīr* Safdar Ali Khan about his peaceful intentions by a notice written in Persian that was immediately sent after entering Hunza territory.[97] After three days he met the *tranphá* Mahomed Rajab from Aliabad who was on duty as a messenger of the *mīr* to China; the poor state of his attire was described. Rarely was a traveller met on the route that was known as the Kanjut Valley with Chupursan and Shimshal as tributaries. Only traces of grazing grounds and temporary settlements were found here.

Bronislaw Grombchevsky realised that the Misgar route was so difficult that only human porterage could traverse it; a pack animal caravan was not feasible. This made that

Opposite
A cantilever bridge supported on one side by a natural rock en route between Lupghar and Reminj; close by is a guard post controlling the passage. The photograph was taken by Bronislaw Grombchevsky on 22 August 1888. The rope bridge (gal) on the Hunza Road between Morkhun and Khaiber consists of three plaited and interconnected ropes which makes the passage of persons and loads a tricky affair. The photograph was taken three days later. Source: Bronislaw Grombchevsky (1888: 115–116, 118–120), the images are reproduced with permission of the Scientific Archive of the Russian Geographical Society, font 45, opis 1, file 4, sheets 60 & 61

route section a first bottleneck on his way to Baltit. However, they took their horses along, some died, and others had accidents. The route across Kirmin Pass (4,115 m) was so challenging and rocky that most horses lost their iron shoes. In the absence of available local replacements, Bronislaw Grombchevsky sent a letter to *wazīr* Dadu kindly requesting him to send for horseshoes to be brought from Gilgit for which he paid a silver *yambu* in advance. He learnt that a good road had been made between Kashmir and Gilgit, and that even a telegraph line up to there enabled speedy communication.[98] The state of the settlements and the route was appalling in Grombchevsky's opinion. He lost another horse at a *perien*, and observed how easily this route could be guarded. Although *mīr* Safdar Ali Khan had sent a message saying that they should not pay locally for any supplies, as he would deduct this from the annual dues and taxes, the people were not ready to give anything away. The journey became more and more difficult, as the passages remained challenging, the horses suffered and food was scarce. The situation improved down the road from Khaiber and especially in lower Gojal. As the people refused to equip the expedition with flour and meat Bronislaw Gromchevesky exchanged a number of letters with Safdar Ali Khan in order to find a way to get supplies in the needed quantity. On 27 August 1888 he wrote: "Thanks to the attention and support of Your Excellency we are happily travelling. However, the road to Kanjut is so difficult that we and our horses are completely exhausted; we can continue our journey only after some rest. I conclude this letter with a kind request. Your Excellency have received me as a Russian guest and you ordered your people to gift all supplies to us, but nobody will accept our payment. This state of affairs makes life in our camps difficult; and I beg Your Excellency to skip this order. You perceived my journey as an official trip and have recommended that this journey should not be a burden to our state budget. To eradicate any possible misunderstandings, I hurry-up to declare that I am a private traveller because of my own curiosity, paying for myself and having the permission but do not follow an order by my government. May God bless you."[99] This letter is followed by further exchanges and misinterpretations, but all the time a safe passage is guaranteed without having exchanged a hostage beforehand. Although he confirms several times that he is not on a mission for the Russian Tsar, Safdar Ali Khan calls him a 'liar'. The weak and sick Bronislaw Grombchevsky sends from his

camp in the unwelcoming Bulchi Das a final letter explaining his viewpoint and suggesting that he could not pretend to be an ambassador if he was not one.[100] For *mīr* Safdar Ali Khan it seemed to have been important that he entered official diplomatic exchanges with the Russian Empire to use these new ties for fending off Great Britain and China.

After some recovery in the dreadful Bulchi Das where no water, no grass and no firewood was available, they mastered the final, no less difficult passage full of bottlenecks (*parī*) to Central Hunza. They reached their destination on 1 September 1888 and were met by *mīrzada* M. Nazim Khan and Qairullah Beg, son of *wazīr* Dadu, in Altit who accompanied them to Baltit.

In M. Nazim Khan's autobiography the visit is remembered very differently; he – who would become the loyal supporter of the British colonial authorities – tells in hindsight the story that he had advised his brother *mīr* Safdar Ali Khan not to let the Russian visitor into the country. But it was the mischievous *wazīr* Dadu who was eager to hatch a plot with the Russians.[101] It was noticed that Bronislaw Grombchevsky remarked the bad state of the Hunza Road while otherwise he was perceived as very diplomatic and polite. M. Nazim Khan insists in his narrative that the mission was commissioned by the Russian Tsar and that Bronislaw Grombchevsky was clandestinely discussing the size and scope of a Russian military garrison not only in Bozai Gumbaz in the Little Pamir, but "… pushing forward another post to occupy Baltit. He said that in the event of this last happening, three hundred rifles and two guns would be sent there and a Russian officer would accompany them and be able to train the men of Hunza."[102] In a private conversation the two exchanged pleasantries, the Russian officer saying that the *mīrzada* was the 'only one with any real intelligence' while all other interlocutors were termed as 'fools', but at the same time he underpinned his opinion that 'there was little doubt but that Hunza would, in time, become Russian territory'.[103] Both eyewitnesses' accounts differ: M. Nazim Khan downplays his role in entertaining the visitor and promotes him as a supporter of the British who were closer to him and Hunza as China and Russia were far away. They sometimes contradict each other in this respect and in the estimation of numbers and size of the presents that were exchanged during their meetings.

On Baltit Fort's veranda the final meeting with *mīr* Safdar Ali Khan is held on 7 September during which the *tham* reiterates his wish to become a vassal under Russia's protection by saying: "Look, you see the other bank of the river, those are the possessions of my arch enemy the ruler of Nagar. My ancestors have conquered Nagar three times, but returned it to their descendants. You see that I can live in peace even among intrigues if my enemies are at least the distance of one gunshot away. I can live without this piece of land. I am not here to beg, but please convey my wish to the 'Noble Imperator' that I am willing to transfer my naked mountains and my entrusted people to him, and I shall be his prime slave. I know that he does not need my country and that it would not benefit him, but the name of the 'White Tsar' has crossed the Celestial Mountains and covered great distances, finally it reaches us poor mountain dwellers. The 'Great Imperator' would not refuse our humble wish thanks to his reputation and because all others are terribly frightened of him. I myself have many enemies, but I am calm because my shoulders are leaning against the rock, which is held by the 'White Tsar'. The British visited my late father and they sent emissaries to me. This is not the place to talk about their presents. I am only saying that in his last letter the Viceroy promised that my country would become the arsenal and the treasury of India. I am disgusted by the British and I have kicked them out of my country. I know that they want me to pay for this, but I have no fear, as I could grasp the mantle of the 'White Tsar' now. I shall send along with you Abdullah, my *diwanbegi*, who will deliver my letters to the Russian Consul [in Kashgar], to the Governor-General of Turkestan and to the Foreign Ministry. I do not dare to write to the emperor directly. I know that my representative is experienced, but I trust you for conveying my wishes. I pray for the health of the 'White Tsar', my emperor." The scene ended with a short prayer and 'amen'.[104]

The detailed account of this final meeting in Bronislaw Grombchevsky's diary underpins the serious concern both sides attached to their exchange. The Russian officer made a repeated and final effort to stress the private character of his journey, but the *tham* refused to take any notice of this and let the visitor return via Gulmit, Pasu, Khaiber, Gircha and Misgar towards Mintaka Pass after five weeks in Hunza. Bronislaw Grombchevsky was quite satisfied with his achievements in terms of reconnaissance and diplomatic results, but he was content with the budgetary aspects of his journey as well; he judged his trip inexpensive in comparison with the 'Gilgit Mission'. Colonel David Lockhart's expenses for gifts were judged as ten times higher and the prices for supplies as well. Bronislaw Gromchevsky frowned about the missed chances to make more use of Russia's reputation in Asia.[105] The richness and detail of his diaries contrast with the prevalent narratives about this first encounter of a Russian traveller with the Hunza *tham* of whom he was quite disappointed and found his 'begging unbearable'.[106]

In his 'History of Hunza' Qudratullah Beg endorsed M. Nazim Khan's contrasting views and recorded a long entry about Grombchevsky's visit and his demands "… that the Mir of Hunza and the people of Hunza should establish friendly relations with the Russian empire and become the obedient subjects of Russia. In this way, the Mir and the people of Hunza would/could qualify themselves for the grant of an annual stipend from the Russian government which will be double the amount and even more

Bronisław Grąbczewski was accompanied by twelve Cossacks and local guides during his travels to Hunza. The Russian Consulate in Kashgar was their place for recreation after strenuous travels, for receiving further instructions and permissions. The photograph of the expedition's leader with some of his companions was taken in front of their yurt in the consulate's premises in 1888. The photograph was reproduced by Wawrzyniec Popiel-Machnicki, Adam Pleskaczyński and Konstancja Pleskaczyńska (2017: xix).

Opposite
A difficult road passage (ovring) with wooden support structures above the Chapursan River near Reminj (above). The difficult route passage between Atabad and Bulchi Das had to be negotiated by all visitors (below). Bronislaw Gromchevsky was one of the first photographers on 30 August 1888 who identified and described these bottlenecks on the Hunza Road. Source: Bronislaw Grombchevsky (1888: 115–116, 125); the images are reproduced with permission of the Scientific Archive of the Russian Geographical Society. font 45, opis 1, file 4, sheets 53 & 78

Altit khan is elevated above the fortified village in Altit rat. The photograph was taken by Bronislaw Grombchevsky on 1 September 1888. Source: Bronislaw Grombchevsky (1888: 127), the image is reproduced with permission of the Scientific Archive of the Russian Geographical Society, font 45, opis 1, file 4, sheet 22

than what they were presently receiving from the British. May it be known that British is a miser but Russians are magnanimous and highly generous. Hence, it is but imperative for you that you must sever all relations and contacts with the British government. In such a scenario, a Russian envoy and a representative would be stationed at Hunza and a military garrison or cantonment would be constructed at Hunza. A large quantity of military equipment and arms and ammunitions like cannons, guns and rifles and other weapons etc. would be procured from Russia and brought to Hunza. In addition, military training would also be imparted to the fighting men of Hunza. In case of establishment of strong and long-term friendly relations with the Russian Empire, the British and Dogras would never dare to place their footsteps on the soil of Hunza. Hence, as of now, please send your own men and representatives with me so that at the town/location of 'Marghelan', I am able to convey your wishes and desires to the court of his majesty the Emperor of Russia. ... You may however be informed and be it known to you, that as of the present and ongoing period, the Russian Empire is concentrating its attention and is busy in construction of military garrisons, forts and cantonments in the locations/area of Murghab, Qizil Ribat, Aaq Tush and Tasghurghan etc. once the construction work in these areas is completed. A military garrison would then be constructed in Hunza to impart military training. However, till the arrival and receipt of military assistance and support from the Russian government: please refrain from involving yourself in disputes and battles etc. with the British government and spend the intervening period with utmost caution and in a state of high alert."[107] Qudratullah Beg's history was published in 1962, and in hindsight the visit of Bronisław Grąbczewski was described in the memory of *wazīrkuč* as a clear-cut plan to entice Hunza away from the British and affiliate it with the Tsarist Empire. Whether the offer was that definite cannot finally be established; the diary of the Russian counterpart, Bronislaw Grombchevsky, alleges that it was the proactive wish of *tham* Safdar Ali Khan to find safety under the Tsar's protective shield. The repetition of his contrasting viewpoint regarding the personal nature of his journey and the lack of military interest in Hunza is revealing and convincing. Bronislaw Grombchevsky visited only once and no further activities took place. The memoirs of M. Nazim Khan and Qudratullah Beg's narrative of the great achievements by the *wazīrkuč* are ample proof of their siding with the British, which in hindsight was – in their strong opinion – the only wise option and a successful policy in the long run. Consequently, the views given in their written versions might have been adjusted by subsequent events and experiences. Bronislaw Grombchevsky's diary is the only written account that was prepared on the scene and documents the nature of personal material expectations, the provision of hospitality as a business, and the meddling in unauthorised negotiations. The diary convincingly documents the proactive leaning of *mīr* Safdar Ali Khan and *wazīr* Tara Beg towards Russia which qualified them to become the arch enemies of the *Raj* and provoked a reaction.

British intelligence was not quite clear about the nature of Bronislaw Grombchevsky's offer or the demands by *mīr* Safdar Ali Khan. During Francis Younghusband's expedition from the Oprang to Shimshal, both explorers met amicably as professional intelligence officers in Kahairan Aksai between 23 to 25 October 1889 and spent a couple of days exchanging their views. Francis Younghusband was quite impressed by his counterpart: "Unlike other Russian officers whom I have met in Eastern Siberia, Captain Grombtchevsky is a thorough gentleman, with an open genial manner, to which one takes immediately, and which undoubtedly has also a very good effect upon the natives.

He strikes one, in fact, as being the typical kind of man for the work he is engaged upon, and had he more money at his command would undoubtedly be a very successful explorer. He is, I have heard, considered by the Russian Geographical Society to be their rising explorer, and his journey into Hunza, in 1888, showed him to be possessed of plenty of boldness and daring."[108] The two opponents got along very well and freely exchanged their world views and their estimations of Hunza's ruler and his behaviour.

Bronislaw Grombchevsky repeated that his travel to Hunza had been a private enterprise: "Referring to his visit to Kanjút last year, Captain Grombtchevsky said that he had repeatedly written to the Rájá to say that he had no political mission whatever. On his arrival at Hunza the Rájá intimated that he would receive him in full durbar, but on Grombtchevsky's mentioning that he intended to ride up with his Cossacks, the Kanjúti officials raised objections. The Russian, however, said that, if he could not ride, he would not go at all; and as the Rájá had already arrived in durbar, and given out that an ambassador from the Great White Czar was coming to treat with him, the Kanjútis had to give in, and, although it was only a few hundred yards distant, Grombtchevsky rode up to the durbar with his escort."[109]

Algernon Durand had studied Francis Younghusband's narrative quite carefully. From the subsequent meeting and enquiries with *mīr* Safdar Ali Khan in Gulmit in December 1889, he recorded in his report: "When Captain Gromchevsky was in Kanjut last year some agreement seems to have been come to. The Raja referred to it during a stormy interview with the bearer of the Government presents sent by the British Agent, but unfortunately did not produce it as he threatened. Captain Younghusband seems to think that arms were promised, the Raja says money, and that it was not paid. The value of the Raja's protestations of attachment to the British Government is of course of the smallest, and may be judged by the fact that he seems this year, either to have invited Captain Gromchevsky to his country according to the latter's statement to Captain Younghusband, or at all events to have shown no great desire to keep him out. For from information brought back some days ago by a man in Safdar Ali's confidence and in the pay of the British Agent, it appears that Captain Gromchevsky wrote asking why he had allowed the British Agent and his party to enter Hunza in contravention of the agreement of last year, and requesting the Raja to send an official to conduct him

The photograph of Gulmit kalha was taken by Bronislaw Grombchevsky on 29 August 1988. The upper fortification of Ondra is visible on the hilltop in the background. He recognised the strategic importance of Gircha Fort when reaching the Hunza Road's first major control post down from Kilik and Mintaka passes on 14 August 1888. The village consisted of twenty farmsteads protected by customary fortifications and was the biggest settlement in Upper Gojal. Source: Bronislaw Grombchevsky (1888: 124–125), the images are reproduced with permission of the Scientific Archive of the Russian Geographical Society, font 45, opis 1, file 4, sheet 56 & 76

Group photograph taken by Captain Bronislaw Grombchevsky at the end of his three-day meeting with Captain Francis Younghusband in October 1889 in Khairan Aksai. Francis Younghusband (1890: 67) mentioned in his report that 'there were fourteen different nationalities represented. After this, when the baggage had started, I made the Gurkha escort march up to the Russian and salute him by presenting arms. He was very much pleased at this compliment, and then he also formed up his escort to salute me, and we afterwards took leave of each other with every assurance of friendship, and with hopes that we should meet again." Source: See for details Wawrzyniec, Popiel-Machnicki, Adam Pleskaczyński and Konstancja Pleskaczyńska (2017, Table 41); the photograph was taken on 25 October 1889 (13 October 1889 Julian Calendar). Courtesy of Pahar.in

Spotkanie ekspedycji B. Grąbczewskiego z angielską ekspedycją kapitana Younghusbanda w pustyni Raskemu, w listopadzie 1889 r., gdzie znaleźli się razem przedstawiciele 21 narodów (str. 127).

1. Kap. B. Grąbczewski, Polak, 2. kap. Younghusband, Anglik, 3. preparator Conrad, Niemiec, 4. 6 kozaków — Rosjanie, 5. Kozak Matwiejew, Mordwin, znający język mongolski, 6. Tadżycy, 7. Sartowie, 8. Kirgizi, 9. Kandżuci, 10. Wachańczycy, 11. Żołnierze armji angielskiej z indyjskiego plemienia Gurków, 12. Kaszmirczycy, 13. Baltistańczycy, 14. Tybetańczycy, 15. Hindusi, 16. Mongołowie, 17. Kirgizi z Sarykji, 18. Kirgizi tybetańscy, 19. Kaszgarczycy, 20. Toglicy Pachpu, 21. Sarykoltowie

to Hunza. The Raja is reported to have given the non-payment of the promised Russian subsidy, as the reason for his having made friends with the British Government, and to have refused to send any official, but to have replied that if Captain Gromchevsky chose to come he might. In any case the Raja has of course played us false, but as he acted up to his engagement in giving Captain Younghusband a passage through his country, it would be a mistake to press this point, more especially as we cannot prove his having invited the Russian."[110] Both travellers were accompanied only by a dozen soldiers and had to rely on friendly relations and hospitality, which was granted to them in Hunza. Francis Younghusband stressed the latest Hunza raid (taking twenty-one Kirghiz as captives) that took place on the Yarkand road during Captain Grombchevsky's visit in 1888.[111] All of a sudden the Kirghiz had been selected as allies of the British intruders.

When intelligence and news about the Russian activities reached the British representatives in Kashmir, they raised alarm. In their opinion, an invisible border line had been crossed. A tactic was suggested to keep relations in a limbo: "Possibly if we can remain on good terms with him for a year or two the regular receipt of this increased subsidy may keep him fairly loyal to his engagements in the future, but it is unlikely. He has an inordinate idea of his own importance and strength, and he will most likely raid the Yarkand road as of yore. In this case there will be nothing left after stopping his subsidy, but to give him a sharp lesson for which the force at Gilgit will be amply sufficient. The result of such action unfollowed by annexation would be excellent in these parts. The Kanjut Chief's is a name of dread all over the Pamirs, and his successes against Kashmir in former days have given him a fictitious importance in the eyes of the surrounding States."[112]

The outcome is well known; sufficient reasons were identified to send a military force into the Hunza Valley where a decisive battle was fought in Nilt and Thol. The 'Hunza-Nagar Campaign' was a short affair; after reaching Chalt on 28 November 1891, the assault on the Maiun-Nilt line began three days later; the whole campaign ended on 20 December with the total surrender and the capture of the respective capitals in Baltit and Uyum Nagar. The very effective siege of sixteen British officers and about one thousand sepoys – 188 soldiers from the 5th Gurkha regiment, 28 soldiers from the 20th Punjab Infantry, two guns from the 24th Hazara Mountain Battery, seven Bengal Sap-

Gilgit Fort represented an impressive structure in the winter of 1891 to 1892 when the Hunza-Nagar Campaign sealed the fate of independent principalities and when joint forces of the Maharaja of Kashmir and British India succeeded in establishing their paramount power in the Karakoram. Source: Illustrated London News 9 January 1892, page 47. Reproduced with permission © Illustrated London News Ltd/Mary Evans Picture Library

pers and Miners, 661 Kashmir Imperial Service Infantry, 200 roadbuilders from Spedding's Pathans organised in six companies, and some Puniali levies – led to subsequent defeat of the combined Hunza-Nagar forces.[113] The military defeat and loss of sovereignty reversed the attitudes of the inhabitants of the valley towards the British.[114] Mir Safdar Ali Khan fled to China, his half-brother M. Nazim Khan from a morganatic marriage (*arġúndaro*) of his father with a Wakhi woman, Qurbon Bibi from Bori lineage (*qətór*) in Gulmit, was installed as the new *tham,* and a cooperating *wazīr* in the person of Humayun Beg was appointed; both were the guarantors of cordial relations with the Raj. Less than one year later already, 100 men from Hunza and Nagar supported the British garrison in Gilgit on their march towards Chitral: "In the winter of 1892 at the time of the disturbances in Chitral 100 Hunza-Nagar men were suddenly warned to proceed to Gilgit. The orders were issued at 10.30 one morning and by 10 the next night they were at Gilgit, sixty-five miles distant, and from there proceeded by double marches towards Chitral. They are good mountaineers on rock; they can endure extremes of cold well; and, as they are most of them accustomed to shooting, they ought to be easily trained into good riflemen."[115] After their defeat, conquest and the installation of a reliable ruler, Francis Younghusband articulated his admiration for the people who could easily master difficult roads and endure severe challenges. As a strategist, he saw in them the future local forces that would become the Gilgit Scouts and would help to reduce the cost of administration and border security.

At some distance, Francis Younghusband remembered the pleasures of competing with Bronisław Grąbczewski. In 1926 Younghusband, who was introduced by the president of the Royal Geographical Society as 'the *doyen* of all Himalayan explorers in this country',[116] presented a paper stating: "Such a region must obviously attract geographers. Grombchevski and I have made but a bare reconnaissance of it, and it must be examined more at leisure; for we had to explore it at a time when we were not only at rivalry with one another, but also when both of us had to beware of attack by raiders from Hunza. And I hope when it is thus explored

The pencil and watercolour of the Marquess Curzon of Kedleston was a study for a drawing by Leslie Ward. Source: Vanity Fair 18 June 1892. Courtesy of the National Portrait Gallery

the result will not be merely a map, but such a description of it as will enable that innumerable multitude who cannot go there to see and enjoy its incomparable beauty. All Grombchevski and I could do was to plod along the valley bottoms and ascend the passes."[117]

In the meantime, their findings had been transferred into maps and were discussed in geographical societies. The map produced by Bronisław Grąbczewski and translated by William Barnes Steveni confused its readers about the location of Hunza because 'Kanjut' was on the right side of the river and Hunza on the left.[118] Gottlieb Wilhelm Leitner mentioned the fact, although acknowledging the rest of the map.[119] The cartographic authority James Thomas Walker regarded the Pamir map as a pioneering effort: "Consequently, when Captain Grombchevsky's map was published it was immediately accepted as the first work of a trained geographer in regions which had not yet been fully explored. It was therefore taken as the basis of the map of the Pamirs and adjacent regions, which is given in our 'Proceedings' for last April, to illustrate Captain Younghusband's account of his travels."[120] Bronisław Grąbczewski is still remembered more often because of his meeting with Francis Younghusband on the Pamirs than because of his being the first Russian officer to visit Hunza. The latter visit only contributed as a prerequisite for the hawks of the *Raj* to promote an immediate conquest of Hunza and Nagar.

The section taken from George Curzon's map 'The Pamirs and adjoining territories of Central Asia and India' shows the British sphere of interest in the aftermath of the Hunza-Nagar Campaign of 1891 which was extended westwards towards Chitral where in 1895 the Chitral Campaign was fought. The map was prepared by Henry Sharbau to accompany his three-piece article series in the Geographical Journal in 1896. The map provides great topographical detail from all kinds of available sources and gave its scale in statute miles and versts. It claims to be the authoritative source of its time and presents the geopolitical landscape in its physical properties, orography, infrastructure and settlement patterns. Source: George Nathaniel Curzon (1896). Courtesy of the Pamir Archive Collection

RANGKUL PAMIR
Rang Kul
SAREZ PAMIR
Murghabi
Mustagh Ata
25350
YARKAND
Yangi-Shahr
Kargbalik
Tashkurghan
PAMIR
LITTLE PAMIR
TAGHDUMBASH PAMIR
MARIOM PAMIR
PAMIR-I-WAKHAN
MUSTAGH
KARAKORAM
KUEN LUN
KANJUT
NAGAR
HISPAR GLACIER
BIAFO GL.
GILGIT
Rakaposhi
25550
Khunjerab P.
Mintaka-aksai
Kilik P.
Shimshal P.
Lofty snow clad mountains
Chalt
Nomal
Gulmit
Hunza
Baltit
Bunji
Hot Sulphur Spring
Chakmaktin Kul
13100
Beshtarik
Kugiar
Ak Masjid
Bozai Gumbaz
Khora Bort P.
Irshad P.
Karambar P.

The photograph's title 'hill fort showing entrance destroyed in H.N. bombard' refers to the theatre of war in Nilt along the Hunza Road where the decisive battle in the Hunza-Nagar Campaign took place. Richard William George Hingston (1913: 70–71) observed: "The fort was a human rabbit-warren, a maze of tiny houses and narrow streets without any regularity or any design. Every little village in this district has its fort, in fact the fort and the village are one. Isolated houses were rare along this route and the whole countryside appear[s] to have congregated in the interior of these old mudden forts the protection of which, in olden times, was no doubt a necessity for them." In 1892, Arthur McCormick (1895: 114) 'inspected the fort, the blown-up gate, and narrow, dirty streets' in Nilt; his companion William Conway (1895: 220–221) remembered the 'ruined sangars, and the bullet marks on the rocks'. Source: Richard William George Hingston (1913). Photograph No. H. 31 was taken on 19 May 1913 and is reproduced with permission of the Trinity College Library Dublin (MS 10484/7/31)

After the Great Game – triangulation for a common grid

The successors of these last warriors of the Great Game enjoyed a professional form of cooperation. The mountain area, which had functioned as a buffer zone between British India and Russia, now became an area of common interest and communication. After the Anglo-Russian Agreement of 1907 settled the competition between the two imperial rivals, a resurvey of the trigonometrical triangulations and the linking of both grids was decided upon during the International Geodetic Conference in Hamburg in 1912.[121] The Russian network had reached the foot of the Pamirs, and the Survey of India was to expand its measurement network from Kashmir across the Karakoram towards the Pamirs. The observations had been taken northwards and the Indian network had been expanded up to Gilgit by 1911, while their Russian counterparts had reached Pamirski Post (Murghab) from Osh.[122] The missing link was to be established in the following year. In 1913, the British and Russian trigonometrical triangulations were joined at Sarbulak in the Taghdumbash Pamir. Lieutenant Kenneth Mason was the leader of the British group that approached the Taghdumbash Pamir from the Hunza Valley.[123] Captain Richard Hingston acted as 'surgeon and naturalist with the Indo-Russian Triangulation expedition' from April to September 1913. They perceived the Hunza Valley as a most challenging and complex mountain terrain. Richard Hingston wrote in his 'private journal' after reaching Gulmit on 24 May: "The Hunza gorge, through which our route today lay, is stupendous. The gorges of the other rivers, along which we marched, bear no comparison to it in depth and magnitude. … the little narrow track – it hardly can be called a road – … lay along the sandy or gravelly bed of the river and, in other parts, it rose hundreds of feet high up on the sheer rocky cliffs. The construction of this track did not seem to require much actual hewing of the rock as advantage was taken of every natural ledge which presented itself and the intervals between these ledges were bridged over by beams of wood supported by stout pegs and on the surface of which were placed smaller transverse branches and large flat stones. It was on the whole a primitive structure and was usually lost when followed down into the sand of the river bed while frequently it was completely broken by minor branches of the Hunza river which, on such occasions, had to be forded."[124] During their return journey at the height of the flood season, the stretch between Gulmit and Atabad seemed to have been even more impressive as he wrote on 14 August: "Most of our route lay over the narrow ledge-like track which clings to the sheer granite cliffs. The flooded river dashed and eddied thousands of feet beneath and, in places, the track was so narrow and the cliffs so precipitous that a single false step should hurl one down into the torrent. The construction of this primitive road along the vertical face of the cliff by means of wooden pegs driven into the solid rock, and surmounted by flat stones and branches of trees, is a monument to the ingenuity and industry of this hardy mountain race. It appeals to the mind as a more worthy feat than the broad thoroughfares of our great cities."[125] These travellers were keen and trained observers and significantly contributed to the knowledge about the valley. Their reports described the route condition for adventurous travellers as acceptable; their perception of the road and again its comparison with other traffic arteries were again quite controversial. Just before the outbreak of Wold War I this joint effort in triangulation was achieved, and some experimenting with new technological equipment for geodesy and map-making such as stereoscopic photography was performed.

The war and the October Revolution changed the relationship and terminated any interest in cooperative programmes again. The doyens of Karakoram exploration had to retire. Bronisław Grąbczewski himself was awarded the gold medal of the Imperial Geographical Society in St. Petersburg for his findings in Kanjut. Subsequently, he donated his expedition journals to its archive where they were more or less forgotten until recently.[126] The reason was that he had made a distinguished career before losing respect after the October Revolution, as Francis Younghusband mentioned in his talk to the Royal Geographical Society: "But before concluding I would say a word about my rival but friend, Captain Grombchevski. After I had finished my exploration of the Shaksgam Valley and had discovered the Shimshal Pass I met him in another valley, a small tributary of the Yarkand River, on his way to Shaksgam. … I also met him in 1890 in Yarkand, on my third journey – this time to the Pamirs. On both occasions he showed me the warmest cordiality. But I did not hear of him again – except that he was holding important posts in Manchuria – till last year, when he sent me a book he had written in Polish describing his journey, and from the Polish Legation I heard that he had had the saddest ending to a distinguished career. He had risen to high position in the Russian service, but in the revolution had been thrown into prison and robbed of all his goods. Through the assistance of the Japanese he had been able to get back to Warsaw. But he was completely broken down [and has since died]. I was at least able to assure him of the admiration we British had for him."[127] It took three decades to change the public attitude and opinion towards a highly competitive drive to win the loyalty of the Hunza rulers and to present the results as the winning streak in a gentlemen's game in the Karakoram. The political implications had been controversially discussed in diplomatic and intelligence circles. Finally, the proponents of the control of the 'Northern Frontier' had succeeded in integrating the mountain region into British colonial control limited by international boundaries; consequently, they would promote the implementation of an adequate infrastructure.

A sketch of the British Residency in Gilgit drawn at the time when the successor of John Biddulph, Algernon Durand, held the post of Political Agent in Gilgit. After the successful Hunza-Nagar Campaign the residency represented the modern building style; its verandas and windows were copied by local rulers in their forts. The seat of the highest British representative in the Gilgit Agency became the coordinating office for espionage and reporting, planning of local rulers' appointments and interventions to maintain indirect rule, stimulating enterprises and Central Asian trade. Foreign visitors ended up in its premises; thus the residency became the expatriates' hub and meeting place.
Source: Illustrated London News 9 January 1892, page 47. Reproduced with permission © Illustrated London News Ltd/Mary Evans Picture Library

The photograph titled '23 May the expedition fording the Hunza River' shows the approach road to Gulmit by the Hunza River Valley and was taken by Richard William George Hingston, who worked as surgeon and naturalist with the Indo-Russian Triangulation expedition from April to September 1913. He noted in his diary "Today civilization in its simplest form has been left behind. No telephone poles stand along the hilly pathway dignified by the name of road. No rest houses await us at the end of the day's march. Our letters are received by relays of mail runners but the arrangement is our own. We are away in the deep gorges of the Hindu-Kush, 'far from the madding crowd'. Today's march continued along the right bank of the deep Hunza gorge. Along the deep steep cliffs of schist and limestone hung the narrow pathway which, in numerous places, was lost to view in the flow of debris which is continually pouring down upon it. At one part of the road it was necessary to make an individual rush in order to escape the stones which at any moment might have made an unpleasant acquaintance with us. The Hunza gorge is here deep and wide but owing to the river broken up into numerous smaller streams we had but little difficulty in fording it." Source: Richard William George Hingston (1913: 76–77). Photograph No. H. 39 is reproduced with permission of the Trinity College Library Dublin (MS 10484/7/39)

Colonial competition over accessibility and dominance

The successful conquest of the Hunza Valley in 1891 significantly changed Algernon Durand's assessment of roads: "Such were the roads all through this region five years ago. Now a good mountain road is complete to Gilgit, the Indus is bridged at Bunji, a passable road leads to Chilas, and the communications generally are improving. But if the roads are wild and unpleasant for riders troubled with nerves, the scenery to which they give access surpasses in grandeur any that it has been my lot to admire."[128] With British interference in the region, suddenly accessibility had improved. Such a development was linked to the amelioration based on an early private-public entrepreneurship. Edward Knight, the Times correspondent during the Hunza-Nagar Campaign and allegedly commanding one of Spedding's Pathan companies during the Hunza-Nagar Campaign[129], reported to London: "The Gilgit road begar, already much mitigated, will soon be a thing of the past. A properly organised coolie corps is taking its place and the military road, which Spedding is now constructing, will allow of the use of mules instead of men."[130] Edward Knight felt quite privileged to travel on this extraordinary road. "With two or three exceptions only, the officers of the Gilgit Agency, and Spedding, with his staff, who are constructing the new military road, are the only Englishmen who have been admitted into the country."[131] He was quite scared about what was to be expected on his itinerary: "The Gilgit road up to now has been the scene of successive raids, little wars, and troubles of various sorts. Here one did indeed always dwell in the midst of alarms ... And, again, the transport of grain for the troops, a difficult matter on the Gilgit road even with a good organisation, afforded an easy opportunity to certain officials for robbing the State on a large scale; with the result that a very insufficient supply of food used to reach Gilgit, and that the unfortunate sepoys stationed there were left to starve through the winter on the smallest possible rations of grain, ..."[132] After the success in toppling the resistance in Hunza and Nagar, Edward Knight was even more convinced what had been achieved: 'As a matter of fact, the Gilgit road is now as safe as the road from Brighton to London' which he eagerly communicated to his bosses and editors in his articles and his best-selling book 'Where three empires meet'.[133] His success was due to the above-mentioned public-private partnership of colonial dimensions: "As soon as Spedding was informed that an expedition into the Kanjut Valley had been decided upon, he volunteered to withdraw a number of his men from the Gilgit road work and place them, together with some of his staff, at the disposal of Colonel Durand as a sapper and miner corps. His offer was gladly accepted."[134] Algernon Durand was obsessed with the security provided by a stable road access; therefore, he enjoyed the outcome of stabilising the road access from Kashmir as well as extending it into the Hunza Valley. Prior to the campaign he had estimated the cost of improving the Bandipur-Gilgit Road at close to half a million rupees; the maintenance of the agency would be shared between the Kashmir Darbar (one quarter) and the British Raj and would cost one lakh rupees annually. Additional investments in military installations such as forts and additional roads would amount to a further two million rupees.[135] A permanent British presence in Gilgit would not come cheaply.

Francis Younghusband advised improving the road in the Hunza Valley up to the *darband* at Khaiber. A *darband* is a gate at a narrow passage that is easily controlled; the *mīr* of Hunza had established several to control the movement within his domain: "Gilgit in a similar way would be covered by defensive position at Darkot on the Yasin route, and at some point, probably Khaibar, on the Hunza route. Up to this point it is necessary to have good roads so that our troops could be brought rapidly up to them. Beyond them our strength would lie in keeping the roads in their present execrable condition."[136] The most experienced explorer of 'roads and passes' and bold strategist suggested keeping upper Gojal as a buffer at the border with bad roads.

Shortly after the Hunza-Nagar Campaign and the subsequent instalment of *mīr* M. Nazim Khan as ruler of Hunza as well as the endorsement of Zafar Khan's nominal rule[137] in Nagar, the British authorities perceived the Hunza Valley as a 'pacified' region. Captain George Howard Bretherton, who was among the early game hunters visiting Gilgit, summarised the British achievement less than a decade after the conquest: "Until the campaign of 1891 destroyed their [Hunza and Nagar] independence, these rajas used to sell, or otherwise inconsiderately dispose of, their subjects' property as they thought fit, and no Nathan was found to rebuke their cruelty and rapacity. Now, however, times are changed, and it is expected that in a few years the people of Hunza and Nagar, because of their sudden prosperity and immu-

Algernon Durand is depicted with his dog in his lap at his Gilgit Agency post prior to the commencement of the Hunza-Nagar Campaign. Source: The Graphic 19 December 1891, page 724

Arthur McCormick looked 'west over Nagyr from the housetops' on 14th June 1892: Their expedition reports shaped the perception of Nagar as less impressive than Hunza which entered colonial documentation and assessments. Source: The sketch is reproduced in William Martin Conway (1894a: 234)

Opposite
The section of the map 'Bagrot-Hunza-Nagyr and the Hispar Glacier' is the first detailed map of the Nagar area at the mouth of the Hispar glacier that was mapped by Walter Martin Conway in its entirety and with the continuation along the Biafo Glacier. It shows the settlement structure in Hunza and Nagar by the end of the 19th century when fortified villages were located in the Hunza, Miatsil, Hopar and Hispar valleys. Source: Walter Martin Conway (1894). Courtesy of the Pamir Archive Collection

nity from oppression, will have hopelessly outgrown their Liliputian valley, and extensive emigration will be the only safeguard against starvation. As late as 1891 the Hunza men were noted throughout Central Asia as daring robbers. Now that they cannot loot, they have devoted themselves more than ever to agriculture. They are a pleasant, rosy-faced people, much given to games; but they like to arrange their own ports for themselves."[138] In his assessment he describes the transition from the savage 'Liliputian valley' to the civilised realm, even if they still prefer their own games and sports. The forced road access was justified as a liberation act for the local coolies: "Before British influence made its mark on the country, the roads were so bad as to be practically impassable for laden animals; so that, when transport was wanted by the authorities, the unfortunate villagers were collected and forced to carry the loads on their own backs. This system of compulsory labour was called *Kar-Begar*. The men were paid for their work at scales fixed by authorities; but the hire in nowise compensated the cultivator for being dragged away from the sowing of his little plot of ground, or from reaping of the harvest on which he and his family mainly depended for their year's food. In a military centre like Gilgit movements of troops, with their baggage and food supplies, are frequent, and great hardship to the country people was the result. This evil was accentuated by the fact that corrupt underlings, to whom was confided the duty of collecting the carriers, often exercised their powers oppressively and unjustly. However, the British engineer officer has now made good roads all through different States, and public transport has been organised and is kept up; so that at last the country people can pursue their avocations without fear of the bugbear of *Kar-begar*, the dread of which used to haunt them night and day. The consequence of this is that more time and energy are devoted to developing the natural resources of the country, and the villagers are now prosperous and contended. Thus the extension of British influence to include these distant regions has been productive of nothing but good to the people."[139] The veteran of the 'Siege of Chitral' was impressed by the civilian achievements that had resulted from the conquest, and measured the degree of 'pacification' in terms of required personal presence of colonial officers close to the end of the 19th century: "The tribes comprised in this Agency are not naturally fanatical. It is a remarkable fact that nowhere on the northern or northwestern frontier of India, except in the Gilgit Agency, would it be safe for a white man to wander about alone and without weapons. But British officers travel, without guards and unarmed, all through Gilgit, Hunza, Nagar, Yasin, and Chilas, and receive everywhere nothing but courtesy and consideration from the tribesman. ... There are usually, nowadays, scattered about the Gilgit Agency from fifteen to twenty British officers, who are employed in various capacities. For some years past there have been two or three English ladies also ..."[140] The fact that regional domination could be achieved with comparatively little force in such a short span of time contributed to the mutual trust between colonisers and the local rulers appointed by them. Their partnership shaped the path of internal developments through 'indirect rule' and reinforced the British authority regarding external relationships. The implied remoteness and perceived reclusion disguises the fact that the conquest was motivated by imperial designs rather than intended as a civilising mission. A possible Russian interference and a potential competitor's threat had been kept at bay. The remaining worrying factor was Hunza's relationship with China.

Therefore, they directed their main attention towards Hunza and focused their interest there: "Nagar is of less importance than Hunza as it has no outlet to the north, and no relations with China ..."[141]

The gateway to the Karakoram Mountains was wide open, though on difficult tracks in a challenging terrain. Less than a year after the successful conquest, British travellers started to frequent the valley; one of the early visitors was William Martin Conway. His book on 'Climbing and exploration in the Karakoram-Himalayas' was published in 1894 and contributed to his reputation as the British pioneer of mountaineering in the Karakoram. Immediately upon his arrival in Gilgit he realised the strategic location: "The situation of Gilgit has always rendered it an important point, for so many ancient routes converge upon it. Travellers and traders coming from Central Asia whether from the east or westsides of the Pamirs, are naturally led to this place. The several passes that converge upon Hunza lead ultimately to Gilgit, and so do those that give access from the northwest to Yasin."[142] To make access easier for travellers he was dreaming of a railway line across the high Karakoram passes. His perspective of providing access was governed by the British-Indian idea of establishing a network of railway lines as the most modern traffic infrastructure. Thus, Martin Conway thought of a link connecting the Indian system with the Russian railways and Gilgit becoming a commercial hub with a railway station. He was a daring visionary when he predicted the substantial impact of rail traffic crossing the Karakoram Mountains by the end of the 19th century: "… there should come a period of peaceful industry and of good fellowship between the peoples of Central Asia, Kashmir, and the Indus valley – and who shall say that these happy results may not be brought about, even in the near future? – Gilgit must grow to be an important trade centre, and possibly, … a railway junction on the line from India to Kashgar, where the Samarkand branch will turn off! At present there is practically no commerce passing through the place, and its chief interest to India lies in its strategic importance, which cannot well be overrated."[143] Martin Conway and his accompanying artist Arthur McCormick felt so safe that they did not see any difficulty or harm in looting a cemetery in the night before seeing the *mīr* for an audience in Uyum Nagar. They packed the two stolen human skulls and sent their booty out of the valley to the Gilgit station before paying a courtesy call to Hunza on 16 June 1892. After leaving Uyum Nagar they had to cross Miatsil (aka Mamutsil, *mamú ċhil*) by a rope bridge as well as the Hunza River opposite Altit by the same means to reach Baltit. Not all companions were ready to use these country-style makeshift bridges made from plaited juniper twigs (*gal*) or willow (*śask*) twigs. Some went back to Tashot to use the wooden bridge there, and preferred the long detour to reach Baltit. Even the eminent mountaineer William Martin Conway lacked trust in that rope bridge and needed roped assistance by his Swiss mountain guide Mattias Zurbriggen to be led to Ganesh.[144] These bridges scared many visitors as they very often were in a poor state of maintenance and repair, and posed real obstacles to smooth travel.[145] After Conway and McCormick's visit to Baltit Fort they continued their exploration of the route to Hispar and Baltistan.[146] Martin Conway's book and maps popularised the assemblage of Karakoram peaks and their approach routes for like-minded Europeans.

Other explorers were mainly concerned with practical aspects of travel. The commander of the British troops in Gilgit, Captain Ernest James Medley, ventured a prediction two years later: "The road [Punjab-Gilgit-Yarkand] will in fact become the Grand Trunk road from Central Asia to India."[147] He could see the disconnection between two major road systems. The Grand Trunk Road of the Moghul era formed the starting point for the British road and

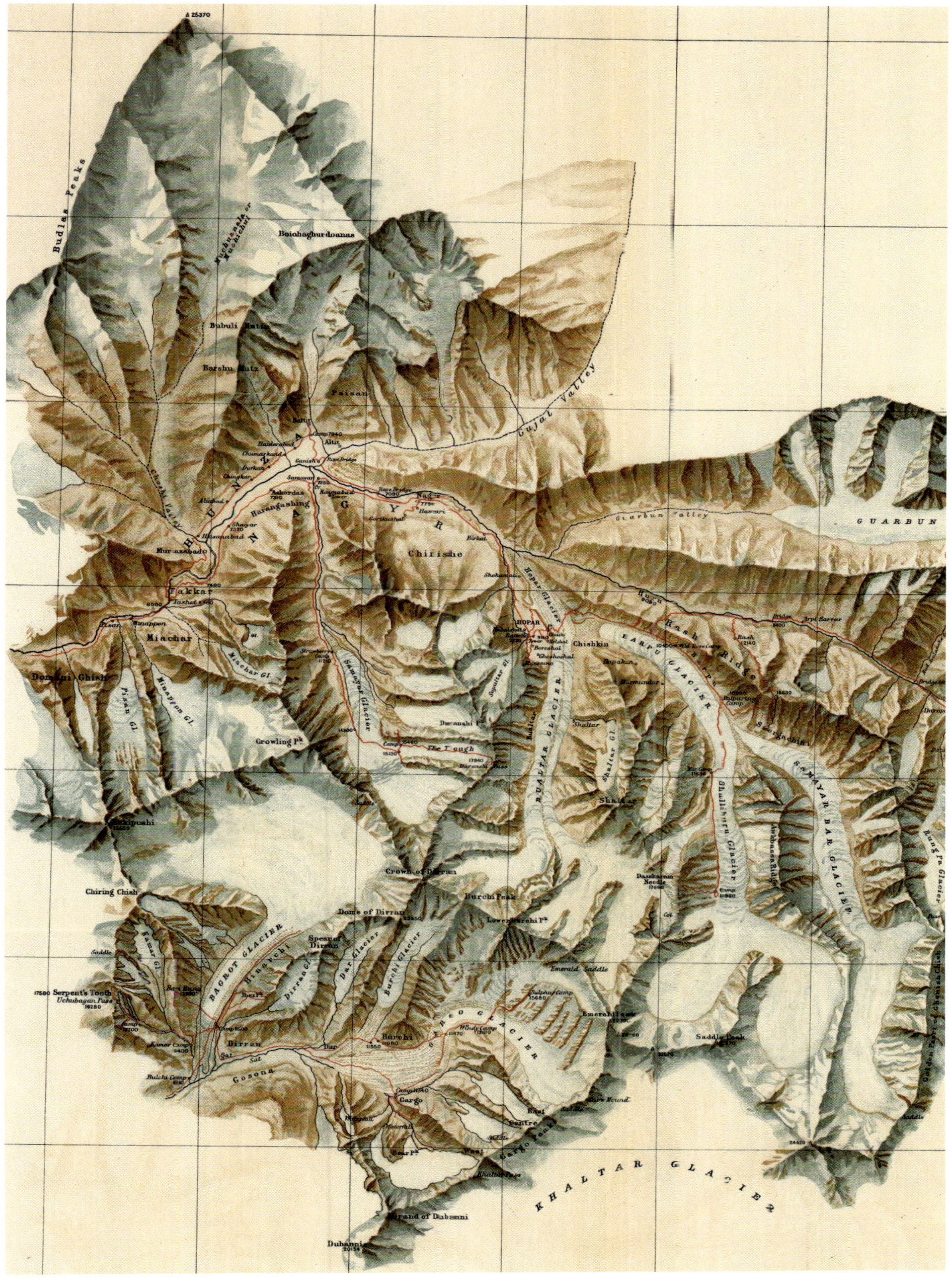

The drawing shows the looting of a cemetery in Nagar by Arthur McCormick and William Martin Conway on 14 June 1892 in search of human skulls that they sent back to England for inspection and measurement. Arthur McCormick (1895: 128) did not mind sketching his and his companion's breach of decency and hospitality during the night before their reception by tham Zafar Zahid Khan in Uyum Nagar. Their theft took place just six months after both principalities were defeated during the Hunza-Nagar Campaign. Source: The sketch is reproduced in William Martin Conway (1894a: 244)

rail network in the northern part of the subcontinent. This was quite detached from the road networks in Turkestan connecting the oases of the Tarim Basin with the Fergana Valley oases towns. From his Gilgit-based perspective, a vision emerged to connect these two networks through a trans-montane artery via the Karakoram Mountains. When Major Ernest James Medley himself ventured on a journey, he realised the discrepancy between vision and reality and assessed the properties of the existing road: "From Gilgit as far as Baltit, the capital of the little state of Hunza, there is now a very fair pack road; but from Baltit up to the British frontier, on the top of the Kilik Pass, the road has been well described as the 'worst in the world'. This badness is due, not so much to the actual condition of the road, as to the fact that, when the pass is open, the valley down which the road leads from the pass is shut, because then the melting snow prevents the valley road being used; while, when the winter comes on and the water in the valley is low enough for ponies to proceed along the bottom of the valley, the pass is then too deep in snow for crossing."[148] After finishing his five-year-long service in the Gilgit Agency he was on his return journey from 'India to England via central Asia and Siberia' in 1900 and realised: "The roads in Asia are not metalled, consequently the going, as long as one is on wheels, is somewhat rough."[149] The imperial travellers' prognosis was taken up much later. It would need other 'tools of empire'[150] to switch from the exhausting travels on foot and/or horseback to wheeled and motorised transport. The physical options were limited as major mountain systems had to be traversed on the way northwards. When Marc Aurel Stein travelled the route from Hunza to Kashgar at the beginning of the 20th century he realised that he was using the 'Gilgit-Transport Road' that would require exact planning because of its special logistical conditions.[151] Percy Etherton assessed the Gilgit route 'as an engineering feat of a high order' in 1909 that crossed an area between Srinagar and Gilgit, which "... is to a very large extent unproductive, so that all the stores and impedimenta of a thousand and one kinds had to be imported. During the construction of the road large gangs of coolies of many different tribes and castes were employed, the management of whom called for a display of that tact and administrative ability in which our frontier officials excel. The road averages some ten or a dozen feet in width, and during the summer months, when the route is open and free from snow, a transport service is maintained from Srinagar to Chalt, two marches beyond Gilgit. The open season extends from May 15th to about the middle of October, after which traffic is discontinued and the mail services are kept up by post runners, who daily risk their lives amongst the dark ravines and steep valleys of the Gilgit route."[152] He attributed the achievement to Algernon Durand who had developed this vital axis as a precondition for controlling and governing the territories beyond Gilgit.

The northernmost British Consulate in Kashgar became a centre of attention as a hub for commercial exchange, strategic interest and political tension. Since 1899, when Andijan in the Fergana Valley was linked with the Russian Middle Asian Railway, the transport costs for bringing European goods via Moscow to Kashgar had become competitive. Among other commodities, chintz, silk, matches and sugar of Russian make entered the Gilgit Agency through this route.[153] The southern routes from British India to Kashgar required forty to nearly sixty days of caravan transport from the nearest railhead to Kashgar. By contrast, the Andijan route covered the distance of 554 kilometres from the railhead via Osh and Irkeshtam in only twelve days and had the additional advantage that the only major pass en route to Kashgar, the Terek Dawan (3,710 m), was a comparatively easy one.[154] Time-wise, Andijan was in pre-revolutionary times the halfway mark on the journey from London to Kashgar.[155] When Clarmont Percival Skrine delivered the first of his two lectures on the 'roads to Kashgar' in the Royal Central Asian Society on 20 March 1925, the connections between British India and Xinjiang were identified as three routes, two leading from Srinagar to Kashgar and Yarkand, and one from Peshawar to Kashgar.[156] After the October Revolution and up to this time the British had been the dominant external force in Kashgar. That would change soon as Soviet commodities arrived again at Kashgar bazaar. Consequently, the British authorities were contemplating another upcoming competition: how to cope with it and what would be the best and safest link road between British India and the Kashgar oasis.

Meeting of route explorers and trade agents for the British Raj. Sir George Macartney, British Consul-General in Kashgar from 1890 to 1918, is sitting on the left with Francis Younghusband (second from right) in 1891, when the latter frequently called at the newly established consulate in Kashgar to meet George Macartney between expeditions and missions. The other two are Richard Beech (with a dog) and Henry Lennard who pretended to be hunters on the Pamirs. Because of Lennard's role in the 'Hunza Campaign' Patrick French (1994: 90) suspected him of being 'an agent of the Indian Government'. Both had accompanied Francis Younghusband on their journey from Leh to Yarkand where they had met with George Macartney and Captain Bronislaw Grombchevsky. The four British travellers and officers met again briefly in Kashgar before Christmas where the photograph was taken (Francis Younghusband 1896: 291–292, 316). Source: India Office Library Mss Eur F197/674(1), reproduced with permission by the British Library

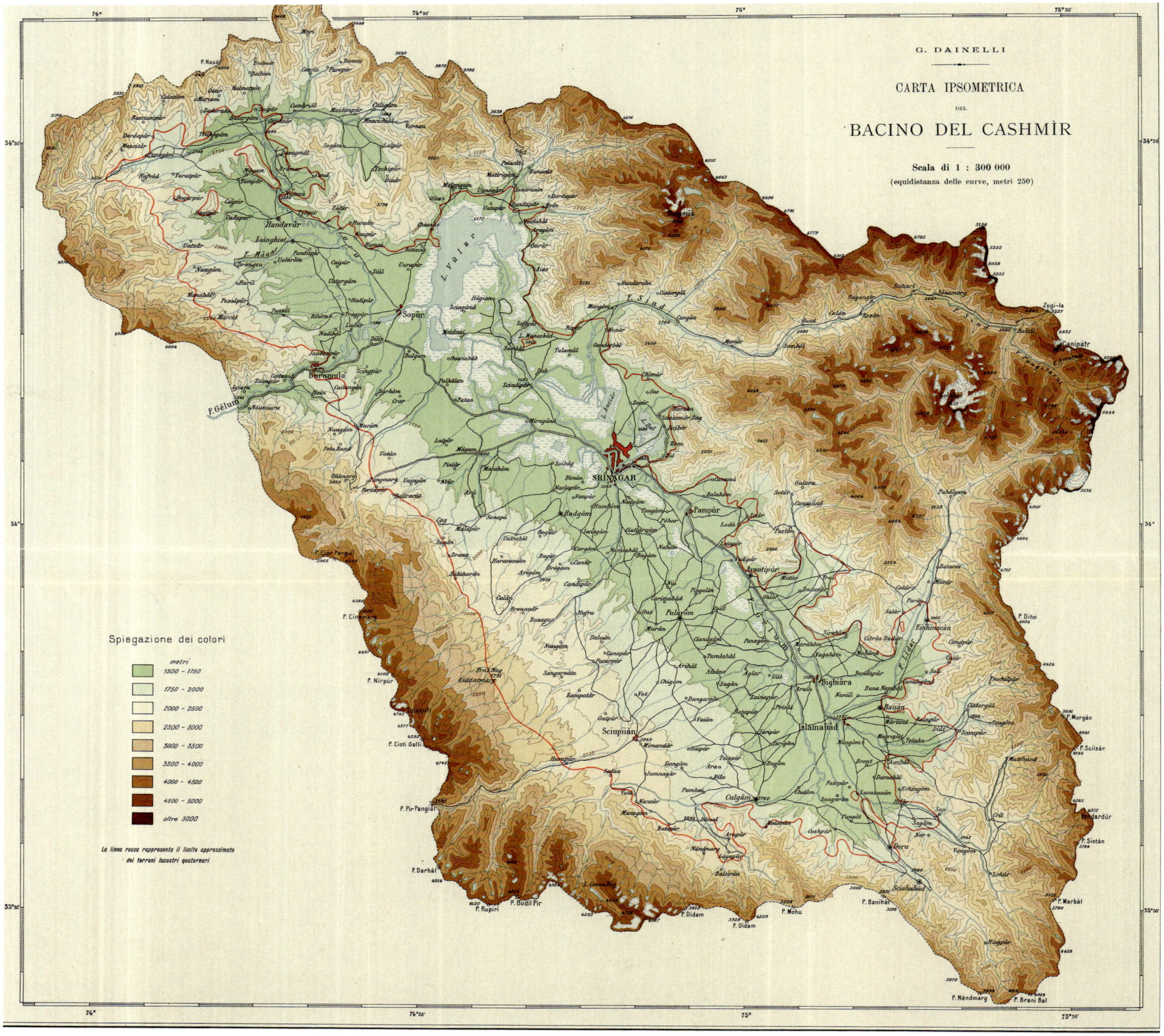
G. DAINELLI
CARTA IPSOMETRICA
DEL
BACINO DEL CASHMÌR
Scala di 1 : 300 000
(equidistanza delle curve, metri 250)
Spiegazione dei colori
metri
1500 - 1750
1750 - 2000
2000 - 2500
2500 - 3000
3000 - 3500
3500 - 4000
4000 - 4500
4500 - 5000
oltre 5000
La linea rossa rappresenta il limite approssimato
dei terreni lacustri quaternari
74°
74°30'
75°
75°30'
34°30'
34°
33°30'
SRINAGAR
Bàramula
Sopùr
Handavùr
Badgòm
Pampùr
Avantipùr
Islamabad
Bigbiara
Sciupiàn
Pulavòm
F. Gèlum
P. Nirpùr
P. Cioti Galli
P. Pir Pangiàl
P. Darhàl
P. Rupiri
P. Budil Pir
P. Didam
P. Mohu
P. Banihàl
P. Marbàl
P. Sintàn
P. Scilsàr
P. Morgàn
P. Ditoi
P. Nàndmarg
P. Brani Bal
Zogi-la
Canipàtr
L. Vular

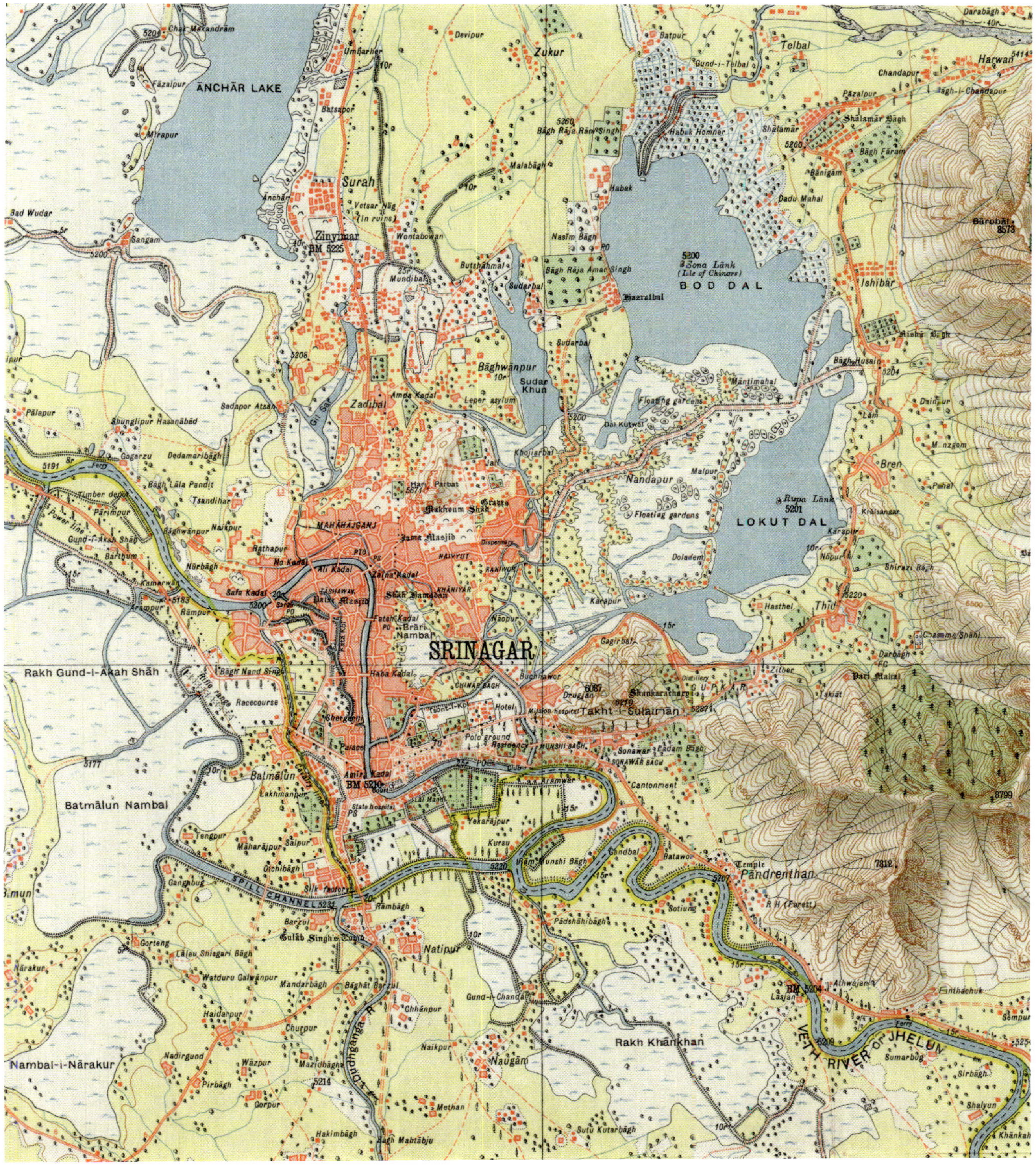

The section of the map 'Kashmir & Jammu' shows Srinagar Town and its environs in 1911. The residence of the Maharaja of Kashmir is watered by the Jhelum River and is located close to the fertile floating gardens of Dal Lake. The Maharaja's Shergarhi palace is located on the left bank of Jhelum, while the British Residency is found just opposite on the right bank, west of Takht-i Suleiman (1,850 m). The present fort Hari Parbat originates from the short-lived Durrani reign and served as a prison for rāja Azur (Uzr) Khan who together with mīr Safdar Ali Khan had fought against the British-Kashmiri troops during the Hunza-Nagar Campaign. After his capture in China and return to Nagar, he was deported to Srinagar as a prisoner and kept there until his death in 1922; see Government of India (1939: 8); Qudratullah Beg (1962: 553). Source: Survey of India (1912). Courtesy of the Pamir Archive Collection.

Opposite
The orographic map of the Valley of Kashmir shows the fertile and wide intra-montane basin with its western accesses from the Jhelum Valley and the Pir Panjal. Northwards the route from Srinagar towards Bendipura and further north to Tragbal Pass (3,638 m) and Gilgit is indicated, as well as the route via the Sind Valley towards Zoji La and Leh in Ladakh. The latter caravan routes are comparatively much more difficult than the western access routes that became the first motor roads connecting Kashmir with the down-country road system. Source: Giotto Dainelli (1934b): Relazioni Scientifiche della Spedizione Italiana de Filippi nell'Himàlaia, Caracorùm e Turchestàn Cinese (1913–1914) Serie II, Resultati geologici e geografici; Vol. 3: Studi sul glaciale. Tavole; Tav. CLXIII. Courtesy of the Pamir Archive Collection

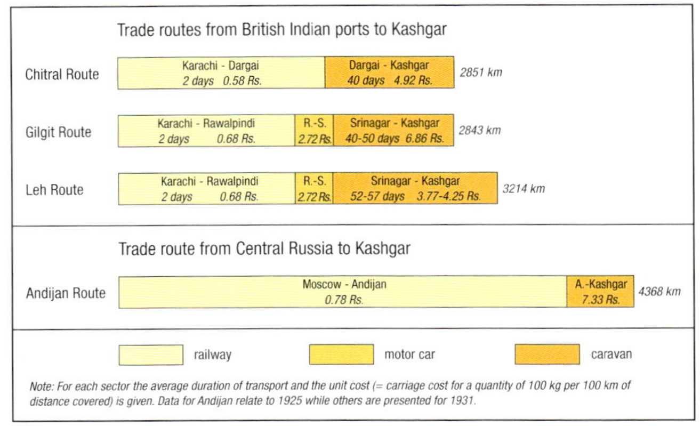

The comparison of cost and time demand for the transport of goods from Karachi to Kashgar reveals that the Chitral and Gilgit routes are shorter than the Leh route by nearly four hundred kilometres. The much longer supply of European commodities from Moscow to Kashgar via the Andijan route has the advantage that the distance covered by caravan is the shortest; the main transport from Moscow to Andijan would use the railway line. Comparing the cost for a standard pack animal load of 2.5 maund (equal to 93.3 kg), on the British Indian approaches, the Chitral route would be the fastest with 42 days and the most cost-efficient with 59.6 Rs; the Leh route would take 56 to 61 days and the transport cost could vary between 66.2 and 73.1 Rs. The Gilgit route would be of intermediate duration with seasonal variations between 44 to 54 days, but would be by far the most costly as transport charges would rise to 83.2 Rs, partly because of the high transit dues in the Hunza Valley. Source: Data for 1931 compiled from Fuad Kazak (1937: 98–99) and General Staff India (1928, 1942;

Three routes compared – Chitral, Gilgit, and Leh

A heated debate began in the 1920s after trade volume had increased, leading to the question which road should be developed as the prime link to Kashgar. Considering distance, the Chitral route was the shortest link from the railhead in Peshawar, followed by the Gilgit route via the Hunza Valley. By far the longest and most physically challenging was the Leh route, despite having a significant trade volume from Kashmir to Xinjiang. Reginald Schomberg had argued in his report from Kashgar in 1927 that the "… most important point … is the opening of the route via Gilgit and the Pamirs. At present it takes 65 days via Leh for goods to reach Kashgar from railhead: and this under favourable conditions. Via Gilgit goods could reach Kashgar from Havelian (N. W. Railway) in 40 days, with only two passes as against six to cross."[157] From his point of view the answer would have been simple and 'Hunza and Nagar would benefit considerably'.[158] About the same time Major Arthur Edward Broadbent Parsons, who later became governor of the North-West Frontier Province, was quite clear about the dual importance of road access. The strategic value of the 'northern frontier' was soberly assessed: "There can be no military threat to India from Sinkiang. Both the Karakoram and Mintaka/Kilik routes are impracticable for a large force in any case, and for a small force if opposed by the forces on the spot. There are no other routes. … But the Sovietisation of Sinkiang would be a source of anxiety."[159] In economic terms he realised British India's disadvantage in competing with Russian goods: "At present, owing to the difficulties of the trade routes, India cannot compete with Russia in anything except dyes and the better class of cloth. It is possible that with the reduction in transport charges India could compete also in the ordinary class of piece-goods and perhaps other articles."[160] He realised that there was an immediate need for action and compared the possible options for improved road connections across the high passes. After providing detailed descriptions Major Arthur Parsons concluded: "… even with the improvement of trade routes Indian trade seems unlikely to be able to compete with Russian, other reasons are not, I think, strong enough to justify much expenditure on the road."[161] Copies of his report were sent to all involved line departments in the colonial administrative network; the various comments and replies stimulated a heated debate in London, Delhi and Srinagar. Between the Department of Commercial Intelligence and Statistics and the Resident in Kashmir letters were exchanged and statements prepared as well as opinions collected from the Consul-Generals in Kashgar, George Gillan and Frederick Williamson, and the Political Agent in Gilgit, Herbert Todd. Major Arthur Parsons maintained his viewpoint which matched that of Evelyn Howell, the Resident in Kashmir, by stating: "The conclusion to which I have come is that it is quite useless to spend much money on improving the road above Baltit. Indian trade can never hope to compete successfully with the Russian trade and the traders themselves are in most cases so unenterprising as to be hardly worth helping. If the dues levied in Hunza and Nagir can be abolished or considerably reduced, the traders will be inclined to travel by the Gilgit route. If this cannot be done they will continue to use the Leh route, even if the road above Baltit should be improved. … The thing is obviously dead. My note [see above] has perhaps been a little misunderstood. My suggestion was that everything except fostering trade was useless to increase our influence in Sinkiang and therefore it would be worth while examining the purely commercial side of the question of improving the trade routes."[162] Subsequently the strategic aspect lost attraction in all considerations, and the Director of Military Operations Brigadier Sydney Frederick Muspratt ended the debate by stating: "On strategic grounds we still hold to the opinion that we do not want too good a road opened up through Hunza. We, however, see no objection to the proposal … of allowing a certain number of traders to use this route …"[163] Afterwards route comparison was made purely on commercial terms. Close to the end of a debate that had lasted more than a decade, a report summarised again the disadvantages and advantages in 1938:

Trade routes across the Hindukush, Pamirs, Karakoram and Himalaya originated in Kashgaria from Kashgar and Yarkand as the oasis centres for entrepôt trade with Afghanistan, Kashmir and British India. All trade caravans had to cross several high passes and met with a variety of environmental, political and economic obstacles. The prominent role of the Pamirs and Wakhan is easily explained as the Baroghil and the Dorah passes are comparatively easy. On the Leh route, five passes above 5 000 metres had to be traversed before reaching the Tarim basin. All three routes significantly differed: the Chitral and Pamir routes had to traverse Afghan territory, whereas the Gilgit and Leh routes originated in Srinagar: thus traders had to comply with Kashmirian dues and rules. Modified from Hermann Kreutzmann (1998: 316)

Chitral route [Kashgar – Wakhan – Chitral – Chakdarra – Peshawar 1,169 km]: "Statistics show that until the last two or three years there has been a regular though small flow of trade along this route. ... It is not clear whether its almost complete cessation during the last two years is due to action by the Sinkiang authorities or to Afghan exactions and restrictions, in Wakhan, but the latter is possibly the chief cause."[164]

Gilgit route [Kashgar/Yarkand-Gilgit-Burzil-Srinagar-Rawalpindi 1,335km]: "The Gilgit route presents probably even greater physical difficulties than the Chitral route, but has the advantage of avoiding Afghan territory and goods pass direct from Sinkiang to India and vice versa. The total value of imports and exports by this route during 1936 was however, less than one Lakh, and any large expenditure on improving it from the point of view of this trade alone would clearly be unjustified. The Government of India propose, however, to consider the possibility of carrying out small improvements on the communication between Gilgit and British India and between Gilgit and the frontier as much from the point of view of the needs of the Agency as from that of Indo-Sinkiang trade."[165]

The Chitral route led through the main bazaar where shops flanked both sides of the narrow lane that was the only thoroughfare for trade caravans coming from the Pamirs on their way to Peshawar via Shandur Pass. The photograph was taken on 17 June 1981

Leh route [Kashgar-Yarkand-Leh-Srinagar-Rawalpindi 1,706 km]: "The Government of India accept the view which is borne out by statistics that the Leh route is the most important of the three routes. ... The physical difficulties are enormous. It is impossible, for instance, to alter the Sasir, bridle the Shyok, and place serais beyond Panamik. The Government of India propose, however, to see if anything can be done to improve amenities of this route at small expense provided the assumption by the Sinkiang Provincial authorities of control over Khotan does not lead to this route being closed indefinitely ..."[166]

During the 1970s the Indian authorities built the Khardung La (5,636 m) road which they claimed to be the highest in the world. The photograph was taken on 24 August 1977

The last route is characterised by the greatest physical difficulties: to span a total distance of 1,170 km it involves at least fifty marches between starting in Srinagar and reaching Yarkand. The notorious Zoji La (3,530 m), Namika La (3,900 m) and Fatu La (4,100 m) have to be traversed to approach Leh from Srinagar. In addition there are five passes to be crossed between Leh and Karghalik such as the Khardung La (5,636 m), Saser Pass (5,249 m), Karakoram Pass (5,575 m), Suget Pass (5,433 m) and Kilian Pass (5,364 m).[167] Several rivers and torrents had to be forded and huge losses in pack animals and goods were caused by these obstacles. For fourteen consecutive marches food and supplies for traders and transport animals had to be carried along. The route was marked by the corpses of travellers and pack animals having lost their lives due to ubiquitous dangers and hazards during their journey. Nevertheless, this route remained through the whole period the most important line for British trade and communication and had prominent advocates.[168] The competition about a future prominence was reduced to the Gilgit and Leh routes as the Chitral route required a transit through Afghanistan.[169]

Although it was by far the longest route, most of the goods were exchanged via Leh. Reasons for this preference are manifold and not primarily linked to environmental

The Leh route connected Kashgaria and Kashmir via Ladakh. The bazaar and oasis of Leh functioned as an important entrepôt for Central Asian traders who contributed to its importance and wealth.
The photograph was taken on 29 August 1977

Collection of transit dues in the Hunza Valley 1920s

Nagar		Hunza			
transit dues	amount (Rs)	transit dues	amount (Rs)	grazing tax[1]	value (*ānā*)
horse for sale	2	horse for sale	5	horse	3
loaded horse	5	loaded horse	4 + 2[2]	cow	3
donkey	3	donkey	2 + 1[2]	donkey	2
cow for sale	1	yak	7	yak	3
sheep or goat	0.5	sheep or goat	1.5	sheep	1
coolie load	1.5	coolie load	2 + 1[2]		

1 The grazing tax collected below the passes to Baltit if animals are released for grazing: horse three *ānā* per night, sheep one *ānā*, donkey two *ānā*, cow three *ānā*, yak three *ānā* (or six *ānā* in the Murkushi grazing grounds). The grazing tax – levied per night – goes to the *mīr* when in uncultivated land whilst that in village lands goes to the village.

2 In addition to the ruler's share the *wazīr* receives his portion of the transit dues directly.

Notes: No taxes are levied by the Kashmir State on goods exported or imported. The rates received the approval of the Political Agent Gilgit in 1924 [one rupee is divided into sixteen *ānā*].

Each merchant gives a present worth four to five rupees to the *mīr* of Hunza's eldest son, similarly a present worth three to five rupees to his second son in Gulmit. Each merchant pays to the headman (*arbāb*) of Gircha, Khaiber, Pasu, Hussaini one rupee each, to the headman of Misgar one to three rupees (according to size of merchandise), to the headman of Ghulkin two rupees and to the one in Gulmit three rupees.

Rates collected from traders transiting the Hunza Valley, approved by the Political Agent in Gilgit. Source: Report on 'Trade: development of trade with India (1 September 1927 – 19 September 1939)' (IOL/P&S/12/2346: 9–11); 'India's concern with Sinkiang' (Government of India (1928); NAI/PR/4001675: 4–5)

conditions. The Gilgit route was almost a quarter shorter in its mountain section where caravans required pack animals and human porterage. On 29 marches between Gilgit and Kashgar it involved fewer ascents and descents of high passes (Mintaka 4,709 m, Ulugh Robat (Subash Pass) 3,900 m) than the other routes. But the Hunza Valley posed enormous difficulties because of its narrow gorge. Difficult passages over scaffolded narrow paths (*parri*, *perien*) allowed transport by porters only instead of pack animals, thus increasing the cost per unit. Numerous river crossings restricted the volume of trade and confined trade caravans to certain seasons. The bad reputation of the former Hunza caravan raiders may have stigmatised this route as well, but more important were the high demands of the greedy rulers of Hunza and Nagar for transit dues such as toll taxes and grazing fees. The *mīr* of Hunza (61 %), his *wazīr* (33 %), the *faráaji* (treasurer 2 %), and the village headmen (4 %) of Gojal (*arbāb*) extracted a total of 2,099 Rs from transit traders between 1892 and 1896. The total income of the *mīr* of Nagar accrued to 571 Rs only.[170] The value of this income can be estimated when it is compared to two other sources of income. The annual revenue derived from caravan raiding by the Hunzukuč was estimated at approximately 1,000 Rs prior to the colonial conquest.[171] After their incorporation into the *Pax Britannica* the payment of annual subsidies was resumed in 1894. The *mīr* of Hunza and Nagar received 3,000 Rs each; the budget was shared between the Kashmir Durbar and the Government of India.[172] These local rulers had managed to secure a significant additional income by controlling the movements through the valley and diminished the prospects of trade despite having been appointed by the colonial administration after their defeat in the Hunza-Nagar Campaign of 1891. They had managed to maintain a certain degree of autonomy in exploiting trespassers in their respective territories.

In Gilgit, the only bazaar of any repute was rather small and mainly operated by outside businessmen. A growing number of Kashmiri shopkeepers had come to Gilgit in the 1860s and started small businesses with goods from Kashmir; around 1895 Punjabi (Hindu and Sikh) had started to open up the 'new bazaar' where in 1897 forty-eight shops were recorded that supplied the British-Kashmiri troops stationed in the garrison.[173]

Since the second half of the 18th century some Kashiro – as the Kashmiri were locally called – had settled in hamlets, which are called even today Kashmiri Bazaar and Kashrote. At the beginning of the 20th century a gazetteer recorded that a prominent position had been filled by the Kashmiri who are "... the largest section of the population in Gilgit proper, but being weavers and carpenters [they] are regarded with some contempt by Shins and Yeshkuns alike."[174] In an assessment five years after the so-called 'pacification' the Political Agent in Gilgit, Stuart Hill Godfrey, stated the potential for trade in his letter to the Resident in Kashmir Adelbert Cecil Talbot: "... received numerous petitions from traders of British India, the Indus Valley, and Yarkand, requesting that arrangements may be made to free the Yarkand route, through Hunza and Gilgit from the obstacles, fiscal and physical, which at present prevent any free flow of commerce, and the establishment of a more extended market in Chinese Turkistan for British and Indian goods. ... principal object is the abolition of local taxes levied by the Mir of Hunza, and the Raja of Nagar in their countries. If arrangements could be made to effect this, the chief obstacle to the development of British Indian trade with Chinese Turkistan through Gilgit will have been removed. ...

The rare photograph of the bridge connecting Hunza and Nagar at the ravine near Donge Das was taken by Captain John Nicholas Price Wood in 1907. Source: Reproduced from 'Military report and gazetteer of the Gilgit Agency and the independent territories of Tangir and Darel' (Government of India 1909: 120)

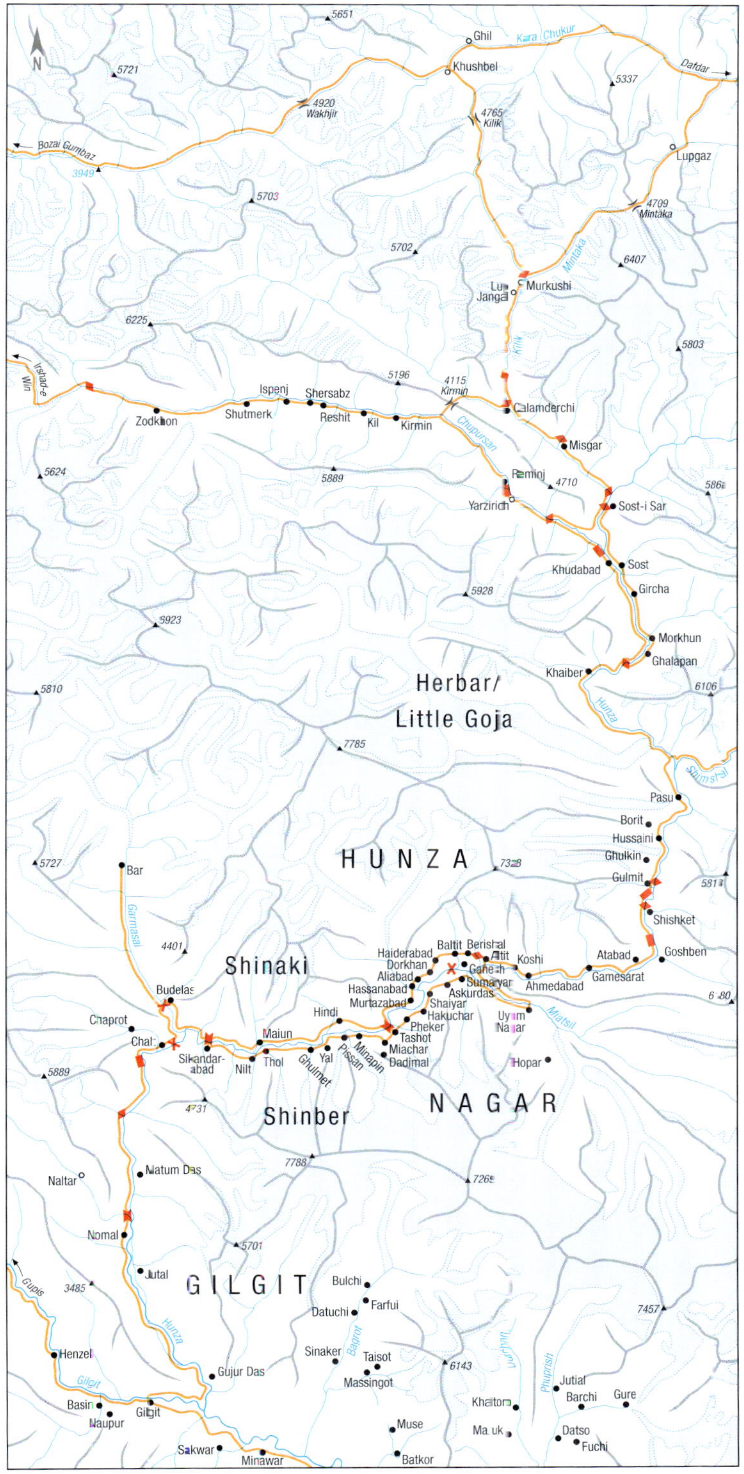

During the 19th century travellers were offered a tightly structured set of available routes for their travel on foot or by horse. Bottlenecks for pack animals occurred at country-specific bridges made of ropes from juniper twigs (gal), birch (tal) or willow (biik) rods. These hampered river crossings as did tricky wooden cantilever bridges (baṣ), easier conditions prevailed in the valley after the establishment of suspension bridges with stable railings. Difficult passages along the road challenged even people on foot where wooden scaffolds supported hanging sections (parri, perien) that hardly allowed a narrow passage along steep cliffs. Travellers were challenged more by these sections than by glacier and river crossings during cold seasons. Mountain passes were regarded as easy traverses in comparison to tracks along rivers. Design by Hermann Kreutzmann

Gilgit Bazaar was the main commercial trading centre in the north of Pakistan, in 1954 it had lost the majority of Sikh and Hindu traders who were replaced by Kashroti and sellers from Gilgit Town's neighbouring valleys; it was not a well-stocked place for purchases, but the only market place for bartering surplus products. From here motor transport was available up to Chalt.
Source: German-Austrian Himalaya-Karakoram expedition 26 May 1954. Courtesy of Gerhard Klamert

Traders here, as in Ladakh, have assured me that if the Gilgit road were opened, British, not Russian, goods would flow into Yarkand. At present there is unfortunately an actual reflux of Russian trade back into Hunza and Gilgit from Yarkand, owing to the Burzil route being monopolized by the Commissariat Department, and the Babusar road not yet fully opened. The few traders who have come down from Yarkand to Hunza this year have brought down Russian cloth and cotton goods exclusively … Moscow, and not Bombay, supply the limited requirements of the Kanjut country, and to some extent of Gilgit also. … There are absolutely no British-made goods of any sort of kind procurable in Hunza and Nagar, … if we exclude a few packages of matches and mirrors brought up by Koliwal traders from the Chilas bazar."[175]

Great potential was attributed to the Gilgit route. Although this thoroughfare could not live up to the vision and projections of Martin Conway, trade between Kashmir and Central Asia via the Hunza valley developed around the British-Kashmiri military headquarters in Gilgit. Edmund Barrow recorded in 1890: "… there is no regular bazar, or any large collection of houses. There are … a few shops close to the fort which supply the wants of the garrison."[176] By the 1920s the small bazaar of Gilgit had grown further and a variety of commercial enterprises were established: "Traders from the Indus Valley districts of Koli and Palas bring up their goods from village to village for sale. The chief articles of their trade are cotton fabrics of white and grey colours, salt in considerable quantity, and also tea, sugar, tobacco and spices. In return for these and in lieu of cash they take grain, gold, ghi and pattu cloths. Pathan traders from Bajaur also bring goods into the Agency viâ Chitral and the Shandur Pass. Musalman merchants from Kashmir established a few shops here and there about 40 years ago, importing their goods from Kashmir, and about 11 years ago [= 1916] these were followed by some Hindu shop-keepers from the Punjab, who opened their shops in Gilgit and whose numbers are slowly increasing. Russian chintz is brought by traders from Yarkand, … while in Yasin and Ishkoman the Wakhis from Wakhan sell ponies, harness, numdas and a little coarse salt in return for grain."[177] Gilgit bazaar was regarded by Reginald Schomberg in his often exaggerated statements as 'the best outside the valley of Kashmir' at the time when dual administration in the Gilgit *wazārat* was terminated in 1935.[178] Five years later Gilgit bazaar had grown to one hundred and fifty shops, which were mainly operated by Kashmiri traders.[179] Very few Burusho and Kho had begun bazaar business as well.

Although the *tham* of Hunza and Nagar were not directly involved in Gilgit's bazaar business they profited from grazing and toll taxes levied from traders crossing the Hunza Valley. The Political Agent had approved the increased rates for 1924; both rulers guaranteed in return a safe passage through the valley to the Chinese border.[180]

TO CHALT.

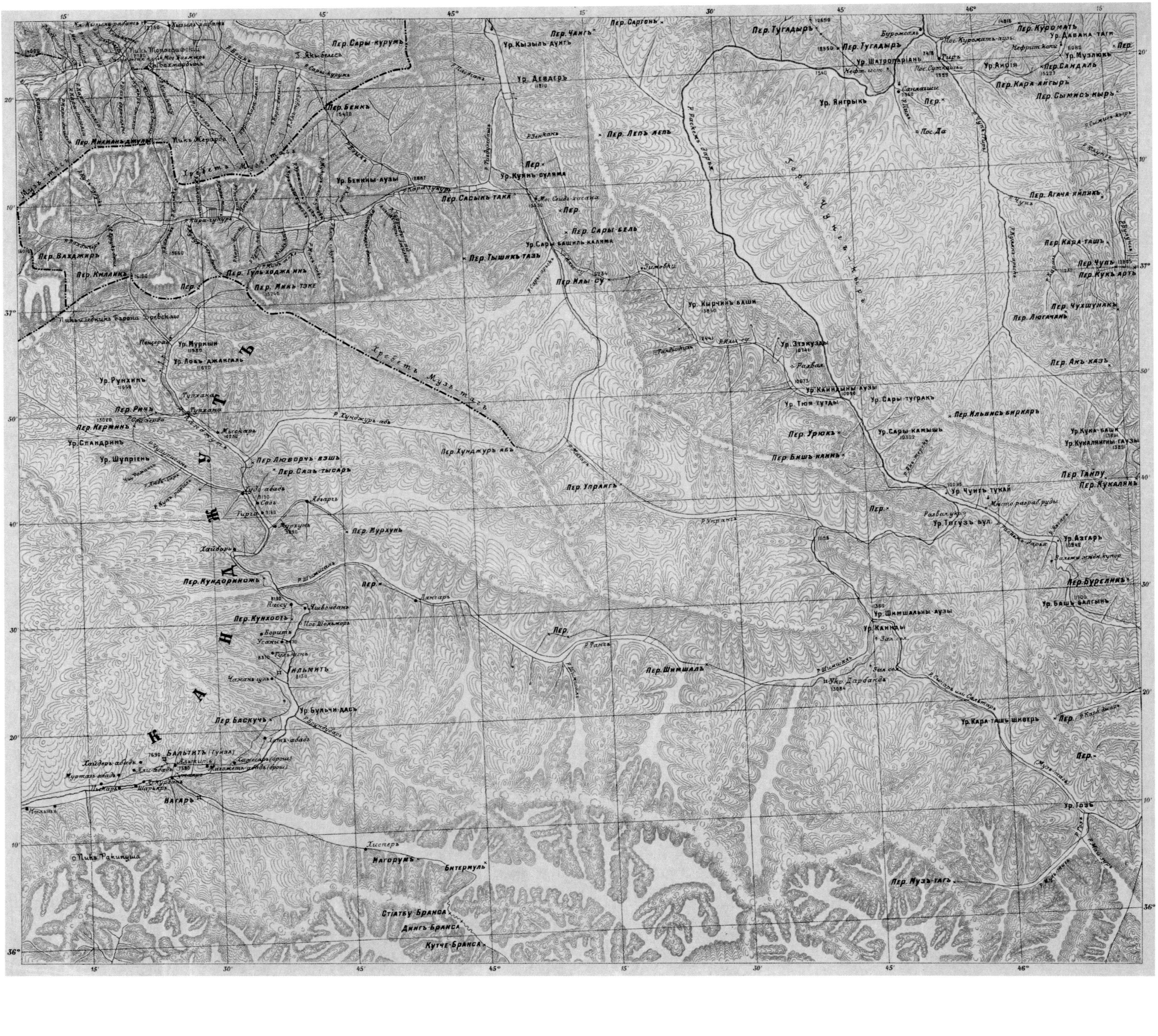
Пер. Сары-курумъ
Пер. Чангъ
Пер. Саргонъ
Пер. Тугадыръ
Ур. Шатропаріань
Пер. Куромать
Ур. Давана-таги
Ур. Музлюкъ
Пер. Сандаль
Пер. Кара-ангыръ
Пер. Сымисъ-ныръ
Ур. Кызыль-Аунгъ
Ур. Ахваеръ
Пер. Бенкъ
Ур. Янгрыкъ
Пер. Лепъ-лепъ
Пер. Михмань-Джулги
Хребетъ Музъ-тагъ
Ур. Беннны-аузы
Ур. Куянъ-суляма
Пер. Сасыкъ-така
Пер. Сары-бель
Пер. Ваxджиръ
Пер. Киланкъ
Пер. Гуль-ходжа-инъ
Пер. Тышинкъ-тазъ
Пер. Илы-су
Пер. Агача-яйлякъ
Пер. Кара-ташъ
Пер. Чупъ
Пер. Кукъ-артъ
Пер. Чухшунакъ
Пер. Люгачанъ
Ур. Кырчинъ-баши
Ур. Мурхши
Ур. Рунхинъ
Пер. Ричъ
Пер. Кермннъ
Ур. Спандринъ
Ур. Шупріенъ
Пер. Акъ-казъ
Ур. Этэкузды
Ур. Каиндыны-аузы
Ур. Тюя-тутды
Ур. Сары-туграхъ
Пер. Ильинисъ-биркаръ
Пер. Урюкъ
Ур. Сары-камышъ
Пер. Бишъ-кинкъ
Пер. Хунджуръ-абъ
Пер. Люворчъ-азшъ
Пер. Сазъ-тысаръ
Пер. Упрангъ
Ур. Чунгъ-тукай
Ур. Тогузъ-бул.
Пер. Тайпу
Пер. Кукалянъ
Ур. Азгаръ
Пер. Мурхунъ
Пер. Кундоринножъ
Пер. Бурсланъ
Ур. Баши-балгынъ
Пер. Кунхость
Ур. Шимшальны-аузы
Ур. Каинды
Пер. Шимшалъ
Ур. Дарбанда
Гильмить
Ур. Бульчи-дасъ
Пер. Баскучъ
Ур. Кара-ташъ-шиверъ
Балтитъ (Гунза)
Нагаръ
Ур. Гозъ
КАНДЖУТЪ
Хиспаръ
Магорумъ
Битермулъ
Пер. Музъ-тагъ
Стіатбу-бранса
Дингъ-бранса
Кутче-бранса
Пикъ Ракипуша

Colonial modernisation of traffic infrastructure

The changes brought about by defeating Hunza and installing a *mīr* of their liking allowed the British Raj to 'modernise' the country according to their own vision. At the turn of the century, Marc Aurel Stein was convinced that the investiture of M. Nazim Khan as *mīr* of Hunza was the right and necessary choice as he regarded him as open to reforms and "… energetic. Road-making, vaccination, and similar Western improvements seem strange as objects of genuine interest in the representative of a family for which intrigue and murder were down to the present generation the main incidents of life. This transformation in its rapidity and evident thoroughness is a striking proof of the results of the pax britannica."[181] However, progress was slow to come; the road retained its challenges in many places.

In terms of accessibility the distinction between the 'backward' and the 'modern' was indicated in the types of bridges and scaffolded, suspended mountain passages (*parīi*) that were the bottleneck for reaching remote places, keeping open a trade route, and reducing travel costs by using pack animal transport etc. Both scared the unexperienced traveller, both feature prominently in travelogues of the 19th and 20th centuries.[182] These trails pass sheer cliffs on support structures of wood or slates that are fixed to the cliff by drilling holes for scaffolding, thus supporting beams and creating narrow passages. They are common in the Pamirs as *avaring* or *ovring* and in the Wakhi-speaking areas as *perien*; other sources mention them as *rafak*, *rafik* or *rafiq*.[183] The famous 'native explorer' Mukhtar Shah aka M-S described *avaring* as a 'hill road along the edge of a precipice'.[184] An observant traveller such as Marc Aurel Stein was quite impressed by them and repeatedly described their construction in challenging locations: "In some places these galleries, or 'Rafiks', as they are locally called, turn in sharp zigzags on the side of cliffs where a false step would prove fatal, while at others again they are steep enough to resemble ladders. To carry loads along these galleries is difficult enough, and for cattle as well as ponies, surefooted as the latter must be in Hunza, they are wholly impassable."[185] These passages made travel not only difficult; they required much higher energetic pursuance than other sections of the route: "Scrambles up precipitous faces of slatey rocks, alternated with still more trying descents to the river-bed; 'Rafiks' and ladders of the type already described were in numerous places the only possible means of getting over the cliffs, often hundreds of feet above the river. The previous five days, however, had accustomed me somewhat to such modes of progress, and it was in comparative freshness that I emerged at last in the less confined portion of the valley above its junction with the gorge of the Khunjerab River."[186] However, he was quite impressed by the enormous speed with which Hunza runners could cover huge distances on difficult mountain paths: "The man had left Hunza on the morning of the 18th, and now he was returning with the Munshi's reply and a considerable load of merchandise which he was bringing back as a private venture. As an illustration of the marching powers of the men of Hunza this feat deserves record. The distance from Hunza to the Kilik is about eighty-one miles, and of the character of the track my experiences so far described will suffice to give an idea. In addition to this and half the return journey, the man had covered twice the route along the Taghdumbash Pamir to and from Tashkurgan, a distance of at least eighty miles each way. Performances of this kind make it easy to understand how the raids of Kanjuti parties could be carried to so great distances, and thanks to the rapidity of their movements, usually with impunity."[187]

On 28 May 1906 Marc Aurel Stein photographed two mail (ḍāk) runners from Hunza (left) at Mirtaka Karaul, the guard post before approaching Mintaka Pass, in company with Sarikoli border guards and their children in front of their yurt. Source: Stein Collection in the British Library. The photograph was previously published by Marc Aurel Stein (1912, I: plate 33) and is reproduced here with kind permission of the International Dunhuang Project Photo 392/26 (57)

Opposite
The so-called 10-Verst map R.IX L8 (1903) is based on the surveys by Bronisław Grąbczewski and was included in the map series that covered the Turkestan borderlands in the Hindukush and Karakoram. 'Kanjut' is mainly known along the Hunza River valley while the eastern and western portions remained unsurveyed. Source: Karta desjativerstnaja Turkestanskago Voennago Okruga, Turkestansk. Voenno-Topogr. Otdel, Pod redakciej Polkovn. Rodionova. 1 : 420,000. Tashkent. Sheet R[ang]. IX L[ist] 8, 1903. Courtesy of the Pamir Archive Collection

During the reconnaissance by the Royal Air Force Intelligence in 1937, a ground survey of accessibility and infrastructure was conducted. In Nagar two suspension bridges and several 'country-style' wooden bridges were used (top to bottom). The upper photograph shows the cantilever bridge across the Chaprot nālā connecting the settlements and village lands in the Chalt oasis; next the 'modern' Chalt suspension bridge becomes visible, connecting the right and left banks of the Hunza River and both parts of Nagar at Harespo. The bottom two photographs show that the reconnaissance group is led over the wooden bridge across the Minapin nālā, and the Hunza Road is reached again after traversing the Tashot suspension bridge. Both suspension bridges were washed away on 7 September 1937 when the Pheker cliff collapsed and caused a major flood.
Source: All photographs by Flight Lieutenant Francis Claude Sturgiss in May 1937; File No. 82(2)-X/39, photos 18, 19, 22, 28; IOL/P&S/12/3288: 175

Local runners were regularly commuting between Hunza and Kashgar and not only managed to cover the distance in a short time, they even made use of their physical strength to bring some commodities from China that were all transported on their backs on these challenging mountain paths, for which feats they were admired and respected. In a later travelogue, Marc Aurel Stein remained intimidated by such passages, and his narrative repeatedly referred to the route obstacles as 'rafaks or ledges of brushwood and stones along otherwise impassable precipices' and 'little galleries built with sticks'; many other travellers shared his view: their ordeal along these sections could only be surpassed by the experience of shaking bridges.[188]

The bridge gained symbolic importance, and its improved version became the visible sign of progress. The only previously known country bridges were those made of plaited juniper twigs (Burushaski *gal*; Wakhi *dut*), consisting of at least three interconnected ropes that gave support for feet and hands to keep the balance while crossing.[189] One of the earliest technical descriptions of a predecessor of 'modern' suspension bridges was provided by Robert Woodthorpe in his 'note on bridges at Gilgit': "The suspension bridges of this country are made of nine plaits of twigs bound together in three. One triple plait forms the footway about 6 inches wide, the other two form the suspension ropes and hand rails, at a height of from 2 to 3 feet above the footway, and are kept apart at intervals by forked sticks. The footway is connected with the upper ropes by further plaits at intervals of about 6 feet. These bridges are capable of bearing from 12 to 20 men on them at once, though, as a rule, 12 is considered the limit of safety. The bridge at Chérkala has a span of 120 yards. They last at least a year. If the footway, instead of consisting of one triple plait only, was formed of two triple plaits kept about 1 ½ or 2 feet apart and lightly planked, and the construction of the bridge slightly altered and strengthened, I imagine the country bridges would answer all purposes, at any rate as a temporary measure, and till such time as the requirements of the new cantonment were fully ascertained, the best site for a bridge decided on, and the necessary materials for a permanent structure obtained. The materials for a twig bridge having been collected by the villagers from the hills above Gilgit, the ropes are plaited in four days, and the bridge is put up in one or two."[190] Robert Woodthorpe seems to have been fearless in crossing these regular features along their route. The bridges across the Hunza River, such

The undated photograph of the rope bridge (gal) spanning Miatsil on the way to Nagar was taken by Reginald Schomberg. The difficulties for load-carriers to cross the shaky and steep gal bridge are clearly visible. Reginald Schomberg (1935a: 110) remarked during his visit in the summer of 1933: "We crossed the Nagir river by a double rope bridge in a bad state of repair, as are so many other things in that state. There was a large rock in midstream and, as the span was too wide for a single bridge, these abominable contraptions were hitched on to either side of it. It took an immense time to traverse, for laden coolies necessarily go but slowly. Copyright Pitt Rivers Museum, University of Oxford; accession number 1998.292.519

The oblique photograph of Gilgit bridge was taken during the flight mission on 8 November 1932. The bridge connects the settlement and garrison of Gilgit – squeezed between the southern terraces that are bordered by mountain ridges and the river bank – with the northern side from where the Hunza Road commences. Konedas is still a deserted place containing only some barracks for the security forces controlling the bridge. Gilgit Fort is situated close to the bridge on the right bank of the Gilgit River. Between fort and bridge the few bazaar shops are located. The substantial residence of the mīr of Nagar is visible as a major structure south of the Punial Road that leads from the Gilgit Fort and bazaar west into Ghizer. The Hunza House seems to be hidden by trees. Source: Aerial photograph titled 'Gilgit bridge and village looking South. 43/I B/1 2' (IOL/P&S/12/1993: 173)

as the ones between Chalt and Harespo, between Gerelt and Sumayar, and between Khaiber and Morkhun, were quite challenging for outsiders.

Another means to cross rivers was by inflated skin rafts (*jáalo*) that were operated by *jaalwán* at places that became fordable during cold seasons.[191] The periods during which glacier melt receded offered much easier passage than the high flood season when even rafts faced difficulties and only bridges allowed the major rivers to be crossed. The local bridges contained no iron or steel parts, being made from juniper twigs or, if bridging narrow passages, from wooden planks based on cantilevers. 'Modern' times were heralded by the steel rope. It could be a single steel cable fixed on both sides in solid rock or a special anchorage of stone heaps. The traveller would sit in a basket (*garaḍí*) that was pulled hand over hand along the rope from one side to the other. These are still found today at places with little traffic to serve small hamlets. The real step forward was the introduction of wire and plank bridges (*šardáṅ, śardáne baṣ, gal, dut*) that allowed laden pack animals to cross. In order to promote progress the Political Agent Major Archibald Duncan Macpherson highlighted in the Gilgit Diary for January 1915; "Communication between Gilgit and Hunza and Nagar threatens to be cut off this summer altogether except by rope bridge as, apart from the fact that it has been found impossible to re-erect the Buladas bridge (near Chalt) which was washed away in the high floods of last summer on its old site (the only one possible for the purpose) owing to a change in the river bed there, the cliff face beyond Chalt, where the road now goes, has crumbled away to such an extent as to render it altogether unsafe. It is proposed, therefore, to take the road in future on the other side of the main river, but this will necessitate the construction of a new bridge for which no arrangements have yet been made."[192] Spanning a bridge across the Hunza River was quite a different affair to crossing its tributaries at certain places. The Political Agent was concerned that communication would remain dependent on country-made bridges that could be washed away at any time. The time was ripe to propose a bigger scheme and to apply modern technology in a bottleneck location as the widened river basin of Chalt and Sikandarabad was perceived.

The Hunza Road would be further improved by constructing some suspension bridges to shorten the passage. To begin with, the 150-m-long Gilgit Bridge was improved to guarantee a safe passage for laden animals. The Resident Engineer Lieutenant-Colonel Chetwynd Henry Haswell inspected Gilgit Fort in August 1921 and saw 'a very fine suspension bridge of 500' span which crosses the Gilgit River and carries the road to Hunza and Nagar'.[193] Smaller wire and plank or cantilever bridges would span the Naltar, Gwachi, and Chaprot *nālā*. After proceeding to Chalt, the Hunza Road would change from the right bank of the river to the left bank above Chalt and after crossing the Garmasai *nālā*. An eighty-metre-long bridge had been constructed between Khizrabad and Sikandarabad in 1915. The road would re-cross the river after the border between Shinber and Nagar Proper via the Tashot Bridge of equal span at Chikas, between Hindi and Murtazabad, before entering the main oasis of Central Hunza. Both bridges were opened for traffic in May 1916.[194] The route was changed in the aftermath of a cliff collapse near Pheker that blocked the river, causing a flood that washed away both bridges, in Sikandarabad and Tashot, on 7 September 1937.[195] Reginald Schomberg commented on the state of affairs eight years later: "Both these bridges have since been washed away and no one misses them. They were useless and extravagant. There is one bridge over the river some six miles from Chalt which is used for traffic to Nagir, and persons who go to Hunza keep on the right bank of the river. The arrangement suits

The bridge linking Gilgit town and its bazaar with public offices across the river in Konedas has lost its importance as the sole crossing. Nowadays several bridges connect both banks and provide the road link to Danyor, Nomal and the Hunza Valley. The photograph was taken on 29 March 2019

everyone, but what is far less suitable is the shocking alignment of the road, long, tedious, exasperating, and indifferent to gradients. It is not easy to strike out a new trace after an alignment has been made, and thus remedy the original mischief of some charlatan or contractor. It is, of course, primarily a matter of money, but it is generally far cheaper and less wasteful in the long run to build a solid track than to tinker eternally with a rotten one. The drawback is that a road that constantly needs repair is a regular goldmine to the contractor (that curse of the modern world and of India in particular), so the hill road never enjoys a good start – it is doomed to be a permanent invalid."[196]

The photograph was taken in the palace garden in Karimabad on 1 September 1930 by Aurel Stein during his journey from Gilgit to an archaeological expedition in Xinjiang. The Hunza levies were assembled in two platoons for a guard of honour. The colonial authorities provided tham M. Nazim Khan with the means to give employment to levies and bodyguards for his own protection. Source: Stein Collection in the British Library. The latter is reproduced here with kind permission of the International Dunhuang Project Photo 392/34 (52)

Preparing and maintaining a sophisticated route

The hazardous event that blocked the river and caused a flood had led to the reopening of the section between Khizrabad and Murtazabad on the right bank of the Hunza River. The advantage was that any crossing of the main river could be avoided between Gilgit and Hunza. The travelling time was shortened by levelling and improving the difficult and precipitous track. In the aftermath, Nagar was connected with the main thoroughfare only by a single suspension bridge.[197] The colonial road-builders had invested to provide the improved bridges leading up to Hunza and Nagar, from where the passage continued on the right bank all the way to upper Gojal, on a comparatively much more difficult track. Reginald Schomberg was not at all content with the efforts of the colonial administrators to improve the road conditions in their area of control. When he returned from Baltit to Gilgit he was complaining: "... we should have to recross the Hunza river, over yet another rope bridge, as detestable and indispensable as its previous fellows. So once more we went down the dismal pack-road again. How often have I wished that the officials who tour in these parts were obliged to walk instead of ride; as, if they did so, many of the superfluous gradients and absurd alignments of the path would disappear. When these great men ride, however, the imperfections of the track do not strike them with the same force as they do poor devils on foot, chiefly coolies, who have to trudge along, cursing the needless ups and downs. It is the contractor who benefits. These roads are invaluable to him, and it is a significant fact that when the road slips it is never repaired, but a new track is made. It was one of our melancholy diversions to comment on the skill of the contractor in devising unnecessary but profitable work on these roads. But after all, contractors, especially in Gilgit, must live, make large profits, and be free from all control."[198] Besides his dislike of contractors (*ṭhekedār*) these comments emphasise the aspect of porterage (*baldákuyo, borwar, muzdūr*). The heavy loads were mainly transported by load-carriers along difficult passages in the Hunza Valley. The *mīr* took such services as part of forced labour, but charged traders for these services that contributed significantly to his income.

The newly established levy post of the Gilgit Scouts in Murkushi took possession by hoisting the Union Jack at this short-lived military station. From here caravans and travellers would proceed to China by Kilik or Mintaka passes. In 1937, the station had been manned from April to mid-August to guard the border owing to the disturbances in Xinjiang. In the following year Murkushi remained abandoned, but two additional military posts were installed in Qalamderchi/Misgar and Darwaza/Shimshal to guard the border region and to watch the ongoing developments in Xinjiang. Source: The photograph was taken by Flight Lieutenant Francis Claude Sturgiss on 20 May 1937; File No. 82(2)-X/39, photo 69; Administration Report of the Gilgit Agency for the years 1937 and 1938 (IOL/P&S/12/3288: 164, 177)

Searching for alternative routes

The comparatively short but challenging section across the Hunza valley became the most expensive part of the whole journey between Rawalpindi and Kashgar. The transit dues were regarded by the traders as a form of special tax payment for the right of passage (*octroi*) rather than a contribution to the maintenance of the route. These extra costs restricted the volume of trade besides physical features such as seasonality of passage or the condition of track and river crossings. In addition, the availability of supplies, porterage and pack animals was limited, just as fodder and fuel were scarce. The unit cost of caravan transport on the Gilgit route amounted to the highest of all routes. This undesirable and unsatisfactory state of affairs remained unchanged for the whole period under consideration, and numerous efforts were made to improve the routes and the conditions for potential traders. Other routes were discussed and proved a certain level of desperation. On the one hand, the debate about a difficult mountain road from Leh to Yarkand in dissociation from the Hunza Road was continued; on the other hand, the search for a motor road went on as well. In a summarising memorandum, the British Consulate-General in Kashgar, Kenneth Packman, took up a suggestion in February 1937 to avoid Hunza and Nagar. Introducing a new route that would be an alternative to the Chitral-Wakhan route, he described one that would lead from Chitral across the Shandur Pass (3,720 m) to Gakuch in Ghizer and via Chilinji Pass (5,335 m) to the Kilik Pass (4,765 m) whence it would follow the conventional route to Tashkurgan and Kashgar.[199] The attraction of the latter route was a combination of the Chitral and Gilgit routes while bypassing most of Hunza

During the process of preparing the lease of the Gilgit Agency the British colonial authorities contemplated various routes for safeguarding their supplies in future. The Srinagar-Gilgit route via Burzil Pass was to be avoided. From the railhead in Havelian the Babusar Pass route via Balakot and the Kaghan Valley was favoured, but had the disadvantage of being open in summers only. Consequently, alternative routes from the railhead in Dargai via Dir and Chitral across the Lowari and Shandur passes and a line via Oghi through the Indus Valley towards Chilas and Gilgit came into focus. Either the northbound trade routes along existing pack animal routes through the Hunza Valley across the Kilik or Mintaka passes towards Kashgar or the Chitral route via Wakhan and the Little Pamir counted as second-best alternatives.
Source: The route alignment is based on various archival records.

Three cantilever bridges were built at narrow and salient points of the Hunza Road in Hassanabad, Khaiber and Sost-i-Sar (top to bottom). These places were at the same time borders for internal administrative sections (maqsòo) in Hunza and control posts for traders on their way from Gilgit to Kashgar and back. In Sost-i-Sar the Hunza River had to be crossed and the bridge was stabilised by fixed steel cables. Source: All photographs by Flight Lieutenant Francis Claude Sturges taken in May 1937; File No. 82(2)-X/39

and Nagar territory and avoiding undue taxation. Nevertheless, it would have to face significant physical challenges and several high passes. It had in common with the Leh route that it would never be feasible to upgrade it to a motor road. The colonial officers of this debate and their counterparts in Kashgar were quite uninterested in introducing modern modes of traffic in High Asia. At this stage, in August 1937, the Political Agent in Gilgit, Major George Kirkbride, reminded the concerned parties in a short statement about the 'development of India's trade routes to Sinkiang' that first of all a ground survey of the Babusar-Gilgit- Mintaka route as well as the Chitral-Wakhan routes would be necessary before any new routings should be suggested. He cautioned: "Whether the best laid road from India will ever be able to compete with the Russian trade road [to] Kashgar is for Consul General in Kashgar to say."[200] Suppressing the fact that the Leh route had significantly lost in importance in recent years, the Kashgar representatives promoted a specific opinion and favoured the route with the longest tradition and overall best performance.

In April 1937, Vice-Consul Michael C. Gillet had investigated the commercial potential of both alternative routes and summarized in his report the achievements and advantages of the Leh route: "The Leh-Yarkand route is used for trade since the Central Asian trade has started and the British Government have done very much to improve this road. We are provided with ease and comfort and difficulties have been removed. Efforts are being made to start Sarikol-Gilgit route but that will harm us very much. ... some of the blessings of Leh route:
(i) Grain is stored in each stage between Srinagar and Saser and it is sold to the carriers for their ponies.
(ii) Sufficient transport can be had on both the sides and goods can be transported easily throughout the year, which is not the case on the other road [Gilgit route].
(iii) Duty paid at the post in India is refunded to us within 3 weeks of the arrival of the goods in Leh and can be utilized again but by the other route it will be paid in India after the goods have been inspected in Kashgar. This takes a long time.
(iv) Leh route has no danger to goods and lives ...
(v) Goods via Leh pay 17 1/2 per cent duty in Khotan area whereas goods via Gilgit for Khotan will have to pay 28 1/2 per cent to Kashgar authorities.
(vi) The carriers who run on Leh route can take for themselves as much rations as they like but on the other route authorities allow hardly as much as is required for the way.
(vii) Goods for export in exchange of import are mostly produced and found in Khotan area. Yarkand produced only Charas [Hashish, *Cannabis sativa*] which has been prohibited by the Government. It is easy for the carriers to take

Khotan goods to Leh but if the same goods are sent via Gilgit it will be a longer route and more costly.
(viii) Khotan goods exported to India pay 12 1/2 per cent to Khotan authorities but when they are sent via Gilgit Kashgar authorities charge another 11 per cent.
(ix) About 2,000 ponies, 1,000 camels and 500 donkeys run on Leh route and this number cannot go by the other route and are not available too.
(x) Goods imported in this country via Leh are mostly brought to Leh via Kullu [Kulu] which is shorter even than Srinagar route and cheaper.
(xi) The export from Turkestan via Leh is consumed in Leh, Kullu, Simla and Lasa [Lhasa], but if they are transported to these markets by the Gilgit route goods will not be then so cheaper hence not paying.
(xii) Goods via Leh are sold in Leh in a great quantity and provide us with money for onward journey hire but Gilgit is not so good a market.
(xiii) Goods via Leh do not pay any duty on way up to Turkestan whereas duty is charged on the other route in two places.
(xiv) Leh route is so wide that camels can go easily on it with bulky bundles of felts but Gilgit route is very narrow for this purpose."[201] The fourteen points were listed in order to show the overall better performance; the colonial administration definitely favoured the Leh route. Trade logistics had been considerably improved (on all routes the cost[202] of transport was reduced by more than a third between 1870 and 1931), taxation seemed to be lowest in this sector, and substantial quantities of goods originated from the Yarkand-Khotan region, which is closer to Ladakh than to the other routes. The volume of trade and the infrastructural requirements of pack animals, fodder, fuel and supplies could be managed favourably only on the Leh route in the opinion of the Consul-General at Kashgar.

Technological innovations would reach the British-dominated mountain regions comparatively late and would end the romanticism that explorers and travellers had connected with the remote mountain abodes, passes and valleys they had 'discovered' and explored. For administrators the challenge was to sustain the flow of supplies at reasonable cost and to safeguard the speedy deployment of troops when borders and security were threatened.

A letter by Sir Evelyn Howell in June 1928 confirmed the governing idea; the 'main considerations were military and

On their way to Upper Hunza the reconnaissance group stayed in several guest houses of different quality (top to bottom). The best facilities were provided in Naltar where the rest-house occasionally functioned as the summer residence of the Political Agent. In Hunza, the first stage after leaving Baltit was Atabad where the mir of Hunza operated a rest-house. The quality standard decreased on the way to Gojal; the stay at Pasu was spent in rather small premises, and the last built-up resting-place at Murkushi before crossing either Kilik or Mintaka pass was comparatively simple and frugally equipped. Source. All photographs by Flight Lieutenant Francis Claude Sturgiss in May 1937; File No. 82(2)-X/39, photos 45, 64, 66, 75

The snow hut on Burzil Pass (4,200 m) is a landmark that hints at the high level of snow that can be experienced on top of the pass between autumn and spring. The hut provided shelter to travellers who were stranded due to weather changes. Source: The photograph was taken by David Lorimer in mid-June 1934. Reproduced with permission © SOAS London

the Bolshevic peril' rather than an improvement of exchange relations based on modern road standards and technology: "At that time it was still our policy to make the Hunza, little Gujhal Road, good enough for a trickle of trade but not so good as to destroy what was then held to be the strategical factor. As a rule, Political Agents have preferred the policy of Sir Robert Walpole and let sleeping dogs lie."[203] The concept of modernisation was overruled by classical ideas of maintaining barrier zones, but a non-terrestrial technology of communication and troop deployment would soon affect strategic reservations.

In the opinion of Major George Kirkbride the whole debate had missed its point by only looking at ground transport. Just before vacating his post in Gilgit he communicated some thoughts he had long cherished : "I believe that if there is ever to be a future for Indo-Sinkiang trade it will be by means of commerce carrying aircraft based on an aerodrome at Gilgit in place of the present excellent landing ground. Merchandise is already carried by air in Canada and with perfection of multi-engined aircraft the advantages of such form of transport between India and Sinkiang via Gilgit will be bound to command serious attention. Even now aircraft could reach Kashgar from Peshawar in a day whereas the land journey will always be limited to the few summer months, and for caravans, can never occupy less than six weeks."[204] In any case a serious situation had occurred for the Hunza Valley by this time. The attempt to bypass Hunza and Nagar would deprive both local rulers of a valuable source of income. By the mid-1930s cross-boundary Central Asian trade had come to a halt, and all entrepôts from Chitral to Leh suffered from significant losses in revenue generation. Apparently the British colonial administration was conducting a debate about the comparison of suitable trade corridors while trade had nearly collapsed. The controversial exchange of arguments about the best possible route was pervaded by departmental interests and regional stakeholders in colonial offices. The preferences of the Kashgar Consulate for the Leh route met with those of representatives in Ladakh when the ties to Kashmir were already dispersing and new administrative setups were under consideration. Nevertheless, the debate about regional traffic alignments continued while technological developments were considered in a parallel strain of arguments.

The advent of motor transport – initially by air only

From the late 1920s onwards, the Central Asian mountain regions had been opened up by the first airplane services to important military stations such as Gilgit and Khorog.[205] William Strang observed for the Foreign Office the developments in Soviet Turkestan and read in the *Pravda* of 29 July 1932 about modern aeronautical communications: "… a new air line has been operating regularly since the middle of July between Stalinabad (Dynshambe), the capital of Tajikistan, and Khorog, in the Badakshan Autonomous Area. The route is said to be one of the most difficult in the world. … The journey, which by road takes about a month, is completed by air in from 3½ to 4 hours."[206] As regards flight connections, British authorities had been quicker than their Russian rivals. During an unofficial visit in 1926 the Air Chief Marshal of the Royal Air Force Sir Edward Ellington had reached Gilgit on foot to explore possible landing grounds. The date would later be remembered as 'the beginning of aeronautical development in the most remote outpost of the Indian Empire'.[207] The first planes on a military mission – four Wapitis – had to land in Chilas on their way to Gilgit in February 1929; following the virgin flights from 1931 onwards annual flight operations took place. To counter the Soviet achievements in the Pamirs a spectacular flight by five Hawker Hart planes was organised under the command of Flight Lieutenant Frederick Harry Isaac; after a brief stay in Gilgit they approached the Hunza Valley: "They flew in formation past the peak of Rakaposhi up the Hunza river valley to Nagir, where the Mir ordered the firing of a saluting gun at intervals as the aircraft passed overhead. From the air every house-top was seen to be crowded with people, who demonstrated their enthusiasm by waving their chogas, and throwing their hats in the air. The sight of the north side of Rakaposhi, … brought gasps of admiration from the flyers, and some magnificent photographs were taken."[208] Somewhat surprisingly, the Times of India recorded: "Over Baltit the fliers noticed little enthusiasm amongst the population. But over Nagar they were entertained by the signs of intense excitement below."[209] Similar flights were operated in the Punial Valley and the command concluded: "… once again aeroplanes had proved up to the hilt their ability to operate safely and effectively over the hill country, ability which has largely contributed in the past few years to the tranquillity of what were formerly among

The reconnaissance team of the Royal Air Force Intelligence landed with four Hart aircraft (top) of Squadron No. 39 (Bomber) from Risalpur in Gilgit aerodrome on 6 May 1937. The team leader was Flight Lieutenant Francis Claude Sturgiss who conducted the survey. For the Gilgit residents it was one of nine occasions in 1937 to watch airplanes land and depart. After about a month in the Hunza Valley the group returned with Valentia aircraft (bottom) to their station in Risalpur. Source: All photographs by Flight Lieutenant Francis Claude Sturgiss on 6 May and 10 June 1937; File No. 82(2)-X/39, photos 9, 10, 76; IOL/P&S/12/3288: 173

The aerial photograph of the Gilgit landing ground in Konedas was taken on 8 November 1934 during a flight mission between Gilgit and Risalpur testing the aeronautical conditions and physical state of the local arrangements for regular flight connections. The present airport was subsequently transferred to the right bank of the Gilgit River, enabling a longer runway. The location shown here is on a terrace between the confluence of Gilgit and Hunza rivers and has been occupied by the Karakoram International University.
Source: Aerial photograph titled 'Gilgit landing ground looking North East. 43/I B/1 5574' (IOL/P&S/12/1993: 177)

the most turbulent provinces on the frontiers of Empire."[210] The domination of air space had to prove the ability to control the imperial borders by means of modern technology. Aerial photographs were introduced as a "… graphic record of a notable piece of pioneer work, both in aviation and in geographical observation. … Most of the photographs show three of the five aeroplanes against the mountains, and the gauge of proportions thus secured is most impressive. The fertile gorges shown shelter small communities so remote from the world that access to them is obtainable only by mule pack or (in modern times) by aeroplane. Air Force men gather strange experiences when they drop down from the skies to such places. The men of the mountains … are not curious about the engines of an aeroplane. What puzzles them most is where propeller 'goes to' when it is set spinning. Its seeming disappearance at high speed is to them just magic".[211] Imperial superiority was shown by controlling and demonstrating modern technology while the isolated dwellers in remote mountain valleys were exposed to the feats of modernisation. Aviation was seen as a replacement for road-building in such locations. After the flights of November 1934 under his command, Flight Lieutenant Mackay recommended a rescheduling of the timings as the Political Agent George Kirkbride had suggested that 'aircraft should visit Gilgit in March at the time of Jhalsa when all the local chiefs are gathered there for their annual meeting'. The display of technological leadership should be made meaningful as 'a visit of aircraft would be of great political significance'.[212]

Up to November 1935, forty military flights were operated to Gilgit aerodrome.[213] The respective down-country airports and supply stations were in the garrisons of Chaklala/Rawalpindi in Punjab and Risalpur/Nowshera in the North-West Frontier Province from where Royal Air Force's Vickers Valentia, Hawker Hart and Westland Wapiti aircraft would take off for Chilas and Gilgit.

Anglo-Soviet competition and different air routings became obvious when in the summer of 1935 an outbreak of 'pneumonic plague' occurred in Kashgar. The British and Russian consulates offered their medical support to the city administration by sending for 'anti-plague serum'. The British authorities flew the serum from India to Gilgit airport from where runners took fifteen days to deliver it at Kashgar. This rivalry hit the newspapers when Peter Fleming reported that the British serum had arrived four days later than the Russian medicine and specialists. The latter were flown to Kashgar from Moscow via Tashkent and had the easier access by avoiding the high and steep Himalayan and Karakoram passes.[214] It became obvious that the distance between Gilgit and Kashgar would need an intermediate airstrip for emergency landing and flight control.

In preparing the ground for regular civilian communication by air travel an alternative route to the Indus Valley was inspected the following year; the route that followed the Kaghan Valley and included the crossing of the Babusar Pass was perceived as probably more suitable for civilian air traffic in case of emergency forced landings.[215] The first civilian flight in the Karakoram took place on 16 January 1936 when an Avro X plane operated by 'Indian National Airways' arrived and it was recorded that 'Sir Eric [Teichman] breakfasted in Gilgit and dined the same day in Delhi'. The pilot seemed to have been impressed as he "… expressed

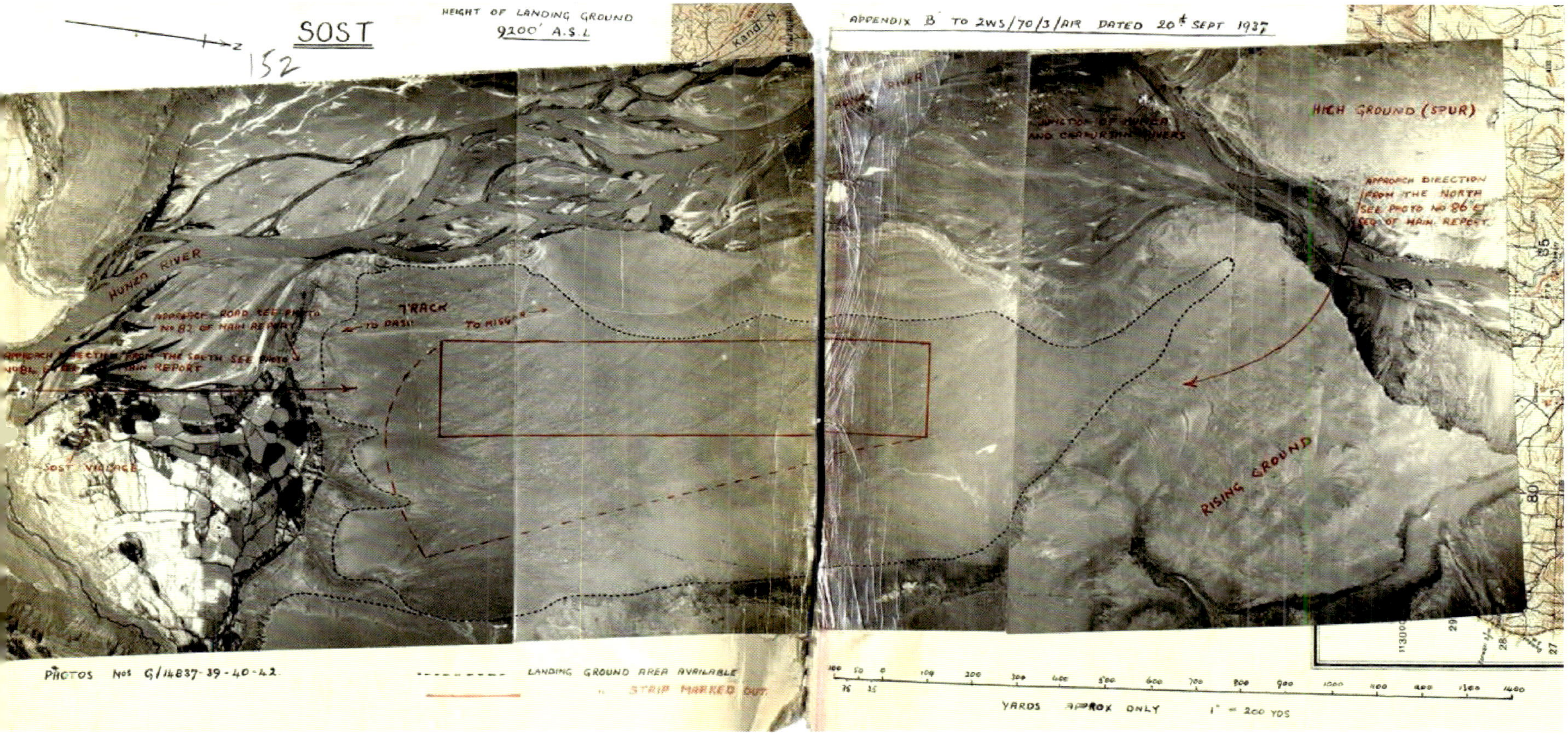

Pasu and Sost were identified as possible landing grounds between Gilgit and Kashgar. Preference was given to Sost as construction costs appeared to be lower; the plan has never been implemented to date. Source: All photographs by Flight Lieutenant Francis Claude Sturgiss in May 1937; File No. 82(2)-X/39, photos 39, 40, 42

surprise and pleasure at the condition of the aerodrome, the extension of which last September, must be regarded as providential in the light of the subsequent visit of this three engine monoplane and of the Valentia troop carrier."[216] In the aftermath of this achievement, grand plans were designed to extend the flight connections to China and cover the distance between Gilgit and Xinjiang, which would have required the establishment of additional landing strips along the route through the Hunza Valley and Sarikol.[217] It soon became obvious that air traffic was too costly and, because of detrimental weather conditions, too unreliable.

Nevertheless, several attempts were started to explore the possibilities of establishing a regular air route between India and China. A possible route to avoid Afghanistan was from Peshawar via Gilgit and the Hunza Valley to Kashgar or Yarkand. Early reconnaissance for landing grounds in the Hunza Valley was commenced on 6 May 1937 when four Valentia aircraft under Squadron Leader Harriss from 39 (Bomber) Squadron Risalpur landed in Gilgit aerodrome. The Royal Air Force Intelligence mission sent a ground survey group to inspect the terrain in the Hunza Valley. Flight Lieutenant Francis Claude Sturgiss collected the necessary information and submitted a detailed report. He was accompanied by Lieutenant Thomas Moreton, Royal Engineers, and the Gilgit Agency surgeon Major Montague Parker Atkinson who went on his regular tour to inspect the agency hospital in Aliabad and the two small dispensaries in Chalt and Uyum Nagar.[218] Spending a full month in the Hunza Valley, they experienced the Hunza Road and explored potential air strips and sites for emergency landing.[219] Among all the potential sites visited only two emerged as feasible with respect to location, accessibility and cost: Pasu and Sost. Their report mentioned that Nagar is 'quite impracticable' for any landing strip.[220] The ground survey suggested

Major George Kirkbride was appointed as Political Agent in Gilgit from 1934 to 1937. A number of reforms were introduced and the lease of the Gilgit Agency implemented during his period in office. He was ardently arguing for infrastructure improvement in the Gilgit Agency. Source: The photograph was taken from the display in the Gilgit Municipal Library

to select Sost as the construction cost of Rs 36,000 would be one thousand rupees lower than in Pasu. For emergency landing they suggested looking at nine sites, starting with Nomal, Jutal and Naltar in the vicinity of Gilgit; the first two were immediately rejected as being too close to the Gilgit aerodrome. The site at Naltar close to the newly established summer residence of the Political Agent was much preferred for 'forced landing' as the emergency strips were termed in the report.[221] In Nagar, two suitable spaces on either side of the Chaprot *nālā* were identified for the reconnaissance team: "Both sites are thus not particularly suitable and were finally ruled out of count on the grounds that they were cultivated areas of extreme value to the local inhabitants whose very existence depends upon them."[222] The *mīr* of Nagar was keen to have a landing ground in his dominion; consequently, another, rather remote place was inspected in Petingish Maidan, across the Hopar Glacier, that could be reached only on foot. After seeing it, the verdict was 'totally unsuitable'.[223] Ghamesarat, Hussaini and Gircha were inspected in Hunza; all sites were located in stretches where the Hunza River had widened its bed considerably and the latter offered scree slopes at Zarabad and Sarteez. The northernmost emergency air strip was suggested for Murkushi, at the junction of Kilik and Mintaka rivers; this was identified as 'quite impracticable'.[224] The conclusion of the ground survey was not convincing for all participant decision-makers in the headquarters of the Royal Air Force and the External Department. Consequently, to test the sites for flight practicability they sent three Harts aircraft of the No. 11 (Bomber) Squadron of the Royal Air Force base in Risalpur to explore potential landing grounds in Pasu and Sost on 11 September 1937.[225] According to their observations and their vertical photographs from the sky, Flight Lieutenants William Stephen Gardner, Francis Claude Sturgiss and Roy Halliday agreed that Pasu would be the most convenient landing strip in terms of options for manoeuvring and accessibility. However, the site in Pasu and its proneness to flooding made their Wing Commander in Risalpur suggest that Sost would be the better choice in terms of reliability and should be given priority due to safety considerations.[226] The debates continued for the rest of the year as to which site would be the best. In any case, the cost could be significantly lowered if the project was commissioned to the *mīr* of Hunza. It was estimated that it could be reduced to less than one quarter of the previous calculation. The *mīr* would use forced labour (*rajaáki*) and still pocket most of the money for himself. The discussion was terminated when the Air Marshall Sir Edgar Ludlow-Hewitt from the Indian Royal Air Force headquarters stated in a letter to the Defence Department in New Delhi dated 25 January 1938: "I consider that the construction of a landing ground in the Hunza Valley might be regarded as of military necessity in the distant future; for the present there are many other works which are of more immediate necessity."[227] The beginning of World War II the year after led to its indefinite postponement.

The successful visit by Air Marshal Sir Edgar Ludlow-Hewitt led immediately to a revaluation and reassessment of the strategic position of Gilgit and Hunza. The Political Agent Georg Kirkbride went into raptures: "If the reconnaissance results in a landing ground at Sost the political consequences will be of first importance. The loyalty of future Mirs will be bound even more stronger to the British Government in a, to them, most depressing manner for no longer will they have a back door for flight. The exit of which Safdar Ali availed himself in 1891 will be closed. If the old strategic barriers are breaking down under the development of aircraft then these valuable northern oases of plain must be denied for ever to an enemy and in what more simple manner than by occupying them ourselves. Not for much longer will there be a strong man in Hunza and Little Gujhal, but the problems of the future will be simplified if it is possible to transport troops into the heart of that State at short notice instead of by a weary and easily opposable march that in the best conditions must take many days."[228] Obviously, the trust in Hunza's loyalty was still limited, and here an instance is shown where technological superiority and cutting-edge military equipment convinced the acting officers of their dominance. The reappraisal seemed to have improved the negotiating position of the British authorities with their neighbours in Xinjiang.

The British desire to support the Guomindang Government in China via Northern India became more important with the success of Mao Zedong in China, with the separatist tendencies in Xinjiang in the mid-1930s, and especially during World War II. In June 1942 it was the experienced traveller and diplomat Eric Teichman, residing in the British Embassy in Chungking, who was involved in making connections between China and British India to survey 'new air routes via Sinkiang'.[229] Ten days later a report was received about the 'Survey of Trans-Himalayan route Peshawar – Yarkand' from the 'Headquarters U.S. Air Forces in India' who had sent a B-24D airplane from Peshawar on 22 June 1942 to sur-

On 22 May 1954 the expedition team landed at Gilgit Airport and was received with military protocol by the Gilgit Scouts band whose lucky mascot was a mountain goat. Source: German-Austrian Himalaya-Karakoram expedition 1954. Courtesy of Gerhard Klamert

vey the potential air route on which DC-3 type aircrafts were supposed to ply.[230] Several obstacles hampered the attempt, beginning with the small size of Peshawar airport preventing the bomber from starting with full fuel load. More severe was the refusal of permission to land in Xinjiang. Three American pilots and the British Air Commodore Manning – who was the commanding officer in Peshawar aerodrome – took off on 24 June 1942. The whole team was impressed by their flight across the Hunza Valley after they had assessed the flight route from Peshawar to Gilgit as easy. The peaks of the main Karakoram range and the narrow passages at Kilik and Mintaka left their mark: "These passes are very narrow and bordered on each side by tremendously steep high mountains. A transport plane attempting to go through them at low level would be in very grave difficulty due to the reduced visibility, turbulence and proximity to the mountain wall. After leaving the Karakoram range we approached the Pamir range, whose summits are more rounded than those we had just passed, and whose heights were a good deal less." Colonel Emmett O'Donnell concluded in his report: "From what could be seen of the passes and defiles, it seems impracticable to even consider running a transport line equipped with DC 3 type airplanes through this stretch. Their service ceiling with full load is not more than 16,000 feet and they have to get over passes which are at 15,500 feet. The defiles are so narrow that few airplanes could turn having once entered at low level. A pilot suddenly encountering snow or cloud in one of them, would find it impossible to turn, equally impossible to fly blind, and would almost certainly crash into the chasm wall. Air Commodore Manning claimed that this was the first time an airplane ever flew the complete breadth of the Himalayan range."[231] The report put to rest most hopes to open up an alternative and cheap supply route for their Guomindang allies in China and to alleviate costly ground transport.

In general, air lifting with the state-of-the-art technology of the first half of the 20th century could not fulfil the demands of material transport required for the supply of administrative and military stations, and for trade and commerce. In awareness of the immediate demand for modern traffic in the mountains, the quest for roads on the ground continued as imperial demands for territorial control grew and political confrontation in Central Asia increased. Flight connections did not take off on a big scale until today. The two operating airports in Northern Pakistan – Gilgit and Skardu – annually served between thirty to forty thousand civilian and military passengers at the beginning of the 21st century.[232] The hope of replacing ground transport by air services never materialised.

Innovations and constraints in Central Asia during the 1930s

French entrepreneurs and their Russian counterparts, succeeded by Soviet planners, had a vision of a motorised future, even for remote and marginal locations, for deserts, steppes and mountain regions. The time had come to experiment with new approaches and innovations. The 1930s should be remembered as a period of significant changes and limitations. The story of the Hunza Road cannot be limited to technological challenges in the valley, which in themselves were very severe; sometimes the mountains and passes seemed to be unsurmountable by motor vehicles. Beyond non-existing infrastructure and technical limitations, the period is characterised by political upheaval in China, administrative reforms and shifting jurisdiction in British India, collectivisation and social reforms in Soviet Central Asia, and reforms and amelioration programmes in Northern Afghanistan. All these events contributed to a regular stream of refugees arriving at the northern passes of Hunza. Refugees from the Soviet Union arrived after April 1932, and more 'ex-Tsarist Russians' fled Kashgar or, together with 'Indian subjects', were driven out from Xinjiang later on. Between October 1932 and December 1933 more than two hundred persons belonging to both groups arrived in Hunza.[233] The escape of the Ismaili *pīr* from Soviet Wakhan to Afghan Wakhan marks the beginning of political and religious persecution that continued during the 1930s.[234] More people attempted to change sides and asked for asylum in British India; comparatively few opted for the reverse direction. The authorities tried to be selective; refugees of high social standing were welcomed, while peasants were treated as 'economic migrants' and were supposed to be refused entry into the Gilgit Agency. Incoming refugees were not allowed to remain or settle in the border region. Misgar had become their first camp in Hunza; subsequently they were transferred to Gilgit and asked to proceed to Srinagar or to Peshawar via Chitral.[235] A Gilgit Scouts post was established in the summer of 1934 to prevent further Russian refugees from entering and to hinder the spread of 'Sovietization' that was feared for Xinjiang. Growing ideological competition and deliberate isolation affected border regimes and commodity exchange. In a period when infrastructure development enabled modern traffic to reach the mountain abodes, international boundaries were closed and cross-border trade ceased to exist between the Soviet Union and Afghanistan, British India and China by the mid-1930s.

The reasons for these developments are not restricted to ideological differences or the preliminaries of World War II and the subsequent Cold War. Within the region, territorial control was expanded to the limits of boundaries that had been laid down mainly by the end of the 19th century. Expanded control regimes, infrastructure and development activities went hand in hand. The Afghan King implemented early joint venture schemes to introduce modern agriculture, processing plants and industries with support from the Weimar Republic's experts for value creation in Qataghan and Badakhshan.[236] The attempts of warlords to autocratically rule Xinjiang revealed the weakness of the Chinese State, resulted in turmoil and attempts to create independent Turkic republics in Ili and Khotan.[237] The most significant and accomplished restructuring was experienced in Soviet Central Asia, where – after the creation of autonomous socialist republics – local communities were transformed according to communist ideas of economies of scale in collective and state farms and cooperatives justifying significant investments in housing and infrastructure schemes.[238]

In Kashmir and the Northern Frontier the British authorities managed to tighten their grip on the 'turbulent frontier' with Afghanistan and China which led to the lease of the Gilgit Agency in 1935. Consequently, a streamlining of administration connected with the search for new sources of revenue generation took place. The areas south of Gilgit on the left bank of the Indus were left in the jurisdiction of the Maharaja, while British control extended to the border areas. As the Times reported: "The change however will not sever the links between the Agency and Kashmir." The cost of governance was always in the forefront, the correspondent saw a reduction of cost in delinking responsibilities for the security of the frontier region: "Henceforth an increased and reinforced irregular force, called the Gilgit Scouts, will garrison the country and watch the passes leading over the Pamirs and the Karakoram to Russian and Chinese territory."[239] Infrastructure development for supplying garrisons and offices in Gilgit was required and it was expected that the taxation of local agricultural yields, the exploitation of timber, and toll taxes from trade would pay for the investments. For the whole period from the lease signed by Maharaja Hari Singh and the Resident in Kashmir, Lionel Edward Lang, to its end prospecting licenses were specifically

Portrait of Maharaja Hari Singh. The bromide print was taken at Bassano Studio on 7 June 1944. Source: Courtesy of the National Portrait Gallery

Opposite
Gilgit Airport has become surrounded by urban dwellings and is part of the town. The runway is too short for present-day demands, the planned extension did not materialise yet; moreover a shift to a different location outside the agglomeration is under discussion. The photograph was taken on 15 March 2019

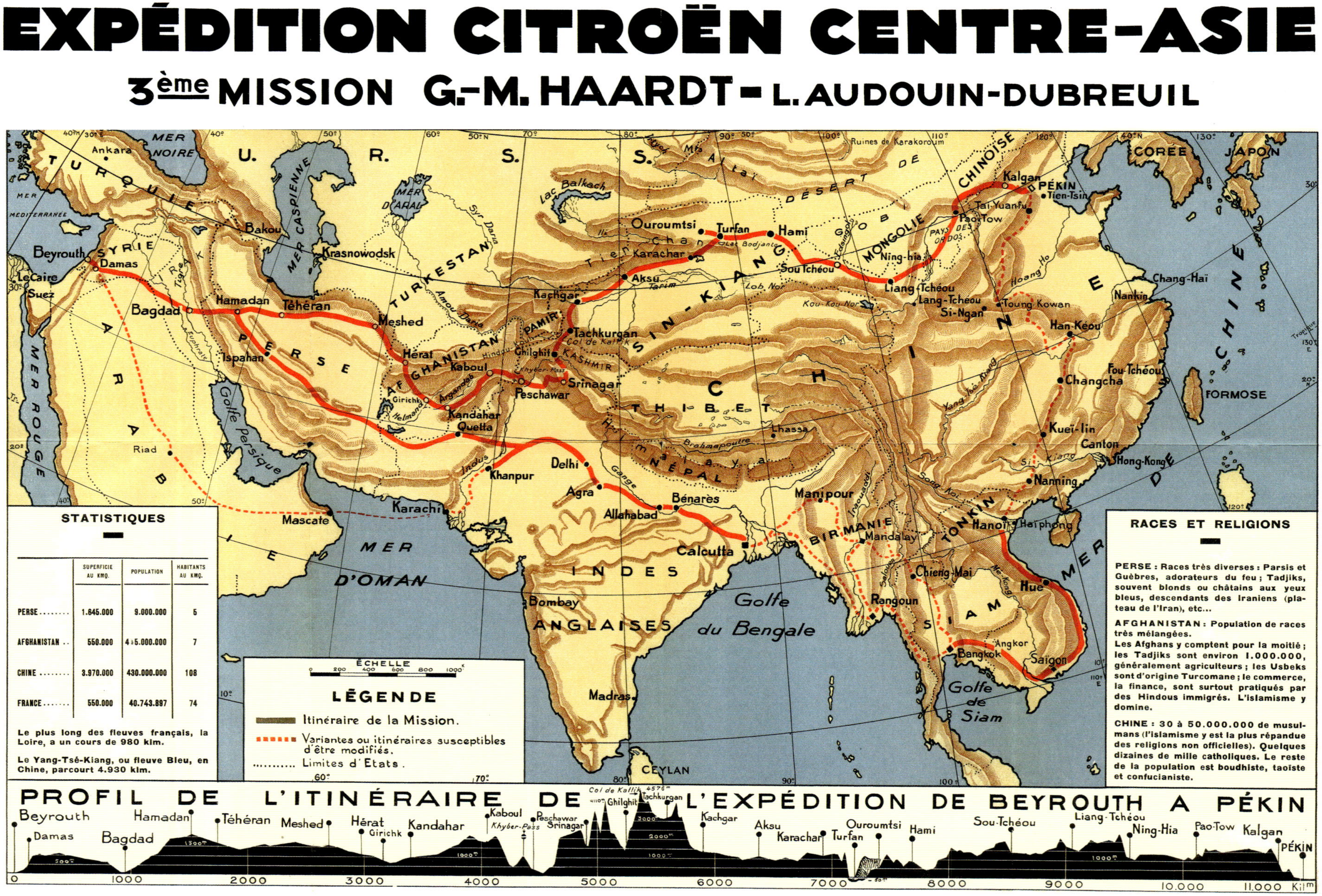

The map shows the proposed route of the 'Expédition Citroën Centre-Asie', popularly known as the 'Croisière jaune' with its two starting points in Beirut and Beijing connecting the main course. It shows the expedition's plan to cover the whole distance by motor transport on the main traverse across Central Asia. The proposed itinerary climaxed in the Kilik Pass where motor transport subsequently failed and where horses and pack animals would become the mode of transport. The Himalaya-Karakoram-Pamir region features prominently as the altitudinal summit region in the centre of the elevation profile that is displayed at the bottom of the illustration. The map was distributed among all Citroën branch offices to advertise the expedition, and small flags were planted on it indicating the progress of the two groups towards the meeting point in Central Asia. Source: The map is contained in a booklet that was distributed after the end of the mission (George Le Fèvre 1933).

banned and mining leases excluded.[240] The Survey of India had already made a new ground survey of the Kaghan Valley in 1930; the following year the Gilgit Agency was mapped again for planning the next steps.[241] The reorganisation of the Gilgit Agency included a realignment of supply lines. The former lifeline between Srinagar and Gilgit via Burzil Pass would not be the route of the future as the territories of the Maharaja of Kashmir were to be bypassed. Major discussions about finding the suitable routing came up after the *Croisière jaune* had attempted to cover the traditional link between the Valley of Kashmir and the Karakoram Mountains. Expectations had been high that motorised land transport was feasible when they started off on their ambitious attempt to achieve the hitherto impossible, i.e., to drive motor cars on a pack animal route across high passes and ford major rivers.

The Burzil Pass (4,200 m) posed a major challenge for the 'Croisière jaune' on their way from Srinagar to Gilgit. After Bandipur the expedition experienced the ascent to the Himalayan passes that detrimental weather conditions had covered in deep snow. An advance party with a long caravan of porters managed to cross on foot before the motor vehicles set off in the second half of July 1931. Source: Oil on canvas painting by Aleksandr Yakovlev (1931); reproduced with kind permission © courtesy of Sotheby's

Croisière jaune – a terrestrial attempt to cross the Karakoram

The 'yellow cruise' to China - conceived and led by Georges Marie Haardt and Louis Audouin-Dubreuil - was a follow-up expedition of their trans-Sahara expedition of 1922 and the *Croisière noire* – the black cruise - through Africa two years later.[242] They were accompanied by a number of experts, among them the painter Aleksandr Yakovlev whose expressive drawings made the expeditions well received and won him a reputation as one of the most prolific 'ethnographic' painters of his time. The African traverse was the role model for connecting antipodes such as Algeria with the Cape. Conceived as promotion tours for Citroën motor cars under the patronage of the French Geographical Society, these cruises accessed remote peripheries. In the aftermath of the Great Depression, they showed how the outcomes of the world economic crisis could be lessened and how potential for future regional development could be tapped and benefit European dominance in colonial settings. Both expeditions proved the feasibility of traversing inaccessible lands by modern technology-inspired modes of transportation. Their very heavy vehicles resembled a combination of motor cars and armoured tanks that were perceived as the cutting-edge inventions for challenging terrains. Connecting the Levant with China was seen as an even greater achievement, as in Asia more attention was given to the benefits of European engineering, thus underpinning colonial dominance in an invaluable setting. The *Croisière jaune* was split into two groups who approached the route from the antipodes. Seven vehicles of the China group started from Kalgan, a prominent location at the Great Wall near Beijing, on 17 May 1931, while the Pamir group had commenced the expedition with six vehicles in Beirut on 4 April 1931; both teams were supposed to meet in Kashgar. From Beirut to Kashmir the journey progressed smoothly; when the Himalaya, Karakoram and Pamirs had to be traversed, however, the path to China proved to be more challenging than anything the organisers had anticipated and experienced to date. Starting the difficult journey from Srinagar on 2 July 1931, they followed the conventional route from Kashmir to Gilgit via Tragbal Pass (3,600 m), Gurais, Burzil Pass (4,200 m) and Astor. Only two vehicles – the *Scarabée d'Or* and the *Croissant d'Argent* - were selected to make the journey across the mountains. Numerous rivers had to be crossed, sometimes on shaking bridges designed for porters and pack animals, sometimes through challenging river fords. The vehicles had to be dismantled on more than one occasion and the pieces had to be packed into loads for which additional local porters were hired. In addition to the expedition members, 300 porters and several hundred pack animals were engaged to

Aleksandr Yakovlev contributed a number of landscape paintings to illustrate the terrain covered by the expedition. The 'foothills of the Karakoram Mountains' indicate the steep slopes and the rugged mountain range emerging from a desert-like structure with scanty vegetation cover. Source: Tempera on canvas painting by Aleksandr Yakovlev (1931); reproduced with kind permission © collection particulière française, courtesy of Sotheby's

transport the technical equipment and food supplies from Kashmir to Gilgit.[243] From the perspective of Herbert Todd, the Political Agent in Gilgit, it was a matter of difficult road access; the entry in his diary reads: "The second echelon with M. Haardt and Colonel Gabriel brought two Citroen Caterpillar cars. The party consisted of 12 Europeans. Astor was reached in nine days from Bandipur, doing one stage per day. No great difficulty was experienced, as the road is supposed to average 10 feet in width. The pace was slow, however, and considerable man-handling was necessary at most corners. At two bridges the cars had to be dismantled. Beyond Astor the recent abnormal rains had greatly damaged the road, and this delayed the cars. Eventually, as the road widening process was so slow, the cars were taken in pieces and carried over the worst places. By the evening of the 31st July the cars were in the neighbourhood of Ramghat Bridge, just below Hattu Pir. … the results so far achieved on the Gilgit Road, seem to testify less to the efficiency of the cars than to the endurance, determination and patience of M. Haardt and his companions."[244] No mention was made of their helpers who in great numbers had to support the expedition at every place along the route. They had to navigate the cars around bends, across bridges, and had to convert the tracks made for pack animals into roads for four-wheel traf-

The 'Croisière jaune' required substantial porterage in their attempt to cross the Himalayan passes; the Burzil was the first major challenge. Aleksandr Yakovlev depicted a halting place for porters to prepare a meal en route to Gilgit. Source: Aleksandr Yakovlev, 8 July 1931

fic. Often the cars had to be disassembled and then carried by coolies across the obstacles.

It took the Pamir group more than a month to reach Gilgit where the last arrived on 4 August 1931. The Citroën motor cars were the first mechanical vehicles ever seen in the Karakoram after a 290-km long journey across the Himalaya. More challenging was the Hunza Valley; here the expedition came to a halt and surrendered to the obstacles posed by the mountain path. Herbert Todd remarked in his diary: "One car was taken on to Nomal on 6th August, taking some 15 hours to cover the 17 miles. From Nomal the car was returned to Gilgit and has been sent back in pieces to Srinagar. The other car was left in Gilgit and has since been presented to Government."[245] Gilgit was the place where capitulation had to be admitted, bearing in mind the challenges of the Hunza Road. Three days after arrival in Gilgit the decision was taken that the second car, the *Scarabée d'Or*, would be dismantled and transported back in pieces along with the returning expedition mechanics to the Citroën Museum in Paris.[246] Inevitable decisions had to be taken, and it was determined that the Hunza Valley would be the last – because insurmountable – destination of the cruise. The British liaison officer Vivian Gabriel expressed his great disappointment. The French mechanics regretted all their previous efforts in making this cruise work when they were informed in Gilgit about the changed plans and the termination of their efforts. Valuable time was spent in dismantling and re-assembling the heavy vehicles; now the final packing of the pieces took place. The garrison town and seat of the Political Agency had received an attractive curiosity with the first ever car reaching the outpost; the *Croissant d'Argent* would remain in Gilgit on display in the polo ground. The parting skeleton expedition group was entertained by *mīr* M. Nazim Khan who invited his exotic guests (*oóśo*) initially to his guesthouse (*oóśo basá*) in Gilgit and afterwards to Baltit.[247] The stay in Hunza gave ample time to Aleksandr Yakovlev – who had travelled to Baltit ahead of the motor cavalcade – to capture in his drawings the scenery, characters and leading personalities of the principality. Herbert Todd was quite impressed by Aleksandr Yakovlev's work, describing him as "… a genius at portrait sketches in crayon, and is drawing local types met along the road. Arrangements were made to collect some of the typical local characters at each stage of the journey. The results are indeed very fine."[248]

Aleksandr Yakovlev portrayed an 'old man of Hunza' during his stay in Baltit on 29 July 1931. Source: Sanguine and pastel on paper by Aleksandr Yakovlev; private collection

After more than a month resting in Gilgit and Hunza, the remaining expedition members proceeded northwards and reached Misgar where the whole group was reunited on 28 August 1931. At the end of the telegraph line from British India communication was possible with the expedition camp in Srinagar from where the latest news about the border-crossing was received.[249] Four years later when Peter Fleming came to Misgar he found "… the rest house, whose walls have been decorated by M. Iacovleff, of the Haardt-Citroen Expedition, with a series of admirable studies in still life."[250] Aleksandr Yakovlev decorated a wall of the rest-house with a mural that showed Georges Marie Haardt, Louis Audouin-Dubreuil, and the artist himself as silhouettes.[251] The expedition had left a mark on the village of Misgar which had inspired the artist to a number of remarkable paintings of the local scenery. Some further delay occurred

Aleksandr Yakovlev's oil and canvas painting of a house and its surrounding landscape in Misgar shows the Burusho house-building style with a room with a view (sarké baldí) on top and storage space for the winter. The landscape in the background is characterised by lacustrine and mudflow deposits that have been deeply incised by fluvial erosion on their way to becoming earth pyramids. These arid and rugged slopes contrast with the flat irrigated fields in the foreground. Source: Oil on canvas painting by Aleksandr Yakovlev, August 1931 in Misgar. Collection Louis-André Citroën; see Emmanuel Bréon (2004: 147)

Aleksandr Yakovlev was among the first expedition members who arrived in Misgar in August 1931 for a stay of two weeks. On joyful events polo matches were celebrated for guests; every village featured a polo ground where the rough variety of Hunza polo was displayed as the most favourite pastime. Source: Tempera on cardboard painting by Alexandre Evgenievich Iacovleff (1934; plate 29), reproduced with kind permission from Gallica. Bibliothèque nationale de France

Beyik (Paik) is located at the bottom of the Kara Chukur Valley and is an important seasonal habitation for Kirghiz nomads in the Taghdumbash Pamir who make use of the fertile pastures among arid and rugged mountains. The clay walls belong to a Chinese post (karaul) controlling the access route to Wakhan. Source: Tempera on canvas laid down on board by Aleksandr Yakovlev, September 1931; reproduced with kind permission © collection particulière française, courtesy of Sotheby's

due to unrest in Xinjiang before the border into China was opened for the remaining expedition's members. Some had spent about a month in Hunza, by now fully convinced that no motor vehicle would have been able to overcome all obstacles here. The only non-motorable stretch of the expedition's itinerary is reported with several images of difficult passages, porters with heavy loads, pack animals and long camel and yak caravans.[252] All travellers had put renewed trust in caravans on their way across the Kilik Pass (4,765 m) to Khushbel. About two hundred Hunza porters and animal drivers had to return from this point, as the Chinese authorities did not allow them beyond the pass and further into settled regions. Only Hunza mail runners delivered their letter bags and files at Kashgar and brought news on the way back; thus, urgent messages could be communicated from Misgar by telegraph to Gilgit and Srinagar. From Kushbel Kirghiz herdsmen had to be hired with their yaks to cross the Taghdumbash, Sarikol and Kara Köl Pamirs towards Kashgar.[253] Despite troubles in entering China, the expedition met with the Chinese *amban* in Tashkurgan who received them for a three-day intermezzo and played the *Marseillaise* on a phonograph for them.[254] After a comparatively fast three-week journey on horseback the Pamir group reached Kashgar on 19 September 1931. They paused a little before proceeding to Aksu where the twelve remaining expedition members were received by four motor cars sent by the China group to their relief.[255] The section between Gilgit and Aksu across the Karakoram and Pamir remained the unsurmountable, non-motorable stretch of the expedition's route; no motor car could be brought as a whole or in pieces on those difficult paths. Thus, the *Croisière jaune* failed to accomplish a continuous motor journey between Beirut and Beijing. The Hunza Road was not yet in a state to cope with modern motor-based technology.

In the 'Geographical Journal' of 1933, Latham Blacker realised the importance of this expedition and acknowledged that Georges Marie Haardt and Louis Audouin-Dubreuil were 'pioneers who foresaw the use and value of the new weapons'.[256] Those weapons were the more or less appropriate motor cars, which Latham Blacker admired a great deal. Moreover, he summarised the importance of the expedition and the intentions of its leaders: "They designed to bring 'the major portion of Asia more within the sphere of the world's knowledge in the same systematic and complete way that the Trans-African expedition had done' in another continent. In the sphere of archaeology they proposed a reconnaissance of ancient remains in Syria and Iraq, and more especially in the monasteries and sanctuaries of Chinese Turkistan. They planned an ethnological study of the native races and their cultures, to be illustrated by paintings, films, photography, and collections of handicrafts. The geographical objects were to map out routes, to record gradients, and to study important land and air lines of transportation. They projected research in short-wave wireless, and planned to collect economic information in the larger commercial centres. Besides this, collections were to be made for museums, and the artists of the expedition were to use brush and pencil for the benefit of science as much as of art. The recording equipment included apparatus for colour photography, and for sound pictures as well as silent films. Haardt believed that in our time the whole face of Asia is rapidly changing, and that Western ideas and customs are spreading everywhere with surprising rapidity. He wished therefore to cross the continent and to preserve for the future a picture of the Asia of to-day by means of photographs, sound and moving pictures, paintings, and by placing data of all sorts within the reach of every one."[257]

Colonel Vivian Gabriel realised the challenges and hardships encountered by the *Croisière jaune* and its adventurous members. After the Soviet competitors had made leeway in their project of building the Pamir Highway, which soon would be opened to vehicular motor traffic, the French expedition met another challenge, this time in its own Indian

After crossing the Kilik Pass the members of the Croisière jaune arrived in the Taghdumbash Pamir; the Chinese magistrate (amban) received the team in his residence in Tashkurgan on 9 October 1931 and played the 'Marseillaise' for them on a phonograph. Georges Le Fèvre (1935: 220) reported: "Twelve Chinese, the entire Celestial colony of Tashkurgan, came into the room. They looked wonderingly at the little group of Frenchmen, who sat uneasy, shocked and silent." Source: The photograph was reproduced in Caroline Haardt de La Baume & Jean-François Ruchaud (2012: 51); see as well Maynard Owen Williams (1932: 353–354)

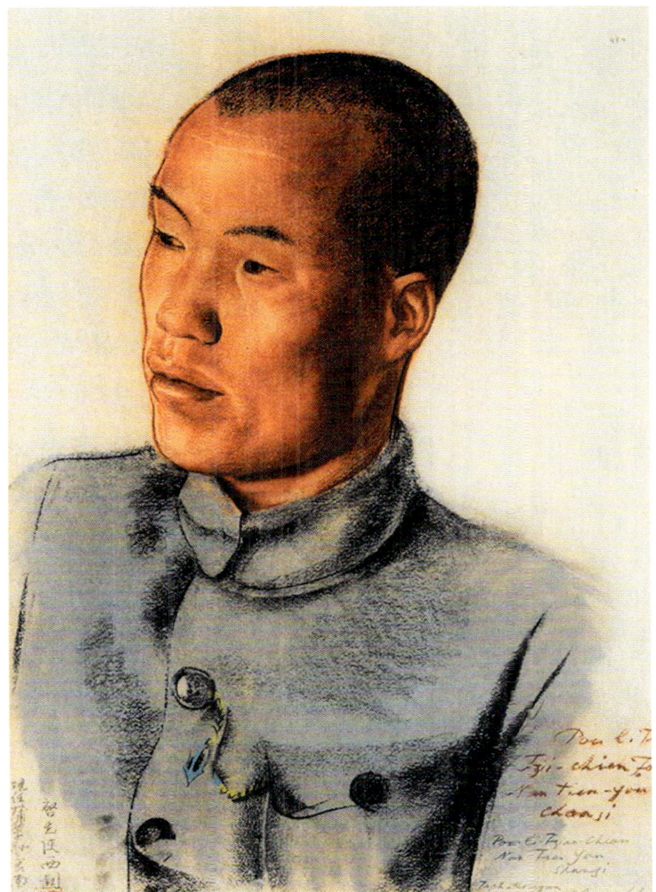

Aleksandr Yakovlev portrayed the Chinese magistrate (amban) in Tashkurgan County (Pou-li-Tzian) on 9 October 1931. Source: The painting was reproduced in Caroline Haardt de La Baume & Jean-François Ruchaud (2012: 51, 79)

After crossing the Kilik Pass (4,765 m) on horseback, the expedition members met with the first Kirghiz nomads in Beyik (Paik) in the Taghdumbash Pamir. After they had left the deeply incised arid Karakoram valleys, the landscape appeared as a fertile late-summer pasture ground in the Pamirian plateaux where glacial melt-waters irrigate the valuable vegetation cover. Two equestrians are on their way down the slope towards the valley bottom, where two yurt encampments are visible, while in the foreground two Kirghiz women, with their substantial white headgear, and a boy are posing in front of a cairn that is decorated with little flags commemorating a person's grave. Source: Oil on canvas laid on board by Aleksandr Yakovlev (1931); reproduced with kind permission © courtesy of Sotheby's

empire, where superior technology was promoted, tested and implemented. A decade earlier, the Citroën factories had begun designing vehicles for cross-country travel in difficult terrain.[258] While the Pamir Highway was proof that roads for motor cars could be constructed even across the highest mountain ranges, the British-Indian Empire had not much to counter the claim of technological and engineering supremacy. André Citroën regarded their achievement as a pioneering adventure into a hidden world that from now on was exposed to the innovations of modernity: "... our cars penetrated into regions where the sound of a motor had never been heard before, where the inhabitants of the humble villages, hidden away in the gigantic folds of the highest chain of mountains in the world, not only had never seen a motor-car but were ignorant even of the existence of a wheeled vehicle. There for centuries all transport had been carried by men or by pack-animals."[259]

Following the *Croisière jaune* some repetition of challenges and obstacles happened when the British diplomat Eric Teichman set out from Beijing to travel by motor car to Kashgar in the autumn of 1935. Like his predecessors, he encountered similar political difficulties in Xinjiang that complicated his journey. His two Ford V-8 30-cwt. trucks – one had previously belonged to the Swedish explorer Sven Hedin who travelled with it in Xinjiang in 1934 – resembled the Russian light trucks that were plying on the Pamir Highway, the Завод имени Сталина, the trucks from Stalin's factory.[260] They were regularly seen in Kashgar at this time. Nevertheless, he was aware that his eagerness to 'complete my mission and get through to India before the passes closed' would not be feasible by lorry.[261] The motor journey ended in Kashgar, but these obstacles did not prevent him from continuing his journey to Gilgit by 'pack animal'.[262] Eric Teichman summarised his experiences and insights: "The gorges of the Karakoram thus defeated, as they will continue in my lifetime to defeat, the attacks of motor-cars. Going the other way, through Russian territory, it should, political obstacles apart, be feasible, with proper preparation, to drive a motor vehicle from Europe across the breadth of Asia to Peking."[263] The Karakoram Mountains posed a greater physical challenge than other potential passages with regard to providing road access and thoroughfares. To approach the mountain arc that separated the Inner Asian plateaux from the Indian subcontinent seemed to be much easier from a northern direction than from the Gangetic and Indus basins.

The Gilgit Road in the wake of the lease of the Gilgit Agency in 1935

In the mid-1930s, a number of serious considerations arose about what should happen to motorable roads in the approach of Gilgit and the Karakoram valleys. Sir Eric Teichman's considerations and reservations were heard and seriously discussed. But when it came to the lease of the Gilgit Agency from the Maharaja of Kashmir for a period of 60 years, various changes needed to be discussed. As the supply lines for the Gilgit garrisons and the Agency were no longer to lead from Srinagar via Burzil Pass to Gilgit – political tension with Kashmir was anticipated – an alternative route had to be developed. The nearest railhead was in Havelian, and under the given circumstances the road connection to Gilgit was planned via Babusar Pass (4,173 m), which was about 400 kilometres compared with 680 kilometres between Gilgit and Srinagar via Burzil Pass.[264] Dated 30 November 1935, the report by the Resident in Kashmir, Lieutenant-Colonel Lionel Edward Lang, estimated the cost of the lease and contained a chapter on 'improvement of communications': "Under the terms of the agreement the British Government have agreed to discontinue, except in cases of emergency, the use of the Bandipur Gilgit Route for the purpose of sending troops to Gilgit or of supplying the garrison stationed there. This entails the use of the Kaghan Valley route for all military purposes. The Kaghan route enters the Gilgit Agency by the Babusar Pass 13,580 feet. The Pass is open from about the middle of June until the middle of October and presents very little difficulty for pack transport. … As this route is now the main line of communication from rail head at Havelian to Gilgit and will, it is anticipated, in future be largely used by Kashgar traders increased importance attaches to the necessity of maintaining it in good repair and keeping it immune from raids. At the present time the road from Havelian as far as Balakote is suitable for motor traffic, and I gather that there is a possibility of its extension as far as Battakundi. From the Babusar Pass to Chilas the road is maintained by the Agency Engineer. In the past this has been in charge of the Kashmir State Engineer and the condition of the road especially in the Indus Valley is far from good. I have asked the Political Agent to arrange for the re-alignment of the road where necessary and for repairs to be carried out as to ensure no interruption to traffic."[265] The state of the road section down from Babusar Pass to

Several administrative reforms preceded the lease of the Gilgit Agency (top); in 1934 the 'Administration Report' remarked "… the transfer to the political control of the Agent of the Governor General North-West Frontier Province, of those almost unknown and practically independent communities which are to be found situated on both banks of the Indus adjoining the District of Hazara and the Malakand Agency. On the right bank of the river they comprise Kandia, Maliar, Razika (Seo), the three Patan Valleys and the Duber Valley and on the left bank Harban, Shatial, Sazin, Shumar, Jalkot and Nila (Palas); the whole Tract being more generally known as Indus Kohistan." After the lease of the Gilgit Agency in 1935 (bottom), the Kashmiri administration (wazīr-i-wazārat) shifted to Bunji while the British colonial administration remained in Gilgit in a specific district that was referred to as Gilgit Tehsil or subdivision. The remaining territories under the domination of the British Raj were listed under the Gilgit Agency, which was reduced from 19,825 to 17,375 square miles. Source: Administration Report of the Gilgit Agency for the year 1934 (IOR L/P&S/12/3288: 207); Administrative boundaries and denominations based on the map 'Kashmir & Jammu and Gilgit Agency', published by the Survey of India in 1940. Design by Hermann Kreutzmann

Babusar Pass (4,173 m) was covered in snow on 5 September 2009; the track leading down towards Chilas is recognisable. The road is regularly closed between October and June due to snow cover.

Chilas was described in more outspoken and specific terms in the administration report for 1935: "This section is not fit for camel transport and in parts is difficult for laden mules. It has been definitely neglected and appears to have been favoured with few if any visits from an Engineering Officer in recent years."[266]

It seems that not only seasonal obstacles hampered the caravan traffic, but also political raids by 'tribes' sometimes led to the closure of the Kaghan route. Obviously, significant investments had to be undertaken to make that route a stable line of supplies. Here the future alignment of the 'Indus Valley Road' could become the lifeline of the Gilgit Agency, at least for four months in summer.

Following the road perception in the status quo report, it seems that there was only one feasible option to link Gilgit with the down-country terminus of the railway at the foot of the mountains in Havelian. Nevertheless, the debate about 'improvement of communications between India and Gilgit in order to maintain and even improve India's trade with Sinkiang'[267] was more complicated and diverse, offering further options. Reasoning about 'an all weather motor road' not only addressed the physical obstacles en route to Gilgit. Territorial control and domination of the inhabitants, safe passage and security of travel were other considerations, besides the overriding cost factor. The Srinagar-Burzil Pass route having been ruled out, three other passages entered the debate. Besides the Kaghan route – some favoured it as the most direct route crossing only one pass, but being snow-bound for many months did not recommend it for a motor road – there was the option of an Indus Valley road bypassing the Kaghan Valley and following the course of the Indus River towards Oghi in Hazara District, where the existing motor road terminated. Here the reservations were mainly whether it would be possible to control and 'pacify' the 'black mountains tribes'. The third option was the combination of the Chitral and Gilgit route via Shandur Pass (3,700 m) and Lowari Top (3,117 m) leading to Malakand Agency.

The Political Agent in Gilgit, George Kirkbride, is quoted by Lieutenant-Colonel Lionel Edward Lang, the Resident in Kashmir, as saying that he "… considers that the cost of constructing a road *via* the Shandur would be less than by the other routes and that the Shandur Pass could be kept open during the winter months by means of motor ploughs. Assuming that the railway is extended to the south side of the Lowarai Pass, he is of opinion that the Shandur route would bring Gilgit 6 miles nearer to railhead than by the Indus route with its railhead at Oghi. Kirkbride, however,

Shandur Pass (3,700 m) resembles a wide plateau with extensive pastures close to its lake. The passage is normally affected by snow cover from autumn to spring. The photograph was taken on 13 June 1991

feels that the Indus Valley route would be a far greater achievement in that it would result in the pacification of a handful of backward tribes and allow of the development of the natural resources of the country. In regard to the Chitral Shandur route, I consider that apart from its being a far longer route it has the further disadvantage of traversing two passes, the Lowarai 10,250 feet, and the Shandur 12,250 feet. The Lowarai is said to be closed to pack transport from November to April and the Shandur though traversable at all times by pack animals could only be kept open for motor traffic during the winter months at considerable expense."[268] The major advantage of this route would be to avoid traversing the Indus Valley, which was not really thought of as a secure passage. George Kirkbride articulated quite stern views on these areas: "The tract of country lying on either side of the Indus below Bunji to Lahtar Nadi where the Indus takes it final, great bend to the south is known comprehensively as Shinaka. … It comprises all Chilas with its connected valleys; Darel and Tangir; Harban, Shatial and Sazin. The last three named have since 1934 come under the orbit of political control of the North-West Frontier Province. Darel and Tangir compose roughly a third of the whole. Geographically the position of both tribes is, from their point of view deplorable. On the north and east is the Scylla of the administered areas of the Agency; on the south, the Charybdis of the Indus, while on the west lies Kandia with whose people Tangir have an ancient and deep seated blood feud. … The system of government is communism in its purest and worst form. Save in a common danger and then not always, each village is independent under its elected headmen (Jashtero). All forms of autocratic government is abhorrent to them. No big man has ever emerged for any length of time. The Khushwaqt Raja Pakhtun Wali managed to establish his rule for a few years but was murdered in 1917. The village headman is the servant, more than the master and has no influence outside his own village. Land is mostly held in 'Shamilat' and this principle extends to the forests and results in inter-village squabbles over timber and grazing rights that vary in intensity. All these tracts are also torn in two by factions known as Shins and Yeshkuns. The affinity of Darel and Tangir with Chilas and their abhorrence of autocratic rule must be an important factor in our future dealings with them."[269] George Kirkbride is convinced that this is a very bad precondition for 'pacifying' the valleys. It would not be an easy match as with Hunza and Nagar. His description carries all the aspects that have been repeated since by generations of anthropologists about the anarchic and segmentary structure in connection with the xenopho-

Rafts are used for transporting felled trees from the lefthand valleys to the right bank of the Indus Valley. The photograph was taken on 26 March 2003

bic behaviour of the inhabitants in the Indus side valleys south of Gilgit. These ascriptions have survived until today and are built on the reputation of their status as *jangli* people or 'backward tribes'.[270] He continues his narrative at length by estimating the number of arms, ammunition and fighting power of the area. The reason for their purchasing power is given as well: "Darel and Tangir have only come into touch with the world outside the Gilgit Agency by the exploitation of their very fine forests by the Attock Trading Company and others through the medium of Mians of Ziarat Kaka Sahib. This is important as out of this arises the legend of Darel and Tangir wealth."[271] Consequently, George Kirkbride is quite convinced that it would be favourable "... to take over the tribes ourselves and administer them in the same loose form we now employ with success in Chilas. ... It could also be pointed out that the making of the road would bring them employment, though the Darelis and Tangiris do not themselves perform cooly work."[272] He suggested a tactic that would slowly bribe the village headmen and other leading personalities who are related to certain *rāja* families by marriage in order to incorporate them into the set of loyal subjects. Nothing comes without a cost and George Kirkbride estimated that 'a round sum of 22,000 rupees would [not] be excessive if it bought the contented allegiance of these tribes."[273] But this investment could be recovered over time: "I should hesitate to say that any substantial revenue would come our way from the forests for some years but they do present the most attractive side of control on the Chilas model, a side that would be lost under the rule of a Raja. ... Any extraction of timber by Government would have to be taken up slowly and it would have to be made quite clear that the owners would enjoy a proper share in the profits. Taking the long view I feel that Government would be justified in spending liberally in the beginning in order to bring about the quick pacification of the tribes. The quicker it is done and the more satisfactory the financial side of the settlement to them the sooner would operations on the forests be possible. A most important factor to be borne in mind is that if the policy of Government results in a road from Oghi to Gilgit the pacification of all the riverian tribes will have been accomplished and that the bugbear of the timber merchant, the tribal chief with his thievings and extortionate tolls, will disappear and it will be possible to float logs from the Gilgit Agency to Khairabad without let or hindrance. Not only the pacification but if necessary the forcible subjugation of Darel and Tangir might be justified on the timber question along, for the control of those forests is not only in the interests of the people but would enable the Agency to assume part of the increasing burden of cost of its administration."[274] He then mentions strategic matters concerning the appropriate time to attack: "There is one very important point. At the time when the political officer and his escort first enter the country I consider it would be most valuable to have a flight of the Royal Air Force temporarily stationed at Chilas. It is unlikely that they would be called upon to bomb these primitive folk but the moral effect would be of immense assistance to the people on the ground. It might just turn the scale in favour of peaceful penetration."[275] Although George Kirkbride suggested an inspection of all roads he could not hide his preference for 'peaceful penetration' into the area of Indus Kohistan where additional wealth could be harvested that would pay, or at least contribute to, the cost of administration, road-building and its maintenance. His views were adopted by the Resident in Kashmir, Lionel Edward Lang, and communicated in his final statement: "To summarize my proposals are: (a) Gradual annexation and inclusion within the Gilgit Agency of the tribal country of Tangiris and Darelis. (b) A survey of the Indus Valley route by a Political Mission with a competent road Engineer. (c) Preference of the route from Havelian to Oghi and thence over the Nadihar Allai tract to Thakot-Kizar-Jalkot and Chilas."[276] The various goals were to be tackled in a combined fashion: revenue generation through local resource exploitation and cross-border trade; administration and security provision at a low cost. The former was a seasonal business, while the latter was an all-year-round affair. George Kirkbride had formulated: "What the trader wants, or should want, is not so much an infallible all weather road as a motor road at the earliest possible point on the section between Gilgit and the Indian (Hunza) Frontier. ... a motor road from Baltit, the Hunza Capital, to the northern Frontier does not, in the opinion of all who have seen that section, enter into practical politics. Little is impossible to an engineer but in this case the cost is obviously so enormous as to rule it out."[277] All euphoric sentiments of having the most modern trans-montane traffic artery were deflated in view of the major physical obstacles and enormous environmental challenges that were lying ahead.

During the winters felled trees are brought down to the Karakoram Highway from where they are transported down country. The timber trade was already well established in colonial times; certain easily accessible forests were completely depleted, a practice that was accelerated afterwards. The forest owners receive their share in form of royalties. The photograph was taken on 19 January 2007

When Kenneth Mason was reflecting about the Himalaya as a 'barrier to modern communication' in 1936, he must have had in mind the steep mountain wall and its narrow, deeply incised valleys.[278] British approaches to traffic development had focused on rail tracks such as the narrow gauge systems from Kalka to Shimla which had already been opened in 1903, following the Hardwar to Dehra Dun (1900) and Siliguri to Darjeeling (1880-1886) lines that are remarkable and rare cases in point. They opened up the mountain barrier for the visitors to a growing number of hill stations, including the Indian summer capital Shimla, thus connecting military and administrative stations with the centres of power.[279] By 1909, the western sector was equipped with three railway lines that had their railheads in Pathankot, Jammu and Dargai. In a strict sense, there were no mountain

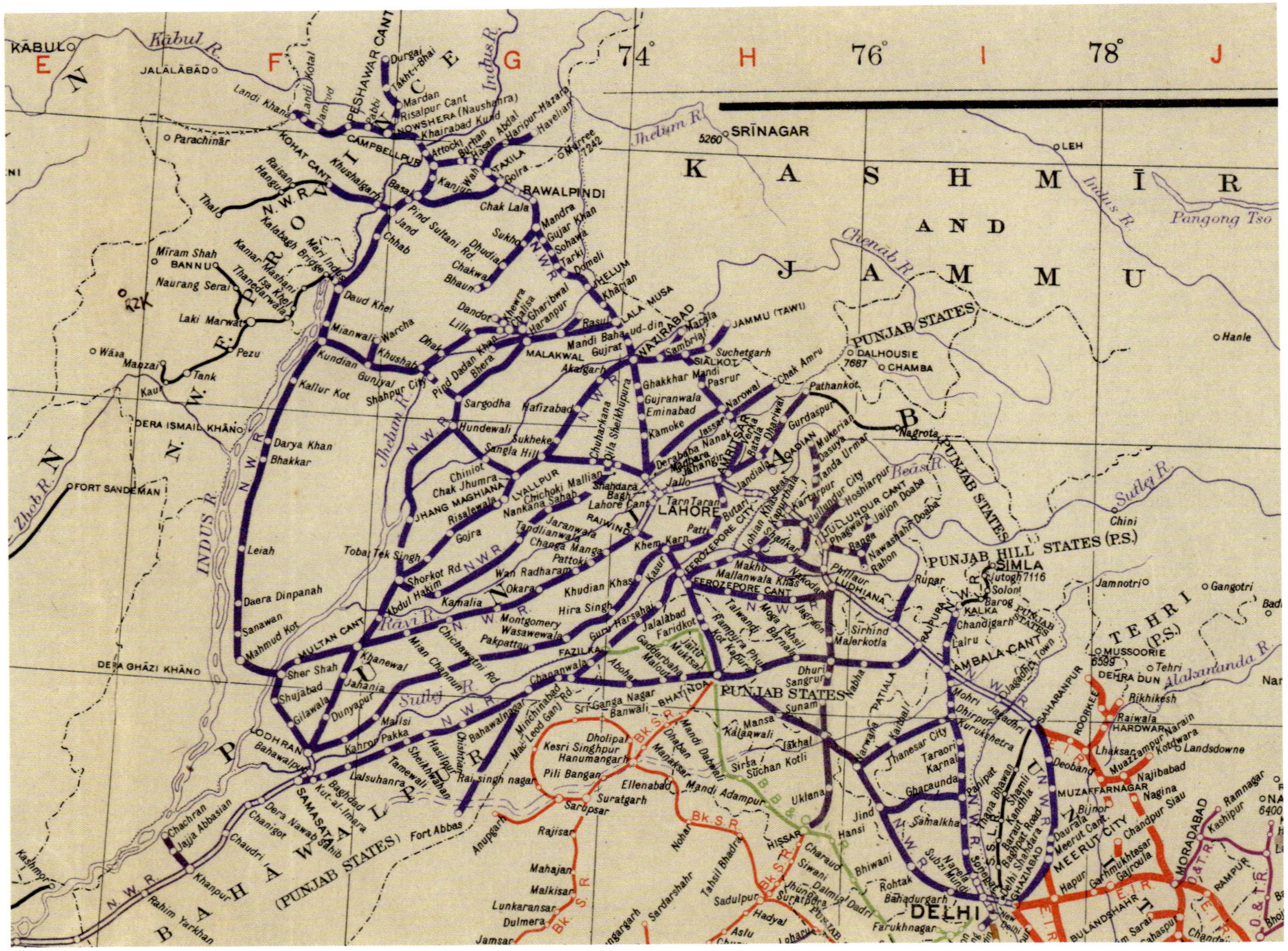

The section of the map 'India. Showing railways open and under construction on 31st March 1945' shows the north-western railway network in British India with the terminal stations in Dargai and Havelian as the closest connections to the Gilgit Agency and Chitral. The Kashmir capital Srinagar was serviced from Jammu-Tawi station. All railway lines ended in the foothills of the mountainous arc. Source: Survey of India (1945). Courtesy of the Pamir Archive Collection

railways. Road development was rather neglected given the economic importance of Kashmir and the strategic importance of the northwestern mountains in the Afghan borderlands. Only Srinagar had good road links that were motorable by 1922 and connected Kashmir's capital with Jammu and Rawalpindi, respectively.[280] From there roads took off to various destinations such as Gilgit, Ladakh, and Zanskar. The classification as a 'road' presupposed that it was suitable for pack animals.

A year after the lease had materialised by 1 August 1935, the first accounting revealed that the administrative cost had increased nearly fivefold from the initial one and a half lakh rupees that had been spent mainly for the upkeep of British officers.[281] Nevertheless, besides other new obligations, roads and routes had to be maintained, and the inventory of communications was expanded to the Hunza Valley. In the company of Major Robert S. Haslett, 17th Queen Victoria's Own Poona Horse, the Political Agent, Major George Kirkbride, travelled the full length of the Hunza Road during June and July 1936 to inspect the road conditions: "From Baltit to Murkushi, 77 miles, the road is a mule track maintained by the Mir of Hunza for which he receives an annual maintenance allowance of a thousand rupees. During July the volume of water in the Hunza river is heavy, fed as it is by snow water from the side nullahs. This necessitates almost continual use of the summer path which climbs a series of precipices. The track is in admirable condition and save for about 400 yards 4 miles south of Gulmit …

A wooden bridge based on one cantilever and fixed in the mountain wall (top) enabled caravans with pack animals to enter the Kilik Valley on their way from Sost to Misgar. In certain sections the Hunza Road was comparatively well built and maintained (middle). Since 1918 the Misgar post-office (bottom) had been the northernmost telegraph station (tār-garh) of the British Raj on the way to Kashgar and contained a post and weather station. Source: All photographs by Flight Lieutenant Francis Claude Sturgiss on 19 May 1937; File No. 82(2)-X/39, photos 54, 55, 60

The Kashmiri telegraph officer in Misgar as portrayed by Alexander Yakovlev in August-September 1931. Four years earlier the British Resident in Kashmir Evelyn Howell stated after officially inspecting the Gilgit Station in 1927 about the pundits: "… a class which does not lack intellectual subtlety but they are in general devoid of the qualities which we expect of our frontier officers. As I can testify from my own experience, most of them are a figure of fun on horseback, and are quite incapable of a day's walk over the hills to look at the scene of a grazing dispute or the source of an irrigation channel. Nor did I meet one who really spoke and understood the local language (Shina), though all, or nearly all, claimed to be able to do so. Officials of this class are necessarily wholly out of sympathy with the population with whom they have to deal. At best they are regarded with tolerant amusement." For any telegraph officer the remote station in Misgar (3,120 m) was not an attractive place, especially during the prolonged cold season. Source: Report of the Resident in Kashmir on his tour in the Gilgit Agency, 1927 (IOL/P&S/3288: 260–263)

Frederick Marshman Bailey left Baltit on 15 May 1918 and the scene shows how travelling was organised. Team leaders enjoyed the journey on horseback, while the subalterns had to walk (top). Not much would change until the mid-1930s when first considerations about improving road conditions and future aspirations for motor transport were envisaged. The photograph (below) shows Qudratullah Beg's house and family mosque; the path below was described by David Lorimer as Baltit's 'main road'. Source: Bailey Collection in the British Library Neg 1083/44 (29); Lorimer Collection in the School of Oriental and African Studies, University of London (PP MS 66), Box II. 36; reproduced with kind permission

and the Batura Glacier ... is fit for laden animals throughout. It takes 50 minutes to cross this Glacier on yaks, but a detour could be made on the left bank by the construction of two bridges. The whole track is maintained by forced labour and is naturally at its best when likely to be used by Government officials. If labour were paid the annual allotment would probably meet the cost of one ten mile stage. At every stage the Mir has built primitive rest houses, fitted with gauze, but without glass. To improve the track as a mule road could be accomplished at no great cost. Its present condition reflects the greatest credit on the Mir while the manner in which some sections are carried over formidable precipices compels one's admiration. ... From Murkushi ... the track branches to the Mintaka and Killik Passes. Both these branches are for most part extremely difficult for laden animals and yaks are in frequent use for riding."[282]

After the early route descriptions by Edmund Barrow and Francis Younghusband, this was an update for the mid-1930s; no mention whatsoever was made of a potential motor road to China. It was still a pack animal route for caravans, although trade had practically become negligible along this route. The only technical mention of motor transport was the idea of building an aerodrome in Pasu with emergency landing grounds in Sost, Misgar and Murkushi.

Besides considerations regarding the passage through the Karakoram valleys, other challenges had come up with the end of dual control in the Gilgit Agency that had existed since 1889. The new administrative setup had led to the withdrawal of 1,210 soldiers and their sutlers from the Kashmir garrison and the Royal Indian Army Service Corps; the latter had provided seventeen service stations along the supply routes where travellers could purchase fodder, grain and stores.[283] Travelling expenses increased in the aftermath. This loss of personnel and amenities was hardly compensated by 500 new British colonial administrative and military posts in Gilgit Town. The new constellation would require improving the living conditions of the local population. The change had affected service purveyors in Gilgit Town and its vicinity to such an extent that the Political Agent George Kirkbride demanded compensational measures. He exemplified possible options in a letter to the Resident in Kashmir in February 1936: "It is inconceivable that there will never be a motor road from British India to Gilgit though even with any impetus that might arise from trade with Sinkiang it seems unlikely that it will be constructed within the next ten years. But in some form or other the future of Gilgit is assured. Either it will become a great mart on a trans-Asiatic trade highway; or in Naltar will enjoy a hillstation rivalling Gulmerg; or it will become an important Air station, both for Military or Civil purposes. In any of these events it cannot be opened up to the world only to reveal a wretched population living under no sylvan conditions but in a state of appalling squalor. I would urge therefore both on humanitarian grounds and for the good name of the British Raj that early steps should be taken to remove a reproach which at present does not lie at our door ...

Hunza and Nagar have benefited by the change. They have never been able to trade in grain, Nagar being only just self-supporting while Hunza cannot feed all its people. On the other hand the increase in the pay and service of the Scouts has benefited them more than anybody. Chilas remains what it is and need not be considered in the present reference. My proposals are however so aimed as to bring a general improvement so that if they succeed the whole Agency benefits eventually ... The proposals are summarised below:

(a) An experimental farm at Jaglote at the mouth of the Sai Nullah opposite Bunji. The farm would be stocked with one or two stud bulls, donkeys, stallions and Merino rams. It would experiment in varieties of wheat and other staple crops and also in fruit trees ...

(b) The installation of an hydraulic ram. If this were successful it might do away with the need to build miles of kuhls [= irrigation channels] at great cost and labour.

(c) A grant for the immediate colonisation of the Pari lands ... An important subsidiary object of this colonisation is to provide an outlet for the surplus Hunza population. ...

(d) A grant for the construction of a kuhl in the Bagrote nullah to irrigate and colonise 2,000 acres of land. There are possibilities in this locality both on account of the excellent condition of the alluvial fan and because the water supply is almost limitless.

(e) A grant to enable the Naupore kuhl to be reconstructed ...

(f) A discretionary grant for the purpose of afforestation near Gilgit for the future supply of firewood and for the planting of road side trees, the supply of farm implements, blasting powder, etc.

(g) A grant to set up a Notified Area Committee in Gilgit

The map 'Afghānistān, Gilgit Agency and Kashmīr & Jammu. Baltit (Hunza), No. 42 L' was published in 1930 and shows the state of topographical knowledge about the Hunza Valley. The areas between the Chinese border and the Karakoram main range seem to be well surveyed in the Misgar and Chapursan valleys. Such detail is found in the remaining map only in the valley bottoms. The well-known sections of Hunza are identified as Shināki, Hunza (Kunjut) and Little Gujhal, while Nagar consists of Nagir and Shen-Bar. Source: Survey of India (1930). Courtesy of Pahar.in

In the aftermath of the Hunza-Nagar Campaign colonial authorities introduced new crops such as potatoes, different wheat varieties and fodder crops, tomatoes and vegetables in the Hunza Valley; half a century later maize was added to the range of cultivars. Two different irrigation practices emerged and have been applied in the fields since: furrow irrigation for tubers such as potatoes that are grown on ridges, and flat parcel inundation for grain crops, alfalfa and kitchen gardens. The photograph was taken in Karimabad on 10 May 2015

proper as to improve the sanitation, etc., of the group of villages in the oasis.

(h) A grant to set up a hand-loom weaving school in Gilgit to improve the local weaving industry which already produces some quite good tweeds.

(i) A grant to set up an industrial school in Gilgit to train youths in carpentry, masons work, iron work, etc.

(j) A grant to improve the Gilgit-Gupis road so as to make it possible for wheeled traffic. With this in view the employment of carpenters and wheelwrights in Gilgit capable of making two-wheeled carts will be necessary.

(k) The provision of loans on very easy terms to enable the zamindars to purchase carts and agricultural implements.

(l) Grazing grounds in the Political Districts are better than anywhere else in the Agency. If the quality of the wool is improved … there is a possibility of establishing a market in Gilgit. At present all woolen cloths used in Gilgit are imported … If Gilgit is eligible for a grant from the Government of India's grant for Rural Uplift and if any still remains over from the reserve of 1935–36 it would seem that Gilgit has a very real claim on an allotment from this source. The only possible difficulty that occurs to me is that the grant apparently has been devoted entirely to British Indian provinces and administrations whereas Gilgit is not part of British India but only an administered area."[284]

Assessment of the local conditions revealed that the little economic activity that was observed on the supply side with agricultural goods, fodder, fuel and porterage had been significantly affected and led to a decline in all prices by a fifth to a third of their previous values; the purchase price for wheat had halved. Major George Kirkbride's comprehensive plan with detailed suggestions and linked to a demand of 50,000 Rs can be regarded as an initial development package under colonial circumstances in which he mentions a 'moral obligation of the Government of India to remedy the economic distress caused by the change in administration'. By emphasising that in the 'matter of Rural Uplift Gilgit is virgin soil' his report marks the beginning of rural development activities that are not far off from measures and packages that are discussed in recent programmes. He is aware of and stresses the opaque political situation of the territory that clearly is under dominance by the 'Imperial Government', but which is incorporated into neither British India nor an assemblage of independent entities. The Government of India allocated 'a single non-recurring grant not exceeding Rs. 30,000 from the unallocated reserve of the Rural Development Fund'; this is only two thirds of the amount allocated in the 1935 budget for the repair and maintenance of roads and buildings by the Public Works Department.[285] The message had been heard, and some measures were taken in an unstable and increasingly insecure cross-boundary political environment.

The necessity to improve living conditions in the Gilgit Agency was not only due to the changed administrative set-up. In Hunza and Nagar the Political Agent had observed that 'Soviet sugar, matches, cotton piece-goods, carpets' were available; but he was consoled by the fact that 'there is very little ready cash in those States so that it is unlikely that a trade post at Murkushi would reveal any startling figures'.[286] Certain activities were amplified to increase local productivity. On his last journey to the Hunza Valley Colonel Reginald Schomberg reflected on the effects of British administration in his ironical criticism: "British rule has brought many benefits, and being human, probably a number of disadvantages in its train, but the advent in Gilgit of despised and misunderstood imperialism, brought the potato with it. This has been a precious boon to Hunza, as it was to Britain in the two great wars. The potato is some compensation for the horde of Pathan and down-country officials who infest Gilgit, since the lease of the Agency to the Government of India."[287] The road had brought agricultural innovations that only much later contributed to the local economy as surplus products. The side-effects were related to Pashtun enterprises in the bazaars and in the marketing of surplus products.[288] Their role as innovative entrepreneurs, itinerant traders and potato dealers has increased since. The potato became a leading crop that changed nutrition patterns in the Pamirian Crossroads and the Himalayas irrespective of British or Russian/Soviet interferences.

Both imperial powers pursued similar ideas in fostering local agricultural production; in terms of transport infrastructure they were heading in different directions. After the October Revolution in the neighbouring Soviet Union the commencement of reforms and infrastructure development posed significant ideological and political challenges in an environment where British India was not leading technological advancements, especially when it came to motor roads and railways.

Potatoes have become the bulky and valuable cash crop since the opening of the Karakoram Highway. Annually Hazara or Pashtun contractors, locally called Paracha, purchase potato bags in the villages along the transport arteries and transfer them to down-country markets by the truck-load. The photographs were taken on 11 September 1990 and 17 September 2013

The first motor vehicle reaching Khorog on the Pamir Highway is displayed on a pedestal there. The reading of the plate remembers the opening of the Pamir Highway with a blurred date of June 1891 (??), probably changed from 1931. The photograph was taken on 13 August 1999

From Pamir Highway to Stalin Road

The 1930s were the period when new projects to enhance terrestrial motor traffic were initiated and implemented. The challenge and response game was on and initially won by the Soviet Union. Their efforts in constructing the Pamir Highway became an engineering feat that surprised their competitors and enemies. In their lectures in the Royal Central Asian Society, travellers who had seen the infrastructural changes in Soviet Central Asia praised them as 'constructive socialism' and admired the modernisation effects and improved communication.[289] More specific and disturbing was the news reported by William Strang: "... successful progress is being made in the construction of a road from Osh in the Eastern part of the Fergansk valley, via Pamirski village, to Khorog, across the Kyzyl-Art (Zaalaisky Range). The road is 702 kilometres long and will cross four mountain ranges. It is stated to be the first metalled road to be constructed in the Badakshan area."[290]

The Soviet Union had entered the path of modernisation in the remote Pamirian Mountains. The first motor vehicle reached Pamirski Post in May 1934. The Tashkent-based correspondent of the 'Pravda' highlighted the new era of transportation with the headline 'A car at an altitude of 5 kilometers'.[291] The following year regular traffic began; travellers and writers were full of praise for such an achievement that was well in tune with a European admiration for the creation of a new society in Central Asia.

The British observers were quite disturbed by the speed with which Soviet engineers mobilised their workforces in accessing formerly remote mountain regions. Military strategists realised that the initial outpost in the Eastern Pamirs – Pamirski Post/Murghab – and the subsequent administrative centre and military headquarters of Badakhshan – Khorog – were now connected to the important supply lines in Kyrgyzstan through the most modern technology available and, as a side-effect, offered new opportunities to all kinds of adventurers and travellers. Wakhan got its first road link in 1939 when the Pamir-Ishkashim road via Khargosh linked Rajon Ishkashim with the Pamir Highway network. For a long time this would remain the only road link to the Panj Valley and the Afghan border.

British concerns grew even further when they learned that the Sary Tash-Irkeshtam road was to be connected with the traffic system at the same time. From Khorog a road to the north was planned in order to reach two new motor roads in the near future: Stalinabad-Kurgan Tyube and Stalinabad-Garm. The road planners had recognised that this road was even more difficult to construct despite lower passes to be traversed.[292] It took another five years for Khorog to be seasonally connected with the Dushanbe-Garm road via Tawildara and Khaburabot (Sagirdasht Pass 3,252m).[293] On 11 September 1940, the New York Times reported the completion of the Stalin Road, a 350-mile-long road from Stalinabad to Khorog with the headline: "Soviet opens mountain road".[294] Seven years later work began to improve the Stalin Road to become an all-weather artery that had previously been closed for four months in winter.[295] The Soviet Union had definitely taken the international lead, and the construction of the Pamir Highway and airfields had shown the world that high mountain regions could be accessed by motor traffic and air links. The Irkeshtam road posed an exception to the other roads that connected remote corners within the Soviet Union. The Irkeshtam road was the outlet towards China and the Kashgar oasis at a time when the internal political situation in Xinjiang was deteriorating and led to the subsequent closure of borders. The 'Russian Trade Agency in Kashgar (Sovsintarg)' finally closed down by the end of May 1943 and the 'Kashgar Weekly Newsletter' remarked: "Trade between Sinkiang and Russia which has flourished for many years, has now completely stopped. Although this trade ensured Sinkiang of an adequate supply of manufactured goods it was mainly profitable to Russia, who acquired the whole of Sinkiang's valuable primary products at very advantageous rates."[296] The Soviet Union had changed the trade policies to termination of cross-boundary exchange and adjusted its infrastructure development to build a road network connecting the socialist republics. The final years of the global wars would change the constellation outside the Soviet Union, and new supply routes to Xinjiang and China were investigated.

The map illustrating the alignment of the Pamir Highway shows the red line from the Andijan's railway junction in the Fergana Basin to the Pamirian terminal at Khorog, close to the Afghan border. The road crosses four major passes; Koi Tesek Pass poses an obstacle on its way from Khorog to Murghab, and onwards the road finds its climax in Akbaital Pass (4,655 m). Never before had a motor road been built at such an altitude, from where its course continues at a high level via Kyzyl Art (4,336 m) and Taldyk Pass (3,589 m) towards Osh. Along the road there, two 'autostations' for car services in Alichur and Jelondy are indicated outside the main settlements. The map is based on geodetic surveys in 1932 and 1934 and was prepared by the cartographers of the Tajik–Pamir expedition N. Jegorov, A. Potulov and G. Skrebej. The eyewitness Gabriel El-Registan (1936: 3) narrates that this achievement was due to the collective 'Pamirstroj' under the leadership of the Bolshevik Federmesser. The Pamir Highway acts as a symbol for the feats of modernisation that are changing every aspect of life on the Pamirs. Source: UdSSR im Bau, No. 12, December 1936. Courtesy of the Pamir Archive Collection

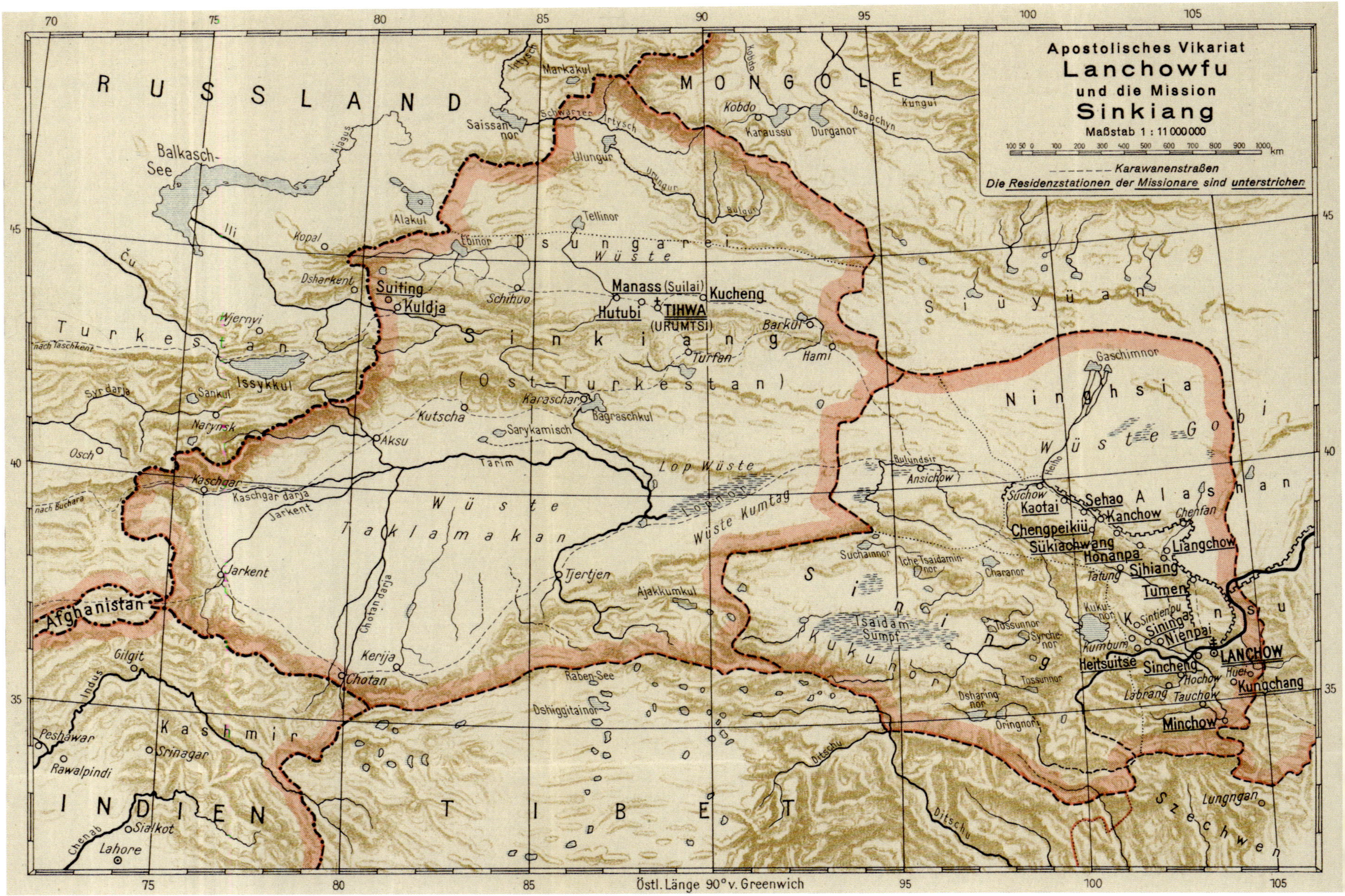

The 1930s are characterised by several transformations in Xinjiang. The map indicates a defined boundary between China and British India. Any connectivity or relations across boundaries between Gilgit and Yarkand are omitted. Source: The map is appended to Johannes Thauren (1931) who described the activities of the Divine Word Missionaries in China. Courtesy of the Pamir Archive Collection

A camel caravan on its way from Kara Köl Lake to Gez Gorge is passing by the famous sand dunes near Bulunkul. Here Aleksandr Yakovlev found his motif on 13 September 1931 at Bulunkul which he described as Bashqul. Oil on canvas laid on board; reproduced with kind permission © collection particulière française, courtesy of Sotheby's. The photographic scene is depicted in Ariane Audouin-Dubreuil (2002: 50); see as well Hermann Kreutzmann (2015a: 462)

Support axis for General Chiang Kai-shek – a road through the Hunza Valley

The 1930s developed into a period of turmoil and rebellion in Xinjiang. During this exciting time, the trade routes between British India and Xinjiang were often interrupted or closed. Trade from Gilgit to Yarkand more or less ceased to exist; so-called Indian traders – Shikarpuri, Pathan, Kashmiri and sometimes Afghans – lost their properties, were expropriated and expelled by Xinjiang's Soviet-backed warlord Sheng Shicai.[297] Regularly refugees would cross the Kilik and Mintaka passes on their exodus from Xinjiang and find their first shelter in Hunza before proceeding to Gilgit.[298] The British authorities had urged *mīr* M. Nazim Khan to terminate his loyalty to the Chinese authorities in Yarkand and to abandon their annual gift exchange.[299] The reasoning was not much concerned with Hunza itself, but they feared that in the probably inevitable conquest by Soviet troops of Xinjiang, including the Taghdumbash Pamir, a confrontation could occur that was to be avoided by all means. The maintenance of buffer zones between British India and the Soviet Union would then collapse. Therefore Olaf Caroe, when planning future external affairs, raised again the question where and how to delimitate the northern boundary, and whether the *mīr* of Hunza could serve two masters – the British and the Soviets who he envisaged would replace the Chinese.[300] Although clarification of alignments was due, the Resident in Kashmir suggested 'the wisest policy is to leave aside the question of frontier delimitation'.[301] Giving up Hunza's claims for grazing taxes beyond the Kilik and Mintaka passes would be a small price compared to the cost of conflict with neighbouring powers or strengthening the defence of the northern frontier. If *mīr* M. Nazim Khan were to give up his claims to levy pasture taxes in Taghdumbash Pamir and Sarikol and to cultivate his own land in Raskam, he would be compensated with cultivable land in Oshikandas and an increase in his subsidy. Major George Kirkbride announced 'the end of a very old chapter in local history' in the 'Gilgit Agency Political Diary for the month of April 1937'. To abolish this cross-border relationship was the strong desire of the colonial administration, although some property was claimed and would remain with the *mīr* of Hunza: "It has long been felt that in the present conditions that exist in Sinkiang a continuance of his relationship was undesirable and in fact had become increasingly difficult for the Mir to assert his grazing rights. The Mir of

In Kashgasu (4,130 m) the felt mats (ijīn) are still produced that in olden times were part of the grazing taxes. In Kuibaland, the former name for Kashgasu above Pirali, more than fifty households use the pastures that immediately border at Hunza territory in the vicinity of Khunjerab Pass. The photograph was taken on 5 August 1991

Hunza has therefore agreed to sever all these connections with Chinese Turkistan and to receive certain compensation from Government. This includes an increase in his subsidy of 3,000 rupees and a valuable jagir of 312 acres at the mouth of the Bagrote Nullah ... His delight with the jagir is very apparent and his men are already working on the land in a manner which puts the local Gilgitis to shame. He was little inclined to cavil at the size of the increase in subsidy but when he perceived that such attitude was likely to be unpopular he remarked with a smile; 'Well a child will always cry to its mother for as much milk as it can get'. ... These arrangements have deliberately ignored the Mir's estate in Yarkand and it will be interesting to see what will happen to it when authorities in Urumchi are fully apprised of the new situation."[302] After Hunza's special cross-border relationship ended, the situation for 'British subjects' and traders in Xinjiang did not improve for the next five years. For *mīr* M. Nazim Khan it was the completion of a cycle that began with his instalment in the aftermath of the Hunza-Nagar Campaign, opening the Hunza Valley for British domination, and terminating his loyalty relationship based on annual gift exchange with China. One year before his death, all these events had fallen in his reign and he remained the faithful subject of the British authorities. The gift exchange with Kashmir had long ago lost its relevance, especially after the lease of the Gilgit Agency to the British.

The situation in Xinjiang did not improve as the position of foreigners had been weakened since the mid-1930s. Targets were often so-called 'British subjects'. Two entries in the Gilgit Diaries of February and May 1941 may suffice to explain the anti-British sentiments: "53 more deportees from Sinkiang arrived in the Agency during the month, including R. S. Rattan Singh, British Aksakal of Yangi Hissar, Abdul Karim, British Aksakal of Goma, and Abdul Rahman, an influential merchant of Khotan. The majority of these claim to be British subjects. They appear to have been treated with a little more consideration than those deported in November and December, for most of them have been able to bring a part of their moveable property and their horses with them. Three or four are destitute. These will be maintained and repatriated at Government expense and the remainder permitted to stay in Gilgit at their own expense until the passes open. But although more fortunate in the matter of personal property than earlier deportees their suffering has been greater. A woman and a child died of exposure on the

Mintaka Pass and R. S. Rattan Singh and one other man died from the effects of frost bite on arrival at Misgar."[303] As the Babusar Pass was closed until summer, the refugees had to stay in Hunza and Gilgit for quite some time before leaving for down country; some managed to leave Gilgit via Shandur Pass for Chitral by the end of February. The loss of significant property seems to have been a regular effect of expulsion: "The most recent deportee from Sinkiang to arrive in this Agency is Hans Raj, an Indian trader, who reached Gilgit on the 28th May. This individual was permitted to bring his property with him and left Kashgar with 25 pony loads of goods, having apparently paid the pony hire as far as Misgar in advance. He was accompanied by two other Hindus, Kishen Chand and Narain Dass. At Lupgaz, on the Chinese side of the Mintaka Pass the Muleteers dumped the goods of the party and decamped. Hans Raj and his companions therefore arrived only with what they could carry. Subsequently it was reported that Chinese soldiery removed the goods from Lupgaz."[304] Relations between Xinjiang and British India had become hostile and affected not only affluent traders and shopkeepers; even British appointees, officials and staff were not spared. On his transfer from Gilgit to Kashgar in October 1940, the designated Consul-General Eric Shipton was "... detained at Dafdar for 4 days ... and Mr. Shipton and party were treated with gross discourtesy, amounting to hostility, by the Chinese officers there."[305] The tides would change soon.

The mīr of Hunza collected grazing taxes in the Taghdumbash Pamir until 1937; Kirghiz, Sariqoli and Wakhi were obliged to deliver their dues on an annual basis. The Wakhi pasture settlement of Sikhtikar (3,920 m) is characterised by three yurts (xirgó) high above Mazar in the Taghdumbash Valley; see map in Hermann Kreutzmann (2015a: 394). The photograph was taken on 3 August 1991

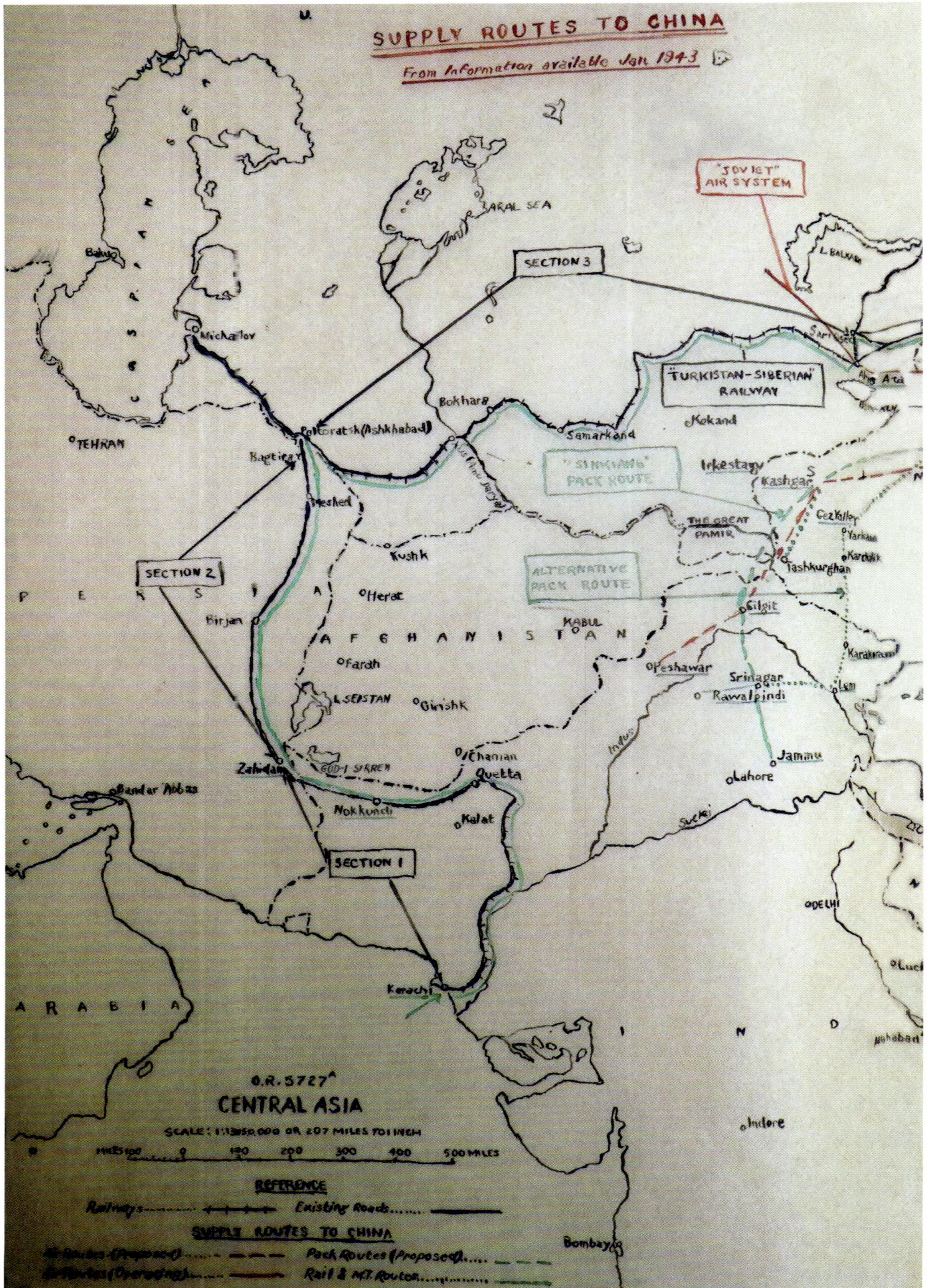

Reconfiguration in Xinjiang

The political constellation was reconfigured when the powerful Xinjiang leader Sheng Shicai severed his cordial relations with the Soviets in Xinjiang and turned towards the Republicans by joining the Guomindang government under General Chiang Kai-shek in Chongqing to fight against the communists under Mao Zedong and against the Japanese. A new perspective for supporting these forces and reopening trade relations between Gilgit and Kashgar appeared and led to bilateral negotiations.

In August-September 1942, the outgoing British Consul-General in Kashgar, Eric Shipton, and his successor Michael Gillett had confidential discussions with their National Chinese counterpart Mr. Pan, deputy of Sheng Shicai, about the possibility of a motor road across the Karakoram to connect Southern Xinjiang with British India. The shortage of military supplies for fighting the Japanese was so severe that Mr. Pan saw only one solution and stated that 'they have decided on a motor road'. Eric Shipton, the author of 'blank on the map'[306] had significant previous experiences as a mountaineer and surveyor in the mountainous interface between Central and South Asia; his colleague Michael Gillett was an accomplished traveller himself so he tried to explain to his counterparts how challenging and difficult the terrain was. His ground knowledge enabled him to paint a realistic picture of what would be required to build such a road: "I pointed out, as objectively as I could, the enormous difficulties of such a project, more particularly on the Indian side, in a country where scarcity of labour and almost complete lack of supplies added to physical and climatic difficulties: I then took them over the country from Kashgar to Gilgit as well as I could from memory and without maps. I gave it as my opinion that the Kilik pass would probably provide a slightly easier alignment than the Mintaka, suggested that if labour were available the Kashgar-Gez-P'uli (Tashkurghan) route would probably be better than the Yarkand-P'uli route, and said there was a 5' mule-transport road from Srinagar to Gilgit that crossed the difficult Burzil pass. I told them I had already talked to Chang Chia-ngao about the matter and showed him a couple of albums of photos of the country."[307] His Chinese counterparts were strongly determined to have that road and pursued the case further. The British representatives had already stated 'how very interested we and America were in getting supplies to China'; internally it was made clear: "The scheme has great possibilities, particularly if most attention is directed to the traditional Karakoram route and the Gilgit road with its potential local porters as well as animals is used to supplement the other route. But the full co-operation of the Sinkiang Government must be secured, and this only can be secured through the Central Government"[308] The outcome of the bilateral talks was that the Chinese partners asked for advice how to proceed. Eric Shipton and Michael Gillett suggested commissioning a survey party to get first-hand experience about the proposed road alignment. These talks were clandestinely held, and exchanges were strictly confidential. Only the India Office in Whitehall and the Embassy in Chongqing were negotiating with the Republican Chinese Government – that had relocated to Chongqing from Nanjing – and the provincial warlord Sheng Shicai. In the following years, even the confidential Gilgit Political Diaries contained only passing remarks about survey teams and visitors, while the local rulers in Gilgit, Hunza and Nagar were left in the dark.[309] The 'History of Hunza' does not mention any major road project that was negotiated in the course of more than two years.[310] This bilateral road project remained reserved for high-level discussions totally excluding local affected communities and rulers of principalities.

On the ground in the Gilgit Agency, other issues were discussed such as the summer flood of 1941 that had destroyed part of the Hunza Road, requiring significant repair. The Political Agent Gerald Charles Lawrence Crichton reported in his monthly diary for July 1941: "An abnormally high flood was caused in the Hunza river on 15 July by the bursting of an ice dam which had been forming for some months, by glacier action, in the Shimshal (Shingshal) River. … At noon that day the flood reached the junction of the Hunza and Gilgit rivers at Danyore … where it rose to a maximum of about 40 feet above the usual summer level. It subsided rapidly after four hours. Large portions of the road between Nomal and Passu have been washed away and extensive repairs are necessary and as is known no lives have been lost in the flood, but a few cattle are reported drowned. A country suspension bridge across the Shimshal river and the Danyore Ferry boat (in Gilgit) were washed away."[311] The Hunza Road was severely affected by these recurring events of glacier lake outburst floods. The damage was caused in the bottleneck passage through the Hunza Valley; no alternative routing was available. Communications for pack animals and muleteers were significantly affected; they posed major challenges for the maintaining agencies. Major Gerald Crichton entered another entry in his diary a month later: "The Resident Engineer, who is now in Hunza reports that the Shimshal flood of July severely damages the road between Baltit and Passu particularly in the vicinity of Atabad, where two miles of realignment and 450 yards of rock blasting will be necessary to reopen the road to animal transport. There has been practically no interruption to pedestrian traffic since the flood as the Mir of Hunza has constructed new footpaths, but these are too hazardous for animals or for heavily laden coolie. The state of the road has delayed delivery of heavy material for the construction of Kalam Darchi Scouts post."[312] While the ground staff was committed to maintain the existing state of the road after any kind of hazardous event, the superiors in faraway locations were more concerned with the grand design of modern needs and future modes of transport.

The Central Asian road connection was meant to link the battle fields of General Chiang Kai-shek with British supply routes. The aim was to counteract Mao Zedong's Red Army. The western terminal fortification of China's Great Wall at Jiayuquan has remained an attractive spot for all kinds of visitors to date. The photograph was taken by Sabine Felmy on 8 September 1989

Opposite
The sketch map on 'supply routes to China' indicates the envisaged new motor road from Peshawar via Gilgit and the Hunza Valley to Kashgar and onward to Hami in order to connect with existing road networks. In addition a 'pack route' from Jammu to Kashgar and Hami via Gilgit was proposed. Source: (IOL/P&S/12/4616)

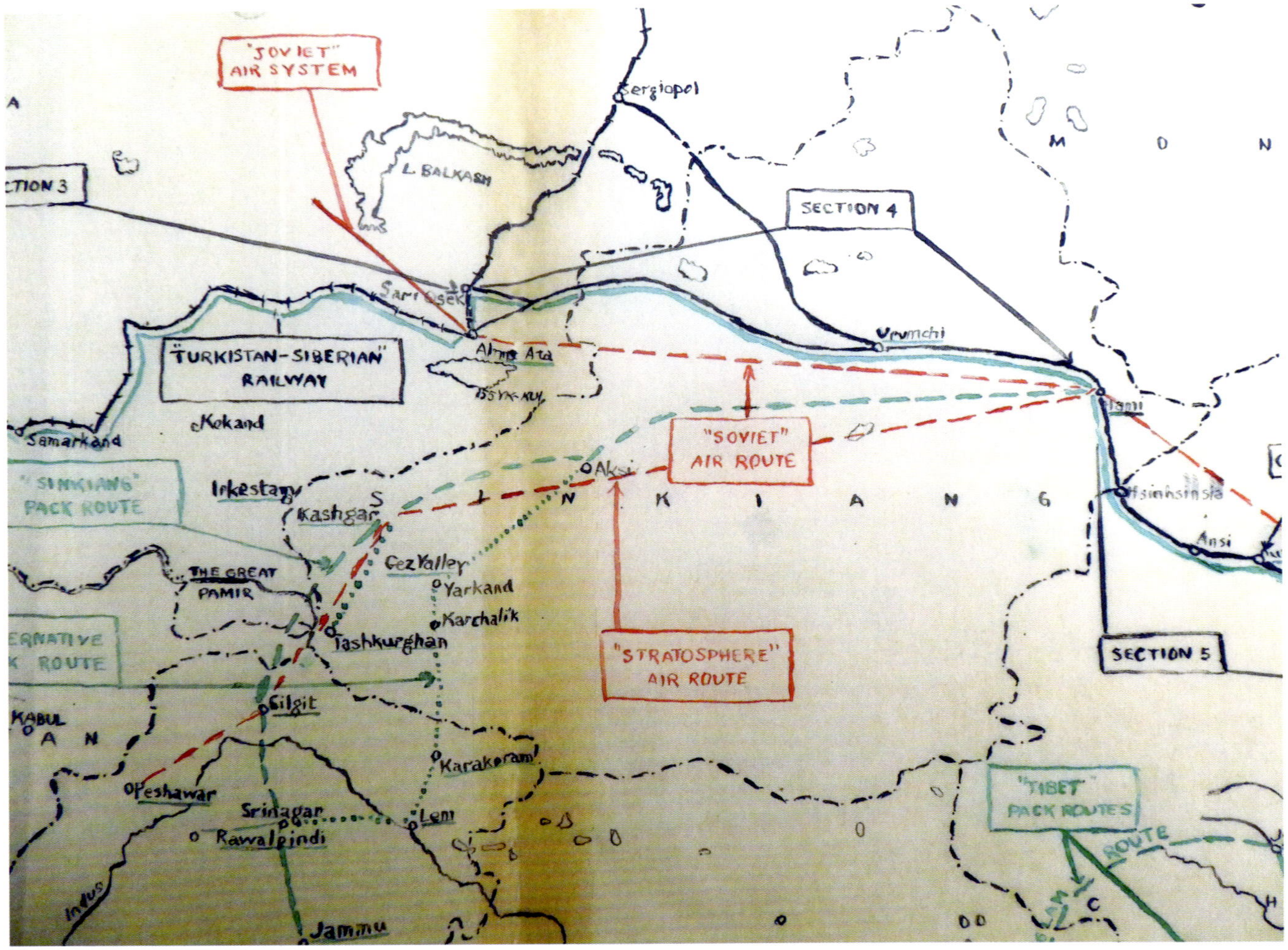

All supply routes from the western Central Asian sector and British India were supposed to lead to Hami and Sinsinsin (Xingxingxia) at the boundary between Gansu and Xinjiang Provinces. The sketch map classified as 'most secret' represents the information and planning stage of January 1943. Source: India Office Library, Departmental Papers: Political & Secret Internal Files & Collections 1931–1947 – Collection 46: Routes Central Asia (IOL/P&S/12/4616)

Improving the Hunza Road for motor transport

The following months in 1942 were directed to the potential of the trade route rather than to its obstacles. The vision was focused on estimating the capacity of the existing supply routes. The 'transport capacity of China-India caravan route' which described two sections – the 38-day and 1,179 kilometres between Lanzhou and Sinsinsin [Xingxingxia] on hand-carts, mules, camels, and the 3½ months on the 2,685 kilometres-long Sinsinsin-Kilik section on horse-carts, yak-carts, in the higher elevations camels, mules, donkeys and yaks – was estimated as barely 1,000 tons per month for 'government supplies'.[313]

On 2 March 1943, the India Office in Whitehall communicated its reformulated Xinjiang policy to the Secretary in the External Affairs Department in New Delhi. This was summarised by Roland T. Peel: "The objectives of our policy in Sinkiang would thus appear in the main to be to ensure: (1) that Sinkiang remains in name and as far as possible in substance a part of China, in order thus to prevent or at least to delay the establishment of complete Russian control over it. At the same time we should be on our guard against any tendency by the Kuomintang to use it as a base for political infiltration towards India; (2) that the Indian community in Sinkiang receives fair treatment; (3) that some resuscitation of trade between India and Sinkiang takes place; (4) that supplies can be passed to China proper through Sinkiang and (5) that we retain in Sinkiang a listening post from which to gather intelligence about Central Asia."[314] Regarding communications, the India Office favoured the 'Karakoram pack route' but had reservations whether the Chinese Government would be able 'to contend against the strong autonomous tendency in the Province'; for its own policy, the danger of Soviet interference, the separatist tendencies and estranged relations between province and centre should be carefully balanced. Trade and supplies would help to stabilise the position: "Even a trickle of goods in short supply, which would give promise of more ample quantities later on, might have considerable political value …"[315]

The issue of constructing a motor road was now embedded in a comprehensive strategy to backstop Guomindang rule in Xinjiang and fend off separatist Turkish movements. Consequently, the plan of having a trans-mountainous motor road matured.

In the following months two options were discussed; in the wake of the reopening of the Leh-Yarkand route, two reconnaissance teams were commissioned to investigate the potential of the Leh to Yarkand and the Gilgit Road for upgrading to a motor road. In an internal office memorandum in the 'Political Department' some hope was expressed that the reconnaissance parties would experience all possible obstacles and consequently would let go the idea of constructing a motor road. Their Chinese counterpart, 'Mr. Tseng [Yang-fu], the newly appointed Minister of Communications, a determined and successful road builder', was quite enthusiastic and should be brought to think about down-to-earth considerations: "This is probably the best and only way to convince the Chinese that pack transport via Leh is the only practicable way of getting goods into China by this route. As Sir H[orace] Seymour has said, the realities of the situation will be brought home to them by practical experience. It is a pity, however, that they will be travelling during the easiest months and that they will not have experience of snow. … It is to be hoped that the survey party will not return with rosy reports as to the possibility of constructing roads over the Karakorams as we know only too well that obstacles which to us are insuperable, bear no such terrors for the Chinese whose ideas are more theoretical than practical. It is possible that the party may have a nuisance value but the risk is outweighed by the value of the gesture of letting the Chinese try to help themselves."[316]

Until the 1990s camels and horses played a prominent role in public transport. Goods for the re-established Sunday market were brought from cultivation areas within the Yarkand oasis by camel cart; customers used horse taxis for urban traffic; while heavily loaded lorries brought market visitors back to their hamlets.
The photographs were taken on 7 and 8 September 1991

A couple of months later Sir Eric Teichman had another conversation with Minister Tseng Yang-fu and discussed the road question again. It was obvious that he preferred the Gilgit route as the passes to be traversed were less high. During the conversation the question of cost was raised for the first time. The Chinese government expected their British

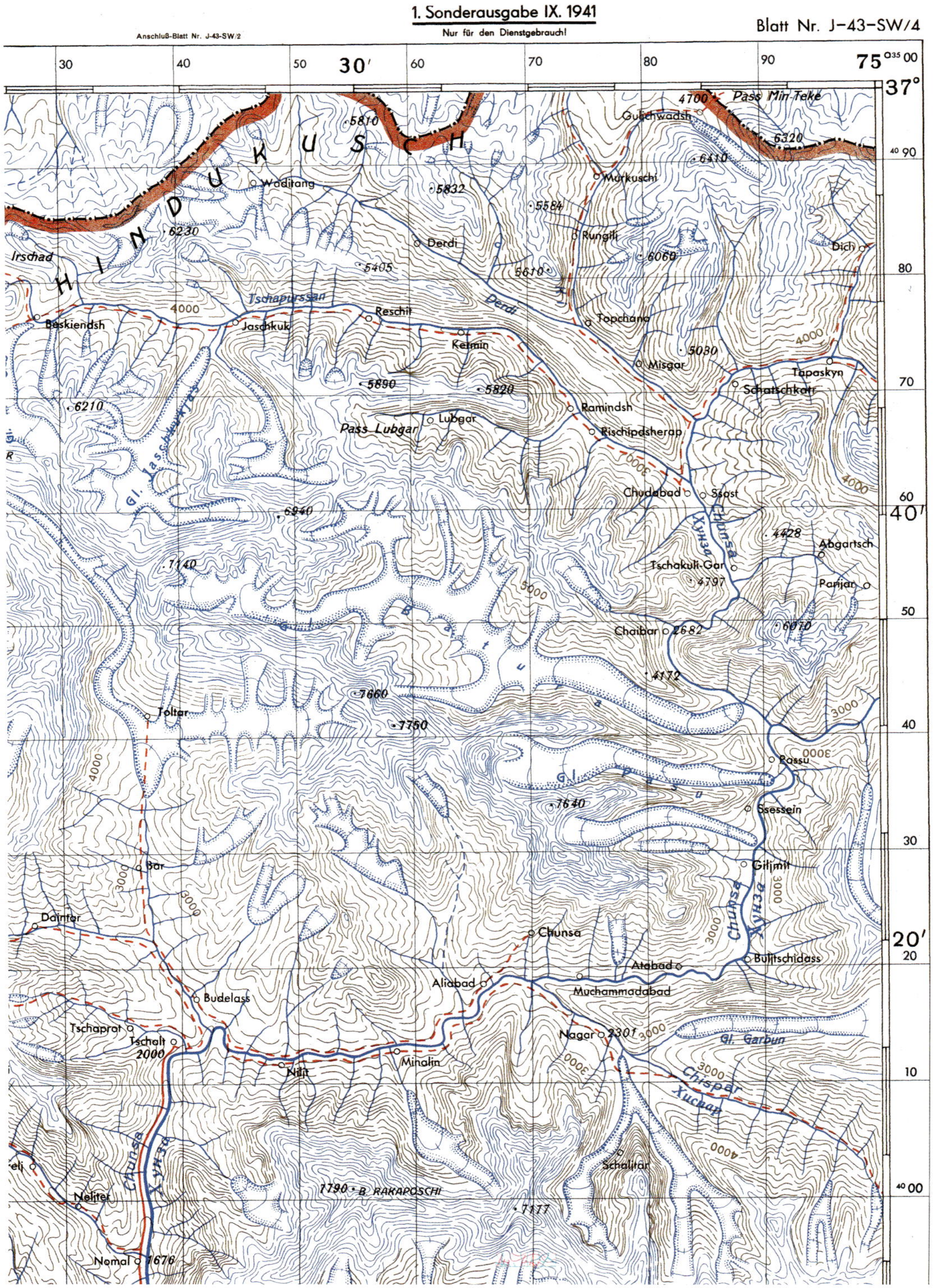

counterparts to bear the cost, and even if the war were to end soon the road would be a necessity.[317] Minister Tseng Yang-fu persisted and in a letter summarising the discussion with Sir Horace Seymour, which contained a map and a memorandum, stated "... this highway should not present any serious difficulty. On our side, i.e. from Yarkand to Killik Pass – 404 km, we are fairly sure that we can finish it within six months. Although the distance on the Indian side, starting from Killik Pass to Balakote is 560 km, it already has a pack animal road of 320 km. which only needs widening. The actual construction work on the Indian side is therefore less than that on the Chinese side. With the help of machinery, perhaps you can possibly finish the construction in a shorter time."[318] The appended report on the 964-km-long 'Proposed Sinkiang-India Highway' contained a full description of the four road sections: "Balakote to Gilgit. The present 8' wide pack-animal road can be improved easily into a motor road. The Babussar Pass will require mechanical snow clearing equipment from October to May. Sloping gently down from the Pass to the Indus River, the road becomes an easy wide track over sandy stretches with very little rock cuttings. Crossing of the Indus presents no difficulty (present 280' span can be shortened). Timber for construction purposes is abundant for the first half of this section. ... Gilgit to Misgar. The road ascends very gradually (4 % to 5 %) by following the valley of the Hunza River. As there is little rain or snow fall, work can be carried throughout the year. Along the Hunza River valley for about 120 km, the road cross[es], or, runs at the foot of, high steep loose rock slopes; hard rock is met in the narrow gorges which total about 30 miles. The road is very narrow (2 ½') along most of the way. Numerous side streams fed by glaciers have to be crossed by small bridges or by fordings. ... Misgar to Puli. There is an easy wide slope throughout. The Kilik Pass is covered with 20' snow in winter and early spring; it is an easy pass. The whole section is barren except for grazing grass. Construction in this section is very easy. ... Puli to Yarkand. Following the Puli river eastwards, the whole section is easy to build; work can be carried on throughout the year. The road goes over the flat sand bank of the river and rock cutting[s] in the gorges do not total more than 32 km. There are numerous villages; timber and food supplies are available."[319] It seemed to be comparatively easy and possible to implement the whole scheme in less than one year. This was the Chinese view, probably influenced by wishful thinking.

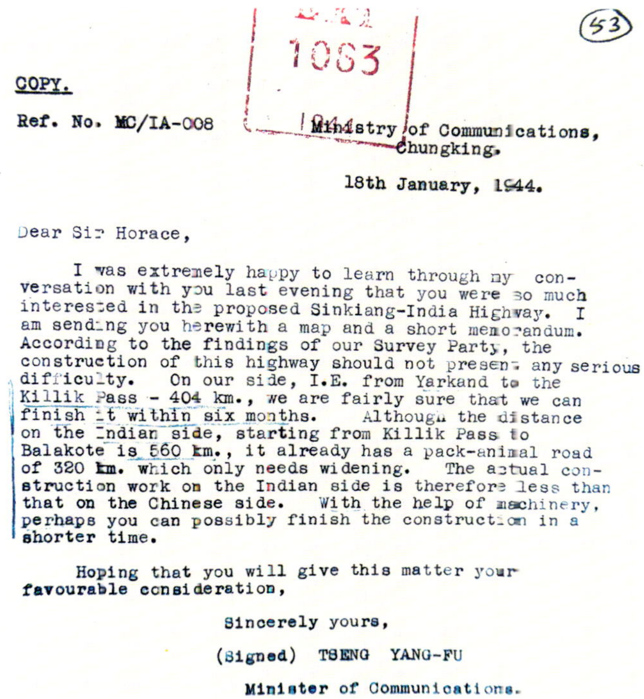

COPY.

Ref. No. MC/IA-008

Ministry of Communications,
Chungking.

18th January, 1944.

Dear Sir Horace,

I was extremely happy to learn through my conversation with you last evening that you were so much interested in the proposed Sinkiang-India Highway. I am sending you herewith a map and a short memorandum. According to the findings of our Survey Party, the construction of this highway should not present any serious difficulty. On our side, I.E. from Yarkand to the Killik Pass - 404 km., we are fairly sure that we can finish it within six months. Although the distance on the Indian side, starting from Killik Pass to Balakote is 560 km., it already has a pack-animal road of 320 km. which only needs widening. The actual construction work on the Indian side is therefore less than that on the Chinese side. With the help of machinery, perhaps you can possibly finish the construction in a shorter time.

Hoping that you will give this matter your favourable consideration,

Sincerely yours,

(Signed) TSENG YANG-FU

Minister of Communications.

The letter by Tseng Yang-fu, Minister of Communications for the Guomindang government, was addressed to Seymour Horace, the British ambassador to their government in Chungking (Chongqing). The minister is convinced that the motor road project of more than one thousand kilometres could be implemented within six months. Source: India Office Library & Records: Departmental Paper: 'Road routes via Karakoram Mountains to Sinkiang' (IOL/P&S/12/4609: 53)

Sir Olaf Caroe, External Affairs Department in New Delhi, supported the reservations made by the British Ambassador in Chongqing, Sir Horace Seymour. The former was familiar with the terrain that was to be traversed; he described the Hunza Valley as 'probably the most stupendous cliff-like formation through which any river passes in the world'.[320] The trustworthiness of the Chinese assessment was not questioned, although grave suspicions remained: "It is fairly clear that the survey party were told to report as they did to maintain the psychological courage of the Chinese, and that they have in fact merely reported what they were expected to report. It is probable also that the prompt suggestion (made before any indication of our views had been offered) that the Government of India' opposition, if it developed, would be based on strategical considerations and the proximity of such an alignment to Russia and Afghanistan, has deliberately been put forward in order that at a later date it could be used to save face when, as the survey party must know, the construction of this road proves to be impracticable."[321] Olaf Caroe saw the opportunity to retreat from the ambiguous project grounding his reservations in strategic aspects.

Owing to delay in translation and misplacement, the 'Memorandum on the proposed motor road from North-west India to Sinkiang via Gilgit' by Dr. K. Chaucer Wu based on the expert's survey of the Chinese Ministry of Communications reached the British Embassy in Chongqing on 22 June 1944 and the External Affairs Department in New Delhi much later.[322] The enthusiasm articulated earlier for having a motor road that would provide the crucial link between the port of Karachi and the Chinese road network in Lanzhou had entered the narrative. A 'programme for the construction' with calculations of construction time, required manpower and machinery was delivered: "The populated and cultivated lands being only at both ends, work must start as early as possible from Balakote and from the Sinkiang side. The most difficult section is in the Hunza Valley which is scantily populated. Man-power and food supplies are abundant in Sinkiang as well as in the plains of the Punjab, Kashmir and North-West Frontier Province of India. Most of the manpower and food supplies for the Hunza Valley, can be obtained from the Sinkiang side."[323] Labour constraints were a big issue as under the existing time pressure an army-like workforce would be required. Only the 240-km-long stretch from Gilgit to Kilik Pass was marked as 'difficult section', whereas the other parts were considered as 'caravan road to be improved' or the part from Kilik Pass to Tashkurgan as an 'easy section'. From Balakot to Gilgit 8.4 million man-days were estimated to "... accomplish the work in about 3 to 4 months, 150 sets of machinery should be used, 30,000 common and 10,000 skilled laborers should be mobilized."[324]

In the Hunza Valley the calculation summed up "... about 9,000,000 man-days. Assuming half of the work to be carried out by machinery for about 4 months, and half by man-power for about 8 months, then 20,000 common and 10,000 skilled laborers are required. It is estimated that about 10,000 men could be mobilized in the Hunza and Nagar States, and Sinkiang Province can send over about 20,000 men. The work should be started as early as possible in the spring of 1945."[325] The calculations are highly exaggerated; in 1941, the total male population of Hunza and Nagar was about 15,000 persons; two thirds of them would have been required as full-time workers during the proposed scheme.[326]

From Kilik to Tashkurgan: "About 1,800,000 man-days will be required; 10,000 men can finish the whole section within 6 months."[327] For the final stretch between Tashkur-

Opposite
The 'Russland (Turkestan) 1 : 200,000' map represented here with its section on the Hunza Valley is a specimen of a World War II series of German cartography in preparation for warfare in Asia. The series is based on the enlargement and translation of Russian maps. The toponymical terms lack accuracy, and the ground terrain is a compilation from previous maps. These war maps were prepared in 1941; within less than a year the Chinese Guomindang government and Great Britain would begin negotiating the construction of a supply road from British India to Xinjiang via the Hunza Valley. Source: Russland (Turkestan) J-43-SW/4. Courtesy of the Pamir Archive Collection

The map shows the route between Kashgar and Srinagar in 1944; the only motorable sections are between Kashgar and Yangi Hissar in the north, and from Srinagar to Bandipura in the south. The suggested alignment follows the conventional route from Srinagar to Tashkurgan; from there an alternative routing is followed via three passes – Chichiklik, Tari Art and Kashka Su – and then down to the southern Silk Route bordering the Takla Makan Shamo where the motorable road is met. Source: Survey of India 1945 Courtesy of Pahar.in

gan and Yarkand: "About 7,700,000 man-days are required, 30,000 common and 15,000 skilled laborers can finish the whole section in about six months."[328] In total it was calculated that 70,000 road-workers requiring 10,000 tons of food supplies and 5,500 tons of engineering supplies plus 800 tons of equipment would suffice to build the road and would be supported by 150 sets of dozers, graders, scrapers, compressors and crushers, 15 sets of snow-ploughs, 310 trucks and 30 jeeps. The supplies would be transported by 2,400 pack animals from British-India and 3,000 from Xinjiang.[329] For the Chinese experts – among them Wang P'eng, the leader of the team, and his deputy K. T. Yung, Assistant Engineer-in-Chief, Indo-Sinkiang Reconnoitering Party, Ministry of Communications – this project was feasible and could be implemented by spring of 1946. They termed it easier than the construction of the Burma Road which suited a similar purpose.[330] This early vision of a Karakoram Highway did not meet with the same enthusiasm on the British side of the negotiation table. Initial reservations were articulated by Horace Seymour, the British Ambassador in Chongqing, who sought some backing from Olaf Caroe in the External Affairs Department in New Delhi. He was convinced that "… while the construction of this road may be impracticable, I think the Ministry of Communications here honestly believe that it can be made, and I think it is important to show by receiving their experts a cooperative and sympathetic spirit towards the project, even though the enclosed memorandum seems to ignore many of the difficulties rather than to solve them."[331] The cost estimate which was supposed to be listed in 'Appendix 4' is absent from the file; thus it cannot be established what financial volume was calculated at the time. These considerations governed the response of the British-Indian representatives who realised that the Guomindang were prepared to shoulder a substantial part of the cost in the Gilgit Agency.

As Olaf Caroe was on leave, his deputy Robert Richardson Burnett answered on behalf of New Delhi on 5 October 1944.[332] The 'Government of India's Road Engineers' examined the memorandum and concluded that 'as an engineering project the section of some 350 miles at the Indian end is not impossible': "As a war project of the first priority to be rushed through at all costs, it might be possible, they consider, to open some sort of road in two years, but this could be done only by devoting labour, engineers, mechanical equipment, steel, and so on, to an extent which is not

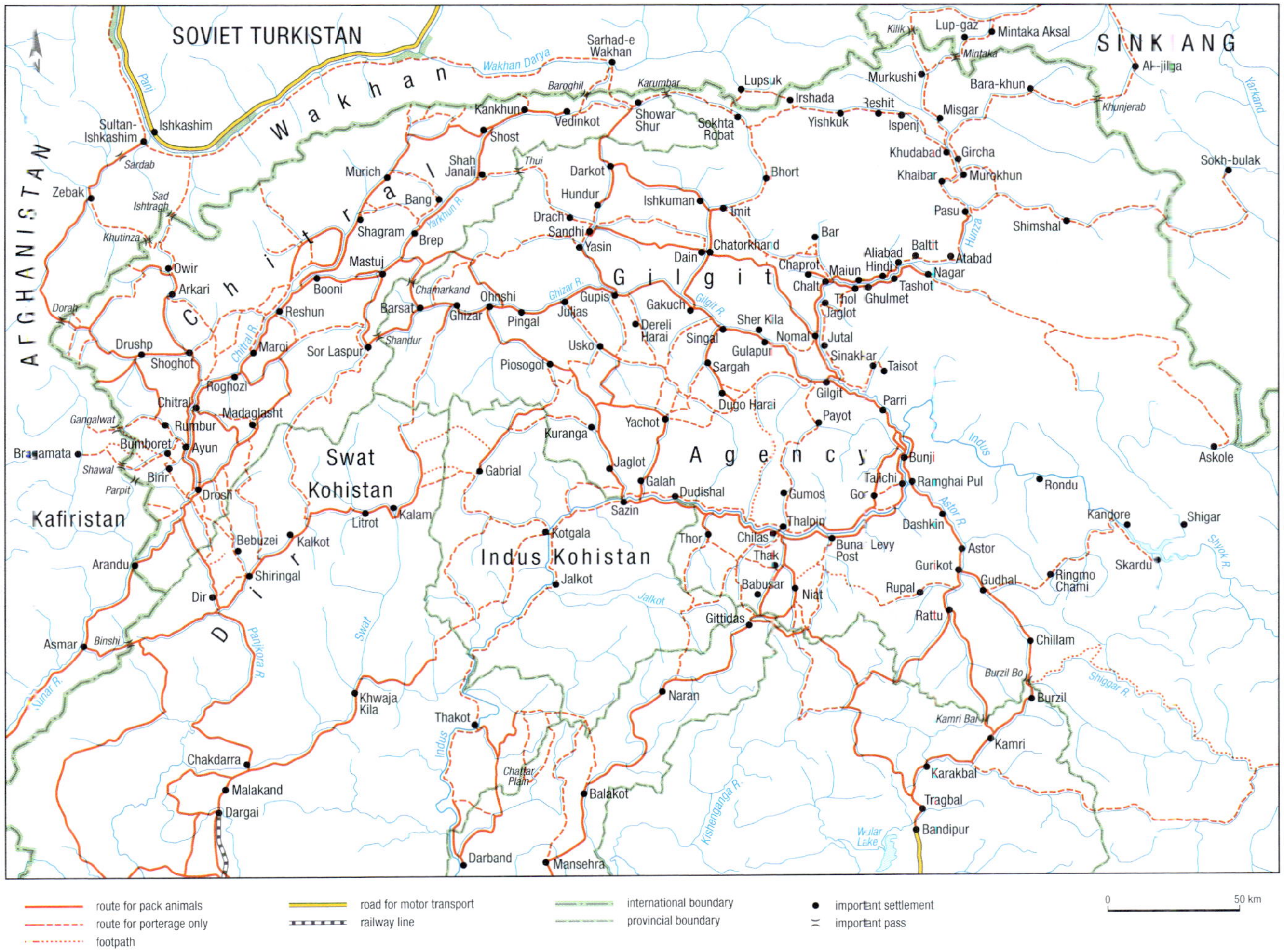

The General Staff India needed an overview map of existing routes and their quality in 1942. It distinguishes the relevant categories: main and other routes are separated from camel or mule routes and coolie roads as well as foot tracks. The map shows the high importance of pack animal transport on most of the supply lines from Malakand Agency, Hazara and the Vale of Kashmir towards the Afghan, Soviet and Chinese borders. The state of affairs was the planning tool for the mutually envisaged and desired motor road linking British India with Xinjiang and China via Gilgit and the Hunza Valley. Source: Representation based on 'Index map to routes in Chitral, Gilgit and Kohistan', published by General Staff India in Simla, July 1942. Design by Hermann Kreutzmann

possible owing to other demands of an operational nature which inevitably have to be accorded the highest priority. Moreover, great difficulties of rock cutting and bridging might be encountered which would lead to serious delays in the completion of the road, and it is not unreasonable to anticipate that no substantial advantage to the prosecution of the war would arise even if the road could be completed within a period of two years."[333] The reservations projected by the engineers were built on giving secondary or tertiary priority to this road, although they might deem it feasible from a construction point of view.

Economic considerations could not cover up other deficits, on the contrary: "The benefit to be expected is also, in the opinion of our road experts, likely to prove ephemeral, for the road can never be a commercial success, freight rates would be prohibitive, and with the revival of more accessible routes which will follow the defeat of the Japanese, the value of this route is likely to decline rather than to increase as time goes on. … In fact, to put it plainly, the whole project is regarded here as chimerical, and certainly the reverse of a contribution to the winning of the war …"[334] The effort of building the road was seen only as a military and strategic route for the support of the Republican Guomindang Government in China.

In the end, the road was never built, construction did not even start, and the file was closed with the repeated statement that the 'Government of India think the idea chimerical, in view of difficulties of labour and terrain'.[335] As a predecessor

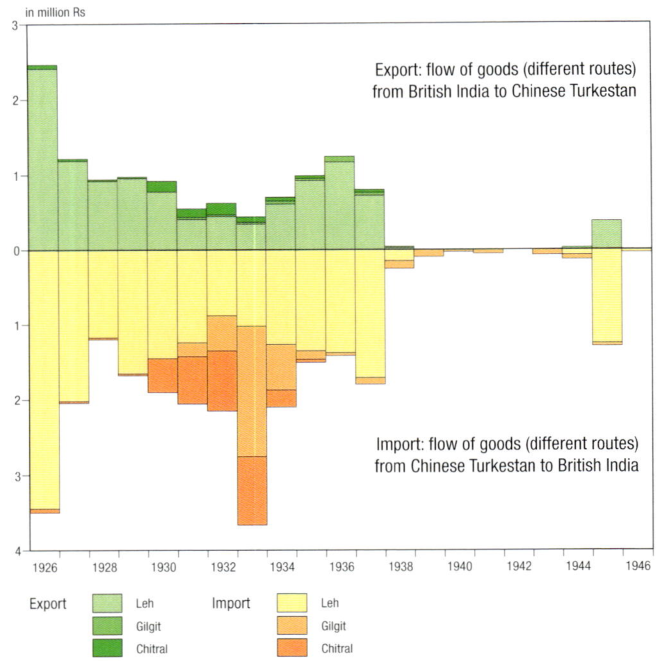

The trade between British India and Chinese Turkestan flourished with some alterations along three routes until 1937. With the exception of 1945, it never reached its former volume again. The trade figures show that the Leh route was the most prosperous of all, while the Gilgit or Hunza Road showed significant imports from China between 1931 and 1934 only. Source: Diagram modified after Hermann Kreutzmann (1989a: 26; 1998b: 318); trade data are compiled from entries in IOL/P&S/12/3285; IOL/P&S/12/3288; IOR/2/1076/232. Design by Hermann Kreutzmann

of the Karakoram Highway this project was fated to remain a non-starter. Somehow the 'proposed motor road from Northwest India to Sinkiang via Gilgit' has totally evaporated from collective memory. Between the Political Agent in Gilgit and the *mīr* Ghazan Khan no infrastructure-related projects for the Hunza Valley were discussed. The *mīr* not being of the same calibre as his father, *mīr* M. Nazim Khan, the British officers in charge were mainly concerned with his drinking problems, and his seven-year-rule ended on 7 May 1945.[336] The internal British reporting system reveals that the administrators were mainly concerned with stability in the Hunza Valley; any kind of disturbances – be it a personal disorder of the rulers, a sectarian problem between Shia and Maulai[337] believers, or an unrest among the Wakhi inhabitants who felt badly treated by the *mīr* – would be recorded. The major project of modernisation – to bring motor traffic on a cross-border road through the Karakoram – did not find any mention in the bilateral negotiations and entries of the monthly diaries.

Contemporary considerations connected with the Guomindang-inspired motor road have illuminated subsequent debates about vital issues of implementing such a major project. The strategic consideration of the road's proximity to Soviet terrain would affect the future alignment of the Pak-China Friendship Highway and would route the road through the Khunjerab Valley up to its pass, which was selected because it was further away from the Soviet border than the more suitable Kilik and Mintaka passes.[338] The reservations about economic feasibility would be repeated on and off, as such a road construction would only be justified by strategic considerations to support military movements. The recovery of the investment for the construction and maintenance of such roads would have been very difficult given the limited exchange of goods that occurred during colonial times. The aspect of infrastructure provision for the communities living along the road was never discussed in any dossier or memorandum. Road construction remained a technical and strategic affair by external actors who considered mainly the difficult terrain conditions for establishing a passage and thoroughfare. The voices of the affected mountain dwellers remained unheard: a constellation that has changed only a little since.

Coinciding with the end of World War II, the 'proposed motor road from Northwest India to Sinkiang via Gilgit' did not materialise and remained a paper tiger. Favourable conditions for the implementation of such a costly and challenging project needed further changes in the structure of regional politics. The major road link between the Grand Trunk Road of South Asia and the Central Asian highways was realised only after Pakistan's independence and the Chinese Revolution. Connecting the Grand Trunk Road with the Southern Silk Road became feasible when Pakistan and China joined forces in a symbolic and strategic effort to fend off a common enemy.

Post-colonial nation-building through traffic networks

Immediately after Pakistan's independence the political map of the mountainous part of the country was under transformation. Two and half months after Pakistan came into existence, an 'Independent Republic of Gilgit' was announced to terminate Dogra rule in an area that had been on lease for a period of 60 years since 1935, but was handed back by the British to the Kashmir Durbar on 1 August 1947 after exactly twelve years of lease.[339] The Gilgit freedom fighters arrested the Dogra *wazīr-i-wazārat* Ghansara Singh and declared the formal end of any relationship with the Maharaja of Kashmir. Their aim was the accession to Pakistan, which obviously was not the intention of the Maharaja of Kashmir, Hari Singh, who planned to build a new motor road connecting Srinagar with the Indian railhead in Pathankot.[340] Such a road was not available to the promotors of Gilgit's independence; their intention to accede to Pakistan was difficult to communicate as appropriate road links did not exist that would enable a speedy beginning of negotiations. The story of Gilgit's short-lived independence has been told many times and does not need to be repeated here; in connection with the Kashmir dispute it is part of the partition deed (*taqsīm nāmah*) that has caused many casualties on both sides.[341] Independence from colonial rule, accession to newly created states, and the separation of communities followed the partition of British India and have affected the remote mountain valleys as well. The outcome of these events is self-evident; the Gilgit Agency including the states of Hunza and Nagar acceded to Pakistan although their status under international law is still in limbo. Hindu and Sikh traders left Gilgit bazaar as well as many professionals (*pundit*) who had been employed as scribes and accountants in the colonial administration.[342] Their reputation had been ridiculed by their employers. These developments resulted in a principal change of entrepreneurship in the bazaars. Only twenty-one shops of Muslim Kashmiri were operated in Gilgit in 1964, but seventy-five shops had been taken over by Pashtuns and Hazara traders. John Staley assessed further modifications besides growth: "Participation in trade by Burusho Yeshkuns and Kho has grown until they now [1964] outnumber outsiders, not only in the recent development of shops in the villages of the principalities, but also in the main bazaars. In relation to the growth in the overall

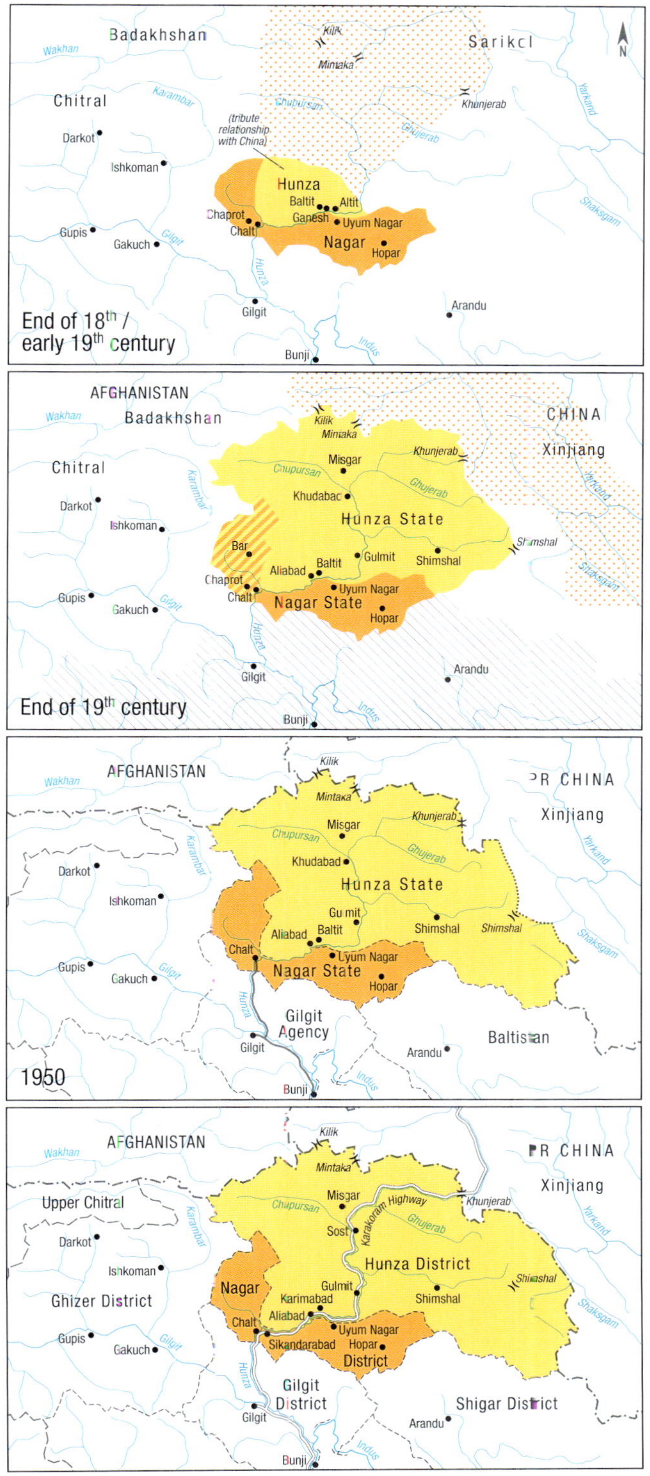

Four phases of territorial development and administrative terminologies are depicted: (1) until the end of the 18th/beginning of the 19th century, Hunza's spatial control was mainly concentrated in the three original settlements Altit, Baltit, Ganesh, and a few other villages in the central oasis and Shinaki. Nagar had already expanded from Uyum Nagar and Hopar into Shinber and controlled Chalt and Chaprot from time to time. (2) Mir Silum Khan (1790–1824) expanded Hunza's sphere of influence significantly and incorporated Herbar/Gojal up to the Karakoram passes. Grazing taxes were even levied beyond the passes in Sarikol where he erected a stone monument north of Dafdar to manifest his claim. A tribute relationship between Hunza and China might have originated from before Silum Khan's reign and lasted until 1937. When the Maharaja of Kashmir expanded his rule north of Gilgit, Hunza and Nagar entered, after some struggle, a tribute relationship with the Dogra rulers, Nagar since 1867, Hunza since 1869 (John Biddulph 1881: 28). Chalt, Chaprot and Garmasai (Bar and Budelas) became contested areas between Hunza, Kashmir, and Nagar. Following the military intervention by British-Kashmiri troops in 1891 the conflict ended and Nagar has remained in control of the area. (3) After the independence of Pakistan Hunza and Nagar acceded to the new country and remained semi-autonomous as 'states' until Nagar State was abolished in 1972 and finally Hunza State two years later. (4) Presently, Hunza and Nagar are administrated as two separate districts within Gilgit-Baltistan
Design by Hermann Kreutzmann

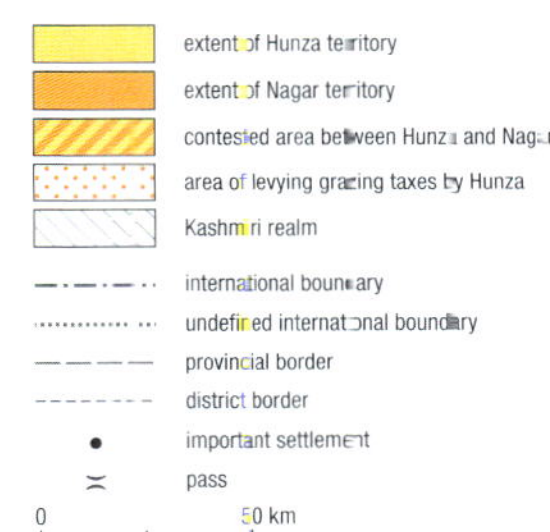

THE HIGH KARAKORAM

ANCIENT STATES AND MODERN COMMUNICATIONS

From Our Special Correspondent in Gilgit

This remote spot in the high Karakoram mountains to the north of Kashmir proper, formerly reached only after 12 days' hard trekking from Peshawar or Srinagar—and that in summer months only—has now been linked with the outside world by air. Every day transport aircraft operating for the Pakistan Government make hazardous 250-mile flights from Peshawar up the Indus river gorges to Gilgit, covering the distance in under two hours and bringing much-needed supplies to the 100,000 people inhabiting Gilgit Agency.

In addition Pakistani military engineers working for the past year have widened the existing 350-mile caravan route from Peshawar to Gilgit *via* Chilas State over Babusar Pass (13,700ft. high) so that jeeps can traverse the whole distance within four days. From Gilgit a branch track leads south-eastwards 100 miles up the Indus river to Skardu, the capital of Baltistan and a strategic centre in the rebel-held territory of northern Kashmir. These routes, however, are blocked by snow for the greater part of the year. The Babusar Pass has only this month been cleared and opened for traffic and jeeps are now bringing military supplies and essential commodities like cloth, salt, sugar, kerosene oil, and matches into Gilgit Agency.

Formerly this was an area of great strategic importance, because it borders upon the Pamirs, or the "roof of the world," where the territories of India, Afghanistan, China, and Russia converge. To-day modern methods of warfare have largely deprived Gilgit of its importance as the northern gateway of the Indian sub-continent; airborne troops could be flown over the tangled mass of mountains where Pamir, Hindu Kush, and Karakoram meet and could land in Peshawar or Srinagar, or even farther south in the plains of the Punjab.

CARAVAN TRADE CENTRE

But Gilgit remains the centre of an age-old caravan traffic from Sinkiang or Chinese Turkestan. Even to-day one meets on the narrow mountain path running southward from the frontier post of Misgar through the rugged but splendid Hunza valley scores of traders and pilgrims from Kashgar, with their ponies and donkeys laden with silk and cotton cloths, carpets, sheepskin coats, and other merchandise which they sell to people starved of such supplies ever since the Kashmir war started nearly two years ago and the natural trade arteries to the southward were cut. Your Correspondent flew here by Dakota aircraft, following the course of the Indus river, with mountains towering on both sides of the narrow valley and with the magnificent peak of Nanga Parbat, 26,660ft. high, providing an awe-inspiring spectacle. Thereafter he trekked eight days through mountains to Baltit, the capital of Hunza State, and back to Gilgit.

Since the partition of India in August, 1947, only two westerners have visited Hunza. One was the British Consul-General returning from Kashgar (where the Consulate has been divided between India and Pakistan) and the other is an American geologist who is still in these parts. No official of the Pakistan Government has gone much beyond Gilgit, and although both Hunza and Nagar States and other lesser units in Gilgit Agency have declared their intention of acceding to Pakistan, Karachi has not formally accepted their accession. This is presumably because acceptance of accession by areas formerly under the nominal suzerainty of the Maharaja of Kashmir (but in practice virtually independent) might prejudice the Pakistan case before the United Nations commission and possibly strengthen the Indian claim to Kashmir proper.

This much is certain—in spite of economic hardships resulting from the Kashmir struggle and shortage of such essential commodities as cloth, salt, and sugar, the 100,000 inhabitants of this completely Muslim region are whole-heartedly with Pakistan. Ever since the bloodless revolution of October, 1947, when the people of Gilgit under the leadership of their own military officers rose and expelled the Kashmir State authorities and army, the populace has been determined to throw in its lot with Islam as represented by Pakistan. In addition to several thousand Gilgitis, Hunzas, and Nagaris serving with the Pakistan Army, more than 1,000 irregulars, known as Gilgit Scouts, have given an excellent account of themselves in the Kashmir fighting and still hold the northern areas of that State as far south as the Burzil Pass and Gurais. They would hotly resent any suggestion that they should cede one inch of this territory to India.

REMOTE "SHANGRI-LA"

Trekking northwards from Gilgit along the narrow gorge of the Hunza river, with its alternating stretches of barren precipitous cliffs and delightful villages nestling among fertile fields in narrow valleys, your Correspondent found that, although the people's economic life had been hard hit by the Kashmir dispute, they were unanimous in their support of Pakistan. This applied equally to the Nagaris, who are mainly members of the Shia sect of Islam, and to the Hunzas, who are Ismailis, or followers of the Aga Khan.

Hunza itself, which has always exercised a strange fascination for British officials and for the few visitors who managed to reach this remote "Shangri-La," continues to live its healthy, happy life under the benign paternalism of its ruler, Mir Mahomed Jamal, Khan of Hunza. His subjects contrive to enjoy their traditional pursuits of polo and folk-dancing without such modern adjuncts as the wireless, cinemas, and electricity or even gramophones. Hunza is also free from such modern curses as Customs (although it borders on China), police, law courts, or politics. A single telephone line links the Mir's house with the Political Agent at Gilgit. Crime is rare—there have been six murders in the past half-century among its 24,000 inhabitants—and all disputes are settled personally by the Mir sitting in daily audience with village notables. There is no death sentence; the worst punishment that can be inflicted is banishment from this lovely land.

Yet Hunza has not remained untouched by world events. Some 700 of its sons served in the Indian Army in the second world war, and many are to-day serving in the Pakistan Army. Their remittances home help to eke out the meagre rewards of husbandry at an average altitude of 8,000ft., and their ideas also make for a changed outlook among the youth of Hunza. But although jeeps now run in Gilgit and aircraft arrive daily, anyone who has seen these fine upstanding Hunza people toiling manually and playing joyously in their mountain fastness must hope fervently that they will not be spoiled by too many outside influences.

The newspaper cutting from the 'Times' is one of the first post-colonial references to Gilgit and announces the advent of a new age of motor transport. Source: The Times 22 June 1949

sizes of the bazaars since 1947 – the number of shops has increased 200 % – their rate of entry has been even more rapid. From less than 10, the number of regional inhabitants with shops at Gilgit has increased to 278, representing nearly 75 % of the present total of 400. The 278 include 133 Hunza Burusho, 107 Gilgitis (mostly Yeshkuns), and 33 Nagar Burusho. Burusho are also present in many of the subsidiary bazaars, and are now the most numerous and widely-spread trading group in the Gilgit Agency ..."[343] New opportunities had come up for mountain-dwellers that had coincided with the exodus of Kashmiri *pundit* and the closure of the Chinese border in the early 1950s. The latter event motivated some stranded Uigur traders to open shops in Gilgit bazaar; twenty were counted by 1964. This feature became a permanent one in Gilgit. Furthermore, the former provision of cross-boundary traders and the local British-Kashmiri garrisons had been replaced by the demands for enhanced supplies within the Karakoram valleys. A stronger commercial link between down-country Pakistan and Gilgit emerged, an amplified trend connected with the growth of military presence during the 1965 Kashmir war.

The former Anglo-Russian confrontation was replaced by an enduring Indo-Pakistan conflict in the name of Kashmir.[344] The Swat State supported Pakistan as well and would be considered as part of the road link to Gilgit as motor traffic seemed to be feasible from Nowshera up to Karora at the eastern foot of the Shangla Pass (2,100 m), from where it was only sixteen kilometres to the Indus Valley junction at Besham.[345] Later ground-checks revealed obvious difficulties in reaching even Karora at all. If proceeding from there further down to the Indus Valley was envisaged it was certainly next to impossible to cover the difficult terrain with road-building equipment.

In terms of road policy, the need for an all-weather motor road linking Gilgit with Rawalpindi via the Kaghan Valley became a pressing issue. In 1948, an intermediate effort by the Corps of Engineers resulted in the upgrading of the Babusar Road to become jeepable for the few months in a year that it was open.

In June 1949, the Times reported that the first jeep had reached Gilgit after a four-day journey: "This remote spot in the high Karakoram mountains to the north of Kashmir proper, formerly reached only after 12 day's hard trekking from Peshawar or Srinagar – and that in summer months only – has now been linked with the outside world by air. Every day transport aircraft operating for the Pakistan Government make hazardous 250-mile flights from Peshawar up the Indus river gorges to Gilgit, covering the distance in under two hours and bringing much-needed supplies to the 100,000 people inhabiting Gilgit Agency. In addition Pakistani military engineers working for the past year have widened the existing 350-mile caravan route from Peshawar to Gilgit via Chilas State over Babusar Pass (13,700 ft. high) so that jeeps can traverse the whole distance within four days. From Gilgit a branch track leads south-eastwards 100 miles up the Indus river to Skardu, the capital of Baltistan and a strategic centre in the rebel-held territory of northern Kashmir. These routes, however, are blocked by snow for the greater part of the year. The Babusar Pass has only this month been cleared and opened for traffic and jeeps are now bringing military supplies and essential commodities like cloth, salt, sugar, kerosene oil, and matches into Gilgit Agency."[346] Air-lifting to the landing strips in Chilas and Gilgit was to be a complementary though costly measure to support the Pakistani presence in Gilgit.[347]

Muhammad Mumtaz Khalid has described in detail how the enthusiasm about having a road link was confronted with high budgetary costs, reluctant administrative and military institutions – that were themselves preoccupied with other pressing issues – and 'red-tapeism' which postponed the beginning of proper surveying and construction for about a decade.[348] The cost for a road link had been estimated in 1952 at 48.9 million Rs and in 1956 at 67.7 million Rs (about 14.2 million US dollars) for an asphalt road including three major bridges.[349] The high cost led to the postponement of the project. It needed the sole authority of Field Marshal M. Ayub Khan as Chief Martial Law Administrator who had participated in his former capacity as Commander-in-Chief in the previous consultations to order the Indus Valley road-building project as an operational task. The road project was to be implemented through the Corps of Engineers and would be interpreted as 'one of the most spectacular examples of nation-building by the Pakistan Army' through infrastructure provision.[350] Nevertheless, the estimated cost was halved, as only a road for light vehicles was to be built; and the task was to be shared between the Pakistan Army and the Central Public Works Department. Ideally, the section in Swat State was to be commissioned to the Wali of Swat on a contractual basis, while the road from Gilgit to Thor was to be improved by the Political Agent in

Gilgit with some professional support from army engineers. In November 1958, preparations for the construction of the Indus Valley Road commenced and reached the implementation stage.

Strategic roads – the Indus Valley Road as a beginning

A year after the construction of the Indus Valley Road was decreed, President Ayub Khan announced on 23 October 1959 that his government would open border talks with China. The previous month a Chinese map had reached the Foreign Ministry in which parts of Hunza were identified as Chinese territory, suggesting an international boundary along the subcontinental watershed from Mintaka Pass via Khunjerab Pass to Shimshal Pass, from where the course of the border continued eastwards towards Tibet.[351] Both events belong to the 'zeitgeist' of nation-building efforts and securing international boundaries in an environment governed by the Kashmir dispute, the undecided status of princely states, and a limbo in which it was not clear whether the *mīr* of Hunza would renew his tribute relationship with China. Rumours had spread right from the beginning of Pakistan's independence that he was still engaged in double-dealing in order to increase his stakes. In a letter to the British High Commissioner in Karachi composed by his deputy Charles Duke, the situation in Gilgit in February 1948 was summarised: "The Mir of Hunza also met the [Pakistan] government representative from here [Khan Sahib Muhammad Alam] and promised support to the Political Agent in maintaining the old system. It appears however that the Mir has resumed relations with the Chinese, which dropped while there was a British Political Agent in Gilgit, through the Amban of Tashkurgan. The Mir has received recognition by the Chinese of his claim to certain disputed grazing grounds on his border with Sinkiang, and has also been granted the right to issue Chinese passports for travel to Sinkiang. The Mir however asserts that if there were any sign of a return of communist influence in Sinkiang he would again sever his connections with them."[352] The 'old system' was the continuation of 'indirect rule' in a post-colonial setting and a termination of loyalty relations with China. In addition, two different strategies conflicted with each other, which was called in Duke's report the 'anomalous position of the Pakistan Government towards the Azad

The precondition for the expansion of motor traffic towards the Hunza Valley was a network of suspension bridges and jeepable roads that would enable the transport of commodities, goods and tools for road construction from the south. Source: German-Austrian Himalaya-Karakoram expedition in May 1954. Courtesy of Gerhard Klamert

Opposite
Babusar Pass (4,173 m) offered the only approach road for motor transport from Rawalpindi to Gilgit during the summer months. The upper Kaghan Valley was regularly visited by Afghan pastoralists who offered transport services by their camel caravans. The 1950s and 1960s were the period when different kinds of camel and mule caravans provided the bulk transport while jeep transport was still more costly. Source: German-Austrian Himalaya-Karakoram expedition 28 August 1954. Courtesy of Gerhard Klamert

Kashmir organization'.[353] While the Pakistan Government tried to maintain a *status quo ante*, Azad Kashmir's interest in the region was deviating from that position. General Tariq, the commander of their forces, had tried to override the authority of the Political Agent and had declared and ordered that the Gilgit Agency would be Azad Kashmir territory 'of which Major Aslam was in sole charge, and the Gilgit Scouts were to be regarded as Azad Kashmir forces'.[354] The confrontation of various interests in Gilgit that has endured until today and that has still found no solution emerged immediately after independence. A decade later, the central interests of the Pakistan Government needed to be streamlined.

The construction of the Gilgit Valley Road offered a symbolic opportunity to show the proper linkages of the Gilgit Agency with Pakistan as no road link whatsoever existed between Azad Kashmir and Gilgit. There was a pressing need for some visible progress on the ground; therefore, it was decided to make it a one-lane jeepable track that would later be widened to a truckable road. The 225 kilometres between Besham and Chilas became the task of the Army Engineers, while improvement of the 137-km-long jeepable track between Chilas and Gilgit was the responsibility of the Political Agent in Gilgit. Cost considerations governed the decision to reduce the width of the road from 5.5 to 3.6 metres. Its gradient was fixed at 1:12 after 1:15 seemed too costly; the turning radius was cut from thirty to eighteen metres, the bridges were designed for light vehicles only, so a new term was invented for its specification as a 'jeepable road on 3-ton alignment'.[355] Despite expectations that this road was to open up isolated areas and that public buses should be able to ply on the dirt track, the 'president's desire of opening up of the area by the quickest possible means initially, without aiming at too high construction standards of the gradient, alignment and width'[356] would become the guiding principle for implementing the still ambitious project.

After a second starting point was opened for roadworks from the north in December 1959, a 'two-pronged attack'[357] was due. From Thor in the Gilgit sector the road should be built towards Sazin, and after a delayed improvement of the landing strip in Chilas, which reduced the airlifting distance from Rawalpindi and saved some transport time to Thor, work continued in both directions. Nevertheless, progress was advancing only slowly; during the first year, only five kilometres of road were completed.[358] As no proper ground survey had been done during the first decade of planning, many surprises occurred. The logistical challenge in setting up the required infrastructure for the Corps of Engineers, supplying them with equipment such as compressors and explosives, and provisions of food and shelter for the workforce had been gravely underestimated. Detrimental geological, hydrological and orographic features, heat waves and snowfall, landslides and flooding hampered the work and led to several unforeseen interruptions. Chilas developed as an important hub for distributing goods towards the construction sites. As Babusar Pass was only open during summer months, necessary goods were stored in dumps in Battakundi in the Kaghan Valley until they could be shipped across the pass in July towards storages in Chilas. In October 1963, after a leaking oil barrel had caught fire, the bazaar was destroyed, and required rebuilding. The airstrip had been made fit for large carrier aircrafts in 1961, so progress in roadbuilding was possible.[359]

A serious problem was posed by the shortage of a local manual labour force to support the army units. First, the Corps of Engineers hired local Kohistani residents for manual labour, but they could not provide sufficient workers. Additional labourers stemmed from neighbouring valleys in the north and south, and, as a last resort, prisoners were included in the workforce, which had grown to 3,000 civilian local labourers, 600 prisoners and 750 army engineers by the summer of 1964 when both working teams had nearly reached each other, being only 35 miles apart. Besides the Karora-Chilas sector, the link from Besham to Abbottabad was roughly completed as a *kachchā* dirt track when the Indo-Pakistan war broke out on 6 September 1965.[360] All work came to a halt, and the artery fell into disuse as no vehicles could ply the route. During the first phase, the achievements were limited, but the ground was prepared for future improvements and the advent of a *pakkā* asphalt road.

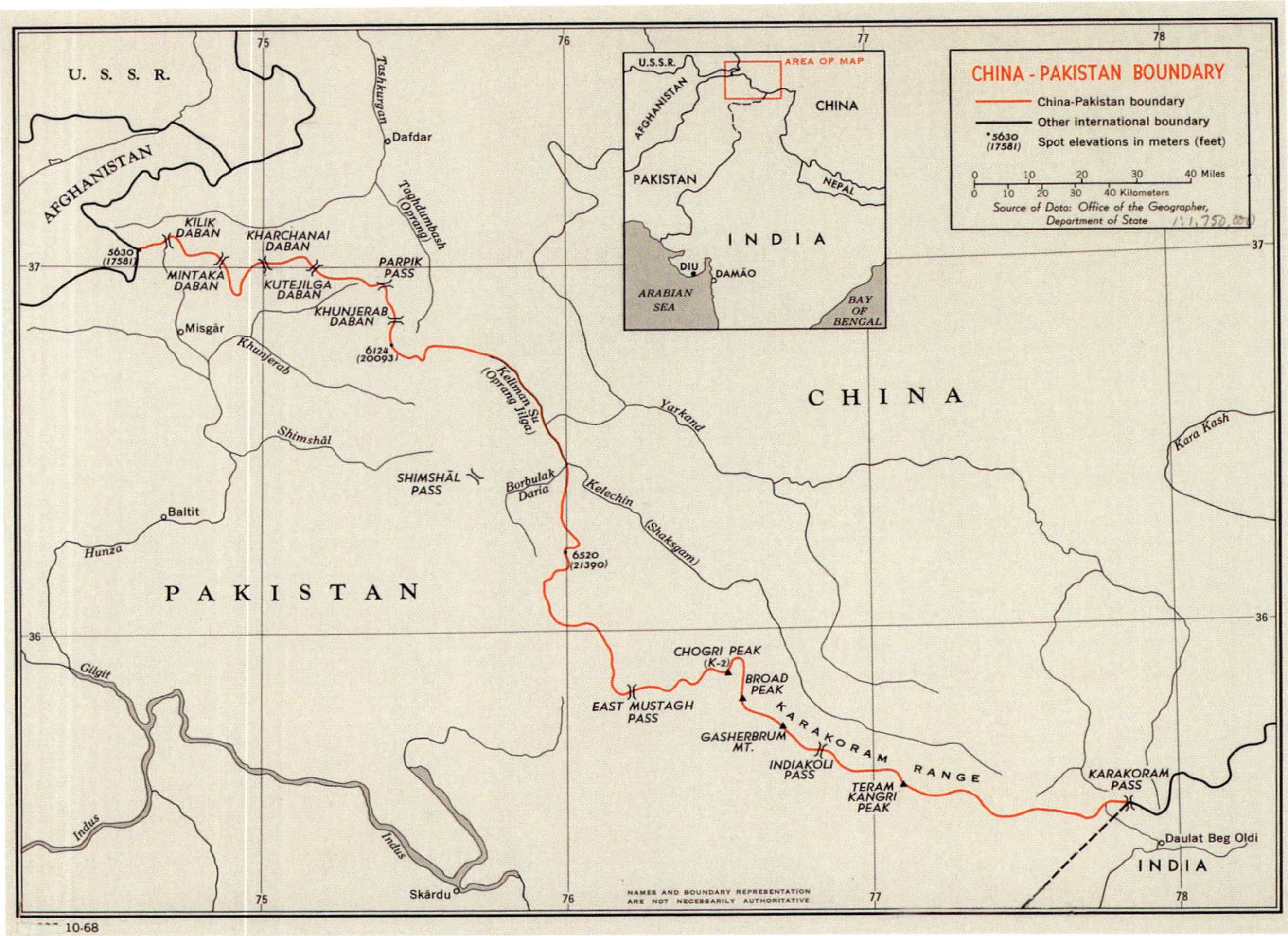

The map 'China-Pakistan boundary' was published by the United States Central Intelligence Agency in 1968. The red line shows the agreed-upon boundary delineation between Wakhan and the Karakoram Pass as in the outcome of the 1963 bilateral Boundary Agreement. Source: Library of Congress, Geography and Map Division, https://www.loc.gov/item/2007627235/

Opposite
The 'Map of the Boundary between China's Sinkiang and the contiguous areas the defence of which is under the actual control of Pakistan' was prepared in the aftermath of the Pak-China Agreement of 1963 under which Pakistan ceded 8,800 km^2 to China and which resulted in the bilateral demarcation of the common boundary (Alastair Lamb 1964b). The section given here shows the border alignment east and west of Khunjerab Pass. In the west, the former trade route runs from Hunza via the Kilik and Mintaka Passes towards the Kara Chukur Valley and Tashkurgan. From its western extension to the Oprang Jilga the boundary alignment follows the mountain crest route, while in the eastern section the Oprang and Muztagh Rivers represent the border line. The riparian boundary is due to the agreement that the Shimshal shepherds were to regain access to their winter pastures north of the subcontinental watershed, which they approached from the Shimshal Pass; see map in Hermann Kreutzmann (2015a: 481) showing the area in question. Source: Survey of Pakistan (1966)

Contextual changes in boundary-making – conflict and war in Western High Asia

The experience of a military confrontation with India firmly reminded the Pakistan authorities of the necessity of having an all-weather road link from down-country garrisons with their troops in the north where the Northern Scouts and Gilgit Scouts were operating on the borders. The context in which these considerations took place had changed since the beginning of road-building.

After the map that claimed Pakistani territory in the Karakoram as belonging to China had been published and Pakistan had offered negotiations about the border alignment, it took more than two years until May 1962 for the Chinese Government to respond and agree to negotiate the border issue.[361] The overall situation had become more complicated. Within Pakistan Ayub Khan's internal reforms had resulted in the implementation of a 'One-Unit plan' in West Pakistan, merging all provinces of West Pakistan, but maintaining the undefined status of Azad Kashmir and the Gilgit Agency as well as continuing selective patronising of Pashtun leaders in the tribal areas where the 'Frontier Crimes Regulations' were still applied.[362] The Hunza-Sarikol boundary had been closed since the early 1950s; by 1958, China had broken off all relations with the Soviet Union, and Xinjiang's international borders were hermetically closed; by 1959, a significant portion of Indian territory had been under Chinese control in Aksai Chin and a road connecting Xinjiang with

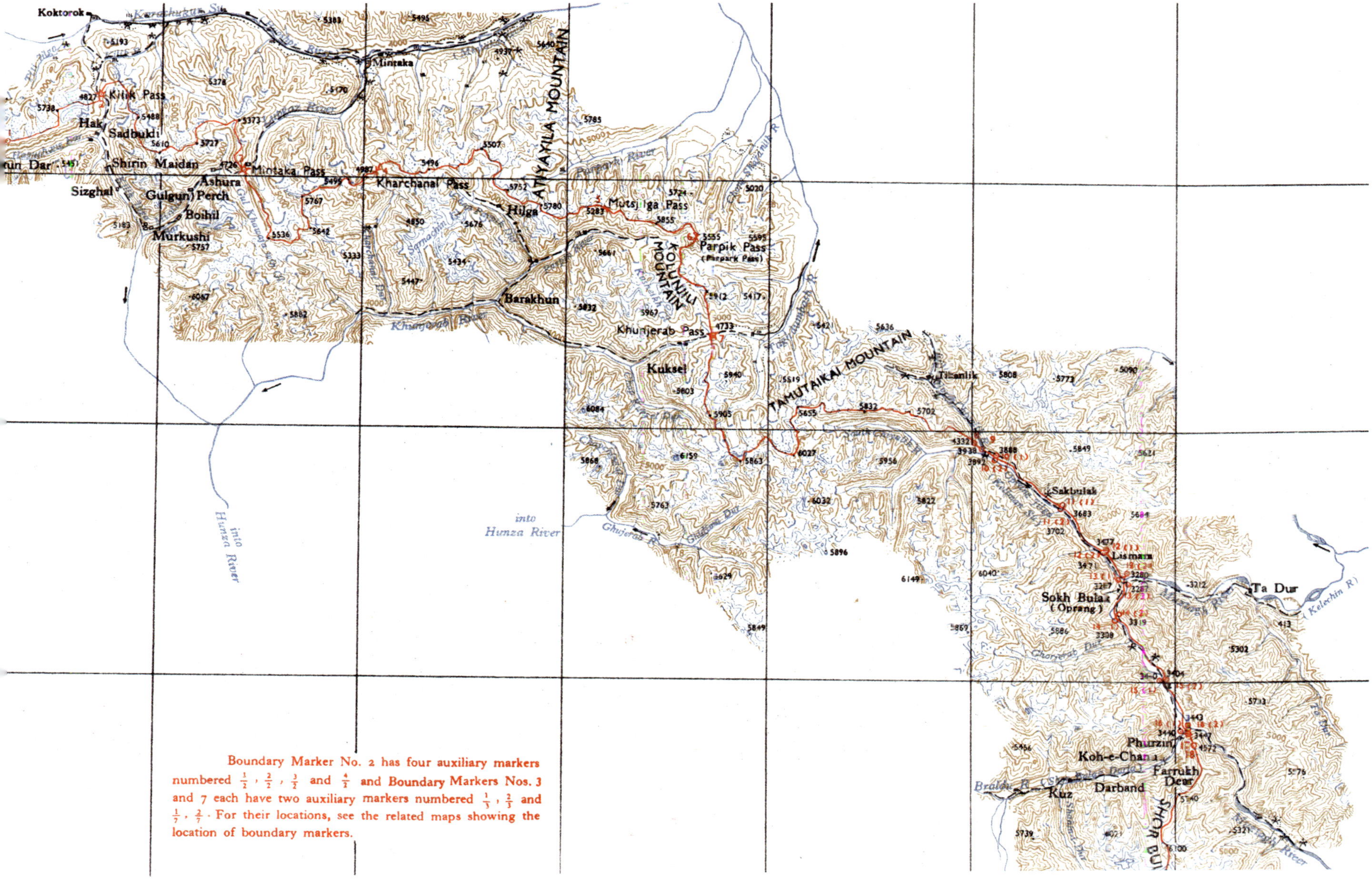

Tibet had been built, largely unnoticed by India's border forces.[363] The stalemate in the Kashmir issue – Pakistan's demand for self-determination in Jammu & Kashmir and a referendum held under the auspices of the United Nations versus the Indian claim of territorial control in Azad Kashmir and the Gilgit Agency – had not been brought forward to any negotiated agreement or durable solution for the inhabitants of Kashmir. Pakistan was somehow bound to strong relations with the United States of America, being a member of different Cold War-specific treaty organisations. Since the mid-1950s Pakistan had been a member of the Southeast Asia Collective Treaty Organization as well as the Central Treaty Organization, which documented its alliance with the West and its interest in anti-Communist regional political stability. In May 1962 President Ayub Khan's government probed a new thrust and announced a more open foreign

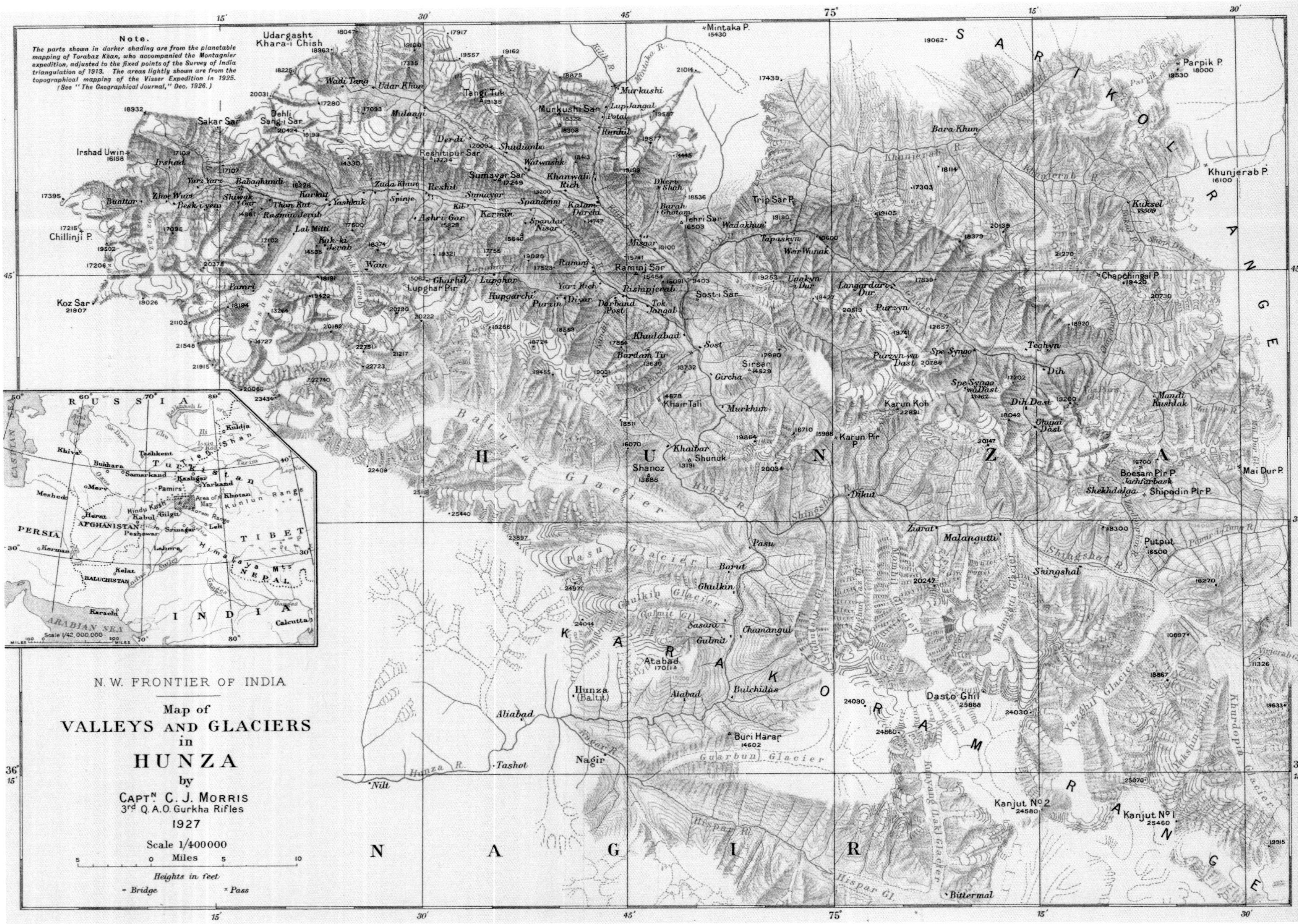

The 'Map of valleys and glaciers in Hunza' was prepared by Captain Charles Morris during his reconnaissance in 1927 and published the following year in the Geographical Journal. Source: Charles John Morris (1928). Courtesy of the Pamir Archive Collection

policy that should not only include strong western ties, but also allow friendly relations with China and the Soviet Union.[364] India and China were on the verge of fighting a war about their international boundaries within the Aksai Chin area. During these times, it took only few months starting from 13 October 1962 to negotiate the Sino-Pakistan Border Agreement.[365] One big issue was the Shimshal Pass area. The state of negotiations was disclosed by the Pakistan side to the British Foreign Office which summarised that "… Shimshal had been recognized as belonging to Hunza until 1936, when as a result of an error by the Political Agent [George Kirkbride] of the time, the Chinese had obtained possession of it. The Shimshalis had thereupon withdrawn south of the watershed [Shimshal Pass], and had lived more or less as refugees in Hunza ever since. The main asset in the Shimshal valley was the salt mines, on which the Hunza area had depended, and recently salt had been as scarce as gold, and could only be obtained from Western Pakistan, by air to Gilgit. The Opran valley was a summer pasture area which, before 1936, had been seasonally occupied during summer by cattle breeding people from the south of the watershed. These two areas represented in all about 700 square miles of territory. They had not been settled by the Chinese, and indeed Opran was not strictly speaking accessible from the Chinese side. The Chinese had had several frontier posts in Shimshal which they would now vacate. They had also built barracks."[366] The Pakistan government interpreted the result as a major success, having gained or re-occupied the grazing grounds in the Shimshal Pamir down to the Braldu and Darwaza, and only giving up territory that was not under their control anyhow.[367] The boundary-making philosophy adopted a plan to follow, where possible, the subcontinental watershed with a few exceptions such as in Hunza's territory where river boundaries were accepted. India rejected the agreement as 'illegal' as this agreement supposedly covered Indian territory, and India disagreed with such a border alignment. Nothing has changed until today; the respective positions have remained the same; all reconciliation efforts to bridge the respective standpoints have failed so far.[368]

In practical terms, the Sino-Pakistan Border Agreement changed bilateral relations considerably as the Chinese government included in the package a possible 'treaty of commerce and friendship and assistance in mineral exploitation and industrial development', which did not become an explicit part of the agreement. The speculation about including road construction in the treaty has sprouted many buds, as some commentators saw a network of cross-border road connections emerging that would have affected strategic positions and triggered immediate threats by the contestants.[369] Both parties explicitly stated in the agreement that a peaceful solution should be pursued for the Kashmir crisis and the Indo-China boundary dispute; nevertheless, both secured each other's support. In late 1962, the course was set for a friendly future beneficial collaboration that would offer technical and financial cooperation. Pakistan became one of the biggest receivers of Chinese aid packages; until the completion of the Karakoram Highway Pakistan had received 620 million US dollars in economic assistance as grants or on very soft loan conditions that could be paid back in kind. By 1970, about a quarter of the loans had been waived and converted into grants.[370] Some of the investments were used to establish industrial enterprises that would produce equipment and materials required for road construction. These favourable circumstances would proactively support the completion of the Pak-China Friendship Highway aka Karakoram Highway within two decades and the implementation of the China-Pakistan Economic Corridor within half a century.

The Sino-Pakistan Border Agreement triggered the construction of the Karakoram Highway. In order to keep away from controversial international boundaries the hitherto less important Khunjerab Pass (4,655 m) was selected as the crossing-point for connecting both countries with a trans-mountain passage. Snowfall is affecting travel conditions during winters. The photograph was taken on 2 November 1985

The most fertile village lands of Karimabad are found in Bul Mal through which the Hunza trade route passed, connecting Gilgit with Tashkurgan and Kashgar. In the middle ground the historical route – known as the middle path (mákućim gan) – is visible and led to the polo ground (šabáran) below Baltit Fort. With the advent of motor transport the jeep road was diverted from Haiderabad Har along the Dalá, the main channel; its sidewalk provided the necessary width and led to the hamlet of Karimabad, close to the new palace where visitors were received. Presently mákućim gan functions as a local connection between different hamlets.
The photograph was taken on 1 April 1985

From Indus Valley Road to Karakoram Highway

The Chinese and Pakistan motor road systems remained separated by a distance of several hundred kilometres when the agreement was signed. In 1958, the first road linked Tashkurgan in Sarikol with Kashgar; further south towards the Pakistan border there was still a cul-de-sac that remained for the next two decades. Similarly, the first jeep reached Hunza on 1 December 1957 bringing Commissioner Kiani from Gilgit. Before that, the jeep track had ended in Chalt. Ian Stephens travelled the 'new road' in April 1952: "The jeep-track, though efficiently engineered for its purpose, was very uneven, and we progressed in a succession of forward rushes and sudden decelerations, with noisy gear-changes and much braking. Occasionally the track took us venturing forth between stone guide-posts among the crimped grey dunes on the valley floor, traversing drifts where wheels might whizz gripless in sand, more often, preferring solid ground, it wriggled along the in-cut lower rim of the great leftward rock-wall, over slabs and shelves, across scree-tips, into small sharp side-nullahs or up steep slopes. At places it held us in thin air, supported over some hundreds of feet of emptiness on out-hanging props of wood or stone."[371] He was much impressed on that short journey from Gilgit to Chalt; it took some four years more to bring the road forward into the 'motorless country' of Shinaki and Shinber that were connected by suspension bridges.[372] Its route followed the narrow footpath that had been sufficiently widened to let a light vehicle pass; instead of a three-day foot journey the travelling time was reduced to eight hours. The *mákućim gan* – the middle path of Central Hunza – was transformed from a bridle and foot path into a jeepable dirt track that later was shifted to follow the wider course and path of the *dalá* or Khul [*gūl*] Samarqand in the main oasis. Baltit and Varshideh, the present Tashkurgan, being a distance of 300 kilometres apart, still required a journey on foot or horseback in difficult terrain that was not yet fit for motor transport. Nevertheless, for both Hunza and Sarikol, their respective road connections contributed to significant change. Sarikol's prime contribution to Xinjiang's economy was from animal husbandry, and it would remain so for more than a generation until the border was opened in 1986. In 1984, only ten percent of livestock and its products left the Autonomous County of the Tajik Nationality Tashkurgan towards Kashgar; there was next to no cross-boundary export from an area that was fully dependent on most of its necessities from Kashgar.[373] It was only in 1989 that yaks started to be exported to Hunza.

Road and rural development

The principality of Hunza had always been food-deficient; especially when rice supplies from China failed to arrive after the border closure in the winter of 1950–1951 The commencement of motor transport made a significant contribution to safeguarding external food supplies. The pressure to alleviate this scarcity was immense. Pakistan's first Village Aid Five-Year Plan in 1956 was extended to the Gilgit Agency three years later.[374] On the basis of a participatory approach village communities could benefit from infrastructure projects; the construction of suspension bridges to span the Hunza river near Danyor and the Gilgit river at Sher Qila were facilitated from these funds.[375] The dire need becomes evident in view of their contribution that consisted of three-quarters of the cost, provision of all unskilled labour required and cutting all the wood for bridge construction from communal forests.[376] This was the first rural development experience in Northern Pakistan when the Central Government covered '75 per cent of all non-recurring expenditure and 50 per cent of recurring expenditure'[377] within the granted packages. Out of the annual Village Aid Programme's budget of Rs 300,000 (approx. US $ 65,000), two thirds were spent on transport alone. Not surprisingly, the transport charges for one *maund* of goods from Rawalpindi to Gilgit amounted to 25 to 35 Rs, whereas the carriage cost on the return trip ranged between zero and eight rupees, highlighting the limited to negligible exports from the mountains.[378] Air transport was no alternative as the cost of bringing a sack of chemical fertiliser from down country to Gilgit – five rupees on terrestrial roads – increased by a factor of twelve to sixty rupees.[379] Interregional transport charges prior to the Karakoram Highway reflect the then difficult access for traders and their goods to Hunza. While carriage cost per *maund* ranged around seven rupees on the Yasin-Gupis to Gilgit and Chilas to Gilgit routes, the jeep charges on the shorter distance between Baltit and Gilgit amounted to twelve rupees per *maund*. It was twice as costly to transport a sack of grain or fertiliser to Central Hunza

In spring the annual cleaning of irrigation channels commences when sediments are excavated. Along Karimabad's main channel, Khul Samarkand or Dalá, the sediments are deposited on the Dalá gan, the first jeepable road of Karimabad since 1960. The photograph was taken on 1 April 1985

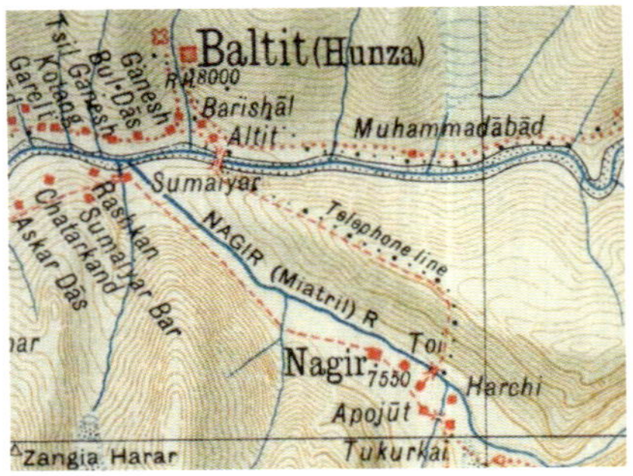

The main obstacles along the shortest road connection between Baltit and Uyum Nagar were the 'rope bridge across the Nagar River' (top) that was made from plaited juniper or willow twigs (gal) and connected Uyum Nagar and Sumayar with the right bank of Miatsil at a place called Apojut. From there the road led to another wooden bridge (bottom) that had been built by mīr M. Nazim Khan to reach the Hunza side at the gorge section near Ganesh and Done Das. The rope bridge across the Miatsil had one peculiarity: it consisted of two sections across a major boulder. Reginald Schomberg (1935a: 110) had described the challenging state of both and the time delay while crossing them with big loads in 1933. Upon his return in 1945 not much had changed: "… another rope one, but in good order, though most unpleasant as it was long with an unpleasant sag, and the usual gymnastics had to be performed. Every time that I cross one of these rope bridges, I dislike the exercise more, and so do my followers. I can quite understand men taking a long detour sooner than face one of these bridges. On the other side of the river, we walked high above the right bank, then over a spur just above the junction with the Hunza river, and so down to the bridge across that stream, which formed the boundary between the two states." (RSAA/SC/SCH/2/47: 58). These two bottlenecks were removed when Gojal was linked with the central oasis by a jeepable bridge in 1960; following this achievement a different approach was feasible to Uyum Nagar. Source: Photographs by Flight Lieutenant Francis Claude Sturgiss on 28 May 1937; File No. 82(2)-X/39, photo 69 and 70; Quarter-inch map 'Afghānistān, Gilgit Agency and Kashmīr & Jammu. Baltit (Hunza), No. 42 L', Survey of India 1930

than to Yasin.[380] This pioneering programme was accompanied by accelerated infrastructure development. From 1957 onwards, there would be a sequence of road extensions into the Hunza Valley. To overcome the initial terminus of the jeep road in Chalt, a new suspension bridge had been constructed to cross the Bar River from Budelas to Khizrabad, the first Hunza village in Shinaki. From there the road continued on the right bank, but its latest construction came to a halt in Hindi. The next obstacle was the sliding area of Chikas where steep slopes regularly released loose material and pebbles that often would shoot down in a short but rather dangerous stretch between Hindi and Murtazabad; the place name relates to the garnets (*ćhíkas*) which are found here in substantial quantities.[381] Although the footpath and pack animal track followed this route, it seemed to be an unsurmountable obstacle for jeep transport; so a 'modern technology' was applied for the sake of clearing this section. An oversized wooden gondola (*garaḍí*) was constructed that was fixed to two steel ropes that led across the Hunza River. The jeep was fastened and secured on the wooden construction, and the gondola was pulled across the river to the other bank. These *garaḍí* have been quite common for the transport of individuals and light goods in the Karakoram; but an *uyúm garaḍí* for the transfer of jeeps was an innovation. Two winch-operated sets of *uyúm garaḍí* were required to cross from Hindi to Tashot and to re-cross less than two kilometres further down the river to Murtazabad at the place where presently the Tashot Bridge is located.[382] Hassanabad and the central oasis were in comfortable reach. The crossing of tributary rivers was comparatively easily managed by wire and plank bridges, whereas the main river was too wide for the affordable technology of the time. The effort applied here for crossing the Hunza River highlights the urgent need to end the cul-de-sac position and to enable motor transport to Hunza and Nagar. One of the earliest narratives about a continuous motor journey to Baltit was supplied by John Edwards in his report on the Batura-Muztagh expedition: "Eventually the road was repaired and the expedition left Gilgit [in May 1959] in nine jeeps for the five-hour trip to Baltit. The road to Baltit is a seven-foot-wide track with no parapet, built on to the cliff face along the Hunza gorge. From our seats we could look down into the foaming brown waters of the Hunza river boiling over the rocks, three or four hundred feet below."[383] The feat of constructing the jeep track had obviously impressed him. It was a great change; the travelling time between Gilgit and Baltit was shortened from a two- to three-day journey to a daytime travel by jeep.[384] In Hunza itself two jeeps were available; both belonged to *mīr* Jamal Khan.[385] He could not drive in his saloon car from down-country Pakistan to his residence in Baltit; that vehicle remained in Rawalpindi.

Upon the first visit of the Aga Khan to Hunza, the road to Altit and Baltit was upgraded by the Aga Khan Voluntary Corps. On 23 October 1960 Karim al-Hussaini, three year after becoming Aga Khan IV, arrived on the front passenger seat in an open jeep, with *mīr* Jamal Khan squatting in the rear. After the Ganesh suspension bridge was widened and stabilised in 1962, the first jeeps could operate up to Gulmit. Previously, all traffic had followed the right bank of the Hunza River; from Altit to Koshi and Ahmedabad, from there to Ghamesarat and Sarat, and then on a difficult track to Gulmit. Shishket could only be reached by a 114-m-long lofty bridge on which small wooden planks, about a metre apart, were fixed and supported by some steel ropes. On the occasion of President Ayub Khan's visit to Gojal, the jeep track was extended to Pasu in 1964; after Altit a second bridge for jeep-crossing had to be installed between Shisket and Gulmit from where the road continued on the right bank. In Pasu the tracks for porters bifurcated: the Shimshal route continued to the pass leading into China; the second route followed the ancient trading route to Sarikol. From Pasu onwards caravans would cross the Batura Glacier to proceed to Misgar; no motor transport was feasible across glaciers.

The old caravan routes offered two options for travellers with porters: either to proceed on the right bank to Khudabad and Tok Jangal, cross the Chupursan and Kilik Rivers by local bridges on their way to Misgar, or to change to the left bank before reaching Ghalapan and continue via Morkhun, Gircha and Sost. After Sost the Hunza River had to be crossed in Belli (Bili) again to reach Misgar. The road-building challenge was to construct a jeep track on the right bank connecting Pasu with Khaiber, from where upper Gojal could be reached after constructing another bridge to the Abgerchi villages. Morkhun became a staging post for Chinese roadbuilders which initiated fresh protest from Indian observers who suspected that more than one road was constructed in the Hunza Valley.[386] From Sost another bridge was required to connect with the Misgar road. In 1969, just in time for the first cross-boundary trade caravans, the project was accomplished and the jeep track reached up to Misgar.[387] Fifty camels had brought loads from China that reached Misgar by August.[388] This highly symbolic act was staged as a visible sign of an enhanced Sino-Pakistan friendship and commodity exchange among equals. For people in Hunza the Kashgar road (*Qaśgare gan*) had been re-opened and followed the old silk route (*šarā-e rešum*). Their view was not only directed northwards: expectations had grown for a closer connection with down-country Pakistan and for development packages, subsidies, and further infrastructure assets.

The short-lived Village Agricultural and Industrial Development Programme (1953–1959) was continued and modified under various names. During President Ayub Khan's time it was known as 'Basic Democracies' (1960-1968); Zulfiqar Ali Bhutto renamed it as the 'People's Works Programme' and introduced the specialised 'Integrated Rural Development Programme' (1972–1977) for agricultural regions.[389] Zia-ul-Haq turned it into 'Local Bodies & Rural Development' that implemented a 'Community Basic Services Programme' (1981-1985); during this phase several link roads in the Hunza Valley were initiated.[390] Benazir Bhutto introduced a 'Social Action Programme' that is well remembered for its school-building packages in even the remotest locations, and Pervez Musharraf continued the implementation of small-scale localised projects from the 'Khushal Pakistan Fund'. All these schemes would be termed rural development and/or poverty alleviation programmes in contemporary parlance. The Indus Valley Road seems to have been a precondition for extending measures for rural uplift from down-country Pakistan into the Himalaya and Karakoram mountains. The road depended on international cooperation sponsored by institutional and private sources as well. In 1981, the Food and Agriculture Organisation in connection with the United Nations Development Programme started a five-year 'Project for Integrated Rural Development in the Northern Areas', shortly after the Aga Khan Rural Support Programme commenced operations in December 1982. During his second visit to Gilgit as an advisor to the Aga Khan Rural Support Programme, the doyen of rural development in Pakistan, Akhtar Hameed Khan, observed: "Recently roads are bringing the Gilgit district into closer relationship with the economy of the plains. The impact of this relationship can already be felt, and it will become more and more significant. It is more than probable that with the growing integration, the same economic and social trends will become dominant here as are already dominant in the southern plains." The grass-roots-oriented ac-

The bridge connecting Konedas with Danyor commemorates the Village AID programme during which this important connection to the left bank of the Hunza River was established. Construction was difficult; thus a tunnel had to be dug out through the river terrace. The photographs were taken on 15 May 1985 and 25 July 1989

The Indus is deeply incised into the alluvial valley filling. Little space is occupied by oasis settlements. Their linkage with down-country Pakistan by the Indus Valley Road initiated a significant transformation of exchange relations for the antipodes. The photograph was taken on 11 October 1985

tivist for participatory approaches feared that 'contractors, traders, tractor owners, who will also, sooner or later, become large landowners' and lead to a 'not very happy kind of development'.[391] Akhtar Hameed Khan's reservations did not emerge as a new social structure in the strict sense, but the process of change had already been accelerated with mixed results. His observation was articulated as a warning for development practitioners. During his subsequent six visits, the 'mentor of the programme'[392] never mentioned the role of the Karakoram Highway again; it seemed to have been already a given frame condition.

Becoming internationally connected

When work on the Indus Valley Road was resumed after the Indo-Pakistan War of 1965, a dual strategy was followed. Initially, the Pakistan Army pursued its task to enable a road link between down country and Gilgit. The state of affairs when the war commenced in September 1965 was that the existing road would enable a truck to go as far as Pattan, and a jeep could proceed further up to Kamila.[393] Progress having been speeded up with a restructured army setup, work on the Indus Valley Road continued in the spring of 1966.

During the same period, the Pakistan Government negotiated with their Chinese counterparts a link of the Indus Valley Road beyond Gilgit to connect with Xinjiang's road network. The idea of a Karakoram Highway materialised, and a new organisation called 'Frontier Works Organisation' was founded to shoulder this road project of quite a different dimension.

The project's history followed the dual approach from now on. First, the existing Indus Valley Road had to be widened from a jeepable to a truckable road. Second, a ground survey was required to identify the road alignment between Gilgit and the Karakoram passes from where it would connect in the Taghdumbash Pamir with Tashkurgan. The survey was commissioned and was carried out between May and August 1966.[394] As both projects developed their own dynamics in a domestic and international dimension, their achievements are sometimes mixed up. The domestically important Indus Valley Road had its breakthrough on 11 August 1967 when a convoy consisting of eight light trucks (*gattoos*) left Kamila to reach Chilas the next day.[395] The goal of widening the road to accommodate bigger vehicles than jeeps had been achieved, although it would take many more years to improve the track and construct sophisticated bridges that were specified for the heavier traffic now envisaged.

The joint Karakoram Highway project would convert the domestic supply line into a cross-boundary artery, thus influencing the quality and size of road work all along the line. As the Pak-Chinese cooperation within Northern Pakistan was kept secret, rumours and speculations thrived. Some authors and journalists announced that more than one road would be built between Pakistan and China.[396] In the literature China was alleged to have built an ambitious road from Skardu across Karakoram Pass to Yarkand. No proof has ever been provided as the Karakoram Pass lies in the to-be-disputed area of the Siachen Glacier combat zone and has been factually under Indian administration.[397] Other speculations discussed the scope of the work alliance as well as the number of Chinese engineers and labourers involved. Again, the developments took various steps, and a variety of achievements was announced. The timeline refers to different instances that were celebrated as highlights or breakthroughs, opening ceremonies on different levels of representation, and local events that opened new venues for commerce and supply.

Resuming the China trade

During the interwar years, most of the work was commenced and important agreements were made. Instances that need to be mentioned are the agreement about bilateral trade 'to reopen the ancient Silk Route between Gilgit and Sinkiang which had been closed during the past two decades' that was signed on 21 October 1967.[398] The joint communique stated: "The agreement is intended for the benefit of the people living in the border areas of both the countries. It provides for movement of goods between these border areas on a limited basis and meets the long-felt need for re-opening of the traditional overland trade route which existed for centuries."[399] The road work was completed by 28 September 1968; external British observers were not fully convinced about the quality of the road and especially its strategic value. The Far Eastern Department in the London Foreign Office cabled to its embassy in Beijing: "Construction of the road is reported to have been difficult and though a road of sorts may have been completed by about this time ... it is certain that in its present form it is of fairly low strategic value and that this is unlikely to be set right under 3-7 years. Some lucky people who have been over part of it describe it as 'an absolutely hair-raising dirt track ... with 600 ft. sheer drops, landslides, 1 in 1 gradients ... The road is only open from 6 a. m. to midday as after that melting snow causes the waterfall to become too powerful.' It is also currently reckoned by Rawalpindi to be 'quite incapable of carrying anything larger than a Russian half ton Gaz jeep'."[400] Beyond strategic considerations the road fulfilled a bilateral symbolic function; a newly-built trans-mountain passage enabled the revival and resumption of cross-boundary trade.

After re-opening the Pakistan-China border for barter trade in 1967 the Gilgit bazaar became a centre for Chinese goods – especially tools, cloth and household crockery – their range would expand in coming years. The photograph was taken on 28 October 1990

The intensified Pak-China cooperation led to the re-opening of the old Mintaka route on 26 August 1969 when two caravans of pack animals met in Misgar, exchanged goods worth two and a half lakh rupees there, and signed mutual trade agreements.[401] After this successful endeavour both governments agreed to formalise this trade arrangement by founding trading organisations. The partner of the 'National Chinese Trading Organization' in Gilgit was the 'Northern Areas Traders Co-operative', with close to three thousand local shareholders who assumed the monopoly of handling the commodity exchange by annual caravans.[402] Nevertheless, only four shops were specialised on Chinese goods in Gilgit Bazaar in the initial years.[403] The barter trade agreement between Pakistan China Trade Cooperative Society and Foreign Trade Bureau Xinjiang was renewed in April and signed on 5 June 1981. The list of commodities from Pakistan included 'cotton and woolen goods, nylon fabrics, cigarettes, razor blades and leather goods. Pakistan would receive, in return, tea, certain types of cotton and woolen fabrics, chinaware, hardware and farm tools'.[404] At this stage, the spectrum of goods was rather limited and the trade volume of eight million rupees for the financial year 1981–1982 not excessive, but the trade partnership seemed to be

In the early years of bilateral trade across the Khunjerab Pass, the former British Consulate in Kashgar's Chini-bagh functioned as a customs clearance station. Pashtun traders are assembling their loads for transport to Pakistan. The photograph was taken on 5 August 1990

based on exchange between equal partners. A contemporary comment on the trade exchange indicated: "The steady and significant rise in imports from China could be attributed to three factors, the most important of which is the growing amount of Chinese aid to Pakistan. The lion's share of this aid is tied to the purchase of Chinese goods. The two other factors are the barter and border trade agreements that encourage the import of Chinese goods."[405] The amount and scope were to change over time, when the still rather limited number of entrepreneurs would diversify and grow. At this stage, the trade volume was more symbolic than of significant economic effect.

It seemed as if the modern highway would follow the ancient caravan route. In physical terms, this would have been the most suitable alignment as Mintaka Pass could be tackled more easily than the neighbouring Kilik or Khunjerab passes. Kilik was too far away: a passage across this pass would have involved an unnecessary detour, while the Khunjerab Valley was notorious for its difficult terrain and high water levels during summers. The narrow gorge with its steep flanks afforded numerous river crossings that posed some difficulties at high flood levels. Finally, the Khunjerab Pass was selected as the entry point; it was preferred on account of being the saddle furthest away from the Afghan and Soviet borders.

By 16 February 1971, road construction had improved to a stage where the Chinese and Pakistani groups met and joined forces. Under the responsibility of the Frontier Works Organisation, out of the 483 kilometres between Thakot and Halegush nearly three quarters had been completed up to the specified width of 8.5 metres; several bridges, protective works and asphalting remained to be finished.[406] The 160 kilometres up to the Khunjerab Pass were constructed by Chinese roadbuilders. A ceremony was held at Shishket stating 'the Karakoram Friendship Highway was declared open and said to have replaced the old caravan route.'[407] The high-level delegation included Mr. Yang Chieh, Chinese Minister of Communications, and the Chief of Staff Pakistan Army, General Abdul Hamid Khan, who jointly performed the opening ceremony, which was attended by 3,000 spectators.[408] This function went unnoticed internationally as Chinese roadbuilders avoided public attention; especially since August 1974, foreign visitors had been banned from visiting the region as the scope and size of Chinese involvement was to be hidden.[409] During our first attempt to visit in July 1977 the road was still not open to foreign travellers; the ban was only relaxed the year after.

The memorial commemorates the victims of road-building between 1966 and 1972 in the Hunza Valley who were working in the Engineer Battalion 103. The photograph was taken on 17 March 2019

Environmental challenges and geo-hazards

The Chinese roadbuilders left immediately after the ceremony and did not experience the significant damage by glacial advance that occurred a few days later. The same happened in other places. A major blow fell further south where the Pattan earthquake of 28 December 1974 killed more than 5,000 people and injured 17,000, more than 30,000 houses were destroyed, and a major portion of the Indus Valley Road lay in shambles.[410] A Pak-Chinese joint effort managed to re-open the road just two weeks later. Reconstruction and relief measures took much longer; the subsequent Kohistan Development Board was a pioneering rural development project on a substantial scale. It became obvious that this road would need constant re-building as several sectors were implemented in areas with very unstable slope conditions.[411] For decades to come, these sections of steep slopes were marked as 'sliding areas' and warned all passengers of immediate dangers. Yal parri in Nagar, the collapsing moraine ridges in Jaglot Gah, or the notorious Tattapani, south of Gilgit, are constant reminders of unstable slope conditions. Edward Derbyshire, Monique Fort and Lewis Owen summarised the results of their detailed investigation: "That part of the highway [between Khunjerab Pass and Gilgit bridge] steeper than 3° lies in the three dominant geomorphological environments (gorges and rock cliffs 17.8 %; sediment fans 22.7 %; and talus/scree slopes 20.7 %). This pattern is almost perfectly reproduced in the sections steeper than 4°, and is closely reflected in the statistics for road surface damage. Almost three-quarters of the highway that has lost 50 % or more of its metalled surface is accounted for by these same three geomorphological settings (22.5 %, 25.3 % and 23.9 %, respectively). Much of the road damage is a result of boulder impact), mass slumping of slope debris, and scree sliding."[412] Even graver was the complete destruction of the Shishket Bridge in connection with the surging Balt Bar glacier on 12 April 1974. The lake that formed behind it extended all the way to Pasu and submerged the newly constructed Friendship bridge under twenty metres of sediment.[413] It took more than three years until 13 July 1977 before the dam broke and the stored water was released. The bridge could not be recovered and was built anew at a different location where, afterwards, the road became prone to annual temporary closure due to snow avalanches in spring.[414] The advancing Batura glacier required an alternative alignment, materialised in two temporary steel bridges circumventing the glacier at the left bank of the Hunza River in 1974. Annually damaging events of varied magnitudes were recorded that affected the road quality; some led to its closure over days and weeks, calling for army action to re-open the thoroughfare. The changing delivery of debris production and deposition at the Ghulkin glacier caused a major challenge to road engineers in 1980 and again in 1986.[415] The prevailing environment presents its challenges, and these cannot be avoided. Edward Derbyshire, Monique Fort and Lewis Owen concluded for the Hunza Valley: "The main geo-hazards are sliding and falling rock and debris, flash floods, fan progradation, collapse induced by fluvial abstraction, and cliff failure in thick deposits of glacigenic debris. Although examples of these may be found in places along most sections of the highway, the vulnerability of the KKH shows considerable variation along its length."[416]

A challenging winch-operated gondola fixed to a steel-rope (ayúm garaḍí) was installed at the site of the Tashot Bridge. It was one of two sets to enable jeeps and other heavy loads to be brought across the one-hundred-metre-wide Hunza River prior to the construction of a sturdy suspension bridge (pakkā pul). Source: Allah Bakhsh Rajput (1962: 51)

The jeepable road between Hunza and Gilgit was regularly affected by sliding material and collapsing support structures. The photograph was taken on 1 August 1969 by Rose Bauer, Berlin

Landslides and rock falls have been regular features interrupting vehicular traffic along the Karakoram Highway. Maintenance units of the Frontier Works Organization have been responsible for keeping the artery open. The photograph was taken on 6 October 1985

Most major trans-montane road projects for motorised transport have been commissioned and inaugurated since the 1950s. Exceptions have been the road link to the Kashmir Valley from the Punjab and the Pamir Highway and its feeder roads. The latest innovation was the crucial link between the Pak-China Friendship Highway and the Pamir Highway across Kulma Pass in 2004, thus connecting major road systems in Central Asia with Xinjiang and Pakistan. Source: extended and updated table from Hermann Kreutzmann (1998b: 23)

Major trans-montane road systems in High Asia

Mountain range	Destination	Name	Pass	Altitude (in m)[1]	Length (in km)[2]	Date[3]
Hindukush	Kabul-Jalalabad-Peshawar	Khyber Road	Khyber	1,067	232	1963
	Kabul-Qizil Qala-Dushanbe	Salang Road	Salang	3,600	497	1965
Pamir	Osh-Khorog	Pamir Highway	Akbaital	4,655	728	1932
	Dushanbe-Kala-i-Khum-Khorog	Novo Pamir Highway/Stalin Road	Khabu-rabot	3,252	536	1940
	Khargosh-Khorog	Pamir-Ishkashim	Khargosh	4,344	280	1940
	Murghab-Tashkurgan	Kulma Pass Road	Kulma	4,365	150	2004
	Kashgar-Tashkurgan		Subash	3,900	294	1958
Pamir/Karakoram	Khunjerab-Tashkurgan	Friendship Highway	Khunjerab	4,655	120	1968–86
Tienshan	Kashgar-Naryn-Bishkek	Xinjiang-Kyrgyzstan	Torugart	3,752	600	1983
Kun Lun Shan	Yarkand-Gartok	Aksai Chin Road	Khitai	5,341	1,200	1956–57
Karakoram	Thakot-Khunjerab	Karakoram Highway	Khunjerab	4,655	735	1964–78
	Gilgit-Skardu	Skardu Road	-	-	210	1950–68
	Abbottabad-Gilgit	Kagan Route	Babusar	4,173	420	1948–49
Himalaya	Khwazakhela-Chilas	Indus Valley Road	Shangla	2,150	278	1959–65
	Rawalpindi-Srinagar	Kashmir Road	Murree	1,980	320	1922[4]
	Jammu-Srinagar	Banihal Road	Banihal	2,196	330	1916/1922[5]
	Srinagar-Leh	Ladakh Road	Fatu La	4,100	435	1962–74
	Leh-Khardung La-Thoise	Nubra Road	Khardung La	5,636	80	1980
	Birgunj-Kathmandu	Tribhuwan Rajpath	Daman		221	1953–56
	Kathmandu-Xigazê	Arniko Rajmarg	Zanglu	5,481	500	1963–67
	Kalimpong-Xigazê		Natu La	4,310	370	
	Phuntsholing-Trongsa	Indo-Bhutan Highway	Pele La	3,390	375	1959–1962

1 altitude refers to highest point of road (pass)
2 length indicates total destination between connected points
3 date describes the period of construction and/or the opening of sealed surface/asphalt roads
4 developed to become the 'normal route' for Srinagar: 'Until 1916 the only way to travel from Jammu to Srinagar by carriage was by way of Rawalpindi and the Jhelum Valley Road which was opened in 1890' (Alastair Lamb 1991: 16); see *Routes in the Western Himālaya, Kashmīr, &c. Volume I: Pūnch, Kashmīr & Ladākh* by Kenneth Mason (1922: 34) in IOR/V/27/60/160: 34
5 construction started in 1910, in 1916 only opened for the Maharaja family; see Alastair Lamb (1991: 16); Kenneth Mason (1922: 58)

Opposite
The China Bridge became the symbol of the Karakoram Highway. The Dasu crossing of the Indus River in Indus Kohistan was an important change of routing by connecting Besham on the right bank alignment with Chilas on the left bank. The photograph was taken on 19 January 2007

Connecting networks to make an international highway

A further drawback was that this road was basically built as an isolated artefact without proper links at both ends to existing traffic networks. The completion of an approach road from Abbottabad via Mansehra, Chattar Plains, Batagram to Thakot was delayed and would remain so for many years.[417] Somehow, the symbolic significance of the Pak-China Friendship Highway released certain synergies in the central sections that were not seriously supported at the ends. On the Chinese side the road between Tashkurgan and Khunjerab Pass was not in an appropriate state; it took another seven years to make the connecting track into a road. By 1975, the first trucks could reach Central Hunza, and the thoroughfare was completed in time for the 1978 inauguration. At this stage, the Pakistan section was asphalted, while the section in the Taghdumbash Pamir took much longer: even at the official opening for international traffic in 1986 it had not materialised.

The project that had started as the narrow Indus Valley Road to connect the Northern Areas with down-country Pakistan had become the Pak-China Friendship Highway. This friendship (*Pak-Cheen dosti*) was tested in the final five years before its inauguration in 1978. As quality standards differed and geo-hazards had interrupted and destroyed completed sections, protective walls and bridges, a solution to the recurring challenges was sought and a joint effort was required. On 15 July 1973, China and Pakistan signed an agreement in Beijing that would involve 9,000 Chinese road builders and 6,000 Pakistani labourers. The dimension of this joint project resembled the size of the Guomindang Road project from the final years of World War II. Nevertheless, the envisaged time frame was more realistic and was fully required to make the road as stable as possible and a reliable artery. The Chinese team was headed by Meng Tungming, an experienced roadbuilder in mountains who had previously been involved in the initial phase of the Karakoram Highway construction. Nevertheless, a certain adaptation to terrain conditions was unavoidable. After the road's completion Edward Derbyshire, Monique Fort and Lewis A. Owen recorded how the roadbuilders proceeded to smooth the physical challenges for road alignment: "The KKH is a very steep road between the Khunjerab Pass and the Gilgit River. Only 5.3 % is more or less horizontal (± 0.5°),

Atabad village is perched on a very steep cliff high above the Sarat basin. The houses of Atabad (opposite) are located on a terrace that is part of the steep slope providing space for the village lands The photographs were taken on 26 April 1985

over one quarter (27.6 %) is inclined at 3° or more, and over 10 % of its length is steeper than 4°."[418] The final product was esteemed as a great engineering feat, but it had its weak spots as well. The roadbuilders had to make several compromises at places where either avalanches, mudflows or rock falls could not be avoided, or where blasting and drilling had destabilised the mountain flanks to such a degree that aftereffects would occur. Besides all obvious bottlenecks and shortcomings, the roadbuilders had succeeded in creating an international highway that connected historical routes and networks in a manner and at a time when the perceived usage was limited to military and strategic considerations and the potential benefit was neglected.

Socio-political reforms along the road

Zulfiqar Ali Bhutto was an ideologist and a visionary, and he attributed a higher importance to the Pak-China Friendship Highway and its benefits as a symbol of Pak-Chinese comradeship. In tune with his drive to make a second attempt at reforming landownership in independent Pakistan, he announced the abolition of feudalism.[419] He decreed the end of the Hunza State and its merger with Pakistan on 24 September 1974, two years after other princely rulers of northern Pakistan had already been deposed. The Indian Ministry of External Affairs reacted strongly: "Premier Bhutto's announcement ... to the effect that the State of Hunza had been abolished and its area merged in the northern areas to be administered by the Resident Commissioner of Gilgit and Baltistan has been noted with surprise. The haste and arbitrariness of this action is particularly glaring in view of the fact that Hunza is a principality of J and K which had acceded to India, and thus at best Pakistan is only in illegal occupation of the northern areas of J and K."[420]

The British High Commission commented upon the delay in the Hunza case: "We thought at the time [1972] that Mr Bhutto might have been moved to spare Hunza out of deference for the feelings of the Agha Khan, who is the spiritual leader of most of the inhabitants of Hunza – they being Ismailis. If this was true Mr Bhutto has evidently now thought better of it and decided that Hunza's status is an embarrassing anomaly which cannot continue; anomalous because it infringes the tenets of the Awami Raj – 'We cannot afford to have states within the State, the country belongs to peasants, labour and workers and is not the Jagir of Nawabs and Mirs'. Mr Bhutto was reported as saying in this morning's press; ... embarrassing because of the presence of some 8,000 Chinese on the Hunza to Gilgit stretch of the Karakoram Highway. So Hunza joins the ranks of former 'merged' states such as Swat, Dir, Chitral, and Bahawalpur, and its former Ruler like theirs, will doubtless retire into pensioned-off obscurity."[421]

The observation that other principalities had been integrated into Pakistan and that Hunza State was spared from abolishment of *mīri* rule mobilised additional energies in the opposition which were displayed in the so-called 'Hunza & Nagar Liberation Movement'. The leaders would not accept the *mīr* of Hunza's declaration that everybody in Hunza was better off with his rule (*thámkuṣ*) than residents of neighbouring areas; on the contrary, their demand was to be freed from land expropriation and forced labour (*rajáaki*) that was applied during the construction of the Karakoram Highway. People were not compensated for their loss of land and had to work for the state without adequate remuneration.[422] Among the many demands, this was an important claim put forward to the Pakistan authorities. Hunza society was split in two antagonistic camps of opponents and supporters of keeping the inherited reign. Pakistan received its first constitution during this period; although there was no clause for princely states within Pakistan, the position of Hunza was not clarified. The President of the 'Hunza & Nagar Liberation Movement' Maula Madad repeated the demand for equal rights and adequate compensation for the bereavements suffered by Hunza people during the process of the construction of the Karakoram Highway.[423] Finally the movement partially succeeded.

Upon the enquiry by *mīr* Jamal Khan whether there had been any wrongdoing on his side, Zulfiqar Ali Bhutto supposedly responded: "I told him that it was not his mistake. Indeed, it was in accordance with our parliamentary system that we thought it necessary to merge Hunza with Pakistani Gilgit."[424] Himself one of Pakistan's biggest landlords in Sindh, he terminated 'feudalism' in the Karakoram by declaring the former local rulers – none of whom owned a minimal fraction of Bhutto's landholdings – to be retired army officers receiving a Pakistani state pension for their upkeep. No need to mention that in Hunza and Nagar no land was redistributed among the landless. The termination of princely power and the growing neglect of hereditary status were instrumental for a societal transformation in which participation in education and economic success would become markers for appreciation and respect. The growing number of out-migrating people clearly shows that certain restrictions and control measures were alleviated. At the same time the provision of subsidised wheat flour (*āṭā*) was announced in tune with his programme of providing sufficient livelihoods (*roṭi kaprā aur makān*) to all citizens. The state would subsidise the transport between down-country Pakistan and the Karakoram valleys; the cost of overland transport had fallen substantially since the opening of the new artery. Zulfiqar Ali Bhutto abolished all direct taxation for people in the Northern Areas and reformed Ayub Khan's 'Basic Democracies' as a root-level development setup by introducing a 'People's Work Programme' which would be relabelled as 'Integrated Rural Development Programme' shortly after.[425] The administrative setup of the Gilgit Agency and Baltistan was transformed into the Northern Areas in 1972; both former districts were split, and the additional districts of Ghizer and Ghanche came into being. The Northern Areas Advisory Council with sixteen elected members became the top regional legislative body, although disenfranchisement prevailed on the national level and a provincial status was not granted.[426]

The Karakoram Highway had become a symbol of nation-building that was achieved with significant and substantial Chinese support. By 1974, these two external efforts had contributed to the most important transformation in Hunza's economy and society. At the same time, the inhabitants were split in two political factions. Long before the opening of the Karakoram Highway migrants from Hunza and neighbouring valleys had reached the factories of Karachi and had become politicised citizens. Strong supporters

The damaged plaque fixed on the exterior wall of the Karimabad palace still shows the flag that has been the most frequent symbol of Hunza's statehood. The lion which used to hold the flag in its mouth was supposedly added by mīr Jamal Khan (Zafar Iqbal 2018a: 565). Its absence symbolises the end of thámkuṣ, the rule of the Ayáśo dynasty in Hunza. The photograph was taken in Karimabad on 22 March 2015.

of the Pakistan People's party had influenced political opinion back home. Protagonists of hereditary rule by the *tham* were fighting with their opponents for majority support in the valley. The abolishment of *mīri* rule supported the antagonists and offered significantly better and new opportunities to overcome autocratic restrictions of movement, to denigrate preferential treatment of upper classes, and to allow excluded citizens to participate in economic enterprises, mobility, skill development and education. The opening of the Karakoram Highway for public transport – the Northern Areas Transport Corporation (NATCO) began its operations in 1974 – and commodity exchange would affect all communities and households in the Western Himalaya, Karakoram and Eastern Hindukush.[427] Since then, Hunza has occupied a special position as the hub for Central Asian trade and entrepôt for China. At this period trading organisations and cooperative societies were founded and started operations in Central Hunza and Gojal; during the first five years 21 cooperative societies are on record as having been established in various villages.[428] Their main objectives were to purchase daily necessities by applying economies of scale, to provide them at reasonable cost for their members, and to operate shops stocked with additional commodities in local demand. Right from the beginning, they organised the out-of-region marketing of local surplus products such as dried fruit and apples. After the abolishment of the Hunza and Nagar States, skilled representatives of the villages had taken the reorganisation of infrastructure development and trade into their own hands. Retired soldiers, preferably addressed as *subahdār*, would become essential as professionals for accounting, book-keeping, and operating shops. Nevertheless, the development of a fringe infrastructure complementing the Karakoram Highway, such as link roads, roadside bazaars and other enterprises, remained in its nascent state for quite some time coming.

At Tashot one of the first suspension bridges was built; in the 1960s a winch-operated gondola was used to transfer heavy loads across the river. With the Karakoram Highway the main artery was built on the right bank of the Hunza River; Dadimal and Miachar villages could be reached via a new suspension bridge at Tashot. The access is via steep tracts. The photograph was taken on 3 October 1989

Two inaugurations – occasions for celebrations and speeches

Whenever a Chinese delegation is visiting Islamabad banners and posters are displayed that emphasise the Pak-Chinese friendship symbolised in the construction of the Karakoram Highway. The photograph was taken on 17 September 2001

Zulfiqar Ali Bhutto could not reap the fruits of his visionary bond with his Chinese comrades and was unable to participate in the opening ceremony. He had been deposed in a military coup and jailed in a death-cell by his successor Martial Law Administrator General Zia-ul-Haq, who staged the inauguration of the Karakoram Highway at Besham and Thakot on 18 June 1978. Thakot bridge was hailed as a monumental achievement symbolising Pak-China Friendship: "Today the Thakot bridge was beautifully decorated. A high streamer at the southern end of the bridge read in both Urdu and Chinese: 'Long Live Pakistan-China friendship'. The national flags of Pakistan and China were hung on the two silvery pillars. A lot of bunting covered the railings on both sides of the bridge and streamers with patterns of the national flags of the two countries fluttered in the breeze. The bridge, bathed in bright sunshine, looked exceptionally imposing and magnificient. … Accompanied by General Ziaul Haq, Vice-Premier Keng Piao walked from the southern end of the bridge to the northern end to the warm welcome of about 1,000 Pakistani and Chinese builders. The Chinese vice-premier cut the ribbon for the opening to traffic of the highway."[429] A special emphasis was attributed to the mutual bond and people's contribution by the vice-premier: "'How the people longed for centuries to have a broad and level road to facilitate trade between East and West, to carry out cultural exchange and to strengthen friendship! Today, the people's thousand-year-old dream has come true with the completion of the Karakoram Highway.' We spoke highly of the heroic road-builders of the two countries who overcame numerous difficulties by doing arduous work and with an unflinching spirit and finally succeeded in turning a natural barrier into a thoroughfare. He deeply mourned for the Pakistan and Chinese martyrs who laid down their lives for the building of this friendship highway and expressed his heartfelt sympathy tor their families."[430]

During the opening ceremony the number of participants in the joint effort was announced as 35,000 engineers, mechanics and road-builders, split between about 10,000 Chinese and 25,000 Pakistani. Some Chinese accounts claim that 15,000 Chinese road workers participated. The number had risen from initially 15,000 persons; later a more realistic 25,000 participants were recorded, of whom 406 persons lost their lives and 314 were injured during construction and maintenance.[431]

The memorial plaque in Besham reads: "The Karakoram Highway (the ancient Silk Route) built jointly by the People's Republic of China & the Islamic Republic of Pakistan to strengthen the historic bonds of friendship between the peoples of China and Pakistan 1958–1978". A smaller marble tablet below honours the victims in a "tribute to those workers of Pakistan and China who laid down their lives for the construction of this highway". A Chinese cemetery in Danyor commemorates the loss of more than one hundred Chinese lives while the Pakistan army termed the deceased victims as martyrs (*shahīd*) for the national cause. General Zia-ul-Haq and the Chinese chief guest, Vice Premier Keng Piao, both used the opportunity to highlight the symbolic value of the Karakoram Highway as an expression of a strengthened bond between the two countries. China Daily (*Renmin Ribao*) reported on this occasion that the 'project is an aid construction provided by China'.[432] Several places along the Karakoram Highway became important for certain events – inaugurations, landslides, disasters, cemeteries, etc. –, but the place that was the meeting point of Chinese and Pakistani roadbuilders, the point where forces and activities were joined and where contracts ended or began, this place evaded public attention and is completely lost from collective memory Upon enquiry no interlocutor we met in the Hunza Valley knew the location of Halegush (*hále ġaṣ*) or, in Pinyin, Haligexi. The barren place between Atabad and Goshben (Ainabad) is further up the valley from the 'sacred rock' Haldeikish, but is not paralleled with it. It is neither a place for commemoration nor a monument along the road; it is a rather unknown place with fox trails (*hále ġaṣ*), to which people bring juniper trees from higher up the mountain. For the reporters of China Daily, Haligexi was a significant place where Chinese and Pakistani roadbuilders

The Chinese cemetery in Danyor was established to commemorate the victims of road construction prior to the opening of the Karakoram Highway in 1978. The photograph was taken on 3 October 1989

With the construction of the Karakoram Highway most villages in the Hunza Valley were linked to the main road. The suspension bridge at Sarat allowed traffic to Atabad on a very steep road. The bridge connected with the Karakoram Highway near Halegush, the landmark spot that divided different sections of road construction. The photograph was taken on 7 August 1985

met; it was identified as the place where different phases of the project were initiated and concluded. From the Chinese perspective the 'project had two phases: from Khunjerab Pass to Haligexi (June 1968–January 1971); from Haligexi to Thakot (June 1974–May 1978)'.[433]

China was lauded as a guarantor for Pakistan's independence and national unity. The Chinese guest underpinned the necessity of an independent development of sovereign states and the avoidance of external imperial interventions. Both supported each other's stand on bilateral and international crisis management. Zia-ul-Haq's grip on the Northern Areas was reflected when they were administered as a separate 'Martial Law Zone'.[434] Representatives of the Northern Areas merely had observer status even in the *Majlis-i-Shora* created by Zia-ul-Haq as the predecessor of the subsequently installed National Assembly. The 510 kilometres of Karakoram Highway in the Northern Areas were the 'backbone to an emerging network of transportation lines' as Nazir Kamal wrote. He calculated that in addition to the highway "... there are 400 km of metalled jeepable roads, 48 km of unmetalled truck road and 2,144 km of unmetalled jeepable roads in the Northern Areas."[435] From a strategic point of view, the link of Skardu with the Karakoram Highway was the next pressing need; the project for the 170 hard-top kilometres linking the Karakoram Highway from Farhad Bridge with Skardu was commissioned to the Frontier Works Organisation at a cost of 12.5 crores rupees and scheduled for completion within three years.[436] Writing an essay for the 'Strategic Studies' at the time of its inauguration, Nazir Kamal denied any military purpose of the friendship road: "The KKH is essentially a natural extension of the nation-building effort which has been prompted by the obvious need for national integration and concrete socio-economic unity. Its military and strategic significance, which are clearly limited and scarcely unsettling, are incidental."[437]

The accomplishment of the road-building effort was seen as the beginning of intensified close neighbourly contacts with mutual interests in commerce and trade. The Karakoram Highway was epitomised as a crystallisation point of such friendly neighbourly relations and declared a real 'Friendship Highway'. A new border trade agreement was signed on the occasion, allowing Pakistan the import of vehicles and livestock, as well as the usual set of household utensils and tools. In the other direction, it was expected that Pakistani cotton products, dry fruit, cigarettes, medicinal plants and drugs would find a market in China. Here it was already mentioned that the Karakoram Highway might function as an axis that would connect China closer with Muslim countries in the Middle East 'thanks to the vision of the Karakorum pioneers'.[438] This statement comes somewhat as a surprise; its rationale focused on modern warfare as airborne and not as a terrestrial confrontation. Such a view was counteracted by most external observers, especially by Indian commentators who used phrases such as 'road of danger' and 'illegal occupation', thus making the connection with the Cold War and the Kashmir dispute. The completion of the 1,224-kilometre-long Karakoram Highway between Thakot and Kashgar triggered substantial road-building in the Indian Himalayas and especially in Kashmir-Ladakh.

A high price had to be paid to keep the thoroughfare open; the constant damage due to environmental hazards required regular repairs and further work to improve the existing road. The missing network links in the north and south had to be brought to the same quality level as the existing Karakoram Highway. Even along its own course, link roads that would connect villages and hamlets with the main road were more or less non-existent. Hopar and Uyum Nagar were linked to the Karakoram Highway by a narrow jeep track in 1978. Baltit was serviced from Ganesh by a steep and narrow jeep road alone; only in 1981 did the Northern Areas Works Organisation complete the first link road through Central Hunza's upper villages and hamlets from Aliabad to Haiderabad and Baltit, renamed as Karimabad two years later. When the regular operation and maintenance of the highway was achieved, a second ceremony was scheduled to take place on the Hunza-Sarikol border. The formal inauguration of Khunjerab Pass as the gateway for international traffic took place on 27 August 1982. The Governor of the Uigur Autonomous Region Xinjiang, Ismael Ehmed, together with the Federal Minister of Kashmir Affairs, retired Major-General Jamadar, first travelled to Islamabad to sign a protocol permitting the movement of passengers and goods. Five days later, the former unveiled a plaque on the top of the pass and cut the ribbon together with Pakistan's Minister of Communications, Mohyuddin Baluch, formally opening the Khunjerab Pass for bilateral civil traffic.[439]

The subsequent visit of President Zia-ul-Haq to Beijing in October 1982 resulted in a further strengthening of ties; several agreements were finalised including 'to promote cooperation in the economic, scientific, technological and trade fields'.[440] By 1980 Pakistan had become China's biggest trading partner; consequently it was little surprising that the National Bank of Pakistan was permitted to open a branch office 'in Beijing in May 1982, becoming the first bank office in Beijing from a Third World country.'[441] A joint ministerial committee was set up to promote further cooperation and exchange. Cross-boundary trade was part of the agreement, but remained restricted to the annual exchange of trade caravans in autumn. In September 1985 the mail service between China and Pakistan via Khunjerab Pass was resumed on a regular basis.[442]

In the aftermath of the opening, border control posts were erected in Dih and in Pirali; the infrastructure was not yet adequate for smooth checks and clearances. At this stage, those foreign visitors who held a travel permit for the Karakoram Highway were initially only allowed up to Altit. In January 1982, special permits for foreigners were abolished, and their free passage was allowed 'within 60 miles of the Chinese border' up to Batura Bridge near Pasu.[443] This decision was taken in accordance with the Chinese counterparts who still did not allow any foreign visitors to travel southwards from Kashgar towards Pakistan. It took another four years, until 1 May 1986, before both governments agreed to open the Karakoram Highway for a complete passage of foreign visitors from Hunza to Kashgar.[444] First travellers experienced the unpreparedness and lack of facilities along the route. It would take another three years until the road

The plaque on top of Khunjerab Pass commemorates the opening ceremony for cross-boundary traffic that took place here on 27 August 1982 honoured by the Chinese Chairman of the Xinjiang Government and the Pakistani Minister for Kashmir Affairs and Northern Areas who formally inaugurated the Karakoram Highway for regular traffic between both countries. The photograph was taken on 26 July 1989

After the opening of Khunjerab Pass for international traffic both countries had agreed that their vehicles were allowed to bring passengers from their respective starting points to either Tashkurgan or Sost. The red buses of the Northern Transport Corporation took passengers to Tashkurgan and had to return empty. The photograph was taken on 26 July 1989

Close to Shishket Bridge snow avalanches regularly blocked the Karakoram Highway at Wuch Shot in spring. A passage had to be dug out by the Frontier Works Organisation to secure a thoroughfare for the public bus. The photograph was taken on 20 April 1990

on the Chinese side from Khunjerab Pass downwards was given an asphalt topping. Only in October 1988 had the road surface condition reached a state that a fourteen-hour-long journey on the unfinished section between Khunjerab Pass and Kashgar was reduced to four hours.[445] The public travel services did not allow Chinese buses beyond Sost, and the journey of travellers from Sost (2,740 m) to Tashkurgan (3,120 m) was only possible on vehicles operated by the Northern Areas Transport Corporation or Pakistan Tourism Development Authority. These had to deliver all passengers at the border control post cum immigration and customs station in Pirali (4,080 m) from where their onward journey would continue on Chinese buses.[446] The immigration and customs posts of both countries were 120 kilometres apart, a distance that all respective border-crossing vehicles had to cover empty on the way back to their home stations.

The impression prevails that a road that was mainly built for military and strategic purposes would only reluctantly initiate additional utilisations and value-generation on both sides.[447] China was not yet prepared to let any civilian traveller use the new road unless on special delegation and equipped with all kinds of consular exemptions and valid permits. For *ḥāǧ* pilgrims from Xinjiang, the Karakoram Highway became a major facilitator to reach Islamabad from where they would proceed to Mecca by air. General Zia-ul-Haq made a point of becoming well known as a facilitator for Muslim pilgrims en route to the holy places of Islam. In July 1986 he himself went to Sost to welcome a group of pilgrims; the leader of the Chinese delegation, Abdul Mehmood, lauded Zia-ul-Haq for Pakistan's hospitality by facilitating their journey and 'highly appreciated the Islamisation process initiated by him in Pakistan' while the president 'presented copies of the holy Quran to every Chinese pilgrim'.[448] These *ḥāǧ* pilgrims were economically quite active on their return journey, bringing Pakistani goods into Xinjiang. Their example stimulated local entrepreneurs to invest in barter trade with China; cigarettes, cotton and printed silk fabrics, dried fruit and a few other commodities were brought in from Pakistan. Pashtun entrepreneurs from Peshawar invested in ancient carpets and falcon smuggling. In autumn large groups of Pashtuns would visit the villages in the foothills of the Western Kun Lun Shan and collect Khotanese carpets on a house-to-house search trying to convince owners to barter old for new carpets. Others met poachers and purchased falcons that were later prepared and packed for a dangerous journey in the luggage of traders. Their peers would reside in Kashgar's Seamen or Chinibagh hotel, where their helpers and supporters would prepare papers, packaging and transport facilities towards the border. The border-crossing was facilitated by helpful and cooperative customs officers who enabled a smooth forwarding of the valuable commodities in exchange for a small share. The added value in the case of falcons could well be twentyfold after crossing the border. Both valuable commodities were smuggled to down-country Pakistan from where the falcons were flown to the Arabian Peninsula and the carpets entered the international antiquarian collector market in the West. Local business was quite limited, and the tourism flow was affected by internal crises in Xinjiang, e.g. the border was closed in 1990 after the armed April uprising in Baren close to Kashgar.[449]

Link roads connect people, villages and the highway

The second inauguration coincided with the beginning of the Aga Khan Rural Support Programme in northern Pakistan in 1983.[450] The first phase was mainly implemented in the Gilgit District, and the Hunza Valley occupied a prominent position in the activities of the private effort operating as a non-governmental organisation. In conjunction with the Northern Areas Public Works Department and Local Bodies & Rural Development, physical infrastructure projects were implemented.[451] The latter followed a top-down approach of expanding and extending roads, building bridges, police stations, hospitals and schools in remote locations in the Karakoram Valleys. The private development organisation implemented a bottom-up approach attempting to fill on an equitable foundation the power void that had appeared after the deposition of local rulers. By establishing village organisations in which the vast majority of households were represented they perceived this structural reform to be the remedy for the existing 'institutional vacuum', as Akhtar Hameed Khan suggested 'to tame their hostile environment'.[452] These village organisations would choose the most pressing infrastructure project that would benefit most, if not all, its members. The majority of projects identified by village organisations were irrigation-related projects for building new channels or extending old ones. As 'there continued to remain an unmet demand for village level infrastructure', a significant share of all projects was devoted to link roads.[453] Thus, the 1980s became the decade of expanding the road infrastructure by secondary and tertiary roads to connect villages and hamlets with the Karakoram Highway. Within the infrastructure development about one fifth of all projects in the first two decades were allocated to road communications such as link roads, pony tracks and bridges.[454] In a second step, tractor roads were built that helped the mechanisation of agricultural field-tilling such as ploughing and threshing.[455] None of the innovations that had improved agricultural production contributed to answering the question about the future of supplies. Muhammad Abdullah had raised this issue in his report about the prospects of agricultural development in the Northern Areas.[456] His findings suggested that the local population of the Karakoram could not be fed from local agricultural resources. Therefore, he recommended a specialisation in fruit and vegetable seed production in order to enable farmers to earn sufficient funds for purchasing the required basic food stuffs from down-country Pakistan. Again, the Karakoram Highway was instrumental in providing the required infrastructure for an additional supply of food items from down country. Muhammad Abdullah's predictions have come true at a fast pace; the dependency on external food supplies has been ever growing, and local production is fading.

The slow process of improving complementary infrastructure within the mountain oases contributed to a significant transformation of rural economies, agricultural and house-building practices, settlement structures, marketing and mobility. In the 1980s, tourism started to become an important source of local income-generation. After the pioneering mountaineers had established the arrival of a regular number of expeditions, trekkers followed in bigger numbers for low-level tourism. The growing source of tourism-based

Goods destined for Pakistan are loaded after customs clearance on the premises of Chinibagh Hotel for transfer to Tashkurgan. The photograph was taken on 22 August 1993

The Karakoram Highway became the major supply line for wheat and meat from down-country Pakistan. Water buffaloes were sent from Punjab to augment the Gilgit meat bazaar. The photograph was taken on 21 August 2003

income became international culture tours that required hotel and restaurant facilities, guiding and transport services. The numbers visiting Baltit increased from a few hundred after the inauguration of the Karakoram Highway to more than 5,000 annual visitors by the mid-1980s.[457] Tourism made ample use of the new road facility and attracted young people back into the mountain valleys as good incomes could be earned on a seasonal basis during summers. This period saw an uplift in terms of new opportunities and jobs in connection with development programmes, tourism and marketing of agricultural products. Industrial development such as mining and marble productions failed conspicuously owing to lack of access and electricity supply.

New highway and old problems

The Karakoram Highway had been established as an international cross-boundary artery with proper linkages to the respective road networks in Pakistan and Xinjiang. However, several legacies from colonial times were fermenting at the same time and becoming part of a continued confrontation between the centre and the Northern Areas. Its connection and involvement with the unsolved Kashmir dispute remained an Achilles' heel of domestic politics and put an end to all efforts to attain equal citizenship rights for the inhabitants of Gilgit and Baltistan. Right from the beginning of Zulfiqar Ali Bhutto's reforms in northern Pakistan, the demand for their own provincial status was put forward.[458] The mid-1970s were a period of strained relations between the Northern Areas and the central government. During Zia-ul-Haq's Martial Law period the 'National Movement for Democracy' had formed in Pakistan in 1985 and inspired stakeholders of the Northern Areas to launch a 'Movement for the Determination of the Constitutional Position of Northern Areas', calling for the constitutional integration of the Northern Areas into Pakistan and electoral rights for the population. They demanded the release of political prisoners and of judges and public officials who had participated in political protests; their other demand was equal pay for civil servants and equal grants for students from the region. Nearly four decades after independence, constitutional limbo prevailed and created a situation where inhabitants of the Northern Areas felt deprived and treated as second-class citizens.[459] A resolution was presented by leading personalities from the Northern Areas to President Zia-ul-Haq in 1987; these demands have been repeated several times since. When Benazir Bhutto came to office the granting of provincial status to the Northern Areas was contemplated again.[460] During the interim government of Moeen Qureshi public unrest and sectarian clashes caused casualties and curfew conditions in August 1993 that resulted in a reform package two months later. The administrative setup was modified, but the constitutional status remained untouched.[461] In the light of rising public unrest and continuing tensions new solutions were explored. During Benazir Bhutto's second term in office, a strong promise was made to alter the constitutional status in favour of the people of Gilgit and Baltistan. A reform package was debated; in February 1994 its implementation was postponed because of tensions between India and

Pakistan in the Kashmir issue.[462] When she visited Gilgit on 13 May 1996 the time seemed to be ripe to explicitly announce 'a legal frame work order, giving right of franchise to the people of Northern Areas'. Her statement was seconded by the Federal Minister of Kashmir Affairs and Northern Areas, Muhammad Afzal Khan who, in his capacity as Chief Executive Northern Areas, stated that 'the Northern Areas Council was fully empowered to make decisions within the legal framework adding it was not less than a provincial assembly'.[463] Benazir Bhutto failed to implement her promises, just as many similar announcements were never fulfilled.

Her second-time follower Nawaz Sharif did not alter the constitutional status of the Northern Areas one iota. In 2002, President General Pervez Musharraf indicated that he would 'delink' the Northern Areas issue from the Kashmir dispute and bring the administered area at par with other provinces of Pakistan.[464] These announcements have become a common feature with various governments, but none has yet been sufficiently audacious to grant an equal constitutional status.

Ganesh Bridge of the Karakoram Highway was built at the same place as the former suspension bridge, whose remaining cables are still recognisable. The narrow gorge section was the preferred place for bridging the Hunza River before its confluence with Miatsil. The photograph was taken on 21 September 1984

Sectarian tensions regularly emerge during religious celebrations such as muḥarram processions including flagellations. The Karakoram Highway provided the arena for such performances of Hunzukuč and Nagarkuč that required representatives from security services and agencies to maintain peaceful conditions. Sectarian clashes occurred mainly in Gilgit Town and outside the Hunza Valley. The photographs were taken on 6 October 1984

Sectarian clashes

The undecided status of the constitutional relationship between the Northern Areas and the centre, as well as several interventions during Zulfiqar Ali Bhutto's regime such as the abolishment of hereditary rule and the banning of the *muḥarram* processions in Gilgit contributed to growing tensions that were finding an outlet and attracting preachers and activists from down-country. The latter incident can be seen as the beginning of what later became known as 'sectarian clashes' between Sunni and Twelvershia factions that have continued until today.[465] In addition, the opening of the Karakoram Highway "... coincided with the rise of religious militancy in Pakistan. While unlocking Gilgit-Baltistan from its physical isolation and ushering in economic opportunity, the highway has also increased Gilgit-Baltistan's vulnerability to new threats, such as the influx of illegal weapons, drugs, and intolerant attitudes from the south, and it has changed the demographics of Gilgit and other towns."[466]

What have been locally identified as 'sectarian clashes' have become a severe law and order issue in northern Pakistan; in May 1988 more than 500 people were killed when a mob attacked a Shia village at the bottom of the Bagrot Valley, causing the closure of the Karakoram Highway. Since then, on and off, incidents of sectarian clashes have occurred and disturbed public order.[467] Several measures have been taken, but contributed only to a temporary calming-down. Members of the Gilgit-Baltistan civil society perceive these developments as a side-effect of further integration into Pakistan's mainstream politics where sometimes cause and effect of sectarian clashes cannot be distinguished.[468] The 1990s have been a lost decade for the citizens of northern Pakistan as structural problems remained at an impasse and certain developments led to positions that became more and more entrenched. For political activists and community stakeholders, an improved physical infrastructure and subsequently introduced interlinked development measures were not sufficient compensation for deficiencies in political representation, their missing recognition under international law, and a perceived neglect in participation in national welfare. The decade leading up to the beginning of the 21st century was a period of consolidation and internal problems. Insecurity was increased by uncertainty about the status and by intensified sectarian unrest, which needed only a spark to release tension. In February-March 1990 such an incident occurred. A minibus with Twelvershiite and Ismaili passengers was kidnapped close to Jaglot; all passengers were murdered in front of a Sunni clergyman. When the news spread that nine persons had been killed and that the culprits and suspects escaped any verdict and punishment, unrest prevailed and led to the temporary arrest of more than 1,500 persons.[469] Between 1988 and 2005, it was estimated by government authorities that 373 persons were killed in the Northern Areas during sectarian clashes; in the following year 100 more persons lost their lives. On several occasions these clashes were triggered by a textbook dispute about the representation of Shia Islam. Other observers see these encounters as part of a conflict constellation that has been extended from down-country Pakistan and Swat into the Northern Areas.[470] A particularly bad year was 2012, when on 18 February eighteen Shia pilgrims were killed in Kohistan district on the Karakoram Highway, followed by twenty more who were killed at Chilas. On 16 August twenty-two civilians were killed by a mob targeting Shia Muslims near Babusar Pass, but also causing the deaths of four Sunni Muslims who protested against killing innocent fellow Muslims.[471] A side-effect of these events was often the temporary closure of the Karakoram Highway, substantially increased roadside security and/or the introduction of escorted travelling on the road when buses and individual vehicles were accompanied by gunmen and convoyed through sectors that were marked as dangerous. Sometimes night travel was banned and curfews were introduced. The police authorities tried to maintain peace at a high cost and to enable smooth conveyance through the Indus Valley. In the aftermath such events episodically occurred. Except for 9/11 no other event reduced the number of international visitors as drastically. From one day to the next, tourism collapsed as a source of income and was not compensated for by domestic tourism. Between 2007 and 2014 it was reported that 71 'terrorist attacks' had occurred between Diamir and Gilgit alone; the vast majority were attributed to sectarian clashes, while sixteen were instigated by outside militant groups.[472] The highest international media coverage was given on 23 June 2013 when, in the vicinity of the Karakoram Highway, nine mountain climbers and two local companions were murdered in a Nanga Parbat base camp. A splinter group of the *Tehreek-e-Taliban Pakistan* by the name of *Junud-e-Hafsa* claimed responsibility, thus documenting that terrorist acts in response to international events had reached the Indus Valley.[473] The failure of the state authorities to arrest the culprits and their local supporters did not contribute to public trust in the safety of travel on the Karakoram Highway. There was a gap of nearly fifteen years before tourism would thrive again and would return as an unpredicted boom in domestic leisure travel.

Ali Masjid is the Shia mosque that was built in Aliabad after much controversy in the 1980s. It is the major place of Twelvershia worship in Hunza. The photograph was taken in Aliabad on 22 March 2019

In the aftermath of the 2005 Kashmir earthquake many residents from the affected Hazara region sought refuge in Indus Kohistan on the bank of the Indus River where village communities provided space for their tents. The local residents from Pattan and hosts had themselves experienced a severe and destructive earthquake in December 1974. The photograph was taken on 29 March 2006

Domestic infrastructure development and trade

Especially after the Cold War and the independence of former Soviet Central Asian republics, there commenced a reorientation and reinterpretation of the function of the Karakoram Highway as a salient artery connecting Pakistan with these new countries. Defence considerations and trade potential became the driving forces earmarking the Karakoram Highway as a priority project in 'upgradation and rehabilitation'. During her visit to Gilgit, Prime Minister Benazir Bhutto announced on 18 May 1996 a three-year-long improvement scheme that was to prepare the infrastructure for implementing a transit trade agreement between China, Kazakhstan, Kyrgyzstan and Pakistan that had been signed in Gilgit the year before.[474] Great expectations were anticipated from this new form of regional cooperation.

During Pervez Musharraf's rule, certain road projects were revived. Keeping in mind the Kargil War of 1999, the linking of Astor to the Karakoram Highway via a substantial bridge for heavy traffic was realised three years later.[475] Part of this project was to establish an alternative approach road to Baltistan via Deosai Plateau. The hydro-electrical project of building Basha Dam would cause a substantial reconstruction of a major portion of the Karakoram Highway in the area inundated by the storage reservoir upriver of the dam. In connection with the disturbed parts of the Karakoram Highway, the idea was put forward to upgrade the jeep track from Thak Nullah up to Babusar Pass and further on to Balakot and Mansehra into a fully asphalted truck road.[476] At least seasonally this alternative road could relieve traffic in the Chilas-Thakot-Mansehra sector where disturbances from 'dacoits' operating from the Allai Valley on the section between Thakot and Battagram had affected traffic security. Other hot spots were in the Chilas area and around Dubair. In any case, the Babusar Road would be necessary to relieve the sector that would be affected by construction of Basha Dam. The road had to be improved to such a standard that container trucks could make the journey via Babusar Pass. The Kashmir earthquake of 2005 reminded all authorities concerned how vulnerable the connection was between Gilgit and down-country Pakistan. Although the Northern Areas were least affected by this earthquake, Balakot and Battagram experienced huge devastations. A further alternative road from Gilgit to down-country Pakistan has been the route via Ghizer and the Shandur Pass to Chitral. This project was expanded by asphalting major parts of the road into Gupis and Yasin after the Kargil War of 1999. In the name of the Yasini soldier Havildar Lalak Jan, who had died in the battle and was posthumously awarded the highest military order (*Nishan-e-Haider*) for extraordinary gallantry, the road to his native village of Hundur was improved. The Gilgit-Chitral road has not become a real alternative route as the opening of the Lowari tunnel was postponed. The tunnel project was commenced in 1975, then abandoned for thirty years and resumed afterwards. It would facilitate the delivery of external supplies to Chitral. As an alternative route from Gilgit it is a comparatively long detour to down-country Pakistan via Dir, Malakand and Peshawar. Essentially, the Karakoram Highway has remained the only lifeline for Northern Pakistan's connectivity with China and down-country Pakistan.

During Pervez Musharraf's rule infrastructure was developed to such an extent that the network of secondary and tertiary roads was expanded, and a great number of roads received asphalt toppings. In the Hunza Valley, all villages could be reached by vehicular traffic after the most remote village Shimshal was connected with the Karakoram Highway in 2003.[477] In Nagar, the remote village of Hispar is reachable by jeep traffic as are all other villages; and even a number

The memory of the 1999 Kargil War is kept alive in Yasin. From his village Hundur the Havildar Lalak Jan participated and found his death. To commemorate his bold action he was posthumously honoured with the 'Nishan-e-Haider', Pakistan's most prestigious gallantry award. Such an acknowledgement is accompanied by compensatory measures for his family and other assets for the village; a new road and school are located close to his memorial. The photograph was taken on 16 March 2019

of summer settlements such as Boiber above Morkhun can now be accessed regularly. The Karakoram Highway has developed into the main supply line for wheat flour, vegetables and chicken from Hazara, live animals and other food items from the Punjab, and it has been used for the export of potatoes, apples, dry and perishable fresh fruit. Since the beginning of the 21st century, about 68,000 to 85,000 tons of potatoes were annually shipped from the Karakoram valleys into Punjab.[478] Besides being a centrepiece of the domestic food chain, the Karakoram Highway served as the export line for Kohistan timber that was partly floated down the Indus before being loaded on to lorries at suitable landing places. At an early peak in 2005, the Karakoram Highway served as a corridor for about 4,000 Chinese heavy-goods vehicles laden with containers that arrived at Sost for customs clearance before the goods were shipped to Gilgit and down the Indus Valley by about 12,000 Pakistan-style lorries. The annual import volume of goods from China amounted to about 49.5 thousand tons valuing at 2.4 billion rupees; in the opposite direction it was only 1695 tons worth 131 million rupees. Over a span of five years the Chinese volume was sixteen times higher than that of goods exported from Pakistan to China.[479] Thus, the Pak-China trade had become a one-sided affair and developed into an import channel into Pakistan.

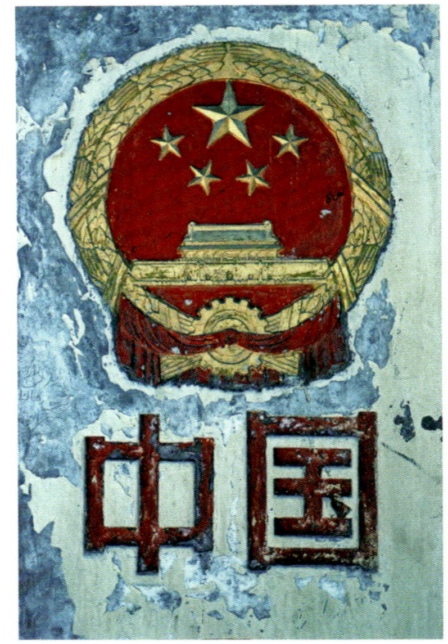

The Karakoram Highway was initially built as the Indus Valley Road to connect Gilgit with down-country Pakistan via Shangla Pass and the Swat Valley. Later on, a Sino-Pakistan joint-venture developed the artery into an international highway that was linked in the south from Besham and Thakot via Mansehra and Abbottabad with the colonial Grand Trunk Road near Taxila. In the north, the Hunza Road was connected across Khunjerab Pass with the Taghdumbash Pamir, from where a road was built to Kashgar via Tashkurgan. The Karakoram Highway was rebuilt four times and in its latest edition has become a thoroughfare for container trucks as part of the China-Pakistan Economic Corridor. Since 2004, a new road has connected this artery with the pioneering Pamir Highway in Tajikistan via Kulma Pass; thus, the first motor road across the Pamir is now part of a network that connects the Central and South Asian transport lines with the Chinese road system of Xinjiang. Source: Modified and updated on the basis of Hermann Kreutzmann (2015a: 120). Design by Hermann Kreutzmann

Karakoram Highway Improvement Project – the second phase

Infrastructure developments in China and Pakistan have followed different rationales. In the beginning the connection of the remote borderlands of the Taghdumbash Pamir and the Karakoram was begun at about the same time as an integration strategy in the 1950s. With the completion of the first Karakoram Highway, Pakistan pioneered in having a standard road connection, whereas the Chinese sector had to wait until 1988 to have an asphalt road. In the aftermath, the paces of development differed significantly in both countries. The Chinese speed of infrastructure expansion and improvement led to another rebuilding of the Tashkurgan to Kashgar sector, which was completed by 2008. The new road resembled the high-quality tarmac road leading up to Khunjerab Pass where a new border control post and pre-checking station had been established along with a barbed wire fence that stretched across Khunjerab Pass and contained a gate that could be closed off. Both countries had established several checkpoints along the route between Sost and Tashkurghan where the new immigration and customs clearing station is known as 'Khunjerab Port'. The unequal velocity of infrastructure utilisation in both countries is partly related to China's introduction of the 'Great Development of the West' (*xibu da kaifa*) campaign at the turn of the century, whereas Pakistan was exposed to enormous challenges and vast losses in the aftermath of 9/11. The Chinese part of the Friendship Highway was already well equipped and embedded in a variety of economic enterprises when considerations began to intensify bilateral cooperation.

The capacity of the Karakoram Highway seemed to have reached its limits at certain stages in Pakistan where the existing road had deteriorated over time. Terrestrial trade across Khunjerab Pass had increased, but public and private traffic grew to an even greater degree. The high numbers of buses and private vehicles became obvious when road blockages led to traffic jams and delays. China and Pakistan negotiated about terms to improve their exchange relations. Certain facilities to ease the bilateral goods exchange such as the Sost Dry Port were registered in 2004 as a joint venture, operated by Sinotrans and the Silk Route Dry Port Trust.[480] By June 2006, when the Sost Dry Port had become operational, the National Highway Authority –which now supervised the Karakoram Highway – signed a memorandum of understanding with the Chinese Assets Supervision and Administration Commission to widen the Karakoram Highway to seven and a half metres and to improve its alignment for an upgradation and faster conveyance.[481] The vision laid down in this agreement was much more extensive. In future not only a sophisticated highway for vehicular traffic should connect both countries but also a glass-fibre cable for communications, a pipeline for conveying fossil energy resources, and a trans-Karakoram railway line connecting the Pakistan railhead in Havelian with the newly established Urumchi-Kashgar railway in Southern Xinjiang.[482] To commemorate this new phase of cooperation and friendship, a new fenced-in concrete arch was constructed on the top of Khunjerab Pass. The latest addition to the monuments features a gateway through which all vehicles have to pass, equipped with camera surveillance, sentry vehicles, guard posts and observation pulpits. The arc with the golden letter inscriptions of People's Republic of China (*zhonghua renmin gonghe guo*) was erected just behind the boundary stones with the state emblems of China and Pakistan close to the memorial tablet commemorating the Karakoram Highway's inauguration in 1982. Slightly to the side, the Pak-

Thakot bridge has been rebuilt for the fourth phase of the Karakoram Highway and remained an important security and traffic checkpoint when entering from Hazara District into Besham. The photograph was taken on 7 May 2015

In the aftermath of the Atabad Landslide, communication of Gojal towards the south was blocked by a two-kilometre long dam that had to be traversed to reach Atabad Lake from where boats could ply. The boats were brought from Tarbela Lake to start immediate operation after a difficult passage. Initial efforts to work on the recovery of the Karakoram Highway soon proved that a grand scheme was required to reopen the road. The photographs were taken on 30 & 31 July 2010

istan army has erected a comparatively simple column with a metal plate depicting a longish text as a 'tribute to pioneers'. It is mainly a lauding of the gallant 101 Road Construction Batallion: "The Khunjarabees nevertheless trudged along merrily hacking a track in the somber mountain fastness for the glory of the nation. ... The fact that they were in competition with the Chinese kept their hearts on fire. ... This humble beginning was the first successful step on a long odyssey of road construction by Chinese and Pakistani road builders which finally culminated in 1978 in the shape of Karakoram Highway: the eighth wonder of the world." In recent times the army contribution has dwindled.

The China Road and Bridge Corporation became the prime contractor for the extension and widening of the Karakoram Highway in the 335-km-long northern sector between Raikot Bridge and Khunjerab Pass. The section was to become the first chapter of the new Karakoram Highway and was subdivided into four operating areas where simultaneously 18 open-cut tunnels, 1,110 culverts and one million cubic metres of retaining walls were constructed, thus

After three years Atabad Lake had become a regular feature and boats were plying and ferrying people and goods between Atabad, Gulmit and Shishket. The photograph was taken on 26 August 2013

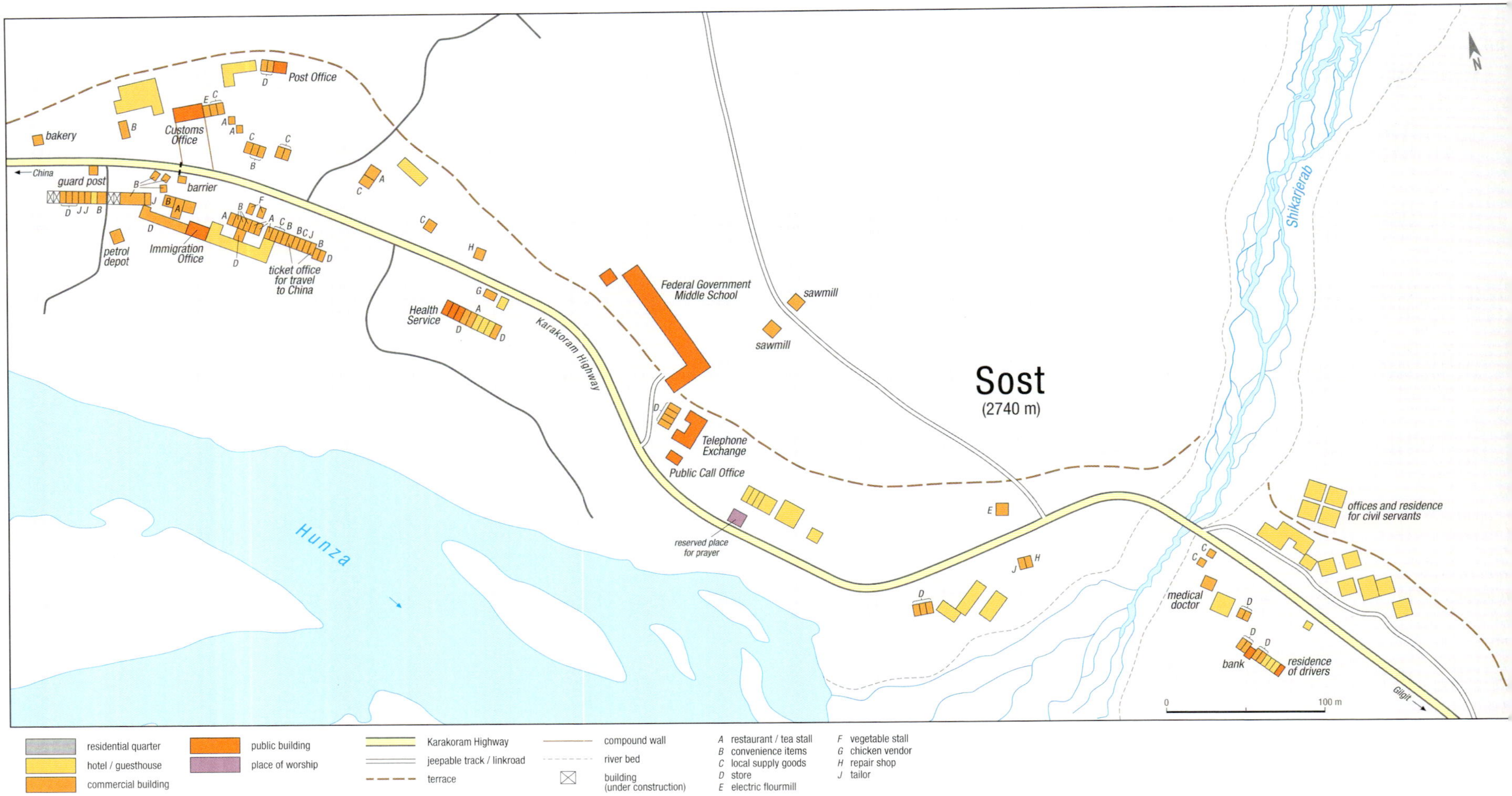
Sost
(2740 m)
Post Office
Customs Office
bakery
China
guard post
barrier
petrol depot
Immigration Office
ticket office for travel to China
Health Service
Karakoram Highway
Federal Government Middle School
sawmill
sawmill
Telephone Exchange
Public Call Office
reserved place for prayer
Hunza
Shikarjerab
offices and residence for civil servants
medical doctor
bank
residence of drivers
Gilgit
0
100 m
N
residential quarter
hotel / guesthouse
commercial building
public building
place of worship
Karakoram Highway
jeepable track / linkroad
terrace
compound wall
river bed
building (under construction)
A restaurant / tea stall
B convenience items
C local supply goods
D store
E electric flourmill
F vegetable stall
G chicken vendor
H repair shop
J tailor

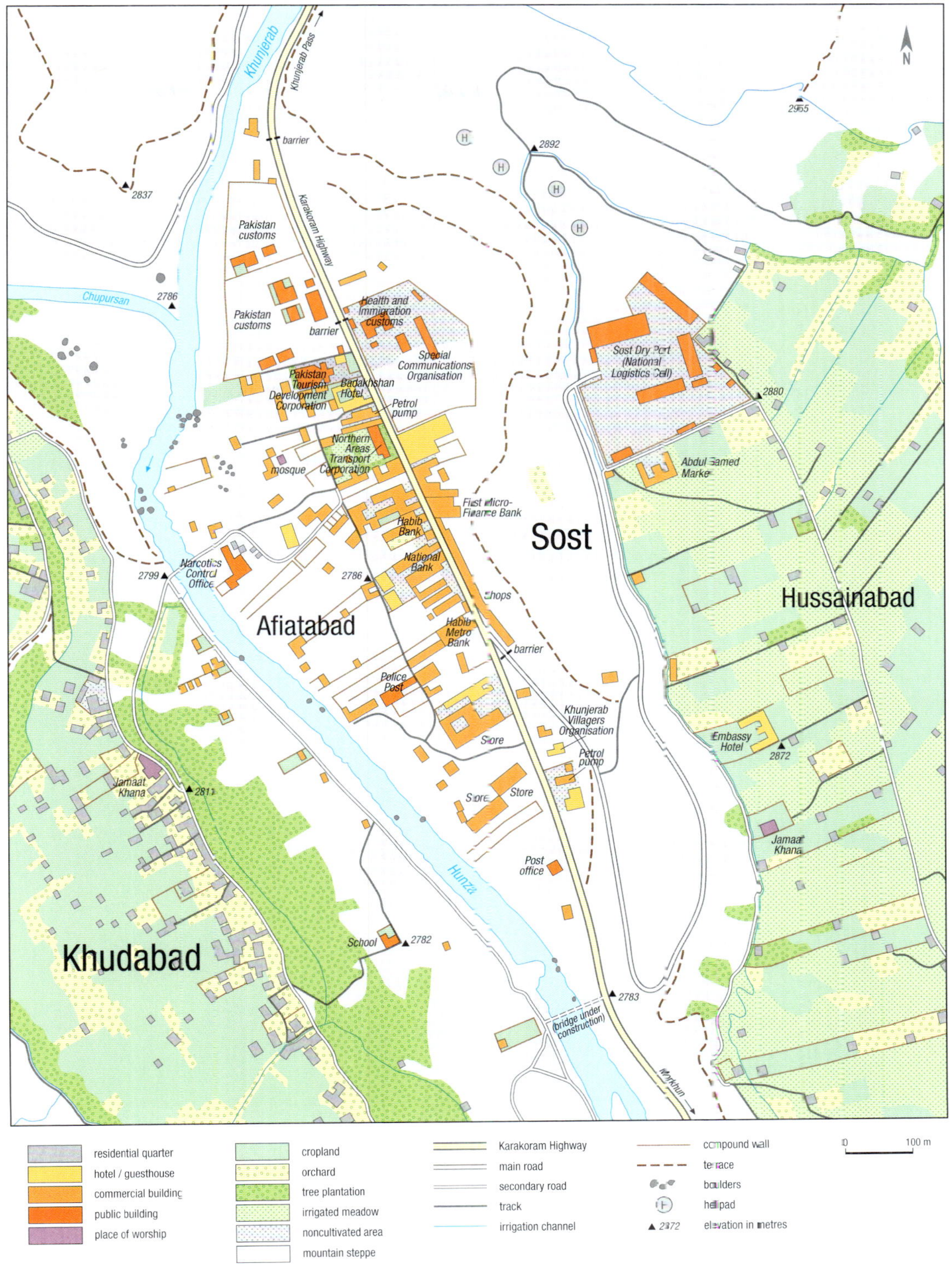

The border bazaar and customs post in Sost shifted after its initial establishment in 1986 from the eastern part towards Afiatabad. The 1990 mapping (left) shows the new agglomeration near the barrier and immigration office where forwarding agencies, traders and other businesses have formed the nucleus of the last village in Hunza before crossing the Khunjerab Pass. Only state-owned vehicles were allowed to ply between Sost and Tashkurgan; passengers from Pakistan to China used Northern Areas Transport Corporation vehicles (top) while travellers from Tashkurgan to Sost boarded Chinese buses; all vehicles returned empty. Sost Bazaar had become an established commercial centre by 2017 (bottom left). Preparations for transforming Sost into Pakistan's entry and exit station along the China-Pakistan Economic Corridor are visible in significant building projects (right). The Sost Dry Port was taken over by the National Logistics Cell in 2015. Customs and immigration facilities were expanded; stores and warehouses, forwarding companies and auxiliary services established their infrastructure in the Afiatabad steppe. The hamlet is located along the Karakoram Highway and provides ample space for the required establishments. On the right bank of the Khunjerab River Khudabad has continued to be a village with agricultural activities and residential quarters. The same applies to the hamlet of Hussainabad in Sost; both villages are located high above the valley floor and provide irrigated agricultural land on river terraces with farmsteads nearby. Source: Survey of 1990 is modified on the basis of Hermann Kreutzmann (1996a: 306); update was prepared by the author's own survey in March 2019; the photographs were taken on 25 July 1990, 5 August 1993 and 30 July 2017

With the opening of a series of tunnels above the blocked Hunza River the road connection between China and Pakistan was re-established five years after the Atabad disaster and two years after the conclusion of the Karakoram Highway Improvement Project. The photograph was taken on 26 March 2019

removing existing bottlenecks and improving thirteen sliding areas by state of the art protection.[483] The work began on 1 August 2008, and less than two years later the collapse of the Atabad slope, the massive rockfall and forming of a rock-filled dam and the subsequent creation of the Atabad Lake led to the inundation of about twenty kilometres of Karakoram Highway between Sarat and Gulmit. The slip occurred on 4 January 2010, the new barrier was overtopped by May, and the lake formed with a maximum depth of two hundred metres and a length of twenty-five kilometres.[484] All terrestrial connections between Gilgit and Kashgar were interrupted, a substantial part of the improved highway was under water, no equipment and goods could be transported overland any more. All attempts to open or break the dam had failed. It took several months until the summer before the first boats were brought from Tarbela Lake to initiate a ferry transport on the lake. Approach roads and landing places had to be laid out; the changing water volume shifted them. The Atabad disaster – as it was termed – paralysed all existing plans; it changed nearly everything in infrastructure development, in cross-boundary commodity exchange and in the supply situation of the Gojali people. The Chinese Government introduced a relief programme for Gojal to supply households with wheat flour, cooking oil and other necessities. For the first time, the Chinese Government was identified as an international donor in disaster relief and rural development.

Re-planning of the new Karakoram Highway had become a necessity, and a cost-efficient solution was needed that would enable heavy container vehicles to drive on the road in future. The engineering solution called for tunnel constructions; in the end it took only five years after the disaster to build five Pak-China Friendship tunnels between Atabad and Shishket. The new alignment is high above Lake Gojal as it was acknowledged that the dam and lake might be stable features with a longer future.[485] The road improvement was concluded by 2013, but it took another two years until five tunnels were opened for regular traffic on 14 September 2015 and lauded as a significant contribution to 'promote interconnectivity between China and Pakistan' at a cost of more than a quarter billion US dollars.[486] The Karakoram Highway Realignment Project bypassed the costly bottleneck of boat transfer via Atabad Lake and provided a much safer and cheaper passage from Central Hunza towards Shishket, from where a new bridge with a higher carrying capacity connected to Gulmit and upper Gojal. To date the Karakoram Highway occupies more than one third of all asphalted roads in Gilgit-Baltistan, thus underlining its primary importance.[487]

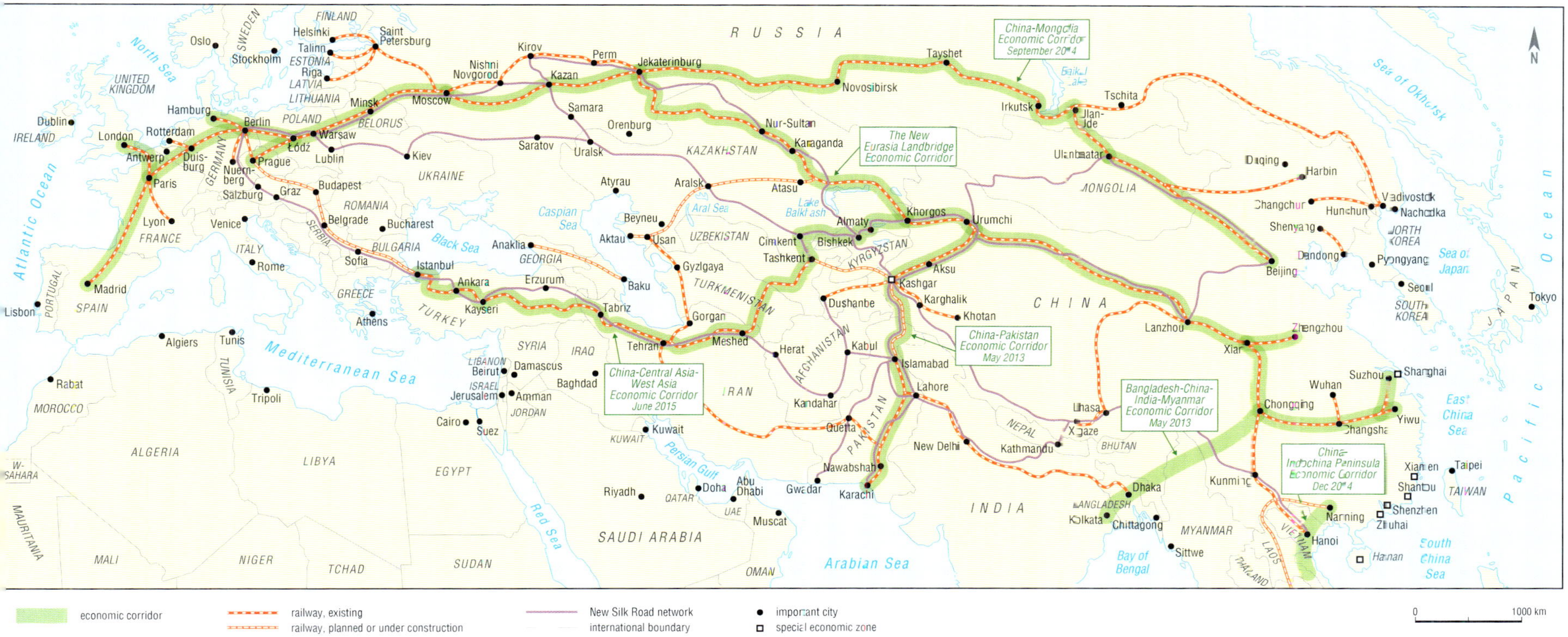

Under the banner of the 'New Silk Road' Chinese-initiated infrastructure projects labelled as 'as one terrestrial belt and one maritime road' are connected so as to open a number of thoroughfares connecting China with Europe via Central Asia. Design by Hermann Kreutzmann

A web of mountain arteries in the age of global silk roads

The reopening of the Karakoram Highway coincided with the commencement of the next step in China's Eurasian strategies. In Astana in September 2013 President Xi Jinping announced a new initiative that is now known under several names: Silk Road Economic Belt, One Belt, one Road Initiative (OBOR), Belt and Road Initiative (BRI). Since then, an acceleration and streamlining of activities in China's relations with Europe, Central and South Asia can be observed. Existing initiatives and projects have been subsumed under this umbrella, and new plans have been made for prospective and enhanced connectivity of existing infrastructure networks and/or additional roads and railways that incorporate important regions into a network of links and trans-continental arteries. For the Central Asian mountain areas and especially for Pakistan this initiative has significantly increased the importance of Kashgar. At the terminus of the Karakoram Highway, the single road at its outskirts has been renamed 'Pamier Dadao'. Kashgar has been identified as the commercial hub for Central Asian and Pakistan trade. In 2010, Kashgar became the first special economic zone of China that was not located in the coastal East. Significant constructions of offices, warehouses and factory space took place in two locations: Kashi Special Economic Zone, sponsored by Shenzhen, and Guang Zhou New City, sponsored by Guangdong province. The latter is offering a 'multinational frontier trade area' and a 'bonded zone of border trading for eight countries in international commercial street'.[488] Traders from Central Asia were invited to open shops and warehouses in Kashgar. Trade with neighbouring countries had received a new artery financed by the Islamic Development Bank and built by a Turkish consortium. It connected Kashgar with Dushanbe via Khorog when the Kulma Pass Road was opened in 2004. In the aftermath commodity export across Kulma Pass (4,363 m) and cleared at China's highest custom station in Karasu increased tenfold within one decade and reached nearly two billion US $ in 2014.[489] With just seventy kilometres length the Kulma Pass Road connected two existing international trade routes. The Pamir Highway connected with the Karakoram Highway and

The gateway to China marks the highest point of the Karakoram Highway on Khunjerab Pass (4,655 m). As part of the China-Pakistan Economic Corridor a special CPEC patrol was set up that belongs to Gilgit-Baltistan police and is using vehicles donated by China Aid. The photograph was taken on 10 July 2017

opened a new network linking Central, South and East Asia. Several executives, traders and organisations from Pakistan were invited to participate here and were offered space in Kashgar either free or at nominal rates.

All these developments that took place after 2010 have led to another reconstruction of the third phase of the Karakoram Highway that had been completed on the Chinese side by 2008. Now the Karakoram Highway 4.0 is under construction and documents in its appearance and shape another step forward. The recent rebuilding has changed the previous alignments in favour of a road built on concrete, with steel-reinforced stilts and open-cut tunnels. After previously making a dirt road between Tashkurgan and Kashgar in 1958, asphalting the route by 1988, and realigning and broadening the highway by 2008, the latest outlay is a remarkable and very costly investment in a cross-boundary artery that is now labelled the 'China-Pakistan Economic Corridor'.

For Gilgit-Baltistan – which received this new appellation in September 2009 when 'an order to provide greater political empowerment and better governance to the people of Gilgit-Baltistan' was issued by the central government – nothing changed with respect to the constitutional limbo.[490] Gilgit-Baltistan continues to be disenfranchised in national elections. Although certain fiscal responsibilities were transferred from the centre, the Ministry of Kashmir Affairs and Gilgit-Baltistan remained the steering committee for Gilgit-Baltistan. Some hope arose when the thirty-first amendment of the constitutions led to the merger of the Federally-Administered Tribal Areas with Khyber-Pakhtunkhwa province in May 2018. In contrast to this abolishment of a colonial legacy in the former North-West Frontier Province, the new government order abolished the Gilgit-Baltistan Council, which had functioned as the second parliamentary chamber, and renamed the Legislative Assembly into just Gilgit-Baltistan Assembly. Full veto rights that can supersede all decision-making and legislation by the Gilgit-Baltistan Assembly were transferred to the Prime Minister of Pakistan.[491] Besides shifting authority from the president of Pakistan to the prime minister and cutting the legislative powers of the assembly, a greater shift occurred in other sectors. The prevention of qualified judges of Gilgit-Baltistan from attaining the highest judiciary position was confirmed. Furthermore, the citizenship law was amended. Ibn Abdur Rahman, the well-known human rights activist, highlighted the change: "All Pakistanis recognised by the Pakistan Citizenship Act of 1951 as well as residents and those holding GB domicile will be citizens of the area (the only word used to describe GB, as the word 'province' is never used for it). This definition of citizenship will give Pakistanis from outside GB significant advantages over the natives."[492] The amendment will have significant effect in property acquisitions along the China-Pakistan Economic Corridor. Previously, outsiders were not permitted to purchase property in the Hunza Valley; on several occasions earlier attempts were stopped. With this move outside investments in property, mining and resource exploitation are feasible.

The latest move is again a reflection of a political limbo and the inertia of status quo situations. During the Cold War, constitutional status was denied to the Northern Areas as the Kashmir conflict parties and their international backers failed to implement resolutions passed by the United Nations in 1949 already. The limbo has remained as India and Pakistan never came to terms to solve the Kashmir conflict by negotiating a solution and signing a peace treaty. For representatives of Gilgit-Baltistan's civil society, at least two major stakeholders are less interested in a change: the army and the Azad Kashmir lobbyists. Both could lose their advantages and leverage for maintaining their positions in Pakistan politics.[493] The present situation has remained the same; only the Chinese pressure to increase stability in the China-Pakistan Economic Corridor has resulted in a domestic rollback. Certain powers that had been transferred to Gilgit-Baltistan in 2009 were taken back nine years later. The rationale for this lies in security considerations. The central government has become aware of the bottleneck situation provided by the Karakoram Highway as the centrepiece of the bilateral economic corridor. No alternative routing is possible, and the present highway leads through disputed territory that is only referred to in the governance order as 'area'. A general move to tighten control and security measures included not only the familiar disenfranchisement in national elections and the non-representation in federal institutions but also China's equipping the Gilgit-Baltistan police with a new 'CPEC patrol', the Gilgit-Baltistan authorities not being considered in a position to guarantee public security in the area. The police vehicles were supplied and sponsored by 'China Aid', a development organisation that had previously helped to supply remote communities with food items after the Atabad landslide had cut them off from the roads to Gilgit. Consequently, it is not surprising that non-sympathising observers claim a growing expropriation of local residents in terms of decision-making, economic participation and benefiting from the China-Pakistan Economic Corridor. The latest infrastructural asset might be the most influential and consequential change the Hunza Valley has experienced so far.

In economic terms the one-sided exchange patterns have continued. After operations began at the mutual joint venture Sost Dry Port in 2006, the amount of goods shipped from China to Pakistan got a short-lived boost and reached a worth of about six billion Pakistan rupees in the following year, but goods from Pakistan exported to China via Khunjerab Pass reached only a value of two hundred million rupees.[494] The latter figures have not changed much in recent years and doubled by 2018. Chinese exports slumped after their first peak in 2007, were severely affected by the Atabad Lake disaster, and were worth only one third in 2011. After the opening of the Pak-China Friendship tunnels the former exchange pattern emerged again, with Chinese exports to Pakistan varying between six and eight billion rupees by 2018; in terms of truckloads 3,818 containers reached Sost from China that year.[495] The value of Chinese goods compared with Pakistan's exports to Kashgar fluctuates between factors of fifteen and thirty times higher than the imports. In comparison with the overall trade-based exchange between Pakistan and Chi-

Customs clearance and immigration take place at the outskirts of Tashkurgan for incoming and outgoing passengers arriving from or bound for Pakistan. At an altitude of 3,120 metres the office has been named Khunjerab Port. The photograph was taken on 6 September 2016

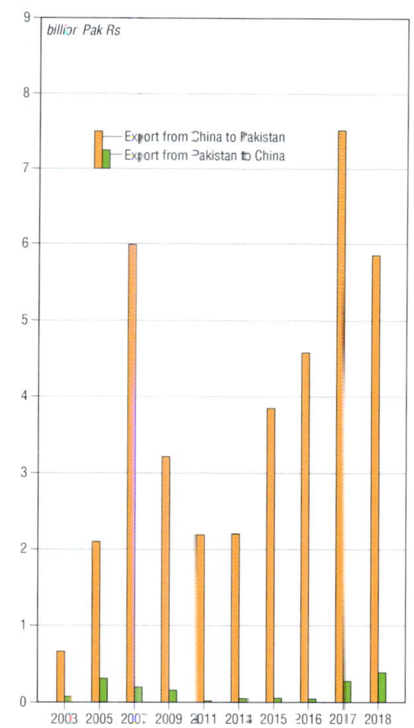

The commodity exchange between China and Pakistan via Khunjerab Pass is a one-sided affair. Substantial Chinese exports reach Pakistan while insignificant amounts of goods pass the customs posts in the opposite direction. Source: Data derived from Pamir Times 3 July 2019. Design by Hermann Kreutzmann

na, the land-based trade in commodities ranges at the level of about a third of a percentage point and is quite insignificant compared to sea trade.[496] The China-Pakistan Economic Corridor still has a long way to go to live up to its title. Presently the political importance and symbolic value dominate the general discussion while its impact on Gilgit-Baltistan and the Hunza Valley in particular has been tremendous.

The monument commemorating the joint Pak-China efforts in constructing and improving the Karakoram Highway is a landmark sometimes also used for political demonstrations such as election campaigns since the completion of the third phase of its reconstruction was inaugurated in 2013 after five years. Now a better road connection between the Pak-China border and Raikot Bridge exists and has reduced travelling time. The China Road & Bridge Corporation functions as the general contractor in partnership with Pakistan's National Highway Authority. The photograph was taken on 24 May 2015

Overshadowing existing settlements a new alignment of the highway is being constructed on stilts between Battagram and Thakot. In future congested areas and traffic bottlenecks will be by-passed. The photograph was taken on 12 March 2019

ENDNOTES OF PAGES 58–208

1 'Khunjerab Pass opens to civilian traffic. Another milestone in Pakistan-China cooperation' in Dawn Overseas Weekly 6 (36), 3–9 September 1982; see as well Dawn, 25 September 2000; The Muslim, 19 May 1996; Pervez Musharraf (2006: 56).
2 People's Daily 18 June 1978.
3 Quoted after N. N. (1978: 42); Sally Wriggins (1980: 25).
4 Nazir Kamal (1979: 20).
5 Ahmad Hasan Dani (1983: 9; 1989); see Dawn, 29 July 1968 and 10 October 1970. The Wakhi version of the term 'silk road' is derived from Persian *xāh-rāh-e abrīšam*.
6 James Scott (2009) has shown escape strategies in his book on 'the art of not being governed' for Burma/Myanmar.
7 Gonzalo Aguirre Beltran (1979).
8 Title of the book of Mahnaz Ispahani (1989) that discusses the 'politics of access' in Central Asia.
9 See Nigel Allan (1986); David Butz and Nancy Cook (2011); Hermann Kreutzmann (2004c); Rinki Sarkar (2010); Thomas Wilbanks (1972).
10 A statement first made by Robert Chambers in 1980 and repeated several times (1983: 13; 2006: 17).
11 Jason Neelis (2012: 13–14); see as well Nayanjot Lahiri (1992: 401).
12 The journeys and routes of famous Chinese pilgrims across the Pamirs were presented in Hermann Kreutzmann (2015a: 22); see as well Samuel Beal (1884) and for more recent records Michael Gillett (1965); Mishi Saran (2005).
13 Jason Neelis (2012: 16); see Ditte Bandini-König (2003); Ahmed Hasan Dani (1985); Gérard Fussman (1978); Heidelberger Akademie der Wissenschaften (1994); Oskar von Hinüber (1989); Karl Jettmar and Volker Thewalt (1985); Harald Hauptmann (2007); Jason Neelis (2011).
14 Jason Neelis (2012: 18).
15 Jason Neelis (2012: 17, map).
16 Sven Hedin and Albert Herrmann (1922: 3); for the context of *tsungling* see Hermann Kreutzmann (2015a: 69).
17 See the Qianlong maps prepared by Jesuit cartographers and their Chinese counterparts in Hermann Kreutzmann (2015a: 47; 2017a: 24–27).
18 See the map 'Turan oder Türkistan', published in 'Atlas von Asien 3. Lieferung 1852'. To illustrate C. Ritter's Erdkunde, Buch III. West-Asien, Theil VII, designed by Heinrich Kiepert. The relevant section of the map is reproduced in Hermann Kreutzmann (2015a: 26) West-Asien, Theil VII, designed by Heinrich Kiepert.
19 See for Adolph Schlagintweit and Chokan Valikhanov's travels Hermann Kreutzmann (2015a: 71–77); for the fate encountered by the former in Kashgar see Hermann Kreutzmann (2015c).
20 James Thomas Walker (1873: 137).
21 Patrick Alexander Vans Agnew (1915: 300).
22 Patrick Alexander Vans Agnew (1915: 300–302). Lieutenant Ralph Young prepared a rather simple sketch map that is based on local explanations in its sections of Hunza and Nagar above Budelas. A section of the map is reproduced in Wolfgang Heichel (2010: 29).
23 Patrick Alexander Vans Agnew in his diary entry of 10 September 1847.
24 John Biddulph (1876) described his experiences in his personal diaries titled 'Trip to the Chitral Passes' 25 April to 16 November 1876. The chapter on Hunza is comparatively brief (pages 104–118) as his main objective was to explore Chitral and the north-western passes and passages. For his career and performance see Hermann Kreutzmann (2017a: 75–81).
25 A second source for John Biddulph's visit in Hunza is the autobiography of *tham* M. Nazim Khan that was recorded by his grandson Jamal Khan and translated from Persian into English by Khan Bahadur Muhammad Maseh Pal. Mimeographed copies were occasionally distributed locally. A printed version was published in the personal records of David Lorimer (1979: 312–376); the relevant episode is mentioned on page 313. A more recent, poorly edited version was released by Zahoor Hussain Khan (M. Nazim Khan 2014).
26 John Biddulph (1876: 104).
27 John Biddulph (1876: 109).
28 John Biddulph (1876: 114). Beehoo could be Beko who functioned as a representative of the Hunza *tham* on other occasions as well; see Mohomed Nazim Khan (1936: 2–3).
29 Gottlieb Wilhelm Leitner (1889: 73).
30 John Biddulph (1878: 240; 1881: 28–29); Frederic Drew (1875: 447, 450, 457). The sources differ sometimes about the dates when Chalt and Chaprot changed hands and when tribute relationships with Kashmir commenced (Stuart Colvin Bayley 1892: 3; John Biddulph 1880: 23; Arthur Henry McMahon 1898: 4; George Ogilvie 1929: 6–7; Reginald Schomberg 1935a: 151–152). Decades later, when the lease of the Gilgit Agency was at stake, Reginald Schomberg (1935a: 97) still regretted that Hunza had lost its direct access to Gilgit because of Nagar's dominance in Chalt and Chaprot.
31 John Biddulph (1880).
32 On the 'Gilgit Mission' see William Stephen Alexander Lockhart, and Robert Gosset Woodthorpe (1889); see as well the comments of *mīr* M. Nazim Khan (1936: 12–14) and an account of the mission's findings in Hermann Kreutzmann (2015a: 96–99; 2017a: 86–93).
33 Edmund Barrow (1888); part II is devoted to Dardistan and contains his recorded information of the Hunza Valley. An earlier version contained a separate volume on Dardistan (1887b).
34 Edmund Barrow (1885: 2). The surveyors were trained at the civil engineering college founded by James Thomason in Rorkee in 1847
35 Edmund Barrow (1885: 1).
36 Edmund Barrow (1889: 6).
37 Bakshi Mulraj was described by William Lockhart and Robert Woodthorpe (1889: 393) being 'the first Hindú who has yet vis-

ited Hunza; the Kashsmír officials sent here are always Mussulmáns'.

38 Edmund Barrow's route description is taken from his personal diary titled 'Chitral-Hunza Diary 1886'. The following quotes about the road condition are taken from the first 26 pages of his diary that began at New Year of 1886. See as well for the Hunza part of the 'Gilgit Mission' William Lockhart and Robert Woodthorpe (1889: 383–402).

39 Edmund Barrow (1886: 9).

40 Edmund Barrow (1886: 8). See for a description of 1892 and reflection Oskar Eckenstein (1896: 80–82).

41 William Lockhart and Robert Woodthorpe (1889: 382); see as well Robert Woodthorpe's (1886: 281–282) 'note on the bridges at Gilgit' with a detailed descriptions of the configuration of a twig bridge.

42 Edmund Barrow (1886: 8).

43 Edmund Barrow (1886: 9).

44 Edmund Barrow (1886: 10).

45 Edmund Barrow (1886: 11–12). The obstacle encountered remained such for a long time afterwards; Henry Haughton described the Chechar Pari between Gwachi and Chalt during his traverse on 13 April 1912 in a similar manner (IOL/Mss Eur/D531/5).

46 Edmund Barrow (1886: 14).

47 Edmund Barrow (1886: 14).

48 William Lockhart and Robert Woodthorpe (1889: 387).

49 Edmund Barrow (1886: 15–16).

50 Letter of William Lockhart to Mortimer Durand, sent from Hunza, dated 24 April 1886 (William Lockhart and Robert Woodthorpe 1889: 389–390).

51 Edmund Barrow (1886: 16).

52 Colonel William Lockhart wrote letters to Mortimer Durand on a daily basis while in Hunza; he approached the Maharaja of Kashmir, Pratap Singh, in the Chaprot case as well and played for time and indecisiveness (William Lockhart and Robert Woodthorpe 1889: 389–394).

53 Edmund Barrow (1886: 21–22).

54 Edmund Barrow (1886: 22).

55 Edmund Barrow (1886: 22).

56 Edmund Barrow (1886: 25).

57 William Lockhart and Robert Woodthorpe (1889: 401–405).

58 Edmund Barrow (1886: 17).

59 Edmund Barrow (1885: 24).

60 Francis Younghusband (1926: 231).

61 Edmund Barrow (1889: 5).

62 See Hermann Kreutzmann (2017a: 92).

63 Edmund Barrow (1885: 110).

64 See John Biddulph (1880; 1881); Munshi Abdul Rahim (1885); Hermann Kreutzmann (2017a: 89–97).

65 N. N. (1922). John Biddulph and Reginald Schomberg devoted their post-intelligence service life to ornithology and botany (IOL/Mss Eur/C371 and F/157/249).

66 Secret letter from India, No. 103, dated 15th July 1881 (Henry Wollaston 1881: 18).

67 William Lockhart and Robert Woodthorpe (1889: 110–111).

68 John Biddulph (1878: 242–243); see William Lockhart and Robert Woodthorpe (1889: 83–83); therefore, the mission's report seems to be partly inconsistent.

69 Francis Younghusband (1895a: 4).

70 Francis Younghusband (1895a: 5).

71 Francis Younghusband (1895a: 5–9).

72 Henry Lansdell (1893, II: 18); George Nathaniel Curzon (1896: 26).

73 Bronislaw Grombchevsky (1888/2015: 107–108).

74 Henry Hugh Deasy (1901a: 109).

75 Edmund Barrow (1889: 5).

76 George Kynaston Cockerill (1895; 1896) was sent to the Hunza Valley in 1892 to correct Edmund Barrow's data and published two appendices to his gazetteer.

77 Details of his order are given in the file 'Reestablishment of a British Agency at Gilgit, 1889' (IOL/P&S/7/57/19-20).

78 Quote from Algernon Durand's memorandum reproduced by Stuart Colvin Bayley (1892: 8) in his 'Note on the complication with Hunza' on 25 January 1892, shortly after the British victory in the Hunza-Nagar Campaign (IOL/P&S/18/A 83).

79 Algernon Durand (1894: 6–7).

80 In his report on the 'present position in Gilgit' (IOL/P&S/7/57); Algernon Durand (1888: 45).

81 Bronislaw Grombchevsky (1888/2015: 108–110, 111, 121–122).

82 Bronislaw Grombchevsky (1888/2015: 113–115).

83 Stuart Colvin Bayley (1892: 4)

84 Stuart Colvin Bayley (1892: 11).

85 Stuart Colvin Bayley (1892: 11); see as well IOR/2/1079/245.

86 IOL/P&S/18/A 83: 8.

87 John Biddulph (1876: 116).

88 William Lockhart and Robert Woodthorpe (1889: 275).

89 'Memorandum on the question of acquiring Gilgit, and erecting there a British cantonment' Gilgit, 25 February 1886 by Colonel William Lockhart (William Lockhart and Robert Woodthorpe 1889: 276–280).

90 Mikhail Bashkanov (2018: 269).

91 This enterprise made him a pioneering hero who was recently named 'Dervish of the Hindukush'; Mikhail Kazbekovich Bashkanov, Alexander Kolesnikov and Maria Matveeva (2015) edited and published Bronisław Grąbczewski's expedition diaries under this title. On 12 April 1888 he was elected a member of the prestigious Imperial Geographical Society (Mikhail Bashkanov 2018: 268). Mikhail Bashkanov (2018: 267) estimated his budget as sufficient in comparison with other contemporary travellers for five people travelling for a period of four and a half months.

92 Francis Younghusband (1892: 222).

93 Bronisław Ludwig Grąbczewski (1891); William Barnes Steveni (1892); Francis Younghusband (1892; 1895a; 1896: 268); see Grąbczewski's map and his itinerary to Hunza reproduced in Hermann Kreutzmann (2015a: 91–93).

94 Bronislaw Grombchevsky (1888/2015: 108); see as well Mikhail Bashkanov (2018: 275). The diaries of Bronisław Ludwig Grąb-

czewski or Bronislav Liudvigovich Grombchevsky (1888/2015) were recently published in St. Petersburg under the title 'Dervish of the Hindukush'. The following references are related to this latest edition of 2015.
95 Bronislaw Grombchevsky (1888/2015: 106–108).
96 Bronislaw Grombchevsky (1888/2015: 107). Language experts have rejected this alleged connection as unfounded.
97 Bronislaw Grombchevsky (1888/2015: 108).
98 Bronislaw Grombchevsky (1888/2015: 118, 133). Eventually he received the horseshoes in Baltit, the charge of one silver yambu plus fourteen rupees amounts to more than one rouble per horseshoe and is estimated as very costly. The telegraph line did not function as rumoured; only in 1892 the direct line between Srinagar and Gilgit via Dras became functional (Siegfried Weber 2007: 90).
99 Bronislaw Grombchevsky (1888/2015: 121–122).
100 Letter from Bulchi Das, dated 29 August 1888 (Bronislaw Grombchevsky 1888/2015: 124–125).
101 The comments on the visit of Bronislaw Grombchevsky are recorded in Mohomed Nazim Khan (1936: 26–31) who got the dates wrong as he postponed the visit to 1890.
102 Mohomed Nazim Khan (1936: 28).
103 Mohomed Nazim Khan (1936: 29–30).
104 Bronislaw Grombchevsky (1888/2015: 134–135).
105 Bronislaw Grombchevsky (1888/2015: 148–149).
106 Mikhail Bashkanov (2018: 276) interprets the character of the mission a bit differently as he understands Bronislaw Gombchevsky's part as an official representative of the Russian Tsar as a mere survival strategy.
107 Qudratullah Beg (1962: 510).
108 Francis Edward Younghusband (1890: 59). Bronislaw Grombchevsky received for his Hunza expedition a special mention by the Vice-President of the Imperial Geographical Society, Pjotr Semenov, who "… put together a brief summary of Grombchevskii's work during the expedition in Hunza and noted that the officer 'carried out the entrusted mission brilliantly', having himself been 'convinced from conversations with the traveller'. The Council decided – 'whilst waiting for the results of the materials obtained by Grombchevskii, to award him for the successful completion of his mission a small gold medal'" (quoted from the archival materials by Mikhail Bashkanov 2018: 277).
109 Francis Edward Younghusband (1890: 64).
110 The quotes from Algernon Durand's 'Gilgit Agency Report for 1889' are taken from Stuart Colvin Bayley (1892: 5–6); see Francis Edward Younghusband's (1890) original report with detailed descriptions.
111 Francis Edward Younghusband (1890: 49).
112 Stuart Colvin Bayley (1892: 6).
113 Hugh Lewis Nevill (1912: 137); see as well Algernon Durand (1899: 254–255); Wilfred Malleson (1907: 12–28).
114 Hugh Lewis Nevill (1912: 137); see as well Algernon Durand (1899: 254–255); Wilfred Malleson (1907: 12–28).
115 Francis Younghusband (1895a: 7). The Hunza levies were trained by George Cockerill (1922: 99) who was Director of Military Intelligence, a propaganda branch in the Military Intelligence Department of the War Office in 1916.
116 Francis Younghusband (1926: 230).
117 Francis Younghusband (1926: 228).
118 William Barnes Steveni (1891); the map was initially 'published in the 'Izvestia', VoL XXV. (1889) of the Russian Imperial Geographical Society' as stated by James Thomas Walker (1892: 780). Robert Woodthorpe designed a fairly accurate map of the area visited by the Gilgit Mission, reproduced in Hermann Kreutzmann (2015a: 97).
119 James Thomas Walker (1892: 780).
120 St. George Littledale (1892: 34).
121 Richard William George Hingston (1913: 4) Hermann Kreutzmann (2015a: 54; 2017a: 51); Kenneth Mason (1955: 143).
122 Kenneth Mason (1955: 144–146).
123 Brief entries in the Gilgit Agency Political Diaries could be found by the Political Agent, Archibald Duncan Macpherson, for the months of May and September 1913 without referring to any details (IOL/P&S/10/826: 198, 206).
124 Richard William George Hingston (1913: 78–79).
125 Richard William George Hingston (1913: 165).
126 Mikhail Kazbekovich Bashkanov, Alexander Kolesnikov and Maria Matveeva (2015) published his diaries. His photographs are kept at the Scientific Archive of the Russian Geographical Society in St. Petersburg; a selection was published by Mikhail Kazbekovich Bashkanov, Alexander Kolesnikov, Maria Matveeva and Alexander Ivanovich Glukhov (2017).
127 Francis Younghusband (1926: 229–230).
128 Algernon Durand (1894: 7).
129 Algernon Durand (1899: 254–270); Wilfred Malleson (1907: 12–28); Hugh Lewis Nevill (1912: 136).
130 Edward Knight (1893: 69).
131 Edward Knight (1893: 184).
132 Edward Knight (1893: 282).
133 Edward Knight (1893: 250).
134 Edward Knight (1893: 356).
135 Algernon Durand (1888: 49).
136 Francis Younghusband (1895a: 9–10). A description and a photograph of the Khaiber *darband* are reproduced in John Nicholas Price Wood (1910: 27; photograph following page 22) and Philips Christiaan Visser (1928: 112; photograph following page 88). The Gwachi gate (*darband*) was still visible on the Hunza Road, controlling the entry to Nagar, when the members of the *Croisière jaune* passed by in August 1931 (Georges Le Fèvre 1935: 205). Henry Haughton mentioned a *darband* at Rishipjerab between Khudabad and Reminj in the Chupursan Valley (IOL/Mss Eur/D531/5). See for the mention of *darband* Frederic Drew (1875: 410); General Staff India (1942: 150); Henry Haughton (1913: 233); Gottlieb Wilhelm Leitner (1894, vi: 8); Reginald Schomberg (1936a: 77, 88, 213–216); Francis Younghusband (1896: 259).
137 Factually, his son *mīr* Sikandar Khan ruled Nagar. see Jürgen Frembgen (1985b: 213–214).

138 Captain George Howard Bretherton (1898: 874) participated in the 'Siege of Chitral' and was impressed by the friendliness expressed towards British officers in the Gilgit Agency.
139 George Howard Bretherton (1898: 875).
140 George Howard Bretherton (1898: 876).
141 Stuart Colvin Bayley (1895: 2). This statement was often repeated since; see Algernon Durand (1899: 139); Reginald Schomberg (1935a: 158) who mentions that Nagar has "... played an insignificant rôle in history; even the States of Punyal and Yasin have been more in evidence. ... Nagir is such a weak partner [for the British] that it has rather been ignored."
142 William Martin Conway (1894a: 144).
143 William Martin Conway (1894a: 144).
144 William Martin Conway (1894a: 249); Arthur McCormick (1895: 119–120); Oskar Eckenstein (1896: 137–138). This apprehension has not ceased later; John Staley (1982: 30) reports that it befell a ruler of Chitral and a president of the Alpine Club as well.
145 See photographs by Resident Engineer Chetwynd Henry Haswell of 1921 (IOR/2/1083/280: 59–60), Filippo de Filippi (1912: 115–116, 325), Robert Kappeler (1950: 75).
146 Arthur McCormick (1895: 126–127); see Laurence Duckworth (1894) for the subsequent 'scientific' handling of the skulls; see for similar techniques of acquisition Moritz von Brescius (2018: 148) providing the example of the skull collection of the Schlagintweit brothers in India in mid-19th century.
147 Quotation from Captain Ernest James Medley, commanding the troops at Gilgit (IOR/2/1064/45: India Office Records: Crown Representative's Records – Indian State Residencies: Kashmir Residency Files: Indo-Chinese Turkistan trade. Report by S. H. Godfrey, Political Agent Gilgit, to A. C. Talbot, Resident Kashmir, Gilgit 17. 12. 1896).
148 Major Ernest James Medley (1901: 281).
149 Major Ernest James Medley (1901: 286–287).
150 Daniel Headrick (1981).
151 Marc Aurel Stein (1903: 5).
152 Percy Thomas Etherton (1911: 21).
153 Administration report of the Gilgit Agency for the year 1935, 1936 (IOL/P&S/12/3288).
154 Hermann Kreutzmann (1998a: 291); Erich Thiel (1934: 201–203). Presently, vehicular transport bypasses Terek Dawan and prefers Taldyk Pass (3,541 m).
155 Pavel Stepanovich Nazaroff (1935: 18); Clarmont Percival Skrine (1925: 235–236); Ella Constance Sykes (1916: 16–18); Ella Constance Sykes and Percy Molesworth Sykes (1920: 21–38).
156 Clarmont Percival Skrine (1925).
157 Reginald Schomberg. Note on Chinese Turkistan. 1927. File No. 583-X (NAI/PR/4001671: 11).
158 NAI/PR/4001671: 11.
159 Note by Major Arthur Edward Broadhent Parsons, dated 10 September 1925 (Government of India. Foreign and Political Department. 1928. India's concern with Sinkiang. NAI/PR/4001675: 1).
160 NAI/PR/4001675: 1.
161 NAI/PR/4001675: 1.
162 Note by Major Arthur Edward Broadhent Parsons, dated 14 November 1928 (Government of India. Foreign and Political Department. 1928. India's concern with Sinkiang. NAI/PR/4001675: 6).
163 Note by Brigadier Sydney Frederick Muspratt, dated 18 March 1929 (Government of India. Foreign and Political Department. 1928. India's concern with Sinkiang. NAI/PR/4001675: 12).
164 India Office Library & Records: Crown Representative's Records – Miscellaneous Files (Confidential): Development of trade between India and Chinese Turkestan. File No. 22-C of 1938 (IOR/2/1076/232: 19).
165 IOR/2/1076/232: 19-20.
166 IOR/2/1076/232: 20.
167 See map of the trade route in Hermann Kreutzmann (2015a: 215), descriptions of challenges and obstacles are provided by Edmund Barrow (1890: 97–109); Henri Dauvergne (1892); Eliezer Joldan (1985); Arthur Neve (1979); Clarmont Percival Skrine (1925); Philips Visser (1935: 158).
168 See report on Vice-Consul Gillett's tour to Keria via Yarkand, Karghalik and Khotan. 1st February to 5th April 1937 (IOL/P&S/12/2336); Clarmont Percival Skrine (1925: 229–230).
169 Hermann Kreutzmann (1998a: 298).
170 Indo-Chinese Turkistan trade. Report by S. H. Godfrey, Political Agent Gilgit, to A. C. Talbot, Resident Kashmir, Gilgit 17. 12. 1896 (IOR/2/1064/45: 4).
171 Crown Representative's Records – Indian States Residencies – Gilgit, Chilas, Hunza and Nagir Files (Confidential): Hunza and Nager Subsidies (IOR/2/1079/251: 8).
172 Crown Representative's Records – Indian States Residencies – Gilgit, Chilas, Hunza and Nagir Files (Confidential): Increase of subsidies of certain Chiefs and Governors of the Gilgit Agency 1926 (IOR/2/1083/284: 59).
173 Stuart Hill Godfrey (1898: 43).
174 General Staff India (1928): Military Report and Gazetteer of the Gilgit Agency and the independent territories of Tangir and Darel (IOL/MIL/17/13/73: 168); see Frederic Drew (1875: 433).
175 Indo-Chinese Turkistan trade. Report by S. H. Godfrey, Political Agent Gilgit, to A. C. Talbot, Resident Kashmir, Gilgit 17. 12. 1896 (IOR/2/1064/45: 1–2).
176 Edmund Barrow (1890: 327).
177 IOL/MIL/17/13/73: 31–32.
178 Reginald Schomberg (1935a: 18).
179 John Staley (1966: 249).
180 Letter from the Political Agent Gilgit to the Resident Kashmir, dated 30 April 1928 (NAI/PR/4001675: 16–17).
181 Marc Aurel Stein (1903: 38).
182 William Martin Conway (1894a: 249); George Nathaniel Curzon (1926: 190); Oskar Eckenstein (1896: 137–138); Percy Thomas Etherton (1911: 39); Filippo de Filippi (1912: 115–116,

325); Jack Gannon (1932: 82); Bronislaw Grombchevsky (1888/2015: 77–185); Henry Haughton (1913: 210); Karl Jettmar (1983: 515); Gottlieb Wilhelm Leitner (1892b: 246); Francis William Long (1948: 163); Arthur McCormick (1895: 112, 119-120); Jean Shor (1955: 275-276); Marc Aurel Stein (1903: 38); Eric Teichman (1937: 173); William Lockhart and Robert Woodthorpe (1889: 14–16, 19–20, 281–282, 376, 382–389).

183 Photographs of difficult passages are reproduced in Jean Shor and Frank Shor (1953: 490, 496); Clarmont Percival Skrine (1926, following pages 16 and 30); Marc Aurel Stein (1903: 41); Maynard Owen Williams (1932: 331, 338, 343). John Clark (1956: 34) spells them as *pahari*, Gottlieb Wilhelm Leitner (1894: vi, 8) as *refáq*.

184 Mukhtar Shah (1883: 83); see Hermann Kreutzmann (2017a: 83-86).

185 Marc Aurel Stein (1903: 41).

186 Marc Aurel Stein (1903: 49–50).

187 Marc Aurel Stein (1903: 48).

188 Marc Aurel Stein (1928, II: 856, 865). Most travellers were quite impressed by the vertiginous constructions of narrow paths; most travelogues contain references to them; see e.g. William Martin Conway (1894a: 251, 306, 446, 465); Arthur McCormick (1895: 112) who sketched them; Kumara Menon (1947: 24–45); Clarmont Percival Skrine (1926: 16, photograph opposite); Wolfgang Pillewizer (1960: 95, photograph following page 160).

189 John Biddulph (1876: 97); Reginald Schomberg (1936a: 14, 163). Sometimes juniper was replaced by willow or birch twigs.

190 William Lockhart and Robert Woodthorpe (1889: 282).

191 William Lockhart and Robert Woodthorpe (1889: 304); John Staley (1966: 196). Skin rafts (*jáalo, jhula*) were more common in Baltistan, especially at the crossing from Khapalu to the other bank of the Shyok, the usual approach towards the Hushe Valley. In Ghizer an important *jhula* was operated at the junction of the Ishkoman and Ghizer rivers opposite Gakuch until recently.

192 Gilgit Agency Political Diary for the month of January 1915 (IOL/P&S/10/826: 165).

193 IOR/2/1083/280: 13; the 'Military report and gazetteer of the Gilgit Agency and the independent territories of Tangir and Darel' gave 532 feet as the length of the Gilgit bridge (Government of India 1909: 5).

194 Gilgit Agency Political Diary for the month of May 1916 (IOL/P&S/10/826: 116). The locations of the bridges are marked on the map 'Afghānistān, Gilgit Agency and Kashmīr & Jammu. Baltit (Hunza), No. 42 L (1930); see for an early reference on these bridges Marc Aurel Stein (1903: 28, 33–34). The 'Military report and gazetteer of the Gilgit Agency and the independent territories of Tangir and Darel' of 1909 described the earlier versions as a 335-feet-long bridge between Khizrabad and Sikandarabad, and the Tashot bridge as 300 feet (Government of India 1909: 5).

195 Gilgit Agency Political Diary September 1937 (IOL/P&S/12/3288: 7).

196 Reginald Schomberg (1945: 9–10) in his unpublished 'Karakoram finale' (RSAA/SC/SCH/2/47).

197 See the detailed description in 'General Staff India 1942: Routes in Chitral, Gilgit and Kohistan' (IOL/MIL/17/13/58: 143–149). Wolfgang Pillewizer (1960: 59) regarded the Sikandarabad suspension bridge as the only stable bridge of the Hunza Valley in 1954.

198 Reginald Schomberg (1945: 61; RSAA/SC/SCH/2/47).

199 Demi-official letter from British Consulate-General Kashgar, Kenneth Packman, to Sir Herbert Aubrey Francis Metcalfe, Foreign and Political Department in New Delhi, dated 18 February 1937 (IOL/P&S/12/2382).

200 Demi-official letter Political Agent Gilgit, Camp Naltar, to Lieutenant Colonel L. E. Lang, Resident Kashgar, dated 31 August 1937 (IOL/P&S/12/2382).

201 Report on Vice-Consul Gillett's tour to Keria via Yarkand, Karghalik and Khotan. 1st February to 5th April 1937 (IOL/P&S/12/2336).

202 See for a detailed account of the development of transport costs Fuad Kazak (1937: 98).

203 The quote refers to Sir Evelyn Howell's confidential letter No. F.-11-C/26, dated 11th June 1928, which is given as an example for contemporary reasoning in the 'Confidential demi-official letter from Major G. Kirkbride, Political Agent Gilgit, to Lieut.-Colonel L. E. Lang, Resident in Kashmir, No. 48-T, dated the 7th August 1936'.

204 Demi-official letter Political Agent Gilgit, Camp Naltar, to Lieutenant Colonel L. E. Lang, Resident Kashgar, dated 31 August 1937 (IOL/P&S/12/2382).

205 See the picture of an early airplane reaching Khorog by Pavel Luknitsky reproduced in Hermann Kreutzmann (2017a: 215).

206 The quote is taken from a report by the British Embassy in Moscow to Sir John Simon in the Foreign Office in London dated 2 August 1932 (IOL/P&S/12/2276).

207 Report on visit of Air Marshal Sir Edgar Ludlow-Hewitt … Air Officer Commanding the Royal Air Force in India to Gilgit, October 21st to 24th 1936 by George Kirkbride (1936: 1–4).

208 Within the 'Aviation notes' of 19 January 1933 a chapter was published titled 'Above the roof of the world' (IOL/P&S/12/1993: 145).

209 The undated article in the 'Times of India Mail Edition' was titled 'Flying the Himalayas. Aviators' thrilling experience' (IOL/P&S/12/1993: 146).

210 IOL/P&S/12/1993: 145.

211 Quote from an article titled 'Air pictures of the Himalaya' that was published in the Birmingham Post on 18 January 1933 (IOL/P&S/12/1993: 144).

212 'Recommendations for future flights' resulting from the 1934 flight to Gilgit (IOL/P&S/12/1993: 140–141).

213 George Kirkbride (1936: 4).

214 The Times, 10 December 1935.

215 'Memorandum on Gilgit Flight 1936' (IOL/P6S/12/1993: 106–107).

216 Major George Kirkbride mentioned this event in the 'Gilgit Agency Political Diary for the month of January 1936' as 'the first visit of civilian aircraft to Gilgit (IOL/P&S/12/3285).
217 Air navigation in the Gilgit Agency. Report on the Hunza valley ground reconnaissance (IOR/2/1085/298); Gilgit Agency: Political diaries January 1931 to May 1947 (IOL/P&S/12/3285).
218 The Aliabad hospital was built in 1928 and opened the following year (ILO/P&S/10/973: 64).
219 Report on the Hunza Valley ground reconnaissance 6th May – 8th June, 1937 (NAI/PR/4002107: 45).
220 NAI/PR/4002107: 75.
221 NAI/PR/4002107: 73-79.
222 NAI/PR/4002107: 76.
223 NAI/PR/4002107: 75.
224 NAI/PR/4002107: 73-79
225 Major George Kirkbride referred to the reconnaissance flight in the 'Gilgit Agency Political Diary for the month of May 1937 (IOL/P&S/12/3285: 231) and a special report (George Kirkbride 1936: 252). One aerial photograph taken during this mission showing Pasu village is reproduced in Hermann Kreutzmann (2012e: 60).
226 NAI/PR/4002107: 145–150.
227 NAI/PR/4002107: 24.
228 Major George Kirkbride (1936: 252).
229 Letter by Eric Teichman from Embassy Chungking to Foreign Office P. L. No. 169 of 17 June 1942 (IOL/P&S/12/4610).
230 Headquarters U.S. Air Forces in India, Tenth U. S. Air Force report to the Commanding General Headquarters 10th U. S. Air Force, New Delhi, dated 27 June 1942 (IOL/P&S/12/4610).
231 IOL/P&S/12/4610: 2–3.
232 World Bank (2011: 151).
233 See the entries in the Gilgit Diaries from April 1932 onwards (IOL/P&S/12/3285: 232–345).
234 Hermann Kreutzmann 2015a: 267–286, 364, 432; 2017a: 202–203).
235 Gilgit Agency Political Diary for December 1932 (IOL/P&S/12/3285: 345).
236 See William Kerr Fraser-Tytler (1953); Max Klimburg (1956); Hermann Kreutzmann and Stefan Schütte (2010; 2012); Geoff Watson (2009).
237 See Ildikó Bellér-Hann (2008); David Brophy (2016); Andrew Forbes (1986); Justin Jacobs (2016); James Millward (2007).
238 Hermann Kreutzmann (2015a; 2017a); Hermann Kreutzmann and Teiji Watanabe (2016).
239 The Times, 29 March 1935.
240 'Confidential letter from the Resident in Kashmir No. D. 686-C/35 dated Srinagar, the 9th November 1935' containing the report about the 'Future of the Gilgit Agency' and the cost of communications (IOL/P&S/12/3287: 77–81); the Administration report for the Gilgit Agency of the year 1935 (IOL/P&S/12/3288: 199) contains the brief, five-article long lease. See as well Ainslee Embree (1977); Hermann Kreutzmann (1989a; 1991a; 2015a); Magnus Marsden and Benjamin Hopkins (2013).
241 See the map 'Surveys in Swāt, Chitrāl, Gilgit and neighbouring territories, 1925 to 1931' published by the Survey of India in 1934.
242 Ariane Audouin-Dubreuil (2002: 8); Latham Blacker (1933: 53).
243 See Ariane Audouin-Dubreuil (2002: 51–59).
244 Entry No. 58 by Herbert Todd in the 'Gilgit Agency Political Diary for the month of July 1931' (IOL/P&S/12/3285).
245 Entry No. 76 by Herbert Todd in the 'Gilgit Agency Political Diary for the month of August 1931' (IOL/P&S/12/3285).
246 Georges Le Fèvre (1935: 204); Maynard Owen Williams (1932: 341).
247 Peter Fleming (1936: 373) reports that the Citroën car still on display was one of the few attractions in Gilgit; see as well The Times, 11 December 1935.
248 Entry No. 58 by Herbert Todd in the 'Gilgit Agency Political Diary for the month of July 1931' (IOL/P&S/12/3285).
249 The telegraph line existed since summer of 1918 when the line from Gilgit to Misgar was established (IOL/P&S/10/826).
250 The Times, 11 December 1935; see as well Kumara Menon (1947: 44).
251 The photograph is reproduced in Jean-François Ruchaud and Caroline Haardt de la Baume (2012).
252 The photographs of Maynard Owen Williams (1932: 338–362) are a vivid testimony of the changed circumstances of travel.
253 Georges Le Fèvre (1935: 213–216); see route map in Hermann Kreutzmann (2016b: 26).
254 Georges Le Fèvre (1935: 217–220); Maynard Owen Williams (1932: 354).
255 Both groups reunited in Urumchi on 26 October 1931; see Ariane Audouin-Dubreuil (2002: 49–61); Latham Blacker (1933: 57); Georges Le Fèvre (1935: 232).
256 Latham Blacker (1933: 53).
257 Latham Blacker (1933: 54–55).
258 André Citroën (1935: 9–10).
259 André Citroën (1935: 14).
260 These trucks are better known abbreviated as SIL from the oldest Russian manufacturer (Завод имени Лихачёва) founded in 1916 and renamed as Завод имени Сталина (SIS) in 1931; see Eric Shipton (1969: 133) for Eric Teichman's vehicles.
261 Eric Teichman (1937: 29).
262 Eric Teichman (1937).
263 Eric Teichman (1937: 181).
264 Administration report for the Gilgit Agency of the year 1935 (IOL/P&S/12/3288: 195).
265 'Confidential letter from the Resident in Kashmir No. D. 686-C/35 dated Srinagar, the 9th November 1935' containing the report about the 'Future of the Gilgit Agency' and the cost of communications (IOL/P&S/12/3287: 77–81).
266 Administration report for the Gilgit Agency of the year 1935 (IOL/P&S/12/3288: 195).
267 'Confidential demi-official letter from Lieut-Col. L. E. Lang,

Resident in Kashmir No. F. 13-C/34 dated Srinagar, the 10th November 1936' (IOL/P&S/12/2382: 1–4).
268 'Confidential demi-official letter from Lieut-Col. L. E. Lang, Resident in Kashmir No. F. 13-C/34 dated Srinagar, the 10th November 1936' (IOL/P&S/12/2382: 4).
269 'Confidential demi-official letter from Major G. Kirkbride, Political Agent Gilgit, to Lieut.-Colonel L. E. Lang, Resident in Kashmir, No. 48-T, dated the 7th August 1936' (IOL/P&S/12/2382: 6–7).
270 See Aasim Akhtar 1997; Fredrik Barth (1956a, b); John Biddulph (1880); Amar Singh Chohan (1998: 215–218); Karl Jettmar (1960; 1961, 1982b; 1983); Lincoln Keiser (1991); Adam Nayyar (1986); John Staley (1969; 1982); Mohammad Manzar Zarin and Ruth Laila Schmidt (1984).
271 'Confidential demi-official letter from Major G. Kirkbride, Political Agent Gilgit, to Lieut.-Colonel L. E. Lang, Resident in Kashmir, No. 48-T, dated the 7th August 1936' (IOL/P&S/12/2382: 7).
272 IOL/P&S/12/2382: 9.
273 IOL/P&S/12/2382: 10.
274 IOL/P&S/12/2382: 10.
275 IOL/P&S/12/2382: 10.
276 'Confidential demi-official letter from Lieut-Col. Lionel Edward Lang, Resident in Kashmir No. F. 13-C/34 dated Srinagar, the 10th November 1936' (IOL/P&S/12/2382: 4).
277 'Confidential demi-official letter from Major G. Kirkbride, Political Agent Gilgit, to Lieut.-Colonel L. E. Lang, Resident in Kashmir, No. 48-T, dated the 7th August 1936' (IOL/P&S/12/2382: 5).
278 Kenneth Mason (1936).
279 Hermann Kreutzmann (2004c: 194).
280 Alastair Lamb (1991: 16); Walter Lawrence (1908: 69); Kenneth Mason (1922: 34, 58).
281 Administration report for the Gilgit Agency of the year 1936 (IOL/P&S/12/3288: 185).
282 Major George Kirkbride in the 'Gilgit Agency Political Diary for the month of July 1936 (IOL/P&S/12/3285: 260).
283 Administration report for the Gilgit Agency of the year 1935 (IOL/P&S/12/3288: 195).
284 Letter No. 13-T of 1936, dated Camp Chilas, the 21st February 1936 by Major George Kirkbride to the Assistant to the Resident Kashmir (IOL/P&S/12/3294).
285 Letter No. F. 257-X/36 from the Additional Secretary to the Government of India in the Foreign and Political Department, Kumara Padmanbha Sivasankara Menon, dated Simla, 13th August 1936 to the Resident Kashmir, Lieut.-Colonel Lionel Edward Lang (IOL/P&S/12/3294); Administration report for the Gilgit Agency of the year 1935 (IOL/P&S/12/3288: 198).
286 Administration report for the Gilgit Agency of the year 1935 (IOL/P&S/12/3288: 196).
287 Reginald Schomberg (1945: 13); unpublished manuscript (RSAA/SC/SCH/2/47).
288 Matthias Weinreich (2009).
289 Abdul Qadir Khan (1930).
290 Pravda of 2 August 1932. The quote is taken from a report by the British Embassy in Moscow to Sir John Simon in the Foreign Office in London (IOL/P&S/12/2276).
291 Pravda, No. 139, May 22, 1934, p. 6.
292 Moses Markovich Levin (1933: 215).
293 IOL/P&S/12/2276.
294 New York Times, September 12, 1940.
295 A 'secret' message from Raymond Johns from the Economic Intelligence Department dated 31 January 1948, which was based on Sotsialisticheskoe Zemledelie of 28 December 1947 (IOL/P&S/12/3303).
296 Kashgar Weekly Letter of 3 June 1943 (IOL/P&S/12/2403/782).
297 See Gardner Bovingdon (2010); Andrew Forbes (1986); James Millward (2007). They provide the context for the crucial developments in Eastern Turkestan/Xinjiang by the mid-1930s.
298 The Gilgit Political Agency monthly diaries of the 1930s (IOL/P&S/12/3285) contain regular entries of refugees arriving on the high passes on the Hunza border; see the chapter in Hermann Kreutzmann (2015: 280–285) with detailed records on the emigration and refugee movement from Xinjiang.
299 See the exchange of letters between Gilgit and Kashgar in 'Claim of Chinese administrator in Kashgar over sovereignty over Hunza' in 1935 (FO 371/19335); further details are documented in Irmtraud Müller-Stellrecht (1978); Hunza's income from grazing rights in Sarikol is listed in Hermann Kreutzmann (1996a: 358–361).
300 Secret letter of Olaf Caroe, Foreign and Political Department in Simla, to Lieutenant-Colonel Lionel Edward Lang, Resident in Kashmir, dated 22 June 1935 (FO 371/19335: 208–213).
301 Demi-official letter by Lieutenant-Colonel Lionel Edward Lang, Resident in Kashmir, dated 30 July 1935 (FO 371/19335: 225).
302 IOL/P&S/12/3285: 233.
303 Major Gerald Charles Lawrence Crichton in Gilgit Agency Political Diary for the month of February 1941 (IOL/P&S/12/3285: 123).
304 Gilgit Agency Political Diary for the month of May 1941 (IOL/P&S/12/3285: 121).
305 Gilgit Agency Political Diary for the month of November 1940 (IOL/P&S/12/3285: 129).
306 Eric Shipton (1938).
307 Extract from Confidential Report from British Consulate-General, Kashgar No. 926/H/7/H, dated 2nd October 1942 (IOL/P&S/12/4609: 146–147), prepared by Michael Gillett; see as well Eric Shipton (1969: 133–137).
308 IOL/P&S/12/4609: 145.
309 The 'Gilgit Agency Political Diaries' for August and September 1943 mentioned that the 'Chinese survey team' consisting of 12 members and 16 servants had arrived in Gilgit by 11 August. Furthermore it was reported that 'Mr. Chu, Chief of the Chinese survey party with four assistants returned from Mintaka Karaul and left for Delhi' on 14 September 1943 (IOL/P&S/12/3285: 66, 68).

310 Haji Qudratullah Beg (1962).
311 Gilgit Agency Political Diary for the month of July 1941 (IOL/P&S/12/3285: 114).
312 Gilgit Agency Political Diary for the month of August 1941 (IOL/P&S/12/3285: 115).
313 IOL/P&S/12/4609: 126–127. Sinsinsin or in Pinyin Xingxingxia, on the road between Hami and Anhsi, was spelled by Marc Aurel Stein as Hsing-Hsing-Hsia and identified as the boundary between Gansu and Xinjiang Provinces (Marc Aurel Stein 1916: map following page 225).
314 The letter was filed under the heading 'Karakoram pack road' Ext. 496/43, Fast air mail. Secret. India Office Whitehall, dated 2 March 1943 (IOL/P&S/12/4609: 114).
315 IOL/P&S/12/4609: 115.
316 IOL/P&S/12/4609: 79.
317 Memorandum by Sir Eric Teichman of interview with Mr. Tseng Yang-fu, Minister of Communications, dated 5th August 1943 (IOL/P&S/12/4609: 67–68).
318 Copy of a letter, dated 18 January 1944, Ref. No. MC/IA-008 (IOL/P&S/12/4609: 53).
319 Memorandum on the 'Proposed Sinkiang-India Highway' (IOL/P&S/12/4609: 54–55).
320 Demi-official letter No. 509-X/44 (IOL/P&S/12/4609: 40).
321 IOL/P&S/12/4609: 40.
322 IOL/P&S/12/4609: 6–14.
323 IOL/P&S/12/4609: 9.
324 IOL/P&S/12/4609: 10.
325 IOL/P&S/12/4609: 11.
326 Muhammad Masih Pal (1934); see Hermann Kreutzmann (1994: 347) for population figures for 1941.
327 IOL/P&S/12/4609: 12.
328 IOL/P&S/12/4609: 12.
329 IOL/P&S/12/4609: 13-14.
330 Letter from the British Consul in Urumchi, Geoffrey R. Tural, to the British Embassy Chungking, dated 28 November 1943 (IOL/P&S/12/4609: 34). Mr. Yung was a railway engineer by training who had participated in the Burma Road project.
331 Letter of Horace Seymour, British Embassy Chungking, to Olaf Caroe, External Affairs Department New Delhi, dated 26 June 1944 (IOL/P&S/12/4609: 5).
332 Demi-official letter No. F. 435-CA/44, dated 5 October 1944 (IOL/P&S/12/4609: 16-17).
333 IOL/P&S/12/4609: 16.
334 IOL/P&S/12/4609: 16.
335 IOL/P&S/12/4609: 3.
336 Gilgit Agency Political Diary for the month of May 1945 (IOL/P&S/12/3285: 28).
337 The term *maulai* has been used as a designation for Ismaili Muslims from the mountain region; see Hermann Kreutzmann (2015a: 137–139); for an early reference Gottlieb Wilhelm Leitner (1894, App. VII).
338 Seymour Topping (1979); see the minutes on 'Kashmir: possibility of Chinese acquisition of Aksai China' by Philip Priestley in the Foreign Office, dated 20 April 1970 (FCO 37/694).
339 Quoted from an article titled 'The October Revolution in Gilgit. Maharaja repudiated: British officers avert bloodshed', published in The Statesman 16 August 1948 (IOL/P&S/12/3303) which is obviously based on the report by Major W. A. Brown of December 1947 (IOL/P&S/13/1860); William Alexander Brown (1998); see as well Sherullah Beg (1976: 8); Fida M. Hassnain (1988: 151).
340 Report by Major W. A. Brown of December 1947 (IOL/P&S/13/1860); reprinted in William Alexander Brown (1998: 270).
341 See Yaqoob Khan Bangash (2010); Sherullah Beg (1976); William Alexander Brown (1998); Ahmad Hasan Dani (1989); Sikandar Khan (1975); Charles Chenevix Trench (1986: 269–279).
342 See John Staley (1969: 241; 1982: 260).
343 John Staley (1966: 259–260).
344 For the effects of the Kashmir conflict on Gilgit-Baltistan see Antía Mato Bouzas (2012; 2017; 2019); Hermann Kreutzmann (2008a; 2016c); Anita Raman (2004); Martin Sökefeld (2005); Chitralekha Zutshi (2004).
345 Muhammad Mumtaz Khalid (2006: 100).
346 The Times (22 June 1949).
347 Muhammad Mumtaz Khalid (2006: 88–90).
348 Muhammad Mumtaz Khalid (2006: 99–112).
349 Muhammad Mumtaz Khalid (2006: 1108–109); Raymond Moore (1979: 143).
350 Muhammad Mumtaz Khalid (2006: 112) quotes from the General Headquarter letter of 23 December 1958. Raymond Moore (1979: 142) prominently discusses this nation-building effort.
351 Anwar Hussein Syed (1974: 82–83); see the maps in Hermann Kreutzmann (2015a: 165, 481) where the watershed line and the disputed territory are identified.
352 Secret letter No. 451/48 by C. B. Duke to Sir Laurence Grafftey-Smith, Peshawar, dated 2 February 1948 (IOL/P&S/12/3303).
353 Secret letter No. 451/48 (IOL/P&S/12/3303).
354 Secret letter No. 451/48 (IOL/P&S/12/3303).
355 Muhammad Mumtaz Khalid (2006: 155–156).
356 Muhammad Mumtaz Khalid (2006: 155).
357 Raymond Moore (1979: 146).
358 Raymond Moore (1979: 147) reported about eight miles as the achievement. A detailed account of interruptions, shortcomings and unexpected challenges during the first phase is given by Muhammad Mumtaz Khalid (2006: 113–132).
359 Muhammad Mumtaz Khalid (2006: 288).
360 Nazir Kamal (1979: 19); Raymond Moore (1979: 150–151).
361 Anwar Hussein Syed (1974: 84).
362 David Dichter (1967); Robert Nichols (2013).
363 Neville Maxwell (1972: 78–95).
364 Ghulam Ali (2017: 55); Anwar Hussein Syed (1974: 86).
365 Anwar Hussein Syed (1974: 87–89). The wording of the agreement and the joint communique are reproduced in Mujtaba

Razvi (1971: 292–301); see as well Alastair Lamb (1964a, b); Dorothy Woodman (1969: 279–311).

366 Minutes by Terence Garvey of the British Foreign Office after a talk with the Pakistan ambassador who disclosed the state of 'Sino-Pakistan border negotiations', dated 5 January 1963 (FO 371/170676).

367 In 1958, according to inspection reports by the Northern Scouts Chinese troops tried to substantiate their claim of territory from an advance military post near the Oprang Pass towards Darband and other areas claimed by the *mīr* of Hunza; since 1953 it had been impossible to occupy pastures there, *mīr* Jamal Khan avoided to mark any boundary in the area as he feared to lose more territory (FO 371/133389).

368 See Hermann Kreutzmann (2015a: 481, map of the Shimshal Pamir; for the Kashmir stalemate Hermann Kreutzmann (2008a; 2013a; 2015b; 2016c); Alastair Lamb (1991: 274–277).

369 Chad Haines (2012: 66) misinterpreted the Mintaka and Khunjerab Pass routes as two different systems as many others did. In fact, the Khunjerab alignment superseded the Mintaka caravan route. Ghulam Ali (2017: 54–55) and Fida M. Hassnain (1978: 19) mention even three; Shashi Bhushan (1978: 2–3) identifies four, but, like some Indian newspapers, shows a map with two roads (Anwar Hussein Syed 1974: 85); see as well FCO 37/694.

370 Yaacov Vertzberger (1983: 644–645).

371 Ian Stephens (1953a: 154-155).

372 Ian Stephens (1953a: 156). Information given by Imam Yar Beg Sani in October 1984; by then he had been the 'road inspector' in Central Hunza for 28 years. See George Band (1955: 79); Barbara Mons (1958: 78–79); Ian Stephens (1953a: 156–160).

373 Hermann Kreutzmann (1996a: 219).

374 L. P. Clark (1960: 21); Akhtar Hameed Khan (1985: 49).

375 A petition for a suspension bridge spanning the Hunza River to connect Gilgit with Danyor had been repeatedly presented during the British lease period; nevertheless, it materialised only after independence (RSAA/SC/SCH/2/47: 68).

376 L. P. Clark (1960: 22); see as well Hermann Kreutzmann (1995b: 221; 1995c).

377 L. P. Clark (1960: 21).

378 Elizabeth Staley (1966: 102); John Staley (1966: 211).

379 L. P. Clark (1960: 21).

380 Elizabeth Staley (1966: 102); John Staley (1966: 210).

381 Reginald Schomberg (1935a: 111).

382 See the photograph in Wazir Mohammed and Ashraf Khan (1969: 86); Allah Bakhsh Rajput (1962: 51).

383 John Edwards (1960: 49). A year earlier Allen Banik had made the journey by jeep to Baltit; some photographs of the road condition were provided (Allen Banik and Renée Taylor (1960; photographs between pages 96 and 97).

384 In another report about the pioneering jeep journeys of the British and Italian mountaineering expeditions for which nine jeeps were available the travelling time was given as seven hours by John Edwards (1961: 108).

385 John Tobe (1960: 446).

386 See Times of India 27 June 1970.

387 Information given by Imam Yar Beg Sani in October 1984; see Hermann Kreutzmann (1989a: 36).

388 Seymour Topping (1979).

389 Mahmood Hasan Khan (2009: 21–30); Mohammed Qadeer (1977: 4–14).

390 See for more detail Hermann Kreutzmann (1989a: 197–207).

391 Akhtar Hameed Khan (1985: 249).

392 Shoaib Sultan Khan (2009: 359); see as well Mahmood Hasan Khan (2009: 50–51).

393 Muhammad Mumtaz Khalid (2006: 293).

394 Muhammad Mumtaz Khalid (2006: 295–301).

395 Muhammad Mumtaz Khalid (2006: 317).

396 See Ghulam Ali (2017: 68–69); Anwar Hussein Syed (1974: 128).

397 See Shashi Bhushan (1978: 3); Anwar Hussain Syed (1974: 136–137); Yaacov Vertzberger (1983: 643); and the maps in Anis Ali Syed (1985: 29); Hermann Kreutzmann (2015b: 281).

398 Mujtaba Razvi (1971: 185).

399 Dawn, 22 October 1967; see as well Dawn, 29 July and 14 November 1968.

400 Confidential letter by John Boyd to George Walden, dated 10 October 1968 (FCO 21/71: 24).

401 The first trade agreement between both countries was signed on 5 January 1963 and did not refer to overland trade which only commenced six years later. In comparison with overseas exchange this was a negligible quantity of less than one percent of bilateral trade; see reports from Associated Press of Pakistan 21 May 1970 (FCO 37/694). See Dawn, 14 November 1968 and 8 May 1970; Mujtaba Razvi (1971: 186). Afterwards the border trade agreement was renewed annually (Dawn, 8, 12 & 15 May 1970; FCO 37/694; Ikram Baig 1992: 1; Yaacov Vertzberger 1983: 638).

402 Ikram Baig (1992: 2) recorded 2,945 share holders in 1992; see Dawn Overseas Weekly, 3–9 September 1982; Hermann Kreutzmann (1991a: 725); Anwar Hussein Syed (1974: 135–136).

403 Observation by an unknown traveller during a visit to Gilgit in summer 1974 (FCO 37/1502).

404 Dawn Overseas Weekly, 12-18 June 1981; see as well Yaacov Vertzberger (1983: 638).

405 Yaacov Vertzberger (1983: 642).

406 Muhammad Mumtaz Khalid (2009: 273–274). In Chinese reports Halegush was spelled as Haligexi (Renmin Ribao 19 June 1978).

407 Anwar Hussein Syed (1974: 136).

408 Muhammad Mumtaz Khalid (2009: 259–262) who reprinted the wording of the respective speeches held in Shishket.

409 Rob Collister (1977: 137); Muhammad Mumtaz Khalid (2009: 344-347); FCO 37/1502.

410 Muhammad Mumtaz Khalid (2009: 344–347).

411 The 2005 Kashmir earthquake affected the same area again and caused major destructions and interruptions.

412 Edward Derbyshire, Monique Fort and Lewis Owen (2001: 66)
413 Edward Derbyshire and Monique Fort (2006: 91); Kenneth Hewitt (2006 : 65).
414 Salamat Ali (1977).
415 See Edward Derbyshire, Monique Fort and Lewis Owen (2001: 62); Andrew Goudie (1981); Andrew Goudie, David Jones and Denys Brunsden (1984); Wenying Wang, Maohuan Huang and Jianming Chen (1984); Xiangsong Zhang (1984).
416 Edward Derbyshire, Monique Fort and Lewis Owen (2001: 69.
417 Muhammad Mumtaz Khalid (2009: 273–275, 287–318).
418 Edward Derbyshire, Monique Fort and Lewis Owen (2001: 66); see Muhammad Mumtaz Khalid (2009: 343).
419 Zulfiqar Ali Bhutto (1972: 171); see Akbar Zaidi (2015: 19–24, 38–39) for a discussion of feudalism in Pakistan's context and the effectiveness of Bhutto's reforms.
420 Quote from a statement issued by the Ministry of External Affairs from 25 September 1974 (FCO 37/1514: 26).
421 Letter from the British Embassy Islamabad to the Foreign Commonwealth Office, dated 25 September 1974 (FCO 37/1498: 23).
422 Qurban Jan, Secretary of the Hunza & Nagar Liberation Movement, in a press statement on 13 September 1972 (Zafar Iqbal 2017: 535–537).
423 Maula Madad in a press conference on 11 April 1974 (Zafar Iqbal 2017: 537–539).
424 Quoted after Bansi Lal Kak (1977: 90).
425 Hermann Kreutzmann (1989a: 198).
426 Sherullah Beg (1976).
427 Government of Pakistan (1988: 263).
428 Information provided by the Local Bodies and Rural Development Office in Gilgit in November 1984).
429 New China News Agency 18 June 1978.
430 New China News Agency 18 June 1978.
431 Dawn Overseas Weekly, 24–30 June 1978, 21–27 June 1980; Nazir Kamal (1979: 21); Muhammad Mumtaz Khalid (2009: 357); World Bank (2011: 148). Other observers accounted for less than 8,000 Chinese workers during the final phase (FCO 37/1502).
432 Renmin Ribao, 19 June 1978.
433 A separate article devoted to the history of Karakoram Highway summarised the achievements (Renmin Ribao, 19 June 1978, p. 5).
434 This status prevailed during the period of Martial Law between 1977 and 1985; see Ahmad Hasan Dani (1989: 422).
435 Nazir Kamal (1979: 20).
436 Dawn Overseas Weekly, 21–27 June 1980; The Pakistan Times 12 December 1974. According to Ayesha Siddiqa (2007: 116) the Frontier Works Organisation is the 'largest contractor in the country for constructing roads and collecting tolls'.
437 Nazir Kamal (1979: 30).
438 Dawn Overseas Weekly, 24–30 June 1978.
439 Dawn, 27 & 28 August 1982; Dawn Overseas Weekly, 3–9 September 1982; India protested immediately against the opening (Tehran Times, 26 August 1982).
440 Dawn Overseas Weekly, 28 October – 3 November 1982.
441 Yaacov Vertzberger (1983: 643).
442 The Muslim, 11 August 1985.
443 Dawn Overseas Weekly, 5–11 February 1982.
444 The Muslim, 1 May 1986.
445 The Pakistan Times Overseas Weekly 1 November 1987.
446 See a photograph of the Pirali station in Hermann Kreutzmann (2015a: 396). Only in 1992 were immigration and customs shifted to new premises in Tashkurgan, which are now called 'Khunjerab Port'.
447 Several observers highlighted the strategic aspect only and neglected its economic potential; see for example Basil Leonard Clyde Johnson (1979: 184); Anwar Hussein Syed (1974: 137).
448 The Pakistan Times Overseas Weekly, 27 July 1986.
449 James Millward (2000: 129).
450 See for rural development in Northern Pakistan Arif Hasan (2009: 1–26); Akhtar Hameed Khan (1985); Mahmod Hasan Khan (2009: 94–146); Mahmood Hasan Khan and Shoaib Sultan Khan (1992); Shoaib Sultan Khan (2009).
451 The Northern Areas Public Works Department was labelled as Northern Areas Works Organisation until 1985 (Hermann Kreutzmann 1989a: 199).
452 Akhtar Hameed Khan (1985: 249).
453 Abdul Malik, Ali Effendi and Muhammad Darjat (2006: 200).
454 Abdul Malik, Ali Effendi and Muhammad Darjat (2006: 202–203).
455 See Tariq Husain (1987); Benno Pilardeaux (1995; 1998).
456 Muhammad Abdullah (1972).
457 Hermann Kreutzmann (1989a: 195; 1989c; 2005a; 2013d).
458 Zulfiqar Ali Bhutto (1972: 173).
459 The Muslim, 31 May and 20 June 1985; see Izhar Ali Hunzai (2013); Hermann Kreutzmann (2008a: 209–210).
460 The Frontier Post, 3 and 5 December 1989.
461 Aziz Siddiqui (1994); Dawn, 9 February 1994; The Muslim, 28 August and 5 October 1993.
462 Dawn, 9 February 1994.
463 Mehmood Khan, Chief Secretary Northern Areas, in The Muslim, 18 May 1996.
464 Dawn, 24 October 2002.
465 See Nosheen Ali (2009); Muhammad Feyyaz (2011); Izhar Ali Hunzai (2013); Hermann Kreutzmann (2012f; 2015b); Martin Sökefeld (2005); Georg Stöber (2007).
466 Izhar Ali Hunzai (2013: 5).
467 Izhar Ali Hunzai (2013: 5–7); see a table that lists incidents in Hermann Kreutzmann (2008a: 216–217).
468 Nosheen Ali (2009; 2010); David Butz and Nancy Cook (2016); Izhar Ali Hunzai (2013); Hermann Kreutzmann (2008a; 2015b); Anita Raman (2004); Martin Sökefeld (2005).
469 Hermann Kreutzmann (2008a: 216).
470 Muhammad Feyyaz (2011: 15–19); Izhar Ali Hunzai (2013: 3–7).
471 Hermann Kreutzmann (2012f: 240–242); Peer Muhammad (2014: 98–99).

472 Safdar Sial (2014: 34).
473 Safdar Sial (2014: 35).
474 Three billion rupees were earmarked for the improvement scheme, and delegations from China, Kazakhstan, and Kyrgyzstan were present (The Muslim, 19 May 1996).
475 Dawn, 24 October 2002.
476 Dawn, 11 and 25 September 2000.
477 David Butz and Nancy Cook (2011: 356); Nancy Cook and David Butz (2011: 31).
478 Data collected at Chilas Basari checkpoint and kindly made available by the late Abdul Wahid, Director of Agriculture in the Northern Areas Administration in Gilgit on 24 December 2008.
479 The data cover the fiscal years from 2002 to 2007 and were collected from the Central Board of Revenue at its Sost office in 2007 (Muzaffar-ud-din 2009: 214)
480 World Bank (2011: 66).
481 World Bank (2011: 68).
482 The railway to Kashgar started operations in 2000; meanwhile the Southern Xinjiang network has been expanded to Yarkand and Khotan.
483 Just for comparison, the original Karakoram Highway featured 85 bridges and 1,392 culverts (Renmin Ribao, 19 June 1978).
484 Nancy Cook and David Butz (2013; 2015); Muhammad Jawed Iqbal, Fida Hussain Shah, Anwar-ul Haq Chaudhry and Muhammad Naseem Baig (2014); Hermann Kreutzmann (2010; 2012e).
485 Keith Delaney and Stephen Evans (2017).
486 Pamir Times, 28 September 2012. The tunnel work started in July 2012 and took just three years to completion; see homepage of China Road and Bridges Company http://www.crbc.com/site/crbcEN/381/info/2015/3057889.html; accessed 17 May 2018; see as well Pamir Times, 14 September 2015; 24 June 2019.
487 The Karakoram Highway runs 460 kilometres through Gilgit-Baltistan, while its total length of asphalt roads is 1,394 kilometres; (Government of Gilgit-Baltistan 2018: 9).
488 Reading of the denominated areas in the respective special economic zones, 27 August 2016. The eight countries are Russia, Mongolia, Kazakhstan, Kyrgyzstan, Tajikistan, Afghanistan, Pakistan and India.
489 Shanghai Daily, 8 May 2015; see Hermann Kreutzmann (2016b: 27; 2017a: 235).
490 Government of Pakistan (2009); see Caylee Hong (2013); Hermann Kreutzmann (2015b: 283).
491 Government of Pakistan (2018).
492 Dawn, 17 May 2018.
493 See Hermann Kreutzmann (2015b) for various comments.
494 Trade figures are based on Zulfiqar Ali Khan's diagram; the revenue from customs levied at Sost has nearly quadrupled since 2009 and peaked at 2.9 billion rupees in 2017 (Pamir Times, 24 June 2019).
495 Pamir Times, 24 June 2019.
496 Government of Pakistan (2019: 128–135); Pamir Times, 24 June 2019.

Atabad Lake seems to have become a stable feature of the Hunza Valley. It functions as a sedimentation trap for finely suspended matter originating from Gojal's substantial glaciers. At the same time it has become an attractive venue for travellers on the Karakoram Highway. The photograph was taken on 26 March 2019

The photograph by Inayat Ali, who was present on the spot on 4 January 2010 when the Atabad landslide came down to the Hunza River, recorded the sheer masses of loose material and dust that filled the valley. Courtesy of Inayat Ali

Resource perspective – Spatial patterns in a diverse environment

Local assets and external resources: Constraints and coping

A massive rockfall blocked the Hunza River valley at Atabad on 4 January 2010, causing multiple damage and initially claiming nineteen human victims. About twenty people lost their lives, having ignored the warnings not to return to their village lands in Atabad where cracks in the terraces and slopes had occurred much earlier and widened over time. Substantial masses rolled down the slopes, and massive rocks and debris fell into the river valley, resulting in a huge dust cloud. Visibility was badly impaired, and it was some time before the disastrous outcome could be seen. First fears came up that this event had changed the course of life for a far greater number of people who had become more and more dependent on the lifeline of the Karakoram Highway for their commuting and migration, for regional exchange and international trade across the border with China. Gojal's one and only supply line for dearly needed goods from down-country Pakistan had been cut.

It took nearly five months for the lake to be formed by filling up the basin that had come into existence due to the blockage of the valley. About twenty kilometres of the Karakoram Highway were inundated; village lands and habitations of Ainabad, Shishket and Gulmit sank into the lake and were gone for ever. The major concrete bridge between Shishket and Gulmit where the Karakoram Highway changed from the right to the left bank of the Hunza River was never seen again. The lost road infrastructure was accompanied by collateral damage that had its side-effects on the valuable lands of the Hunza mountain farmers in Gojal.

This singular event was of a dimension that has affected all considerations about safety and security for habitations, impacted on activities in road-building, communication infrastructure and supply lines, and created a challenge for public disaster management institutions that seemed unprepared for such events and have been blamed and cursed for their disastrous performance since. Public demonstrations and general strikes have been symbols of despair and trauma in the light of such a major disaster that detrimentally affected daily life in Gojal. Refugees found temporary shelter in Central Hunza; 314 households (U. *čūlhā*) in Lower Gojal had lost their abodes and shifted mainly to villages of their close immediate or affine relatives where some have become permanent residents by now. Their helplessness was sometimes articulated in demonstrations against government institutions. The damage was estimated at a loss of property in the order of more than ninety-seven crore (*karor*) Rs. Finally, the Prime Minister announced a 'relief package' and authorised the Deputy Commissioner's Office to identify 457 needy and eligible households that received a sum of between six and eight lakh (*lākh*) rupees to cover the housing losses plus 30,000 Rs for food support.[1] The nineteen households of the fatal victims plus one household of a volunteer who had lost his life during the rescue operation were compensated with twenty lakh rupees each. The demonstrations caused two fatalities and the burning down of the Deputy Commissioner's Office in Karimabad where all but one of the 33 files relating to the Atabad disaster caught fire and were destroyed.[2] The targets of this ire were administrative and specially established institutions such as the 'National Disaster Management Authority', while on the other hand 'China Aid' was praised for its support of the affected households north of Atabad that could easily be reached from Tashkurgan. Substantial relief packages were distrib-

The analysis of available records and sources for the Hunza Valley provides a list of incidents that led to tragic losses of lives and land, destruction of infrastructure or built-up assets in village locations of the valley. The list is qualified by the type and nature of event and evidence in records: snow avalanche (A), glacier-induced (G), mudslide (M), rockfall (R), thunderstorm (T), wind action (W). The entries have been collected from contemporary reporting systems and hint at the scope and size of recurring disastrous events.

Natural hazards events in the Hunza Valley

Year	Event	Locality	Remarks/extent of damage
1830/1833	G/M	Chupursan	damage to all settlements in Chupursan[1]
1841	G	Badang	blockage of Hunza River near Garmasai[2]
1847	G	Bar	glacial surge leads to abandonment of fields[3]
1858	R	Sarat	loss of village lands due to undercutting of terrace; flooding of all villages from Sarat to Pasu[4]
1860	G	Gwachi	damage to settlement and village lands due to flood[5]
	G	Matum Das	irrigation channel destroyed by major flood[6]
1873	G	Batura	damage to buildings at Pasu and Matum Das[7]
1882	G	Shimshal	large flood caused in Upper Shimshal in September[8]
1884	G	Shimshal	damage to terraces in Ganesh[9]
1888		Gojal	unexpected high snowfall in mid-summer destroyed several crops[10]
1890	G	Nomal-Gilgit	route destroyed by flood[11]
1892	G	Nomal-Chalt	all bridges destroyed, road damaged[12]
1893	M	Matumdas	a mudflow destroyed irrigation channel and settlement in Pratapsinghpura [Matum Das, Rahimabad][13]
	G	Shimshal	damage to terraces in Altit[14]

The access road from Shishket to the Karakoram Highway's bridge spanning the Hunza River is hardly visible as the road and bridge are submerged under water; boats provided the only access for nearly five years after 2010. The photograph was taken on 10 September 2013

1 Reginald Schomberg (1935: 225); see as well Kenneth Hewitt (1982: 261) who relates the advent to Yashkuk Yaz in 1833.
2 Qudratullah Beg (1962: 350–351).
3 Patrick Vans Agnew (1915: 300–302) visited Garmasai Valley in September 1847 and was told that agricultural practices were affected by a glacier advance that caused lower temperatures, thus detrimentally affecting plant growth.
4 Bronislaw Grombchevsky (1888/2015: 137) recorded thirty years later that an earthquake might have released the masses and created a dam with the effect that 'water rose to five hundred feet above the usual level and stood there for eight months, flooding and destroying the villages of Passu, Usani, Gilamit, etc., then the water broke through the dam and was discharged. These villages were rebuilt again, but as we noticed, all the trees here are young." Henry Haughton dated the event as having taken place three years earlier: "… in about 1855 big landslip on left bank formed lake for 9 months which washed away all the lower part of Ghulmit + other cultivated areas." (Henry Haughton in his diary of 20 January 1912; IOL/Mss/Eur/D531/5). See Qudratullah Beg (1962: 151); Frederic Drew (1875: 419); Kenneth Mason (1929); Karlheinz Paffen (1956a: 14); Herbert Todd (1930: 173–174).
5 General Staff India (1928: 143): "A great deal of the ground formerly cultivated was swept away by an inundation."
6 General Staff India (1928: 182); "Part of the village-lands was also swept away by an inundation. This was probably caused by the landslip at Ghammesar … or possibly it was due to floods from the Verigeraf [Virjerab] glacier in the Shingshal valley"; see as well Thakur Singh (1917: 27).
7 Kenneth Hewitt and Jingshi Liu (2010: 534); Kenneth Mason (1929: 20–21): "In 1913 I was informed by the head man of the village of Pasu that about forty years before that date, the Batura glacier had advanced and blocked the Hunza valley, causing a lake to form north of it. I was told that this obstruction eventually burst, and that the consequent flood carried away the old village of Pasu, and caused destruction down the valley." Kenneth Mason doubts the explanation, opined that this event might have been the 1858 disaster and not caused by an advance of the Batura Glacier. Upon further enquiries Herbert Todd (1930: 174) could not find any evidence from his interlocutors: "The people say that the Batura glacier has never caused a flood. It has been known to move across the river, but the water has soon found its way underneath or over the top."
8 Kenneth Hewitt (2014: 254).
9 Herbert Todd (1930: 174): "The old Hunza records show that a lake in the Shingshal valley burst about 1884, and caused considerable damage to lands at Altit and Ganesh, near Baltit."
10 Bronislaw Grombchevsky (1888/2015: 113).
11 IOL/P&S/7/63: 498.
12 IOL/P&S/7/67: 701.
13 Thakur Singh (1917: 27): "Its irrigation channel remained in a precarious condition, and was swept away by a flood in the Nulla immediately after settlement, and the land was deserted by the Assamis [*asāmī*, cultivator]."
14 Kenneth Mason (1929: 21): "Much may be done by careful watching and a sound system of warning. In the Shingshal valley men are despatched each year from the village of Shingshal to report on the positions of the glaciers, and as long ago as 1893 warning of a glacier burst was given in time to prevent loss of life." Herbert Todd confirmed with *mīr* M. Nazim Khan who: "… confirms that a flood from the Shingshal did damage to lands at Altit."

For five years commodities had to be transferred via Atabad Lake on their way from Kashgar to Gilgit; there they had to be reloaded at Atabad shore from the boats on to lorries. The photograph was taken on 1 September 2015 during the final days of this procedure before the road was reopened along a new alignment above the lake.

uted three times among the affected households in Gojal, helping to refill the depleted stocks for times of emergency. In 2010, about 3,500 tons of wheat flour and cooking oil were provided by China Aid which distributed 2,600 tons again in the following year and 2,300 tons in 2013.[3] International relief agencies such as Focus Humanitarian, Pakistan Relief and the World Food Programme stepped in as well as the Aga Khan Rural Support Programme to safeguard the residents and support them during the times when they were cut off from regular communication until an affordable boat service on Atabad Lake was installed. In all respects, the cost of transport and travelling times to bridge the disaster location had increased manifold.

Living in a harsh environment

In Hunza's collective memory the Atabad slide was the most disastrous event caused by a sudden though foreseeable collapse of a slope structure that was previously perceived as sufficiently stable to allow continuous human dwellings. However, such events are not unique incidents; about 152 years earlier, a similar severe rockfall had occurred a few hundred metres away from the Atabad site and blocked the river.[4] The natural dam collapsed, causing a substantial flood down the valley with damage and losses of lives as far as Attock down country. These two closely located and related events are separated by one and a half centuries during

The Shishpar Glacier is characterised by 'black ice' when flowing out of the accumulation zone. The glacier is heavily crevassed up to its confluence with Muchu Har and poses great difficulties for shepherds and animal herds to cross. The photograph was taken on 14 June 1985

Year	Event	Locality	Remarks/extent of damage
1894	M	Hispar[15]	
	T	Baltit	damage due to rainfall, channels destroyed[16]
	T	Sumayar	*jáalo* (ferry) carried away[17]
	T	Gojal	several *jáalo* and bridges destroyed[18]
	T	Ratal	destruction of hamlet, all channels and fields; 23 persons were killed[19]
	G	Hasanabad	bridge destroyed[20]
1895	G	Ganesh	loss of cultivation at Shamets[21]
1897	A	Hakuchar, Chalt	blocked river caused damages to terraces, orchards and tracks[22]
	A	Nilt	avalanche kills two women[23]
	G	Barpu	formation of lake[24]
1899	G	Hasanabad	bridge destroyed[25]
1900	G	Nilt	bridge destroyed[26]
1901	R/M	Chalt	Nomal-Chalt track destroyed[27]
	G	Shimshal	breaking of dam[28]
	G	Shimshal	bridge at Ganesh destroyed[29]
	G	Hasanabad	damage to bridge[30]
1902	G	Hasanabad	damage to bridge[31]
	G	Hasanabad	bridge destroyed[32]
	G	Hasanabad	glacier advance and threat to cultivation[33]
1903	G	Hasanabad	glacier surge (30 m), destruction of canal[34]
	A	Miachar	avalanche blocked road[35]
1904	G	Ahmedabad	glacier surge, destruction of canals[36]
	G	Hasanabad	glacier surge, destruction of canals[37]

15 Martin Conway (1894a: 323).
16 IOL/P&S/7/75.
17 IOL/P&S/7/75.
18 IOL/P&S/7/75.
19 IOL/P&S/7/75.
20 IOL/P&S/7/76.
21 IOL/P&S/7/81; PRO/FO65/1506:282.
22 Gilgit Diary 17 April 1897: Avalanche in Hakuchar (Nagar) on April 9th, destroying the road, damage in orchards and fields, blocking of the river for one hour. In addition, avalanches near Nilt: death of two women, damage in orchards, same in Chalt (IOL/P&S/7/92/539).
23 IOL/P&S/7/92/539.
24 Gilgit Diary 24. April 1897: Three miles above Uyum Nagar a lake was formed, blockage of Hispar River following an advance of Barpu Glacier. The lake was four to five miles long, 100 metres wide, and 60 to 90 metres deep (IOL/P&S/7/92/539).
25 IOL/P&S/7/115/803.
26 IOL/P&S/7/126/967.
27 IOL/P&S/7/134/821.
28 IOL/P&S/7/135/903.
29 IOL/P&S/7/136/1021.
30 IOL/P&S/7/136/1009.
31 IOL/P&S/7/145/886.
32 IOL/P&S/7/146/1051.
33 Gilgit Agency Diary 17 May 1902: "The Hassanabad glacier is reported to have advanced 2 miles during one month and 20 days. It is feared that the Ganesh village cultivation may suffer." (IOL/P&S/7/145/856)
34 IOL/P&S/7/152/468.
35 IOL/P&S/7/154/789.
36 IOL/P&S/7/165/1025.
37 IOL/P&S/7/165/1094.

which the Hunza Valley has undergone a transformation in terms of population, extent of village lands, glacial activities, and expansion of communication infrastructure. The extent of the damage caused by both events is closely related to the more than tenfold growth of the population inhabiting the valley.

Between these two events a high number of small events were recorded which might be only an approximation and are dependent on the sources that can be used for grasping the episodically recurring challenges for human habitations and installations in a harsh environment. The 'History of Hunza' refers to an even earlier event that not only led to loss of land in the upper valley but caused the creation of land lower down the valley. Qudratullah Beg recorded: "It has been narrated through oral tradition that the 'Hunza River' was blocked for almost eight months as a result of a large land slide in which the entire hill/mountain located at a place 'Buddung' had slid down into the river Hunza and had formed a natural dam on the river. This had happened during the era of rule of Mir Ghazanfar Khan (Probably in the year 1841 A.D.). A lake was formed in the river course which extended upstream up to the bottom of settlement of Khyber in Gojal valley. This place up to which the tail of this lake had extended is, therefore, called 'Sarrey Mush' (Sare Musk) meaning the 'head or tail of the lake' in Burushaski language. During this period a religious scholar and a preacher called 'Pir', by the name of Khawaja Shah Dad who was from amongst the 'Pirs' of Sariqool, had arrived at Gojal to educate his disciples in Hunza. Hence he was requested to write a 'Taweez' or amulet and say some sooth saying prayers to cause the natural dam on Hunza River recede and save them all from its destructive effects. It was feared that the rising water could completely inundate the villages of Fasso [Pasu] and Gulmit and there was a real danger of these villages getting submerged into this ever rising lake. On their ~~this~~ request Pir Khawaja Shah Dad wrote a sooth saying amulet or 'Taaweez' and gave them along with a fist full of loose earth on which he had also blown his prayers. The people took this 'Taaweez' and the soil and threw them on the slide blocking the river. By the grace of almighty Allah the river water started to overflow over the dam and started to rapidly erode and cut/wash away the natural dam. Soon the dam was busted and washed away and a huge flood was caused. This flood was so large that it completely washed away the lower portion of village Ganish which is called 'Shamess' [Shamets]. This flood also washed away the upper portion of village Mayun further downstream. However this flood took such a course at the junction of Mayun Nullah that it added a large chunk of land on river bed to the lower portion of Mayun village by shifting it from the side of village Nilt located opposite in Nagar territory."[5] Upon his own consultation of geological records Qudratullah Beg suggested that this event might have been caused by an earthquake in 1841, which at the same time is made responsible for the collapse of a cliff in the Nanga Parbat Range and the blockage of the Indus River (Burushaski *Abaasíin*) close to Ghor.[6] Most authors provide evidence only from the blockage of the Indus; Qudratullah Beg is the only writer to correlate the same event with the blockage of the Hunza River.[7]

Hassanabad's village lands are clearly defined by the flat cultivated fields on top of the carved out glacial terraces that are deeply incised by the Hassanabad nālā, a tributary to the Hunza River. The Karakoram Highway leads to Aliabad which is visible in the background. The photograph was taken on 23 October 1985

Frederic Drew, who prepared the first thorough inventory of resources in Jammu and Kashmir for the Maharaja, devoted some attention to the floods in the Indus that had occurred outside the immediate area of control. Two events

The Sarat basin is partially sealed by the major moraine on which Ghamesarat is located. In the foreground lacustrine deposits indicate the previous existence of a lake. The Hunza River has lowered the basin floor; the northern terraces are used for Sarat's cultivation. The photograph was taken on 26 April 1985

Year	Event	Locality	Remarks/extent of damage
1904	G	Khurdopin	damage to terraces in Shimshal after emptying of two-year-old lake (1902)[38]
1905	R/M	Hasanabad	two flocks of goats covered by rocks and mudslide[39]
	G	Shimshal glaciers	damage to Chalt Bridge and Gilgit-Chalt track destroyed; loss of fields in Pasu and Shimshal[40]
	G	Shimshal	damage of 7 houses in Shamets and to Hunza-Nagar bridge at Ganesh[41]
	G	Pasu	loss of cultivated land[42]
	G	Gilgit-Hunza	formation of glacial lake[43]
	G	Gilgit-Hunza	flood damage to bridges[44]
1906	T	Nomal	road destroyed, one woman and livestock killed[45]
	G	Shimshal, Chalt	damage to bridges in Askurdas, Tashot and Chamogah.[46]
	G	Shimshal	fields, houses and bridges destroyed in Pasu, Hussaini, Gulmit, Ganesh[47]
	G	Shimshal	damage to houses, harvests and village lands[48]
	G	Pasu	damage to houses and village lands[49]
1907	G	Khurdopin	local alert system proves successful[50]
1908	T	Minapin/Miachar	damage to road and bridges[51]
1910	W	Sikandarabad	new bridge destroyed by high winds[52]

38 Fitzstephen Henry Bridges (1930: 172); IOL/ P&S/7/167/1319.
39 IOL/P&S/7/181/1586.
40 IOR/2/1084/289: 153; IOL/P&S/7/180/1426; Herbert Todd (1930: 174): "The usual Shingshal valley lake, formed by the Khurdopin glacier blocking the upper Shingshal valley, caused a heavy flood in the summer [2 August] of 1905. The bridge at Chalt was washed away, and a rise of thirty feet was recorded at Bunji. The Gilgit-Chalt road was extensively damaged."
41 Arthur Neve (1913: 164–165).
42 IOR/2/1084/289: 153; IOL/P&S/7/180/1426; Herbert Todd (1930: 174).
43 Gilgit Diary 8 April 1905: "... owing to the masses of ice breaking off the snout of a glacier, a large lake has formed about 5 miles above the village of Imit" (IOL/P&S/7/177/849). Karumbar and Sukhtay [Sukhtarabad] lakes are mentioned as well as previous events of 1893 and 1902; the former damaged two bridges in Gilgit as an effect of a dam breakage and flood.
44 Gilgit Agency Diary 24 June 1905: Damage of Gilgit Hunza bridge on 14 June as an effect of the Ishkoman flood. The water level increased by 20 feet in the span of a few minutes. Even in Sharot ice blocks were seen in the Punial River, fish were recorded on the fields of Chatorkhand. It was the second time that the Gilgit bridge was destroyed in the span of one year (IOL/P&S/7/179/1229; IOL/P&S/7/180/1426).
45 IOL/P&S/7/193/1654.
46 Gilgit Agency Diary 15 August 1906: "The bursting of the glacier dam in the Shingshal [Shimshal] valley caused a very high flood in Hunza river on the night of 11th August. ... The flood was no doubt the most serious that has occurred since the Agency has been established ... the water in the Hunza river rose 50 feet above summer level and 25 feet above last year's flood. At Bunji the Indus rose 30 feet, and the water was only 16 feet below the road way of the big suspension bridge known as the 'Partab-pul' (across the Indus 7 miles above Bunji). ... The road between Nomal and Chalt has been literally wiped out ..." (IOL/P&S/7/192/1618); see as well Kenneth Mason (1929: 21); Herbert Todd (1930: 175).
47 Thakur Singh (1917: 7); IOL/P&S/7/193/1654.
48 Gilgit Agency Diary 29 August 1906: "... recent Shingshal floods has carried away 35 fields with standing crops, 3 houses, 3 water-mills and 1 orchard. A considerable quantity of wheat, grain, which was lying on the threshing-floors, and fodder has also been swept away" (IOL/P&S/7/193/1654).
49 Gilgit Diary 22 August 1906: "... flood of 11th August caused the following damage: seven cultivated fields and one house were washed away at Pásu, five cultivated fields were washed away at Suseni [Hussaini]. Two fields and rope bridge were washed away at Gilmit, six cultivated fields and some fruit trees were washed away at Ganish." (IOL/P&S/7/193/1654); see as well Thakur Singh (1917: 7). Herbert Todd (1930: 175): "In 1905 the obstruction was not completely cleared, or the glacier again closed in across the Shingshal. A flood considerably higher than that of 1905 resulted on the night of 11–12th August, 1906. Askurdas, Tashot and Chamogah bridges were all washed away. At Chilas a rise of 36 feet was recorded. At Chalt the river rose fifty feet above the normal summer level, and 25 feet above the flood of the previous year."
50 Herbert Todd (1930: 175): "In 1907 a flood was again expected, and an elaborate system of bonfire warnings was arranged from the head of the Shingshal valley to Baltit. Actually however the water found its way over the top of the obstructing Khurdopin glacier, and, cutting a trough down gradually through the ice, dispersed quietly in eleven days."
51 IOL/P&S/7/221/1898.
52 IOL/P&S/7/238/723.

entered his record as most severe: the already mentioned 1841 and 1858 floods. The latter cause is definitely located in the Hunza Valley: "The following is the description of it. At 5 A.M. on the 10th August, 1858, the Indus at Atak (Attock) was very low; at 7 A.M. it had risen 10 feet; by half an hour after noon it had risen 50 feet, and it continued to rise until it stood ninety feet higher than in the morning. The fall was very slow; during the 12th August it returned very much to the position it occupied before the flood came."[8] This sudden increase of Indus water levels down river concerned the authorities who made enquiries about their roots. Several discussions were devoted to the safety of the river banks for camping military troops. Intelligence units found out that the mighty water flow might be a sudden flood event but that the blockage of a river such as the Hunza and the Indus would take several months to fill a lake behind the barrier before the overflow would release the flood wave. Authorities were sensitised to watch out for such events of unknown magnitude; an early warning system would increase safety and would make flood events calculable. Since the occurrence of those two major floods it became part of the Political Agent's reporting requirements to mention in their fortnightly or monthly diaries any physical peculiarities, rockfalls, landslides and glacier advances that could create havoc downstream. Based on George Cockerill's observations and enquiries in 1894, the General Staff India report located the 1858 flood's origin at Ghamesarat: "Here some 40 or 50 years ago a serious landslip occurred, which blocking the river for several months, caused a lake to form. This extended up to and somewhat beyond Pasu. When the obstacle gave way, a vast wave swept down the Hunza Valley, carrying away several forts in Hunza and Nagir and considerably widening the ravine between the two states."[9] This report mentions mainly the effects within the Hunza Valley. In 1859, Major John Reid Becher was sent to Kashmir to enquire about the location and the effects: "I have carefully questioned men from Gilgit, Hoonza, Nuggur, and Ghor ... All accounts agree that the late stoppage was of the river of Hoonza, about a day's journey Northward of Gilgit. It was caused by the subsidence of a mountain called 'Phungurh' on the left or Nuggur bank, from the action of rain and snow above, and of the stream below, in the winter of 1858. The vast fragments of rock, and earth, and trees and drift wood, dammed up the narrow bed of the river for 6 months; when the waters swollen from melting of the snow in the mountains, burst a sudden passage in August, and destroyed many forts and villages and lands of the Hoonza and Gilgit districts. The lake mentioned by Wuzeer Poonoo was probably the back water of this huge dam. ... the bed of the Hoonza is still somewhat encumbered by remaining rocks, and that a guard is maintained there by the Nuggur Raja to prevent the crossing of the hostile people of Hoonza by throwing over a bridge of rope or timber. ... The recent cataclysm of August 1858, is I think proved to have been that of the Hoonza branch, which at least affords an example that the site can vary; this is 300 miles from Attok."[10]

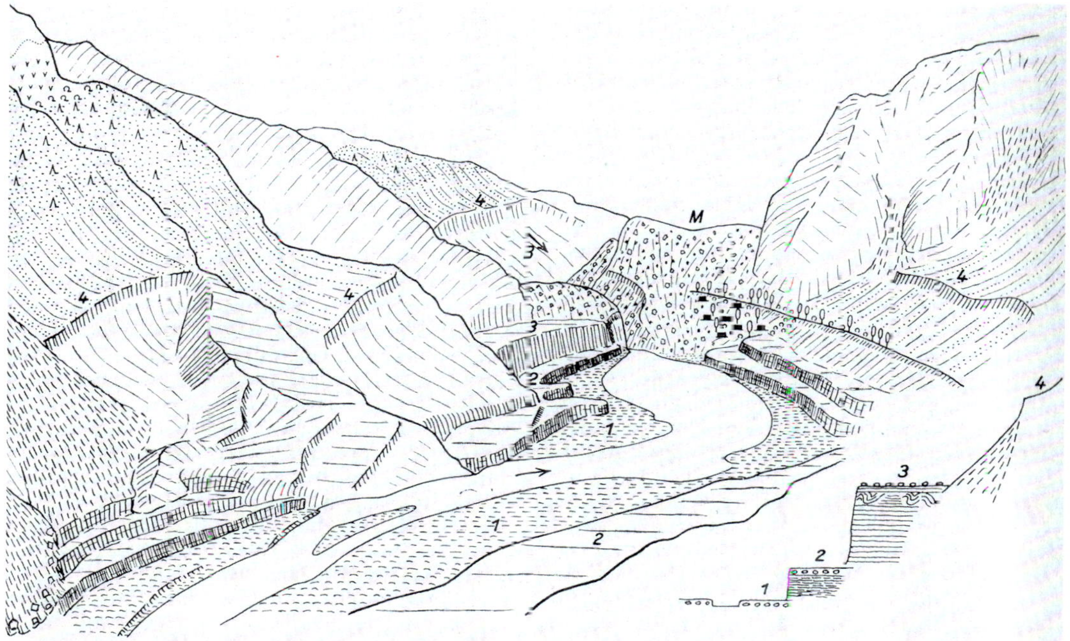

Abb. 3: Das Becken von Saret Hunza-abwärts gegen W (Zeichnung *KH. Paffen*)
M = späteiszeitliche Endmoräne des vereinigten Nordseiten-Gletschers; B = Stelle des Bergsturzes der 1850er Jahre u. wahrscheinlich auch früherer Bergstürze, welche das Becken zeitweilig abriegelten; 1 junge Schotteraue, 2 ca. 6 m mächtige, ungeschichtete Seetonfüllung des vorigen Jhs. mit dünner Schotterauflagerung, 3 um 20 m mächtige warvenähnliche Feinsand-Tonschichtung mit stellenweiser Schlierenzone im ob. Teil und dünner Grünsandauflagerung; 4 letzteiszeitliche Schliffgrenze mit trogschulterähnlicher Verflachung. Punktiert Artemisia-Steppe, im oberen Teil mit Baumwacholder, darüber Birken und Weiden.

The Sarat basin was the arena of several events that affected the livelihoods down river. Karlheinz Paffen prepared the sketch highlighting the scene of the 1858 rockfall (B) blocking the valley close to the lateglacial moraine (M) of Ghamesarat. Source: Karlheinz Paffen (1956a: 14)

Seventy years later, Kenneth Mason began to systematise information on floods and published the relevant sequence of Major John Reid Becher's account to which the Political Agent of the time, Herbert Todd, responded. Referring to his own interviews with surviving eyewitnesses seventy years later, Herbert Todd summarised: "The Second Great Indus Flood, 1858. I do not think that there is room for any doubt as to the origin of this flood. There are old people still alive in Hunza who remember it. They say that there used

The Shishpar Glacier is the main feeder for the Hassanabad Glacier. Its accumulation source is a major basin to which two steep side glaciers lead; snow avalanches are significantly contributing to its accumulation of ice masses. The photograph was taken on 15 June 1985

Year	Event	Locality	Remarks/extent of damage
1910	R	Shayar	Shayar bridge destroyed[53]
	G	Ahmedabad	Gurpi glacier advance, loss of irrigation channel heads[54]
	G	Batura, Pasu	damage to 3 houses, 20 terraces and fruit trees[55]
	G	Sikandarabad	bridge carried away[56]
1911	M	Pheker	human casualties due to landslide[57]
	G	Batura	advance of 300 metres since October, nearly blocking the valley[58]
1913	G	Hassanabad	flood destroyed bridge[59]
1914	G	Raskam	bridge destroyed[60]
1922	G	Shimshal	loss of farms in Shimshal[61]
1923	G	Shimshal	formation of glacial lake[62]
1925	A/R/W	Minapin	Pressure wave destroyed all fruit trees in Minapin[63]
	G	Murkushi/Misgar	flood destroyed bridges[64]
1926	M	Shinaki/Khanabad	five casualties due to collapse of river terrace[65]
1927	G	Khurdopin	damage to bridges and farms in Shimshal[66]

53 IOL/P&S/7/241/1050.
54 Captain Douglas Oliver reported about the general feature of advancing glaciers and the detrimental effects in Ahmedabad (IOL/P&S/7/243/1477: 6).
55 IOL/P&S/7/241/1118.
56 IOL/P&S/7/243/1477.
57 Gilgit Agency Diary December 1911: "An unfortunate accident took place near Fikar village in Nagar where five men were buried by a landslip. Their bodies have not yet been recovered" (IOL/P&S/10/826: 249).
58 Henry Haughton in his diary of 21 January 1912 (IOL/Mss Eur/D531/5).
59 Gilgit Agency Diary August 1913 "The temporary bridge over the Hasanabad stream in Hunza was carried away by a flood on the night of the 29th July" (IOL/P&S/10/826: 199).
60 Gilgit Agency Diary August 1914: "… the Kirghiz crops have now ripened, but that, as was feared, little can now be hoped for from the Hunza crops, a portion having been carried away by the river and the remainder having failed. … The rope (? bridge) across the Mustagh River carried away, but has been re-erected. No further floods are anticipated now in the rivers, as, owing to the lateness of the season, the volume of water in them is beginning to decrease" (IOL/P&S/10/826: 175). Gilgit Agency Diary July 1914: "Shortly after getting the supplies across, the rope bridge over the Mustagh river was seriously damaged by a heavy flood and it was expected that it would be carried away altogether. In this case Raskam will be entirely cut off from Hunza. … The floods in the rivers and *nalas* in both States continue. In addition to the damage already reported (paragraph 92) the Mir of Hunza's country-made cantilever bridge, between Khaibar and Murkhun in Bujhal [Gojal] has only escaped being carried away by being dismantled, the suspension bridge at Baladas (Chalt) is seriously threatened, while the main road, at a point two miles short of Nagar, is in a very bad state owing to the subsidence of the hill due to erosion of its banks by the Nagar river" (IOL/P&S/10/826: 179)
61 Gilgit Agency Diary May 1922: "A big lake is said to have formed in the Shimshal Nullah of Gujhal …" (IOL/P&S/10/973: 226); see as well Philips Visser (1935: 48).
62 Gilgit Agency Diary June 1923: "It is reported that a lake is again forming this year in the Shimshal Nullah of Gujhal. The lake is now about half full. The Mir of Hunza is detailing a guard to report its condition daily" (IOL/P&S/10/973: 191). Gilgit Agency Diary July 1923: "The water in the Shimshal lake succeeded in cutting its way through the dam and escaping to a very appreciable extent, but the passage seems again to have become blocked as the water in the lake is once more gradually rising" (IOL/P&S/10/973: 188).
63 Gilgit Agency Diary October 1925: "One man and a considerable number of sheep and goats are reported to have been killed at Minapin by a violent wind caused by the fall of an enormous mass of snow from Rakaposhi. Nearly all the fruit trees in the villages of Minapin and Pison have been destroyed by the wind, which will cause much hardship for many years" (IOL/P&S/10/973: 126-127).
64 Gilgit Agency Diary June 1925: "The country bridge at Murkushi and one between Murkushi and Misgar have been washed away by floods. The Kashgar mail is being sent along a track on the left bank of the stream. The bridges will be re-built by the Mir of Hunza when the water sub-sides" (IOL/P&S/10/973: 132).
65 Gilgit Agency Diary August 1926: "Five persons (1 male and 4 females) were proceeding in a party from their village, Khanabad, to collect fuel along the bank of the Hunza River and an overhanging bank suddenly gave way and buried the party. Two bodies (1 male and 1 female) were recovered, but the rest of the bodies have not been found" (IOL/P&S/10/973: 106).
66 Kenneth Mason (1929: 22): "In 1925 a considerable lake was observed above the Khurdopin glacier, under which the waters then found an outlet. It seems probable that this outlet became choked, probably by pressure from behind, and that the catastrophe, which was predicted by the Vissers [Jenny Visser-Hooft 1926: 128] in 1925, occurred two years afterwards, though the damage does not appear to have been severe." Charles Morris (1928: 525) heard in Misgar that 'one of the glacier lakes above that place [Shimshal] burst and caused heavy floods in the valley'; Herbert Todd's (1930: 175) enquiries with the ruler resulted: "The Mir of Hunza declares that there was no flood this year, nor is there any official record. If there was one, it must have been very small." Kenneth Mason followed-up this remark and quoted from Charles Morris's diary: "It seems that after our departure from Shingshal on 25th June one of the smaller glacier lakes at the head of the valley broke, thus allowing a large volume of water to escape down the valley. The rope-bridge at Shingshal was washed away and several of the more low-lying houses were partly submerged." The effects might have been only locally felt: "… it seems probable that the glacier barrier did not burst, the water probably reaching the top of the ice and cutting a trough comparatively harmlessly through the glacier. It is also possible that the glacier was not cut through to its base and the lake only partly drained as happened in 1907." Quoted from a 'Note by Editor' [Kenneth Mason] appended to Herbert Todd (1930: 175).

to be a small catchment area forming a lake on the top of the hill forming the right ridge of the Hunza valley, some nine miles above Baltit and about a mile below Atabad. The lake was called 'Shaikalti Phari', and was opposite the place now shown on the map as Ganesar (really Ghammesar). The bank of the lake suddenly gave way and down came a tremendous mass of the hill-side into the Hunza valley, below, completely blocking the river to a height of many hundreds of feet. The river was blocked for nine months and a lake formed, stretching right back to the mouth of the Shingshal river and the Batura glacier. The old grey-beards of Hunza say that when the obstruction eventually gave way, the air remained misty for a week. Enormous damage was done to all cultivation near the river. At the present day the remains of this obstruction are very noticeable, in fact, the stream has only cut a narrow passage through the debris, the track to Kashgar has to climb over the obstruction, and a very steep climb it is. Upstream of the obstruction the river, right back to Pasu, has only a gradual fall, and the usual huge boulders of the Hunza river are here buried under fine shingle. On each bank, some distance above the high-flood level, can be seen distinct traces of a sandy deposit, which might indicate the bed of the lake."[11] It seems that this event had become part of Hunza's collective memory.

The effects of glacial surges and subsequent outburst floods with high destructive potential motivated various observers, scholars and travellers to contribute to a better understanding of cause and effect. Kenneth Hewitt recorded thirty-one glacier surges in the span of one and a half centuries in the Karakoram, of which the Hunza Valley contains eight known and five suspected glaciers; in addition, about sixty-two known glacier advances blocked the Karakoram valleys, fifteen of which were located in the Hunza Basin.[12] The abundance of glacial phenomena made it worthwhile to focus on further investigations. Applying the methods of geomorphology and local enquiry, Karlheinz Paffen relates an event in the 1850s on the left bank of the Hunza River at Sarat, where a rockfall or slide blocked the river for six months and created a lake that had reached up to Gulmit.[13] He provided a geomorphological sketch map of the location and identified the sliding site down the slope. The valley bottom was still filled with major boulders that were identified as having been dislocated during the event.

The local chronicler Qudratullah Beg was more concerned with the loss of land and potential gains from the

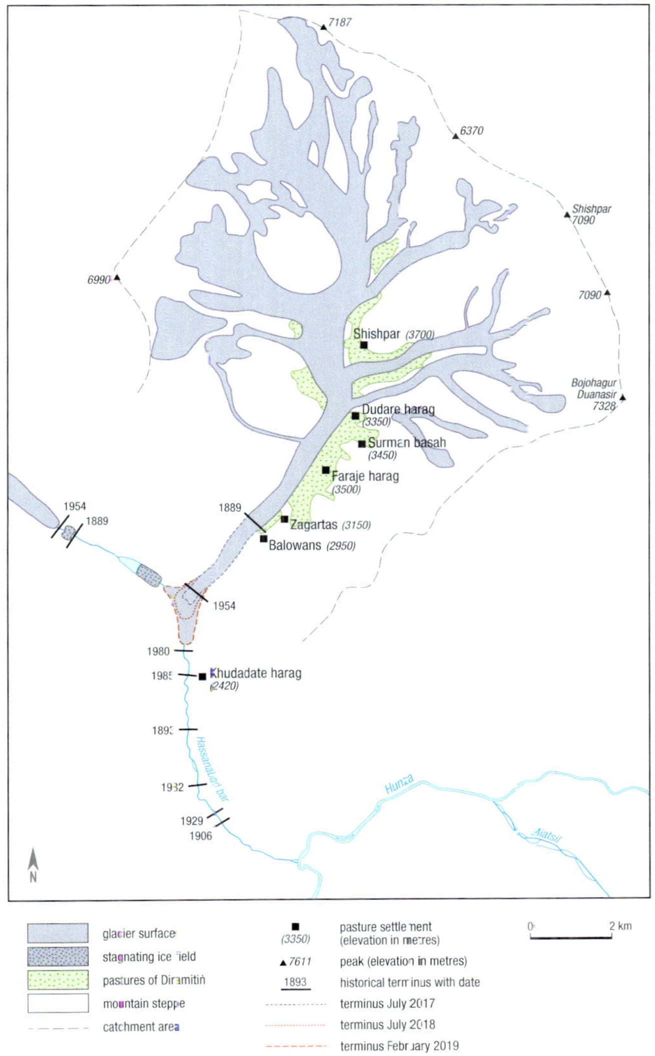

The Shishpar Glacier finds its meltwater outlet in the Hassanabad bar. The glacier terminus has been episodically measured since 1889 and has been used as a prime indicator for glacier fluctuations in the Hunza-Karakoram. In 1984, the western Muchu Har Glacier and the northern Shishpar Glacier joined before their common terminus close to the shepherds' hut at Khudadate harag; presently the western branch appears to be dispersed in ground moraine that contains some stagnating ice fields. In 2019, the main Shishpar Glacier advanced and temporarily destroyed the channel head of Aliabad góčil and Hassanabad bridge. The Shishpar catchment is one of the hereditary pastures of the Diramitin of Central Hunza. Small patches of valuable pasture are found between the various arms of the glacier. Source: Adapted and updated from Andrew Goudie, David Jones and Denys Brunsden (1984: 430); Hermann Kreutzmann (1989a: 68); Hans-Jochen Schneider (1959; 1962). Design by Hermann Kreutzmann

disaster and added another effect: "Sarat slide and natural of Hunza river damming diverted the course of Hunza river below Mayan [Maiun] resulting into shifting of a large chunk of Nilt lands along river bed towards the banks on Mayan side. This added a sizeable piece of cultivable land to the lands of Mayan. Hence, land holdings in Mayan got expanded and increased. This newly added chunk of land added as a result of flooding in the river, is called 'Dunyal'. Mir Ghazanfar Khan later divided this newly added tract of land and distributed it among the tenants of his personal lands."[14] The 1858 flood contributed to the undercutting of river terraces and the subsequent accumulation of sediments where

A state-of-the-art suspension bridge was established at Tashot after the opening of the Karakoram Highway; several attempts had previously failed to construct a sturdy bridge for a lasting vital connection between Nagar and Hunza where the river gorge is narrow, but surrounded by loose glacial deposits. The photograph was taken on 19 May 1985

Year	Event	Locality	Remarks/extent of damage
1928	G	Kilik	local flood in the valley originating in Kilik Pass[67]
1930	G	Shimshal	loss of habitations and orchards in Rezgine ben (Farmanabad)[68]
1934	G	Ganesh	damage to cultivated lands at Shamets (Ganesh)[69]
1935	R	Pheker	damage to terraces at Pheker and Hakuchar[70]
1937	M/R	Shinber	collapse of cliff on 5th July, blockage of river, 150 m high dam that collapsed on 7th, washed away Tashot and Chalt bridges, 23 people were drowned[71]
1938		Budalas	flood destroyed bridge[72]
1941	G	Shimshal	abnormally high flood on 15 July, bursting of ice dam in Shimshal that was building up for some months[73]
1944	G	Shimshal	damage to terraces in Pasu[74]

66 Kenneth Mason (1929: 22): "In 1925 a considerable lake was observed above the Khurdopin glacier, under which the waters then found an outlet. It seems probable that this outlet became choked, probably by pressure from behind, and that the catastrophe, which was predicted by the Vissers [Jenny Visser-Hooft 1926: 128] in 1925, occurred two years afterwards, though the damage does not appear to have been severe." Charles Morris (1928: 525) heard in Misgar that 'one of the glacier lakes above that place [Shimshal] burst and caused heavy floods in the valley'; Herbert Todd's (1930: 175) enquiries with the ruler resulted: "The Mir of Hunza declares that there was no flood this year, nor is there any official record. If there was one, it must have been very small." Kenneth Mason followed-up this remark and quoted from Charles Morris's diary: "It seems that after our departure from Shingshal on 25th June one of the smaller glacier lakes at the head of the valley broke, thus allowing a large volume of water to escape down the valley. The rope-bridge at Shingshal was washed away and several of the more low-lying houses were partly submerged." The effects might have been only locally felt: "... it seems probable that the glacier barrier did not burst, the water probably reaching the top of the ice and cutting a trough comparatively harmlessly through the glacier. It is also possible that the glacier was not cut through to its base and the lake only partly drained as happened in 1907." Quoted from a 'Note by Editor' [Kenneth Mason] appended to Herbert Todd (1930: 175).

67 Kenneth Hewitt (1982: 261).

68 Aga Khan Agency for Habitat (AKAH) Disasters Record (2018).

69 Local knowledge.

70 Sabine Felmy (1986: 19).

71 Gilgit Agency Political Diary September 1937: "On the 5th a collapse of cliff at Phikar in Nagar occurred completely blocking the Hunza river ... Several cultivated fields also collapsed and 23 persons are reported to have been drowned. The dam so formed was 150 feet in thickness. It burst on the 7th and first washed away the Tashot suspension bridge and then the Chalt suspension bridge 23 miles below Tashot. The road from Chalt direct to Hunza on the right bank of the Hunza river which was abandoned many years ago has now been put in order as temporary measure" (IOL/P&S/12/3285: 219; IOL/P&S/12/3288: 7). The damage caused a loss of Rs 30,000 in infrastructure assets (IOL/P&S/12/3288: 175). The fear of recurring events affected the choices for a landing ground either in Pasu or Sost as Flight Lieutenant Francis Claude Sturgiss noted in his report: "In view of this latest land slide the possibility of a repetition of the flooding of Pasu must be considered. ... Some thirty years ago [1906] due to a large landslide lower down the Hunza River the river was blocked, and the maidan [*maidān*, here probably the polo ground] at Pasu was under water for a considerable period. It appears possible but improbable that this will recur" (IOR/2/1085/298: 6, 24).

72 Gilgit Agency Political Diary August 1938: "The new Budalas bridge (Chalt) constructed last year was washed away by a flood in the Nullah in July last" (IOL/P&S/12/3285: 194).

73 "At noon that day the flood reached the junction of the Hunza and Gilgit rivers at Danyore ... where it rose to a maximum of about 40 feet above the usual summer level. It subsided rapidly after four hours. Large portions of the road between Nomal and Passu have been washed away and extensive repairs are necessary and as is known no lives have been lost in the flood, but a few cattle are reported drowned. A country suspension bridge across the Shimshal river and the Danyore Ferry boat (in Gilgit) were washed away." (IOL/P&S/12/3285: 114). Gerald Crichton entered on 11 August 1941 into the Gilgit Agency Political Diary for the month of August 1941 the following: "The resident Engineer reported that the Shimshal flood of July severely damages the road between Baltit and Passu particularly in the vicinity of Atabad, where two miles of realignment and 450 yards of rock blasting will be necessary to reopen the road to animal transport. There has been practically no interruption to pedestrian traffic since the flood as the Mir of Hunza has constructed new footpaths, but these are too hazardous for animals or for heavily laden coolie. The state of the road has delayed delivery of heavy material for the construction of Kalam Darchi Scouts post" (IOL/P&S/12/3285: 116; IOL/P&S/12/3288: 64).

74 Frank Saunders (1983: 107).

Road access to one of the oldest villages in Nagar is quite challenging as the jeep route from Tashot Bridge ascends a steep cliff in zigzag turns to reach the plateau where Pheker village is located. The photograph was taken on 19 November 1984

the braided river slowed down. The existing colonial records are the single source for 'minor' events that did not make it into the 'History of Hunza'. Nevertheless, the damage and destruction reported first of all focused on roads and bridges; in addition the events had severe impacts on habitation, village lands, and irrigation channels and last but not least on numerous human lives that were lost during such calamities.

In the 1920s, the Dutch diplomat Philips Christiaan Visser and his wife Jenny Visser-Hooft devoted their attention to the Karakoram glaciers and made long and specific investigations and observations. She commented in view of the living conditions in a high mountain desert among glaciers: "Human life in these barren Kara-Korum valleys is a continuous struggle for existence: a struggle not only to force the sterile soil to yield enough to keep them from starvation, but also a struggle against the terrible menace of Nature. On every side these men see the devastating phenomenal forces at work: the avalanches thundering down, the treacherous stones falling, rocks crumbling where they have stood for ages, rivers suddenly flooding their habitations and the fields their patient labour has wrested from the cruel desert."[15] In her opinion there remained only a narrow space for survival in a harsh environment, and that was episodically threatened.

It is revealing that a settlement hardly ever had to be completely abandoned after such an event; Atabad in 2010 might be one of such recent cases where resettlement would not be advisable and feasible. The earliest event in our record was the devastating glacier lake outburst flood that, according to a local legend, destroyed all but one of the settlements in the outwash plain of the Yishkuk Glacier in the Chupursan Valley in the 1830s.[16] It took about a century before resettlement was considered and systematically implemented from the 1920s onwards. The first channel was built in 'Irshad Valley' in 1918; a continuous process of settlement commenced that has lasted until today.[17] The village of Yarzirich

The braided river at Pasu has undercut the river terrace on which the village lands are situated. The photograph was taken on 31 August 2015

Year	Event	Locality	Remarks/extent of damage
1944	G	Gircha	river undercutting damaged ten out of Gircha's fifteen houses[75]
1945	R, T	Atabad	blocking of river, damage to terraces in the valley bottom[76]
1947	R		stone avalanche destroys flock[77]
	M	Nomal	Gilgit-Chalt track damaged[78]
1950	G	Masot	river bank under-cutting led to loss of village lands[79]
1952	G	Altit	agricultural lands in Yavari Das lost due to under-cutting of river[80]
1954	A	Gwachi-Chalt	avalanche destroyed road[81]
	G	Budelas	bridge washed away by flood[82]
1956	G	Hussaini	advancing Ghulkin glacier destroyed channel heads of Hussaini[83]
1957	G	Sumayar	river bank under-cutting due to flood from Shimshal River[84]
	G	Shimshal[85]	
1959	G	Shimshal	damage to irrigation channel in Nomal; destruction of Sikandarabad bridge[86]
	G	Pasu	Shimshal outburst flood destroyed a substantial share of Pasu's village lands in August[87]
	T	Minapin, Uyum	destruction of houses and road infrastructure in Nagar and damage to the palace in Uyum
		Nagar	Nagar due to heavy rains in June[88]
1960	G	Shimshal	damage to bridge in Pasu[89]
	R	Maiun	damage to orchards[90]
1960-1964	G	Khurdopin	loss of farms and terraces in Shimshal in consecutive years[91]
1962	G	Shimshal	damage to bridge in Pasu[92]
1963	T	Sikandarabad	damage to irrigation channel[93]
	G	Sumayar	river bank under-cutting due to flood from Shimshal River[94]
1970	A/T	Atabad bala	orchards and four irrigation channels damaged[95]
1972	G/M	Batura	damage to bridge[96]
	A/T	Miachar	tree plantations damaged[97]

Opposite
Hussaini and Pasu are located in a narrow cultivable space between major glacier environments and the deeply incised canyon of the Hunza River. Both villages have expanded towards neighbouring scree slopes where they established new village lands that have often been a reaction to losses of valuable terraced lands due to under-cutting by the braided Hunza River. Kharamabad and Zarabad are reached by challenging footbridges across the Hunza River, while Janabad Das is the new Pasu settlement serviced by the Karakoram Highway. Both villages have access rights to pastures along the Batura Glacier that pose a valuable resource for combined mountain agriculture. Source: Topography based on Hunza-Karakoram, scale 1:100,000; Quickbird data, 11 May 2007; modified representation based on Hermann Kreutzmann (2012e: 54; 2015a: 471)

75 Kumara Menon (1947: 38-39); a flash flood in July had caused the collapsing river terraces.
76 Kumara Menon (1947: 33) was told about a severe rockfall blocking the river, no proper date was given: "Some time ago, a whole hill, with rocks and boulders as big as dinosaurs, came down during the rains and blocked the river for quite a fortnight. So effectively was the river blocked that, to the amazement of the villagers, the river which had been flowing towards Baltit flowed back towards Passu! The cultivation in Atabad was altogether ruined. Now they are beginning to cultivate their lands again." Most likely he referred to the village lands of Sarat.
77 Hans Gyr (1949: 68).
78 Hans Gyr (1949: 72).
79 AKAH Disasters Record (2018).
80 AKAH Disasters Record (2018).
81 Wolfgang Pillewizer (1960: 27).
82 Wolfgang Pillewizer (1960: 59).
83 Wolfgang Pillewizer (1960: 97).
84 AKAH Disasters Record (2018).
85 Christian Charles (1985: 369).
86 Richard Finsterwalder (1960: 787-788); Wolfgang Pillewizer (1960: 100); Gertraud Repp (1963: 209).
87 Gerhart Klamert (expedition diary for 24 August 1959) recorded at Gilgit a five-metre-high flood released from an overflowing glacial lake in Shimshal destroying bridges, roads and cultivated lands on its way down the valley; Wolfgang Pillewizer (1960: 100-101).
88 Hans-Jochen Schneider (1961a: 92-93).
89 John Clark (1960: 22).
90 Local knowledge.
91 Local knowledge.
92 Local knowledge.
93 AKAH Disasters Record (2018).
94 AKAH Disasters Record (2018).
95 AKAH Disasters Record (2018).
96 Andrew Goudie et al. (1984: 389).
97 AKAH Disasters Record (2018).

settled by people from Gulmit is one of the latest additions. The Sarat rockfall of 1858 led to the withdrawal from the juvenile settlement of Altikuċ there; a resettlement took place only in the 1930s.[18]

Temporary abandonment occurred in such cases as Matum Das where, shortly after a new irrigation channel was built and the village was renamed Pratapsinghpura in 1893, a flood from a side valley (*nālā*) destroyed the irrigation infrastructure and the habitations of settlers from Jaglot and Jutal who gave up the place and withdrew. Matum Das, nowadays known as Rahimabad, was resettled in 1904 after the *mīr* of Hunza got permission from the British and Kashmir authorities to repair the irrigation system and to recultivate the village lands in order to settle Hunzukuċ there.[19] The recurring mentions of damage in Shamets – 1858, 1895, 1905, and 1934 – are because these village lands were located in the floodplain of the Hunza River and were prone to episodic inundation. It has been a gamble with time whether tree plantations and alfalfa fields there would reward the farmer.

Other recurring events are floods caused by glacial lakes that formed when Shimshal glaciers such as Yazghil, Khurdopin and Virjerab blocked the valley.[20] The subsequent flood events have destroyed more than half of the former compact village of Shimshal itself. Pasu has been most affected further down the valley. In the list of events and in the 'History of Hunza' Pasu is the village that suffered more than any other from terrace under-cuttings leading to the loss of habitations and village lands. The records show repeated mention of Pasu, where in 1910 three houses, twenty field parcels (*wundr*) and many fruit trees were destroyed. Another flood occurred in 1944 and caused similar damage.[21] A more severe series of glacial floods is connected in the collective memory with the years 1960 to 1964. In fact, it began already in the summer of 1959 when surges of the glaciers located in the upper Shimshal Valley – Khurdopin, Virjerab, Yazghil – caused large ice-dams to close off the valley and allowed a major lake to form. A major outburst flood was expected and Hunza's early warning system – operated by light signals across the mountain ridges between Shimshal and Pasu and from there down the valley by telephone – had informed *mīr* Jamal Khan and the Political Agent in Gilgit about the imminent risk.[22] The latter ordered the dismantling of most bridges in the Hunza Valley. Richard Finsterwalder summarised the sequence of events leading to the outburst flood: "… on 21 August with the sudden burst of a

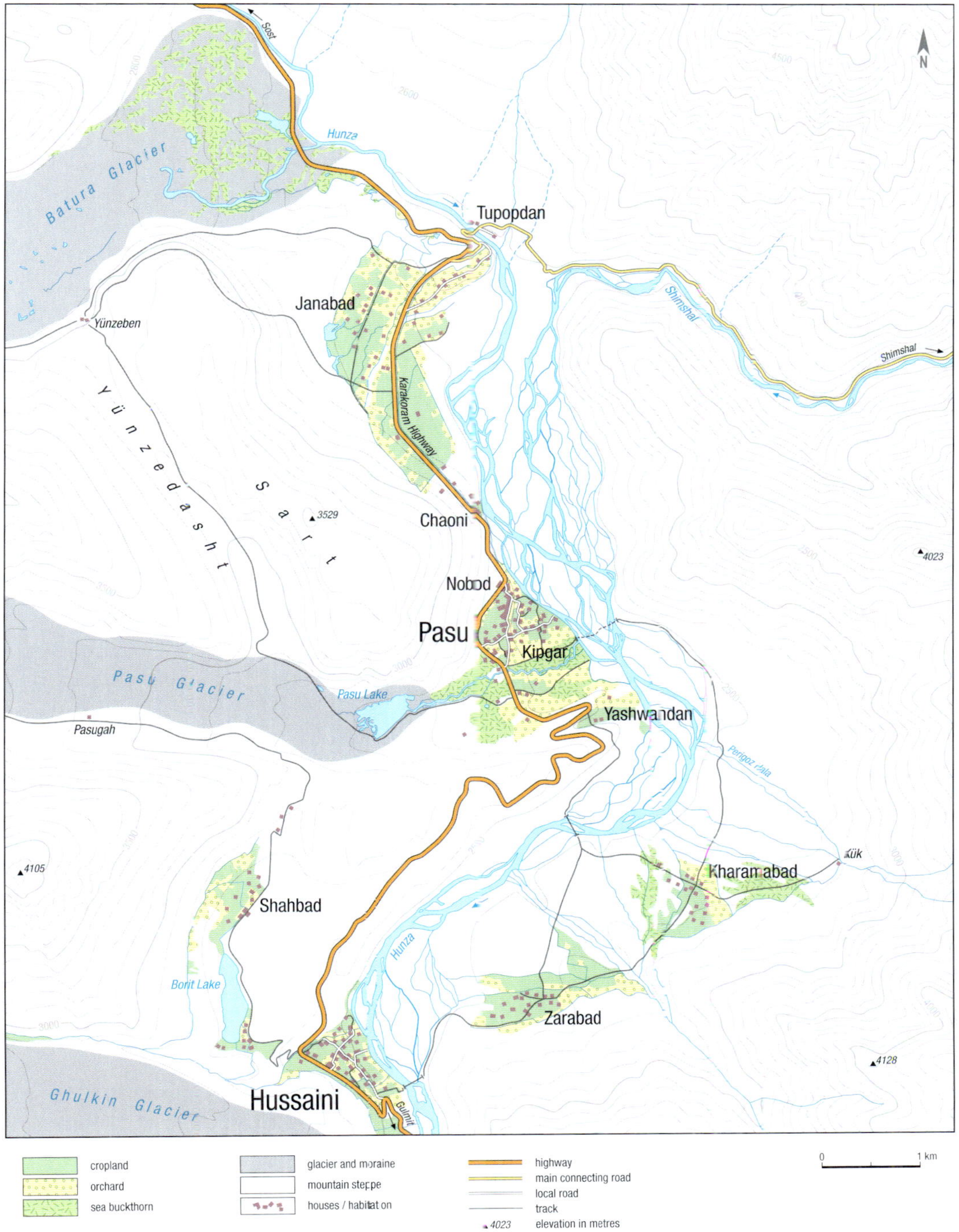

Along with Virjerab and Khurdopin, the white Yazghil is one of three major glaciers that are prone to block the upper Shimshal Valley and to episodically release major floods down the valley causing damage and terrace undercutting in Gojal. The photograph was taken on 17 July 2017

Year	Event	Locality	Remarks/extent of damage
1973	M	Momhill	damage to bridge[98]
1974	A/T	Altit	one house damaged and more than 100 trees destroyed due to flash flood in Ultar Bar[99]
	M	Balt Bar	damage to bridge, road, and plantations; six people died[100]
	G/M	Batura	damage to bridge due to river bank under-cutting[101]
	M	Halegush	damage to road-building on Karakoram Highway near Sarat[102]
1974-1993	A	Gulmit	annually covering/blocking of Karakoram Highway with snow/ice[103]
1975	T	Dadimal	village lands and orchards, fifteen houses partially damaged[104]
	G	Balt Bar	damage to bridge[105]
1976	A/T	Shimshal	agricultural lands and orchards in Baturkushi damaged[106]
	M	Yal	damage to road-building on Karakoram Highway near Sarat[107]
1976-1979	M	Shishket	glacier surge[108]
1978	M	Haiderabad	several irrigation channel heads destroyed in Dutang Shung (*ṣuṅ*)[109]
	G	Shimshal	damage to terraces in Pasu[110]
1980	G	Ghulkin	damage to Karakoram Highway[111]
1982	M	Koshi	one house and four irrigation channels damaged[112]
1983	R	Atabad	damage to irrigation channel in Atabad[113]
1985	A/T	Atabad bala	orchards and four irrigation channels damaged[114]
1986	G	Ghulkin	damage to Karakoram Highway[115]
	G	Pasu	six houses damaged in centre of Pasu[116]
1988	A	Jaglot	snow avalanche blocked Karakoram Highway[117]
	M	Ainabad	blocking of Karakoram Highway[118]
1990	A/M/R	Jaglot	blocking of Karakoram Highway[119]
	M	Sarat	blocking of Karakoram Highway[120]
	R	Multanza	rockfall killing traveller[121]

98 Nazir Kamal (1979: 24).
99 AKAH Disasters Record (2018).
100 AKAH Disasters Record (2018); Wenying Wang et al. (1984: 76).
101 AKAH Disasters Record (2018); Andrew Goudie (1981: 310); Andrew Goudie et al. (1984: 389).
102 People's Daily 18 June 1978 reported: "The glacier avalanche near Halle-Gush caused major landslide in 1974, dumping five million cubic metres into the Hunza River."
103 Own observation and confirmation.
104 AKAH Disasters Record (2018).
105 Salamat Ali (1977: 66).
106 AKAH Disasters Record (2018).
107 People's Daily 18 June 1978 reported: "The landslide in Yal in 1976 brought down 60,000 cubic metres of earth at one fell swoop."
108 Kenneth Hewitt et al. (2017); Frank Saunders (1983: 107).
109 AKAH Disasters Record (2018).
110 Andrew Goudie et al. (1984: 389).
111 Edward Derbyshire, Monique Fort and Lewis Owen (2001: 62); Andrew Goudie (1981: 310).
112 AKAH Disasters Record (2018).
113 Frank Saunders (1983).
114 AKAH Disasters Record (2018).
115 Own observation.
116 AKAH Disasters Record (2018).
117 Edward Derbyshire and Monique Fort (2006: 92); own observation.
118 "In July 1988, the road was cut for three days by an extensive debris flow, composed of granodioritic boulders in a dense, coarse sandy matrix. ... The debris-flow triggering factor in the 1988 case was heavy and continuous rainfall during the preceding 3-4 days, a situation that can occur from late spring to autumn and which is liable to occur at least once a year" (Edward Derbyshire, Monique Fort and Lewis Owen 2001: 62).
119 Own observation.
120 Own observation.
121 Own observation.

The Shimshal River has undercut the main settlement terrace. The former fortified village was mainly washed away; only some houses remain at the edge of the river terrace. Since the 1960s the houses have been relocated within the village lands and newly established hamlets. The photograph was taken on 9 May 1985

lake dammed by the tongue of one of the big glaciers in the very remote Shimshal Valley … It is assumed that it was the Malangutti or the Yazghil Glacier … The flood caused by the burst had a depth of about 30 m at the junction of the Shimshal with the Hunza, about 40 km from the assumed position of the lake. The flood destroyed the village of Pasu near its junction, and further bridges over the Hunza River were swept away, causing heavy losses. The bursting of the lake had been expected three months before. Perhaps the bursting of the lake was caused by the extraordinary precipitation which had occurred seven weeks earlier."[23] The last sentence documents that the causes of glacier surges in the Shimshal Valley had not been investigated and were not well understood. The frequency of such events was alarming and the damage to settlements and infrastructure substantial.

The major suspension bridge that connected Hunza and Nagar near Sikandarabad and several minor ones were destroyed as well as part of the jeep track.[24] A bridge in Pasu was damaged, but more severely huge tracts of cultivated land collapsed into the river owing to its under-cutting of the main terrace. Those years are remembered in Pasu as the 'danger years'. Some village elders had a vivid memory of these four years of loss. Hassan Khan's garden (*bağ*) was the first to vanish; on a daily basis people could observe the collapse of the cliff and the loss of their valuable property. The Virjerab Glacier became a synonym for destructive natural forces. Immediately plans were made to shift a substantial share of the population to Janabad Dasht where irrigation channels (*wod*) had to be made and to the higher-lying hamlet of Nobod (new village) above the old settlement nucleus (*yadiyor*) where 18 households found an abode.[25] The force of the outburst floods had been so strong that all efforts to control the undercutting of Pasu's river banks by establishing protective bunds failed. The process of land loss could not be stopped. Comparison of aerial photographs and satellite images from 1937, 1966 and 2007 has revealed the size

During summers Pasu women cross the Hunza River twice daily on a shaky 140-metre-long suspension bridge that allows them to work on their fields in Kharamabad and to look after their livestock. The photograph was taken on 30 June 1985

Year	Event	Locality	Remarks/extent of damage
1991	A/M/R	Jaglot	blocking of Karakoram Highway[122]
1992	A/M/T	Gilgit-Sost	blocking of Karakoram Highway at several places after heavy rains and snowfall[123]
	M	Minapin	one person died, one house destroyed[124]
	M/R/T	Hunza	destruction of bridges and paths[125]
	A	Hispar	damage to three irrigation channels and village lands[126]
	T	Shimshal	damage in pasture settlements and village lands[127]
	R	Shimshal	road blockage due to rockfall[128]
	A/M	Chupursan	damage in pasture settlements and village lands[129]
1993	M	Murtazabad	blocking of Karakoram Highway[130]
1995	A	Nilt	snow avalanche blocked Karakoram Highway; six houses partially damaged and agricultural terraces[131]
	A	Dadimal	one person killed, cattle-sheds and orchards damaged due to snow avalanche[132]
	R	Shayar	two houses, irrigation channel, ten cattle-sheds and orchards damaged[133]
1996	M	Faizabad	timber orchard destroyed[134]
1997	T	Chaprot	three houses, village lands destroyed[135]
1998	R	Masot	shop, irrigation channel and main road damaged[136]
	M	Ghalapan	two houses destroyed[137]
	R	Kirmin/Ispenj	damaged terraces[138]
1998-2005	G	Haiderabad	regular river bank under-cutting led to the loss of agricultural land and orchards[139]
1999	R	Misgar	damage to houses[140]
2001	A	Shimshal	link road and irrigation channel in Wodthur damaged[141]
2004	G	Chaprot	ten houses lost due to river bank under-cutting[142]
2005	A/T	Miachar	twenty houses and two mosques partially damaged, as well as cattle sheds and several irrigation channels damaged[143]
	A	Karimabad	five animals killed in Ultar pasture[144]

122 Own observation.
123 Local information.
124 AKAH Disasters Record (2018).
125 Local information.
126 AKAH Disasters Record (2018).
127 Local information.
128 Lasafam Iturrizaga (1997: 311–312): "After an unusually intense precipitation event from 07.09–09.09.1992 the whole of the new Shimshal Road was largely destroyed ... The Karakorum Highway (KKH) ... was also not passable for motor traffic in all of its length for two weeks because of numerous mass movements. ... Entire road sections were swept away by landslides. Also the route was overrun by many mudflows. Thus the routeway seems to be the most dangerous part of the landscape under these weather conditions. The man-made cuts into the cliffs for the road remove support from the cliff sections."
129 Local information.
130 Own observation.
131 AKAH Disasters Record (2018).
132 AKAH Disasters Record (2018).
133 AKAH Disasters Record (2018).
134 AKAH Disasters Record (2018).
135 AKAH Disasters Record (2018).
136 AKAH Disasters Record (2018).
137 AKAH Disasters Record (2018).
138 AKAH Disasters Record (2018).
139 AKAH Disasters Record (2018).
140 AKAH Disasters Record (2018).
141 AKAH Disasters Record (2018).
142 AKAH Disasters Record (2018).
143 AKAH Disasters Record (2018).
144 AKAH Disasters Record (2018).

Opposite
The Hunza River has undercut the major terraces and scree slopes between Hussaini and Pasu, thus reducing the village lands. The wide river bed can only be crossed via a long suspension bridge that is hardly visible from afar. The photograph was taken on 24 September 1990

of damage that affected the agricultural and settlement base of Pasu.[26] The riverbed has been widening and the Hunza River is meandering in its expanded valley. It is a recurring feature of the Hunza River valley that narrow stretches are followed by wide basins that in many cases have been formed by outburst floods.

In a similar manner, Hussaini lost valuable land and its unusual location is a prime example for the widening valley. Both villages have reacted by cultivating additional lands on both banks of the Hunza River or by creating filial settlements elsewhere.[27] A generation after the major flood event when the lake had reached all the way up to Pasu, about forty households were living there during the 'Gilgit Mission' that recorded the village size in 1886.[28] Despite out-migration the present number of households is fourfold and has built houses in safer locations such as Janabad.[29] The only settlement that was completely abandoned was Abdullah Khan Dasht[30] in the Borit area. It was founded as Sholemal and supplied with irrigation water from the southern bank of the Pasu Glacier in the 1930s. After the channel head (*sar*) had been destroyed by glacial movement the alfalfa cultivation in Sholemal and the hamlet of Jalalabad were given up, and all subsequent attempts to re-establish a water supply have failed so far. Other small hamlets close to Borit are still struggling today to safeguard their water supply from Ghulkin Glacier. The village of Gircha in Upper Gojal shifted from its original location on a river terrace close to the cliff further upslope towards Sarteez, which is nowadays regarded as the main settlement of Gircha that had lost habitations and village lands owing to undercutting in the aftermath of high floods. This process is continuing. The general impression supports the thesis that older villages are supposedly in safer locations than younger affiliations. Nevertheless, exceptions are possible, as in the case of Miachar where since 2005 the threatened village has been losing valuable habitations and lands and where a serious discussion whether to evacuate this settlement location is unavoidable.

The majority of recorded events are linked to direct or indirect action, interrupting communication infrastructure – roads and bridges – that in their function of connecting places have to be routed through dangerous and hazardous locations. For the colonial traveller the scaffold constructions for suspended mountain paths (*parii* or *perien*) were regarded as prime challenges in managing the narrow

The Minapin Glacier is fed from a major basin that stretches from the eastern flank of Diran (7,269 m) to Rakaposhi (7,788 m). The photograph shows the Diran section and was taken on 21 July 1985

Year	Event	Locality	Remarks/extent of damage
2005	T	Shimshal	orchard in Barkut damaged[145]
since 2005	G/R	Sumayar	regular damage to link road due to river bank under-cutting[146]
2005–2010	M	Sikandarabad	damage to irrigation channels and village lands[147]
	M/R	Ahmadabad	damage to village lands and link road[148]
2006	G	Pisan	bridge to Minapin destroyed[149]
	T	Minapin	one person died, one link road and bridge, two irrigation channels and several houses partially damaged[150]
	T	Miachar	about five houses and one mosque totally and a number of buildings partially damaged, link road and irrigations channels interrupted[151]
	R	Hispar	orchards, cattle-sheds, and village lands damaged[152]
2008	G/T	Gilgit-Hunza	the Gilgit-Hunza bridge (*China pul*) partly collapsed due to high August flood in Hunza River[153]
	T	Minapin	interruption of link roads; several houses damaged[154]
	G	Hussaini	damage to village lands due to massive water release from Ghulkin glacier[155]
	T	Gulmit	bridge and agricultural land destroyed[156]
	G	Pasu	massive water release from Pasu Glacier damaged houses and property[157]
2009	M	Aliabad	one house destroyed[158]
	A	Gulmit	snow avalanche blocking the Karakoram Highway at the notorious point of Wuch (*wuč*) Shoth, close to Shishket Bridge[159]
	G	Pasu	nine houses, agricultural fields and orchards damaged[160]
	M	Khaiber	bridge (*mīr skord*) between Ghalapan and Khaiber collapsed due to massive landslide[161]
	M	Shimshal	water-mills in Aminabad damaged[162]
2010	M	Danyor	seven houses destroyed[163]
	T	Nomal	one house destroyed[164]
	T	Jutal	25 houses destroyed due to thunderstorm[165]
	T	Chaprot	one house and one mosque, village lands partially destroyed[166]
	G	Chaprot	three houses and village lands destroyed due to river bank under-cutting[167]
	R	Chalt	damage to village lands[168]

145 AKAH Disasters Record (2018).
146 AKAH Disasters Record (2018).
147 AKAH Disasters Record (2018).
148 AKAH Disasters Record (2018).
149 AKAH Disasters Record (2018).
150 AKAH Disasters Record (2018).
151 AKAH Disasters Record (2018).
152 AKAH Disasters Record (2018).
153 Pamir Times, 17 September 2008.
154 AKAH Disasters Record (2018).
155 Pamir Times, 22 May 2008.
156 AKAH Disasters Record (2018).
157 Pamir Times, 6 January 2008.
158 AKAH Disasters Record (2018).
159 Pamir Times, 28 February 2009.
160 AKAH Disasters Record (2018).
161 Pamir Times, 19 March 2009.
162 AKAH Disasters Record (2018).
163 AKAH Disasters Record (2018).
164 AKAH Disasters Record (2018).
165 AKAH Disasters Record (2018).
166 AKAH Disasters Record (2018).
167 AKAH Disasters Record (2018).
168 AKAH Disasters Record (2018).

track, as the overhanging structures afforded some sense of equilibrium and reduced acrophobia. Nevertheless, many dangerous locations were found between these man-made structures where avalanches, pebble and rockfalls, land and mudslides as well as sudden collapses of morainic structures, terraces and slopes could expose the traveller to danger. The travel risks multiplied during and after blizzards and thunderstorms. In an arid mountain environment, precipitation is an unwelcome trigger of mass movement. Caravan leaders and drivers were respected for their ability to organise safe travel under the detrimental conditions that often occur in mountain areas. The Karakoram Mountains offer a contrasting picture of physical challenges. The Hunza Valley is characterised by deeply incised valleys and steep slopes from where the glaciated areas are visible only in the distance. The northern landscapes are less harsh on the other side of the Karakoram passes such as Kilik, Khunjerab or Mintaka and have Central Asian features of plateaux-like structures. Here, the glaciated regions are less extensive; the slopes are less precipitous; the undulating steppe region characterises

The fertile village of Minapin is located on a protected river terrace on the northern slope of Rakaposhi. On several occasions natural hazards occurred that effected the cultivated lands and infrastructure. In 192[illegible] a pressure wave destroyed all fruit trees and killed livestock; sometimes torrential rains damaged the road infrastructure and houses. The photograph was taken on 20 July 198[illegible]

Atabad Lake changed Gulmit's perspective on its property downstream. Access to Gaush pastures on the opposite slope was affected by the loss of Shishket Bridge and the drowned village lands of Shishket and Ainabad; the latter's old name Goshben is the Wakhi expression for 'lower Gaush'; only the high-lying pastures were not inundated. The photograph was taken on 8 September 2013

Year	Event	Locality	Remarks/extent of damage
2010	M/R	Hussainabad	Shinaki-Hunza link road destroyed, two people injured, loss of land, damaged irrigation channels[169]
	T	Pisan	bridge to Minapin destroyed[170]
	T	Dadimal	village lands and orchards, fifteen houses partially damaged[171]
	T	Shayar	road, irrigation channel and village lands damaged[172]
	A	Aliabad	snow avalanche uprooted 50 trees[173]
	M	Haiderabad	several irrigation channel heads destroyed in Dutang Shung (*ṣuṅ*) and Haiderabd Har[174]
	G	Hispar	river bank under-cutting destroyed the link road, loss of irrigated lands and orchards[175]
	M/T	Altit	Duiker road interrupted due to flash flood in Talmushi Bar, agricultural terraces destroyed[176]
	R	Atabad	major rockfall blocking the Hunza Valley, forming a lake behind the dam that has persisted since[177]
	M/R	Gojal	blockage of Karakoram Highway on several places between Gulmit and Sost; formation of Atabad Lake causing inundation of infrastructure, habitations and village lands in Ainabad, Shishket and Gulmit[178]
	M	Shimshal	water channel damaged in Wodthur; several access roads and orchards affected[179]
	G	Chupursan	six bridges near the glacier terminus were destroyed in April[180]
2011	M	Chalt	landslide between Chalt and Jaglot blocked Karakoram Highway[181]
	M/T	Miachar	one person injured, five irrigation channels and two jeep roads damaged[182]
	M	Altit	damaged terraces and orchards in Talmushi Bar[183]
	A	Chupursan	snow avalanches affected pastures[184]
2012	T	Pisan	bridge on link road destroyed[185]
	T	Minapin	one person died, link road damaged[186]
	M	Maiun	several landslides destroyed water channel, twenty households had to be evacuated[187]
	M/R	Yal	at the Yal Pari sliding area two Chinese road workers were killed[188]
	R	Shayar	six cattle-sheds and orchards damaged[189]
2013	R	Atabad Lake	two people were killed due to rockfall[190]
	G	Pasu	under-cutting of river terraces[191]

169 AKAH Disasters Record (2018); Pamir Times, 9 October 2010.
170 AKAH Disasters Record (2018).
171 AKAH Disasters Record (2018).
172 AKAH Disasters Record (2018).
173 AKAH Disasters Record (2018).
174 AKAH Disasters Record (2018).
175 AKAH Disasters Record (2018).
176 AKAH Disasters Record (2018).
177 Kenneth Hewitt (2014); Hermann Kreutzmann (2010); Pamir Times, 4 January 2010.
178 AKAH Disasters Record (2018); Pamir Times, 31 May 2013.
179 AKAH Disasters Record (2018).
180 AKAH Disasters Record (2018).
181 Pamir Times, 11 September 2011.
182 AKAH Disasters Record (2018).
183 AKAH Disasters Record (2018).
184 AKAH Disasters Record (2018).
185 AKAH Disasters Record (2018).
186 AKAH Disasters Record (2018).
187 Pamir Times, 29 September 2012.
188 Pamir Times, 1 October 2012.
189 AKAH Disasters Record (2018).
190 Pamir Times, 16 July 2013.
191 Pamir Times, 31 August 2013.

the Taghdumbash Pamir. The latter posed comparatively fewer challenges for making a road access and is less prone to some of the geohazards mentioned above.

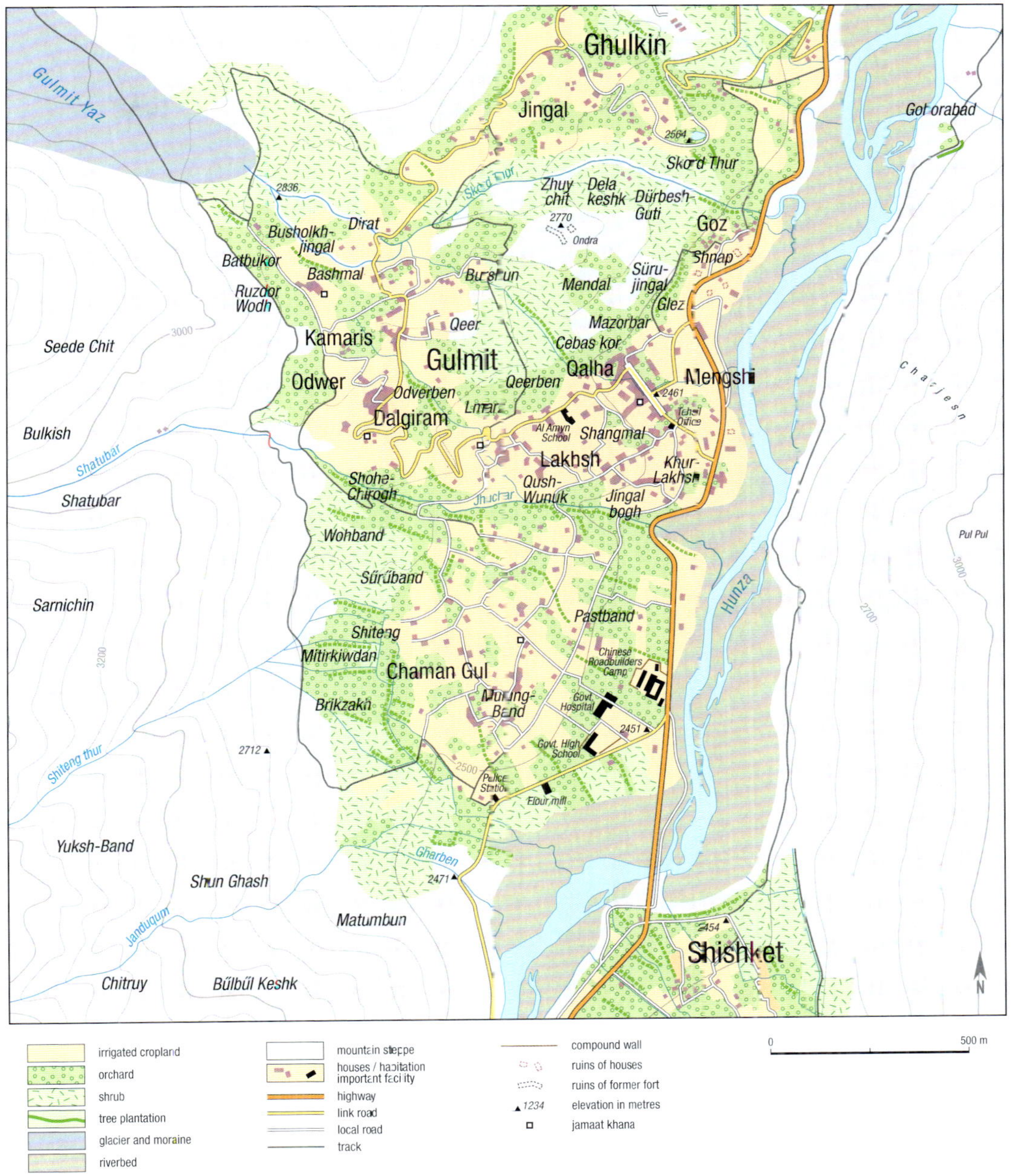

The largest Wakhi village is characterised by the remnants of a residence of the mīr of Hunza that was visited in former times when the tham spent some time in Gulmit in autumn for hunting. The houses of the leading families are found in its vicinity. From this nuclear settlement an expansion into filial hamlets has taken place. The new settlement of Chaman Gul is the latest and fastest growing hamlet, featuring a huge jamáat khana that was inaugurated in July 2017. Many households from Gulmit have built houses and cultivated new terraces. Some people refer to Chaman Gul as the 'New Gulmit' where affluent young families have found a place to live. After the regulation of Atabad Lake, the reconstruction of the Karakoram Highway along a slightly changed alignment and a new major Chinese bridge connecting Shishket and Gulmit, some of the lost land could be reclaimed as is shown in the map. Own design based on 'Hunza-Karakorum' map, satellite images, and Hermann Kreutzmann (2015a: 469).

Regularly roads and tracks are washed away after hazardous events. The Naltar Road was interrupted for long periods in summer of 2017. The photograph was taken on 5 August 2017

Year	Event	Locality	Remarks/extent of damage
2014	M	Ghulmet	portion of the village lands collapsed and fell into the river bed, 25 households evacuated[192]
	G	Miachar	evacuation of residents from village lands, precautionary measures were taken when cracks in river terrace widened[193]
	R	Shikas	a rockfall blocked the Karakoram Highway[194]
	G	Yazeyupk	Ghulkin glacier changed its terminal moraine and blocked the Karakoram Highway[195]
	G	Hussaini	irrigation channels destroyed by glacial movement[196]
2016	M/R	Rahimabad	landslide and rockfall threatened 25 houses[197]
	A/M/R	Atabad-Shishket	the newly opened Pak-China friendship tunnels were blocked after severe landslides in spring
	M/R	Gojal	major landslides block new tunnels above Atabad Lake[198]
2017	T	Shimshal	Shimshal road blocked after river bank undercutting[199]
2018	M	Misgar	landslide blocked the Misgar Road[200]
2019	G	Shishpar	advancing glacier destroyed the intake for hydroelectric plant and destroyed the channel head of *Aliabad gočil*[201]

*) Abbreviations:

A = snow avalanche G = damage caused by outburst floods from glacial barrages. These barrages were formed through glacier advances blocking a valley and thus creating lakes of different sizes, which eventually feed and propagate the outburst floods.
M = mudslide/landslide R = rockfall/rockslide T = thunderstorms causing flash floods W = wind action

192 Pamir Times, 3 October 2014.
193 Pamir Times, 11 February 2014.
194 Pamir Times, 21 October 2014.
195 Pamir Times, 4 July 2014.
196 Pamir Times, 7 May 2014.
197 Pamir Times, 17 April 2016.
198 Pamir Times, 5 April 2016.
199 Own observation.
200 Pamir Times, 19 May 2018.
201 Own enquiry in March 2019.

The occurrence of landslides and rockfalls had continued for several months and increased at the site of the Atabad disaster by the end of July 2010; after significant amounts of rainfall in the Hindukush, Karakoram and Himalayas in the following days the water masses accumulated to a wave that left the upper parts of the Indus watershed and inundated major populated parts of Sind Province and caused the flood disaster there. The photograph was taken at Atabad on 31 July 2010

Following page
The view from the west into the Hunza Valley presents major oases in the lower section of the picture such as Sikandarabad, Chalt and Chaprot in Nagar. The string of irrigated oases on both banks in the valley bottom reaches up to Atabad Lake in the centre of the picture. Just below the lake the major oases of Central Hunza and Nagar are visible flanked by Rakaposhi on the left bank and Ultar peaks on the right bank of the main river. Both highest elevations of nearly 7,900 metres are part of the Karakoram main range through which the Hunza River has cut its course at the valley floor of 2,100 m in the central oases. The deeply incised valley reaches from the desert and steppe at the bottom to the glaciated zone at the top. In between, the irrigated oases and pastures are located. Two broad depressions border the Disteghil Sar (7,788 m) massif and show the Hispar Valley and its glacier originating from Snow Lake in the east with Hispar Pass (5,150 m) at the extreme right of the picture; the glaciated depression northwest of Atabad Lake belongs to the 56-kilometre-long Batura Glacier and its neighbours. The Gojal villages towards the north are hidden in the narrow valley leading to Khunjerab, Kilik and Mintaka passes on the Chinese border. Source: Satellite image processed and provided by Christoph Hormann

Hunza Valley – from risk to potential

Various approaches have been applied to explain and understand living conditions in challenging high mountain environments such as the Karakoram Mountains. In early phases of mountain research these areas were mainly characterised by their physical dangers; at the same time limited accessibility to the resource potential of mountain ranges reduced active exploitation. Mountain passes would be crossed in order to get from one lowland to another. The human perspective emphasised the backwardness and remoteness of mountain valleys that contributed to the maintenance of 'traditional life' and preservation of cultural traits.[31] The study of archetypal patterns of behaviour, cultivation and herding practices, the search for the origin of plant species, unusual specimens and archaic languages: all these topics justified mountain-related research. Subsequently an enhanced specialisation in geographical research topics focused on a variety of more or less isolated aspects. A perspective informed by hazards and risk research emphasised the dangers posed by a mountain environment where verticality and exceptional elevation ranges can mobilise relief energy and become triggers for mass movements. A threatening environment would be classified in danger zones and exposed to hazard mapping. Highland-lowland interdependencies and linkages were perceived in a more comprehensive and holistic assessment of mountain processes. Geomorphologists and glaciologists highlighted the effects of glacial accumulation and ablation, avalanche nourishment and para-glacial phenomena on a landscape that is significantly affected by related action and provides the hydraulic resources for human habitations. The retreat of shrinking glaciers has survived as the main objective of investigation in freeze-thaw environments.[32] Nearly half of the listed events that have affected communications, habitations and village lands are related to glacial surge phenomena and outburst floods only. Climatologists and soil specialists judged the potential of this high mountain desert as a mixed blessing where the edaphic and weather conditions found in the valley bottoms allow crop farming, while the higher altitudes provide the hydraulic resources for irrigation purposes. The juxtaposition of various resources at different altitudinal layers and their connection by human action has been one of the main interests of human-geographical research in mountains.[33] Agronomists, botanists and vegetation experts were impressed by the transition from desert conditions in the valley bottoms to the sub-humid conditions just below the nival zone.[34] A precipitation gradient that increases by one order of magnitude from the valley bottom at about 2,000 m up to levels above 4,000 to 5,500 m correlates with the transition from desert-steppe plant species to areas that feature valuable grass and shrub resources.[35] This humidity gradient contributes to a differentiated use of available plant distribution at various elevation layers. It enables a combined mountain agriculture that is split into crop farming and aboriculture in the lower areas where permanent settlements are located and seasonally utilised higher areas where hunting and pastoral practices prevail.[36] At the same time, the horizontal variation between oasis, steppe and desert has been observed. Slopes with different exposure and aspect provide quite a spectrum of options in a sensitive area. Marc Aurel Stein observed the difference between Hunza and Nagar during his travels in the valley: "While the Nagir side of the valley shows a cheerful succession of villages, the opposite side, which belongs to Hunza, is here for the most part a rocky waste. The difference is easily accounted for by the increased supply of water which Rakiposhi provides."[37] A few years later Henry Lawrence Haughton observed "… a difference between

View from Malikamu Shikari towards Hopar and Hispar. The glaciated mountains tower high above the glaciers which flow in deeply incised valleys. The photograph was taken on 21 September 1985

Flat irrigated terraces are characteristic for the village lands of Central Hunza. The apricot blossom indicates the abundance of fruit trees in Karimabad. The photograph was taken on 3 April 1985

the formation of the fields in Nagar and those in Hunza. Although in both cases the cultivation is more or less terraced, the surface of the Hunza fields is made level … The reason for this difference is, I believe, firstly, because the hills on the Nagar side of the valley are for the most part of a much more gentle slope than those on the Hunza side; and, secondly, the difference is a matter of irrigation. In Hunza many people are all dependent upon one large kul for their water, so that times have to be allotted to them for watering their fields, village by village, and kom [*roóm*] by kom, which means that it always falls to some of them to turn out and water their ground by night, attending to each field separately. In Nagar, where there are a larger number of small kuls [*gūl*] flowing from various directions, the people are seldom obliged to irrigate by night. But even if obliged to do so, a Nagar man can water his ground without having to sit up all night to watch it; for the fields being upon a gentle slope, he can just turn the water on to one and let it find its own way down into the next, without fear of the retaining wall being carried away, as would probably occur in a Hunza field if it were left flooded for any length of time."[38]

For transport researchers, the marginality and remoteness of these mountain valleys counted as a locational disadvantage due to the physical challenges of a steep and high mountain system that could be perceived as a 'barrier to modern communication'.[39] The comparative approach of seeing in mountain areas only the infrastructural deficits not generally featured in the lowlands has informed most colonial studies that emphasised the asymmetry of assets. This negative reflection has been a driving force in development theory for the explanation of marginality and remoteness as backward, non-modern and societies caught in tradition and cultural traits.

All aspects and viewpoints mentioned so far are mainly informed by disciplinary reduction, although it seems obvious that human-environmental interaction may be the only key to explain and/or probably justify attributions such as barriers, hazards, risks and threats in specific contexts. Consequently, the conspectus presented here introduces environmental resources as assets for human use by highlighting constraints and their impact on coping and utilisation strategies. A recently introduced approach to perceive agricultural change by conceptualising farming systems as 'assemblages of diverse actants' has revealed the interrelatedness of 'factors and actors' and provided convincing evidence of the observed effects.[40] The Hunza Valley will initially be introduced in its physical factor endowment by keeping in mind specific challenges for actors. The separate introduction of certain properties leads to a more comprehensive and equal treatment of a number of contributing factors for shaping the living conditions in the Hunza Valley.

Aleksandr Yakovlev contributed a number of landscape paintings to illustrate the terrain covered by the expedition. The 'foothills of the Karakoram Mountains' indicate the steep slopes and the rugged mountain range emerging from a desert-like structure with scanty vegetation cover. Source: Oil on canvas painting by Aleksandr Yakovlev (1931)

Karakoram or Muztagh – life among boulders and glaciers

In the early 20th century the distinction between Hindukush and Karakoram was debated, and Sidney Burrard and Henry Hayden suggested that the watershed between the Gilgit and Hunza Rivers should be the dividing line as they had rightly observed that both mountain ranges are 'different sections of the same crustal fold."[41] Looking from afar on the Karakoram Mountains, the observer sees a landscape of stark contrasts that is reflected in the translation of its conventional name as 'black gravel' or in the alternative ascription Muztagh, a term that literally refers to 'snow mountains'. During the 'Karakoram nomenclature' debate in the 1930s some protagonists suggested the latter as an alternative toponym for the whole mountain range.[42] Both appellations from Turki languages describe the concurrence and contrast of ruggedness, gravel and extensive glaciation. Its exceptional position in the Greater Himalayan Region, which recently was more commonly referred to as the 'third pole', locates the Karakoram as a highly glaciated mountain area surrounded by arid lowlands. Since the mountains of High Asia have been rediscovered as a 'visible indicator' in the climate change debate, the 'third pole' discussion has devoted scrutiny to the unique interactions among the atmosphere, cryosphere, hydrosphere and biosphere with huge impacts on its glacial environments.[43] The Karakoram – 71°-79° E, 35°-36° N – occupies a niche in the southwestern corner of this area which has encountered significant environmental changes during the last century.

If the total land area of the Karakoram is estimated at 60,000 square kilometres, about one third can be related to perennial snow and ice cover.[44] The meltwaters leading out of the high elevated glacier basins drain through the dendritic river systems into the Indus River. Steep slopes connect the elevated mountain peaks with the deeply incised valley bottoms. The main Karakoram Range stretches for 650 kilometres from the Afghan border towards Ladakh and Tibet. Kenneth Hewitt has recently emphasised that verticality might or should be regarded as the 'primary Himalayan dimension'; among all descriptions that reflect the spatial-

The canyon of the Hunza River is deeply incised into it moranic filling. The village lands and the old khan settlement of Miachar are located on the river terrace where a comparatively flat ground is available. The photograph was taken on 19 May 1985

ity and extension of geographic phenomena, the third and vertical dimension occupies a special position in the given environment.[45]

About one quarter of the Karakoram is taken up by the Hunza Basin. Of its 14,100 square kilometres land area, 4,450 square kilometres are covered by glaciers located in the main Karakoram and Rakaposhi ranges.[46] The terminology applied by Kenneth Hewitt has replaced older and differing versions when the main Karakoram Range was identified as 'Batura-Muztagh' or 'Hunza-Karakorum' and the Rakaposhi Range as 'Saltoro-Karakorum' or 'Kailās Range'.[47] Despite all terminological variations most observers have admired the sheer force with which the Hunza River has cut its bed across the main ridges and thus created very steep cliffs and deeply incised valley sections.

The course of the Hunza River is unique as it is the only valley cutting across the entire range, having created a canyon-like valley with bordering flat river terraces, outwash fans and scree slopes.[48] In global comparison, some of the elevational extremes are met in the Hunza Valley where the difference in altitude between Rakaposhi (7,788 m) and the valley bottom of the Hunza River near Ghulmet (1,830 m) of about six kilometres is less than eleven kilometres apart in horizontal distance. Similar steepness of gradients is encountered between Bojohaghur Duanasir (7,350 m) towering above Altit and the Hunza River below (2,050 m) where the slope inclination averages about sixty percent.[49] Such an environment results in a high relief energy that can mobilise gravitational movements along slopes leading to debris accumulations close to the valley bottom. The Karakoram in general and the Hunza Basin in particular feature some of the longest glaciers outside the high latitudes, among them the Batura with a length of sixty kilometres and the Hispar which is about fifty-three kilometres long. The Batura has its highest point at 7,795 m, and its tongue reaches down to its terminus at 2,540 m. Their extensive debris covers in the ablation zone have helped to protect and stabilise glacier tongues at altitudes as low as 2,300 m. The ten largest glaciers are responsible for over half of the Hunza Basin's ice cover that is completed by numerous smaller glaciers and smaller ice masses.[50] The big glaciers significantly contribute to the Indus waters but are less valuable for village irrigation in the cultivated oases. The scattered village oases located on river terraces, outwash fans and scree slopes mainly depend on the contributions from small ice masses that can be tapped in the side valleys. The vertical regime between glacier accumulation and ablation zones is another distinguishing feature of the Hunza Basin as its large glaciers 'descend much lower than those in the rest of the Himalaya'.[51] A number of glaciers reach far below 2,500 m in the Batura and Rakaposhi ranges; those valuable hydraulic resources can be tapped close to the compact oases.

Utilisation of verticality in a high mountain desert

The criteria for establishing human habitations in such an environment are the accessibility and availability of sufficient quantities of water as well as suitable edaphic and thermal conditions with favourable soils for cultivating crops within an annual cycle that allows crops to ripen. Given the locational arid conditions of the Karakoram, the Hunza Valley represents a typical subtropical steppe environment in a mountainous setting with distinct altitudinal zonation of vegetation cover. The classification of vegetational belts begins at the valley bottom with the arid conditions in the desert steppe where Chenopodiaceae (*Haloxylon thomsonii*) are omnipresent and followed by artemisia (*Artemisia fragrans*) in the colline belt that reaches up to approximately 2,100 to 2,700 m, in some exposed locations significantly higher. Both zones are the major settlement regions for permanent habitations. The artemisia shrubs (B. *móiṅ*) are a valuable fodder and fuel resource. Following the slope gradient upwards, humid-temperate stretches are found in northern aspects where patches of remnant coniferous woods may occur locally. This zone, which used to be characterised by juniper (*Juniperus excelsa*), is nearly extinct in most of Hun-

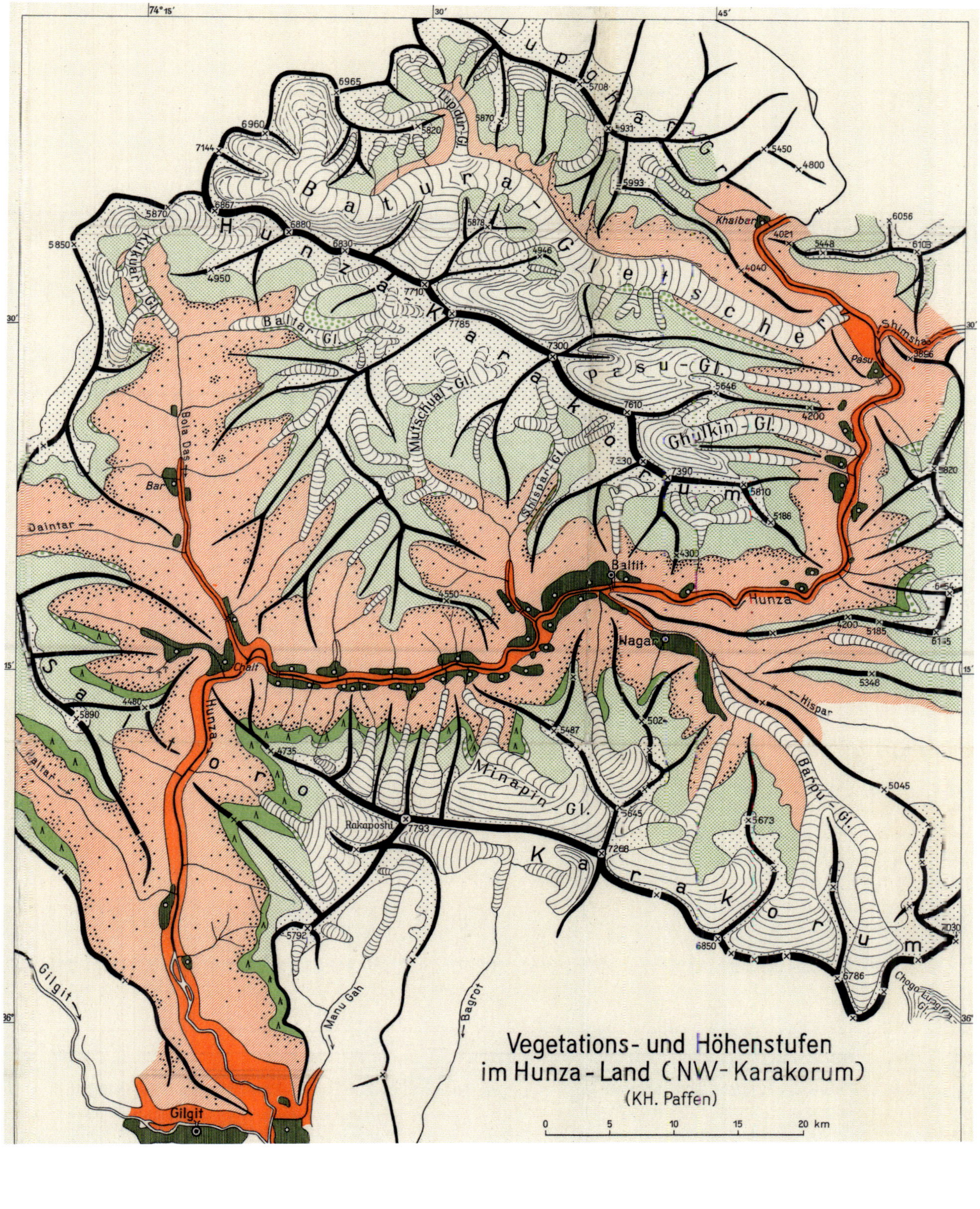

The spatial distribution of vegetation cover in the Hunza Valley as recorded by Karlheinz Paffen in 1954. Its altitudinal distribution is highlighted by bordering mountain ridges, glaciers and rivers. Source: Karlheinz Paffen, Wolfgang Pillewizer and Hans-Jochen Schneider (1956, supplement)

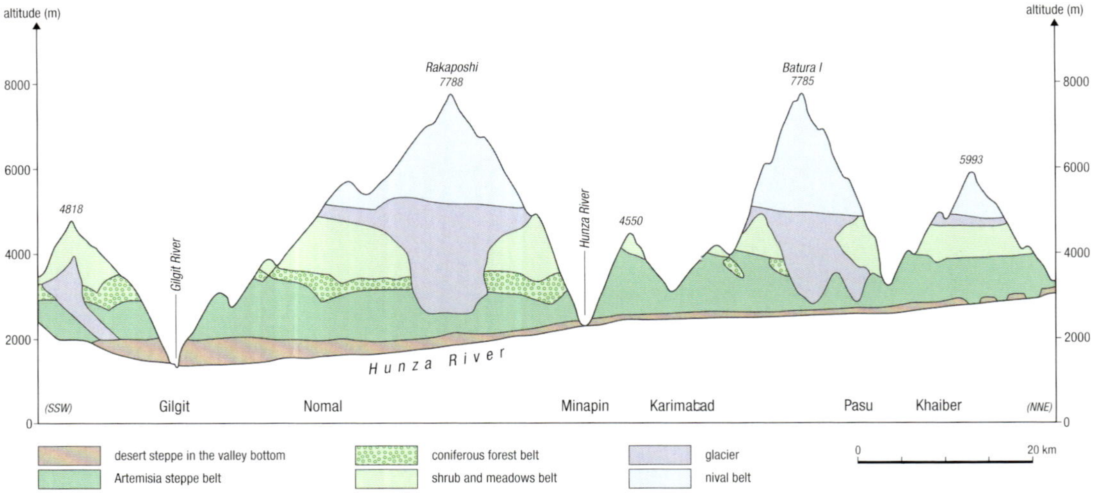

The cross-section through the Hunza Valley shows the steep slope gradients (five times elevated in the graph) and the various layers of vegetation cover. In the valley bottom where the irrigated oases are located, the characteristic environment features vegetation formations that belong to desert and steppe. Higher up, there is an extended zone of Artemisia steppe that is bordered in places by a narrow belt of coniferous forest, followed by shrubs and meadows that provide a valuable fodder resource in the vicinity of glaciers and below the nival belt. Source: Adapted from Karlheinz Paffen, Wolfgang Pillewizer and Hans-Jochen Schneider (1956, supplement) and Hermann Kreutzmann (1989a: 76).

za, but there are patches left in Nagar.[52] Economically more important is the next belt whose meadows and shrubs can be utilised for summer settlements in which additional cultivation of grain crops formerly took place. Most of these temporary settlements have been given up or remained as intermediate resting stages on the way up to the high pastures that follow and reach up to the zone of glacier moraines, perennial snow and ice.

The staged structure of vegetational belts offers a compact space for combined mountain agriculture where the desert-steppe conditions of the valley bottoms are suitable locations for oasis cultivation if water can be guided to the scree slopes and river terraces that offer suitable edaphic and thermal conditions. The intermediate stage offers wormwood shrubs (B. *móiṅ;* W. *təbəsk*) for fuel; these are augmented by irrigated plantations of sea buckthorn (*Hippophae rhamnoides*) that are widespread especially in Gojal (B. *ćhaṣ*, W. *zax̌*) and are a prime source for cooking and heating. The former function as a summer settlement with cultivation, spring and autumn pasturing having diminished in most places. Where accessibility and irrigation are available, these areas have been converted into valuable potato cultivation fields. The upper zones form the second pillar of combined mountain agriculture and offer sedges (*Kobresia, Carex*) for forage in the summer pastures.[53] Especially the regions close to glaciers and their ablation valleys are of prime importance for seasonal settlements and all-year-long pastures for non-milk-giving yaks. Significant rainfall in high altitudes contrasts with the arid conditions in the villages.

Temperature variation in a mountain desert

Climatically the Karakoram mountains form a barrier between the monsoon-dominated lowlands of the Indian subcontinent and the arid belt of Central Asia with its huge desert basins of the Tarim and Fergana. The Karakoram valleys are thus affected by a monsoonal climatic regime as well as by westerly depressions forming a transition zone. In the vertical dimension, extreme differences of precipitation conditions have been recorded between arid, desert-like valley bottoms and the humid nival zone, thus separating potential settlement regions from those where sufficient humidity is available.

Until the mid-1980s total annual precipitation in the Hunza Valley at Karimabad was recorded at as low as 145 mm on average and for Gilgit at 131 mm; consequently both valley areas range well below minimal requirements for rain-fed (*barani*) cultivation. More recent data recordings for a new station in Gerelt give 194 mm, and for the last three decades in Gilgit an average precipitation of 143 mm has been recorded.[54] On the other hand, measurements of ablation and related calculations suggest a maximum precipitation of approximately 2,000 mm at 5,000 metres altitude. This significant gradient explains the desert conditions in the villages and the enormous glaciation in the upper elevations: the typical environmental feature of verticality symbolised in snow-covered Rakaposhi (7,788m) and Disteghil Sar (7,885m) towering above the arid valley bottoms. The 'scope of vertical atmospheric gradients is critical for the area-altitude distribution of the ice cover'; while the average precipitation for the Hunza Basin was calculated as 816 mm, we find the accumulation of four fifths of hydraulic resources bound in the altitude range between 4,000 and 6,000 metres.[55] The summer meltwaters are guided from the glaciers and snowfields by irrigation channels towards the village lands in the valley bottoms where edaphic and thermal conditions are favourable and allow cultivation when water can be made available.

Mean average temperatures vary seasonally with maxima in July-August and minima in January with an amplitude of twenty-five degrees centigrade. In a glaciated region like the Karakoram these variations in temperature determine and actuate the volume of available meltwater for irrigation in the valley bottoms. The Hunza River offers fifty-two times more water during summers than winters.[56] The period of seasonal meltwater release relates to the climatic conditions and, in connection with the altitudinal location of settlements, determines the length of the cultivation period for crops in irrigation oases. The average duration of annual growing cycles ranges from 307 days for Gilgit (1,450 metres above sea-level), 260 days for Karimabad (2,300 metres), to 195 days only for Misgar (3,102 metres) where semi- and humid conditions only prevail in November to March when snow cover is likely and no plant growth takes place. Overall, these irrigation oases and cultivated areas cover less than one percent of the surface of the Karakoram Mountains, but remain the focal points for human settlements and agriculture-based livelihoods. Combined mountain agriculture has the advantage of simultaneous fodder production in the permanent homesteads for herds which are grazed in the high-lying pastures during the summers. The limiting factor for animal husbandry is the necessity to provide seven to nine months of feed which has to be produced on private or common property village lands or purchased. About half of the area of Northern Pakistan is covered by extensively utilised pastures, which are only available during three to five months in summer. Consequently, the usual structural settlement setup and the familiar agricultural production system have productively used the salient features of the given mountain desert environment. Such a technical approach to the environmental and economic potential neglects other diversities that are linked to human agency and historical processes.

The Kilik River valley is dominated by the village of Misgar. At an altitude of 3,102 metres the cultivation period is two months shorter than in Central Hunza and allows only single-cropping of wheat and potatoes. Its ecological setting in a high mountain desert is easily detectable outside the irrigated oasis. The photograph was taken on 23 September 2000

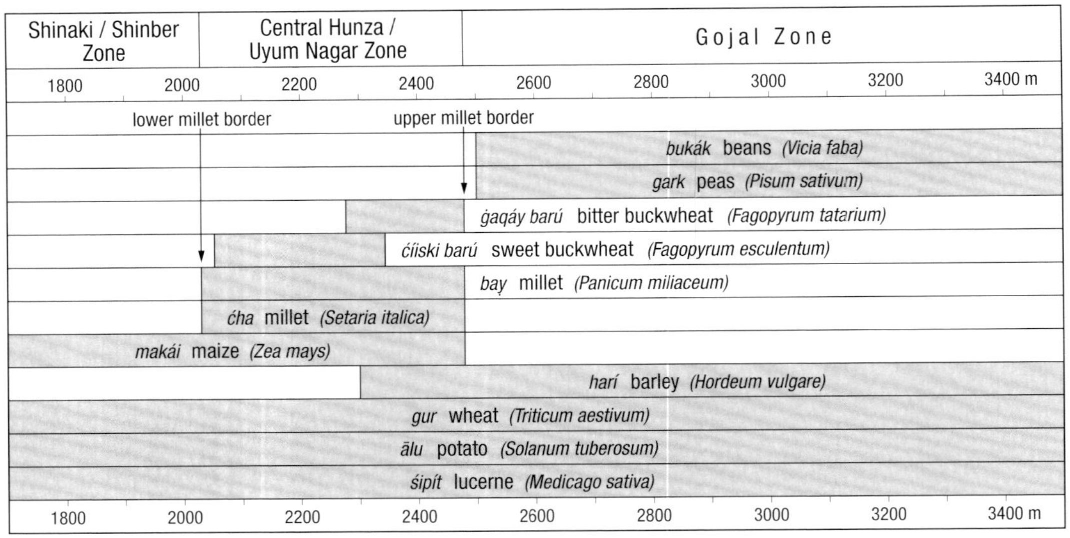

The three distinguishable cultivation zones of the Hunza Valley are separated by the upper and lower limits of millet and buckwheat as sequential crops in the intermediate zone, thus placing a more complex grain production between single- and double-cropping areas. Source: Modified from Hermann Kreutzmann (1989a: 101; 2006: 333)

Opposite
The agro-ecology of the Hunza Valley is characterised by the location of oases in the valley bottom. In the lower section between Gilgit and Thol these villages feature a full double-cropping zone (do-faṣlī). In Uyum Nagar, Hopar and all of Gojal single-cropping (ek-faṣlī) is dominating the cultivation patterns. The village oases in between – Central Hunza, parts of Shinaki, Shinber and Nagar – are located in a transition zone where, by applying certain crop sequences, double-cropping might be accomplished under favourable weather conditions.
Source: Modified version based on Hermann Kreutzmann (2006: 332).

Ecological differentiation along the Hunza Valley

The major characteristic of the village lands of Hunza and Nagar is the transition from a double-cropping zone in the low elevations (Khizrabad and Chalt at approx. 1,800 m) to an area of only single cropping (up to the highest settlements of Shimshal and Hispar at approx. 3,500m). Such sequences are mainly attributed to a change of agro-climatic conditions with altitude, although the selection of crops, choices and preferences and agricultural management need to be taken into account as well. It is evident that these are references to previous cultivation patterns and have been modified in recent years when new crops were introduced along with faster-growing varieties. In the inherited system of differentiation the central oases of Hunza and Nagar are distinguished from the higher and lower-lying valley sections. The dominant position of the densely inhabited central oases is correlated with the attempt to cultivate two crops (*do-faṣlī*) per annum. The sequential crop grows faster than the basic food crops of barley and wheat. These staple grains are followed by millets and buckwheat which are only cultivated within the limits of this intermediate zone. Consequently, these sequential crops could be identified as the distinguishing markers within the cultivation patterns of the Hunza Valley. They do not occupy such a prominence in the local dietary patterns. It might be permissible to assume that the sequential crops identify the grave shortage of food stuffs that has prevailed for generations in the central oases. Since external supplies of cereals have augmented the households' storage the importance of millets and buckwheat has been diminished further.

Thus, three dominant zones, separated by elevation levels, may be distinguished in the course of the Hunza Valley. Incidentally, they somehow correlate with local terms derived from settlement patterns. Another coincidence is that we find Shina-speakers only in the lower section of the valley, and Wakhi settlements only in the upper valley in Gojal. The sections are named according to local regional distinctions.

(1) Shinaki or Shinber – double-cropping (*do-faṣlī*) zone
Here two crops are regularly cultivated in the course of a growing season with a sequence of maize (*Zea mays*, B. *makái*) following winter wheat (*Triticum aestivum*, B. *gur*, W. *ỹədím*). The thermal conditions allow both crops to ripen. The critical point is reached at altitudes of 2,000 m – Thol and Khanabad – where maize is often replaced by millets, either foxtail varieties (*Setaria italica*, B. *ćha*) or panicled ones (*Panicum miliaceum*, bur. *bay*). This border coincides with the lower limit of millet cultivation in the Hunza Valley.

(2) Central Hunza and Uyum Nagar zone:
intermediate zone of single/double cropping
Here the transition between double and single cropping takes place and is strongly dependent on the weather conditions of each year. In some years it is feasible to harvest two sequential crops of summer barley (*Hordeum vulgare*, B. *harí*, W. *yirk*) followed by maize or potatoes (*Solanum tuberosum*, *ālu*). A second sequence is composed of winter wheat followed by millets or buckwheat (*Fagopyrum esculentum* or *F. tataricum*, B. *ćiiski barú*, *ġaqáy barú*). It regularly happens that the crop sown in summer after the bread grains (barley, wheat) have been harvested cannot be brought to full ripening. Then it has to be fed to the livestock as additional fodder. The sequences have been diversified in recent years owing to efforts by development institutions and changing markets.

(3) Gojal zone: single-cropping zone
Although only lying a little higher than the intermediate zone, here single crops of wheat, barley, potatoes, beans (*Vicia faba*, B. *bukák*, W. *bəqlá*) and peas (*Pisum sativum*, B. *gark*, W. *ṣ̌ax*) are harvested without any stress of a limited

vegetation period as no sequences are sown apart from some fodder crops.

Prevalent explanations for three distinctive zones of cultivation strategies are presented mainly from an agro-ecological logic. Strong winds in Gojal (*bād-e Gojal*) resulting in slightly more adverse environmental conditions are regarded as a signal for regularly occurring difficulties in the case of rotation cycles of two crops planned. Proof for such a micro-climatic reason is still pending and unlikely. Elizabeth and John Staley tried to explain these phenomena in the 1960s already from a population-pressure-dominated point of view.[57] They concluded that in Hunza the highest elevations for double cropping are due to high population density although their empirical work was restricted to Nagar. Historical records exist from periods when the population of Central Hunza was less than a third of today's, and already then double cropping was the cultivation strategy although higher emphasis was put on barley.[58] Landholdings are comparatively larger in Uyum Nagar, but the same strategies apply for crop cultivation. Gojali farmers claim until today that their landholdings are bigger than further down the valley and that this is the reason for their adherence to single cropping. They regard cultivation practices in Central Hunza and Uyum Nagar as those of kitchen gardens because of scattered and split fields and small plot sizes. More light can be shed if we acknowledge that village communities have set rules for themselves in an irrigation system which makes it complicated for members to drop out, that other agricultural and off-farm activities need to be considered, and that a strongly community-controlled system of cultivation is being modified and in some places even disappearing. Traditional practices of crop cultivation, drawing irrigation water from channels with long-established user rights and adhering to commonly accepted rotation cycles prove to be measures of low input and substantial output, while all forms of so-called 'modern' farming require higher investments and afford less secure results.

The correlation between settlement zones and cultivation practices needs to be further analysed and evaluated. In popular understanding there exist ecological boundaries that would refrain Burusho to move for settlement to Herbar. The upper valley had the reputation of being unsuitable for orchard cultivation. As the apricot tree was identified as a life-saving asset for household consumption popular writers such as Emily Overend Lorimer narrated that "…

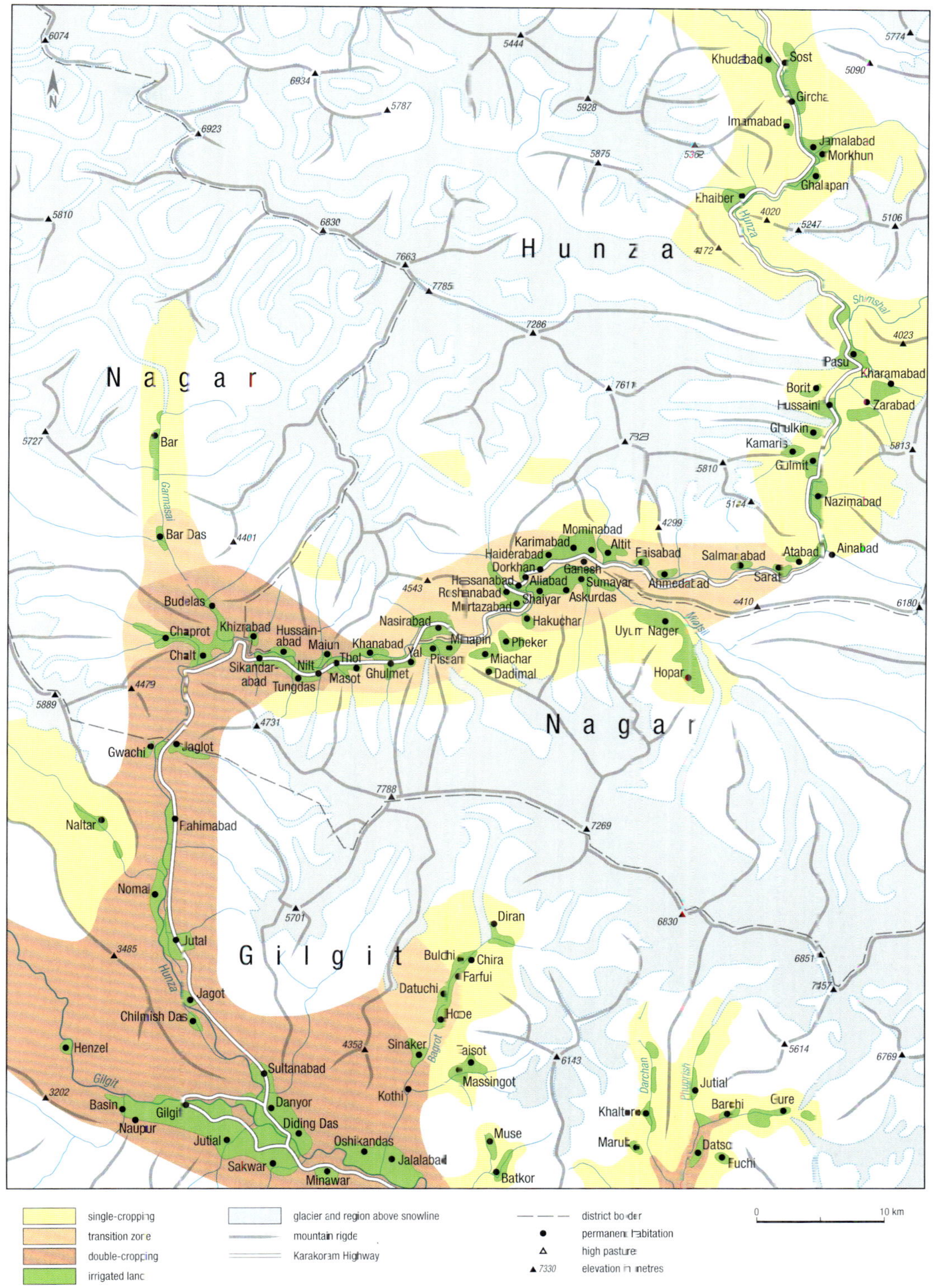

The Wakhi villages of Gojal have significantly changed since Reginald Schomberg's visits in the 1930s. The oasis of Ghulkin is surrounded by tree plantations in which poplars and fruit trees dominate; sea buckthorn is cultivated as a major fuel. The Pasu group towers in the background behind the mighty moraines of the Ghulkin Glacier. The photograph was taken on 26 August 1993

no Hunza women will go to live above the height where apricots will ripen. So some of the higher Hunza villages at 10,000 and 11,000 feet are inhabited by Wakhis who have come in from the north and, I suppose, find even the ground of Hunza better than their own homes in Wakhan."[59] Nowadays Burusho villages in Gojal such as Misgar, Nazimabad and Khudabad are proud of their fruit cultivation and generate a significant crop from orchards. While Emily Lorimer was impressed by the Burusho of Hunza in contrast to all neighbours, other visitors compared a wider range of valley settings. Reginald Schomberg, an experienced botanist and traveller in the Karakoram and Pamirs tried to characterise landscapes and inhabitants in a decisive manner. About the Wakhi who had settled in Gojal and adjacent regions he noted that they 'seemed content to live under miserable conditions that no nomad would tolerate'.[60] Reginald Schomberg seemed obsessed with that 'nomad' notion and described the Wakhi settlers of Pasu twenty years later: "They have settled on the land, but one of their nomadic instincts has survived – a strange indifference to trees. All the Wakhi cares for is one tree for shade in front of his house but at this village even that embellishment was wanting. Indeed the whole plain seemed neglected. The few scattered trees were well grown, and it seemed to me that the Wakhis might have done more. The plain was poorly cultivated although there was abundant water. The excuse given was that the water came too late for newly-sown barley and dried up too soon for the crop to ripen. The truth is not quite so simple. The Wakhi is a poor cultivator, and his agricultural labours are limited to the bare minimum of what is needed. Indeed, he seems still to be essentially a nomad, tamed perhaps by having lived a settled life for generations in a house, but nevertheless far happier on the hillside with his flocks than busy on the soil."[61] The strong views of Reginald Schomberg have been contradicted by other achievements; the inhabitants of Pasu were threatened by a number of natural hazards in their risk-prone environment. Nevertheless, they have managed to maintain a reputation for growing tasty apples. It seems to be worthwhile to have a closer look on the actors and factors that contributed to the emergence of the cultural landscapes of the Karakoram valleys. Tracing the process of settling the harsh environment in consecutive phases will produce further insight into the complexity of decision-making and adaptation practices that migrants have developed in challenging ecological settings.

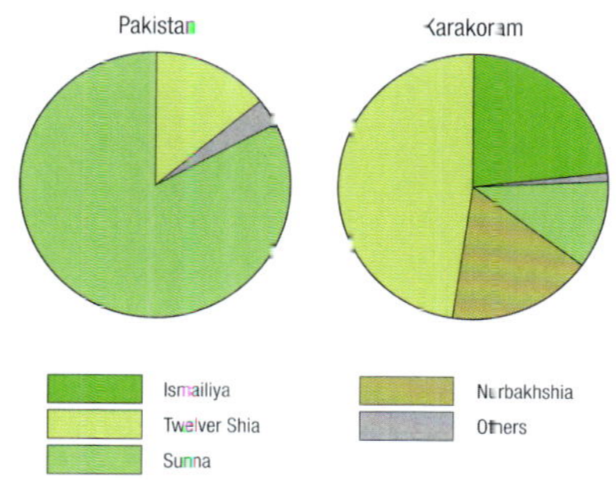

In denominational terms the Karakoram valleys significantly differ from the whole of Pakistan where the vast majority of people reside down country. There the Sunni majority is augmented by Shia and other communities who constitute less then a fifth of the total. In the Karakoram valleys the Twelver Shia represent roughly half of all believers, a quarter being Ismailis and a fifth Nurbakhshias. The Sunni represent the smallest segment and are mainly concentrated in the urban bazaars as well as west and south of Gilgit. Source: Own design based on Hermann Kreutzmann (1995a; 2005a).

Human diversity in the Karakoram

Early ethnographers and geographers realised that the Karakoram Mountains are an area of significant ethno-linguistic diversity and referred to these aspects when describing the communities.[62] Whenever it came to linking these diversities with human practices of agriculture, settlement and resource utilisation, a certain neglect of human diversity seemed to prevail. The 'mountain farmer' featured as a representative of a backward and non-modern approach to tilling the land and keeping livestock. Jack Ives and Bruno Messerli attempted a shift of geographical paradigms in mountain studies to honour the contribution of farming households. They referred to 'the mountain subsistence farmers, men and women of the Himalaya, [as] the best hope for resolution of the dilemma'.[63] The academic debate was characterised by generalisations: "All hopes for solving the 'Himalayan Dilemma' were projected on to the mountain farmers, although – and that is the surprising fact – comparatively little attention was given to the complex living conditions of the people. The problem of neglect was symbolised in the personification of the mountain farmer and in addressing 'his' needs as being somehow representative. The perception of common practices and problems among mountain dwellers in developing countries lacked at least sufficient imagination of the high degree of variation between regions, groups and households and their members. As there seems to be no single person representing the average European alpine farmer or the nomads of the world, there is definitely no such a person in existence who resembles the Third World mountain inhabitant."[64] The conventional approach followed a dual principle in attributing societal stagnation to 'tradition'. In a first approach, it did not distinguish among mountain people between poor and rich, between landed and landless, between members of different denominational groups with their own communal setups and structures. Even a classification into various language groups was mainly highlighted as a distinction between mountains and plains. In a follow-up, it described some archaic agricultural and settlement practices that had failed to receive and adapt an external momentum for modernisation in the colonial as well as in the post-colonial realm and that had failed to enter the development era. Often the backwardness was related to being cut off in terms of accessibility and infrastructure. Group Captain Shah Khan, son of *mīr* M. Nazim Khan, hero of the days of independence and author of an autobiography opened his chapter on the 'liberation of Gilgit and Baltistan' by describing the area as 'geographically isolated' and with 'no contact with what was then India' in 1947: "Means of communication did not exist. Excepting the Mirs and Rajas, none had a radio set. Still worse, none knew about the newspaper. Only a few newspapers reached this area and those were meant for the government officials. As far as the general public was concerned, no information was available for them and very little did they know of the world."[65] Shah Khan was referring to the lack of communication with the new capital of Pakistan in Karachi during a crucial time of independence fighting. Since then, the topic of isolation has been repeated numerous times despite the fact that the Karakoram Highway and many connecting link roads have been built and an institutional network of education, telecommunication, electronic and print media has been established. Many other interconnections and exchanges have become a regular asset of the region.

Consultancy reports and policy papers have often ignored and played down these changes and often take considerations of isolation as their starting point. The 2011 economic report on Gilgit-Baltistan represents the school of thought that emphasises isolation rather than exchange relations that have existed over centuries. A description of isolation is quite surprising and seems out of date in an environment where people have been traders and migrants for centuries. Some people have become bureaucrats, guards and night-watchmen, postal runners and mercenaries in the colonial era and have continued to work as civil servants and army personnel until today. The mountain region has produced and released thousands of students who have successfully attended classes in down-country colleges and received degrees from universities in Pakistan. Under the heading 'coping with isolation', the generalisation in the assessment report becomes obvious: "Firstly, remoteness and scattered populations impair livelihoods in myriad ways; most farmers in GB [Gilgit-Baltistan] are only able to sow one crop during the short growing season at higher altitudes, for example, increasing the risks they face. Similarly, the nascency of the private sector means that the opportunities for employment in the formal part of the economy and hence access to social security, are quite limited. … Marginalized segments of society find it difficult to get help in the face of shocks – a supply side constraint. These groups include

'On the road to Gilgit' shows the challenges for approaching the mountain valleys even for Gurkha soldiers in the British-Indian Army. Source: The Graphic 12 January 1891, issue 1260

the nonproductive poor (such as people who are disabled and elderly widows), and others who are excluded for social, ethnic and religious reasons. In other cases, some who would be deemed to have a legitimate need for assistance do not avail of existing support – a demand side issue."[66] A number of dimensions that could be insightful are omitted, the existing diversity is overridden and the socio-cultural and spatial spread of communities is grossly neglected. In fact, such descriptions are a general feature of national and international reports, which tend to highlight a backward farming community that needs external incentives to take a step forward. The 1997 report on 'Mountains of the world. A global priority' highlighted the 'cultural persistence' as part of 'their character as peripheral areas' which includes a 'psychological and cultural distance'.[67] The absence of urban centres was made responsible for marginality: "Peripheral remoteness corresponds to the fact that mountains are obviously rather unsuitable for the development of urban centres with extensive hinterlands."[68] Here the classical approach of modernisation theory was applied to the mountains and has survived in the minds of quite a number of planners and policy makers. Remedy could only be brought from outside; development was perceived as an allochthonous process with external actors as advisors. The 'global priority' report continued: "Two of the most important requirements for modernisation are modern transportation and communication systems."[69] Those were attributed only to European mountain systems.

In mountain research, it took well into the 1990s before empirical studies sensitised the academic community to a more comprehensive approach in perceiving non-European mountain communities. The neglect is partly explainable by the fact that the leading authors originated mainly from backgrounds in Western natural science and development economics. The attempt to describe mountain communities in Pakistan and Central Asia in more detail and comprehensively might be a first step in creating equal terms for communication about mountain societies.

Sensitivity to these diversities could provide some insights into how certain communities have come to the place where they settled and from where they have spread in to new locations. The reconstruction of the settlement process offers an opportunity to trace the steps of spatial appropriation, the emergence of borders and boundaries, and the coping strategies with human and environmental risk and disasters. Retracing migration processes and power games could contribute to a reading of the map of language and denominational distribution and to a reconstruction of the placement of human habitations in a challenging socio-political environment.

One of the most popular myths that prevail in the Hindukush and Karakoram is that Burusho and Kalasha might be the descendants of soldiers who were left behind by Alexander the Great.[70] This narrative is found from Anatolia to India and has contributed to popularise linkages and suggest long-distance relations. In the case of Burusho and Kalasha they have led to competition between the Greek and the Macedonian governments for their relatives in the Hindukush and Karakoram.[71] Although to date there is no archaeological evidence of any traces of Alexander's troops going north of Besham in the Indus Valley, the saga has been adopted and used for a long time. Only recently have new methodological approaches combined genetic and language studies and convincingly proved that there are no connections whatsoever that could support the myth, which still remains an entertaining story.[72] Nevertheless, the complex history of migration over millennia and centuries in South and Central Asia poses a major challenge for any kind of reconstruction over long periods. In our context, it is more important to explain the distribution of language groups, their relationships and communication patterns.

The 'Map illustrating the languages spoken beyond the North-Western Frontier of British India' was published by George Abraham Grierson in the early phase leading to his opus magnum of compiling and editing the 'Linguistic survey of India' (1908–1928). Here clear-cut borders are given for vernaculars separating distribution areas of Khowar, Shina, Burushaski and Balti from the Pamirian languages in the north. The language borders follow watershed lines that are similar to political boundaries, leaving for example Wakhi in the Afghan and Russian realms, and Sariqoli within China. Source: George Abraham Grierson (1900, map following page 510). Courtesy of the Pamir Archive Collection

A mosaic of languages – autochthonous enclaves and emerging patterns across the Hindukush-Karakoram

The mountainous interface between South and Central Asia is characterised by a significant language spread and variety belonging to Indo-Iranian, Altaic, Sino-Tibetan and isolated language groups. In different historical contexts, reaching from colonial/imperial times to postcolonial nation-building and the age of globalisation, the use of local vernaculars has been challenged by a multilingual landscape. In the Pamirian Crossroads allochthonous languages such as English, Russian and Chinese have functioned as outclassing and suppressing languages; most administrative and census reports are still written in these languages as well as the vast majority of articles, books and reports; most of the relevant electronic media are using them. The dominant colonial languages have represented administration, bureaucracy and external powers. While the leading and monopolising role of these languages can be attributed to boundary-making and the reaches of colonial empires and subsequent nation states, it should be acknowledged that local vernaculars are distributed across such borders and have functioned in those borderlands as bridges in communication and trade. A borderland can be perceived as a region "… bisected by the boundary line between states, which in comparative per-

The 'rope bridge at Gilgit' posed a communication bottleneck and was the only connection towards the Hunza Valley from the Gilgit garrison towards Nomal. Source: The Graphic 19 December 1891, page 720

spective is presumed to encapsulate a variety of identities, social networks and formal and informal, legal and illegal relationships which tie together people in the areas contiguous to the borderline on both of its sides."[73] Borders functioned as ordering systems as well as enabling structures for cross-boundary communication.

Consequently, language use, commonalities and distinctions in its communities should play a major role. English, Russian and Chinese served colonial purposes, and it might be due to an absence of a dominant *lingua franca* that in the Karakoram and Eastern Hindukush English still supersedes Urdu or any widespread vernacular in Pakistan. In the Pamirian Crossroads the colonial separation into Eastern and Western Turkestan marks the boundary of Chinese in Xinjiang and Russian in Tajikistan. Similarly, George Abraham Grierson marked clear-cut borders in his 'Map illustrating the languages spoken beyond the North-Western Frontier of British India', which was published in 1900 and distinguishes three groups – Iranian, Indo-Aryan and Non-Aryan languages – within a fine structure of appellations and distribution of local vernaculars.[74] It suggests spatial congruence between language distribution and territory in a period of the Great Game when imperial desires were aiming to claim and justify territories as 'spheres of influence' in order to establish 'scientific' and/or 'natural' boundaries between empires. The outcome of imperial boundary-making has created borderlands that are mainly characterised by territorial division rather than by ethno-linguistic distinction. Present-day boundaries clearly separate territories of colonial languages and divide community spaces in which vernaculars have developed on their own.

In awareness of dynamic socioeconomic processes and fluid behavioural patterns, it can be observed that more than twenty different languages were spoken within the area then and still are today. Without essentialising and constituting communities on the basis of specific traits and practices, certain group constellations fill the local arena, led by dominant and well-connected leaderships. Difference, multiplicity and patterns of heterogeneity are common features in the Pamirian Crossroads and elsewhere. On the mountain scene, place-based communities that identify themselves as members sharing the same vernaculars are representing certain claims and interests against the dominant external and authoritative language users from the urban centres and capital regions. Being insiders and outsiders in a world of cosmopolitanism and globalisation becomes a blurred feature given the dimension of flows and variations in mobility. The contradictions of self-definition and external labelling become obvious when conflicting interests and demands for linguistic representation are negotiated. Neil Brenner has stated that "... globalization researchers have begun to deploy a barrage of distinctively geographical prefixes–e.g. 'sub-', 'supra-', 'trans-', 'meso-', and 'inter-' – to describe various emergent social processes that appear to operate below, above, beyond, or between entrenched geopolitical boundaries."[75] The results are often opaque ascriptions and blurred pictures of a dynamic process that seems to be difficult to grasp and fix. Neil Brenner argues that, by overcoming a state-centrist perspective, a new global perspective leads to 'the production of new configurations of territoriality on both sub- and supra-national geographical scales'.[76] Colonial and post-colonial re-figurations of state territory, administrative unit reforms and renaming processes of incorporated spatial entities are characteristic of the Pamirian Crossroads.[77] The same applies to studying language communities that have encountered a re-scaling of their supra-local relations in the age of migration and enhanced telecommunication. The cases presented here are, first, an example for creating communities and social classes in early colonial encounters by applying all kinds of markers and, second, an interpretation based on empirical material from a remote mountain area, representing a complexity in local settings that is embedded in trans-regional networks and communication structures. Approaching such diversity patterns, a historical interpretation of migration and mobility could form the basis for an analysis of contemporary mosaics of highly dynamic speech communities in multi-local settings that have achieved a communication practice across political boundaries that has been enabled by globalised media and mobile phones.

The positioning of small languages challenged by globalisation, language and minority policies

A survey of comparatively small languages in remote regions poses a great challenge when mobility in all directions is a common feature. Remoteness is a relative category as these languages are centred in state peripheries. Members of small language groups are very often neglected and excluded

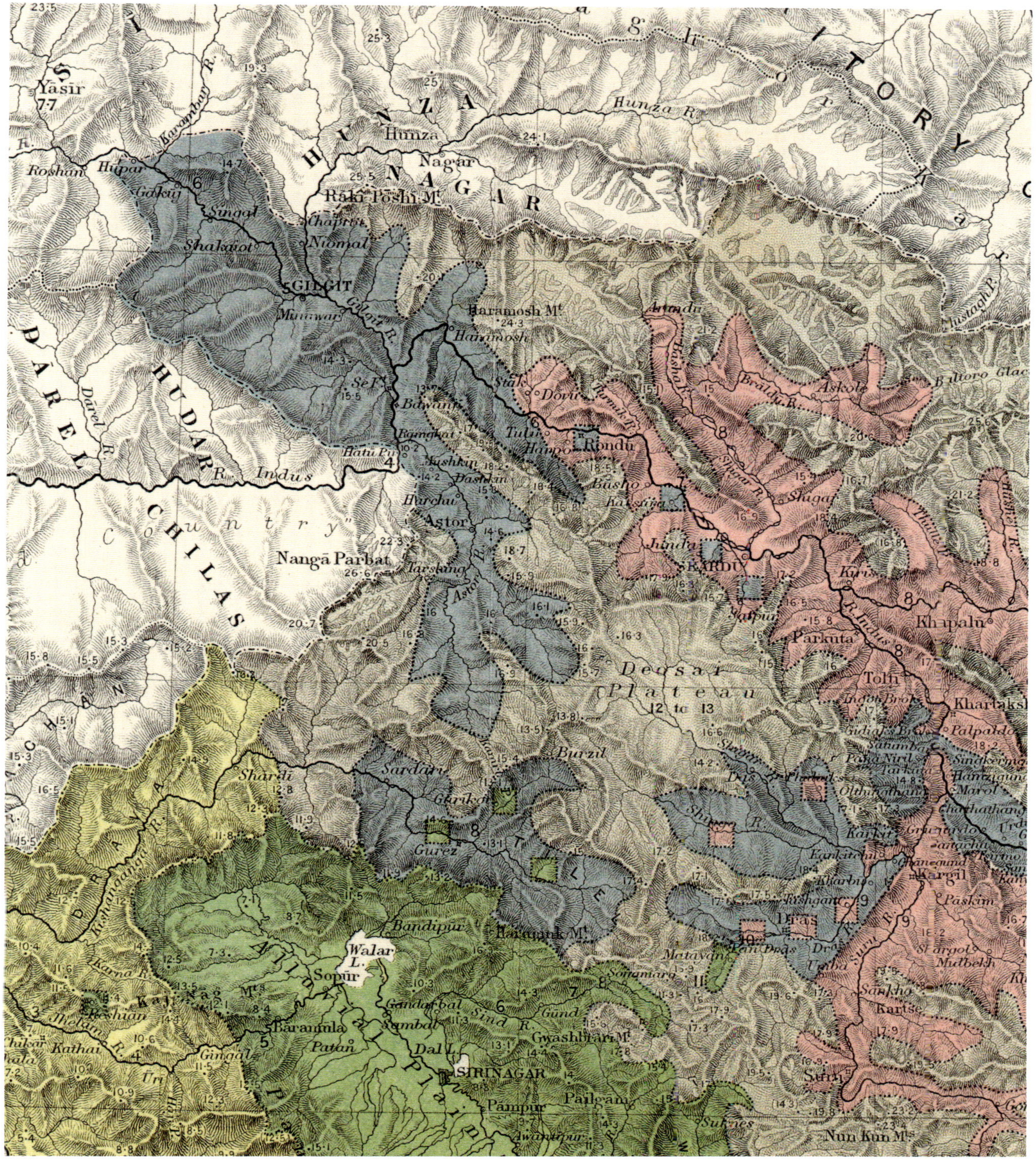

The 'language map' in Frederic Drew's account of the 'territories of the Maharaja of Jummoo and Kashmir' identifies only two language groups in what is now Gilgit-Baltistan. The Gilgit area is attributed as 'Dard', which emphasises the Shina-speakers. Baltistan is connected with the Tibetan subgroup of Balti and contains some Dard pockets. Source: Frederic Drew (1875, appendix). Courtesy of the Pamir Archive Collection

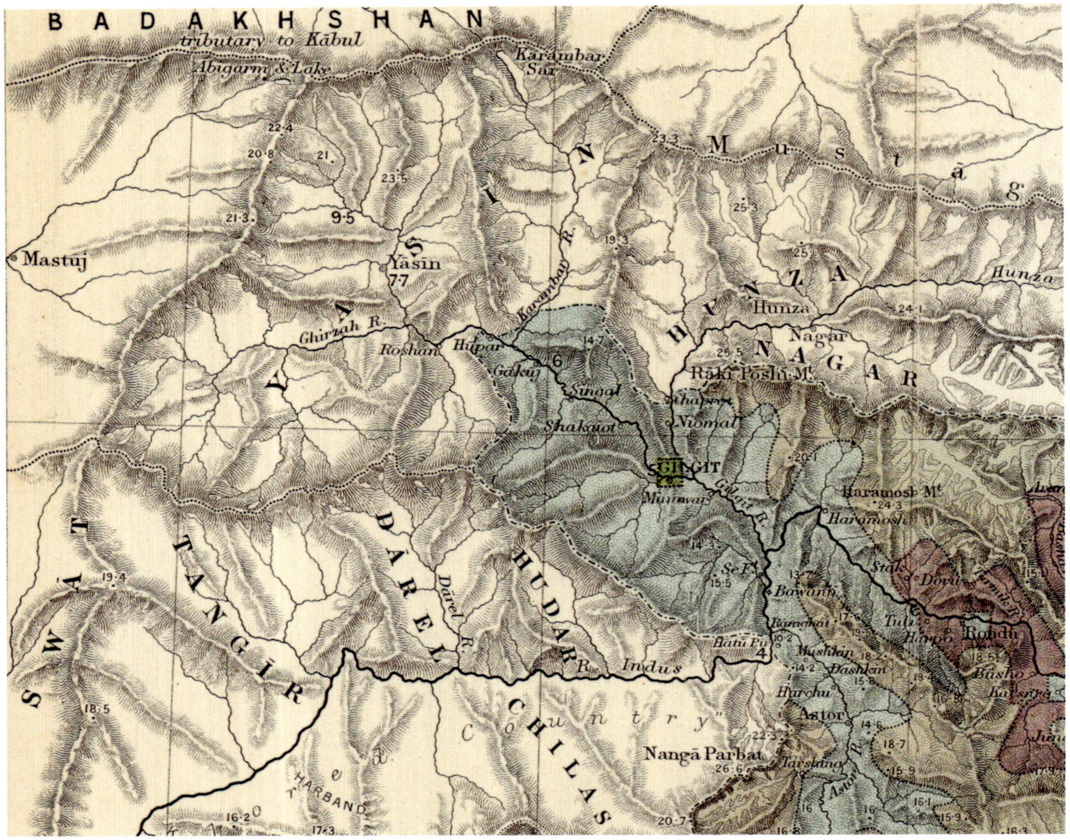

The 'race map' in Frederic Drew's account of the 'territories of the Maharaja of Jummoo and Kashmir' depicts Gilgit garrison as 'Kashmiri', and the 'Governorship of Gilgit' is included in the 'outlying governorships' of the Maharaja as 'Dard'. Hunza and Nagar as well as Chalt and Chaprot are excluded, as are Ishkoman, Yasin, and Chilas with the Darel, Hudar and Tangir valleys. Astor and Haramosh are shown as Dard areas, while eastwards only a few Dard pockets are found in Baltistan. Source: Frederic Drew (1875, appendix) Courtesy of the Pamir Archive Collection

by national census authorities, which concentrate on majority groups and omit minorities that are labelled as 'others' only in a residual grouping.[78] A multitude of languages remain unrecorded and/or are hidden in areas of ambiguity. Comparisons with historical language descriptions and colonial records reveal a peculiar dynamism in terms of mobility in such regions. This dynamism is regularly negated when people are described as stagnant and less interested. A first encounter between members of local speech communities in the Hunza Valley and British and Kashmiri colonial officers commenced in the second half of the 19th century. Early explorers such as John Biddulph, Frederic Drew and Gottlieb Wilhelm Leitner served their clients, had a keen interest in understanding the diversity among people, and were searching for parameters to categorise sub-groups along markers such as caste, religion, language, profession and physical appearance. Their material entered confidential reports and easily accessible books for the interested public.[79] Their findings were utilised for administrative purposes especially after the Hunza-Nagar Campaign of 1891 when local resistance was broken, compliant rulers were installed, and a Political Officer was appointed in Hunza. The curiosity shift from a conqueror's to an administrator's perspective is reflected in the reports, handbooks and gazetteers that were subsequently published. During the first decade of British domination the seminal works of the above-mentioned explorers were augmented by ground surveys conducted by staff from the Political Agency in Gilgit. The Political Agent Stuart Hill Godfrey wrote the initial 'Report on the Gilgit Agency and Wazarat and the countries of Chilas, Hunza-Nagar, and Yasin including Ashkuman, Ghizr, and Koh 1896-97' which is characterised as an inventory of historical background and local conditions.[80] His substantial report serves the purpose of establishing an 'adequate' colonial administration and sympathises with cooperating partners from the ruling elites. The gazetteer addressed language and religion only in passing and did not make a big issue of social class distinctions. The state of nutrition was mentioned besides references about the elites. With the turn of the century a different approach becomes visible. It might be due to the different personalities of Stuart Hill Godfrey and his successors (of whom Bertrand Evelyn Mellish Gurdon deserves special mention) that the attitude of enquiry and the interpretation of the material changed significantly. Every subsequent gazetteer and military report is based on its predecessor; as a rule it corrects, updates and augments the existing information.

A decade after Stuart Hill Godfrey's confidential report, a new version with updated information collected by Bertrand Evelyn Mellish Gurdon in his 'report on the Gilgit Agency' from 1906 was published titled the 'Military report and gazetteer of the Gilgit Agency and the independent territories of Tangir and Darel'.[81] Again the colonial authorities meant this to be a confidential handbook for administrators and officers to get their bearings in a rather unknown environment. The Assistant Quarter Master General Colonel Wilfred Malleson, from the Intelligence Branch in Simla, issued the new version that contained quite a different approach to describing the people in terms of caste, religion, language, profession and physical appearance. The first two chapters were devoted to geography and ethnography. The colonial administration of Gilgit was searching for a paramount term to describe the people in the Agency and adopted the catch-all category coined by Gottlieb Wilhelm Leitner who attrib-used the term 'Dard' to the speakers of Dardic languages such as Kashmiri, Khowar, Kalasha, and Shina as well as to the inhabitants of Dardistan in general.[82] In contrast to Gottlieb Wilhelm Leitner's good intentions and his later continuous advocacy of the people, the reader of the 'military report' would not find in the narrative any favourable and sympathetic terms for the description of Dards who are perceived to be the residents of the Gilgit Agency: "As a whole, the Dards are lacking in energy and adaptability, unwilling to strike out new modes of life, or to employ themselves otherwise than with agriculture. They appear contented with the same poverty that satisfied their forefathers, and their want of cohesion or enterprise seems to show that they are doomed to be absorbed by more vigorous races. In disposition they are tractable, good tempered, and indolent to a degree; unambitious and unwarlike, neither cruel nor quarrelsome, fond of rejoicing and merry-making, and submitting readily to constituted authority."[83] The colonial classification highlighted the perceived hierarchical social classes of Rono (*mīr* and *wazīr*), Shin (self-assessed aristocracy), and Yeshkun (Burusho) and linked them to 'ethnic' categories. Apart from those basic foundations, *sayid* as a respected religious class are distinguished as well as – at the opposite end of the social ladder – the artisan classes of Kamin (millers and potters), Ḍom (blacksmiths, musicians, and cobblers), and Shoto (leatherworkers).[84] Again separately treated are Kashmiri and Gujur who are regarded as recent immigrants. In describing the people and their behaviour, the report opted for categories related to physical anthropology and highlighted appearance and cleanliness. A vernacular-based classification which includes language choices for the upper classes is attached to the report: "Chitrali is spoken by all the better classes in Gilgit and Punial and by everybody in Yasin, but apart from this the population may be grouped philologically as follows:

4 classes, speaking Shina or Dard language – 1. Shin. 2. Yashkuns (mixed). 3. Kamin. 4. Ḍom and Shotos.

3 classes, speaking Punjabi – 1. Gujars. 2. Kaghani traders. 3. Saiyids.

2 classes, speaking the Afghan language, Pushtu – 1. Koliwals. 2. Mullas and their pupils.

And apart from these is the most interesting language of all. Burishki, spoken by the pure Yashkuns of Hunza, Nagir and Yasin. Its foundation has been identified as of Turanian origin, and it cannot be classed with any other Dard language."[85] The classification shows a rather selective approach in attributing language use to social classes despite some grave misperceptions of vernaculars spoken in different settings. The follow-up report from 1928 realised that the term 'Dard' might be less appropriate: "The term 'Dardistan' has been applied by Doctor Leitner to the whole of this area, though it is not quite clear how he arrived at the name, which appears to be unknown to the people themselves."[86] In terms of language description not much had changed and the same tabular structure was repeated.[87] Language attainment seemed not to have raised attention from the colonial administration until the Political Agent David Lorimer made it his personal interest during his duty in Gilgit and after his retirement in Hunza.[88] Prior to their departure to Hunza in spring 1934 Emily Lorimer wrote to her publisher Gilchrist Thompson: "Colonel Lorimer and I are off in March 1 on a crazy, middle-aged adventure, back to the Hindu Kush and Karakoram, to anthropologize amongst our beloved mountain people of Hunza (Gilgit Agency), whose language he was the first white man to learn."[89] In an article published after their return she wrote "… my hus-

GROUP OF DARDS, &c.

Mirza Beg, an Astóri, Sunni Muhammadan). — Gulám Muhammad, a Ghilghiti, (Shiah Muhammadan). — Dr. Leitner's Hindustani Munshi. — Dr. Leitner. — Kázim, from Skardo, (Little Tibet). — A. Chilási, (standing) Batshu, (a Báshgeli Kafir, sitting). — The man in the white dress is Málek, also a Kalásha Káfir, subject of Chitrál; the last man on the right is Gharib Shah, a Chilási.

Gottlieb Wilhelm Leitner is sitting in an armchair among his companions and informants from Astor, Chilas, Chitral, Gilgit and Skardu whom he would identify as 'Dards'. He popularised the term 'Dardistan' as a comprehensive category for the inhabitants of the Eastern Hindukush and Karakoram in order to distinguish this intermediate region from the two major entities of the Himalayas and the Pamirs. The photograph was taken in September 1866. In his critical remarks, Kai Friese (2000: 8) commented on the photograph: "A pugnaciously ugly man, seated, with a row of fiercely whiskered retainers standing on either side. The man directly behind him holds a fan over his master." Gottlieb Wilhelm Leitner became one of the fiercest advocates for 'his' Dards and the most prominent anti-colonial commentator of his time. Source: Gottlieb Wilhelm Leitner (1877, following page 8). Courtesy of the Universitätsbibliothek Heidelberg

The 'dance at Ghilghit' is taking place in the ground in front of the Gilgit Fort. Local dancers are performing in front of Gottlieb Wilhelm Leitner who is sitting on an elevated veranda. Source: Gottlieb Wilhelm Leitner (1877, Part III, following page 30) Courtesy of the Universitätsbibliothek Heidelberg

band was the first white man – and I, of course, the first white woman – to have learnt their language, and no British officer had ever settled down amongst them to get to know about their daily life."[90] About this time Reginald Schomberg frequently visited the Hunza Valley and neighbouring regions. His assessment and categorisation of residents is strongly linked to language groups and behavioural characteristics and does not eschew racist ascriptions. In his last book manuscript on the 'Karakoram Finale' he summarises his prejudices based on his own observations starting with the Wakhi: "They are of Persian origin, and originally nomads. They have settled on the land, but one of their nomadic instincts has survived – a strange indifference to trees. All the Wakhi cares for is one tree for shade in front of his house ... he seems still to be essentially a nomad, tamed perhaps by having lived a settled life for generations in a house, but nevertheless far happier on the hillside with his flocks than busy on the soil."[91] He developed a greater admiration for Hunzukuċ which might be a result of his regular conversations with his 'followers' who mainly came from Hunza: "The men of Hunza have a far greater practical knowledge of glaciers and their habits than do the scientists and travellers who write pontifically about them after a brief visit."[92] This statement contrasts significantly with his opinion about others: "The Balti is essentially a fool. He emigrates in large numbers every year to neighbouring places, and far down into India to work during the winter for a small wage. There is no need whatever for this migration, but he is too stupid to realise it. ... This criticism is by no means all mine. Indeed most is the considered opinion of my followers, who were highly competent to judge. ... I had forgotten how insignificant the Balti was. I was rather taken aback when I saw these eight little chaps, dirty, scrubby and undersized. There was nothing of the 'homo alpinus' about them, and they compared very poorly with our Hunza and Wakhi followers. There was not one of them over 5 feet 6 inches in height, and their physique seemed poor."[93] In the lower part of the Hunza Valley Reginald Schomberg visited Matum Das, a settlement of Burusho that puzzled him as being located in a Shina-speaking environment: "The Hunza colonists had developed the land in a wonderful way, with the skill and labour peculiar to their race. They did not look very healthy, and declared that it was the fault of a wild bean which they could not eradicate from their crops, which became mixed up and eaten with the wheat, and which, amongst other nasty effects, made the people wobble at the knees. Another cause of weakness was the huge consumption of alcohol, made from the abundant and excellent grapes that grew everywhere, and of which we received large quantities. The Hunza folk are fond of strong liquor. They make first-rate wine, and are particularly attached to a very potent spirit which is very easily made from mulberries, but the manufacture of which is forbidden by the Mir."[94] Further down the valley he visited Juglot, a village of Shina-speakers: "They were all Shins, who were probably the first invaders of the Gilgit valley, and who, though no longer regarded as different race, still consider themselves the aristocracy of the country. They are all Mohammedans of the Shiah sect, and were observing the yearly fast of Ramazan with great zeal, but no enthusiasm, as no one can feel that emotion on an empty stomach in the harvest field and the hot sun."[95] Not much impressed, he continues with deliberations about cleanliness: "They never washed, in spite of thinking themselves people of rank and position. Their houses stank to the high heavens."[96] Confusion came up when these 'races'

The undated photograph most likely shows Reginald Schomberg with his travel companions at a resting place. Source: Schomberg Collection (RSAA/SC/Sch/1/99/2). The image is reproduced with permission of the Royal Society for Asian Affairs in London.

mixed, in this case marriage between Burusho and Shin: "It was the more surprising therefore to hear that forty women of the neighbouring Hunza settlement of Matum Das had married these Shins. It was explained that there was nothing else for them to do. When I expressed the hope that the children of these unions would be less apathetic and gutless, I was told that no decent offspring could be expected from such fathers. This was unduly hard. The Shins are not as bad as that. I was asked how Hunza women would produce manly sons with fathers like these Shins? If this be so, then the future of the race must indeed be bleak."[97] Finally, Reginald Schomberg made the comparison with his favoured porters from Shimshal: "But we had been spoilt by our Shimshalis, with whom the poor Shins were bound to compare unfavourably both in energy and cleanliness. It is interesting to note that in the Gilgit Agency, there are three races who are active, capable and progressive – the people of Hunza, Punial and Yasin, and they can be grouped in the order given. It is true that the inhabitants of Chilas might be included if they were not so lawless and intractable; they are like the zebra or African elephant, impossible to domesticate. It is a pity that they remain untamed savages with a predilection for murder which so far the administration, forbidden to inflict capital punishment, has been unable to overcome. Imprisonment is no deterrent to people of this kind. For them years in jail, where they are decently clothed and adequately fed, are no penalty in comparison with the joy of killing an enemy."[98] The narratives of Reginald Schomberg are full of pejorative statements and prejudices; his remarks represent a classification scheme that attributes certain qualities and deficiencies to whole communities. For him it is quite clear whom to trust and whom to believe. The only consolation might be that he sometimes contradicts himself. Overall, he represents the 'zeitgeist' in categorising people along visible and easily gathered attributes such as cleanliness, language and religion. The colonial urge to classify people in their environments is ubiquitous and functions as a thread running through his writings. Reginald Schomberg referred to the term 'Dard' only once in this text when describing a Brokpa: "He used to be called a Dard, but that very useful description has been frowned upon by the scientific pedants who probably find it too simple."[99] No bigger contrast could exist between the methodological approaches of David Lorimer and Reginald Schomberg in understanding the Dards. Whether too simplistic or too com-

During Reginald Schomberg's visit in 1931 he was entertained by a band of Dom musicians in the Karimabad garden where mīr M. Nazim Khan received his guests. Source: Schomberg Collection (RSAA/SC/Sch/1/8/7). The undated image is reproduced with permission of the Royal Society for Asian Affairs in London.

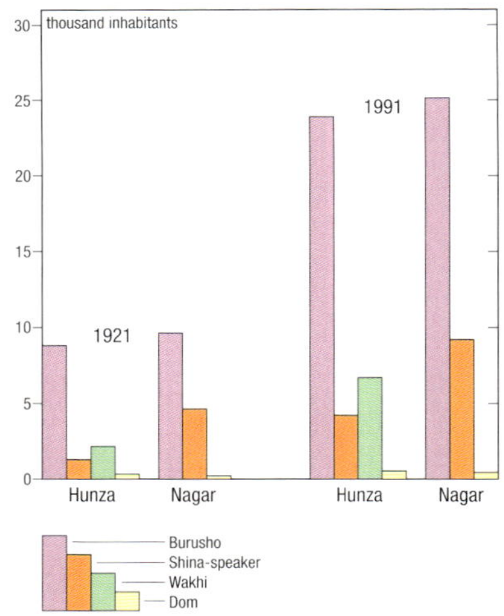

The language affiliation in Hunza and Nagar has not significantly changed from 1921 to 1991. Burushaski has remained the dominant vernacular in both entities. Source: Own design with data based on Muhammad Masih Pal (1928) and Hermann Kreutzmann (1996b: 177).

plex an approach depends on the viewpoint; the desire to find a common term for such a small, but obviously highly diversified population in such a vast and remote mountain area has remained on the agenda. Pejorative ascriptions were quite common; a member of the German-Austrian Karakoram expedition termed the Shin as 'schmutzig, verlaust, kropfig. Kontrast springt sofort in die Augen'.[100]

Anthropologists seem to have been confused in a similar manner as the colonial visitors. Irmtraud Müller-Stellrecht categorised the Dards in a mixture of linguistic and ethnographic terms by counting the speakers of Dardic languages and including Burushaski and Werchikwar speakers, but excluding Kalasha who speak a Dardic language. Her emphasis was on the inhabitants of a geographic space rather than on language, thus adopting Gottlieb Wilhelm Leitner's approach who explained his view: "In a restricted sense the Dards are the race inhabiting the mountainous country of Shinaki, ... but I include under that designation not the only the Chilasis, Astoris, Ghilghitis, Dureylis, &c, but also the people of Hunza, Nagyr, Chitral and Kaffiristan."[101] John Staley preferred to label them 'Kohistani', mountain dwellers between Kafiristan and Baltistan, and north of Indus Kohistan towards the borders of Wakhan; this is a term strongly disliked and repudiated in Ghizer, Gilgit, and the Hunza Valley.[102] Neither Dardistan nor Kohistan have much significance in daily practices today; but the distinction along language barriers is omnipresent.

The perception of languages has always been a marker for structuring the population in the Hindukush-Karakoram. Very often the language use is linked to heritage and immigration. The mountainous region is rich in language diversity that has attracted several generations of researchers to figure out the meaning of autochthonous and allochthonous languages and the nexus between language affiliation and migration.

Mobility and language diversity

Migration processes can often be linked to linguistic groups and resulted in particular cultivation and settlement patterns. Present-day multi-locality, schooling and in-migration add to a growing language complexity that is aggravated by restructurings through governmental changes and the impact of development organisations sponsored by European, North American and the Arab World–based donors. The Hindukush-Karakoram was classified as a big 'linguistic museum' by the nestor of language studies Georg Buddruss who continues as follows: "But the Northern Areas are not only a museum of languages now extinct but also of living vernaculars hitherto very imperfectly known to the linguistic world."[103] In order to deepen the insight into the polyglot communities of the Pamirian Crossroads and their distinctions, a selective approach is applied in the following. Six attributions related to migration processes will be distinguished, of which the first four are related to historical mobility and help scholars understand settlement processes, immigration and mobility within the mountain area. The latter two more recent developments focus on an enhanced process of intra-mountainous migration within the Pamir-Hindukush-Karakoram-Himalaya interface, on the one hand, and mobility between the predominantly rural high mountain areas and the urban centres in the forelands, on the other. Thus, mobility in a multitude of migration corridors connects the region to a globalised world and enhances its importance at the hinge within Crossroads Asia.

Little information may be derived from official statistics when the complex linguistic pattern of the Hindukush, Karakoram and Pamirs is at stake; for instance, the 'Tajikistan Republic law on language' only acknowledged in 1989 the existence and special status of Pamir languages, but did not give 'official permission for writing'.[104] Other countries are not necessarily concerned with giving such permission or developing a language policy. The Population Census Organisation of Pakistan includes in their questionnaire a column in which eight language options are given for the selection of the respondents. The 'mother tongue' is asked for as the 'language spoken in the household'.[105] The eight stated languages are qualified as national idioms as they represent the majority of speakers. Identifying a single idiom in Gilgit-Baltistan as a regional language significantly differs from mainstream Pakistan where each province features at least one or two of those languages with a significant share. In Gilgit-Baltistan and in Chitral only three to five percent of all households stated one of those highlighted national idioms as their mother tongue during the last census, while more than 95% belonged to the category of 'others'. The application of this term sheds some light on the distinctiveness of the region concerned. In recent years, various activities to initiate a legislative process for

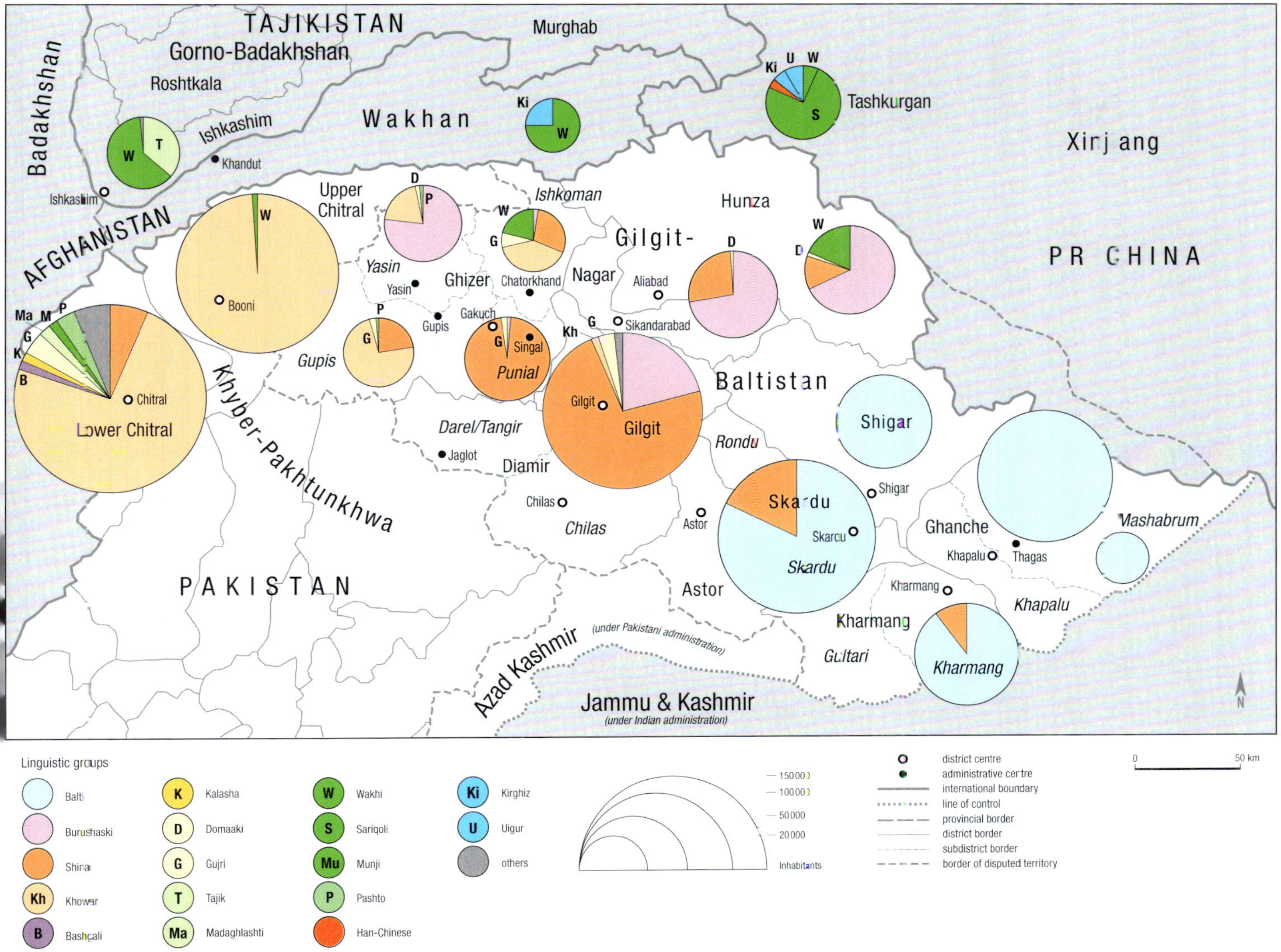

Chitral and Gilgit-Baltistan feature a high diversity of languages. In the north, Pamirian languages are represented; only Wakhi is found in all four countries and has a majority in Gojal. Burushaski as an isolated vernacular dominates in Hunza, Nagar, Yasin, and has a fair share in the urban population of Gilgit; Shina speakers dominate the central part, while the other major Dard language is Khowar that is found in Chitral and Ghizer. Tibetan idioms are found only in Baltistan. Source: Data are derived from the language survey by Hermann Kreutzmann (1995a, 2005c). Design by Hermann Kreutzmann

a new language bill in Pakistan can be observed. The Shina Language and Culture Promotion Society has taken the lead in Gilgit-Baltistan.[106] Although suggestions were made to include Balti and Shina in the list of national languages along with Baluchi, Brahui, Hindko, Pashto, Punjabi, Seraiki, Sindhi and Urdu, none of the recent amendments of the constitution have implemented the move yet. The quest for recognition of the languages of Gilgit-Baltistan is part of a movement to broaden the category of official languages beyond English and Urdu as fixed in article 251 of the Constitution of Pakistan since 1973. The article claimed that Urdu would replace English 'for official and other purposes within fifteen years' which would have been by 1988.[107] Besides Urdu selected regional vernaculars were identified, and Provincial Assemblies were allowed to 'prescribe measures for the teaching, promotion and use of provincial language in addition to national language'.[108] For Gilgit-Baltistan - which does not enjoy a provincial status to date - English has remained an important language in administration, development and education although Urdu is quite prominent

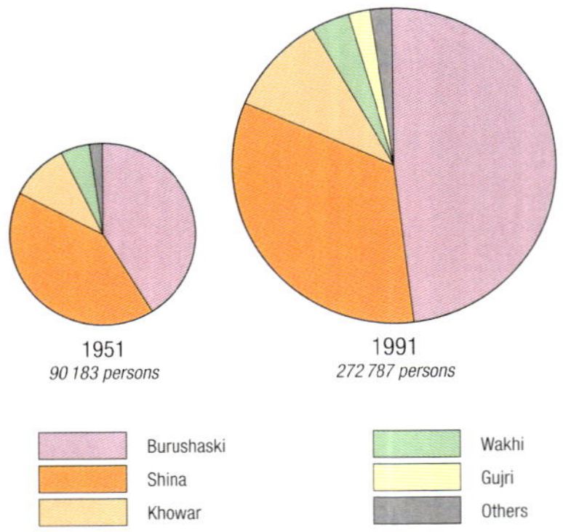

The first census in independent Pakistan that featured language data was conducted in 1951. Its findings for the Gilgit and Ghizer districts are compared with our survey four decades later. The share of Burushaski seems to have slightly grown, while the other proportions remained similar, Shina appearing as the second-most important vernacular. Khowar is mainly spoken in Ghizer in those regions that were previously governed by Chitral and settled under the Khushwaqte guśpúr who attracted new settlers. Source: Own design; data based on Government of Azad Kashmir (1952); Hermann Kreutzmann (1995a; 2005a)

as well in the army, in bureaucracy, and in schools. Local vernaculars have retained their respective importance for individual speech communities.

At the same time the prevalent census category 'other' disguises a substantial number of languages not to be found in other parts of Pakistan, e.g. Burushaski, Shina and Balti. Nevertheless, some are the *lingua franca* of formerly independent microstates or principalities – such is the case in Chitral, Gilgit, Hunza, Nagar and Shigar – or of the regional bazaar towns such as Khowar, Shina, Burushaski and Balti. Despite the neglect in official statistics, a number of linguists have devoted their interest to a constellation that is unique because of its high degree of linguistic diversity.[109] These patterns do not stop at borders. Similarly the minority nationalities' (*minzu*) languages of the neighbouring Uigur Autonomous Region Xinjiang within the People's Republic of China are comparatively small with fewer than 50,000 speakers, as in the case of Tajik.[110] The Eastern Iranian languages that are incorporated in the 'Tajik' *minzu* belong to the Pamirian group, as it would be distinguished in neighbouring Tajikistan. Tajik in the Chinese context includes Pamirian/Eastern Iranian vernaculars such as Sariqoli, Shughni and Wakhi.[111] These would not be identified as Tajik in Afghanistan, Tajikistan and Pakistan because this term is exclusively reserved for Western Iranian languages. Here the distinction as Pamirian languages is important.[112] In both countries Pamirian language speakers are mainly found in autonomous districts with distinct legislative frameworks: Gorno-Badakhshan in Tajikistan and Tashkurgan Tajik Autonomous County within Xinjiang. In both areas we find Kirghiz as a Turkic language as well, for which in Xinjiang a separate Kizilsu Kirghiz Autonomous Prefecture was created sixty years ago. All languages are excluded as media of instruction from schools and institutions of higher learning. The applied concepts of regional autonomy had a spatialisation of ethno-linguistic affiliation and territorial distribution in mind.

Depending on the respective viewpoint of observers, not all aspects of either affirmative action or discrimination by exclusion and stigmatisation are implemented by applying language parameters in Pakistan where discrimination along vectors of gender and religion predominates. However, vernaculars from the Hindukush and Karakoram that are sidelined in any respect nevertheless play an important role in shaping community perceptions.

A survey of linguistic and denominational diversity

Two parameters of distinction that characterise houses, hamlets and villages in people's perceptions are language and religion. Although most people command more than one language, mostly one vernacular is attributed to them when meeting others. These ascriptions are well known and used in conversations, political debates and classifications in daily life. Nevertheless, owing to marriage and migration households can be mixed with regard to denomination and language. These affiliations play an important role in terms of community and stigmatisation, as they are one level up from direct kinship relations. Most people would feel obliged or even prefer to purchase goods from their immediate and affine relatives, or they would ask for help and support from these groups. That practice prevails in village contexts, but in urban bazaar settings regional affiliation, belonging to the same language group and/or religious community may become important for decision-making or for turning to someone for advice and assistance. Consequently, it was comparatively straight-forward to record this information on a household basis and to aggregate the data on a village basis to get a rough idea of spatial patterns of distribution. Somehow surprisingly, neither before our own survey was conducted in the early 1990s nor afterwards was any effort made to gain a more precise picture of the denominational and language diversity in northern Pakistan. From our village-based survey of the languages in the Eastern Hindukush and Karakoram, some striking patterns emerge on first sight within the mountain region.[113] The western and central part is dominated by Indic languages, in the centre we find two valleys where the only dialects of the isolated Burushaski language are to be located, while the eastern part is dominated by the Sino-Tibetan Balti language. Speakers from the Altaic and Iranian language groups are found in the northern border areas to Afghanistan and Xinjiang (China). More than twenty-five different languages belonging to the four above-mentioned groups have been recorded in the Eastern Hindukush and Karakoram. On this level, the even more complex spatial distribution patterns are still unrevealed, but in an initial step some linkages between the origin and migration history of members of different language groups can be established. Basically four groups could be distinguished:

(1) *Autochthonous languages in compact settlement areas*
Mountain exceptionalism is represented by the isolated language of Burushaski, which is confined to this mountain region. No link to any other language group could be established so far.[114] Karl Jettmar summarised the "… evidence suggesting that this other group [the Burusho] goes back to an antecedent stratum of immigrants or even the original inhabitants."[115] While the meaning of original might be doubted or accepted as a relative category, it is quite probable that the two Burushaski-speaking valleys of today - Hunza and Yasin - were once connected via the Gilgit Valley and that Shina has superseded and replaced Burushaski there.[116] In this category of autochthonous languages a similar role can be attributed to the Nuristani idioms which are mainly to be found in a compact area of diffusion in the Eastern Hindukush.[117] Likewise, Balti has to be added as the dominant language of Baltistan; together with Purik and Ladakhi, it forms the westernmost exponent of an archaic dialect of Tibetan. Traces suggesting an expansion, contraction or displacement of the distribution areas of these language groups have been presented from toponymic incongruencies and from narratives describing migration processes. In the group of autochthonous languages the Pamirian languages should be mentioned, including the dominant ones of Shughni, Roshani and Wakhi. In the Pamirian core, the languages bearing the name of the mountains occupy a contiguous area that is mainly divided between Afghanistan and Tajikistan.[118] The autochthonous languages are still characterised by a rather compact and easily identifiable regional segment of spatial pockets.

(2) *Indic languages of early migrants*
Scattered information is available about the initial immigration of Prakrit speakers. Evidence is based on oral traditions and linguistic analysis.[119] Most probably, immigration started about a millennium ago and resulted in the occupation of the lower parts of the valleys by the ancestors of the present-day Khowar, Maiyã and Shina speakers. The area of their distribution is bordered in the south by Pashto immigrants in Indus Kohistan, and by Pashto speakers in the Dir-Swat area. Gilgit and Chitral became their political centres, from which further settlements spread into the side valleys. Kho people are spread over a wide territory in Chitral and Ghizer. It was suggested that the Kalasha speakers previously dominated a much wider area in Chitral into which

On their way from Hunza to Kashgar the expedition's artist Aleksandr Yakovlev portrayed personalities en route and invited sitters from all language groups. Hunza's envoy Daulat Shah was on his mission to Kashgar to deliver Hunza's annual tribute of fifteen miskal of gold dust. He was a trusted confidant of the mir of Hunza. The portrait was painted by Aleksandr Yakovlev on 9 September 1931 when the Pamir group of the Croisière jaune had arrived in Tashkurgan. Unfortunately, less than a month later 'Daulat Shah died in the Consulate Hospital [in Kashgar] on the 5th October, from blood poisoning resulting from a kick on the leg by a horse' as was reported in the Gilgit Agency Political Diary for the months of September and October 1931 (IOL/P&S/12/3285: 370). Source: Alexandre Evgenievich Iacovleff (1934; plate 28), reproduced with kind permission from Gallica. Bibliothèque nationale de France

the Kho immigrated "submerging the indigenous Kalasha language in all but a few side valleys".[120] Along with these migrants, Ḍomaaki speakers arrived in the mountain belt and became prominent as the professional groups of musicians and blacksmiths.[121] As professionals providing services they have been mainly settling with Khowar-, Shina- and Burushaski-speaking groups. Their own language has been replenished with loan words from the prevalent vernaculars. As the Ḍom are a minority in every village where they settle, they are very often neglected in perception although they provide vital services to the respective communities.

(3) *Later immigrants and refugees from Eastern Iranian and Altaic language groups*
During the last two centuries scattered groups of refugees and migrants settled in various valleys of the Hindukush-Karakoram. In general, they were allocated cultivable

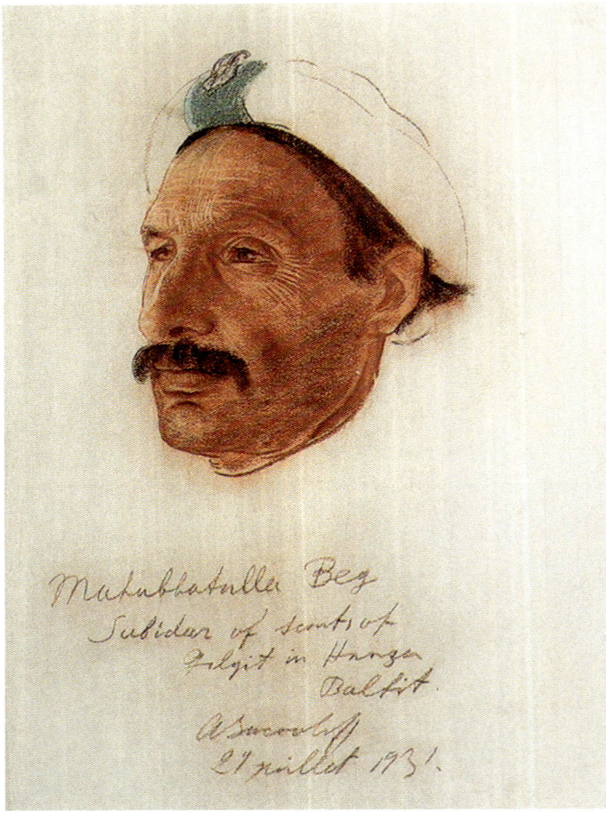

The portrait of Muhabbattullah Beg, son of M. Reza Beg and father of Qudratullah Beg, is a rare specimen portraying a member of the wazirkuċ. Muhabbattulah Beg (left) is here represented as a sūbedār of the Gilgit Scouts, and not as one of the collectors of Hunza's historiography. Sanghi Khan (middle) from the village of Maiun was identified as 'numberdar' by Aleksandr Yakovlev. In David Lorimer's list of office-bearers he was named as a new tranphá in 1934. Mahmud Sekand, a Shina-speaking levy serving in the Gilgit Scouts, was portrayed at Maiun in Shinaki. Source: Mixed media drawings on paper by Aleksandr Yakovlev, Maiun and Baltit 25 and 27 July 1931. Reproduction with kind permission of Kunstbibliothek der Staatlichen Museen zu Berlin

land at the upper limit of settlements and have been instrumental in the expansion of the ecumene by converting pastures into cropped land. From Badakhshan, speakers of Iranian idioms such as Munji, Madaghlashti, and Wakhi have to be mentioned as well as Turk refugees from the northern fringes such as Uigur and Kirghiz who found a temporary or permanent abode in these valleys.[122] The flow of immigrants was not necessarily unidirectional; depending on economic and political conditions in the respective countries there were quite some movements backwards and forwards.

(4) *Immigration of Kashmiri artisans and Gujur nomads fom the Indus Basin*

The sizeable Kashmiri population in Gilgit Town originates from immigrants during the time of Ahmed Shah Abdali in the second half of the 18th century. They are known as 'Kashiro' and are artisans such as 'weavers, goldsmiths and carpenters' and 'form the largest section of the population' at the beginning of the 20th century in Gilgit.[123] Following the transformation of vast areas in the Punjab into canal colonies and controlled areas by the British colonial power, the grazing grounds of Gujur nomads shrank. A significant number of Gujur nomads migrated to the mountain rim in search of new abodes and pastures in consequence of these developments, which commenced in the second half of the 19th century on a big scale. This process of lowland-highland migration has continued until today. Some Gujurs settled in Chitral and Gilgit-Baltistan. In Ishkoman an oral version of local memory was recorded which traces the migration history of Gujurs back to an event in which a Syed Ismail Shaheed was defeated by a Sikh army under the leadership of the famous Punjabi ruler Ranjit Singh at Balakot in 1840. The defeated Gujurs dispersed and sought refuge in the mountains in the aftermath. It is most likely according to this account that the Gujur settlers of Chilas and Gilgit are said to be descendants of those refugees. From there new grazing grounds were occupied in neighbouring valleys such as Ishkoman and Yasin. The original name of Sultanabad in

the Hunza Valley is Gujurdas, their winter station on their cyclical migration to the high-lying summer pastures in the Naltar Valley.

The classification of established language groups in the Eastern Hindukush and Karakoram distinguishes autochthonous settlers and extra-mountainous/extra-territorial immigrants rooted in certain valleys. In addition, two more groups should be mentioned which have been important for more recent migration processes:

(5) *Intra-montane migration*

Significant migration within the mountain belt has taken place during the 20th century in search of cultivable land and grazing grounds and is still continuing at a slower pace today. New settlements were established in previously unoccupied territory either on barren terraces (*das*, *dasht*) through irrigation or by converting temporary summer and/or pasture settlements into permanent villages. Shina and Burushaski speakers from the Hunza Valley migrated down the river towards Matum Das, Jutal, and Nomal. Nowadays irrigation colonies are found near Gilgit Town and as far away as in Punial, Ishkoman, and Yasin. A comparatively recent development is the migration of households to the commercial and administrative centres of Chitral and Gilgit-Baltistan. Individuals and households temporarily and sometimes permanently migrate to new locations in search of off-farm employment in government offices and development agencies.

(6) *Immigration and temporary population exchange between lowlands and highlands*

Persian and Urdu had entered the mountainous regions as court and legal languages. Urdu replaced Persian in the Kashmiri court administration around the same time in the second half of the 19th century when Kashmiri influence was more strongly felt. This shift had an impact on very few local residents, as schooling was a privilege of the elite. Schools were in their infancy and mainly based on Persian as the medium of instruction. Talking about contemporary Urdu

Sharafat from Gulmit is here depicted with a Wakhi cap (skid) in a Pamirian tradition. The village represenative (lambardār) of Pasu, M. Adab, is wearing a local cap made from homespun sheep wool (phárčin). The respected gentleman on the right in Aleksandr Yakovlev's painting is Mirza Dolik, the village headman (arbāb) of Gircha in Gojal. Source: Sanguine and pastel on paper by Aleksandr Yakovlev in Gulmit, 31 July, in Pasu, 1 August and in Gircha, 3 August 1931

The headman (traṅphá) of Misgar, Sage Ali, had inherited his position from his father Ali Gohar and represented the mīr in the village where Aleksandr Yakovlev spent some time waiting for the other expedition members to arrive. Sage Ali is wearing a Hunza cap (phárčin) made from white homespun sheep wool, a jacket of the same material and wide cotton trousers. The leather boots are of local make. Source: Sanguine, pastel, charcoal on paper painting by Aleksandr Yakovlev in Misgar, 4 August 1931

speakers, we are faced with quite a different development of temporary migration. Pashto is the only down-country language that has a longer and continuous tradition from the time of some ambulant traders and a few settlers in the second half of the 19th century until today.[124] Today the quota of Urdu, Punjabi, and Pashto becomes statistically significant only in the few urban centres of Northern Pakistan. There the percentage of households can rise up to fifteen percent of the resident population, while in the rural areas on average only two percent stated one of those languages as their mother tongue during the last census.[125] This trend has amplified in recent years: in the 1998 census seventeen percent of all immigrants were from the Pashto speaking areas.[126] Most of these temporary immigrants are either officers on duty, posted bureaucrats, or entrepreneurs in the bazaars. This process will be further intensified as the China-Pakistan Economic Corridor attracts investments and generates employment in Gilgit-Baltistan. In the other direction an increasing number of out-migrants seeks education, employment and business opportunities in the urban centres of down-country Pakistan. Taking into account the seasonal or temporary character of these migrations, the unique and persistent position of this linguistic region is underlined.

The qualitative classification based on vernacular affiliation and migration history has produced a community stratification that appears quite often as a distinctive measure when competition, resource distribution and access to resources are at stake. Linguistic, denominational and regional ascriptions play a dominant role in everyday life although quite a substantial share of people speak more than one or two languages and are part of distinctive networks. The retracing of the meaning of language community affiliation has shown the range of different communities of speech and their dominant areas of distribution as result of settlement history and migration processes.

Quantitative analysis of the survey data supports the statement of regional linguistic centres of gravity. Balti is the dominant language of all subdivisions in Skardu and Ghanche Districts. More than a quarter of a million mother tongue speakers were identified during the survey in the 1990s. Thus, this Tibetan dialect outnumbers Shina, which is the dominant idiom of Gilgit and Ghizer Districts. More than 150,000 inhabitants stated Shina which is predominantly spoken in northern Pakistan; only a small pocket of their speakers is found in adjacent areas across the 'line of control' in Baramulla and Ladakh.[127] In Hunza and Nagar, Burushaski dominates with more than two thirds of its habitants while Wakhi is spoken by nearly one fifth in Hunza, and Shina by more than a quarter in Nagar, and just an eighth of the population in Hunza. Yasin-Burushaski speakers (Werchikwar) trail behind the number of Khowar speakers in Ghizer District, a fact that underlines the function of this region as an intermediate zone between Chitral and Gilgit. However, in the dominant area of Werchikwar in Yasin, more than three quarters of the people speak this variant of Burushaski. Ghizer was under Chitrali rule and revenue schemes over long periods, resulting in the presence of members of the hereditary leadership and of settlers from

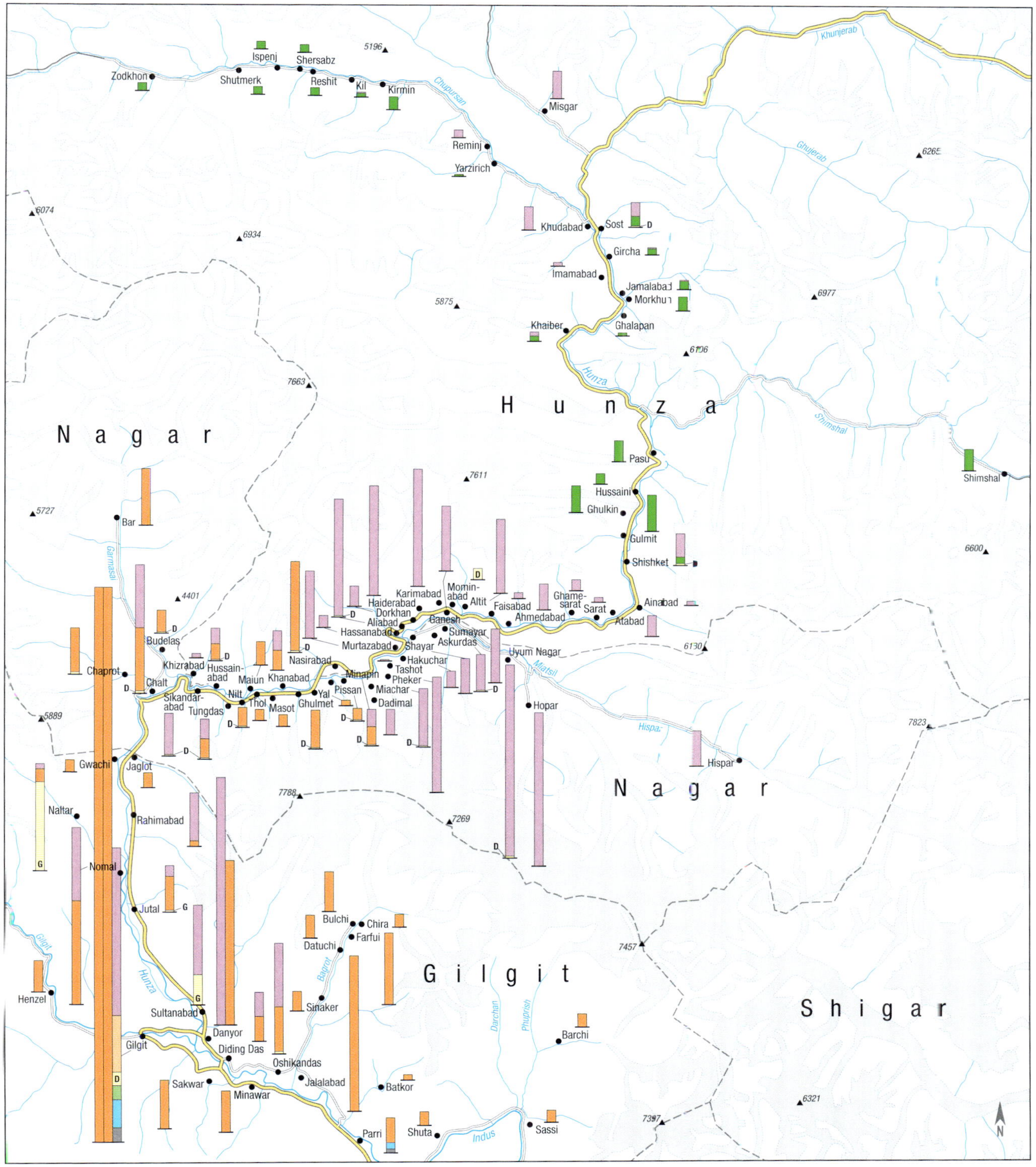

The representation of language groups at village level shows the variation within the Hunza Valley. In the lower section Shina dominates as the major language, in the middle section Burushaski is the prevalent vernacular spoken in Central Hunza and Nagar, while Wakhi is the dominant language of Gojal. Domaaki as the vernacular of the Dom is found in certain villages where artisans are located; the only exclusively Dom village in Hunza is Mominabad. Source: Own design with data based on Hermann Kreutzmann (1995a; 2005a).

The inaugural meeting of the Wakhi-Tajik Culture Association took place in Gulmit. Activists and stakeholders gathered in order to establish the association, to discuss possible scripts for Wakhi language and fields of mutual interest for promoting Wakhi heritage. The photograph was taken by Sabine Felmy on 28 April 1990

Chitral. Four out of five persons in Chitral are Khowar speakers. In neighbouring Badakhshan three quarters of the population speak Wakhi in Wakhan Woluswali, while across the Amu Darya boundary with Tajikistan in Gorno-Badakhshan nearly two thirds belong to the same language group in Rajon Ishkashim.[128] Sariqoli dominates over Wakhi in the contiguous Tashkurgan County in Xinjiang; both languages are related and belong to the Eastern Iranian branch of Iranian languages. Wakhi settlement regions are found in the upper parts of the Hunza, Ishkoman and Yarkhun valleys. Next follows Gujri with local importance in the Gilgit subdivision and Ghizer. Smaller groups of Ḍomaaki, Uigur, Kohistani and Pashto speakers, each with less than 2,000 persons, reside in the Gilgit and Ghizer Districts.[129] Needless to state that not all languages are exclusively limited to the core area. Speakers are found outside the mountains as well and across international boundaries in neighbouring countries. The Wakhi are an example of a group distributed in the 'Wakhan Quadrangle'.[130]

Language and script

Most of the unwritten vernaculars have often been described as a 'language remnant', as dying and soon to be extinct languages.[131] Various investigations were directed towards these endangered languages with rather limited numbers of speakers because they were regarded as remains of a past without any future.[132] One of those languages is Ḍomaaki for which not only a low number of fewer than 500 speakers was estimated, but also a severe language shift was observed towards Shina and Burushaski.[133] Ḍomaaki is listed with some Nuristani and Pamirian languages as 'definitely endangered' in the Atlas of the World's Languages in Danger, whereas Balti, Burushaski, Khowar and Maiyã are only termed 'vulnerable'.[134] For some reason Shina is missing from the map, but Hakim Elnazarov made sure that Shina and Kashmiri could be considered as not endangered.[135] He opines that the rescue operations seem to be feasible only from outside and concludes: "Given the current geopolitical interests of the countries of Central and South Asia and the instability in some regions, the reinvigoration of endangered languages is unlikely to be a priority. The efforts of international organisations to increase awareness of endangered languages as part of human rights initiatives seem to be the best option in the current situation."[136]

He and others are not convinced that local efforts and access to new media could stabilise the 'survival' of local vernaculars. The US-American faith-based, missionary-inspired Summer Institute of Linguistics paired with the National Institute of Pakistan Studies of Quaid-e Azam University in Islamabad to publish a five-volume Sociolinguistic Survey of Northern Pakistan in 1992. It was the first endeavour of its kind since the colonial enterprise of George Abraham Grierson who edited an eleven-volume Linguistic Survey of India with two volumes covering most of the area concerned by providing a summary of the contemporary state of the art.[137] Surprisingly, most languages are very much 'alive' and in daily use by growing speech communities. The range of language use can be quite selective and may vary from person to person. The senior librarian of Gilgit's Municipal Library, Sherbaz Ali Khan Barcha, described language use as follows a decade ago: "Shina is used for speaking with parents and family; Urdu is used for communication in public, outside the family; English is used for employment in private and government sectors; Arabic is used for religious rituals."[138]

The list could be expanded and modified for individuals in different contexts by adding the ritual use of Persian and by replacing Shina by one or even more languages spoken in households. Mobile telephone communication, internet publishing, document repositories and easier access to information technology have supported a move to expand the language use of Shina and its application in daily routines.

Scripted vernaculars in new media

One of the major challenges for the use of local vernaculars in modern forms of publishing and printing is the lack of scripts for most of the languages. Consequently, all of the printed media started off and still function as Urdu- and English-medium daily newspapers and weeklies. Nowadays the majority is electronically accessible through internet providers as well. The Pamir Media Group pioneered the publication of Wakhi poems in a Wakhi script adapted from linguistic writing systems and modified for their purposes.[139] Burushaski, Shina and Wakhi were the first vernaculars that were allocated radio time for local language broadcasting from Gilgit Town. The programming for a thirty-minute-long broadcast required some structuring based on script. Consequently, different versions of local scripts were used that are – as in the case of Wakhi – either based on the phonetic scripts applied by language researchers such as Aleksandr Leonovich Gryunberg and Ivan Michailovich Stéblin-Kamensky or derived from the Urdu script and individually adapted for the purpose.[140] During the first and founding meeting of the Wakhi-Tajik Culture Association (WTCA) in Gulmit on 28 April 1990, a dispute immediately arose which would be the 'appropriate script' for the Wakhi language. As no consensus or compromise was mediated, no binding commitment could be reached in terms of a single and in future applicable script – either the Arabic/Urdu calligraphy or the predominant international Latin script. This stalemate has lasted until today. The two factions practised their respective versions of an adapted script. Recently, more people have followed the Latin script as different disseminating multipliers adopted it for their publications, such as the Pamir Media Group, the Pamir Times, and adherents of social media with blogs in Wakhi.

Shortly after the Wakhi-Tajik Culture Association was founded, it started operating the *Bam-e Dunya* radio programme; their presenters do use scripts according to personal preference for their preparation of broadcasts. The *Bam-e Dunya* radio programme is a good example of cross-boundary communication as it can be received in neighbouring Afghanistan, China and Tajikistan by Wakhi speakers there. Radio transmission has enhanced communication and supported a tendency towards preserved texts and recordings in adapted art forms which Georg Buddruss and Almuth Degener called the 'meeting place', a collection of radio features in Gilgit's Shina language by Mohammad Amin Zia.[141] Shina broadcasting has operated from Gilgit since 1979; pioneering transmissions from Rawalpindi already began in 1949.

Daily and weekly Urdu newspapers are popular in Gilgit-Baltistan, despite the significant growth of electronic media they have maintained their readership. The photograph was taken in Gilgit bazaar on 29 March 2019

One of the oldest mosques of Hunza has survived in Haiderabad; its inscription hints at 1806, the zenith of tham Silum Khan's rule. The photograph was taken on 10 October 1984

A Khowar programme has been transmitted since the mid-1960s, a Burushaski programme since the mid-1980s. All those programmes require scripts. A pioneer was Shahzada Hussam-ul-Mulk, who founded the *Anjuman-e-Taraqqi Khowar* in Drosh (Chitral) in 1956; he developed a textbook for instruction and published Khowar texts from 1967 onwards.[142] A non-standardised writing system was derived from a modified Arabo-Persian alphabet for the purpose of preparing radio manuscripts; this is used by Shina literati and poets who are organised in the Gilgit-based 'Karakoram Writers Forum'.[143] The latter organisation acts as an umbrella institution of the above-mentioned regional representations of vernaculars and includes as well the 'Balti Culture Academy' in Skardu. Such institutions prepared the ground for the adoption of new media. By now, there are 35 years of experience in radio broadcasting from Gilgit using local vernaculars.

Radio was a pioneering technology in electronic communication and has been followed by television, internet newspapers and blogs. In its initial phase the internet offered services in English only, representing different speech communities. After the passage of time and because of the internet's success story, the different news services that bear the label 'Times' have introduced additional language services. They serve the purpose of articulating regionalised interests, such as Ghizer Times, and political thrusts. The Gilgit-Baltistan Tribune is utilised as a distribution forum for the views of the 'Balawaristan National Front' (BNF). Urdu was added as a second language in order to reach more people. The spectrum of users has expanded again by the introduction of Burushaski and Wakhi blogs. It seems to be only a question of time before blogs in a variety of vernaculars are accessible. These have become useful sources of information and act as exchange networks for migrants who work in the urban centres outside the mountain regions or abroad. A significant number of international migrants enjoy these platforms besides the usual mobile phone communication, social media and Skype services. The latter have been instrumental for a revival of local vernaculars. However, the process continues; the demand of the 'Shina Language and Culture Promotion Society' that local vernaculars should be regarded and respected as 'national languages', taught from primary to university level, and that a Shinalogy department should be established at the Karakoram International University is prominent on the agenda.[144]

Denominational diversity

The distinguishing parameter of language opens up a field of a multitude of vernaculars with regional cores and dialectical variants. The overall picture is quite clear in terms of denomination when the predominant religion of Islam is concerned. In the Pakistan population census reports, no distinction is offered in terms of denominations of Islam. A different picture occurred in colonial reports. The early gazetteers were interested to distinguish between Sunni, Shia and Ismaili Muslims. The Political Agent Bertrand Evelyn Mellish Gurdon took a keen interest in the numbers game, and his 'report on the Gilgit Agency' from 1906 is an early identification of Punial, Yasin, Ishkoman, Ghizer and Hunza as predominantly Maulai or Ismaili districts, while Nagar and parts of Gilgit and Bunji were identified as Twelvershia-dominated districts. Sunni strongholds were Chilas, Tangir and Darel; Kuh was a district with nearly equally distributed Sunni and Maulai populations. The Gilgit Agency of that time had nearly equal shares of each denomination with a slight majority of Twelvershia adherents.[145] This distribution is more or less in tune with the overall perception of the mix in Gilgit Town. If the 'old' districts of Gilgit and Ghizer are compared, the data from 1906 and 1990 show a significant rise in the Ismailiya who compose nearly half of the population and an increase in Sunni believers who are recorded as one fifth, while the Twelvershia have dropped to less than a third.[146] The data would be modified if the Chilas and Baltistan Districts were included in the sample.

In our survey in most parts of Gilgit-Baltistan – excluding Astor and Chilas – close to one hundred percent of all interlocutors claimed to be Muslim. Data interpretation revealed four denominations in the Gilgit-Baltistan sample that can be distinguished and that have regional impact.[147] About half of all inhabitants of Gilgit-Baltistan declared they were Twelvershia, about a quarter belonged to the Ismailiya, one seventh to the Nurbakhshia and one eighth to the Sunni. In Ghanche District three quarters of the population belong to the Nurbakhshia which has its sole pocket in Baltistan and its stronghold in Ghanche. The Twelvershia dominates from Kharmang to Nagar and includes most of Skardu and Shigar. The Ismailiya has been strongly correlated with Hunza and Ghizer.[148] Sunni believers are found in the Indus Valley around Chilas. A special case is Gilgit District with Gilgit Town as an urban area where Sunni, Twelvershia and Ismail-

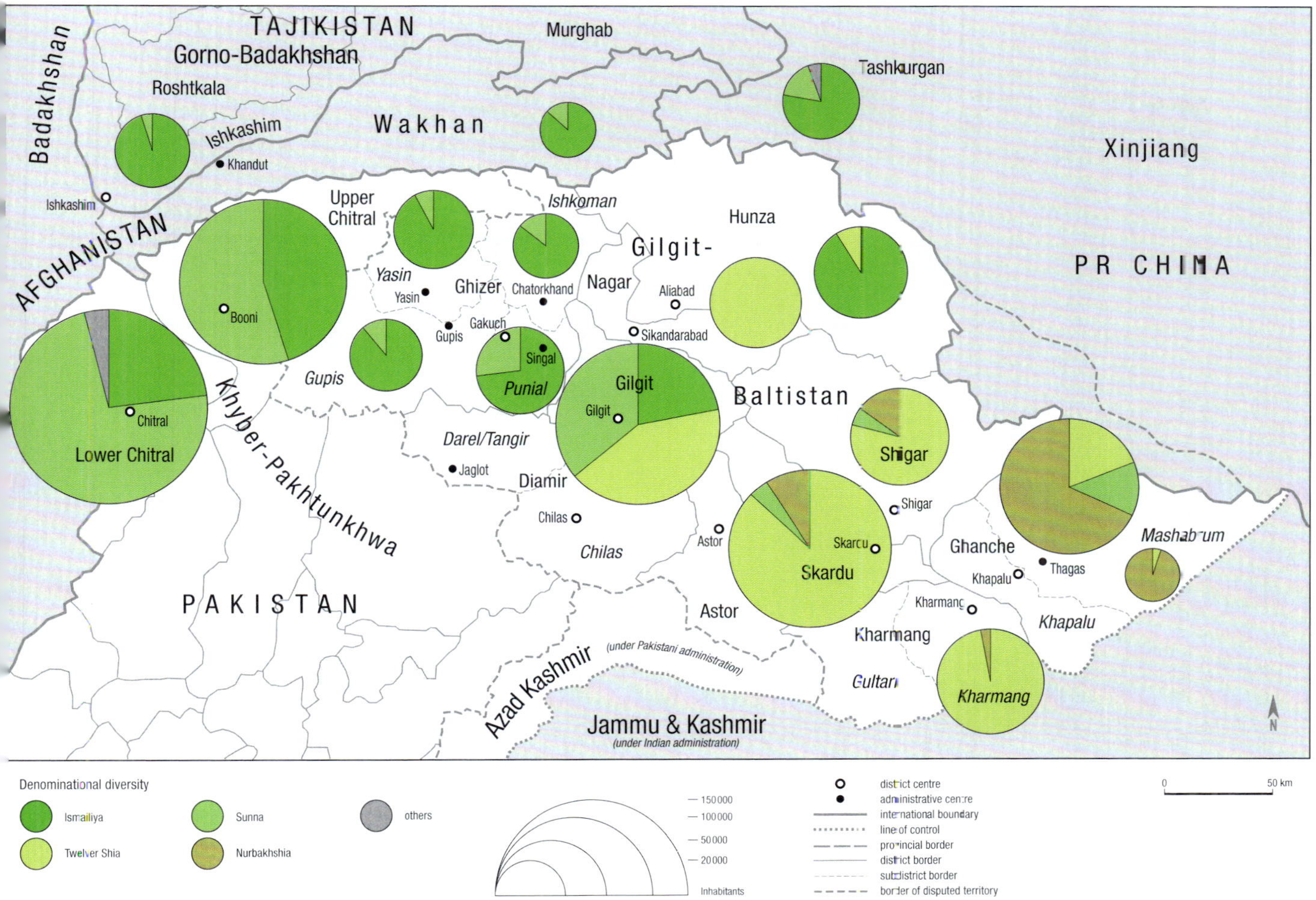

In terms of denominational diversity Chitral and Gilgit-Baltistan display some regional variations. While Sunni believers are dominant in Chitral, they have remained minorities in Gilgit-Baltistan where the western and northern parts are dominated by Ismailis. Twelver Shias are strongly represented in Nagar and Baltistan where in addition Nurbakhshias are dominant in Ghanche and Mashabrum. Gilgit Town appears to be a tripartite arena inhabited by Sunni, Twelver Shia and Ismailis. Source: Map modified and updated from Hermann Kreutzmann (1995a: Fig. 2); denominational data based on author's own survey in 1990–1991. Design by Hermann Kreutzmann

iya are represented in nearly equal shares. Sectarian tensions have developed into clashes and conflicts in recent years, especially in Baltistan, south of Gilgit and lower down the Indus Valley. Migration to the urban centre of Gilgit-Baltistan has increased in the aftermath of the recent economic transformations and modified the denominational split in Gilgit Town. Migration from Hunza and Ghizer has increased the Ismaili population, while cash growth in Chilas District due to timber royalties and compensation allowances for the Basha Dam construction has attracted Sunni settlers to the property market in and around Gilgit Town.[149] Professional migration and investors from down-country Pakistan have amplified this trend. In recent years, the administrative and urban centres of Gilgit-Baltistan – Gilgit Town, Gakuch, Skardu, and Aliabad – have grown faster than their surroundings and have been confronted with a significant change of settlers. Gilgit City, which a recent report called the major urban centre of Gilgit-Baltistan, is estimated to have crossed the one lakh population line by now. The Population Census Organisation of Pakistan projected the urban growth in Gilgit Town twenty years ago as reaching 92,365 inhabitants by 2018; the latest data available suggest the present population to be more than twice as high with 216,760 in the city.[150] The recent impact of higher investments due to the China-Pakistan Economic Corridor has significantly accelerated the growth. The urbanised areas have become the centres of communal conflicts as regards the building of places of worship in certain neighbourhoods, the representation of denominational interpretations of Islam or participation in regional affairs and the allocation of

In the premises of a Shia mosque in Sikandarabad school education is offered. The architectural style has adopted a gabled roof to which a veranda is attached that features wooden pillars. The photograph was taken on 26 October 1985

posts in the administration. Tensions have grown in recent years. The 1928 updated version of the gazetteer reported a rather amicable environment in the Gilgit Agency: "The religion of the whole country is some form of Muhammadism, three different types of which are now striving for the mastery. From the south, *Suni Mullas* have carried their tenets up the different valleys with more or less success; from the east a current of *Shia* doctrine has set in from Skardu, and from the west the curious *Maulai* or *Rafizi* tenets have found their way. It is to be noticed that the subversion of Hinduism and Buddhism by Muhammadanism in the remoter valleys seems to have been extremely gradual and due more to conversion than to persecution. Wherever *Suni* and *Shia* are found living together they seem to practise a mutual tolerance rare in other Muhammadan communities."[151] That observation reflected a general picture although the political diaries contain references to tensions now and then. In Hunza, the minority Twelvershia community had sometimes felt uncomfortable with their treatment by their *tham*, especially after the reforms in Ismaili practices since the 1920s.

The quality and frequency of conflicts have changed during the last generation; the notion of tensions has become a regular feature. Sectarian conflicts in Gilgit-Baltistan are primarily due to communalism-motivated clashes between the majority Twelvershia and minority Sunni groups. Armed clashes with fatalities have recurred frequently since the 1980s, coming to a head in 1988 when many died in the Gilgit District and especially in Jalalabad.[152] Polarisation, with its concomitant identity effects, occurs primarily between these two communities. Many observers claim that these kinds of lobbyism, strategies and group formation were copied in the Gilgit-Baltistan arena from developments in down-country Pakistan that stimulated the growing confrontation. Endeavouring to contain the conflict and to mediate, the administration arranges regular meetings at which representatives of the parties concerned in the shaky peace agreement are called to account. These representatives are the Sunni *Anjuman-e-Ahle Sunnat* and the *Anjuman-e-Imamia,* which acts for the Twelvershia population of Gilgit. Thus, authorised negotiators and members of the administration have contributed to reducing the tensions and supporting the maintenance of civil rules and public order. However, the Gilgit Police registered 117 murder cases that were related to sectarian clashes; other sources estimate the casualties in the order of one thousand.[153] The Jalalabad massacre and the deteriorating security situation in Karachi were two factors that motivated quite a number of people to remigrate into the Hunza Valley and to engage in building facilities such as schools there. The aim was to make accessible in Aliabad and Karimabad infrastructure assets that were previously only available in down-country Pakistan and in Gilgit Town. The exceptional growth of Aliabad is connected to these developments that were further accelerated after 9/11.

The steady flow of immigrants and out-migrants in Gilgit-Baltistan is a comparatively recent phenomenon that was amplified by improved infrastructure and travel arrangements. The Karakoram Highway was the single most important artery that has supported a steady increase of mobility and migration. There are no signs that this process is being attenuated, given the vision of new opportunities linked to the China-Pakistan Economic Corridor.

However, an observed dynamism of population flows is related to a settlement pattern that came into being much earlier and that is shaping and structuring the relationships between communities, villages and households. The reconstruction of these processes for the Hunza Valley offers insights into the historical process of land acquisition, power structures and migration.

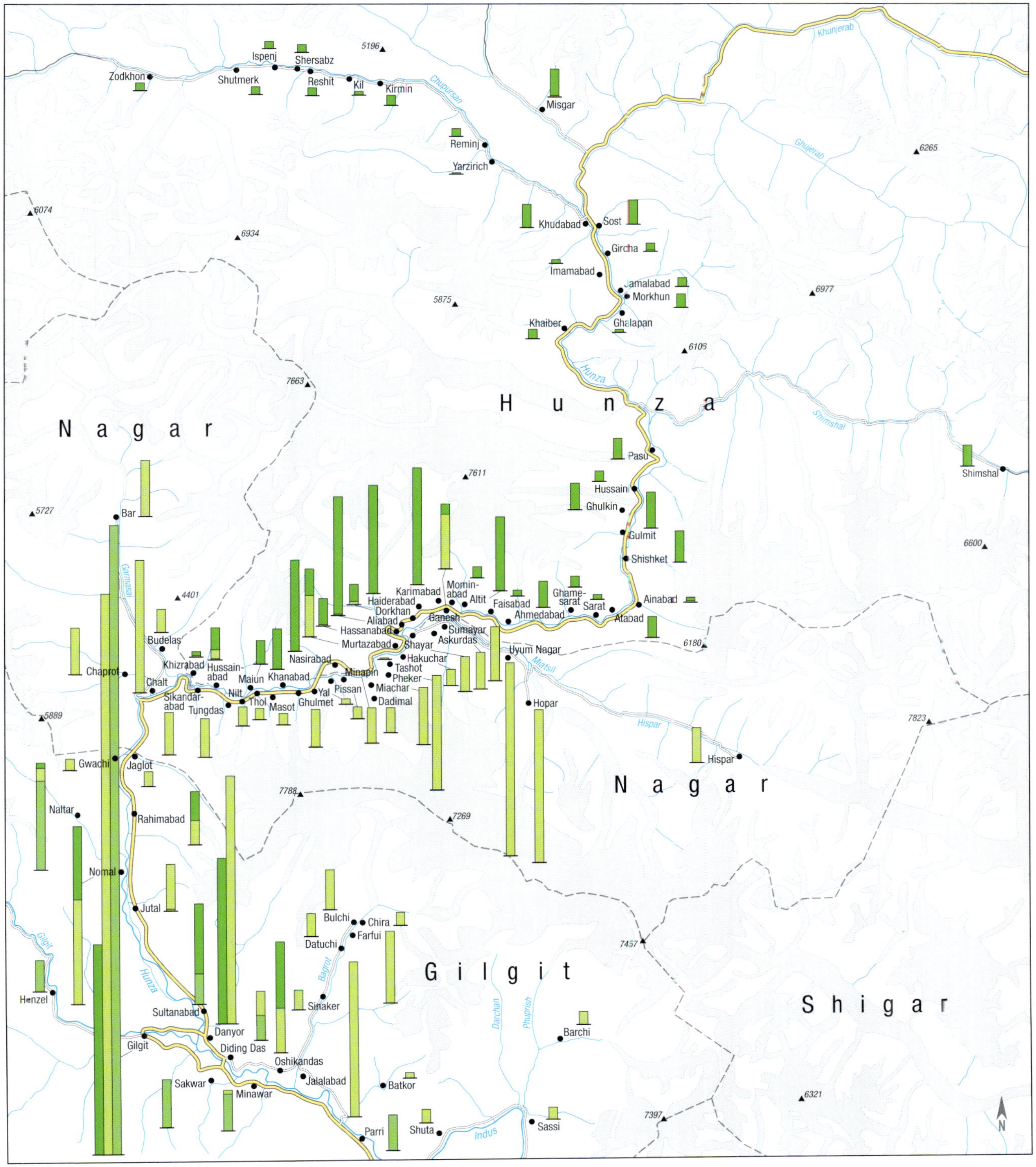

Denominational diversity can be attributed to Gilgit Town where residents belong to the Sunna, Twelver Shia and Ismailiya. In the Hunza Valley nearly exclusively Twelver Shia people are found in Nagar and some villages in Central Hunza. Otherwise Ismaili villages dominate in Shinaki, Central Hunza and Gojal. Source: Own design with data based on Hermann Kreutzmann (1995a; 2005a).

Settlement processes – appropriation, colonisation, immigration and occupation

The majority of permanent settlements are located in the valley bottom where the Hunza Valley appears as a string of irrigated oases.[154] Settlements on scree slopes and river terraces are the main feature of sometimes isolated oases. The highest population concentration is found in the central oases of Hunza and Nagar where the valley comprises a continuous band of interconnected villages on both banks of the river. The size of the villages may vary significantly. Khizrabad and Tashot are very small settlements, whereas Karimabad, Aliabad, Chalt and Uyum Nagar constitute units of several hundred houses and may even contain more than one thousand households. Being the administrative and commercial hub of Hunza District, Aliabad features most functions of a municipality.

The constituting factors and perceptions of a village in the Hunza Valley need to be specified. Administratively the term has no meaning as the bureaucracy addresses union councils as the smallest units, which are constituted as a cluster of neighbourhood settlements. Non-governmental development institutions operate on the basis of village production groups and village organisations (*tanzīm*) which break up the bigger settlements into neighbourhood units (*muḥalla*) and which try to make the cooperatives comparable units of similar size. In recent years they have created their parallel structure in Local Support Organisations that more or less resemble the union councils. The appellation as a village is mainly based on principles that are connected to their founding periods when combined agriculture constituted the basis of any dwelling: water rights and common property in pastures. Their rural appearance indicates that the village as an institution plays a prominent role in the context of society and economy. Villages are not predominantly centres of identification due to statistical features; rather, villages constitute units of institution, kinship affiliations and property rights. The historical genesis looks back on a more or less long tradition. Each village has its own founding history and occupies its specific role in the local administration (*maqsóo*) and respective taxation with common rights and duties.

The arbitrary character of administration contrasts with a structural understanding of a village. It makes sense in our context to define a village as a socio-economic unit which controls common properties, uses some resources on a community basis and capitalises certain assets as a unit. The village lands contain one or several compact settlement nuclei and scattered houses and stables, individually owned irrigated fields close to the valley floors, seasonal settlements and private lands at intermediate elevations, and natural pastures as a common property at high altitudes. By the landmark of the highest irrigation channel (B. *góčil*, S. *gúuço*, W. *wod*) in the permanent settlement oasis, the village lands (B. *íčit*, W. *qišlóq*) are distinctly separated from the rest. Cultivable, mainly unused and deserted lands (*das, dašt*) are easily distinguished by the absence of sufficient precipitation and/or irrigation. The *Artemisia* steppe there is extensively used by roaming livestock and provides the spare ground for the construction of public and private buildings. High pastures (*ter, ayloq*) offer seasonally high-quality fodder and are accessed by shepherds for the prime purpose of exploiting this natural resource. The combined mountain agriculture followed by the farmers of Hunza and Nagar is firmly based on two foundations due to the absence of substantial forest use: irrigated agriculture and animal husbandry in high pastures. Their share in the income structure was very high in former times, but their importance has decreased steadily in recent years as off-farm employment opportunities have grown.

The earliest data on village sizes go back to Patrick Vans Agnew who visited the valley in September 1847 and reported a likely estimate of about seven thousand houses in Hunza and Nagar.[155] This was quite a substantial number as he counted only one thousand houses for Gilgit. He travelled only up to Chalt and did not see the seats of power and the most populated places of the central oases. The first population census in 1911 returned for Hunza 10,126 and for Nagar 13,347 inhabitants.[156] Since then, a steady growth has been recorded. The number of inhabitants has doubled during the forty years between 1931 and 1971 when for Hunza 26,544 and for Nagar 25,631 inhabitants were fixed.[157] All population figures are rough indicators at best. Since then the population size has more than doubled. The results of the 1998 census should be handled with great care; the exercise of 2017 has not been finalised and published because of reservations about data manipulations.[158] Nevertheless, 'projected data' were released; Nagar has about 70,000 and Hunza 50,000 inhabitants to date according to this estimate.[159]

Opposite
View from Baltit Fort towards the central oases of Hunza and Nagar. In the foreground the roofs and smoke-holes of the houses in Diramishal Khan are visible. The roof tops were used for drying fruit and vegetables in the summer of 1954. The three basic forms of irrigated agricultural land use are next to each other: terraced field parcels (hárkiṣ) for crop farming, extensive fruit orchards (basíkiṣ), and the grassy slopes on the left for irrigated meadows (toq). The deeply incised canyon of the Hunza River separates Hunza from Nagar; only a few patches are cultivated in the river-bed, which is prone to episodic flooding. The main cultivation and settlement zone begins about one hundred metres above the flood-bed. Source: German-Austrian Himalaya-Karakoram expedition 1954. Courtesy of Gerhard Klamert

The population in Hunza and Nagar has increased fivefold during the 20th century according to the census data, which are not very reliable but give an indication of the major trends. Source: Author's own calculations based on data provided by John Biddulph (1880: 24–25); George Cockerill (1895); Census of India (1912: 220; 1933: 59; 1943: 49); Census of Pakistan (1998); Frederic Drew (1875: 554); Government of Gilgit-Baltistan (2018: 2); Government of India (1909: 142–143, 171–172); Government of Pakistan (1975: 11); Stuart Hill Godfrey (1898: 85); Walter Lawrence (1908: 83, 107); Anant Ram and Hira Nand Raina (1933: 59); Muhammad Masih Pal (1928; 1934); Elizabeth Staley (1966: Table 8); The Muslim 16.10.1998; The Superintendent of Government Printing (1890: 368, 615); Patrick Vans Agnew (1915).

Opposite
The fertile oases of Central Hunza and Nagar form a compact settlement area on both sides of the Hunza River. Central Hunza on the right bank in the north comprises the old nuclear settlements of Altit, Baltit (renamed Karimabad since 1983) and Ganesh. All other settlements towards Hassanabad in the west and Ahmedabad in the east are younger filial settlements (thuáan khanánċ). On the southern river bank, the former seat of power is located in Uyum Nagar between the old village of Hopar and the major Burusho villages between Sumayar and Pheker. The appearance is less compact; nevertheless, Nagar is endowed with substantial lands and a comparatively higher supply of irrigation water. The north-facing slopes are slightly more humid and offer better fodder potential for the locally kept livestock. Hunza profited from the alignment of the Karakoram Highway that bypasses the main Nagar villages. Thus, Aliabad developed into the major commercial centre in this area. Source: Topographical representation is based on Pleiades image 22/08/2014 (Karimabad); Pleiades image 10/08/2013 (Uyum Nagar.) Design by Hermann Kreutzmann

Population development in Hunza, Nagar and Gilgit District

Year	Hunza		Nagar		Gilgit District	
	absolute	%[1]	absolute	%[1]	absolute	%[1]
1847	7,000 houses in Hunza and Nagar[2]				1,000 h. in Gilgit[2]	
1873	–	–	–	–	25,834	–
1880	6,000	–	10,000	–	–	–
1896	8,868	–	11,633	–	–	–
1901	8,667	–	10,001	–	44,731	–
1911	10,126	1.6	13,347	2.9	57,602	2.5
1921	12,117	1.9	14,188	0.6	66,788	1.5
1931	13,241	1.0	13,672	–0.4	69,861	0.4
1941	15,341	1.2	14,874	0.8	83,655	1.8
1951	15,691	0.2	18,353	2.1	90,283	0.7
1961	21,291	3.1	17,623	–0.4	112.938	2.2
1972[3]	26,544	1.9	25,631	3.3	158,385	2.9
1981	27,797	0.5	45,880	7	228,185	4.3
1998[4]	46,665	3.1	53,000	0.8	363,542	2.8
2017[5]	51,250	0.4	71,000	1.6	450,000	1.3

1 Average annual growth rate during the preceding decade.
2 Patrick Vans Agnew (1915: 300–302) estimated the number of households in September 1847 upon his visit of the Hunza Valley up to Chalt.
3 Tahir Ali (1983: 16–17) mentions that the population census for Hunza was delayed until 1974.
4 The statistical area of the former Gilgit District is identical with the present Gilgit and Ghizer Districts since the administrative reform of 1989. The latest data of 1981 (especially for Nagar) and 1998 are rather unreliable as all the absentees, i.e. former residents, migrants etc., have been returned as permanent settlers to the census enumerators, if not others as well. More inaccuracy occurred when seasonal settlements were double-counted with settlers who had been recorded in their homesteads already.
5 The 2017 data are estimates that were published by the Government of Gilgit-Baltistan (2018).

The impact of this population growth has found its spatial expression in the expansion of settlements within the valley and the establishment of extra-territorial migrant colonies outside the former principalities. The settlement process in Hunza has been reconstructed for the last two hundred years.[160] The sources of information include oral traditions about the extension of irrigation networks and the establishment of filial settlements, colonial reports and records as well as travelogues. The systematic recordings of genealogical affiliations of founders of new settlements and village histories have supported the reconstruction of the expansion process. This epochal growth cycle can be divided into four different phases; for practical reasons the formative phase of Hunza as a microstate is linked to the eighteenth century. In the end a significant transition occurred that kept Nagar more or less at the same size, while Hunza expanded north- and eastwards touching the limits of the Chinese Empire. The nineteenth century is characterised by an expansion within the newly acquired and controlled lands. The founding of new settlements was supported by substantial population growth due to accepting refugees and incorporating migrant lineages into the Burusho kinship system. Consequently, people have been integrated into the Burusho community who have been immigrating from southern or northern neighbouring valleys; some arrived as refugees seeking security and avoiding persecution, others took residence as servants of notables. The knowledge about these adoptions and incorporations of migrants are still alive among local chronologists and genealogists today. There exist certain lineages and extended families who are proud of their relatives in various language groups. The case of Niazi Ali's family from Wakhan might serve as a case in point. His son Banda Ali fostered Asman Pari, daughter of *mīr* Fateh Ali Khan and sister of *mīr* Ali Mardan Shah from Wakhan. Asman Pari was married to the Hunza *tham* Safdar Ali Khan most probably around 1877. Thus it happened that she and her two foster brothers arrived in Sarikol in 1891. Her elder foster brother Nazar Ali would become the founder of a Shersabz line who subsequently settled in Chupursan Valley and have remained Wakhi speakers there until today. The younger brother Din Ali and his offspring became servants of *tham* M. Nazim Khan and permanently settled in Baltit.[161] It obviously took only one additional generation adopting Burushaski as their language. Now, three generations later, nobody would exclude them from the community of Burushaski speakers. Many similar cases could be presented where language change and incorporation into Burusho lineage systems occurred and have become apparent in settlement patterns and landholdings.

Structurally, this period changed shape after the colonial conquest, commencing with the *Hunza-Nagar Campaign* of 1891. Measures of indirect rule, colonial backing of the local *tham*, a regional security setup governed by the interests of the dominant empires, and the further influx of settlers from Wakhan visibly changed the layouts of villages that gave up their fortified, compact hamlets (*khan*) in order to settle within the village lands (*girám*). The incorporation of Hunza and Nagar into the colonial administration of the Gilgit Agency supported migratory processes that allowed people from the Hunza Valley to settle outside the realm of their *tham*, to cultivate barren lands and to take up service jobs in Gilgit. The beginning of the 20th century marks the upcoming opportunities for out-migration that benefit a

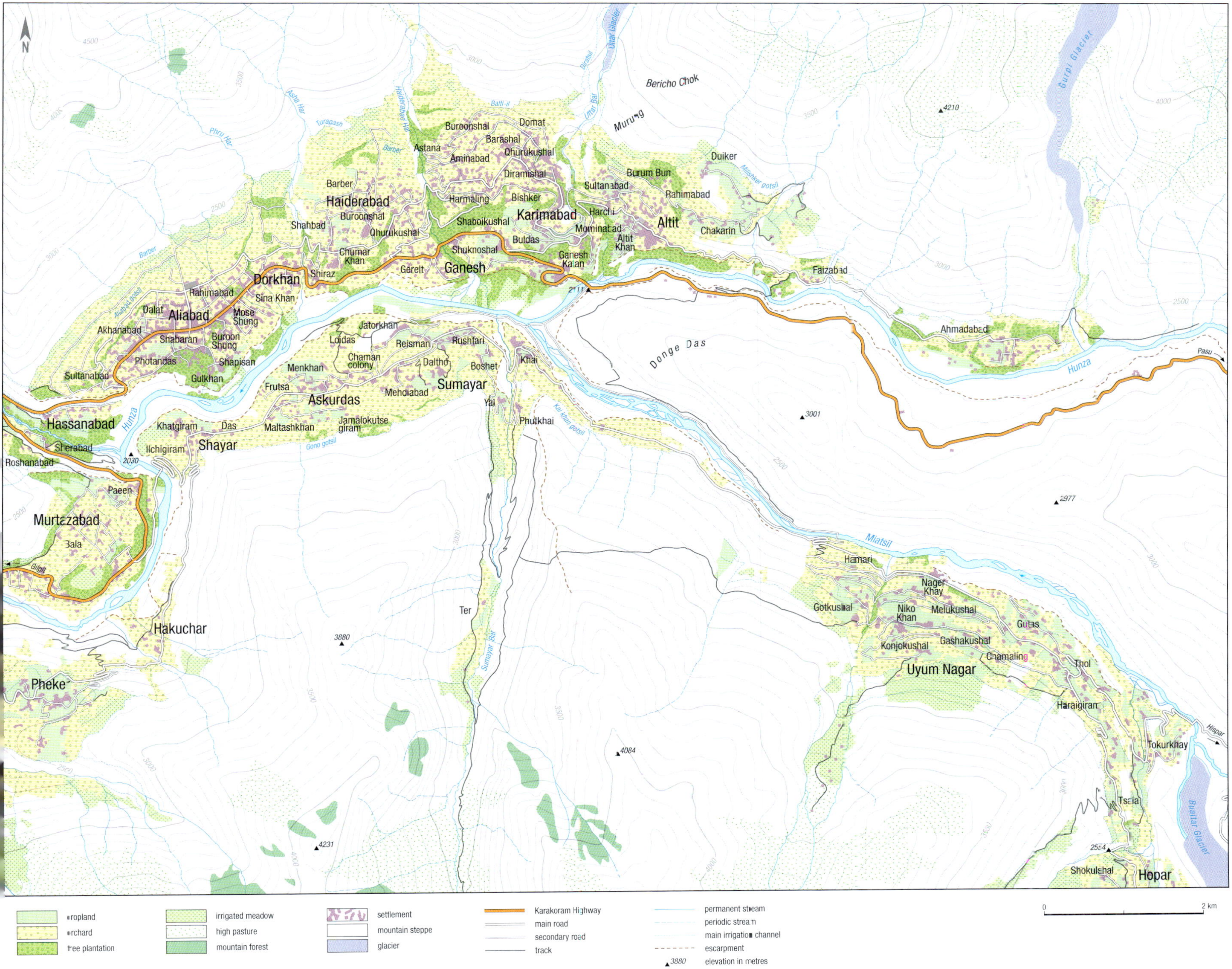
N
Bericho Chok
Murung
Gurpi Glacier
Ultar Glacier
Ultar Bar
Duiker
Burum Bun
Rahimabad
Sultanabad
Altit
Chakarin
Faizabad
Buroonshal
Domat
Barashal
Qhurukushal
Astana
Aminabad
Diramishal
Barber
Haiderabad
Buroonshal
Qhurukushal
Shahbad
Harmaling
Bishker
Shaboikushal
Karimabad
Harchi
Mominabad
Altit Khan
Buldas
Shuknoshal
Ganesh Kalan
Ganesh
Gerelt
Chumar Khan
Shiraz
Dorkhan
Sina Khan
Rahimabad
Dalat
Aliabad
Mose Shung
Akhanabad
Shabaran
Buroon Shung
Photandas
Shapisan
Gulkhan
Sultanabad
Jatorkhan
Loidas
Reisman
Rushfari
Chaman colony
Daltho
Boshet
Khai
Menkhan
Frutsa
Mehdiabad
Sumayar
Askurdas
Yal
Phutkhai
Maltashkhan
Jamalokutse giram
Khatgiram
Das
Hassanabad
Hunza
Sherabad
Roshanabad
Ilchigiram
Shayar
Gono gotsil
Paeen
Murtazabad
Bala
Gilgit
Hakuchar
Pheker
Ter
Sumayar Bar
Donge Das
Ahmadabad
Pasu
Miatsil
Hamari
Nager Khay
Gotkushal
Niko Khan
Melukushal
Gutas
Konjokushal
Gashakushal
Chamaling
Uyum Nagar
Thol
Haraigiram
Hispar
Tokurkhay
Tsala
Shokulshal
Hopar
Bualtar Glacier
Kai Khan gotsil
2111
2030
3001
2977
3880
4084
4231
4210
2544
cropland
orchard
tree plantation
irrigated meadow
high pasture
mountain forest
settlement
mountain steppe
glacier
Karakoram Highway
main road
secondary road
track
permanent stream
periodic stream
main irrigation channel
escarpment
3880 elevation in metres
0
2 km

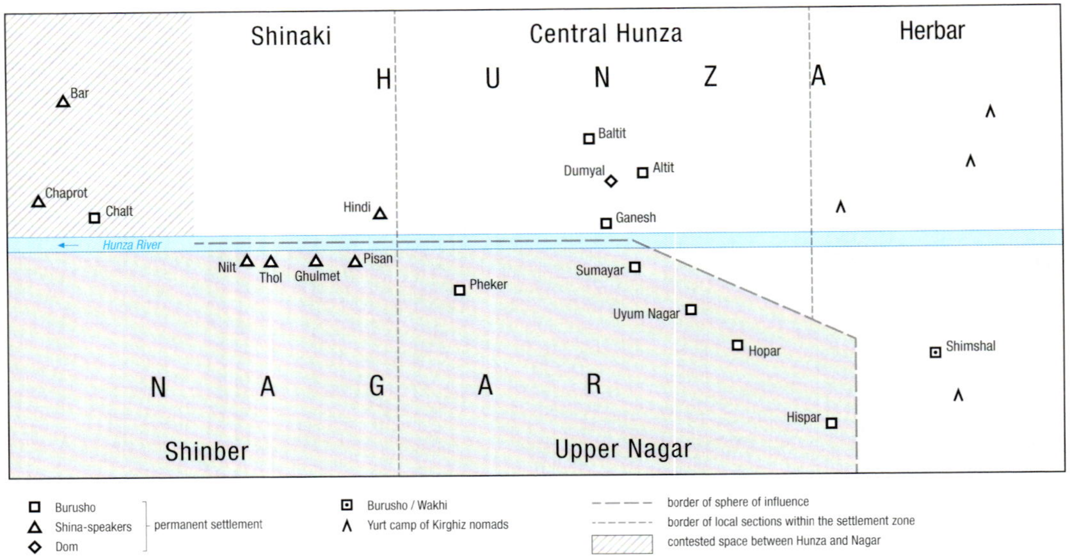

In the 18th century the settlement structure in the Hunza Valley consisted of a limited number of fortified villages (khan) and a compact space of control. The centre of Hunza was concentrated around the three nuclear Burusho villages Altit, Baltit, Ganesh with the artisan caste forced to settle in a separate Ḍom hamlet in Dumyal. In Shinaki probably only Hindi existed, while on the left bank of the river four Shina-speaking villages in Nagar are on record in Shinber. The lower part on the right bank was a contested space between Hunza and Nagar; both tried to control the fertile areas of Chalt and Chaprot. Nagar's centre was located in the upper valley where Hispar and Hopar were far away from the main river. The seat of power was to be found in Uyum Nagar while Sumayar and Pheker are mentioned as well as old Burusho villages. Herbar was seasonally utilised by Kirghiz nomads during summers, while a small Wakhi settlement might have existed in Shimshal from where Burusho caravan raiding expeditions set off to the Leh-Yarkand trade route. Source: Modified from Hermann Kreutzmann (1996b: 179; 2005b: 47; 2006: 261).

substantial number of household members who are initially supported by their kinsfolk and subsequently become independent dwellers in their new abodes. A new dynamism occurred after the end of *mīri* rule in the Hunza Valley and the opening of the Karakoram Highway; both events boosted spatial processes of settlement reorientation and fostered increased out-migration. Remittances and pensions were invested in bazaar shops and residential quarters that have resulted in dispersed and uncontrolled building of houses in an area lacking any kind of land use plan and zoning. This process is continuing and being amplified by a renewed tourism boom and expected new investments in the framework of the China-Pakistan Economic Corridor.

The reconstruction of the settlement process in the Hunza Valley allows some insight into the spatial expression of power structures, community relations and external influences that have contributed to shaping the high mountain oases of Hunza and Nagar.

Period of nuclear villages

Of all villages in Central Hunza still existing today, the oldest seem to be the three original *khan* (fortified villages) of Ganesh, Altit and Baltit, as well as the artisans' settlement of Dumyal or Berishal. The Ḍom still provide services as musicians and blacksmiths to the Burusho farming communities of the three 'original' villages in the main irrigation oasis of Central Hunza.[162] The *khan* is not a specific structural form that is limited to the Hunza Valley; fortified settlements – Urdu *kot*, Pashto *qal'a* – are familiar buildings in the Hindukush-Karakoram and reflect the urge for security and protection against outsiders and invaders.[163]

A similar setup for Nagar is on record for the Burusho settlements; the oldest village seems to be Hopar followed by Uyum Nagar, Sumayar, and Pheker.[164] Some Nagarkuċ regard Pheker as the earliest fortified village (*khan*) of Nagar. Chalt, Chaprot and Hispar might have been out of reach for the Hunza and Nagar *tham* during this period.[165]

The remnants of the old nuclei suggest a close relationship between site selection and defence purposes while safeguarding access to water supply and agricultural lands at the same time. Hindi constituted the only *khan* in the lower region of Shinaki; in Shinber Pisan, Thol, Nilt and probably Ghulmet existed already. It seems as if Nagar had a higher population – or less concentrated communities – than Hunza in the central and lower valley sections. The upper part of the Hunza Valley (*Herbar*) was seasonally occupied by Kirghiz nomads who used the high pastures there during summers. The system of fortified villages and their structural elements was quite common all over the Hindukush-Karakoram region, giving protection in times of threat by outside intruders.

The precondition for establishing fortified villages was a regular water supply. Thus, the narrations linked to the expansion of the irrigation systems contain clues about the settlement genetics. Ganesh is regarded as the oldest *khan* in Central Hunza because the *Hamáći* channel is accepted as the first irrigation line that tapped Ultar Bar to supply the *Hamáćatiṅ*, the earliest settlers of Ganesh. The subsequent *Altit góċil* diverted water from the same source to Altit, which is regarded as the older seat of power (*thámkuṣe diś*) occupied by the Hunza ruler compared to Baltit.[166] However, one of the oldest irrigation channels might be the *Balti íl* that tapped the Bululo snow meltwaters, which were led all the way into the fortified Diramishal Khan.[167] During this period, Hunza and Nagar were rather small chiefdoms, as was Gilgit. However, the fortified villages document that external threats existed, albeit the major threats were between the two direct neighbours that claim their origin from two brothers who separated and became arch enemies. The ruling dynasty in Nagar claimed descent from Moghlot, the

Hunza *Ayáśo* dynasty from his brother Girkis.[168] The story has often been told in many variants to document and maintain the enmity between the two very similar – in terms of language, old belief systems, and resources – but competing neighbours. This enmity has been celebrated for centuries now. It was surprising for the colonial authorities and worth an entry in the 'political diaries' many decades later that a marriage was arranged between a *sayid* from Nagar with a farmer's daughter from Hunza.[169] The narrative of adverse interests contributed to fixing the origin of the old settlement and the ruling class in the central oasis. Hunza was a rather small entity as its territory was limited to the central oasis and parts of Shinaki. Soon that would change.

Pre-colonial phase of oasis expansion and internal colonisation

The settlement at the beginning of the 19th century was the fortified *khan* where all houses were located in places that could not easily be irrigated. Each household possessed a *girám* outside in the village lands, where the cattle-sheds were located within or close to the orchards (*basíkiṣ*) that occupied patches within the cultivated terraced and irrigated fields for crop cultivation (*harkiṣ*). Apparently the population of the fortified villages increased, and so did the demand to cultivate new lands. It became necessary to expand into newly developed land parcels and to establish filial settlements there with ever greater distances from the agricultural resources. A boost in this direction must have occurred at the end of the 18th and beginning of the 19th century when *tham* Silum Khan III initiated an amelioration scheme within Central Hunza and Shinaki by founding filial settlements in connection with an expansion of the irrigation system. New channels were built into the new fortified settlements (*thuáaṅ khanánċ*); all *Báltikuċ* participated in the construction of the *Dalá*, which became the central water lifeline between Baltit and Aliabad. The setup of building new irrigation networks is the strongest clue, in connection with kinship configurations, to explain the settlement process. Along the *Dalá* we find *Báltikuċ* settlers; the area of *Bimal Das* became Haiderabad, whose origin could be around 1806 according to an inscription in the oldest local mosque. Seven years later the *Dalá* is supposed to have reached the barren lands covered by a white mudflow (*burúm mos*); henceforth they became known as Aliabad.[170] From Altit, expansion took place towards Koshi and Ahmedabad where *Áltikuċ* live. Ganesh expanded westwards all the way to its filial settlements of Gerelt, Dorkhan up to Murtazbad where settlers are found that are affiliated with *Gániśkuċ*. In all villages where new land was cultivated, a substantial share remained in the hands of the *tham* and *wazīrkuċ* while the remaining field parcels were divided among the persons from the original village who had participated in the construction and establishment of the village.[171] Kinship group affiliations are a prime indicator for finding the groups that participated in this process leading to an intensification of land use in the central oases.

The enterprising ruler *mīr* Silum Khan III became *tham* of the chiefdom in 1784 and followed a dual strategy.[172] He extended Hunza's sphere of dominance to the north towards Sarikol and Wakhan, expelled Kirghiz nomads from *Herbar* and allowed immigrating Wakhi settlers to found villages there. The local historians have made this a narrative of brave Burusho who expelled Kirghiz and invited Wakhi to settle. Qudratullah Beg tells the story in his 'History of Hunza': "He [Silum Khan] also settled Gojalis in these localities and kept his herds of goats and sheep and transport animals, like horses, mules and donkeys etc., and those of Gojalis under the supervision of these men from Gojal. However the Kyrghiz nomads carried out frequent raids and created trouble and problems for the new-settlers.

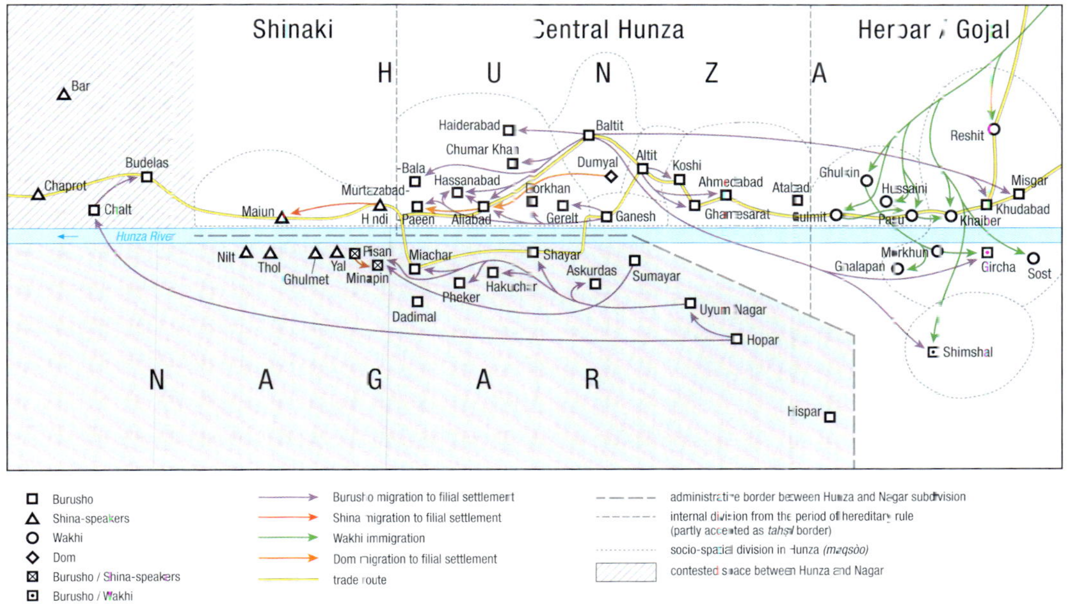

During the 19th century settlement expansion took place through the enlargement of existing oases and the creation of filial settlements; especially in Central Hunza and Nagar this process contributed to the creation of huge and compact irrigated oases. The settlement process in Central Hunza is exemplified in the description by Tahir Ali (1982: 160–166) about the foundation of Murtazabad. In Shinber and Shinaki the process of meliorisation was much slower. In Gojal a major transformation occurred after Kirghiz nomads were driven out of their customary grazing grounds and Wakhi immigrants were offered space for founding new villages. Gojal around Gulmit, Ghulkin, Hussaini and Pasu formed the earlier part (Gojal I), while Abgerchi and the Chupursan Valley (Gojal II) followed somewhat later. The hereditary rulers of Hunza had subdivided their territory into nine sections (maqsoo) that enjoyed different rights and responsibilities. Their claims on Chalt and Chaprot were not realized. Source: Modified and expanded from Hermann Kreutzmann (1989a: 52).

The view towards the East is opening up a perspective on Hunza. The Kirghiz yak-driver Naseeb from Andemin in Little Pamir is resting and holding our horses' bridles on top of Irshad-e Win Pass (4,979 m) which connects the Chupursan Valley with Wakhan. Several Wakhi immigrants took this route before settling in Gojal. The photograph was taken on 15 June 2000

Mir Salim Khan as a punitive response sent a strong force of chosen brave men of central Hunza under the command of Wazir Dara Beg son of Dervish to punish and evict these troublesome nomadic people. Wazir Dara Beg took his force to Chupurson and with his military skills and abilities defeated these nomads and killed many of them and had made heaps of their dead. He was finally able to push the remaining Kyrghiz over the passes and across the mountains and evicted them once for all. This force had killed a large number of Kyrghiz men hence many graves be found in the areas located between Misgar and the passes of Kilik and Mintaka even to this day."[173]

Thus, the upper valley became a Burusho *cordon sanitaire*, now called Little Gojal, and attracted Wakhi migrants and refugees from Big Gojal, which is Wakhan.[174] The northern passes – Irshad-eWin, Kilik, Mintaka, Khunjerab – were guarded from the Burusho-controlled *khan* of Misgar, while Chupursan was marked from Reminj. The eastern section with the Shimshal Pass would follow soon, and Khudabad was founded during *mīr* Ghazan Khan's reign as a second *khan* representing Burusho control of the passes by settling *guśpúr* Nafees Khan here.[175] While passes were the lines of control in the north, the Hunza River was an open flank towards Nagar, and, in the south, Maiun formed an important defence line towards Gilgit. Here narrow passages along the route (*darband*, *parī́*) offered easy control of access. The trust in these outward defence systems must have been limited as additional measures to protect the settlements were taken. All new villages of this period were designed as compact settlements with fortifications in the traditional *khan* style. Henry Haughton described the protective function of *parī́* and *darband* at vital strategic sections in the Hunza Valley: "Chéchar Pari [between Gwachi and Chalt] … The valley here becomes very narrow & is flanked on either side by huge cliffs which thus formed as it were the 'gate' into Hunza and Nagyr in the old days & one too which could be very easily defended. … At Khaibar there is a 'Derband' against raids from Kirghiz or Chinese. It is a very strong position & the present sangars were apparently by the Sahib commanding the British detachment which remained at Khaibar for about 2 months after Hunza and Nagar were taken in 1891. The fortified gateway on the only path across the nullah is the fancy part of the original defences. There is also a watchtower on a spur on the far (or left bank) of the stream."[176] Even within Hunza, a system of obstacles and barriers (*darband*) had been introduced along the routes to control the movements of inhabitants and travellers between the settlement areas of different sections (*maqsòo*) of Hunza. The village of Gulmit was locked and guarded at a gate where Burusho and Wakhi kept vigil.[177]

In every village there were either aristocratic relatives (*guśpúr*) or one to two persons of trust (*traṅphá*, *arbāb*), often linked to the *wazīrkuč*. For instance, during Bronislaw Grombchevsky's visit the stepbrother of *mīr* Safdar Ali Khan in Khudabad, the *guśpúr* M. Nafis Khan, appeared on the scene as a go-between. The trustworthy representative in Gojal was Mirza Hasan in Gulmit, a cousin of *wazīr* Tara Beg; both were regarded by the visitor as the two strong persons next to the *tham*. In Aliabad the *wazīr* made his brother Nasir the village representative.[178] For David Lorimer, the *traṅphá*, whom he sometimes called *traṅphá wazīr*, were the strong representatives of the *tham* in the villages.[179] The number of villages increased from five to twenty-five during this period of internal colonisation.

Similar processes took place in Nagar where the new villages of Askurdas and Shayar were established in the main oasis of Nagar, nowadays called Sas Valley which is an acronym representing Sumayar, Askurdas and Shayar; all are located on the left Hunza River terrace.[180] Hakuchar, Dadimal and Miachar are new villages in the Burusho realm, while Minapin, founded in the first half of the 19th century, became a mixed village where some Shin from Pisan settled along with Burusho from Sumayar. Yal and Masot came into being in Shinber, and Nagar's interest and claim on Chalt was underlined by sending some Burusho settlers from Hopar there after reconquering it from Hunza in December 1875.[181] The Bar or Garmasai Valley has functioned as an additional asset and valuable northward extension for Chalt; the whole valley has augmented Nagar's possessions and acts as a buffer between Gilgit and Hunza.

Settlement concentration processes under colonial supremacy

Hunza and Nagar constituted neither a remote microcosm nor the main arena during the Great Game when Anglo-Russian rivalry led to decisive moments of boundary-making in the Karakoram valleys.[182] The short but effective Hunza-Nagar Campaign[183] of 1891 had a lasting influence on the region's society, economy and settlement patterns. The structural element of *khan* ceased to be the dominant form of constructing habitations. The changed security situation, the cooperation with the British and Kashmiri administrations in Gilgit, and their backing in their relations with other neighbours seems to have brought a feeling of less vulnerability towards external attacks. The *khan* slowly disappeared as an architectural construction principle; it was brought to an end by dismantling the fortifications and expanding new buildings beyond the walls. A process of populating unfortified hamlets (*girám*) within the village lands started in places that were formerly reserved for cattle-sheds and storage facilities. The movement out of the *khan* into the *girám* had the welcome effect of reducing the distance between habitations and agricultural fields and orchards.[184] The sprawl of farmhouses into hamlets saved energy and time as most transport was done by human porterage (*baldá, sarúu*).

With the accomplishment of the British 'pacification' of Hunza and Nagar, a narrative was rejuvenated that had once given the pretext for conquering the valley: the tale of caravan raiding as a threat to regional security of movement and trade. It was obvious for Martin Conway during his visit in the early 1890s that combined agriculture would not suffice to feed the people of Hunza. He explained caravan raiding as a necessity to survive: "When population in Hunza increased to the bare subsistence level a bad season involved either starvation or a raid."[185] The cooperating trustee of the colonial forces, *wazīr* Humayun, justified the raids in the same manner: "We were hungry. The herds and the flocks of the Kirghiz on the Pamirs were ours for the taking. It was not our warriors who went; it was just the lads. They went in bands when they pleased, and captured caravans. Most of the plunder went to the Thum [*tham*], but they kept bullocks and sheep for themselves. How else could they live? The country is too small for us to graze our cattle."[186] In Hunza and Nagar the caravan raids were providing needed supplies for a wider range of the society and their livelihoods, while the slave-trade was an upper-class phenomenon.[187] A side-effect of their activities had been the expansion of Hunza territory at the expense of Kirghiz who lost access to their summer grazing grounds in Gojal, Raskam and Shimshal. The latter was regarded as a periphery where enterprising settlers with an emphasis on animal husbandry would find ample resources. Shimshal and its Nagar equivalent Hispar have been enclosed since the first half of the 19th century and treated in a number of narratives as penal colonies to which the *mīr* would expel unwelcome settlers and local trouble-makers.[188] The very notion of this makes little sense as people could easily have escaped from here; it seems more likely that the settlers in these places were more heavily taxed and had more forced labour obligations than others. Although many tales were told about these penal settlements, little evidence was presented. Their location at the foot of mighty glaciers and their remoteness and distance to the centres of power might have contributed to their reputation.

After the conquest of Hunza and Nagar, the narrative of caravan raiding and expansion into the northern headwaters changed into a story of Hunza as a land without sufficient space to feed its people and where food security could not be achieved. Further amelioration schemes and other measures were explored. The colonisation process of barren lands was stimulated in the peripheral regions of Shinaki and Gojal in connection with the construction of new irrigation channels (B. *góčil*, S. *gúuçc*, W. *wod*). A number of new villages were founded especially during the forty-six-year-long reign of *mīr* M. Nazim Khan until 1938. He and his *wazīr* – from Humayun Beg to Shukrullah Beg and finally with Zarparast – were the prominent figures during the second phase of internal colonisation. The expansion of channel networks and the amelioration of cultivable tracts led to the intertwining

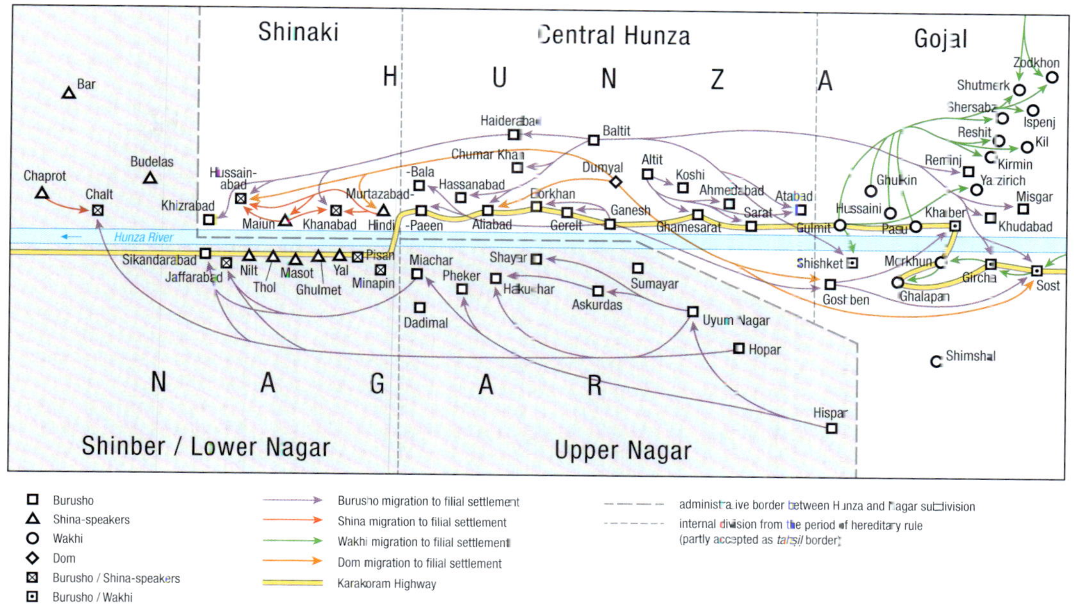

The most recent phase of settlement transformation is characterised by an expansion between and within existing villages; households shift their abodes from a nuclear settlement or a hamlet into their own agricultural lands wherever space is provided. In Gulmit a substantial relocation and shift was observed towards Chaman Gul, which is part of Gulmit's village lands. A few new settlements were founded during the 20th century, most notably Sikandarabad in Nagar, which subsequently acquired an important administrative function owing to its location on the Karakoram Highway. Several new villages were established at the territorial margins in Shinaki, Shinber and Chupursan. Source: Modified and updated on the basis of Hermann Kreutzmann (1996b: 180; 2005b: 47; 2006: 261)

When Altikuč began to found filial settlements one of the early attempts was made in Koshi up the valley. The compact village lands were established for less than a dozen households high above the Hunza River on a steep cliff. Even today the approach to the renamed village of Faizabad is somewhat challenging. The photograph was taken on 31 May 1985

of the compact network of contiguous major oases in Central Hunza based on available technology.

Within Shinaki the barren lands of Motos (Murtus) became one of the first new settlements when the irrigation channel was established in 1904; four years later the first settlers of Khanabad arrived from Hindi to develop terraces and to build houses there.[189] Along the route to Gojal, Sarat was developed along with the hamlets by the name of Nazimabad in Shishket and Sost. Henry Haughton called Nazimabad 'New Ghulmit' and dated its beginning to 1903.[190] As Gircha was losing land owing to undercutting of the settlement terrace by the Hunza River, a new sister settlement by the name of Sarteez was inaugurated in a safer location. In 1912 'New Gircha' had grown to thirteen households of Burusho settlers.[191] Reminj became an early Burusho stronghold in Chupursan, additional farms were built after the new channel of 1923, while the remainder of the long valley was substantially resettled in five Wakhi villages from the 1920s onwards.[192] During Henry Haughton's visit in 1912 there were thirty households in Reshit, but only five in Kil and two in Kirmin.[193] The *mīr* of Hunza had built two huts for his shepherds close to the *ziarat* of Babaghundi near Yishkük where he kept some flocks.[194] There were setbacks as well; six men were killed while constructing an irrigation channel when a cliff collapsed and buried them in 1928.[195] In the span of thirty-nine years *mīr* M. Nazim Khan claimed to have founded ten villages for 270 households up to 1931.[196] At the beginning of his reign the three 'original' villages had population sizes of a similar order: Baltit with 345, Altit with 120, and Ganesh with 143 houses.[197] From a compact nuclear oasis the settlements had spatially expanded and connected a wide area of common belonging, communication and production.

The major project that *mīr* M. Nazim Khan and his *wazīr* Humayun Beg commenced in Shinaki in 1912 was a long channel leading from Murtazabad to Hindi. This alignment was a major challenge as rough and instable terrain had to be tackled. The availability of abundant cultivable land in Hindi would have given space for one hundred households; more than a thousand people worked on the site in December 1913, but the project failed.[198] His greed for wealth motivated and stimulated *mīr* M. Nazim Khan to experiment with more land development in Hunza's periphery. Not only had *mīr* M. Nazim Khan been active in constructing new channels and establishing new settlements but he had considered for many years a change in the agricultural taxation scheme. With all his *mīri* lands at hand, control was no longer that easy and the trust in his office-bearers was rather limited. He therefore introduced a new system of revenue generation by directly leasing out his lands to peasants for their own cultivation, out of which they would pay him a fixed amount (*khuṭúkal*).[199] Hitherto, tilling the land had been a part of the forced labour (*begar*, *rajaáki*) obligations that did not provide satisfactory returns, especially in newly developed settlements where nearly no spare time and energy was left to the cultivator (*asāmī*). Only after the death of *wazīr* Humayun Beg did *mīr* M. Nazim Khan introduce the *khuṭúkal* as a new approach to levying land tax in September 1916 and laid the foundation for this more reliable income from his new lands that were ever increasing owing to the settlement expansion and attraction of settlers and refugees from Wakhan and other areas.[200]

Mīr M. Nazim Khan and *wazīr* Shukrullah Beg successfully continued amelioration projects in Shinaki and developed the section between Maiun to Seko which became Hussainabad. In addition to the lands reserved for the *mīr* and *wazīr*, lower Seko was developed for sixteen households by constructing a canal in the summer of 1918; upper Seko expanded by thirty-three households in the summer of 1924.[201] Major George Kirkbride was impressed by *mīr* M. Nazim Khan's continuous efforts to cultivate new lands when he toured the Hunza Valley in the summer of 1936: "All along the route, noticeably at Ghammessar, Gulmit (right bank), Gircha and Sost, there is evidence of the Mir's activity in taking waste land under cultivation. At Gulmit he is busy making provisions for his younger sons and to this end large areas have been enclosed and are being carefully planted with trees and cultivated. Four miles north of Misgar he is experimenting with barley."[202] With these projects and activities the present settlement structure in the Hunza Valley was basically completed; later additions could only augment and expand the existing structure of appropriating settlement space in the valley.

The photograph of Reshit Fort was taken by Marc Aurel Stein on 4 September 1913. The wazīr Muhabbatullah Beg is posing in front of the fortified village. Source: Stein Collection in the British Library. The latter is reproduced here with kind permission of the International Dunhuang Project Photo 392/28 (150)

Letting people leave – initial out-migration

From the beginning of the 20th century a dialogue, sometimes a dispute, had developed between the *mīr* of Hunza and the British colonial administration in Gilgit about the carrying capacity of the Karakoram valleys. The *mīr* demanded barren land within the Gilgit Agency for cultivation and the establishment of additional irrigation oases outside of Hunza. Both negotiators, the Political Agent and the *mīr*, would profit from such schemes as the colonial administration would increase local production and its tax revenue and the *mīr* would augment his income and land-holdings. In a region where food shortages occurred regularly and where

Snow-clad Ultar peaks are towering high above Hunza's central oasis, which can easily be reached by a temporary bridge from Sumayar crossing Miatsil in winter. The photograph was taken on 25 November 1984

Opposite
On the southern bank of the Hunza River lies Sumayar, one of the most affluent villages in Nagar owing to its proceeds from mining. The settlement pattern reveals khan-like, compact structures in Jatorkhan, Daltho and Khai; augmented by scattered farmsteads within the village lands. Ganesh and Sumayar were comparatively similar villages prior to the Karakoram Highway. Now the settlement structure in Hunza is denser and even more scattered than in Nagar. The land between the confluence of Miatsil with the Hunza River has remained a desert-steppe as a nearby source for irrigation is lacking. The valley bottoms of the Hunza canyon have been re-cultivated after a major flood destroyed habitations, fields and orchards there in 1934. Source: Land use distinction and topography are based on Pleiades image 22/08/2014 (Karimabad); Pleiades image 10/08/2013 (Uyum Nagar). Design by Hermann Kreutzmann

supplies for the garrisons were scarce, any additional production was needed; in fact, such schemes became the nucleus of a number of colonial and post-colonial development projects. In a file titled 'Possibility of providing land for the surplus population of Hunza in the Gilgit Agency' the Political Agent, Captain John Manners-Smith, suggested at the turn of the century: "As regards land in the Gilgit Agency he [the *mīr* of Hunza] would gladly accept any wasteland provided it seemed possible for his people to bring water onto it and provided the grant of the land did not involve the entire separation of his people from connection with him and his authority over them."[203] Earlier attempts by colonial departments like the 'Cultivation Company of the Gilgit Agency' had failed to construct lasting irrigation channels at Matum Das and Chamogar.

Consequently, the beginning of the 20th century marks the out-migration of Hunzukuč for cultivating barren lands in the Gilgit Agency. These were early opportunities for people to escape the direct grip of their rulers.[204] Those who settled in the outskirts of Gilgit – Jutial and Khomer – would become rich by rising land prices in the future. Different resettlement projects resulted for Hunza farmers who became known as experts for amelioration by implementing irrigation projects in challenging locations. In 1908, the first barren lands were allocated in Matum Das (renamed Pratabsinghpura, nowadays Rahimabad); the second scheme followed in 1912. Initially settlers from Jutal and Jaglot had tried to establish an oasis there; after a flood destroyed the channel they abandoned the one hundred and seventy *kanāl* of cultivated land. On the recommendation of the Political Agent in 1901, Hunza settlers were permitted to repair the channel and re-cultivate the land as an 'experimental measure'; the first ten years were tax-free, the following two decades would see an increase to five and ten percent, respectively. Later on, similar conditions should prevail as in neighbouring villages. The *mīr* of Hunza would have no right to interfere.[205] The Kashmir Darbar provided a loan of 828 Rs while the Imperial Government granted the remainder of the cost of two thousand rupees for free. "The land amounting to 1086 kanals 13 marlas was distributed by the Wazir Wazarat in 40 equal shares amongst the Mir of Hunza, his Wazir, Humayun Beg, and other people of Hunza and they were put in possession of it from Rabi 1962. The total land of the village as measured now comes to 559 acres, of which cultivation is 134 acres."[206] By acquiring new lands in Matum Das the Hunza elite lost their grip on the new settlers there, who have been oriented towards Gilgit ever since.

The British authorities offered barren land in Konedas, opposite Gilgit, for a 'Hunza colony' in 1912. The offer was declined after discussing the issue in a *jirga* with representatives of all villages in Baltit, as it seemed that the effort to build a channel would be too costly in relation to its effect and value.[207] More successful projects would come up later. The channel project of Punal Das, later renamed Oshikandas, provided Hunza farmers with 312 acres of land and the Bagroti landowners with 1,188 acres in 1938-39. Initially fifty-seven households from Hunza settled there; twenty-seven from Hindi, the other from Central Hunza – Aliabad, Haiderabad, Ganesh.[208] *Mīr* M. Nazim Khan's successor, his son Ghazan Khan, added only one project that commenced in 1940; the new irrigation scheme in Danyor attracted more migrants – twenty households from Shinaki and sixty-two from Central Hunza. Each household received forty-five *kanāl* of land and had to deliver one *man* of wheat to the *mīr* of Hunza in annual taxes.[209] Danyor has substantially grown to one of the most fertile oases close to Gilgit Town. Besides the founding of those irrigation colonies, individual farmers acquired agricultural lands there.[210] The villages with migrants from Hunza and Nagar include Nomal, Naltar, Gujur Das (Sultanabad), Jutal, Gwachi, Diding Das (Muhammadabad). Hunza settlers have cultivated the colony of Golodas at the junction of the Ishkoman and Gilgit rivers; in Ishkoman itself, some settled in Bar Jangal. The share of settlers in agrarian colonies outside of Hunza equalled about one sixth of the resident population within Hunza in the 1960s.[211] This emigration process has continued until today, although nowadays the pull effects of Gilgit as a workplace for non-agrarian occupations supersede the attraction of amelioration projects.

Hunza and Nagar compared

In Nagar developments took a slightly different turn. The pressure on space was much less severe, and the *mīr* of Nagar was less interested in cultivating land outside his realm. There were still ample resources in Garmasai and Shinber. The establishment of new settlements by tapping hydraulic resources and constructing long channels contributed to the ambitious project of Sikandarabad that grew to become one of the major oases in Shinber. Two channels were built around 1905 under the guidance of *mīr* Sikandar Khan; five years later Henry Haughton estimated the number of settlers at two hundred households with significant additional potential.[212] Especially after widening the main channel in spring of 1913 the expansion accelerated.[213] After the upper canal had been successfully established, a new channel was constructed to irrigate the Harespo plain between Sikandarabad and Chalt in 1922.[214] In general, Nagar is endowed with better land and hydraulic resources than Hunza. The difference between the two unequal neighbours has been a prolonged topic of debate and prejudices. Already John Biddulph, after his return from the first visit of an Englishman to Hunza in August 1876, discussed this topic with his Kashmiri counterpart, the *wazīr-i-wazārat* Bhai Gunga Singh, who allegedly remarked 'that Hunza would not pay for annexation, as it already suffers from food scarcity on account of over population'.[215] Hunza had notoriously suffered from food insecurity over long periods. Bronislaw Grombchevsky observed significant differences between Hunza and Nagar in 1888: "In terms of bread and gold the Nagar people are wealthier; they are washing gold dust in significant quantities from the river. In contrast, in turn they are poor in terms of dress, weapons and bravery."[216] A quarter century later, Henry Haughton became more explicit: "Climbing up from the river through Ganesh village to Baltit, where the Mir's fort stands imposingly upon a small hill, one of the first details that strikes anyone coming direct from Nagar is the difference between the Hunza and the Nagar men. The latter, as their Hunza neighbours say of them, appear to lack in looks and spirits the brightness which their country lacks, as if the sun, which shines so little on to their side of the valley, shone equally little upon their lives and natures. As a matter of fact, the poor man of Nagar is probably on the whole happier and better off than a man of similar social standing in Hunza ..."[217] Henry Haughton endorsed the argument which had been presented by Algernon Durand when pronouncing the differences between Hunza and Nagar: "There is considerably more cultivation than in Hunza, the land is better watered, and the main slopes of the mountains enclosing the country face north, and are consequently covered with forest. The land has the full benefit of the summer sun, and the crops are splendid. The untold profusion of the apricots, and the quantity of gold which

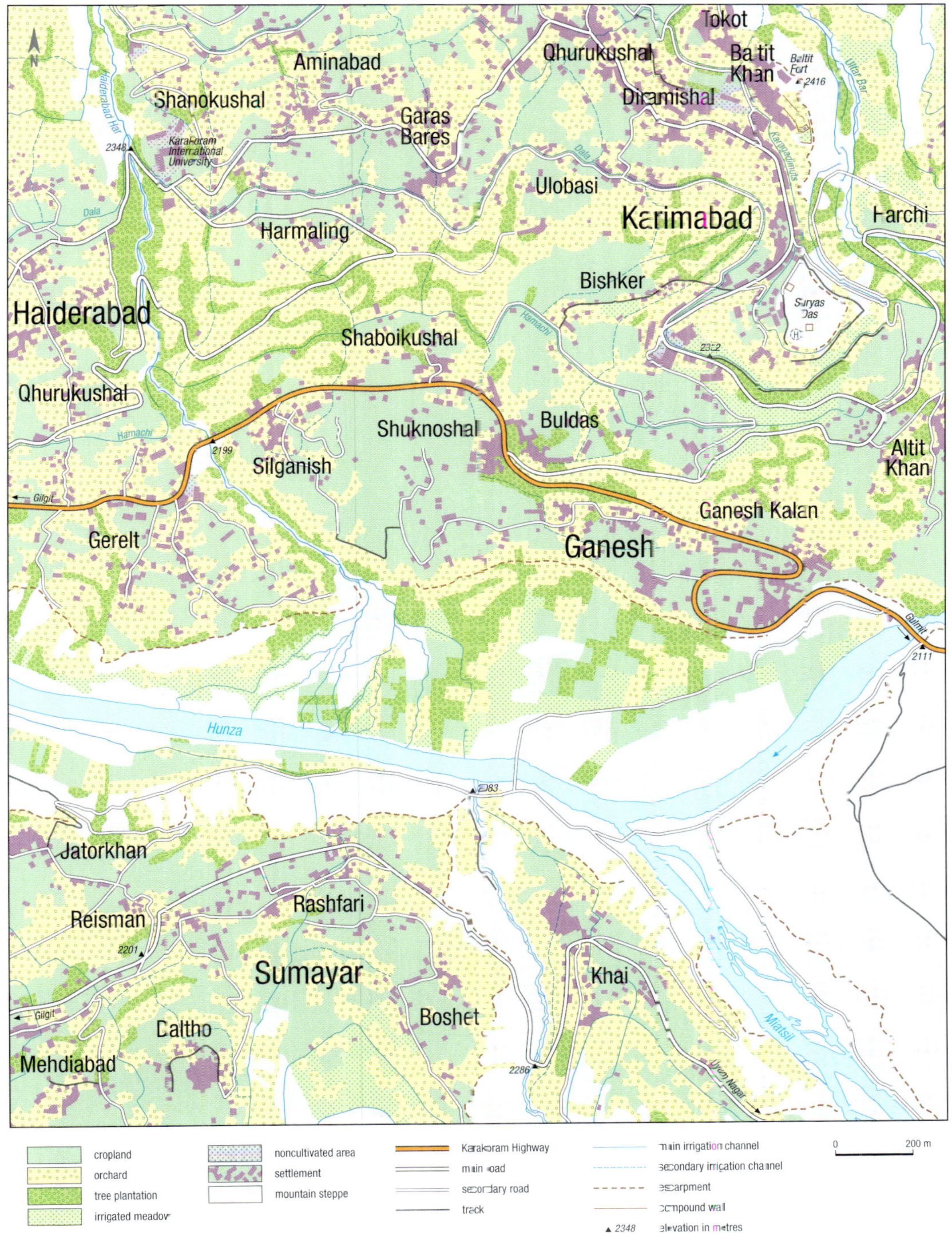

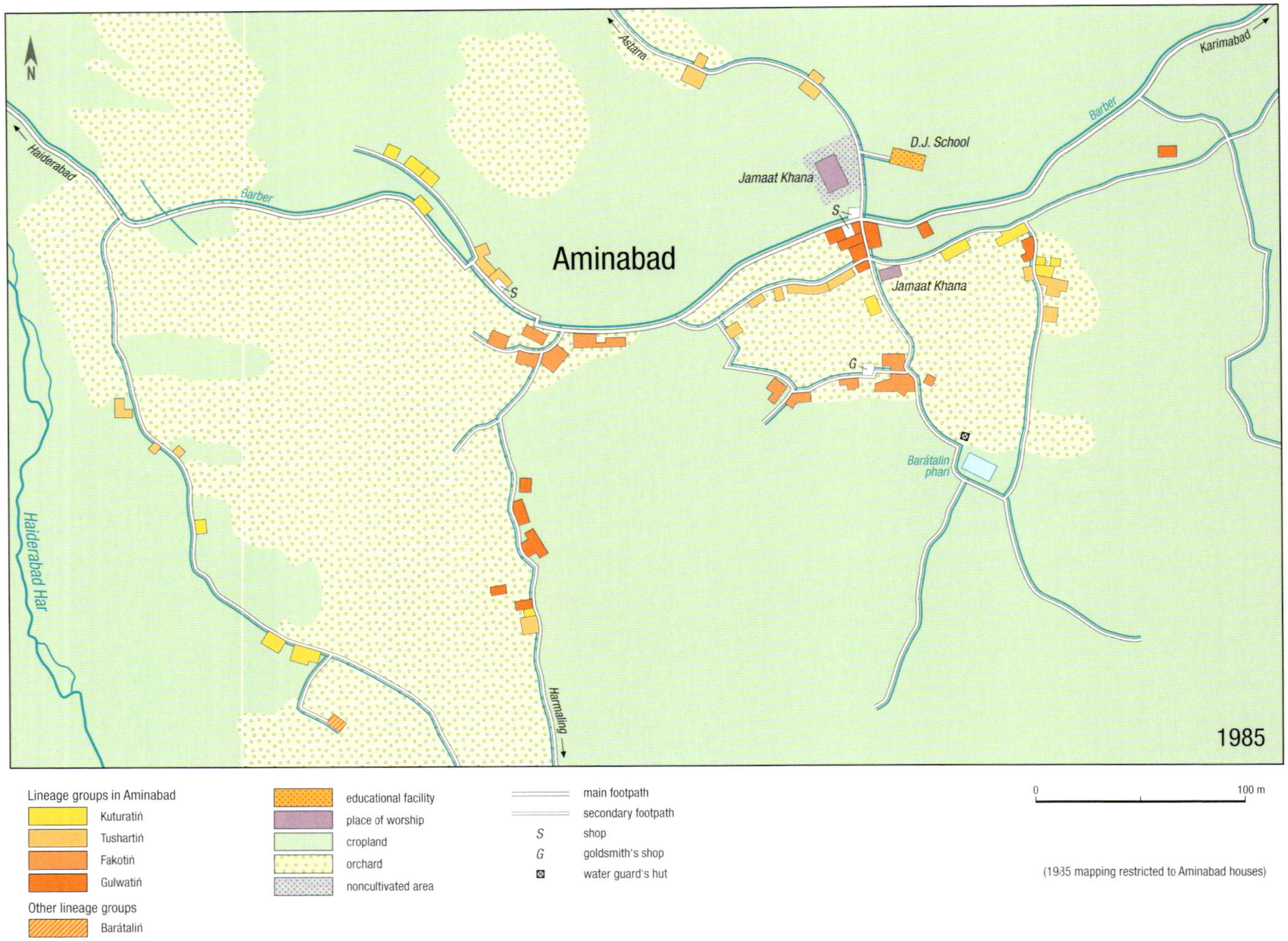

Aminabad (formerly Herra) is presented as an example of lineage group affiliation reflected in its settlement structure in 1985. Four represented guṭí of the Burúoṅ are found in Aminabad; at its outskirts Barátaliṅ houses are located. Source: Data collection supported by Mohmin Hayat, mapping by Sabine Felmy and Hermann Kreutzmann in September 1985; modified representation based on Hermann Kreutzmann (1989a: 59)

Aminabad is a hamlet with inhabitants who mainly belong to Burooṅ lineages. The habitations and village lands represent a typical arrangement of settlement nuclei and a set of well-arranged agricultural terraces on a steep slope. Aminabad is well known as the home of the Fakotiṅ gold- and silversmiths. The photographs were taken on 15 July and 20 September 1985

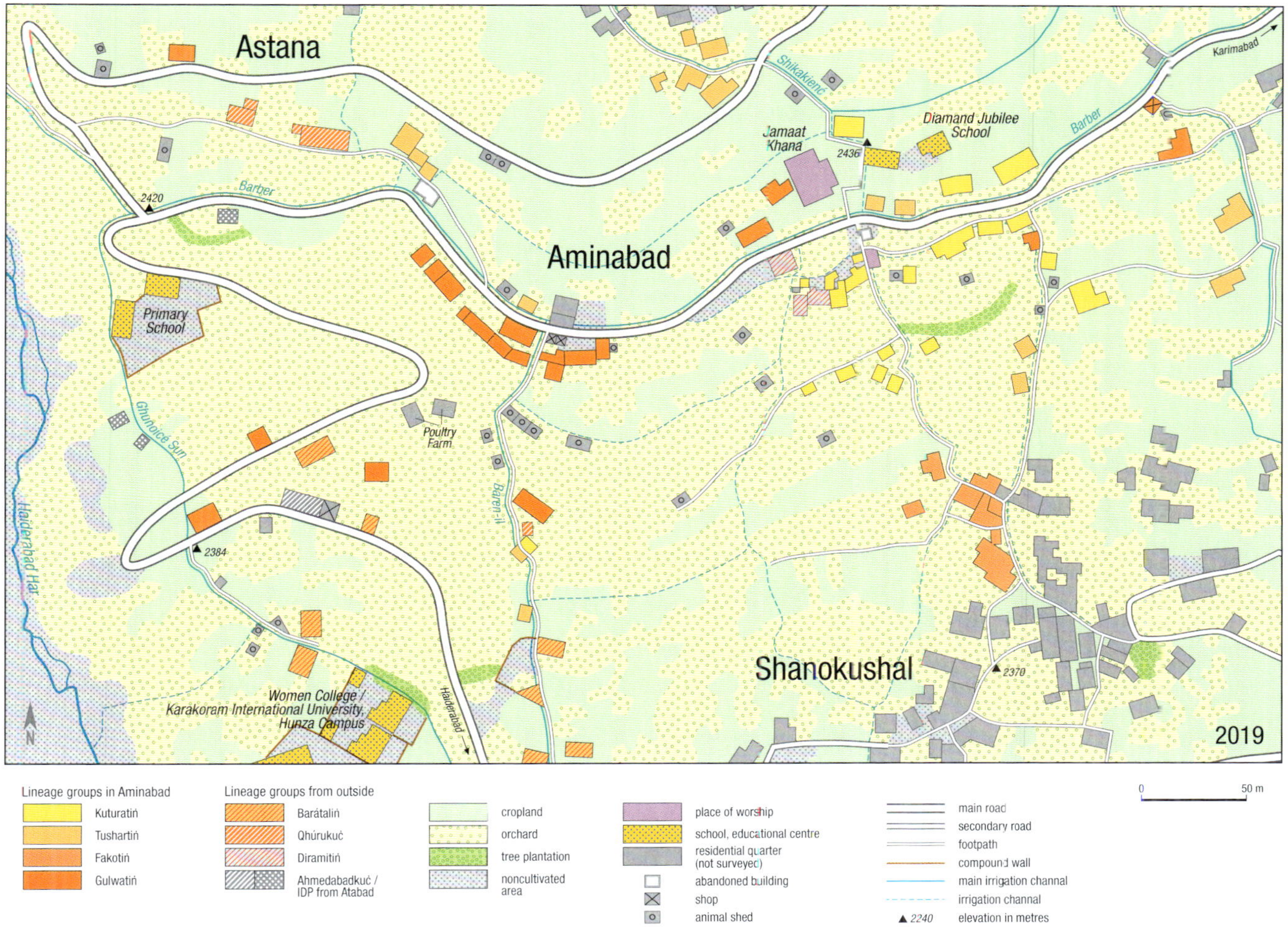

The comparison of Aminabad's settlement structure in 2017 clearly shows the expansion of houses. Not only the former four guṭí are found now but also new settlers have been able to purchase land in Aminabad and to build their houses. The hamlet is now accessible to motor vehicles. Source: Data collection and mapping by Didar Ali, Sabine Felmy and Hermann Kreutzmann in July 2017

The view from Astana towards Aminabad opens up a panorama that includes more recent three-storey buildings. Lower down, commercial buildings and wide roads have been added to the infrastructure. The confluence of the Mictsil and Hunza Rivers is visible in the background, in front of the mountain massif with peaks above 7,000 metres reaching from the Ultar Peaks to Disteghil Sar. The photograph was taken on 26 March 2019

The Hunza canyon separates the central oases of Hunza and Nagar. The barren lands of Shayar in the foreground contrast with Aliabad, which has developed being the commercial hub for Hunza and Nagar since the mid-1980s. The photograph was taken on 23 October 1985

Opposite
The emergence of Aliabad as the commercial centre of Central Hunza is strongly linked to the Karakoram Highway. Most administrative functions of the newly created Hunza District are concentrated here in addition to the main bazaar and dominant shopping centre of the valley; petrol pumps, banks, hospitals, schools and colleges are located on valuable former agricultural fields; offices are concentrated off the Karakoram Highway. Built in 1983, the first link road started in Sultanabad to service Haiderabad and Karimabad; the second connects with the Karakoram Highway in the centre of Aliabad, meeting with the older link road in Rahimabad, where sufficient space was found for the new high school. Although connected by a bridge that allows tractors and light lorries to pass, Shayar and its neighbouring villages in Nagar failed to grow at a similar pace. Source: Land use distinction and topography are based on Pleiades image 22/08/2014 (Karimabad); Pleiades image 10/08/2013 (Uyum Nagar). Design by Hermann Kreutzmann

can be washed out of every stream, has gained the country amongst the Nagaris themselves the name of 'the land of gold and apricots'. The people, though of identically the same stock as those of Hunza, show marked differences. They are less warlike than their neighbours; they were never implicated in the raiding of caravans, owing principally perhaps to there being no outlet to the north or east leading to caravan tracks; they are, on the whole, better off, owing to the better agricultural conditions, and the possession of splendid grazing grounds; and last, but not least, slavery is unknown amongst them – that is, they would not brook their king selling them or their children, and the exile of the present ruler's eldest son is said to have been caused by his attempt to introduce the custom from Hunza. The men are not so bright and cheery as their Hunza cousins; they strike one as more sedate and morose, and their reputation for dash and gallantry is far lower. The reason they themselves assign for the difference, and the one generally accepted, is that in the winter, when the sun is in the south, the great mountains backing their valley shut out all the light and warmth, of which Hunza facing south enjoys the full benefit, and that the depressing effect of passing day after day in the cold and darkness of their uncomfortable houses leaves an indelible mark on the national character."[218] Reginald Schomberg endorsed the view of different military qualities among the inhabitants of the Hunza Valley. In a letter to his fellow intelligence officer Frederick Marshman Bailey, then Resident in Kashmir, he confided: "I wish your scheme for the G[ilgit] Agency would go through. The chief charm of it to my mind is that it gives employment to the men of Hunza. It seems extraordinary that we do not enlist any of that race. Not Nagar but the Hunza men are fine material."[219] The exposure and light aspect was taken up much later by Emily Overend Lorimer to distinguish inhabitants on both banks of the Hunza River in terms of character and lifestyle.[220] Her account popularised prejudices against Nagarkuč and stylised the Hunzukuč as exceptional beings, a point gladly accepted and amplified by generations of esoteric, health-food and organic gardening writers.[221] The perception of a landscape as the expression of human character and behaviour was projected on Hunza and Nagar in a rather simplified manner, thus neglecting existing insights from climatology, geomorphology and agricultural sciences. The formation of the Hunza Valley and its terraces resembled the same postglacial landscapes in both principalities. The solar radiation provides similar energy intakes for crop farming, and agricultural practices do not differ much. Nagar was better endowed in environmental terms. Many observers made this point of less incentive to leave because of a better endowment with natural resources in Nagar. At the same time, they followed a racist impulse that judged character in terms of light exposure, fair skin and cleanliness. Consequently, there seem to have been less need and zeal for out-migration at the beginning of the 20th century, although the oases close to Gilgit Town nowadays accommodate a sizeable Nagar population as well.

Village growth and response to improved communication systems

Since the independence of Pakistan and the improvement of communication lines towards the Indus basin, the dominant factors to alleviate population growth in the Hunza Valley have been the expansion of existing villages, emigration to Gilgit and extra-montane migration further down country. A new labour market could be accessed as young men became soldiers and workers, especially in the factories of Karachi. The process of founding new independent villages has significantly slowed down in the same manner as agricultural practices diminished in importance for income generation. However, some exceptions occurred during the rule of the last *mīr* of Hunza, Jamal Khan (1945-1974). Small new villages such as Sarteez or Gircha *bālā* have replaced the dwindling location of Gircha *pāīn* after 1950. All other villages have experienced a further increase in households, a concentration process of settlements, and an expansion of new and isolated farmsteads into the village lands. Furthermore, two outlets such as Imamabad and Jamalabad came into being in upper Gojal. The latter was founded after 1960 on the initiative of Ghulam Jaffar and Titam from Morkhun; it consisted of twenty-one local Wakhi and three Burusho households resettled from Misgar and was named in honour of the last *tham*. Some irrigation channel projects such as the *Imamyari* in Altit or the *Dilbar* and *Murku* in Karimabad symbolise the attempts to expand the local nutrition base. With the demise of the last *mīr* Jamal Khan in 1974 and the abolishment of the states in Hunza and Nagar, a hiatus and power vacuum occurred at a time when accessibility and traffic infrastructure were changing at a fast pace.[222]

A new boost for settlement expansion occurred with the extension of roads and the provision of regular motor transport to Hunza and Nagar since 1959. Especially the opening of the Karakoram Highway in 1978 amplified these new options. This principal change of traffic and its subsequent expansion of a transport network linked villages and hamlets to the Karakoram Highway. Furthermore, it connected village lands for access of tractors that enable mechanical ploughing and threshing as well as picking up crops for marketing.[223] These innovations that allowed the easy transport of passengers and bulk goods have supported a reorientation of site selection for commercial and administrative buildings in the villages towards access roads where small bazaars have developed. Valuable agricultural lands were converted to commercial sites for the construction of physical infrastructure. New building materials brought in from distant bazaars allowed for cheaper construction of housing. The prices for a selection of building materials such as cement, corrugated iron sheets and wooden beams from the Indus Valley have undercut purchasing and construction costs for locally available and treated products such as dressed stones and timber from fruit trees and 'donkey-trees' such as poplars (*jarpá*, *safidor*, *tərək*).[224]

Settlement concentration and expansion are presently governed by the network of communication lines for motorised transport. Improved accessibility has not only supported the transport of goods but also enhanced the mobility of people. Rural-urban migration towards Gilgit Town has led to the resettlement of a substantial number of households there. Out-migration has contributed to diminishing the agricultural workforce in the villages where episodic and regular tasks require the commitment of household members. Tilling the fields and maintaining the irrigation system are obligations that document the ownership of agricultural resources in an area devoid of cadastral records and written land titles. For each and every household it is a kind of obligation not to give up their inherited resources and not to let valuable land fall fallow. The communities in the Hunza Valley have up to now maintained their oases despite a significant out-migration.

While out-migration is a significant burden on the allocation and division of labour for tilling the fields and orchards, a reversing trend can be observed in house construction. The absent villagers have not severed their relations to their roots. On the contrary, salaried people, soldiers and

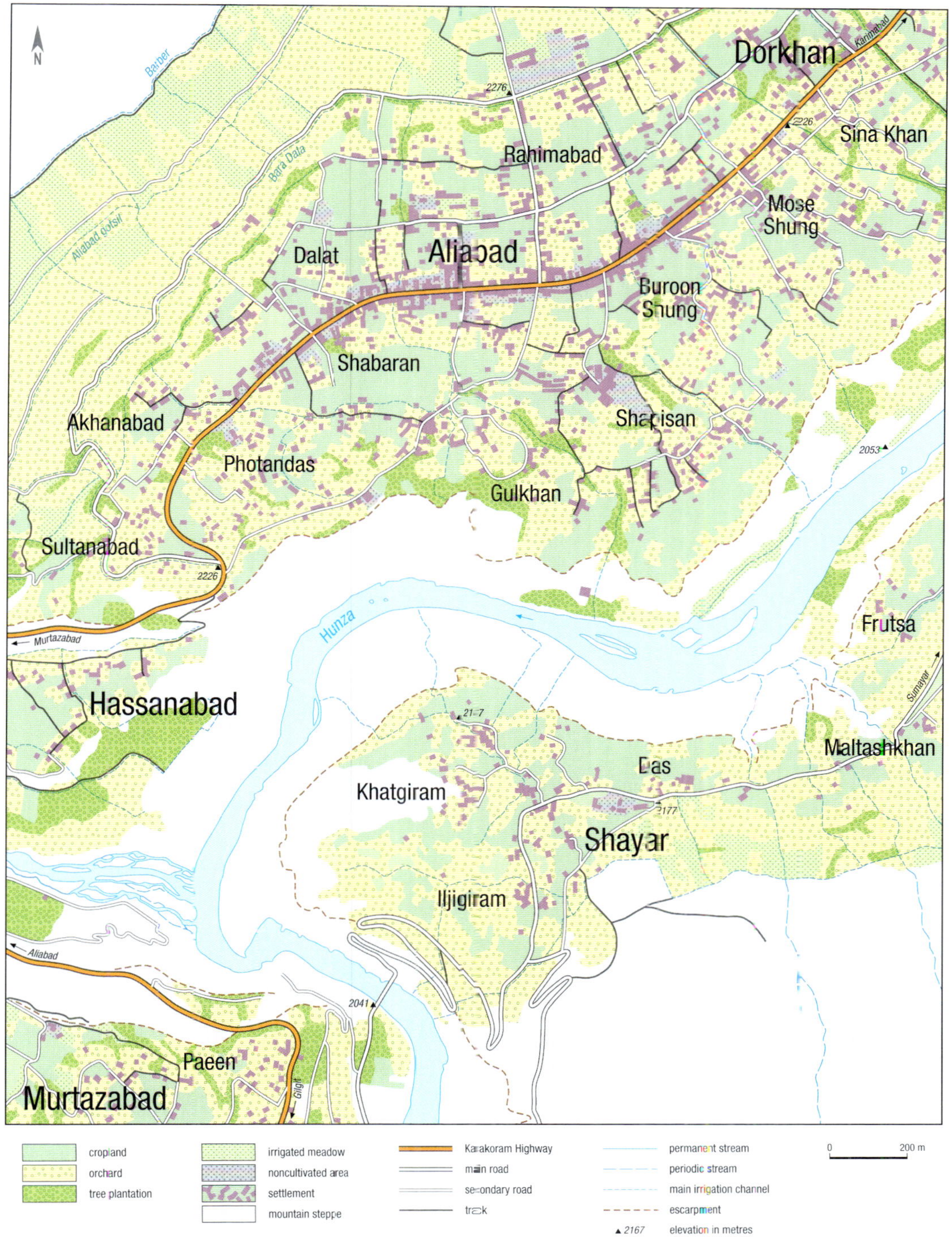

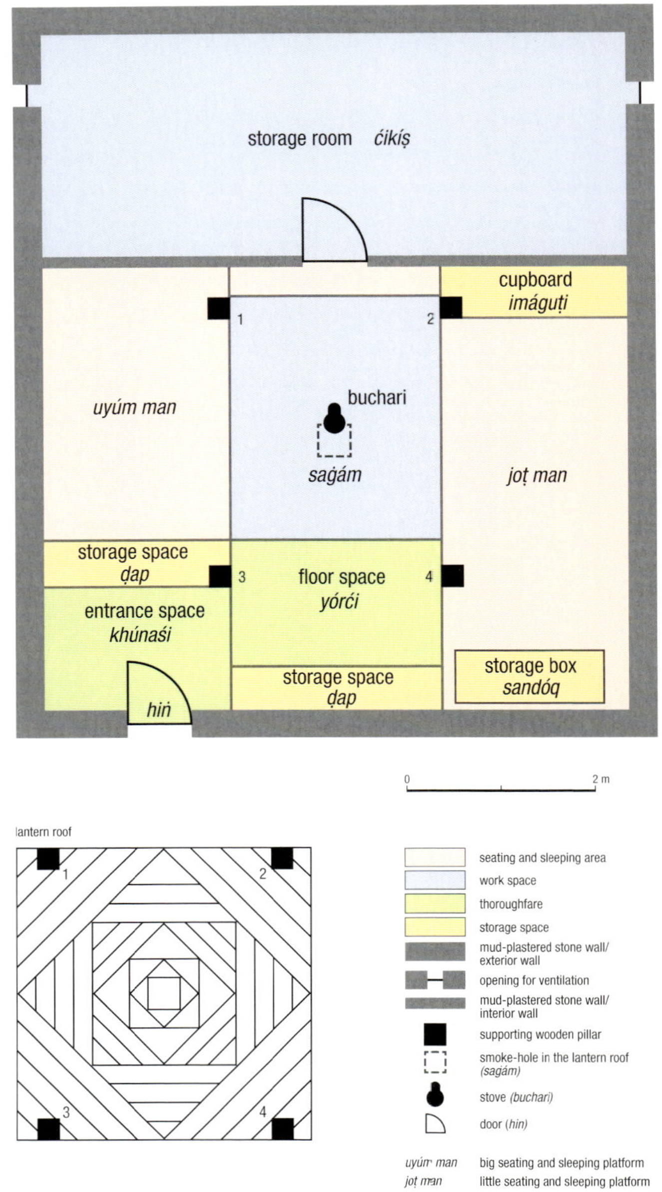

In Baltit khan the one-room houses are attached to each other; the smoke-hole in the roof is the only source of light (top). More spacious homesteads have been built in the hamlets where, in addition to the 'ha', rooms have been attached for guests and additional household members. The photographs were taken on 16 October 1984 and 27 May 1985

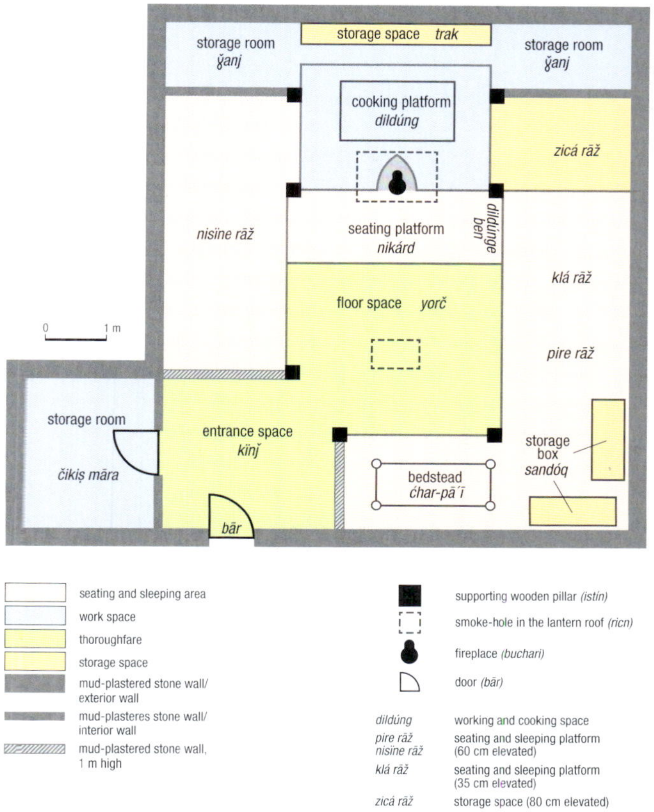

The so-called Burusho house (left) features elements that are found in several house types in the Pamirian Crossroads; the storage room at the back of the house separates food preparation from sitting and sleeping, which take place on the platforms (man). The space close to the entrance is reserved for storage. The only opening for light is in the roof above the fireplace. The lantern roof ceiling is wide as well. In general, the Wakhi house (top) is more spacious and differentiated than the Burusho house. The basic structure is similar, with a storage area opposite the entrance and separate food preparation on an elevated platform. At the sides the sleeping and sitting arrangements can be expanded towards the entrance side. Source: Modified from Hermann Kreutzmann (1989a: 55)

The cooking place is the central feature of a Wakhi house in the village and in the pasture. Dur Jahan is preparing tea in Shuwert (top) during the summer at the 4,620 metre-high settlement. In summer cooking is sometimes moved from the centre to the back of the house in Gulmit (bottom). The photographs were taken on 7 September 2013 and 24 July 2014

After the 1960s floods the jamáat khana was moved to a safer place and built in the contemporary style by Shimshali masons and carpenters with local stone and timber. The photograph was taken on 15 May 2015

labourers have continued to invest part of their income in new housing. In some villages valuable agricultural land has been used for house construction. The small farm sizes have become smaller by construction and separation of joint families. In most villages the space for housing has continuously grown. Strategies to overcome such a land deficit have always been the development of new lands and the establishment of filial settlements. Wherever space was available new hamlets and villages were established. Gojal is probably the most prominent and visible example of such additional lands being newly opened. Villages such as Hussaini, Pasu and Gircha that are well known for land losses due to river undercutting have established new habitations close by. Zarabad and Borit, Janabad and Sarteez are the additional settlement spaces that have been used and made accessible. In Shimshal the major floods of the 1960s destroyed the old *khan*, and initiated a process of resettlement in Dasht (Khizrabad), Sholalakhsh (Aminabad), and Resgineben (Farmanabad) that has continued until today. Nearly everybody who is working outside Hunza has built a new house during the last generation. Similarly, Gulmit has cultivated Chaman Gul, where every household with migrants, either on a weekly basis to Gilgit (*jum'arāti*) or for good, has constructed a new and comparatively big house. In 2017, a new *jamáat khana* was opened in Chaman Gul. The old settlement has lost significant cultivated space for public and community buildings such as schools, hospitals and hotels. In addition, the new central *jamáat khana* in Gulmit is not only a place for worship, it provides office space to more than a dozen community organisations. Space within the old villages has become scarce and very expensive; valuable agricultural lands and orchards are converted into habitations. Other additional lands in Bulchi Das, Gohorabad and Borit are less suitable for this purpose. The investment in new buildings and hamlets is mainly sponsored by villagers who earn substantial incomes far away from home.

The absent workforce has been replaced by elderly household members and women who have tried to compensate for the deficit. Certain seasonal tasks cannot be coped with and have been subcontracted to migrant labourers from Baltistan, Chitral and Wakhan. Mechanisation has played its role, and changing cultivation patterns with less labour-demanding crops have emerged as well. Some crop farming parcels have been turned into orchards or alfalfa plots. These are a sign of limited workforce availability and are affecting water management and its related rules and regulations. They have been the key to understanding authority, community structures, power games and economic strategies. In the next section, the complex setup of making a living in the Karakoram mountain desert will be exemplified.

Opposite
Shimshal village is still characterised by the land loss due to floods and terrace undercutting. In the meantime the fortified village has lost its importance as a centre. Farmsteads and social institutions have moved southward. New hamlets were established nearby on village lands that have significantly expanded to several satellites. Even in the valley bottom some cultivation has taken place, and the high school was established in the flood-prone plain near the river. The photograph was taken on 16 July 2017

Rakaposhi (7,788 m) symbolises the former name – Muztagh meaning snowy mountain – of the Karakoram (black gravel) Range as a highly glaciated massif storing significant ice masses. The photograph was taken on 5 August 2017

Water – lifeline for village dwellers and backbone for oasis agriculture

From a remote viewpoint the oases of the Karakoram have many structural features in common as they are embedded in a similar ecological environment of a high-mountain desert utilising glacier and snow melt for irrigation. A split between two different social systems has been perceived when it comes to societal rules and regulations. John Staley distinguished two principles.[225] First, semi-autonomous and independent principalities represent the dominant feature of hereditary or institutionalised rule in the northern Karakoram valleys. Local *mīr*, *rāja* or *tham* dominated peasant farmers of different social standing in places like Gupis, Yasin, Ishkoman, Punial, Gilgit, Hunza, Nagar, Astor, Skardu and Khapalu. These rulers exercised a certain control over the peasant community, levied taxes and requested forced labour services from the rural households.

Second, acephalous societies or republics have emerged in the mountainous interface of the southern valleys such as Chilas, Tangir, Darel, Gor, Kandia, etc. following the Yusufzai Pashtun conquest of Swat and Indus Kohistan since the 14th century. Their characteristic is the establishment of an egalitarian community to which all landowners belong. They are entitled to participate in the decision-making process of all village activities such as distribution of land, construction and maintenance of irrigation channels. No hereditary ruler would be acknowledged there; every member possesses equal rights. The forum for counselling is the *jirga*, a local assembly basing its decisions on overall consent.[226] Besides the landed members of society there exists a subordinated landless class which works on the fields and as shepherds. They are excluded from all decision-making processes.

John Staley observed distinguishing features between these two societal settings.[227] The cultural landscape of the republics was devoid of any orchards, a trait which has been related to the land rotation system of *wesh*.[228] Agricultural tasks are predominantly performed by indebted wage-labourers from outside, who are tolerated as employees. Intensity of crop farming has been low, and cultivable land abounds. The agricultural resources of the republics are abundant and under-used in comparison with the principalities, where landowners work their small holdings (on average less than one hectare per household) themselves and are engaged in intensive exploitation of available resources. The intensity of land-use is somehow related to the upper limits of certain crops that are significantly higher in the principalities than in the republics. Overall, natural forests have been depleted to a high degree.[229] Scarce water resources are optimised in a highly sophisticated system of water management by ascribing qualified priorities to different crops, orchards and meadows.

An extraordinary position is held by Hunza which has gained a reputation for the development of marginal resources. Hunza professionals managed to cultivate difficult tracts; their expertise in designing and constructing irrigation networks won them assignments in other Karakoram valleys. The Hunza water management system incorporates a number of rules and regulations in this respect which are applied in different configurations in other Karakoram valleys as well. Water management is exemplified by the case of Central Hunza where a high population density correlates with a high demand for water. The set of rules and regu-

The terraces in the vicinity of Buldas hamlet are all green in spring. Winter wheat and summer barley are distributed in similar proportions, other parcels are planted with perennial alfalfa. Very few potato fields are visible in the foreground. The photograph was taken on 25 April 1985

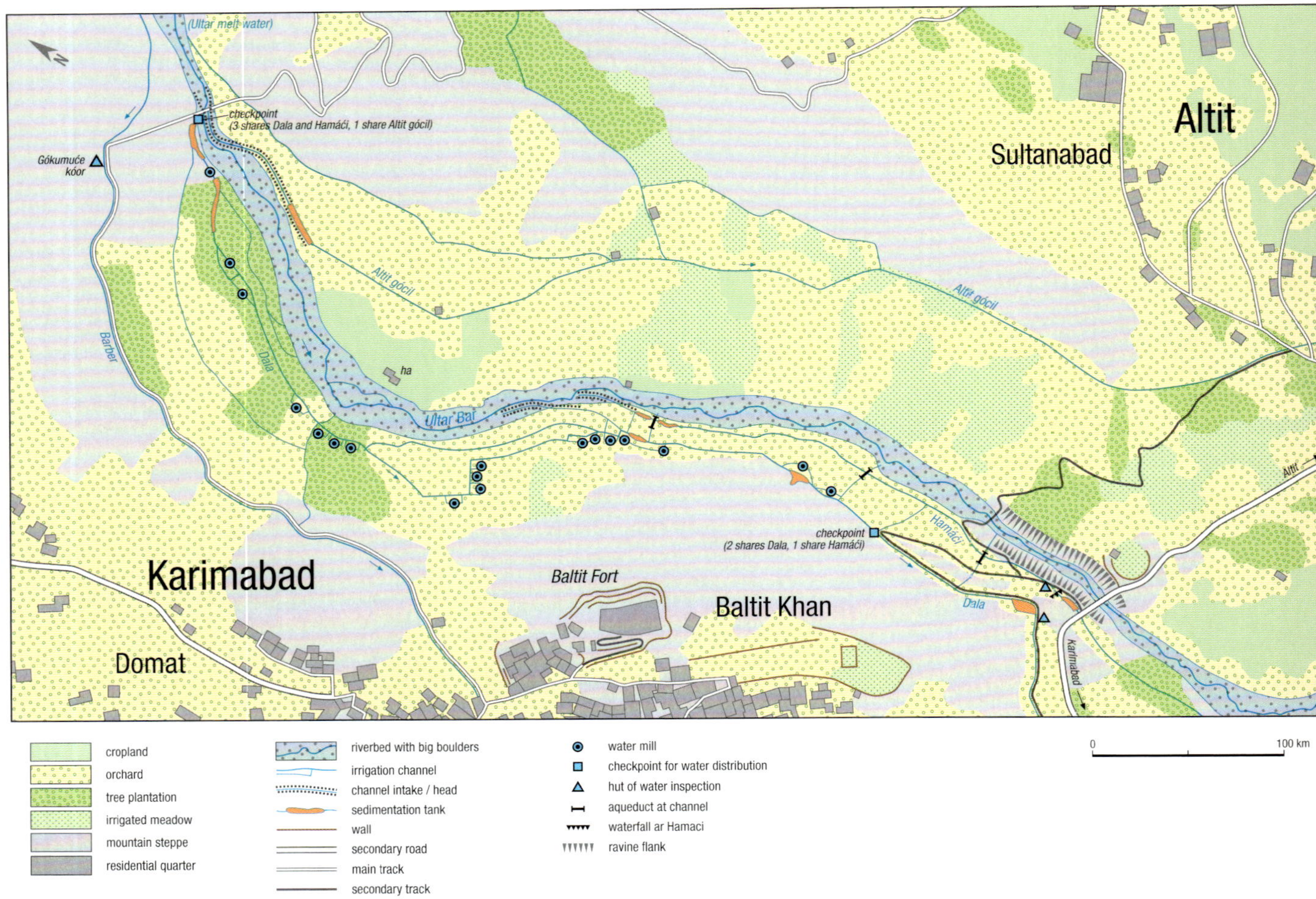

The water supply lifelines for the 'original' nuclear settlements were the meltwaters from Ultar Glacier that were tapped in Harchi Har. In this nālā the water dividers and control posts regulated the distribution between Altit góčil, Hamači, and Dalá (Khul Samarkand). The steep gradient in the valley allowed the establishment of a series of watermills (yáiṅ) that were located mainly along the Dalá, not too far from the settlement. Harchi Har was frequently visited and kept under close surveillance to avoid water theft. The younger Barber channel ran above the three main channels and was managed under a different distribution scheme. Source: Adapted from map in Hermann Kreutzmann (1989a, supplement).

lations is always more complex in scarcity settings; this is the case in Central Hunza, which is located in the transition zone between single- and double-cropping.

Water management in Central Hunza

The vast majority of all irrigation systems in Hindukush, Karakoram and Himalaya use gravity-fed networks to conduct running meltwater through channels which are less steep than the slope gradient and in some cases nearly parallel to contour lines. The same applies to Hunza where the village lands receive their 'glacier milk' (*mamú ċhil, yaze yupk*) from channels which deliver enormous amounts of suspended matter at the same time.

This characteristic surface irrigation system consists of main channels (*dalá*, *gūl/khul* or *góċil*) which start at the head-works (*sar*) with a diversion of water into filter and sedimentation basins. These are necessary to settle huge amounts of suspended silt and sand from the meltwater before sending it down the network. Despite these efforts, sediment deposits still have to be cleared annually from the main channels before the irrigation season starts. The irrigation water is distributed to the fields through a system of secondary channels (*íl*, *ṣuṅ*) that follow the slope gradient. Farmers operate two or more main channels at different elevations. Besides supporting the distribution of water to the field channels, these *íl* and *ṣuṅ* are necessary to divert water from one network to another in case of breaching and to allow for drainage. All main channels below the highest level one are multifunctional as suppliers and drains.

Individual fields receive their allocated share of water through field channels (*dirán*, *ḍir*, *írkiṣ*) which reach all par-

Wheat and alfalfa cover all field parcels of Altit zirám at the beginning of the cultivation season in spring. The irrigated lands are bordered by the steep slopes of Harchi Har in the background and the non-irrigated rock of Burum bun, a perpendicular cliff above the cultivated flat lands of Altit. The photograph was taken on 4 April 1985

Altit village lands in autumn. At the end of the growing season most crops have been harvested; on some fields maize is still standing. Leaves from poplars and apricot trees are changing colour. The photograph was taken on 17 October 1984

Opposite
The cross-section of a typical slope shows the amelioration schemes prevalent in the irrigated oases (íćit) of Hunza and Nagar. Plane field parcels are established supported by massive retaining walls (hurt). The top view reveals the complexity of arranged field parcels (phurúṅ) that are supplied with water from main (góćil, dalá), secondary (ṣuṅ) and field channels (ḍir, írkiṣ). Sluices (ćak, il) are opened and closed for individual fields. The irrigation technique is applied and distinguished in two shapes of fields: furrow-irrigated plots for potato cultivation on ridges, and the inundation of flat parcels for grain crops (hárkiṣ) and vegetables in kitchen gardens (śaníe khúṭkus); see diagram in Hermann Kreutzmann (2015a: 179). A tank (pharí) built with stone walls provides the overnight storage capacity for irrigation water. Source: Modified after Hermann Kreutzmann (1990: 14; 2000a: 98).

cels through the opening and closing of individual field sluices (*ċak*, *il*). Two different types of cultivation beds (*phurúṅ*) are used in Hunza: flat parcels for raising grain crops and leguminous plants in which some small patches for vegetables (*śaníe khúṭkus*) are inserted, and furrow beds for root crops, predominantly potatoes. The flat parcels still account for the majority of croplands (*hárkiṣ*) and receive irrigation water through flooding. The furrow beds are allocated water by opening and connecting these furrows during a single cycle. The cultivation beds meant for raising the staple food crops (*hárkiṣ*) of the village lands (*íćit*) command first priority of water rights and are given preference over orchards (*basíkiṣ*) and irrigated meadows (*ṭoq*). Ranking last, the latter are only irrigated during periods of surplus water. However, during the peak period of glacial meltwaters supplies are sufficient for the upkeep of orchards and meadows. These irrigation techniques and priority measures have been structuring the value appreciation of village lands along mutually accepted guidelines.

Under the prevalent conditions of scarcity, further measures have been taken to optimise the use of the available water. The levelling of terraces reduces water loss due to speedy flow and enables better infiltration. A complex distribution key and substantial measures for its control are observed in regions with a limited supply and intensive cultivation. These measures can vary between seasons of deficiency and surplus. Where and when water availability and supply are abundant, its management requires fewer efforts and less emphasis is placed on setting up rules and controlling consensually fixed regulations. Most villages in Gojal do not face the same severity of water shortages that are notorious from Sikandarabad, Nasirabad (Hindi) and parts of Central Hunza. Attempts to tap further hydraulic resources for augmenting and stabilising the water supply have been a characteristic feature of innovative irrigation practices undertaken by enterprising rulers and village communities.

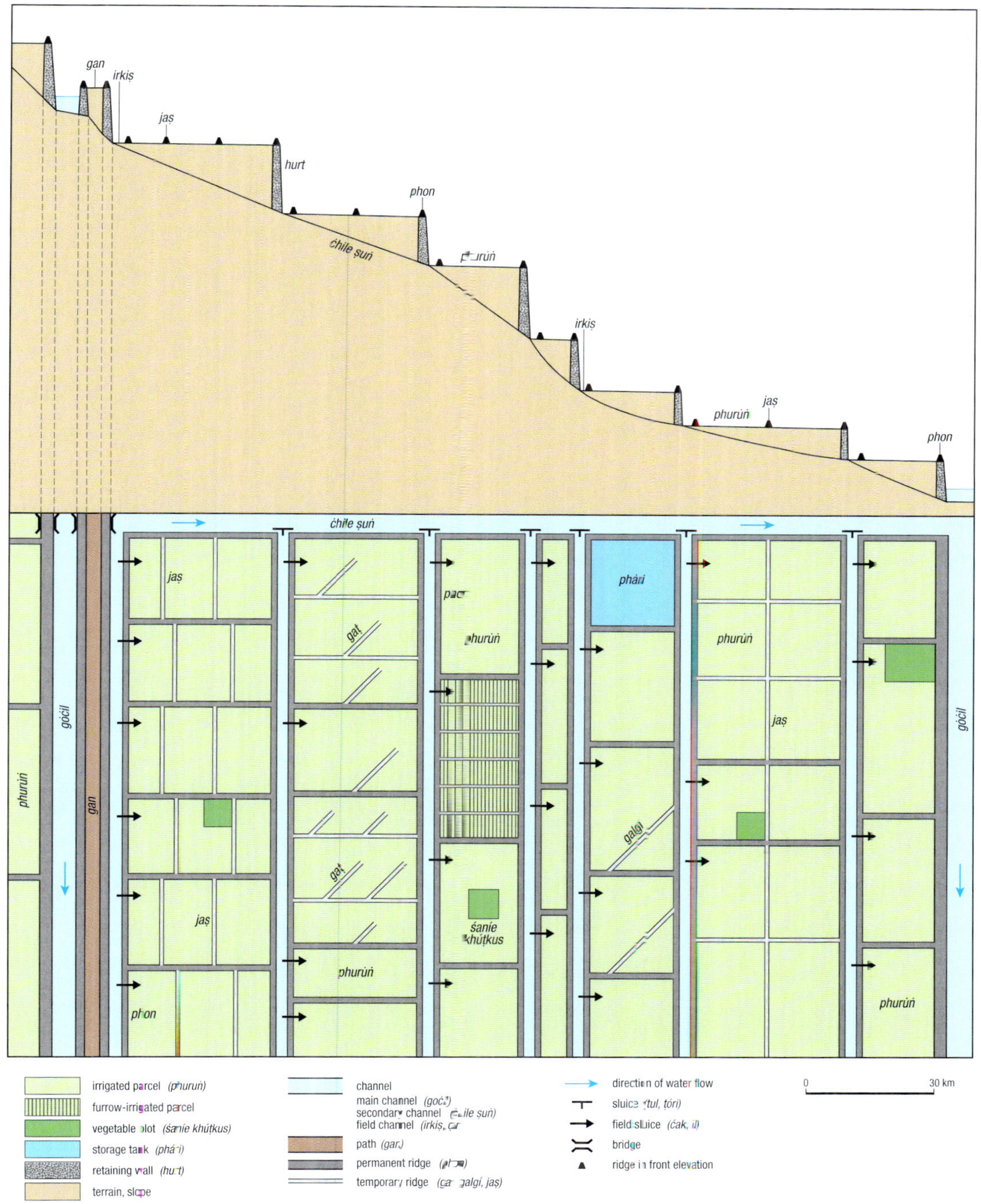

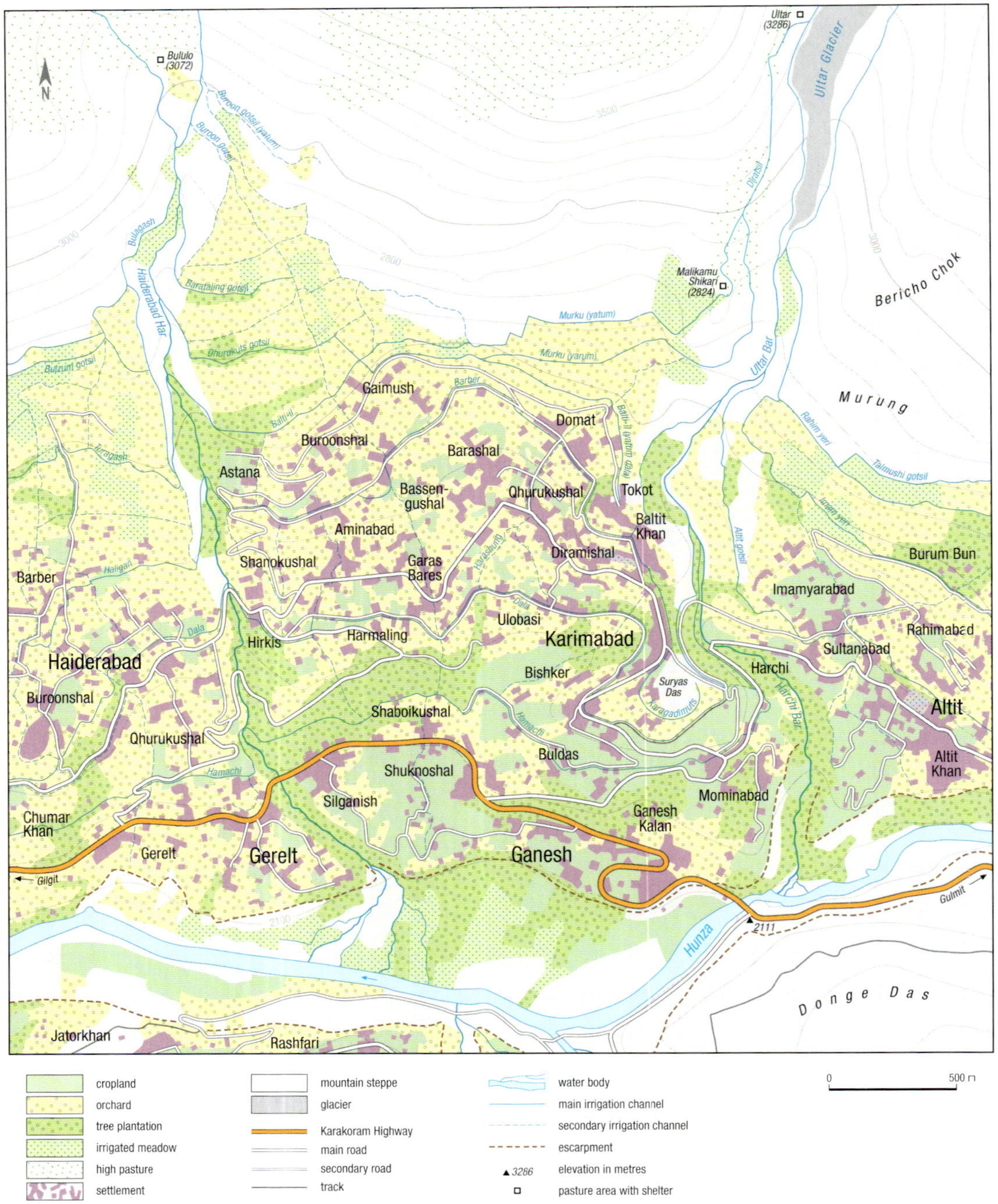

Evolution of the channel network and adaptation of irrigation rules in Central Hunza

The historical dimension in the expanding irrigation system in Central Hunza can be linked to initiatives of different rulers and their *wazīr*. The core of the chiefdom's irrigation system comprised an old structure consisting of the three oldest settlements, which were supplied with water from individual channels leading to their respective village lands. The *Hamáči* channel forms the lifeline of Ganesh, and *Altit góčil* or *Kesari dalá* as it is known in Altit brings meltwater to that village.[230] Both channels tap the Ultar glacier as their main resource. The *Balti íl* links the fortified village of Baltit with the waters of the Bululo snowfields and springs. This – in comparison with glaciers rather meagre – resource basin is drained by the Haiderabad Har which provides the *Balti-íl* with its offtake. The age of these three oldest channels of Central Hunza has not been established, but they could be well over seven hundred years old.[231]

Until *tham* Silum Khan III founded several new villages (*thuáaṅ khanánċ*), no important extensions to the system had been on record. His name is associated with the main channel and the longest lifeline of Central Hunza. The *Dalá* or *Khul (gūl) Samarqand*, as it is locally called, still irrigates the lower portion of the central irrigation oasis between Karimabad and Aliabad. All settlers of Baltit participated in the construction of the new channels, and through this the people of the *thuáaṅ khanánċ*, originating mainly though not exclusively from Baltit, were connected to the *dalá*. Upon commencing its operation, the new channel afforded a revised distribution key for the Ultar meltwaters, dividing half of the available water into the *Dalá* network (two shares) for Baltikuċ, while one share each was allocated for *Altit góčil* and *Hamáči*.[232] The rules that were set at that time have remained the core of water distribution and are regarded as 'old' water rights that will supersede every new addition to the network. The priority accorded to these oldest channels is a valuable asset in periods of water scarcity that occur from autumn to spring.

During summer, when surplus irrigation water above the carrying capacity of the old channels is available, newer channels are eligible for water shares. This newer group is composed of channels constructed under the guidance of Silum Khan's successors. His son and grandson found in *wazīr* Asadullah Beg, acting as *wazīr* between 1847 and 1885,

The meltwaters from Ultar Glacier are divided into four portions. At the upper gate one share is taken out for Altit gočil, at the lower gate two remaining portions are allocated for Dalá and one for Hamáči. At the gates water wardens (čhilgálas) control the observing of rules. The photographs were taken on 17 October 1984 and 2 May 1985

an able local engineer who supervised the construction of the second major channel of Central Hunza bringing water to the upper fields. The *Barber* follows the course of the *dalá* on a higher contour line, thus irrigating fields above the former limit.[233] Members of his own clan respect his name especially as he enabled the Diramitiṅ to exclusively tap the *Diráčhil* spring and command its waters until today.

New channels provided arable land for a growing population and allowed additional settlements in the main oasis. The interest and engagement of the *tham* and *wazīr* were self-rewarded because they reserved substantial newly irrigated lands for themselves; the remainder was distributed among the participating households from various descent groups who operated under forced labour conditions (*rajaáki*). These main channels of Central Hunza were established up to the mid-nineteenth century and have remained the backbone of the present-day hydraulic network. The younger channels could not gain the same importance as they depend solely on surplus water during the peak summer season. Their role might be underestimated as they are important conveyors of water in times of repair and offer alternative diversion channels in case of any kind of damage at other sections in the flow of water. After the boom of channel-building during the first half of the nineteenth century a consolidation phase lasted until *mīr* M. Nazim Khan initiated another phase of channel construction outside the central oasis in Gojal and Shinaki. These young villages command, on average, a good supply of water and require fewer rules and regulations than the core oases.

Seasonal and territorial rules for irrigation

The expansion of the irrigation network and the construction of new channels have enlarged the overall complex codex of water rights in Central Hunza. In old and new areas different rights of water allocation have been laid down for certain channels, both for periods of seasonal water shortage and for surplus times. As a general rule, the irrigated lands are classified into three groups for which strict sequences apply. First priority of irrigation is enjoyed by crop land (*hárkiṣ*) which requires regular watering during the different growth stages and on which the staple cereals are produced. Second priority is reserved for orchards (*basíkiṣ*) that provide the villagers with an additional and valuable nutritional base from fruit trees which require less intensive and less frequent water supply. Third priority goes to irrigated meadows (*ṭoq*) on steep slopes within the village lands: they are entitled to surplus water supply only during non-growth stages in other parcels. This ranking for village lands reflects an irrigation order with preferential treatment of differentiated production zones within similar agro-ecological settings. During the short period after harvest and before sowing the winter crops, no irrigation water is needed for agricultural lands and orchards. Then the irrigated meadows (*ṭoq*) qualify for the full load of available irrigation water.

Seasonality plays a major role in the flexible approach to organising water management in Hunza. The quantity of meltwater released from the Ultar glacier depends on temperature variations and is reflected in the discharge patterns of the tributaries of the main Hunza River. The irrigation calendar for Central Hunza takes into account that periods of surplus alternate with periods of severe deficiency. The basic water supply is precisely distributed during all seasons

Opposite
The present-day oasis of Central Hunza comprises Karimabad with the neighbouring villages of Altit, Ganesh and Haiderabad. The oasis is clearly structured in irrigated cultivated lands (íčit) that are separated from the desert steppe (das) and the scanty pasture lands (ter) located above. On the village lands, prime appreciation and importance are given to agricultural fields for crop farming (hárkiṣ); a secondary position is occupied by orchards (basíkiṣ) and non-fruit tree plantations (jakúr tom), and irrigated meadows (toq) are found as an additional asset. Barren lands and fallow parcels are rare within the catchment area of irrigation channels. In previous times, settlements were located outside the irrigated space; but settlements have spread into the village lands and hamlets (muḥalla) since the end of the 19th century and have occupied a substantial area in the valuable agricultural space. In the aftermath of constructing the Karakoram Highway village communities and public organisations built several link and tractor roads; nearly every hamlet can nowadays be reached by motor transport. Source: Modified and updated map based on Hermann Kreutzmann (1989a: 50)

The ċhilgálas is informing the owner of a land parcel when his turn for irrigation is coming; the farmer is preparing the access to the field parcel and monitors the flow of water. The photograph was taken on 11 November 1984

The synoptic overview contains the structural elements of rules and regulations related to certain water sources, their connected water rights and the distribution and vigilance scheme as recorded in 1985. Subsequently, modifications occurred due to additions, but the core of the rules has been maintained. The irrigation calendar is structured according to the seniority of channels at different water sources. Its distribution key is structured by inherited rights of access and lineage affiliation. Each channel is characterised by a unique schedule and entitlement system that reflects their separate coming into being and the negotiation of shares. Source: Author's own survey and data collection in 1984–1985, and updates for Aliabad góčil. Modified after Hermann Kreutzmann (1989a: 86; 2000a: 105)

Irrigation calendar of Central Hunza

water source	channel system/ priority of supply	water right	duration of cycle (days)	water distribution (one cycle for each system displayed) (* = daytime; + = night-time)	decision about turns/ sequence	number of water guards
Ultar (glacier)	*Altit góčil* (first priority)	one share for Altit (all year)	4½	TUUHHKKS S * + * + * + * + * T = *tham* H = Hakálokuċ U = Usénuċ K = Khanúkuċ S = Susuròotiṅ	annual draw of lots	2
Ultar	*Hamáći* (first priority)	one share for Ganesh (all year)	6	GGGGABCDFFFF * + * + * + * + * + * + Ganesh Giram G = Gamu Ganesh A = Buldas B = Shuknoshal C = Shaboikushal D = Silganish F = Gerelt	fixed sequence overall; day & night turns in Giram	2
Ultar	*Dalá* (*Píire góčil, Mákučim góčil*) (first priority)	two shares for *thuaáṅ khanánċ*	Feb-May: 10 Jun-Sep: no cycle Oct-Nov: 10 Dec-Feb	HHAAAAAAAAAAAAAAAAAA * + * + * + * + * + * + * + * + * + * + H = Haiderabad A = Aliabad (with Dorkhan) *tul* and *ṭóri* system: 2 Haiderabad (outlet hole) 1 Mominabad 1 Baltit 1 Bishker same like Feb-May (see above) drinking water only for *thuaáṅ khanánċ*	annual change from E - W to W - E	3
Ultar	*Karagaḍimuċ* (first priority)	all year round	no cycle	solely for tham	-	1
Ultar	*Barber* (surplus only: first priority)	surplus water from Jun-Sep: day-time: Baltit night-time: *thuaáṅ khanánċ*	7	BHBHBHBDBABABA * + * + * + * + * + * + * + B = Baltit H = Haiderabad D = Dorkhan A = Aliabad	-	2
Ultar	*Imamyari; Rahimyari* (surplus only: second priority)	surplus water for Altit (after *Barber* is filled)	-	according to Altit village key (see above)	-	1
Diráċhil (spring)	*yarum Murku; yáṭum Murku*	Diramitiṅ of Baltit	-	additional irrigation of Dirámitiṅ *íćit*	-	1
Búlulo (snow & spring-water)	*Balti íl* (*yáṭum dalá*) or *Háligan* (Barber) (first priority)	3/8 Haiderabad (= 6 days); 5/8 Baltit (= 10 days)	Feb-Jun: 16 Jun-Dec: 16	DDAAUUQQ1122AAAAUUUUUUQQQQQDDDDD * + * + * + * + * + * + * + * + * + * + * + * + * + * + * + *+ Baltit D = Diramitiṅ 1 = Baskaotiṅ[1] A = Baratalíṅ U = Buroon 2 = Baskaotiṅ[2] Q = Qhurukuċ DDAAUUQQ1122AAAAUUUUUUQQQQDDDDDD * + * + * + * + * + * + * + * + * + * + * + * + * + * + * + *+ Haiderabad | Baltit	four-year cycle in Baltit 1985: AUQD (E-W, top-bottom); 1986: UQDA (W-E, bottom-top)	1
Búlulo	*Buróoṅ góčil; Barátaliṅ góčil; Qhúrukuč góčil* (surplus only) *Bulágaṣ; Turagaṣ; Bulzum góčil* (surplus only)	10 days Baltit (after *Barber* is filled) 6 days for Haiderabad (after *Barber* is filled)	16	distribution in accordance with *Balti íl* distribution in accordance with *Háligan*		1 1 1 1 1 1
Asha Har (snow & spring-water)	*Asha góčil*	6 days for Haiderabad; 1 day Dorkhan	7	HHHHHHHHHHHHDD * + * + * + * + * + * + * + H = Haiderabad D = Dorkhan	fixed sequence: fridays for Dorkhan	1
Hassan-abad nālā	*Aliabad góčil*	4 days each for Aliabad and Haiderabad, two days for Dorkhan	10	AAAAAAAAHHHHHHHHDDDD * + * + * + * + * + * + * + * + * + * + A = Aliabad D = Dorkhan H = Haiderabad	fixed sequence from top to bottom	3

according to a set of rules that is binding to all participants This codex is common knowledge and neither written down nor fixed like in other mountain regions. The contemporary set of rules and regulations is the result of historically acquired water rights of user groups and modifications directed by the hereditary rulers.

The participation of clans and village communities in the construction of new irrigation channels secures their rights to water and lands. The use of individual channels exclusively belongs to certain involved groups. A second criterion for water distribution is the traditional right of access to irrigation water belonging to certain clans or kinship groups. In cases of kinship and spatial congruency, i.e., when the parcelling of village lands follows a group structure, water distribution is highly efficient as certain areas are irrigated at the same time. Seepage and water losses in a dendritic irrigation network are thus minimised.

Karimabad poses an ideally structured example for this kind of setup. Sectors of the village lands are distinguished as belonging to certain kinship groups. The quarters of the four clans of the Diramitiṅ, Burooṅ, Baratalin̈ and Qhurukuċ define the agricultural oasis, only modified by the possessions of the *tham*. The *mīri* lands were traditionally located in every favourable cultivation area, such as Bul Mal, Karimabad (Suryas Das), and Bishker. Since the final years of *thamo* rule the majority of the *mīri* lands have been sold to individual villagers of Karimabad; thus we find there a mixed kinship-related ownership.

A synopsis of water rights, irrigation cycles and rules of distribution for the main channels of Central Hunza reveals the historic evolution and complexity of water management. First priority of access rests with the oldest channels and rights acquired through kinship participation in the construction of channels. The applicability of distribution keys is proved by measurements of water flow in different seasons. The distribution key – the allocation for the *Dalá* alone equals the combined shares of *Altit góčil* and *Hamáči* – is controlled at certain gates where the water flow is diverted to the different villages. In periods of deficiency, extra guards (*ċhíle yáṭkuin*) control the exact adherence to these distribution keys, while during surplus phases control is relaxed.

A second method of rationing is the hourly or daily right of water diversion. For example, the Bululo water is split in a rhythm which allows the inhabitants of Karimabad to use

Water controllers (yáṭkuin) are guarding and inspecting the upper portions of the irrigation system, making sure that the channels are adequately and rightfully filled. During the peak season they reside in the cave day and night. The photograph was taken on 2 May 1985

it for ten days, while the Haiderabadkuċ' share is only six days. The water in the *Barber* channel belongs to the community of Karimabad during the day, while at night-time all water is reserved for the users at the lower end of the *Barber* – Haiderabad-Dorkhan-Aliabad – for the irrigation of their respective fields under this channel's command. Priority rights of the *tham* included a sufficient share of water and the irrigation of his fields only in favourable daylight conditions. Even after the abolition of hereditary rule in 1974 the traditional *mīri* water rights remained valid through the combination of water and land rights.

The user communities distribute the allocated water among themselves according to allocation keys which are laid down by their members. A complex system is applied in Haiderabad. Six days of Bululo water have to be divided among four kinship groups which are the same as in Baltit. The Haiderabadkuċ have formed six user groups of equal size: four groups consisting of Diramitiṅ, Buróoṅ, Barátalin̈ and Qhúrukuċ, respectively, and two groups of Baskáotiṅ (from *bask*, literally meaning surplus, in addition). The members of the Baskáotiṅ are the extra households of the four clans which are numerically in excess of the group size. In Karim-

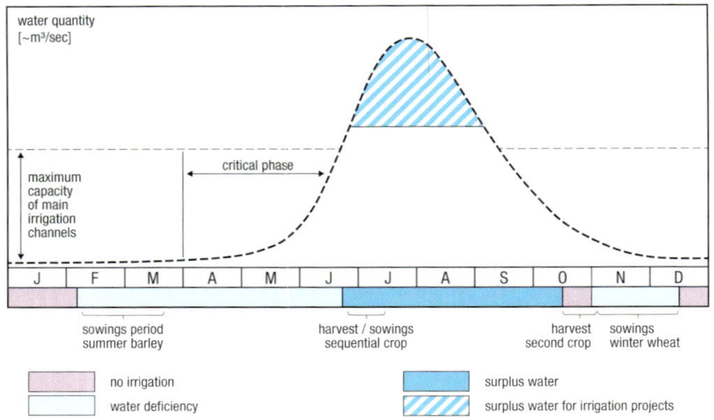

Irrigation and crop rotation in Karimabad, Central Hunza Valley. Source: Modified after Hermann Kreutzmann (2000a: 99; 2011d: 532)

abad a different system applies: the ten-day cycle is distributed among the four *roóm* in a manner that irrigation schedules – consisting of cycles such as two days and three nights or three days and two nights – are allocated to the relevant kinship groups. Every year a different clan begins the irrigation cycle, thus avoiding any preferential arrangement for individual kinship groups. Within these user communities the irrigation sequence varies annually from top to bottom and from east to west. These examples illustrate the wide range of possible communal agreements which have been established in order to safeguard equity for all users.

These cycles of six or ten days are no real indicator for the actual time required to irrigate all the fields of a user group. In periods of water deficiency, e.g. during the sowing of winter wheat (*gur*, *Triticum aestivum*) in November/December or of summer barley (*harí*, *Hordeum vulgare*) in February/March, the total time necessary to irrigate all fields once could amount to forty-eight days. In order to maximise the use of available water, the whole community has adopted a cultivation pattern in which the first crops of winter wheat and summer barley are represented in equal quantities. Both are sown in deficiency periods. By contrast, the second, sequential crops such as maize (*makái*, *Zea mays*), millets (*bay*, *Panicum miliaceum*; *ćha*, *Setaria italica*), potatoes (*ālu*, *Solanum tuberosum*) and buckwheat (*barú*, *Fagopyrum esculentum & tataricum*) are exempt from such regulations. In summer, surplus conditions of no water shortage prevail, and the entire village lands of Central Hunza can be irrigated within a sixteen-day cycle.

Valuable water resources have to be distributed according to systematic management and plant requirements. Top priority is given to cereal crops on irrigated terraces (*hárkiṣ*). After the first watering (*burúui*) of all wheat fields, barley is irrigated. Then follow potato furrows, vegetable plots (*śaníe khúṭkus*) and, last of all, lucerne (*śipíṫ*, *Medicago sativa*). This regulated sequence is repeated at the second watering (*yáṭćil*) and is only relaxed when sufficient water is available in the old channels. With the end of restrictive water use, the first irrigation of orchards (*basíkiṣ*) is permitted. The timing of this date in relation to the summer solstice gives the measurement for the classification; if the channels are full before 21 June, then it is a good water year; if later, a problematic season is at hand and scarcity might be an issue. This complex system of rationing is applied in Central Hunza in those settlements which command limited water sources or have grown beyond their resource capacity. Communities with abundant irrigation water do not need such a highly sophisticated distribution key to safeguard a high probability of sufficient crop production. Villages like Altit, for example, allot the annual sequence of irrigation every year through the drawing of lots, while in Ganesh the rotational structure is fixed.

Different kinship groups (*roóm*) are in command of day or night cycles. Some water rights incorporate the structural inadequacy of the systems, such as in the case of the *Barber* channel, where parties whose parcels are located nearer to the source are given preference. Karimabad as a whole is allowed *Barber* water during daylight hours, while Haiderabad, Dorkhan and Aliabad use the precious resource at night. The farmers of the Diramitiṅ kinship group in Karimabad profit from the location of their agricultural lands in the same manner as the people of Haiderabad and can divert any needed quantity of water to their fields. The difficult location of Aliabad at the lower end of *Barber*, *Dalá* and *Piíre góćil* results in greater water deficiency in this village than in others. Topographical features in combination with settlement history pinpoint the structural injustice. The older settlements at the upper end profit from their seniority, certain descent groups from their relationship with the *wazīrkuć*. As a general rule old age and closeness to power supersede the latecomers and less influential users. An additional factor contributes to the water supply situation, for along the course of the ten-kilometre-long open channels a substantial quantity of water is lost by evaporation and seepage.

Water channels function as multipurpose lines. Besides irrigation, the system provides all washing water as well. Drinking water was taken from the channels in former

Two types of buckwheat – ćiiski barú (Fagopyrum esculentum) and ġaqáy barú (Fagopyrum tataricum) – and two types of millets – ćha (Setaria italica) and bay (Panicum miliaceum) – are the main sequential crops (from left to right) that follow barley and wheat for grain production. The photograph was taken on 19 October 1984

times, but since the 1980s development agencies and village organisations have realised the detrimental influence of open water resources on public health. In most hamlets water-pipes now tap spring water and glacial melt high above the villages at the *íćit* border where sufficient cleanliness can be safeguarded at the intake.

The major administrative and economic centre of Central Hunza, Aliabad, used to be the most dependent village without control of its own water sources. The village community is eager to reduce further losses and has irrigated all fields consecutively: one year the irrigation starts in the east, the following year in the west.

Ecological limitations are responsible for these rules and regulations to a lesser degree than socio-political and historical events. They laid the foundations for certain user groups' water rights, which have to be defended against competitors. Water surplus and deficiency regions are defined by inherited rights of individual groups. Complex irrigation systems reward community efforts during construction and regular maintenance, in which all social groups – household (*ha*), kinship group (*guṭí*, *roóm*), village (*khan*, *mahalá*) – are involved on different levels of participation.

Organisational aspects of water management

The *tham* of Hunza had traditionally taken the initiative for the planning and execution of construction work on irrigation channels. He normally involved his *wazīr* in this enterprise of state importance in his dual function as prime minister and executive engineer for planning and supervision. The respective kinship groups or village communities served as construction workers. Early major channel projects accomplished general construction with simple technical devices and tools such as wooden shovels and ibex horns.[234] Sheer cliffs were mastered with support structures made of wood and stone. The influence and reputation of *wazīr* and workers increased through ingenious channel alignments, and they secured usufructuary rights to the ameliorated lands through their participation in *rajaáki*. For the ruling classes this was one of the opportunities to increase their wealth based on land resources.

Communal work of comparative size is only called for in certain cases nowadays: for example when a new storage tank (*phári*) has to be excavated or when a channel has

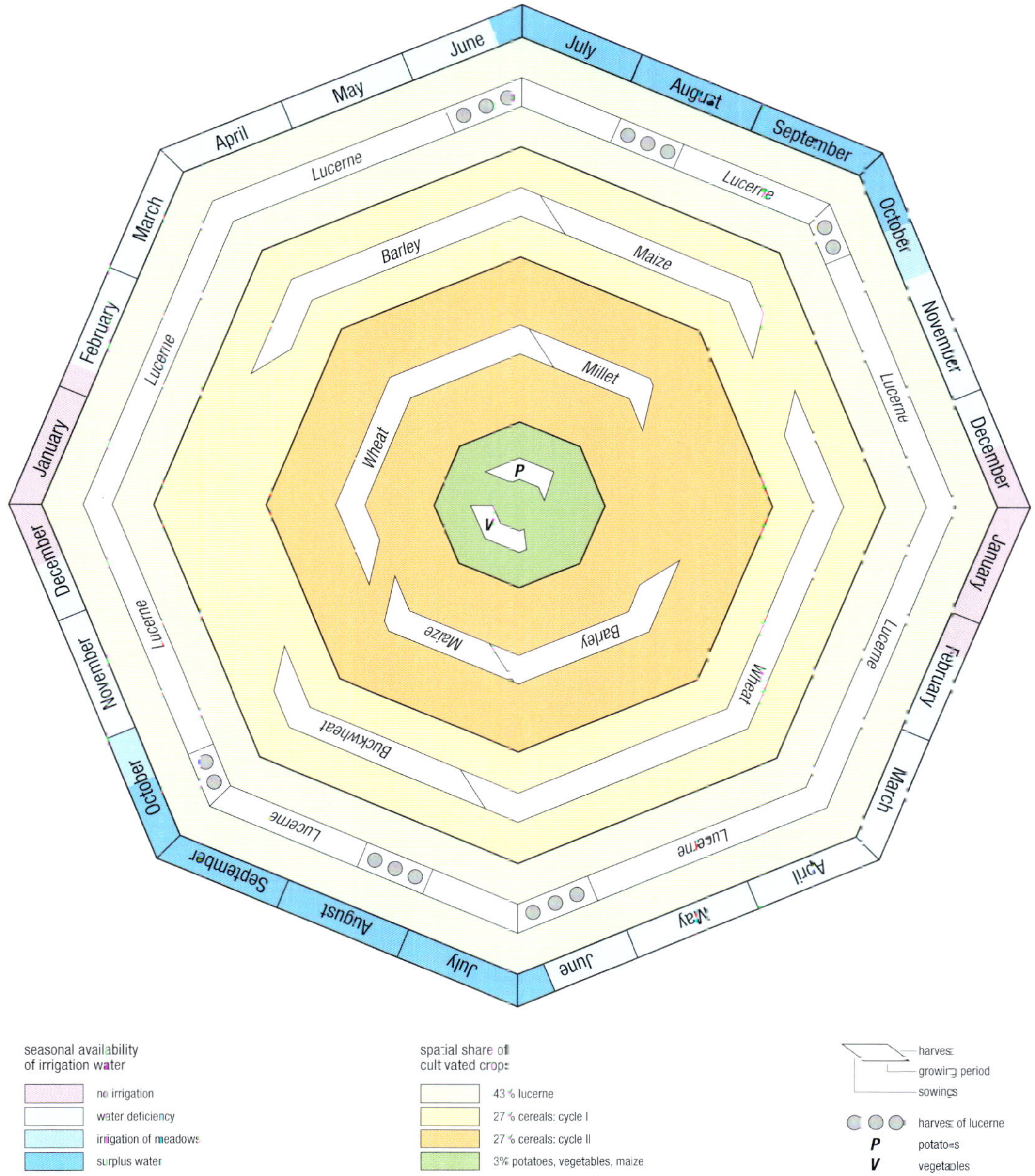

Seasonality of crop cultivation and deficiency of water supply. The diagram shows the relationship between thermal growth conditions for agricultural crops and their inter-linkages with water availability for a two-year growth cycle. Farmers have adapted to the conditions by splitting the sowing of crops between winter wheat and summer barley. Consequently, less water is required for the sowing period when scarcity prevails. Source: Modified after Hermann Kreutzmann (1998c: 196; 2011d: 531)

been breached. In the latter case, all user households have to cooperate until the damaged channel has been repaired or a new course has been excavated. Except for these exceptional or episodic events, fewer individuals are required in the management of channel systems through the regular activities of operation and maintenance. The annual cleaning and repair work starts prior to the first irrigation of crops in spring. Farmers who own land within the channel command area are obligated to remove the deposited sediment of the previous year from the channel-bed. These activities were traditionally controlled by the *wazīr*, who supervised the works and possessed the right to sanction those who did not participate according to the rules.[235] His influence in allocating the various jobs in water management contributed to his esteem as the chief hydraulic supervisor.

Certain offices were abolished in Hunza after 1974 such as the *tham* and the *wazīr* who lost their societal functions. The traditional functions of the *wazīr* were taken over either by the *lambardār* or by village committees. Every year the group of village elders assembles in a *jirga* which discusses the upcoming tasks of control, distribution, and remuneration. The water management posts are allocated among community members for the next irrigation period.

The *jirga* nominates a supervisor and the water workers (*ċhilgálas*) who safeguard the correct distribution schedule, the information about date and timing for the recipients and the allocation schedule of water to individual land parcels. They are responsible for preventing water theft. Minor maintenance and repair work is done by the *ċhilgálas* as well. Non-participating households pay a fixed amount in kind or money which basically pays the salary of the water workers. The number of *ċhilgálas* and the water rates are negotiated anew each year and depend on the size of the available village work force. In olden times the *ċhilgálas* was equalled with the load-carrier (*baldákuyo*) group.[236] Traditionally these salaried duties formed a highly esteemed way of providing an income to those households with insufficient lands to profitably employ their surplus male work force. Given improvements in socio-economic conditions

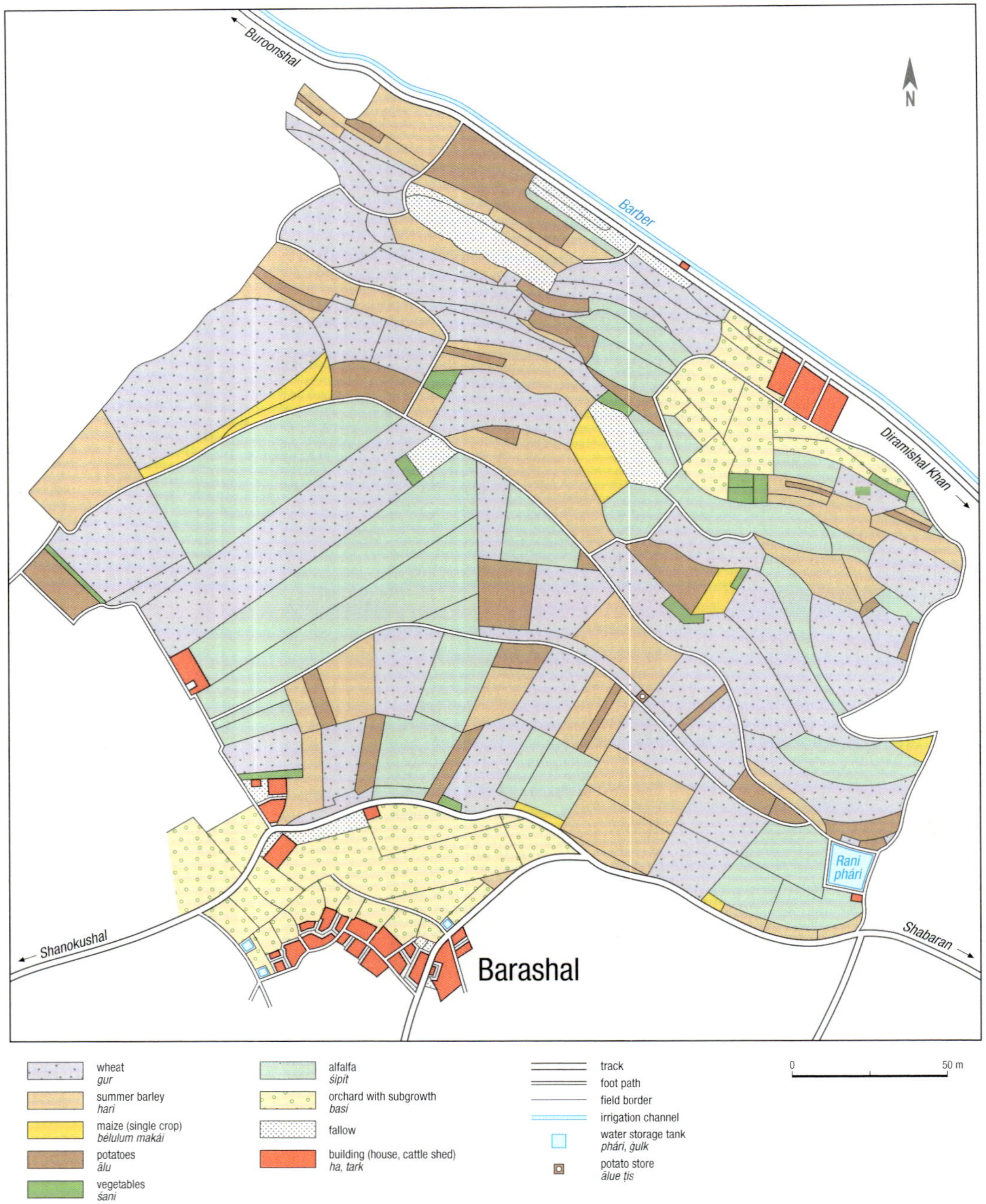

The land use map of Bul Mal shows the structured use of space for settlements, orchards and crop farming in spring 1985. Irrigated by water from the Barber channel, the first grain crop of the year is shared between winter wheat and summer barley; only a few patches are devoted to maize. Potatoes are emerging as a valuable crop for consumption and cash. Quite a substantial space is devoted to alfalfa. Perennial lucerne is planted in order to fix nitrogen and to improve soil conditions. During the 1980s, alfalfa developed as a cash crop for local sale to households that have to augment their fodder basis. Fallowing is rarely observed on small patches. Although limited in space an intensive vegetable cultivation plays an important role for household supplies. Source: Data collection and mapping by Sabine Felmy and Hermann Kreutzmann in April–May 1985; modified and based on Hermann Kreutzmann (1989a, supplement).

By the end of July the summer barley is ready for harvesting while the winter wheat takes a little longer; the green patches are alfalfa fields which are harvested three times per year. At this time there is no shortage of irrigation water, but at the same time there is little need for it. The photograph was taken on 27 June 1985

Most fields have been harvested by mid-October; only some plots with maize are still green. The hamlets are easily distinguishable as they are surrounded by orchards; the trees are changing the colours of their leaves. The photograph was taken on 16 October 1984

with increased labour emigration and more farmers involved in non-agrarian occupations, such traditional jobs in water management have lost their attraction. The difficulty of finding eligible persons has resulted in increased salaries for these irrigation workers in order to secure local personnel and to safeguard the system of water distribution. Though the cash investment in hiring such people remains within the community, the difficulty in obtaining reliable and expert people for these tasks causes a growing problem for some villages. Partly the difficulties are overcome by more than one person sharing a duty function. Nowadays, often even elderly and retired men are appointed as water guards (*ċhíle yáṭkuin*) to supervise the proportional and equitable distribution of irrigation water at the channel heads. These functionaries are elected for one year and safeguard the fair allocation of water to different communities. The *ċhíle yáṭkuin* are responsible for all duties at the channel heads (*sar*) including the cleaning of sedimentation tanks in which suspended sand particles are precipitated. During periods of surplus water, the *ċhíle yáṭkuin* limit the flow capacity of individual channel systems to avoid the breaching of channels. The *ċhíle yáṭkuin* reside in simple huts or the 'cave of the young donkey' (*gókomuċe kóor*) in the Ultar *nālā* and in the vicinity of important channel bifurcations during the entire irrigation period. After the end of the harvest in autumn they are nominally remunerated in kind, collecting one to two kilograms of wheat per channel from every household. In times of water shortage additional supervisors were brought in to support the appointed guards. Settlements like Haiderabad could not afford water loss during the early stages of plant growth when overall water shortage in Central Hunza increased the chances for water theft by individual farmers. Thus, during night time, villagers controlled all ten outlets along the way from Karimabad to Haiderabad in order to protect all *tul* and *ṭóri* from being opened unlawfully. In spite of these measures, water theft along with inheritance quarrels accounted for the majority of disputes in Hunza each year. As the overall importance of crop farming has diminished, the efforts for control and surveillance have been reduced as well.

Traditionally the *tham* fined those found to be stealing water in kind, taking a good share of the penalty for himself. Nowadays local Ismaili Arbitration & Reconciliation Boards are involved in the legal proceedings. They operate cost-free and compensate the winning party with the entire

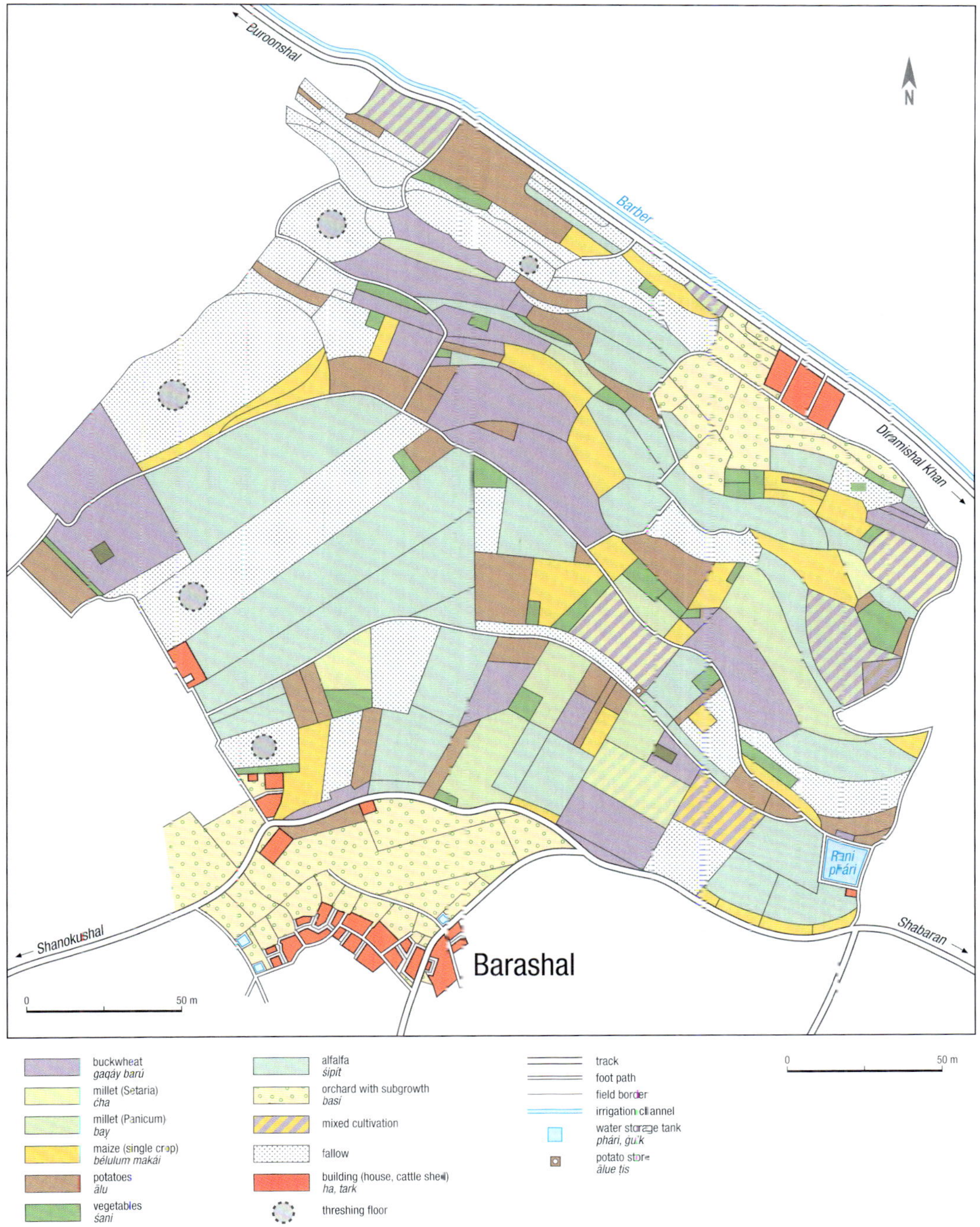

The sequential crops are dominated by buckwheat and millets that follow summer barley. Some wheat fields have been left fallow; in some cases maize has been planted as a second crop Source: Data collection and mapping by Sabine Felmy and Hermann Kreutzmann in September 1985; modified from Hermann Kreutzmann (2006: 336).

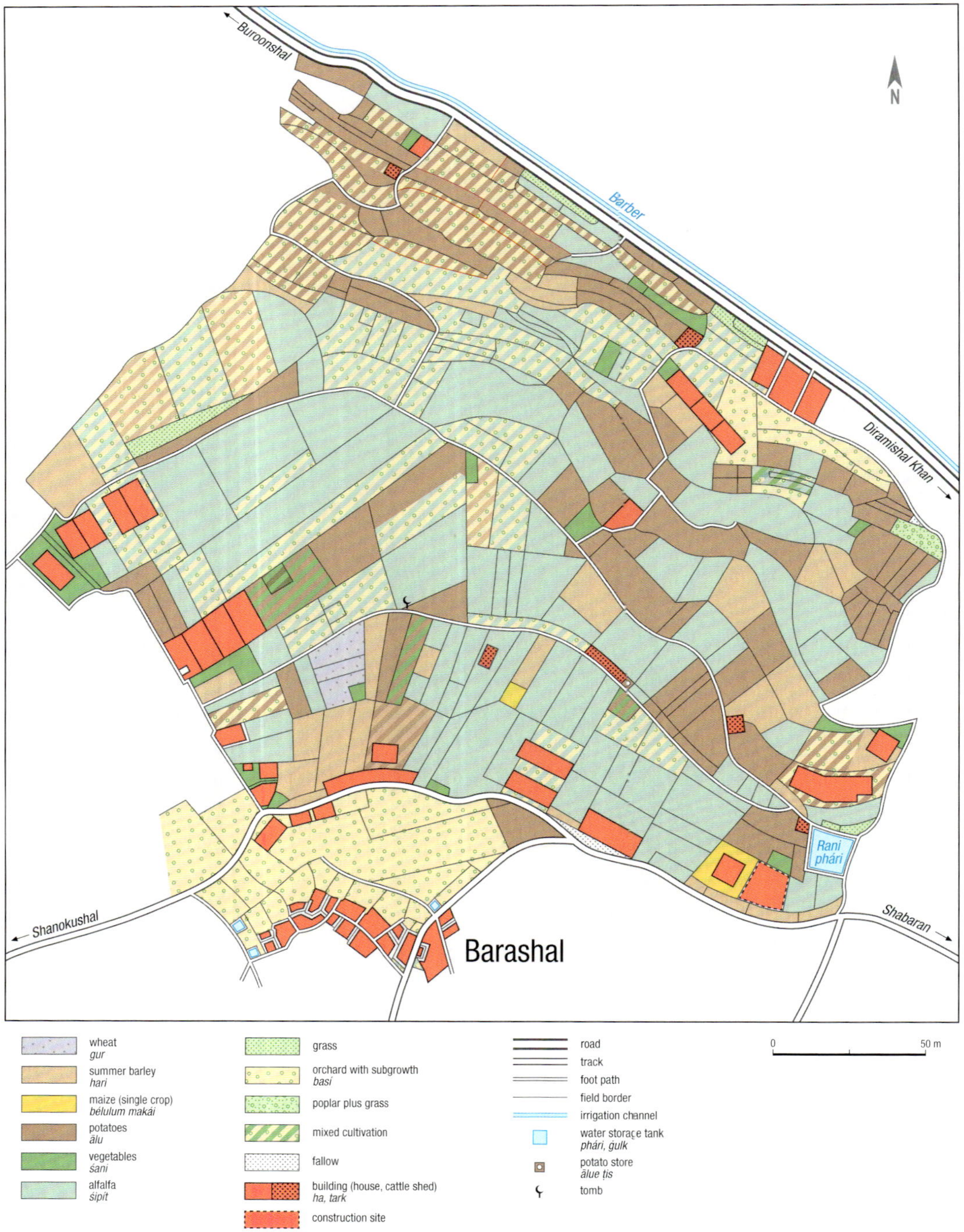

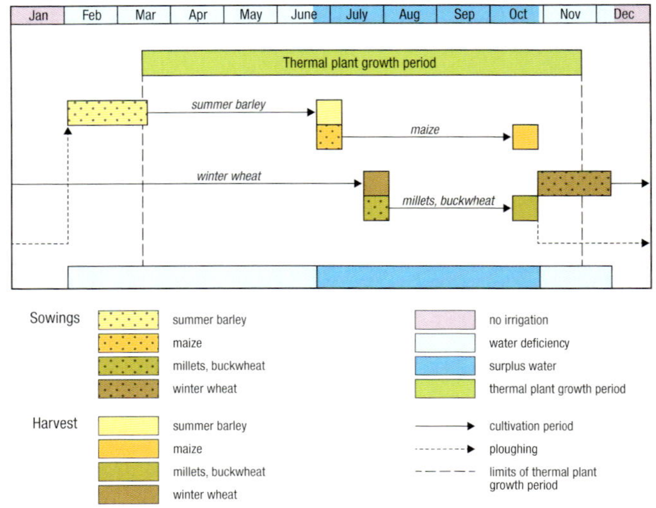

The diagram shows the nexus of seasonality of crop cultivation and deficiency of water supply in Central Hunza. The relationship between thermal growth conditions for agricultural crops and its inter-linkages with water availability are repeated in a two-year growth cycle. Farming practices have adjusted to a scarcity situation by splitting the sowing of crops between winter wheat and summer barley, which results in less water required for the sowing period. Source: Modified from Hermann Kreutzmann (2011d: 531)

By 2015, the field parcels have become much smaller than in 1985. Inheritance and subsequent land division have contributed to smaller landholdings and parcels. The dominant crop is now alfalfa, a marketable commodity for local customers. Several field parcels (hárkiṣ) feature mixed crops as the expansion of orchards (basíkiṣ) into the crop fields has increased. In previous times, fruit and other trees (ja̱kún tom) were banned from the fields if they cast a shadow on the neighbour's parcel. Orchards or single trees used to be practically non-existent in the valuable crop farming areas and were restricted to the space between grain fields, stables and hamlets. Housing now occupies substantially more space as new houses have been built outside the compact dwellings. Source: Data collection and mapping by Sabine Felmy and Hermann Kreutzmann in May 2015.

fine. Some severe cases have even been presented before public courts; normally no end of the dispute was found.

Fair distribution among entitled and involved community members forms the principle in an approach that rejects the separation of water and land rights. The irrigation networks have nowadays become the property of community members who themselves exercise supervision and control on the basis of consent. Scarcity and complicated distribution keys have increased the bureaucratic burden of water management. In periods of deficiency, small but highly valued quantities of water have to be guarded on their way to the fields. This practice requires substantial manpower as sluices and opportunities to leak water are numerous.

Future projects for the construction of new channels have to respect the traditional water rights. This precondition involves certain limitations: excess meltwater can only be utilised during summer surplus periods. In all other seasons no additional water supply is available to be tapped. New channels cannot reduce the ubiquitous seasonal water deficiency. Thus the scope for extending irrigated areas lies mainly in the reduction of seepage and evaporation in existing networks.

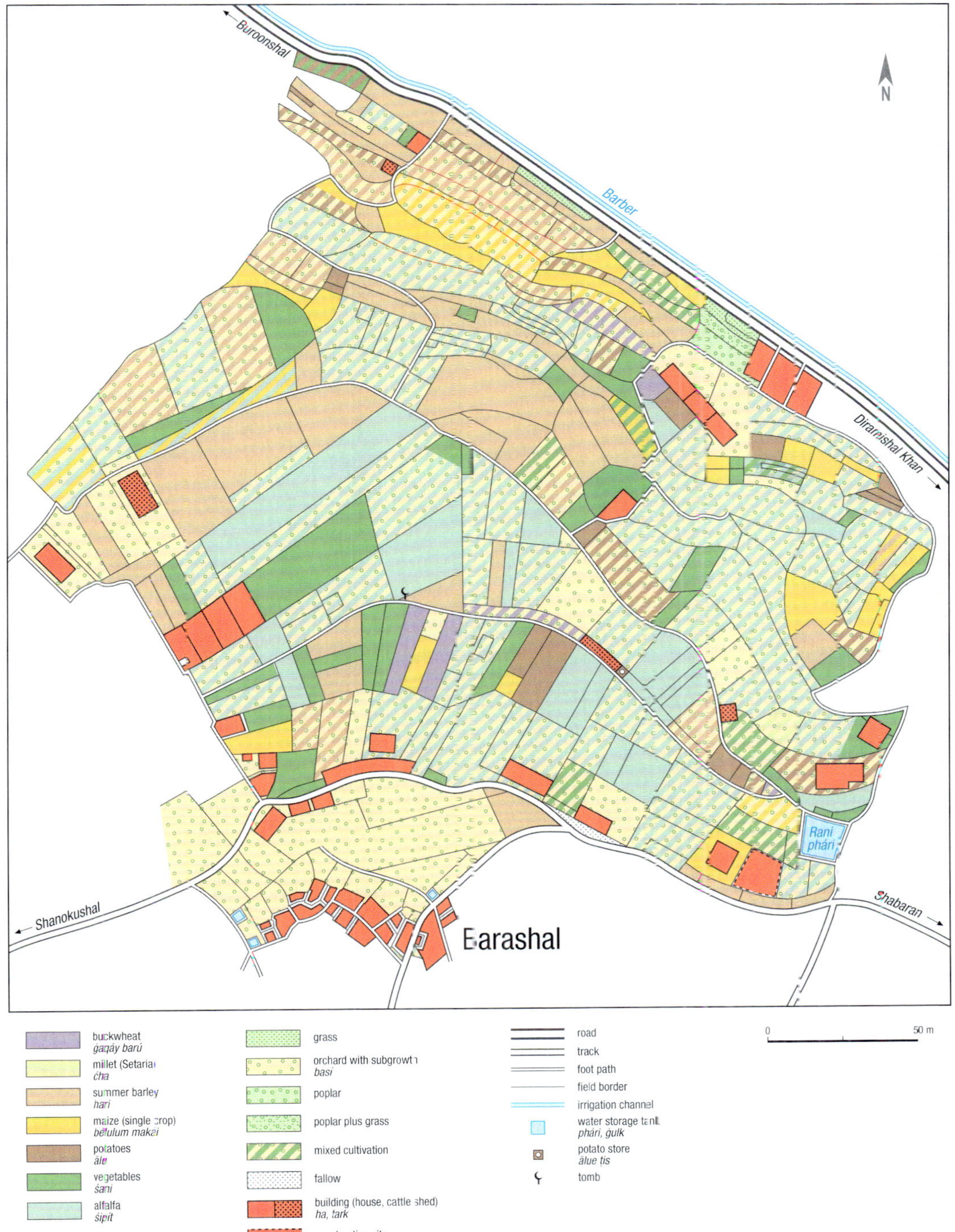

The dominant feature in the autumn of 2013 is a mixed cropping of fruit trees within the former grain-cropping area where grain and other crops are cultivated beneath the trees. As they normally do not mature, they are used as green fodder for the livestock. The increased local demand for fresh vegetables is reflected in the intensively used space (hárkiṣ). Formerly common sequential crops such as buckwheat and millets are hardly found any more. Data collection and mapping by Sabine Felmy and Hermann Kreutzmann in September 2013. Design by Hermann Kreutzmann

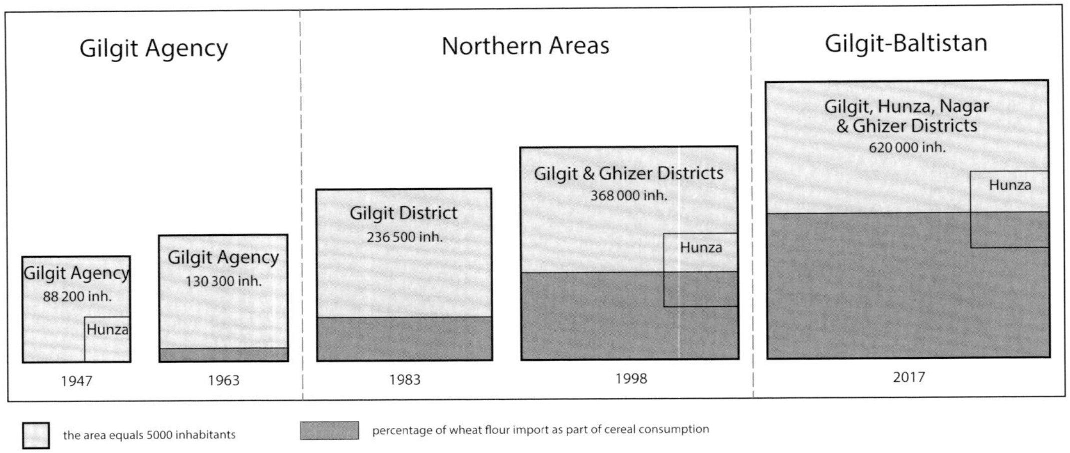

No external food supplies reached the Gilgit Agency prior to Pakistan's independence. First food transfers from down country Pakistan to the Gilgit Agency occurred in the framework of the First Village Aid Programme after 1956. The availability of basic food items from external sources has steadily increased since the Pak China Boundary Agreement in 1963, after Zulfiqar Ali Bhutto's announcement of fully subsidising the transport cost of wheat flour to Gilgit, and after the opening of the Karakoram Highway in 1978. Lower transport costs significantly contributed to the steady increase of wheat flour consignments from the Punjab. By now, about 100,000 metric tons are imported to Gilgit-Baltistan each year, covering about half of all bread cereals. Source: Own calculations and estimates based on Frank Saunders (1983); Elizabeth Staley (1966: 107); John Staley (1966); diagram modified from Hermann Kreutzmann (1995b: 222); recent data based on Gholam Rasul and Abid Hussain (2015: 629); Government of Gilgit-Baltistan (2018). Design by Hermann Kreutzmann

Recent developments in the irrigated oases of Hunza

New channel projects have contributed to increase Hunza's power and economic base during the reign of hereditary rulers and their *wazīr*. The range of such enterprises was not limited to the main settlement oasis of Central Hunza. Territorial claims and gains were manifested through the establishment of new villages in the southern and northern periphery. Such projects could involve groups of twenty to one hundred household members who were supported by their families in the early phase of amelioration. Resettled farmers from Baltit and Haiderabad and refugees/migrants from Wakhan were instrumental for expansive plans of the *tham* in Gojal and founded several villages there. Compared to the other regions, this single-cropping zone abounds with water sources. In Shinaki a limited quantity of meltwater from the Maiun *nālā* was utilised to establish a number of fortified villages in the double-cropping area. The population of Hunza numbered about ten thousand persons when M. Nazim Khan's investiture by the colonial authorities was celebrated in 1892.[237] During his reign as *tham*, a major uplift of the nutritional base took place. At the end of his rule in 1938, the population in Hunza had grown by half. The population has tripled in the sixty years since; a period of two to three generations has experienced the shift from a mainly agriculture-based income to an augmentation by salaries, pensions and proceeds from enterprises. Nevertheless, every generation had a strong interest in expanding the agricultural assets, especially in bringing cultivable wastelands (*das*) under irrigation. The popular belief has remained in Hunza that on average every household is cultivating a minimum of four *kanāl* of land with wheat. Crop farming is still perceived as a kind of responsible duty for every household as the basis for nutrition. In recent years other shifts have taken place when valuable irrigated land was used for house construction or devoted to cash crops such as potatoes and fruit. Fallow land is hardly found in an area where irrigation is esteemed as a life-saving activity and moral duty.

The termination of *mīri* rule in 1974 left a power vacuum which was supposed to be filled by a plethora of government institutions. Their involvement in infrastructure projects significantly increased after the completion of the Karakoram Highway. Within Gilgit-Baltistan a few attempts to establish new irrigation projects were undertaken, while in Hunza itself village communities built their own channels using traditional techniques, but applying motor drills and dynamite for blasting. Village elders were entrusted to supervise the planning and construction of a number of minor projects.

Different development agencies discovered this structural power vacuum and based their target-oriented approaches on self-reliance strategies. Village Production Groups and Village Organisations (*tanzīm*) took over the role of decision-making once held by more traditional institutions such as the *tham* and *wazīr*. These institutions now identified necessary physical infrastructure projects and managed the workload, seeking financial and technological support from external development aid. These agencies have supplied cost-free technical advice to the villages through local engineers in their employ. Other major differences from the past are that the construction cost is now covered from outside sources, man-days are paid for, not supplied through forced labour (*rajaáki*), the channels belong to the community working on them, and all participants are entitled to an equal share of water in the irrigated lands. No extra plots are reserved for members of the former ruling classes or any other high-status individuals any more. Water management and maintenance of channels are delegated to professionals (*ćaukī-dār*) who earn more than any traditional supervisor; development agencies supply modern technical equipment for drilling and blasting at nominal costs as well as engineering expertise.

Under these favourable conditions a number of irrigation projects have been carried out in Hunza, having been identified by the villagers as having top priority. Within the first decade forty-eight physical infrastructure projects were initiated which aim at the expansion and widening of existing *góčil* as well as the construction of new channels. In addition, a few experiments have been conducted with siphon (Nasirabad), pipe (Kamaris) and sprinkler (Ghalapan) irrigation. Several storage reservoirs and sedimentation tanks were also constructed in the productive physical infrastructure section of the Aga Khan Rural Support Programme.

Among these development-agency-led projects were three that reflect that times have changed in the Hunza Valley. Initially, the Food and Agriculture Organisation cooperated with the Agriculture Department in Gilgit and tried to install a prestigious project of high visibility in Hunza. They supported the village of Karimabad in constructing the *Murku góčil* that was bringing water from Ultar Har to the upper *das* and gardens (*basí*) of Karimabad. Without any modesty, the foreign development experts hailed this channel as one solving the cultivation problems of the village by "... adding 380 hectares, almost doubling their area of cultivated land as well as supplying the much needed extra water to existing crops and fruit trees. By providing their own labour, the new lands have cost only Rs. 10 a kanal [equals approximately 500 square metres] in explosives provided by FAO/UNDP resources."[238] They missed the point that additional water resources were not available and that surplus is limited to peak summer season only. This additional channel was eligible only for surplus water and could only be filled at the height of summer. The *Murku góčil* fulfils a desirable function, but not in the expected dimension, let alone solving any problems in food production.

The second project is the prestigious effort to free the water supply of Central Hunza from its total reliance on the one and only Ultar glacier. The first phase of the *Aliabad góčil* – tapping the Hassanabad glacier – was commenced in 1984, and the layout for the first five and most difficult kilometres was accomplished after four years. Covering an easier slope, the length of the channel was doubled and subsequently extended to Haiderabad a year later.[239] It was expected that eventually seven hundred hectares of land would

Upon their completion, the Murku waterfall and channel were advertised as the solution to Karimabad's food production deficit. The Murku góčil is a valuable addition to the existing system and provides an alternative routing for irrigation water in the peak melting season. The steep rock cliffs under Murku góčil are visible in the background; the irrigated terraces are watered by older channel systems. The photograph was taken on 23 June 1985

The Aliabad góčil was built as a community effort in which several village organisations from Central Hunza participated. Workers shared the task of preparing the ground by drilling and blasting a track into the slopes. The photograph was taken on 16 June 1985

The last wazīr in Hunza was Sajidullah Beg who was appointed in 1958 and lost his position when the Hunza State was abolished in 1974. Being in charge of the initial attempts to build a new channel, he was the last in a long line of wazīrkuč who were involved in channel construction. Source: Undated photograph provided by Ejazullah Baig, Karimabad

be watered from this newly tapped source; eventually seventy-five hectares of cropland were added, but significant tree plantations became feasible under the regime of *Aliabad góčil*. This difficult project cost the lives of three local farmers during construction in which all villages concerned between Karimabad to Aliabad participated. The working groups were formed according to lineages (*roóm*) that operated and participated along these well-known lines. During communal work (*rajaáki*) they preferred the conventional and inherited lineage (*roóm*) structure that has a long tradition in irrigation-channel-building despite the existence of village organisations (*tanzīm*) that were formed on a hamlet (*mahalá*) basis at the suggestion of the donors.[240] The *Aliabad góčil* is a great example of a project that could not have been feasible for a village community depending on its own resources. International development aid purveyed a means of experimenting with new techniques and formerly unprofitable projects. A danger lies in sustaining the channel head (*sar*) near the mouth of the Hassanabad glacier. Should the sixteen-kilometre-long glacier advance only a few hundred metres, it would destroy the channel intake which lies in close proximity below the present head. The threat became reality as the Shishpar glacier has been advancing in a surging manner since July 2017. A year later it had crossed the confluence of Muchu Har and Shishpar Valley; further advance movements continued up to spring 2019 and led to the destruction of the channel head. The *Aliabad góčil* has to cover a more than ten-kilometre-long stretch from the source in the glacial valley towards the central oasis and has been a constant project of repair and complex maintenance. Nevertheless, it has managed to alleviate the irrigation situation for all the villages that were at the tail end of Dalá: all *thuáaṅ khananč* such as Aliabad, Dorkhan and Haiderabad. With this additional water source included, the pressure during scarcity periods could be reduced and its functioning has allowed a more reliable irrigation. In order to optimise the use of water, the water rights of Aliabad village were adjusted and modified in such a manner that consecutive top-down irrigation of neighbouring fields is the rule. The farmers have realised that the lineage-based allocation of fields had resulted in scattered field parcels. If the inherited rules of lineage-wise irrigation had been followed further, more water would have been lost along the course of transfer from one field parcel to the other. The inauguration of the *Aliabad góčil* was taken as a turning point of amending the previous rule and introducing the optimal allocation of water on a spatial sequence from top to down. The maintenance work was divided according to beneficiaries: Aliabad and Haiderabad each have four shares of water and are responsible for four days of work each in a ten-day cycle. Dorkhan has only two shares and therefore only half the duties.[241] In Aliabad, where the arable land was about three hundred hectares, an additional seventy-five hectares could be cultivated that were equally divided among the five village organisations. Most farmers devoted the land to orchards that were nearly double in size. The provision of about forty percent more water to the villages has eased the pressure on the most disputed resource which is water.

The third project is the *Dilbar* channel. It was commissioned by *mīr* Jamal Khan during the tenure of his last *wazīr*,

Sajidullah Beg, who took his post in 1958. Efforts were subsequently made to bring additional water from a high-lying channel head to Karimabad. The project was a grand failure. Only in the last ten years could a channel and pipe connection be established that was named *Dilbar*. Funds from Khushal Pakistan were instrumental in funding the equipment and workload to drill, blast and prepare a channel around the cliff above Karimabad. All three examples show that new actors are on the scene; somehow governmental and non-governmental actors have taken over from the village institutions, either before the end of the Hunza State or after. The prime quality that made Hunza's reputation as channel-builders is now dependent on external funding; without donor support no project seems to be feasible.

If sustainable development can be envisaged, these projects will increase the agricultural resource base of this mountain valley. In case of failure, the only asset would be the remuneration of the work force and the communal organisations that have been developed. The tradition of channel-building knows quite a number of failures. Prior to establishing the *Aliabad góĉil*, there were five attempts to bring water from the Hassanabad glacier to Central Hunza up to the end of *mīri* rule.[242] The *Aliabad góĉil* was successful, but heavy rainfall destroyed vast sections of the channel only a few years after operation began.[243] With external funding the repair work could be shouldered; it would probably have been abandoned in earlier times. Ruins of irrigation channels would be neither the first nor the last remnants of attempts to maximise the use of meagre natural resources in the Karakoram.

New irrigation projects attempt to harvest additional waters higher up in Ultar Valley. Diráĉhil and Dilbar are the highest channels that utilise spring and melt water as well as surplus water from Ultar Glacier during the peak melting season; to establish channel heads in such steep and rocky locations is a challenging task. The photograph was taken on 28 June 1985

After harvesting the winter wheat a temporary threshing floor (darč) is prepared on a field parcel by tamping down the soil that is moistened with water. After drying the wheat threshing and winnowing commences; immediately after finalising the tasks the ground is ploughed again and a sequential crop is sown. The photograph was taken in Mominabad on 17 July 1985

Combined mountain agriculture – an ideal model?

The irrigation economy is one of two main pillars of combined mountain agriculture that is effectively augmented by, and strongly interlinked with, animal husbandry. Both are configuring the combined system that depends on irrigation as a highly concentrated and intensive cultivation while animal husbandry is utilising natural vegetation cover in a most extensive way on a seasonal basis. Both are interrelated in the village lands where crop-cultivation provides residues and straw for animal feed, where alfalfa is grown as a valuable fodder, and where certain parts of the village lands are occupied by irrigated meadows (*ṭoq*). In turn, animal husbandry has been the prime supplier of animal manure as a valuable fertiliser. The presence of livestock in the village has posed a number of challenges such as protecting field crops from scavenging and trespassing animals during growth periods. There is a second strong correlation between water and fodder. The embeddedness of the Karakoram's villages in an arid mountain steppe environment puts a special emphasis on water. Pastoralism as the second pillar of combined mountain agriculture is strongly linked to water availability. In the Karakoram Mountains only one percent of the space is used for irrigated agriculture, while half of the area is potentially suitable for pastoral practices. Both spaces are significantly dependent on water. Fodder production in the irrigated oases is strongly linked to irrigation when alfalfa and straw are harvested for livestock consumption. Grass production within the village lands is restricted to irrigated meadows (*ṭoq*). All fodder that is collected outside the *íćit* is found close to glaciers and springs. Most pasture areas in the high elevations are strictly dependent on the availability of flowing water. Consequently, the distance between homestead and pasture area could be quite long and not only linked to hydraulic resources; especially water access is a prime resource where space is a contested commodity. That holds true for most locations in the Hunza Valley and the Karakoram Mountains.[244]

The emphasis on such a strong link was valid when no external inputs had reached agricultural practices. External supplies of chemical fertiliser had entered the production cycles even before the Karakoram Highway was opened. Transport became cheaper in its aftermath and allowed a growth of external inputs from seed to fertiliser and pest control. The value chain is not only driven by inputs; the demand structure has changed as well. It had long been obvious that self-sufficiency in staple crops could not be achieved at any stage. A changing demand structure initiated various shifts in agricultural production. First, options were limited by the cultivation of high-value crops that would thrive under the conditions of a certain mountain climate, specific poor soil conditions, and varying availability of irrigation water. The experiments yielded success in potato and fruit production; to a lesser extent seed and vegetable cultivation succeeded. Potatoes have entered the staple diet of people in Gilgit-Baltistan. In addition, potatoes have become one of the prime cash crops and account for more than three quarters of the crops that are marketed. In 2017, more than 100,000 tons of potatoes left Gilgit-Baltistan; a substantial share originated in the Hunza Valley.[245] Apples lead the list of all marketed fruit: with nearly 11,000 tons they cover more than one third, followed by walnuts, grapes, apricots, pomegranates and cherries.[246] Processing the harvest into dried fruit will add value and durability; apricots, mulberries and walnuts are renowned regional products. The improvement of quality processing and marketing has been transformed from the early stage of village-based cooperative marketing societies that arose with the opening of the Karakoram Highway to private companies that have contracted farmers since 2005 and have successfully offered organic and certified fruit from the Hunza Valley to the world market.[247] A growing number of villagers in Central Hunza and Nagar have

At the end of the cultivation season sequential crops such as sweet buckwheat, millets, maize, potatoes and alfalfa are harvested in Hassanabad. The 1984 pattern of a great variety among small field parcels has dwindled in recent years. The photograph was taken on 25 October 1985

The division of labour has been changing over time. In the mid-1980s the burden of agricultural tasks such as transporting and distributing manure on field parcels, ploughing, threshing and winnowing was mainly assigned to older men and women who had taken over duties from absent husbands, brothers and sons. Weeding of crop fields (bilís) has remained in the female realm, thus providing additional fodder by thinning the ears of barley and wheat. The photographs were taken on 22 October and 11 November 1984, 21 May and 8 July 1985

changed their farming system from a clear-cut division between crops, orchards, and meadows to an inter-mixed system in which fruit trees cover a bigger space in areas where trees were formerly banned. In the undergrowth alfalfa and grain crops are planted and mainly used or sold as fodder. The transformation was recorded on selected field parcels in Bul Mal where the changes have been documented since 1985.[248] The recent pattern reflects the market orientation with cash crops and less labour-intensive crops dominating. Other villages in Gojal and Nagar have concentrated on potato production; every year the purchasers (*ṭhekedār*) from down country travel the Karakoram Highway to load the sacks (*borī*) on to their trucks which have been arranged by their local agents. As yet, few local contractors are part of the marketing chain.[249] The marketing of fresh fruit – cherries have been especially successful – became feasible with improving road conditions and has opened up a new market in down-country Pakistan for perishable goods. In return, the cash earned with agricultural products partly allows the purchase of wheat flour and other staple food items from the plains. Thus, the limited land resources – the average farm size in Gilgit-Baltistan is about three quarters of a hectare – are used in an optimising strategy by cultivating higher-value cash crops.

The ideal model of a combined mountain agriculture has lost its relevance since market access and cash crop production has shifted agricultural practices from low input strategies embedded in local circulation systems to an agricultural production that minimally contributes to household needs in flour, milk products and meat, but satisfies domestic demand for fruit, potatoes and vegetables and generates income from market-oriented surplus production. The strong interlinkages were dissolved not only because of these events; they were strained due to developments in the livestock sector that are connected with a dwindling workforce of shepherds and access to external flour and meat markets.

Apricots are ripe and ready for drying during the main harvesting season of field crops. The burden for female household members and children is extremely high as the fruits have to be prepared for drying during a very short period that requires dry weather conditions. The photograph was taken in Altit on 27 July 2010

In mid-summer most oxen that could be used for ploughing for the sequential crop sowing are outside the village lands on pastures. Before motor-driven ploughs were utilised men pulled the plough. Especially on very narrow terraced fields an ox with yoke would not be able to turn. The photographs were taken on 12 July 1985 and 18 March 2019

Pastoral practices in decline – adaptation and landscape management

On first sight a well-known pattern of pasture use is observed in the Hunza Valley. Livestock herds are seasonally taken out of the villages to high pastures that offer natural vegetation for fodder during the season when crop farming takes place in the villages. After harvest most village lands are open for free grazing (*heṭiṅ*). The time-space relationship seems to be so obvious and practical that this system of combined mountain agriculture was regarded as a well-adapted form of optimising the use of locally available resources.[250] Seasonal migrations of herds of yaks, sheep and goats take place between permanent homesteads in the arid valley grounds and natural high pastures in the vicinity of glaciers. Their schedule reveals a time-space pattern that connects locally available resources at appropriate times of the year. A closer look would reveal that not all animals left the cattle-sheds in the village. The daily milk for tea as the staple diet requires fresh milk in every house. Many households have been keeping a cow or a sheep/goat for these purposes in the village and fed them with greens that were taken out of weeded fields (*bilís*) and all kinds of residue. Fodder was always in short supply, even leaves from the orchard trees were collected as winter feed. Consequently, keeping a horse or a donkey was reserved for affluent households of the ruling elite; only the *tham* had camels as transport animals. These were kept on pastures in Gojal where most of his livestock was grazed. The last camels have been spending their time in Dih at the entrance to Khunjerab National Park.

Transport animals did not play an important role in pre-motor transport days. Most luggage and loads (*baldá*) were carried by porters (*baldákuyo; borwar*); the forced labour (*rajaáki*) regulations had assigned these menial duties to the lower classes. Most households were obliged to provide such services to the *tham* and *wazīr*. Members of the upper class (*uyóṅko*) were exempted from these duties. Oxen were kept or brought back to the villages when required for ploughing (*hárki*) or threshing (*darċ*) in mid-summer and autumn.

The flocks that left the village consisted mainly of sheep and goats (*huyes; kla*). In the case of the Baltikuċ, the closest pasture area is above Karimabad in Ultar and Sekai. Together with their relatives from the new settlements (*thuáaṅ khanánċ*), they shared the pastures on the left bank of the Shishpar glacier that were accessed from the Hassanabad *nālā*. The Burusho rule was that only male shepherds were allowed to access the pastures as the border between the village lands (*íċit*) and the pasture (*ter*) had some significance in purity terms. Access for women in the age-group between menarche and menopause was restricted; the *ter* was the sphere of the shepherds (*huyéltarċ*) and the fairies (*parí*). Irmtraud Stellrecht interpreted a 'gendered landscape' in describing the vertical differentiation of purity and the imposition of restrictions among the Burusho in Hunza.[251] This exclusion of women from access to high pastures is found in all Burusho villages in the Hunza Valley. The Wakhi tradition allowed women to be the masters of the high pastures and in charge of all duties there; men only supported the relocation of herds from one pasture to another, helping to slaughter livestock and guarding the herds on their way from village to pasture and back (*kuč*). An awkward but pragmatic situation was emerging in some combined high pasture settlements that have been shared by Burusho and Wakhi alike.[252] In the Wakhi realm the gendered division of labour allocates the major burden of summer pasturing to the senior woman of the house who is responsible for processing the milk (*δiydor*).[253] Their high pastures are dominated and managed by women who get occasional, partial support from elderly men and their children. Sometimes couples prefer to share the burden and spend the summer together in the high altitudes if sufficient work force is present in the village household to till the fields.[254]

In recent years more and more villages and households have been facing severe shortages of shepherds (*huyéltarč*) and shepherdesses (*δiydor*). A number of reasons are responsible for this scarcity of people who are qualified, willing and able to spend a whole summer in the high pastures. The extremity of the situation can be exemplified in the case of the pastures above Karimabad belonging to the Baltikuč. The prime source for reference is the report of Qudratullah Beg who recorded in Ultar (3,200 m) five to twelve shepherds in the summer of 1935 caring for horses, oxen, sheep and goats, processing the milk into a variety of fresh and durable products, and tilling some fields for grain crops there.[255] Half a century later, we met three to four shepherds for the four *roóm* of Karimabad where oxen, sheep and goats were pastured. Crop cultivation had been given up. Nobody was met with in the summer of 2017 when the hut had been given up and a single male goat was seen roaming on the pasture. The Diramitiṅ kept their sheep and goats in neighbouring Hon where four shepherds took control in 1935; in Sekai (3,600 m) the *wazīrkuč* had reserved a pasture for their oxen and horses: here as well four shepherds were in charge. Both pastures had been given up by 1985. In recent years one member of the *wazīrkuč* experimented with keeping some yaks there. The Diramitiṅ of all the villages in Central Hunza had sent their animals to Shishpar, where forty to seventy shepherds fulfilled the duties of taking care of oxen, sheep and goats and of cultivating some grain crops in 1935. The Burooṅ would send their *huyes* to Muchu Har in the same manner and with an equal number of shepherds. At that time, one shepherd cared for the animals of three to four related households from the same lineage. Fifty years later the number of personnel had fallen to one tenth; the flocks had diminished accordingly. It was difficult to find more than one shepherd in recent years who would willingly maintain the pasture rights of his village by being present there for the whole summer. Many villages in the Hunza Valley face similar problems. In Gulmit and Shishket hardly anyone can be convinced to be present in the pasture settlement of Gaush, which is disputed between the two communities. In neighbouring Ghulkin shepherds from Nagar were hired to attend the herds in Kharachanai. The cost of the shepherds amounted to one lakh rupees in 2013; all milk products – butter (*maltáṣ*, *rúy̌un*) and *qurúṭ* – remain with the shepherds, nothing is delivered to the villagers. They have an income from their pasture from ibex (*yukṣ̌*) hunting licences that were earning more than the cost of shepherding. Even in the most prominent herding community – Shimshal – labour shortage has become a major issue of discussion. Although the local livestock committee is offering substantial wages for shepherds and major shares of the proceeds from milking, it has increasingly become difficult to convince able-bodied persons to take up such duties.[256]

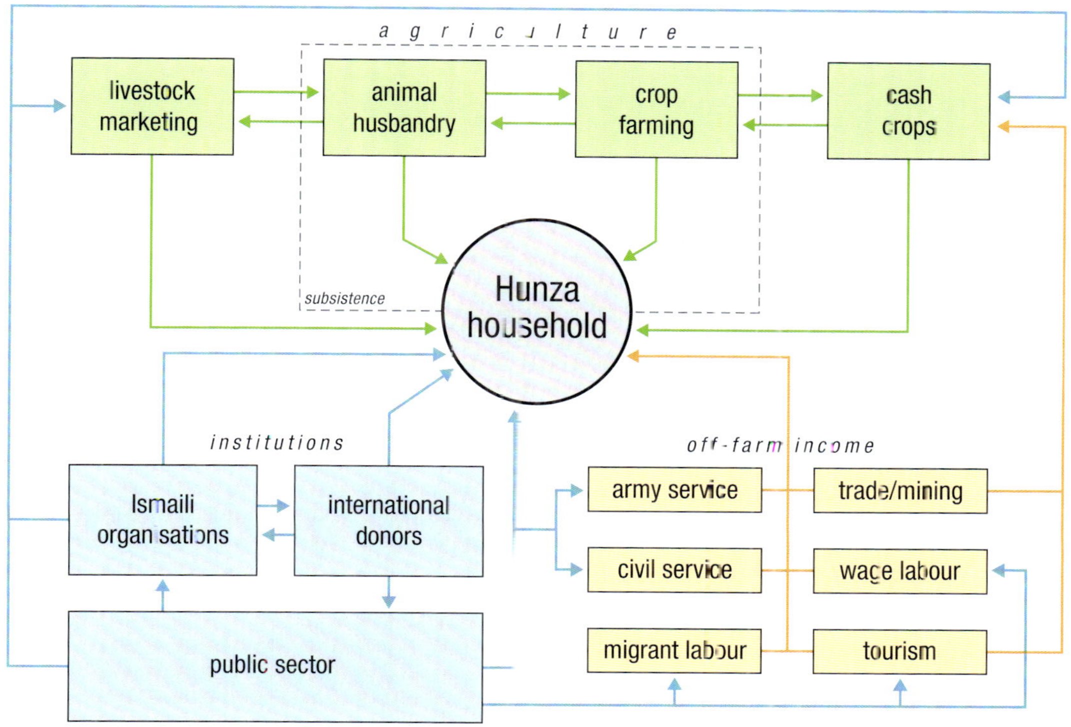

During the first decade after the opening of the Karakoram Highway the diversification of income sources increased. The share of crop farming and animal husbandry for the sustenance of Hunza households diminished in favour of livestock-marketing and cash crop production, such as potatoes and apples. At the same time the share of off-farm income increased due to growing opportunities of employment in various sectors. While army service declined as the former prime source, all other occupations received a boost. Based upon the infrastructural road network, several development packages were implemented by governmental and non-governmental institutions sponsored by higher budgetary assets and international donors. Source: Modified on the basis of Hermann Kreutzmann (1989a: 209; 2006: 330). Design by Hermann Kreutzmann

The pastures between Dafdar and the passes leading into Hunza were in the realm of levying grazing taxes by the mīr of Hunza. Only in 1894 was the village of Dafdar established; from where the shepherds who regularly visit Kashgasu (opposite) originate. Wakhi shepherdesses are milking their goats at Kashgasu (4,130 m). Until 1937 Hunza's rights to levy grazing taxes were maintained and were terminated afterwards. The photographs were taken in Kashgasu on 5 August 1991

Revenue of the *mīr* of Hunza derived from grazing taxes in the Taghdumbash Pamir and Sarikol

Year	revenue in kind				Source
	namdā (felt, woollen blanket)	ropes	*kirpas* (cotton cloth)	*śuqá* (coat)	
1760–1865	grazing dues from Kirghiz[1]				IOL/P&S/10/278: 216
1865–1878	no revenue[2]				IOL/P&S/10/278: 216
1875	looting by Hunzukuč in Taghdumbash[3], extracting goods from Kirghiz				John Biddulph 1876: 116
1878–1887	grazing dues from Kirghiz and Sariqoli[4]				IOL/P&S/10/278: 217
1879–1891	taxes from Shakshu Pakhpu[5]				N. N. 1928: 85
1888	50	–	–	–	IOR/2/1079/251: 8
1889	100	–	–	–	IOR/2/1079/251: 8
1890	90	–	–	–	IOR/2/1079/251: 8
1890	attack of Hunzukuč against Kirghiz nomads in Taghdumbash				IOL/P&S/7/61/52, 53
1891–1895	no revenue[6]				IOL/P&S/10/278: 216
1895–1902	grazing dues of Kirghiz and Sariqoli				IOR/2/1075/217: 40[7]

1 In the aftermath of the conquest of the Taghdumbash Pamir and following the supplanting of the Kirghiz there by the *mīr* of Hunza's actions to drive them back to Tashkurgan his control of the grazing grounds resulted in annual demands for grazing taxes from all users (FO 371/19335: 214). Mir Silum Khan III materialised his claim by erecting a stone monument in Dafdar and he presented scalps of Kirghiz to the Chinese representatives in Kashgar and Yarkand (Stuart Hill Godfrey 1898: 74, Arthur Henry McMahon 1898: 5). Approximately 300 Kirghiz were estimated as having lived there before. The annual tax was supposedly fixed at the rate of one sheep per forty sheep and goats as well as one yak per thirty yaks.

2 During the reign of Yakub Beg (1862–1878) in Kashgaria and his occupation of the Pamir region the *mīr* of Hunza failed to realise any revenue there. The majority of Kirghiz nomads had given up the Taghdumbash Pamir as their grazing grounds during this period and shifted their activities to Aktash. Stuart Hill Godfrey (1898: 74) stated that until 1880 no taxes were collected. Subsequent to attacks by Hunzukuč around 1866/67 the Kirghiz of Taghdumbash retreated to Tagarma (see Thomas Edward Gordon 1876a: 115). The sedentary inhabitants of Sarikol took their place and utilised the grazing grounds in the Pamir.

3 A group of Hunzukuč is reported to have attacked the camps of Kirghiz nomads in Taghdumbash in 1875. They took some Kirghiz as hostages and drove horses and a large flock of sheep away. The *hākim* of Sarikol, Hussan Shah, organised a punitive expedition towards Hunza, liberated the captured Kirghiz and pulled down the fort at Misgar (John Biddulph 1876: 115–116).

4 Around 1880 *mīr* Ghazan Khan permitted Kirghiz nomads (20 tents) under the leadership of Beg Kuchmumabad (Kuch Muhammad) to utilise the pastures in the Taghdumbash Pamir. The taxes were fixed as one *namdā*, one *śuqā*, some rolls of *kirpas*, one rope or one saddle (*jhul*) per tent (*xirgó*). The amount of levies was depending on the size of the flocks. Stuart Hill Godfrey (1898: 74) mentions grazing dues to the amount of one sheep, one *namdā*, one rope and one pair of *paipakh* per tent for this period. Annually Sariqoli shepherds visited the same region since 1883 under same conditions. The competition between shepherds from Hunza and Sarikol led to quarrels in Taghdumbash in 1886, which were pacified by the Chinese representative (Taotai) in Kashgar. According to *mīr* M. Nazim Khan and *wazīr* Humayun Beg he settled the dispute in a local court in Tashkurgan in favour of the Hunzukuč (IOL/P&S/7/66/701; Arthur Henry McMahon 1898: 6).

5 Following a punitive expedition under the leadership of *wazīr* Humayun Beg directed against the inhabitants of Shakshu and Pakhpu (upper Yarkand valley) the annual dues were increased. Until the Hunza Campaign in 1891 the *mīr* of Hunza received annually five *yambu* silver equalling 750–800 Rs (Stuart Hill Godfrey 1898: 74). The taxes are supposed to having been paid partly in kind: Shakshu: 100 sheep, silver being worth 60 Rs, 2 shot-guns; Pakhpu: 10 *namdā*, silver being worth 60 Rs; 2 shot-guns (IOR/2/1079/251). During his escape to Yarkand *mīr* Safdar Ali Khan realised this penalty for the last time. It had been introduced as a making up for the previous enslavement of two Hunzukuč (Stuart Hill Godfrey 1898: 74, Arthur Henry McMahon 1898).

6 No dues were realised in the Taghdumbash Pamir for four years in the aftermath of the Hunza-Nagar Campaign (Stuart Hill Godfrey 1898: 74). This power vacuum was filled by Kirghiz nomads who came back to this region with 200 tents (IOL/P&S/10/278: William Macandrew Marshall 1913). Following a directive by the Chinese *amban* in Tashkurghan the Kirghiz began again in 1895 to pay taxes to the *mīr* of Hunza. The Wakhi settlers of Dafdar (established in 1894) remained exempted from any dues at the same time.

7 The taxes amounted to one *namdā* (length of five metres) per household. Impoverished camps would provide smaller *namdā*, one rope or *paipakh* (socks). In addition, the shepherds transported all dues as far as Murkushi in Hunza (IOL/P&S/10/278). From 1896 onwards the *mīr* was supported by the Chinese administration in allocating his dues (Arthur Henry McMahon 1898: 6). According to the judgement of the Political Agent in Gilgit the revenue of the *mīr* of Hunza from grazing taxes in Taghdumbash accrued to the value of 200-300 Rs annually around the turn of the century. This amount was higher than the tribute paid by Hunza to China and about a tenth of the value of return gifts received by Hunza from the Chinese Emperor (IOR/2/1075/217: 40).

The examples show a tendency of a sustained pastoral policy by the villages to maintain their territorial claims for the future. In economic terms they presently do not generate much profit. None of the communities would give up their rights of access; these are huge spaces of unknown value that are defended against outside intruders and that are supposed to be kept for the next generations. The function of pastoralism for the practice of combined mountain agriculture has been shrinking to a high degree; all its functions as a source of manure can and have been taken over either by chemical fertiliser or by manure from village-based stables.

Given the situation of the overall decline of a practice with enormous adaptive potential, it might be surprising that a detailed account of pastoral practices in the Hunza Valley is given from a historical perspective. Two reasons are important: first, pastoral practices have been an instrument of revenue generation for the ruling classes in the Hunza Valley and are an important leverage in power relations. Second, the pasture lands and interrelated user rights have been a bone of contention for more than a century; in recent years these land conflicts have gained pace. Severe conflicts and land disputes are the result when remote resources become important for the design of national parks, when the access routes to peaks for mountaineers and the trails for trekkers and trophy-hunters lead through such areas and when hydraulic and mineral resources are exploited. A certain aura is connected to these landscapes, not only because they are the abodes of the fairies.

Livestock taxes – a major source of revenue in Hunza State

Before Wakhi settlers immigrated to Gojal, more or less every summer Kirghiz nomads crossed the northern passes of Hunza to the high pastures in Chupursan, Mintaka, Kilik and Khunjerab. They probably migrated even further down the valley with their flocks of yaks, sheep and goats, using horses and camels for transport. Annually they delivered grazing taxes to the *tham* residing in Baltit, or his collectors extracted these directly on the grazing grounds. Pastoral practices were divided between Kirghiz nomadic use in Gojal and combined mountain agriculture in Central Hunza and Shinaki, while it seems unlikely that nomads and transhumant shepherds were visiting the Hunza Valley from the south. The only exception might be a Gujur community

The two Abgerchi shepherdesses are releasing their flocks for the day pasture after the morning milking in Boiber (3,400 m) at the foot of the northern side of the Tupopdan (6,100 m) massif. The photograph was taken on 7 July 1990

Year	revenue in kind				Source
	namdā (felt, woollen blanket)	ropes	*kirpas* (cotton cloth)	*śuqá* (coat)	
1903	22	–	–	–	IOL/P&S/7/157/1277
1904		grazing dues of Kirghiz and Sariqoli			IOL/P&S/7/170/1893
1905	33	31	–	–	IOL/P&S/7/181/1554
1906	45	30	–	–	IOL/P&S/7/205/1602
1907	40	30 (= total value 200 Rs)			IOL/P&S/7/205/1602
1908	40	30	–	–	IOL/P&S/7/222/2027
1909	41	35	2 *kham*[8]	–	IOL/P&S/7/233/1556
1910	50	–	2 *kham*		IOL/P&S/7/243/1407
1911	40	40	10	–	IOL/P&S/7/252/1654
1912–1914	40	30	3–4	–	IOL/P&S/10/826: 173[9]
1914	70	50	9	–	IOL/P&S/10/826:173
1915		regular dues from Sariqoli[10]			IOL/P&S/10/826: 140
1916		regular dues[11]			IOL/P&S/10/826: 107
1917	60	45	15	3	IOL/P&S/10/826: 86
1918		regular dues			IOL/P&S/10/826: 62
1919		regular dues			IOL/P&S/10/826: 32
1920	62	47	17	5	IOL/P&S/10/826: 9
1921	65	32	10	4	IOL/P&S/10/973: 240
1922	60	30	6	–	IOL/P&S/10/973: 217
1923	80	40	20	10	IOL/P&S/10/973: 183
1924	60	40	6	–	IOL/P&S/10/973: 155[12]
1925	52	40	10	–	IOL/P&S/10/973: 129[13]
1926	72	45	4	–	IOL/P&S/10/973: 104
1927	58	42	12	–	IOL/P&S/10/973: 81
1929	55	30	10	–	IOL/P&S/10/973: 36
1928[14]	55	60	10	–	IOL/P&S/10/973: 58

8 The value of coarse cloth (*kham*) amounted to 2 Rs, that of *namdā* to 150 Rs, that of ropes to 20 Rs the value of one *namdā*-sock (*paipakh*) to 0.5 Rs. During this year Sariqoli (77 households) were the only grazing tax (*khiraj*) payers. They demanded their exemption on the same basis as was applicable for Kirghiz (13 households) and Wakhi (25 households). This resistance could take place because of the little authority displayed by the *amban* of Tashkurgan who refused to support the Hunza tax collector Kara Beg. He provided him with 2 *charak āṭā* and 2 *charak* fodder for his horse only while normally a substantially higher provision was allocated: 1 sheep, 1 *charak* rice, 2 *charak āṭā*, 2 *charak* fodder for horses; 2 donkey loads of firewood (IOL/P&S/7/231/1399). The weight measures are converted as such: Five *charak* equal one *man* or *maund* (37.32 kilogrammes).

9 In 1914 the Wakhi settlers of Dafdar (Sarikol) paid grazing dues for the very first time (IOL/P&S/10/826:173). The grazing community was estimated as composed of 40 Sariqoli, 30 Wakhi and 2–3 Kirghiz households at this time (IOL/P&S/10/278).

10 The Wakhi refused to pay any dues, since the Chinese authorities had threatened them with expulsion if they would not oblige the order to discontinue the tax relationship with Hunza (IOL/P&S/10/826: 155: Gilgit Diary March 1915; IOL/P&S/10/826: 143: Gilgit Diary July 1915; IOL/P&S/10/826: 140: Gilgit Diary August 1915). The Wakhi of Kilian paid grazing taxes (in currency) to the Chinese authorities in the same year for the first time. This agreement was negotiated by the British Consul-General Percy Molesworth Sykes (1915: 26).

11 The Wakhi of Dafdar renewed the delivery of grazing taxes to Hunza (IOL/P&S/10/826: 107).

12 Fewer herds were counted in Sarikol, while an increase was registered for Tagarma.

13 Seven households from Mariang (Sarikol) refrained from accessing the pastures in Taghdumbash Pamir.

14 The exchange value of some items in Kashgar in 1928 (IOR/12/50/394):

1 kg *čaras*	5–14Rs	1 horse	65–100 Rs 1 kg
opium	40 Rs	1 donkey	20 Rs
1 *namdā*	2.5–3 Rs	1 sheep/goat	5 Rs
1 carpet	25 Rs		

which had its winter camp in the present-day village lands of Sultanabad (Gujur Das) and spent the summer with its flocks in the Naltar Valley. Nevertheless, the Gujur were out of the taxation grip of the Hunza and Nagar rulers as their power did not extend that far south.

The extra revenue from affluent nomadic communities was bitterly needed for the upkeep of the frugal lifestyle of the Hunza ruler. Mir Silum Khan III succeeded in expanding his territory further north in order to control more Kirghiz pastures. Per forty sheep and per thirty yaks a contribution of one animal had been agreed upon as annual grazing tax.[257]

Relations between the nomads of the Taghdumbash Pamir and the *mīr* of Hunza continued to last for about one and a half centuries. They contributed substantially to his income while he sent his own flocks with Hunza shepherds for summer grazing to the Pamirs as well.[258]

In the aftermath a twofold strategy was applied. The faraway and difficult-to-control Pamir pastures were allocated to nomadic communities such as the Kirghiz and to peasants dwelling in adjacent Wakhan and Sarikol who used those summer pastures in their practice of combined mountain agriculture. Gojal became a settled *maqsòo* within Hunza and featured some fortified villages with Burusho settlers who had migrated from Central Hunza and Wakhi refugees who occupied the single-cropping region in the upper Hunza Valley. As a rule the share of animal husbandry in their combined mixed mountain agriculture was higher than in the lower-lying parts of Hunza, and the livestock taxes (*ilban*) realised in Gojal and in Sarikol were substantial.

As a consequence of this development, Kirghiz nomads insidiously lost their traditional grazing grounds. They were forced to change seasonal itineraries and shift their flocks northwards. Towards the end of the 19th century, British intelligence reports claimed that the competition between outsiders such as Kirghiz and Wakhi and insiders represented by Hunzukuć with their own and their ruler's flocks had led to disputes which were settled by the Chinese representative in Tashkurgan in the *mīr* of Hunza's favour.[259] While in the preceding period the grazing dues were levied in live animals, the production of livestock in Hunza itself seems to have increased substantially. From then on the grazing dues (*khiraǰ*) levied by the *mīr* of Hunza in the Pamirs were paid predominantly in kind such as felts, woollen blankets (*namdā*), ropes from yak hair, cotton cloth (*kirpas*), coarse cloth (*khaám*), coats (*śuqá*), saddles (*ǰhul*), rugs (*śarmá*) and

Hunza rulers levied grazing taxes from Kirghiz, Sariqoli and Wakhi on both sides of Khunjerab Pass – Karajilga (top) in Gojal and Oprang in Sarikol. The photographs were taken on 26 July 1989 and 12 September 1995

The Kirghiz Merghan was portrayed by Aleksandr Yakovlev near the Beyik (Paik) karaul in the Taghdumbash Pamir on 4 September 1931. Source: Alexandre Iacovleff (1934)

Year	revenue in kind: *namdā* (felt, woollen blanket)	ropes	*kirpas* (cotton cloth)	*śuqá* (coat)	Source
1930	61	43	20	–	IOL/P&S/10/973: 9
1931	65	50	25	2	IOL/P&S/12/3285: 370
1932	regular dues				IOL/P&S/12/3285: 329
1933	regular dues				IOL/P&S/12/3285: 329
1934	no dues [15]				IOL/P&S/12/3285: 284
1935	no dues [15]				IOL/P&S/12/3285: 284
1936	no dues [15]				IOL/P&S/12/3285: 284
1937	termination of all tax demands [16]				IOL/P&S/12/3285: 233
1948	attempt to regain pasture rights in Taghdumbash [17]				IOL/P&S/12/2336
since 1985	renewed negotiations about property rights [18]				

15 The riots in Sarikol made the *mīr* of Hunza decide not to send any tax collectors to Taghdumbash Pamir. In addition there were some threats from the direction of the Tajik Pamirs, in the documents 'Tajik Bolshevik' allegedly demanded from the *mīr* of Hunza the evacuation of his herds from the Taghdumbash Pamir (FO 371/19335: 229). The Hunza flocks consisted of three hundred yaks and an equal number of sheep of which thirty sheep were stolen that year (report of Lionel Edward Lang, Resident in Kashmir, on 'Mir of Hunza's rights in the Taghdumbash Pamir and Raskam', dated 21 October 1935; FO 371/19335: 233–238)

16 In 1937 the *mīr* of Hunza gave up all traditional grazing rights for his own herds as well as the right to collect taxes in the pastures of the Taghdumbash Pamir. The average profit for Hunza from the Taghdumbash Pamir revenue and the asset from the grazing of a herd of 300 yaks were estimated in the range of approximately 200–300 Rs annually during the previous three decades (IOL/P&S/12/3292: 319). The British colonial administration compensated the loss of the Pamirian pastures with an increment of subsidies, the provision of barren land in Oshikandas, and offered the *mīr* of Hunza potential grazing rights in Naltar Valley (FO 371/19335: 239–242; IOR/2/1085/296: 23–27).

17 The *mīr* of Hunza took the opportunity to negotiate with the Chinese authorities the option of regaining his traditional grazing rights in the Taghdumbash Pamir. He saw some bargaining potential in respect to its incorporation into Pakistan as the status of Hunza was not clear or defined after the partition of British India. (IOL/P&S/12/2361: Letter from Eric Shipton to Secretary of Government of India, dated Kashgar 5 April 1948). The Chinese Revolution of 1949 and its expansion into Xinjiang terminates these negotiations.

18 Agricultural reforms in China and in Xinjiang entitled the regional authorities to start talks with the heirs of the Hunza ruling family who claimed property in China (especially in Sarikol and Yarkand) prior to the Chinese Revolution. The negotiations led to compensation payments and special allowances for the heirs, especially in the case of Raja Bahadur Khan from Gulmit whose mother originated from Datung in Sarikol.

Opposite
The Pamirian landscape as it was perceived at Beyik (Paik) in the Kara Chukur Valley. Loose material and morainic outcrops cover the arid scree slopes, which resemble desert-like conditions; the Taghdumbash Pamir has its extended fertile pastures in the flat valley bottom. A glaciated range becomes visible only in the background. Source: Oil on canvas painting by Aleksandr Yakovlev, September 1931

socks (*paipakh*). Since the beginning of the 20th century, regular payment of grazing dues has been received in Hunza, amounting in 1931 to sixty-five felts, fifty ropes, twenty-five rolls of cotton cloth and two coats. These goods had an exchange value in the Gilgit and Kashgar bazaars and created a welcome extra source of revenue for the ruler. In addition, the political influence in the border areas and the participation in Central Asian trade as a transit region enhanced the value of authoritative presence in the Pamirs.

Livestock production in the newly developed oases within Upper Hunza had substantially grown during the 19th century. Gojal paid four fifths of all taxes in Hunza although barely one fifth of the population was settled there. In 1894 Gojal delivered three hundred and fifty sheep and goats (*kla*) in addition to fifteen *maund* (one *man* = 37.32 kg) of grain to the tax collectors.[260] This covered basically the regular demand for the Court while additional livestock was available in the personal flocks. Backed by the British authorities in Gilgit, the then ruling *mīr* M. Nazim Khan managed to increase his personal income manifold. Ample evidence is provided by the example of the village of Shimshal, where the best grazing of Hunza is available in the adjacent *pamér*. In the early years of M. Nazim Khan's reign, the livestock dues (*ilban*) amounted to fifty sheep (equivalent of 200 Rs) in 1894; by 1938 the tax collectors had managed to increase the extracted amount to a total of 40 *man* of salt (= 400 Rs), 7 *man* of wheat, 16 *man* of barley, 146 sheep and goats, and two yaks. The value of all goods amounted

The Shimshal Pamir was one of the fertile pasture regions where the mīr of Hunza kept some of his flocks, guarded by local shepherds. This practice ended with the termination of hereditary rule in Hunza. Yak herds are driven into selected summer pastures in the Shimshal Pass area; the photograph was taken on 2 September 2013

to 1,598 Rs, which equals an eightfold increase during his reign.[261] Thus the forty-eight households (about four hundred persons) of Shimshal contributed nearly as many taxes to the Hunza revenue in 1938 as the whole principality of 10,000 inhabitants did in 1894 when a total of 1,800 Rs was collected.[262] The importance of Gojal in providing revenue to the state is obvious when we register that all livestock of three hundred and fifty animals per year in 1891 was extracted from there.[263] The resettlement of Wakhi refugees and Burusho colonisers in Gojal and the subsequent severe taxation practices considerably increased the wealth of the ruling classes. Emphasis on a controlled grazing policy with higher returns from an intensified animal husbandry was reflected in the stocking density all over Hunza. On a reconnaissance tour Colonel Reginald Schomberg visited Gojal and acknowledged the overall importance of animal husbandry in combined mountain agriculture. He regarded the Wakhi pastoral practices within their combined mountain agriculture as equivalent to those of nomads or recently settled nomads.[264]

The exploitative taxation at the climax of *mīr* M. Nazim Khan's reign coincides with the loss of all rights in the Taghdumbash Pamir, such as the levying of grazing taxes as well as sending flocks there from Hunza. The loss was estimated to be equivalent to a meagre 200-300 Rs annually, and the *mīr* was generously compensated by the British authorities. Whereas the Pamir revenue had been stagnating for about three decades, the livestock taxes within Hunza had grown substantially as had the use of grazing grounds.

From a legal aspect the *mīr* of Hunza regarded himself as the sole proprietor of all natural resources including pastures as a document from 1935 explains: "All forests, mountains and pasture lands in Hunza belong to the Mir and have been granted by him to the different communities who take their flocks for grazing to their respective grazing grounds. The Mir is entitled to graze his flocks in any pasture he wishes. The Mir's wood cutters are at liberty to cut wood from every forest. If the Mir betakes himself on a pleasure trip to any of the above pastures, every shepherd should present him with a sheep and one roghan [*rúy̌un*]."[265] The hereditary ruler demonstrated some examples of this when he expropriated the grazing rights at Ultar for his own herds and when he compensated himself for the loss of grazing rights in the Chinese-controlled Pamir by shifting his personal herds to grazing grounds in bordering Kilik, Mintaka and Khunjerab that had been previously used by Hunzukuć peasants.[266] The *tham* himself owned huge flocks that were kept with shepherds in the Kilik Pass region and on the Taghdumbash Pamir; in the summer of 1936 his livestock herds were estimated at four thousand head.[267] In the end he reserved and exclusively maintained the Kilik pastures and the Dih Valley for his flocks while Mintaka could be used by farmers from Misgar, Kharachanai by Ghulkin and the upper Khunjerab Valley by Abgerchi. Giving up pastures meant a significant loss of income in the ruler's annual budget. Pastures have always been perceived as valuable assets that need to be carefully controlled and adequately maintained.

Importance and quantitative decline of animal husbandry in Hunza

In the same year that revenue from pastoral practices had reached its climax, the local chronicler Qudratullah Beg compiled a survey of all pastures in Central Hunza and Shinaki. Augmented by other contemporary sources[268], his work provides us with a reference which forms the basis of comparison with the present situation. Although the number of households doubled in Central Hunza and Shinaki, and quadrupled in Gojal from 1931 until the end of the 20th century, the number of persons involved in animal husbandry had steadily and substantially decreased. In several cases no shepherds at all are recorded today; the workforce has been reduced to nominal representation in other summer settlements. High pasturing as part of combined mountain agriculture has undergone a significant transformation within the last century. In Central Hunza in 1935 every fifth to tenth household usually sent a shepherd to the high pastures, and in Gojal more than three quarters of all households participated in the seasonal migration (*kuč*).[269]

Every evening sheep and goats are brought back to the pasture settlement of Shuwert (4,600 m) for milking and protection during the night. The photograph was taken on 31 August 2013

Half a century later, our initial enquiries revealed a completely different picture; little more than one per cent of the households provided shepherds in Shinaki and Central Hunza. Even in Gojal, the number decreased considerably with the exception of Shimshal where the proportion of households following the difficult tracks to the remote pastures resembled the pattern from fifty years ago.[270] Taking Shimshal as an example, the delayed but significant change in pastoral practices can be estimated. During our enquiries in 1985 the pattern seemed quite stable for more than half a century; but by 2013 the number of shepherdesses who spent the summer in the Shimshal Pass area had halved. During our subsequent visits in the following years, a severe shortage of workforce became more and more evident. The expert shepherdesses had matured but failed to find expert younger successors. During four visits in the Shimshal Pamir up to 2017, the number had varied and declined to fewer than fifteen shepherdesses (*xunedor*) who were expected

The Burusho technique of churning is based on turning sour milk (dumánum mamú) which is filled into an inflated sheepskin for prolonged periods on the shepherd's knees until the butter (maská) separates from the buttermilk (díltar). The photograph was taken on 28 June 1985

On special occasions a piece of butter (maltáṣ) is dug out from an underground store where the butter wrapped in birch bark is sometimes kept for years. The hiding place is kept as a family secret. The photograph was taken on 13 April 1985

Transformation of pasture utilisation in Hunza in the 20th century

Location	Right of pasture	Name of pasture	Number of shepherds 1935	1985	2018	Type of animal 1935	2018	Cultivation 1935	2018
SHINAKI									
Shinaki	Hussainabad, Maiun, Khanabad	Baiyes, above Huruman Har	12		12	H/P	H/O	xx	–
Shinaki	Hussainabad, Maiun, Khanabad	Maiun bar	10	3–4	5	H/O/P	H/O	xx	–
Shinaki	Hussainabad, Maiun, Khanabad	Rui bar	10		4	H/O	H		–
Hindi	Gasuratiṅ	Deinger haráay	3		3	H	H/O	–	–
Hindi	Gasuratiṅ	Chashi haráay	2		4	H	H/O	–	–
Hindi	Taltotiṅ	Proni	4		3	H	H	–	–
Hindi	Photkuċ	Pfulgui haráay	3		3	H	H	–	–
Hindi	Gasuratiṅ	Hundri	5		4	H	H	–	–
Hindi	Machotiṅ	Ghumu ter	4		3	H	H	–	–
CENTRAL HUNZA									
Murtazabad	Murtazabad Kha Khan	Bate khur	3		2	H	O/H	–	–
Hassanabad	Hassanabad	Hachindar	5		2	H	O/H	–	–
Ganesh giram	Ganesh giram	Hachindar	10		2	H	O/H	–	–
Burúoṅ	Burúoṅ	Muchu har	40–70	4	2	H/O	O/H	xx	–
Dirámitiṅ	Dirámitiṅ	Shishpar	40–70	4	3	H/O	Y/O/H	xx	–
Aliabad	Datusingkuċ haráay	Ghowanċ	3			H			
Aliabad	Datusingkuċ Dersam	Chayukuċe	2			H			
Aliabad	Datusingkuċ	Yare haráay	6			H			
Dorkhan	Chaboikuċ	Ghumat	2–3	1	2	H	O/H	xx	–
Haiderabad	Buroon: Mayurukuċ, Hajikuċ	Yate thali	4			H			
Ganesh kalan	Ganeshkuċ, Nagarkuċ, Wakhi	Ganzupar	15–20	–	2	H/O	O/H	–	–
Baltit	Thamo, Baltikuċ	Ultar	5–12	3–4	2	H/O/P	O	xx	–
Baltit	Baltikuċ	Bulen	–	–	2	O	O	–	–
Baltit	Burúoṅ	Bululo	6	–	2	H	O	xx	–
Baltit	Dirámitiṅ	Hon	4	–	2	H	O	–	–
Baltit	Wazirkuċ	Sekai	4	–	2	O/P	O	–	–

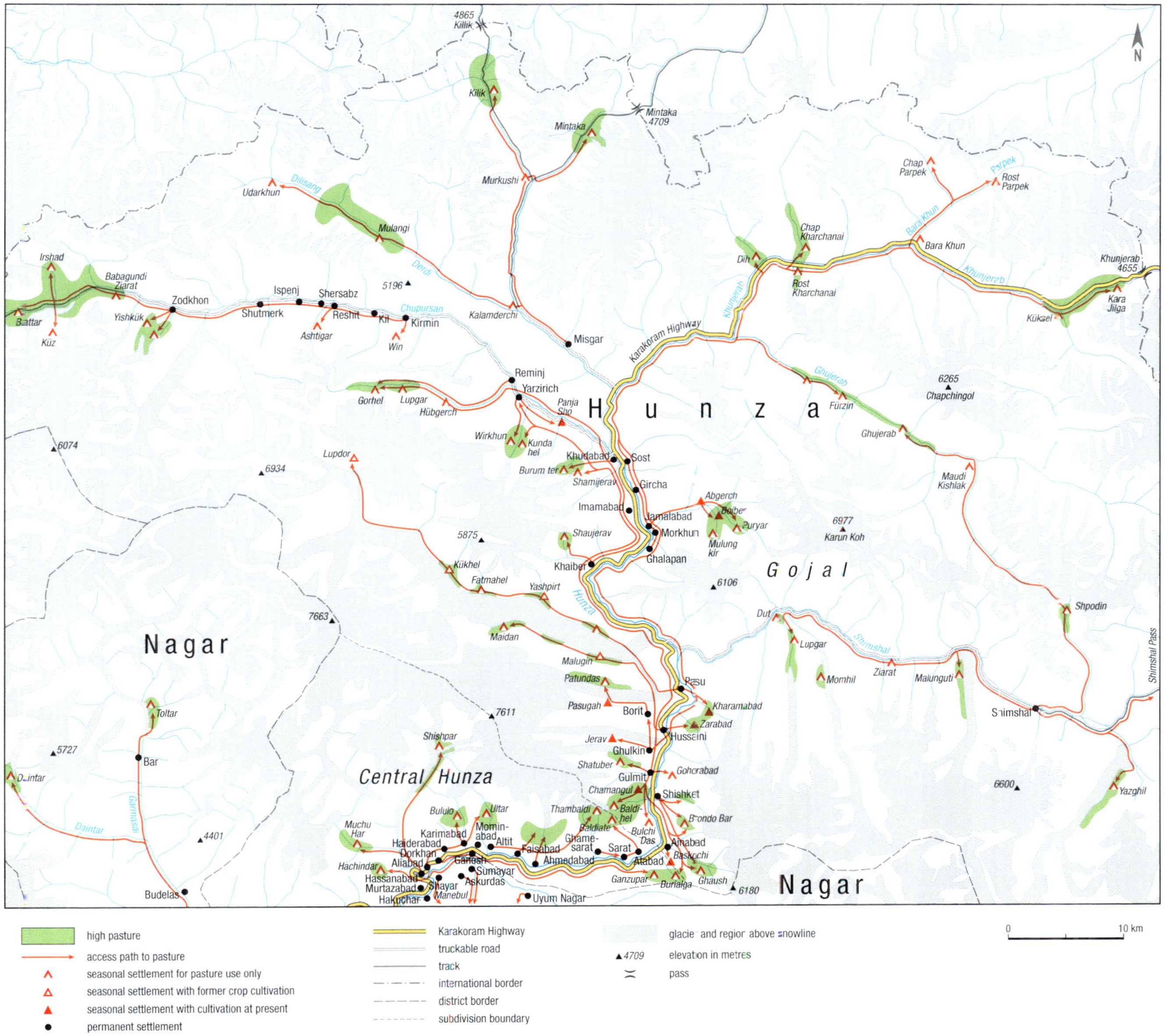

Safida Begum is emptying the wooden butter churner (soyo) after the fresh butter (maská) was removed and the butter milk is put into a cooking pot for the dehydration process in order to produce qurút that keeps well. The photograph was taken on 31 August 2013

Some villages have to cover short distances of a few kilometres from the irrigated oases (íći) to reach their summer pastures (ǰut in Shina, ruṅ and ter in Burushaski, hel, jerav and pamér in Wakhi) in higher elevations between 3,200 and 4,600 m). They cover long distances and have to cross other village lands depending on their access rights. Pasture disputes are common between neighbouring communities although the economic importance of animal husbandry has significantly dropped since more than one generation. Source: Modified and updated from Hermann Kreutzmann (2000b: 93; 2004a: 58)

to fulfil all duties during the summer peak season. The remaining women could not adequately attend to their sheep and goats (*kla*) and yak (*ʒuý*) herds. As a remedial measure the Shimshal Falai Committee concerned with local welfare had introduced a turn (*sur*) system which allocated shifts of shepherding to each Shimshali household that had animals in Shimshal Pamir. The length of each *sur* was according to the size of the household's herds. Thus, supporting hands of men and women from the village were taking shifts. These duty systems work in theory, but there are ample reasons why a person would not arrive in time when the homestead was two days journey away on foot. Despite the introduction of local livestock taxes to pay for shepherding and in spite of having the *sur* system, no appropriate shepherding could be ensured, and the animal losses due to predators and other menaces significantly increased. The process of decline and loss may con-

The buttermilk is boiled down in a process lasting for long hours in Boiber; the final product will be dried on the rooftops or special elevated platforms until the qurút is sufficiently dehydrated and storable. The photographs were taken on 7 July 1990

Location	Right of pasture	Name of pasture	Number of shepherds 1935	1985	2018	Type of animal 1935	2018	Cultivation 1935	2018
Botkuč	Botkuč	Suchash	2	–	2	H	O	–	–
Baltit	Ḍom and Baltikuč	Beričo čok Altitkuče sat	8	2	0	H	–	–	–
Altit	Mamukuč	Móiṅtas	4	2	2	H	O/H	xx	–
Altit	Mamukuč	Talmushi	4		2	H	O/H	–	–
Altit	Altikuč	Khuwate	4		2	H	O/H	–	–
Altit	Altikuč	Tiyash	8			H	–	xx	–
Altit	Bakuč	Ghundoing	2			H	–	–	–
Ahmedabad	Sukuč	Gurpi	–	3–4	2	H	O/H	xx	–
Faisabad	Faisabad	Churd		1		H	–	–	–
Atabad	Gulmitik, Geregatkuč	Baldiate	10	4	2	H	O/H	–	–
GOJAL									
Shishket	Shishket	Gaush, Brondo Bar, Baltin Bar			2	H	O/H	–	–
Gulmit	Gulmit	Baldi hel	8	2	0	H/O	–	–	–
Gulmit	Gulmit	Gaush, Buri alga	2	1	0	H	–	xx	–
Gulmit	Gulmit	Brondo Bar			0	H	–	–	–
Gulmit	Gulmit	Shatuber	–	–	0	O	–	–	–
Gulmit	Gulmit	Shamijerav	12	6	3	H	H/O/Y	–	–
Ghulkin	Ghulkin	Patundas			2		O	–	–
Ghulkin	Ghulkin	Karachanai			3		H	–	–
Hussaini	Hussaini	Batura (South)	21	6–8	2	H/P	O/H	–	–
Pasu	Pasu	Batura (North)	22	13	2	H/Y/P	O/H	xx	–
Shimshal	Shimshal	Shuijerav/ Shuwert	42	80	9	H/Y	Y/H	–	–
Shimshal	Shimshal	Ghujerab	12	6	44	H/Y	Y/H	–	–
Khudabad	Khudabad	Burum ter	6	4	3	H	O/H/Y	–	–
Abgerchi	Abgerchi	Boiber, Puryar, Mulung Kir	18	10	3	H/Y/P	O/H/D	xx	xx
Abgerchi	Abgerchi	Karajilga, Kükhel (Khunjerab)	?	–	2	H/Y	O/H/Y	–	–

Notes and explanations: H = sheep and goats (bur. *huyes*, wakh. *kla*); P = horses (bur. *yabú*, wakh. *yaš*); O = oxen (bur. *har*, wakh. *čat*); D = donkeys (bur. *gókomuč*, wakh. *xur*); Y = yaks (bur. *bépay*, wakh. *zuÿ*) xx = cultivation of grain crops in pasture area; tt = irrigated grasslands (bur. *toq*)

tinue as long as agricultural practices do not generate competitive incomes in comparison with salaried jobs.

This process supports the general impression that the use of seasonal settlements for cultivation and animal husbandry has been greatly reduced, in some cases towards insignificance. Crop farming as a side activity of shepherds in the summer settlements has been given up almost completely. In a few rare cases the cultivated terraces are used for fodder production (grass and alfalfa). The animal mix has remained similar; where pastoralism is alive, only horses are no longer kept. The size of flocks must have diminished considerably although comparative data are difficult to come by. Especially in Central Hunza one or two shepherds are nowadays sufficient to control the herds of their respective communities. The four clans of Karimabad with a population of more than 5,000 plus inhabitants accounted for livestock composed of 2,224 sheep and goats, and 970 cattle in 1992.[271] The average household called fewer than four sheep and goats and fewer than two cows their own. In comparison the mean livestock property in Pasu accounted for twenty-three sheep and goats, and more than seven head of cattle (including three yaks) per household in 1998. Both villages are far from representative for the overall situation. There is a growing tendency to extend cattle husbandry in the permanent settlements all year round. The cow in the homestead meets the household's need for milk and its processed products. In the meantime the decline continues, and a resource that accounts for about half to two thirds of cash income from local agricultural and livestock products fails to be developed as the bottleneck situation of insufficient workforce availability is getting more and more severe.

The decline of pastoralism is a strong indicator for a societal shift from place-based resource utilisation to a diversification of non-local occupations generating substantial shares of household incomes. The shortage of workforce is more strongly felt in shepherding than in other agricultural practices. The decline of animal husbandry is linked to a general shift in economic activities. The contributions from education and migration-based off-farm resources have increased significantly. Jobs in military and civil services or in trade and tourism require employees' availability and presence throughout the year. Education and salaried jobs are not compatible with shepherding. The returns from non-agrarian occupations are higher in general. This holds true especially for out-migrants to Gilgit, Islamabad, Karachi and/or overseas. Even if there are periods with less work load, these time frames do not coincide with the heavy burden in agricultural activities during the summer season. It becomes more difficult for individual households to spare a member as shepherd; even for lineages it is sometimes a problem to find a single person to uphold the pasture claims in the inherited lands.

Every morning and evening the sheep and goats are milked in Shishpar agél; their meagre milk yield is processed into butter. The photograph was taken on 15 June 1985

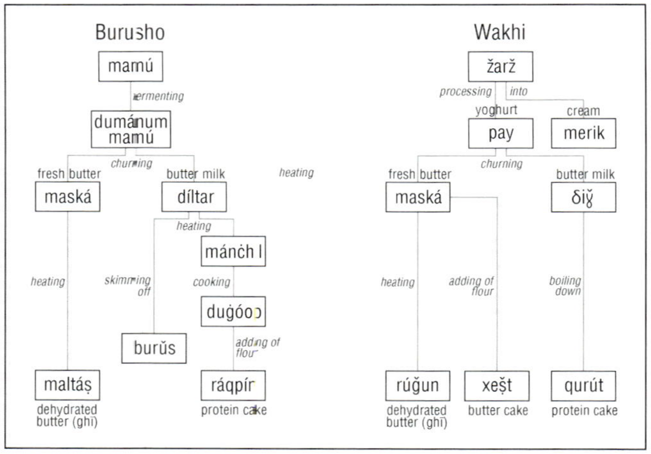

The steps of processing fresh milk into non-perishable products involves various processes that are quite different in Burusho and Wakhi households. Bigger quantities of milk have been processed in most pastures of Gojal, thus resulting in an emphasis on butter and qurút. Burusho shepherds prefer a number of fresh products such as maská and burús, only few long-life butter derivates are produced. Source: Modified after Hermann Kreutzmann (2000b: 99)

*Fresh butter (maská) is the most valuable commodity produced in the high pastures; in its clarified state (rúǧun) it keeps well and is transported to the village homes of the shepherdesses from Shimshal. The photograph was taken in Shuwert on 31 August 201_*

Different strategies are followed to solve the workforce shortage. As off-farm jobs have been predominantly taken up by male household members of a certain age group, the agricultural burden was distributed among the remaining household members: women, elderly men and children. The latter age group is nowadays rarely available for agricultural activities in Hunza as almost all children are registered in schools and attend classes. Many of them become young migrants when they continue their education outside their villages. Boys and young men are seen on the high pastures when they accompany trekking groups who very often follow traditional migration paths of livestock or when they are visiting relatives during vacation after a long stay outside of Hunza. Basically the task of shepherding lies presently in the hands of elderly men and women where community rules permit them to go to the pasture settlements. Traditionally the access of Burusho women was restricted, while it is quite common among the Wakhi. This difference is not revealed by the layout of the combined Wakhi-Burusho pasture settlement in Shamijerav ilga-Burum ter. The dual settlement harbours at the same time the two different milk-processing techniques applied in the Hunza Valley. Wherever livestock is kept in high pastures, durable and highly appreciated consumer goods are produced. The best part of Wakhi animal husbandry in Gojal is controlled and practised by female household members. Where men are involved the age structure of shepherds has changed. Elderly men have to take over duties that were traditionally reserved for the sons of a household for whom it was a privilege to spend the hot summer season in the *Sommerfrische* of the pastures. As the workload is increasing for the permanent household members remaining in the village, a tribute has to be paid to agricultural activities.

The flocks have decreased nearly everywhere. No household is presently able to look after sizeable herds all year long. The extra value from animal husbandry is easily compensated if non-agrarian incomes are available. In the village of Gulmit, 12 % of all 217 households remained without non-agrarian income in 1990; nearly 60 % had more than one additional source. The twenty-six households without non-agrarian resources were mainly small households managed by widows or by elderly people without able-bodied members who would otherwise be engaged in labour, services or other occupations. Private enterprise, public services and educational migration plus labour accounted for the dominant occupations. Dependence on animal manure and products has been reduced since mineral fertiliser has been distributed in Hunza. Fresh meat, poultry, milk powder and cooking oil are regularly on offer in the bazaars, even butter fat (*ghī*) and *qurút* (dehydrated butter milk) are imported from other valleys. The strong relationship on the agrarian subsistence side in the household economy has been weakened by the degree of increased market participation. Furthermore, the pasture rotation system has been simplified owing to the lack of sufficient and qualified personnel. Difficult glacier crossings and passages along narrow and steep tracks require the guidance of experienced shepherds. Easily accessible pastures are used more frequently and for extended periods. This phenomenon exhausts the natural pasture resources in certain areas while giving up additional available pasturage in remote areas. Are we experiencing the final stage of combined mountain agriculture and the end of pastoralism? Does the commercial value of pastures lie only in their fodder, timber and firewood? One aspect needs to be kept in mind. Since the Hunza state and its hereditary rule were abolished in 1974, the proprietary rights of village lands including pastures have been taken over by lineages and village communities. In a region without cadastral surveys and registered property rights, the transfer from ubiquitous *mīri* rule and absolute control over resources to a democratic society with personal landholdings and communal property resembles aspects of agrarian reform. Such a transformation would not necessarily be accomplished undisputedly. The separation of the Hunza State and personal property of the *tham* and his relatives from that of the farmers resulted in different perceptions and prolonged negotiations. Immediately valuable resources such as physical infrastructure, communal meeting places, irrigated lands and cultivable waste were affected, whereas pastures and forests fared less noticeably in the beginning. This perception has changed during the last two decades.

High pastures and their part in the commons

The scenario described above hints at a continuous process of lessening importance for animal husbandry and a dramatically changing socio-economic environment. One would expect that these developments would affect all walks of life. A probable scenario could be the limitation of economic activities in the valley bottom and the permanent settlements, and a fading importance of remote glacier and pasture regions that are only accessible by long moves on foot. That should have severe consequences for agricultural practices. In such a scenario the pasture settlements should look barren and dilapidated; a formerly important section of the commons would be abandoned. This statement holds true for some places such as Shispar along the left bank of the Hassanabad glacier in Central Hunza. While in 1935 forty to seventy shepherds spent the summers in the three habitations surrounding the animal pen (*ǰel*), their number had dropped to four in 1985; by the end of the century only two shepherds (*huéltarć*) from Aliabad inhabited the one remaining building while the other two are in ruins. Nowadays hardly anyone can be convinced to fulfil these duties, and helpers are coming up for support on a shift basis. More people are engaged in driving the flocks up or down the valley. On the right bank of the Hassanbad glacier the pastures of Muchu Har are not used at all. In Tochi the former pasture settlement was converted into an orchard (apple, pine trees and willow), and animals are not permitted to trespass.

In other places such as Shamijerav (Burum Ter) above Khudabad, the Wakhi from Gulmit have sustained an intact pasture settlement which they share with Burusho from Khudabad who have their own corral and housing arrangement. Both groups use their pastures intensively. In 1935 they sent twelve and six shepherds, respectively; in the 1990s the numbers were still six and two. Both communities have a strong interest to document their pasture claims here. Similar developments can be observed in Shimshal where the whole community is actively involved in the summer *kuč* to the *pamér*. Here livestock numbers are still high, and since more than one decade fresh yak stock has been imported from the neighbouring Chinese Tashkurgan County.[272] Yaks always numbered around one thousand head and accounted for about four fifths of all livestock sales until today.[273]

The pastures north and south of the Batura glacier have been used by farmers from Pasu and Hussaini for a long time. In 1998 the 87 households of Pasu sent a herd of 356 cattle, 282 yaks, 1547 goats and 468 sheep to their Batura pastures, while their neighbours from Hussaini sent only

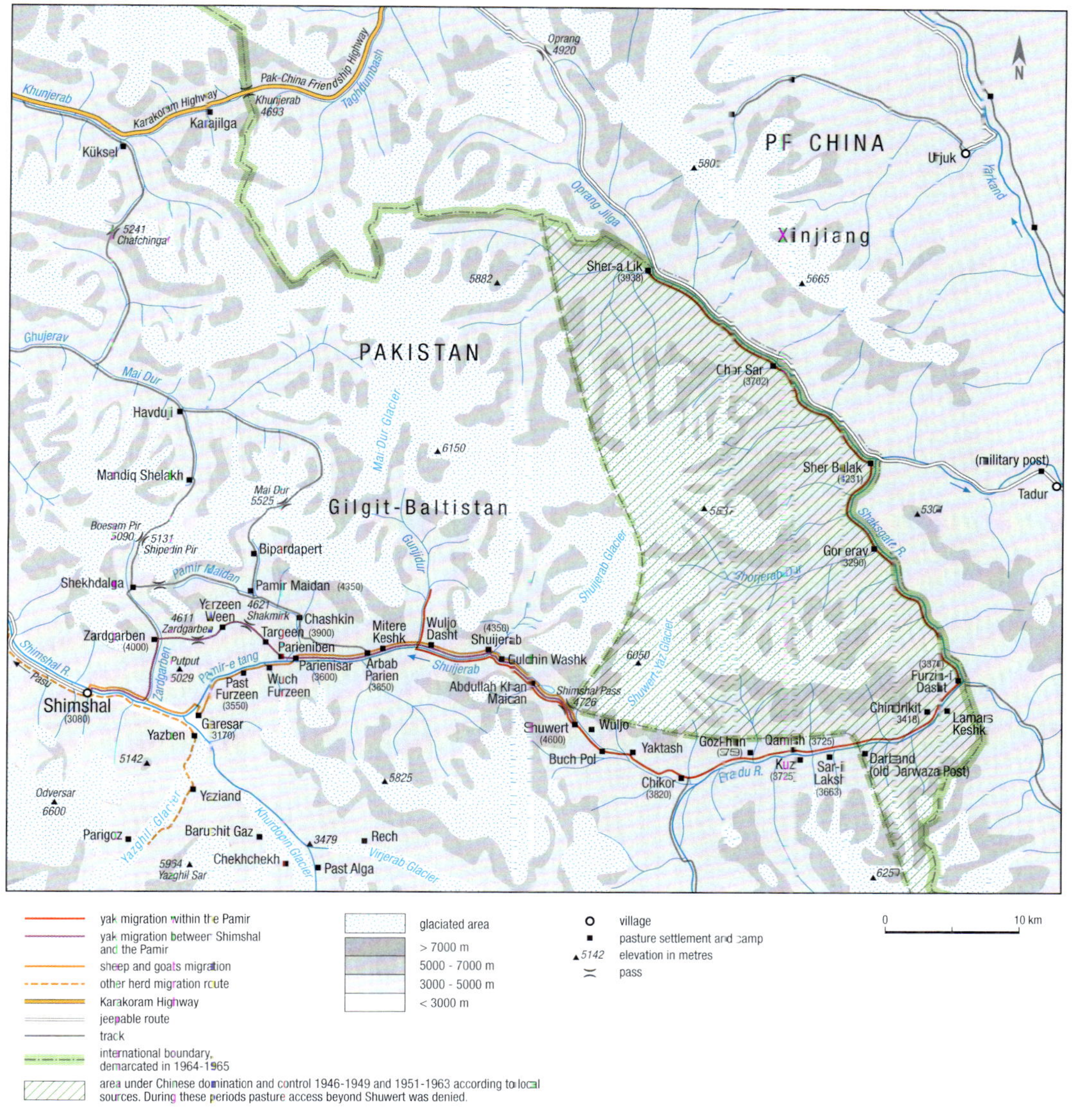

Shimshal was affected by an unclear border delineation between China and Pakistan; as long as the international boundary was not drawn in a mutually acceptable way, a number of territorial disputes occurred. Only after the Pak-China Boundary agreement of 1963 was the shaded area returned to Shimshal, guaranteeing people's access to their inherited pastures. Source: The representation is based on Hermann Kreutzmann (2015a: 481).

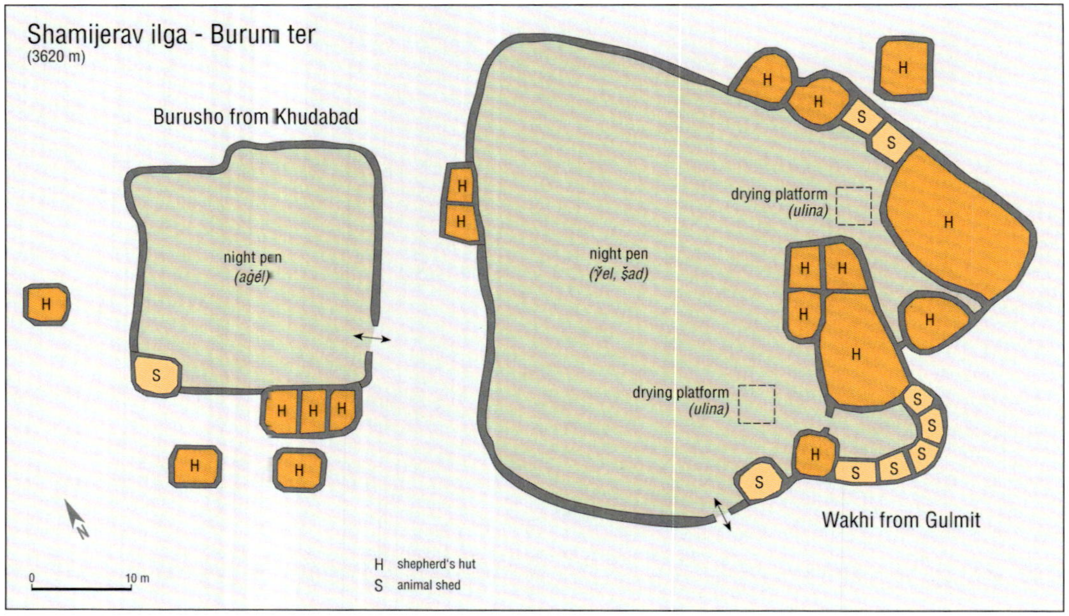

The combined pasture settlement of Shamijerav ilga (Wakh. white river valley pasture) and Burum ter (Bur. white pasture) is a rare feature where Wakhi women spend the summer season with the livestock from far-away Gulmit, and where Burusho men from nearby Khudabad station their sheep and goats (Wakh. kla; Bur. huyés). Both hamlets (3,620m) are economically separated and contain corrals for the livestock, animal sheds and huts for the shepherds who belong to different households and care for the livestock of several households of neighbours and relatives. Design drafted August 27, 1990. Source: Modified from Hermann Kreutzmann (2000b: 98)

47 cattle, 677 goats and 214 sheep.[274] The significant change since the previous survey in 1985 occurs in the increase of yak numbers; all other herds remained quite stable although the number of households had increased as well.[275] Here we find a pattern where a village community depending to a higher degree on off-farm employment than most other villages sustains a system of pasture use through the help of female household members. The productivity of herds has been increased by the introduction of Pamirian yaks.

Labour deficiency has increasingly become a problem for all communities. At the same time attention has shifted again towards the summer pastures. Wherever these pastures are located near the access routes to high mountains or along trekking paths, the communities who share the right of pasture claim the right of guiding and portering in these areas as well. They feel entitled to negotiate the terms for trespassing on their common property. Villages which are affected by mountain tourism such as Pasu and Shimshal have developed a rotation system within their community so all households can participate in this source of income. Shimshal has founded an organisation that functions as the 'Shimshal Nature Trust' and is responsible for abiding by the rules. The combination of pasture use, trekking and mountaineering tourism is an option to bring household incomes at par with salaries generated by other professionals. In a few days a household can earn more cash income than a full summer of shepherding would bring. Thus mountain tourism has led to a new valuation of high pastures.

A second case is provided by the people of Abgerch, i.e., the Wakhi settlements in the Hunza Valley above Khaiber. Their traditional right to two pastures in tributaries of the Khunjerab valley – Kükhel and Karajilga – was revoked when the Khunjerab National Park was inaugurated in 1975. The mountain farmers were promised compensation which was not paid until 1990 when a dispute between the Abgerch people and the Government of Pakistan broke out. Negotiations finally led to a settlement which included preferential provision of jobs to Abgerch people in the national park, control of traffic and hunters, as well as certain access rights to pastures.[276] These mountain farmers faced a dilemma. On the one hand, global interest in the protection of Marco Polo sheep affected their pasture resources; on the other hand, the protection of extremely old juniper trees in the Boiber valley – their second pasture resource – was supported by the International Union of Conservation. The villagers have been compensated for not cutting trees there anymore and for restricted pasture use. For these five villages – Ghalapan, Gircha (Sarteez), Morkhun, Jamalabad, Sost – the access rights to common pastures have become a negotiable quantity in their relations with the regional and federal administration as well as with international organisations. A management plan for the Khunjerab National Park was outlined in 1996, and with the withdrawal of the court case in 2000 its implementation process began. The mountain farmers are permitted to graze about one hundred yaks in the park core zone – compared to 400-500 in the 1990s –, about one hundred sheep and goats in Kukshel, and about eight hundred in Furzindur and Arbob Kuk. In addition, the Khunjerab Village Development Organisation represents the interests of the approximately three hundred farming households. The Khunjerab Village Development Organisation controls the access to certain areas and is engaged in the management of ibex trophy hunting. Thus jobs have been provided, park entrance fees and hunting fees collected, invested for projects of mutual interest and/or distributed among the entitled households.[277] The presence of the Abgerchi at their check post in Belli on the Karakoram Highway has changed the attitude of government institutions and provided a visible symbol of people's participation.

Disputes about the commons

The common property of the upper sections of watersheds tributary (*bar, har, nālā* or *ǰəráv*) to the Hunza River represents one of the last communal resources that have gained importance recently. Never before have more village funds been spent in legal disputes before religious and civil courts. The village of Gulmit is most severely affected of all and presented here as a virulent case. Gulmit's pastures are distributed comparatively far away from the permanent settlement rather than being located just above the homestead. During the 1990s different disputes arose with neighbours about the hereditary rights of pasture use that have not ended yet. In 1990 a severe dispute began with Shishket across the Hunza River. The Bori *qətór* lineage of Gulmit was to be deprived of its right to access Gaush and Bori Alga/Buri Harag. The Ruzdor *qətór* made similar experiences in Bulbulkeshk and Brondo Bar. No solution could be reached for these and other plots close to the Karakoram Highway through mediation by local institutions and negotiations by mutually accepted and respected neutral persons, although kinship and marriage relationships exist between the inhabitants of neighbouring Gulmit and Shishket. The respected religious scholar Alama Nasir-ud-din Hunzai was called to Gulmit and Shishket to solve the dispute in 1990, but failed to arbitrate, like every other authority from the former rulers, Ismaili councils or state authorities afterwards. The whole conflict escalated and became a matter of principle of defending property rights that have not been laid down in written documents. Representatives of public and religious institutions were consulted in vain before the legal procedures took off. During the 1990s more than half a million rupees were spent on lawyers and court fees by Gulmit alone. Similar or even higher contributions were invested by the opponents, not counting all travel expenses and secret meetings of representatives. No solution is in sight besides stay orders issued by the courts permitting both sides to use

Burum ter or Shamijerav. The dual pasture settlement above Khudabad is called Burum ter by Burusho who come from here; Wakhi from Gulmit are entitled to utilise the pasture resources here as well and call the settlement Shamijerav. Both toponyms are translated as white pasture or white valley. Burusho and Wakhi have established their respective living abodes and livestock pens close to each other and operate separately. The photograph was taken on 27 August 1990

The pastures of Gaush are among the disputed areas between Gulmit and Shishket. A compromise was found that the pastures and the livestock pens are divided. Gulmitik shepherds use the left one, while the right one is occupied by Shishket people who can look down towards their village in the background. The photograph was taken on 7 July 1991

Opposite
The dispute between Gulmit and Shishket about respective pasture rights escalated in April 1990 when men from both villages marched on the Karakoram Highway towards Shishket Bridge to push their rights. The stand-off resulted in legal disputes that lasted for twenty-five years and involved high investments in legal and court procedures. The photograph was taken on 29 April 1990

the pastures for a quarter century. The funds spent exceed by far the commercial value from animal husbandry in these pastures for a decade. Another dispute between these two villages occurred about the waterless scrub area of Bulchi Das. All opponents mostly explained the driving force for allocating so much energy on a land dispute as a matter of pride and sometimes interpreted the case as an ethnic dispute. But at the same time there is the hope that potential mineral wealth will be found in barren lands and on pastures and the need to develop irrigated land for future generations, and, of course, those areas are the only land resources left. If a community loses them, the pressure on land is even higher. After twenty-five years of quarrelling Gulmit and Shishket found a reconciliation in 2015; both communities declared Gaush a mutual grazing ground that can be commonly used, corresponding to the customary rights. In the past women had been the shepherdesses in Gaush; since the dispute began, each community had sent one shepherd (*špïn*) each living in the only two separate houses to document their claims. In former times about forty fields (*wundr*) were cultivated in Gaush as land and manure were available; a threshing-place is as abandoned as all field parcels.

In the remaining cases both communities agreed to divide the disputed spaces and separate interests; Brondo Bar and Baltbar would be allocated to Shishket, while the sole ownership of Bulchi Das and Bulbulkeshk was endorsed for Gulmit. So-called 'stay-orders' or temporary injunctions by the respective courts that had kept the disputes on hold and alive were taken back. Here is a case where a dispute between two villages that could not be solved within the communities made its way through Ismaili institutions to Pakistan's civil courts where it was retrieved and resolved by the adversaries.

A similar conflict followed soon about Baldi hel or Baldiate in May 1990, in this case the dispute is between Gulmit and Altit. Again huge funds have been spent for a dispute on a pasture which most of the opponents had never visited. The tradition of one side claims that the pasture was divided in three sections: the western one belonging to Altit, the central one to the *mīr* of Hunza and the eastern one to the Gulmitik. The other side tries to make it a case in point that already during the reign of Shah Ghazanfar (1824-1865) or *mīr* Ghazan Khan (1866-1886) the immigrating settlers from Altit and Baltit were allocated Baldi hel as their pastures. Obviously no eyewitness could be presented for this viewpoint. The abolishment of hereditary rule in 1974 created misperceptions and controversial interpretations of customary rights. The hardship in this enduring case is that one party tried to totally ban the other from access, although the latter had practised animal husbandry in Baldi hel until recently. In such cases knowledgeable and respected village elders were consulted first; in the second instance religious representatives gave their advice before public judicial institutions were addressed. All involved have failed to find a solution. Then the ultimate instance was approached by selecting honourable men from both sides who were requested to take an oath by the Holy Quran and then decide about the property rights. The positions of both parties have not changed as they played off against each other the verdicts from different authorities; in April 2013 the Trial Court in Gilgit gave a verdict, which was challenged by Gulmit six months later because they adhered to a judgement by the Ismaili Arbitration and Reconciliation Board in Hunza. The present situation is that a civil court hearing is again pending which was scheduled for October 2018, but was postponed again. The dispute seems to be a never-ending story of a fight between Burusho and Wakhi.

The third case involved the Gulmitik in a dispute with the farmers of Khudabad about Shamijerav ilga/Burum ter, the high pasture above Khudabad. The customary priority was given to the Gulmitik while the Khudabadkuč were tolerated in the same grazing grounds. The latter attempted to reach a status quo of equal rights in the 1990s, aiming at finally dividing the pastures into two shares. Again huge funds were invested for the legal procedures with only little hope for a mutually accepted solution. In 2002 an agreement was mediated through communal arbitration outside of the civil court that both communities are equally entitled to use the pastures.

These cases have been presented as examples for the need of a settlement in those areas without cadastral surveys. On the other hand, it is quite clear that any settlement could

The flocks are released for the day pasture after the sheep and goats have been milked in the morning in Mulung Kir (3,870 m). The photograph was taken on 8 July 1990

generate more disputes and could become a painful affair for the parties concerned. Currently pastures lead the list of importance in land disputes. Nevertheless, stretches of barren land along roads are coming under dispute in growing numbers. To sum up only briefly the areas of pasture disputes that began in the 1980s and 1990s besides the ones already mentioned: Murtazabad argued with Hassanabad about Hachindar; Ganesh and Shishket/Gulmit about the borders between Gaush and Ganzupar; Pasu and Hussaini about the limits of the adjacent newly colonised areas of Kharamabad and Zarabad in 1980; Hussaini and Ghulkin propose different opinions about the environs of Borit Lake and the lands of the seasonal settlements in Borit from 1978 to 1981; Ghalapan and Kaiber disputed about Dildung kor in 1985; Gulmit and Khudabad about barren lands in Khudabad Sar and Mamla Das. The latter has been resolved by allocating Khudabad Sar to the village and by acknowledging Gulmit's claim to Mamla Das.[278] Kil and Kirmin claimed both stretches of barren land between the two villages in 1978 and 1991; Misgar and Sost tried to claim Belli after it became commercially upvalued owing to its location on the Karakoram Highway in 1990 to 1991. After the construction of a tractor road, Pasu and Shimshal expressed different opinions about the barren stretches of land between Tupopdan and Dut in 1991. The escalation during this period motivated the Gojal Ismaili Students Organisation to start an agitation for communal understanding and Wakhi solidarity in July 1990 when they posted pamphlets in all villages to end the disputes.

Further disputes are expected as infrastructure development along the Karakoram Highway has made formerly less valuable barren tracts now attractive for tourism camping grounds on the shores of Atabad Lake. From the submerged village of Goshben only the higher-lying pasture area of Gaush has survived. From the intensively used permanent settlements where land disputes between the former hereditary ruler and local farmers were dominant in the 1970s and 1980s, the controversies have shifted to the extensively used parts of village lands: pastures and barren tracts. The funds spent on these disputes by far exceed their commercial value at the time of the quarrel and have resulted in huge economic losses for many years. Neither side disclosed any sums that have been allocated for arbitration, lawyers' fees and court judgements. But time has proved the disputers right as nothing has gained higher value than land. Especially after the latest improvement scheme on the Karakoram Highway and its declaration as part of the China-Pakistan Economic Corridor, the situation has changed. Internal disputes between social groups and villages characterised the above-mentioned conflicts; nowadays an external threat is looming. Investors and property-dealers have realised the value of land close to the Karakoram Highway.

Recent changes and future prospects for high mountain pastoralism

The perspectives elaborated on above do not complete the picture of the present state of pastoralism and the estimation of natural grazing grounds. Some observations may hint at a development which would lead to a further decline in agriculture and in pastoralism in general. But at the same time agriculture is very much alive as an important economic resource. In times when the national economy is undergoing difficult crises, when unemployment of skilled people is growing, and when tourism is dependent on global events and affected by market shifts, we could observe that the traditional agricultural resource base of the Hunza Valley has experienced its transformation as well and has adjusted at higher productivity levels when possible. New and valuable cash crops such as potatoes, seeds, fodder crops, cherries and other fruits have been introduced and are becoming profitable. Crop farming no longer depends on the manure supply from animal husbandry, but households prefer to own some livestock for their domestic demands and these animals have to be cared for. In the commercial centre of Hunza, in Karimabad, changes have not only affected the individual growth patterns in irrigated agriculture. The common practice of free grazing (*heṭiṅ*) of all livestock in the cultivated lands between the harvest of the second crop in October and the germination of the next crop in March/April was abolished in 1993. Similar community-based legislation occurred in other villages. As a rule, its enforcement lies with the respective local communities. Private ownership of land was traditionally valid only during the cultivation period when animals were removed to the high pastures. After their return the harvested fields were accessible to everybody's animals until the cultivation period began again in spring. This practice – once regarded by development agencies as one of the most severe obstacles to improved cultivation techniques and an abridged growth pattern – now belongs to the past.[279] The ban is strictly supervised and non-compliance leads to severe penalties.[280] Especially farmers who make experiments with new crops are keen to avoid any losses from livestock predations. Fruit orchards, vegetable and alfalfa plots or seed beds are presently in favour and are devoted to using the maximum vegetation period. Consequently, all sheep and goats are banned from other people's fields all year long. This decision was reached in a consensual village forum (*jirga*) and resulted in the search for a solution how to deal with the remaining livestock. The farmers of Karimabad rejuvenated their previously more or less abandoned pastures. The Burooṅ clan sent six shepherds in 1935 to Bululo with sheep and goats; in 1985 no shepherd supervised the roaming oxen there which were brought up in spring and taken back in autumn. Since 1998 the Burooṅ clan has adopted a turn (*galt*) system in the same way as Baltikuċ (mainly Qhurukuċ) and Ḍom make use of their pasture in Altikuċe sat/Beriċo Ċok. As no household can spare a full-time shepherd, the burden of watching the oxen and milking the few sheep and goats is distributed among all households on a two-day-long shift basis. Every household has to safeguard its participation. Milch cows tend to be kept in stables near the homestead.

The pasture settlement of Patundas (4,130 m) between Batura and Pasu glaciers is located on a high ridge from where the flocks are released in the early morning for grazing at higher elevations. The photograph was taken in Patundas on 7 July 1991. Nowadays no shepherds from Ghulkin – which owns the pasture rights – can be found: hired shepherds from Nagar have taken over this duty during the summer season.

Similar developments took place in Haiderabad and Aliabad and are under discussion in other villages. Reduced numbers of oxen and the absence of horses in Central Hunza have opened up the opportunity to use high pastures differently. Ultar ter (3,600 m) above Karimabad is the best example for a conversion from livestock-keeping to mountain tourism. Abandoned oxen pastures such as Sekai (3,600m) have been stocked for the first time with yaks imported from the Chinese Pamirs.

The question of future prospects of high mountain pastoralism within the context of sustainable development deserves a complex answer, at least for the Hunza Valley. The fodder potential is given, but a workforce that can make an adequate living there is not available. One reason is a seasonal occupation that prevents taking up another salaried occupation. Pastoral practices afford considerable skills, endurance and strength that have been fading in recent years. As long as no higher profit can be made from agricultural activities they will remain a side business for people who are bound to the village for different reasons. While crop farming and arboriculture are more easily integrated into daily routines and village lands are more easily accessible, the challenge of using high pastures is great and people do not have solutions at hand. The use of pastures is mainly discussed in Gojal where animal husbandry has greater importance; practical solutions have so far only increased the burden of women and the elders.

Adaptations and modifications are influenced to a higher degree by political and societal developments than by changing environmental conditions in the region where these practices are applied. Thus, sustainability has to account for all available opportunities under a given set of frame conditions. Consequently, pastoral practices and the use of grazing grounds will continue to play an economic and security-related role in the Hunza economy. The degree of importance for the generation of household incomes may nevertheless vary quite significantly. The importance of political issues in a globalised world could be observed in the aftermath of 9/11. Immediately the tourism industry collapsed, and annual income losses of more than ninety percent from that resource have been recorded for the subsequent years. In consequence dynamic responses to a crisis situation were adapted; young men who previously were employed in tourism temporarily went back to pastoral practices. The composition of household incomes had shifted again to a higher share of subsistence production. This was an intermediate solution before other options came up. Most people who were employed in tourism found new opportunities in non-governmental organisations that acted after the 2005 Kashmir earthquake. Dynamic adaptation to changed circumstances has always been a characteristic of this mountain region.

In early summer the oxen are relocated to high pastures above Gulmit in Bulkish and Shatuber. The approach road is rather challenging as the rock faces are quite slippery and the oxen very weak after a hard winter and spring with limited fodder resources. One men per ox has to be found in order to bring the flock up here for the next three to four months. The scene seems to be of a bygone past when respected household members took a day off from work for the relocation. Nowadays it is difficult to find the workforce for such a task which offers no monetary return and little benefit at the end of the season. The photograph was taken on 19 June 1990

Yaks are kept in a separate night pen (zuy̌e wandan) in Shuwert (4,600 m). Predators have been such a severe menace that preventive security measures not only led to keeping all livestock in night pens but also required the shepherdesses to do night duty. Snow leopards and wolves are held responsible for livestock losses. The photograph was taken on 21 July 2017

Beyond the two main pillars – resources in hunting and prospecting

The Hunza economy – that of the valley and that of individual households – was probably never based on two pillars such as crop farming and pastoralism alone. There were always some persons who replenished their income from proceeds generated by professional skills. Foremost here are the Ḍom, who were stigmatised as artisans (*bériċo*) and professionally worked as blacksmiths and musicians.[281] Some specialists produced woven goods (cloth, mats and carpets), shoes from local leather, wooden beams and planks for house-construction.[282] House-building was an effort in which volunteering neighbours and relatives participated; no special skills were required. Other community members earned some income in kind for providing services to community and relatives in irrigation, supervision, guarding and shepherding. On visits to the pastures firewood was gathered as well as birch bark. The latter was used for wrapping fresh butter before burying it in the ground, and it was a valuable resource for house-building as birch bark was used as an insulation and water protection for the roof. In previous times this was a local commodity for personal use, probably exchanged within the villages. Since the 1980s, firewood and birch bark have been offered on road-side stalls along the Karakoram Highway. Wooden products such as bowls and spoons were produced as a side business as well. Such commodities mainly come from Nagar and are offered to Hunzukuċ and travellers in the bazaars. With the recent increase of domestic tourism, wood processing for souvenir items – spoons and musical instruments – and a growing demand have motivated a few young artisans to open workshops and selling points in Karimabad. Other items that were previously only produced for domestic use such as home-spun woollen blankets (*paṭṭū*), mats (*pálos*) and caps (*phárċin*) have become souvenir items for market sale as well.

Members of Khunjerab Villagers' Organisation regularly observe the movement of wildlife within the boundaries of their assigned area; Farman Raza from Jamalabad is on duty on the checkpoint in Belli near Sost. The photograph was taken on 24 March 2019

Yak herds that are released without daily attendance of shepherds are controlled from time to time; Leili and Noor Ali are on their way to Gunjidur equipped with binoculars and a gun for their own protection. The photograph was taken on 19 July 2019

It has been characteristic for mountain dwellers in general to augment their livelihoods by tapping additional resources that are locally accessible and exploitable. Limitations and specificities characterise these practices. In the context of the Hunza Valley, local hunting for ibex, blue sheep and Marco Polo sheep as well as partridges and migratory birds played an important role in augmenting the nutrition base. Especially in winter and spring when food stores were dwindling, prey from hunting was a valuable additional nutritional resource and could easily be preserved. These practices have changed over time; one reason is the depletion of game, another is the increased protection of flora and fauna in special reserves. Since the inauguration of the Khunjerab National Park in 1975 and the subsequent introduction of protected areas, hunting and poaching have been banned for almost everybody, except for outsiders who are issued costly trophy-hunting licences in certain valleys.[283] For some village communities proceeds from trophy-hunting have become quite valuable, but the majority of households are excluded from this source.

Mineral wealth is another valuable resource that has augmented household incomes over time. Salt has been a rare commodity in Hunza, though of utmost importance for the daily diet. Rock salt was mainly brought into the valley from Koliwal traders who transported it from the Salt Range to Gilgit bazaar from where small amounts arrived in the Hunza Valley; even lower quantities of salt came from Wakhi traders from Badakhshan or from Kashgar. The only exceptions of local production were salt dues brought by Shimshali farmers to the *mīr* of Hunza as part of their taxes. These were gathered from Kuchin and Darwaza in the Braldu and Muztagh valleys.[284] As these areas lay beyond the Shimshal Pass and were under dispute between China and Hunza, the situation became more precarious after the *mīr* of Hunza had given up his claims in 1937. In the aftermath it became known that the *mīr* had annually received forty *man* of salt, and he claimed that this equalled the value of one quarter of his total revenue from Shimshal.[285]

Another mineral wealth of high value was gold dust that was panned in the tributaries and on the banks of the Hunza River as a duty of the low classes (*baldákuyo*) and part of forced labour (*rajáaki*). This obligatory practice faded in colonial times when the lower classes were still forced to pan for the gold dust that had to be delivered by the ruler as his loyalty presents (*gan-giánum*) to China and Kashmir.[286] The latter amount was initially less than one hundred grams when presented on the annual *darbār* in Gilgit, but would be nearly doubled to '25 tilloos of gold' that was an equal weight of '16 tolas and 5 mashas' fixed in the *sanad* of *mīr* M. Nazim Khan; other sources recorded '17 tolas and 4 mashas for Hunza, and 18 tolas for Nagar'.[287] The amount paid to China weighed about fifteen *miskal,* which is less than seventy-five grams. Arthur Henry McMahon reported

In former times gold-panning was part of forced labour duties (rajaáki) in Hunza and Nagar. After the tham relinquished his claim, itinerant specialists (marúć) took over and have found a meagre means of living on the banks of the Hunza and Shimshal rivers. The photographs were taken near Nasirabad on 20 May 1985

that before Yakub Beg's rule in Kashgar the quantity was twenty-five *miskal*.[288] Later on the quantity must have been reduced as the Gilgit Diaries mention that Daulat Shah, Hunza's envoy, was on his way from Baltit to Kashgar to deliver 'the annual present of 15 miscals of gold for the Chinese Government'.[289]

The quantity of up to a quarter of a kilogram of gold dust had to be locally panned by the Hunzukuć of the central oasis and was referred to as the gold dust present from the valley. The monetary value of gold was calculated as twenty rupees per *tolā* in 1916. Later on, gold panning was taken over by itinerant *marúć* who were commissioned by the colonial authorities. Since then they visited the Indus Valley south of Gilgit and paid a nominal amount to the Political Agent.[290] At the latest since the abolition of the Hunza State *marúć* have monopolised certain locations in the Hunza Valley, pitched their camps and engaged in their profession with all family members involved. The meagre proceeds provide a small living from the very fine gold dust panned out of the alluvial deposits of glacier meltwaters.[291]

Mineral exploration and exploitation have older roots in the mining for garnets (*ćhikas*) that were used as bullet cores in Chikas, gemstones – aquamarine, fluorite, sapphire and kunzite – in Hassanabad *nālā*, in Sumayar with aquamarine, tourmaline and topaz. At several other locations success stories about ruby mining were narrated.[292] In former times it happened to be sheer luck when a person gathered the mineral wealth (mainly rubies) of the mountains. This find then belonged to the person extracting it or to the *tham* if he could get hold of it. Since the Pakistan administration took over, government institutions have been involved in prospecting and have issued licences. In theory, all discoveries have to be registered and submitted to the 'Gemstone Corporation of Pakistan Ltd' since 1979; this institution had its headquarters in Gilgit and operated a field office in Aliabad. It is difficult to judge how many persons are involved in informal and micro-scale gemstone mining; the potential wealth of the Hindukush-Karakoram Mountains is acknowledged, and substantial future potential is expected. Since two decades, processing of raw gemstones has been

On the Pamir Triangulation Survey's way up the Kilik Valley to Misgar Richard Hingston noted in his diary for 28 May 1913: "Today we continued our tramp through the sand and water-thorn pebble of the Hunza gorge. The river had to be forded several times, and as the water reached almost as high as the coolies' waists and rushed along with almost torrential force, each crossing caused us considerable amusement accompanied with the fear that some of the retinue or belongings might at any moment be carried away in the fierce current and receive rough treatment in the rapids below. … the coolies stemmed the strong current. They entered the water in batches of three and four, and having joined hands, they advanced in a line parallel to the direction of the stream. The uppermost cooly, by dragging on the next below, prevented him from being washed down while the lowermost, by pushing upwards, produced the same effect. By the combined efforts of all in a batch each rendered the other a mutual support." Source: Richard William George Hingston (1913: 84–85). Photograph No. H. 47 is reproduced with permission of the Trinity College Library Dublin (MS 10484/7/47)

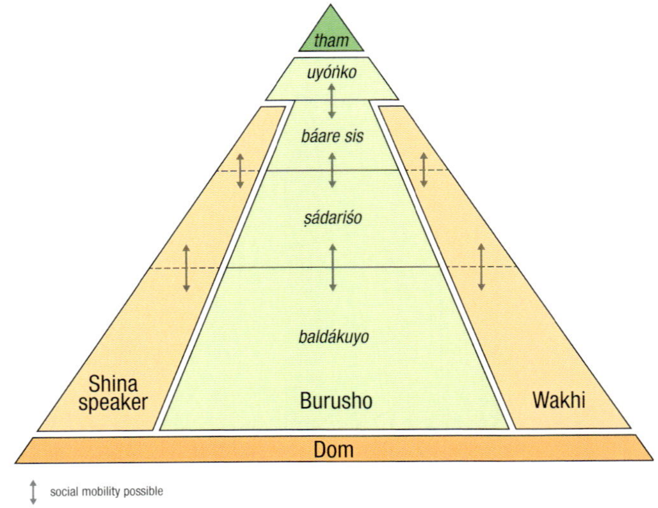

During the pre-colonial Hunza State period, the respective positions of Hunzukuć in their society were clearly perceived. Hunza was well-known for the absence of religious elites such as pīr or sayid. The highest position was occupied by the tham from the Ayáśo dynasty and his family in their secular function. An important role was taken up by the uyóṅko; influential members of the wazirkuć, other office-bearers and advisors of the tham were included in this group. The báare sis and ṣádariśo occupied the middle layer of Hunza society and were characterised by fewer duties and obligations to-wards the tham. They were exempted from load-carrying, which was the menial task of the baldákuyo. The Shina-speakers and Wakhi were more or less sidelined in receiving authority and offices; in respect of tax-paying and other services, they were quite disadvantaged. The excluded group were the professional blacksmiths and musicians, pejoratively called bérićo (their self-appellation is Ḍom). Source: Modified from Hermann Kreutzmann (1989a: 167).

promoted in Gilgit-Baltistan, and with the growth of tourism a local market for these products is thriving.[293]

Another mineral source is linked to dimension stones that are available in great quantities in the Hunza Valley. Several contractors have tried their luck so far and worked on cutting marble slabs, especially since the 1980s in Khaiber and more recently in Yal in Nagar. So far, the bottlenecks have been limited accessibility and energy deficits. All government institutions suggest a high potential for mineral exploitation in the area; a further boost is linked to the projects related to the China-Pakistan Economic Corridor in which the 'Northern Pakistan Zone', besides being a 'border logistical channel', has been allocated the functions of 'ecological conservation' and 'resource exploration'.[294] These prospects have accelerated the fear of local residents of being excluded from benefiting, controlling and decision-making about the exploitation of their inherited resources.

A more classical additional source for income generation is based on human labour. In the Hunza Valley the demand for human porterage was part of the taxation system by the local elites. The *baldákuyo* or *borwar* were categorised as load-carriers. In as much as it was a part-time or seasonal unremunerated profession it was a distinguishing mark of social standing. The layers of Hunza society that were ex-

exempted from load-carrying and any kind of forced labour (*rajaáki*) belonged to the upper classes (*uyóṅko*), who were respected for these privileges. In the pre-colonial setup the next group of higher standing (*báare sis*) was equally exempted from load-carrying, while the servitors (*ṣádariśo*) would be employed as messengers and were serving the upper classes. The bulk task of porterage remained with the *baldákuyo* who in earlier times were responsible for gold washing as well.[295] In Wakhi society the *borwar* were the permanent load-carriers, whereas the middle classes (*tarkhan*, *ʒəqláy žarž*) were exempted from regular duties, but could be called in emergency cases.[296] Firewood, butter, grain, timber and many other goods and commodities had to be relocated within the valley; traders requested the *tham* to arrange coolie porterage which was provided against payment to the *tham* and *wazīr* for whom load-carrying was a source of income. Expeditions and missions of explorers, colonial officers, hunters and travellers requested sometimes hundreds of porters and support staff who were appointed by the ruler. A valuable source of income for individual households emerged from porterage when tourism in the form of mountain expeditions began. Initially guides and porters were brought from the Nepalese Sherpa community to northern Pakistan. Since the 1950s Hunzukuč and Nagarkuč have replaced them; later on people from Baltistan and Chilas joined them to provide services as cooks, guides and porters.[297] Certain villages such as Pasu and Shimshal have produced a significant number of high-altitude porters who are well-known and highly respected.

An even earlier resource for paid human labour was a wide range of functions in the colonial system. Postal runners, guards of the Kashgar Consulate and the Gilgit Political Agency as well as posts as bodyguards and mercenaries in the Gilgit Scouts could augment their households' incomes by salaried jobs.[298] The latter were initially mainly reserved for Burusho as opportunities were given only on the suggestion of the *mīr* who profited from his position all along.[299] The colonial system had established a functioning consulate in Kashgar after 1891; a greater demand for support staff occurred at the beginning of the twentieth century when the Hunzukuč were engaged to keep up communications between the Misgar telegraph office and Kashgar where they found additional employment as trustworthy guards. Their personal performance enhanced the job market; especially in the 1930s, when the lease of the Gilgit Agency was negotiated, Hunzukuč replaced a number of Kashmiri employees in the agency and about 160 men served in the Gilgit Scouts. These connections to the bazaars of Gilgit and Kashgar contributed to small enterprises undertaken by people who got the opportunity to leave the Hunza State and to bring in goods from outside. David Lorimer mentioned that a full year's service in Kashgar would earn a horse and some extra goods to be brought home.[300] A number of people would go as service-providers to Kashmir and Punjab; thus cash income and clothes other than home-spun entered the valley and enabled the purchase of dearly needed food grains from neighbouring Nagar that had a surplus production. Hunza's only surplus crop was apricots that were dried and sold to Gilgit bazaar. These activities augmented the little local exchange with pilgrims and traders crossing the valley. David Lorimer, a true colonial representative, attributed this meagre improvement of exchange relations as an asset and quoted the people as saying: "What anxiety need we have about providing for marriages, and what need to worry about having no land? Government is our father and mother. If they drive us down (i.e. out of the country), Government will give us food and clothing. We owe a debt, as it were, first to God, and then to Government." And he perceived the local appreciation at the birth of sons "They will earn wages and support us", which led David Lorimer to the conclusion "… now by the blessing of the English every day is a fes-

Marble slabs are brought from a nearby quarry in Yal to the roadside close to the junction of the Minapin road where electricity and water are available for tile production. The photograph was taken on 5 August 2017

The abandoned Misgar post office originates from colonial times; from here the Hunza postal runners carried the mailbag to Kashgar and returned with messages from there. Nowadays a new post office is operated by the Pakistan postal system while the old building became the office of Misgar's cattle breeding farm. Misgar has been sidelined, as the postal communication between Pakistan and China follows the route along the Karakoram Highway from Sost to Tashkurgan. The photograph was taken on 23 September 2000

tival".[301] Other contemporary reports and photographs do not document such a joyful state of livelihoods in Hunza. In fact, the 1930s were a period when most visitors reported an appalling state of affairs. The remarks of David Lorimer indicate that – long before motor road infrastructure such as the Karakoram Highway at its zenith came into being – a trend occurred that enabled people to diversify their income structures, generate funds from services of various kinds, and participate in economic activities such as barter trading, purchase and sale of goods of daily and episodic need. Consequently, cash money entered the valley and more people had rupees at their disposal. Jean Bowie Shor fell for the tales of *tham* Jamal Khan in 1952 who presented himself as the benevolent ruler who supposedly was the only cash-owner and 'conducts all imports and exports on behalf of his subjects'.[302] The money-based market economy had reached Hunza already half a century earlier.[303]

The present highly diversified pattern of non-agrarian jobs has emerged out of these occupations; access to educational institutions was the trigger and driving force for professional changes as much as the early opportunities of leaving the valley.[304] The present household income structure is characterised by a significant generation of funds from qualified professional work outside the valley. Presently, remittances in multi-local settings play an important role in sustaining the livelihoods of household members residing in the Hunza Valley.

The idea of combined mountain agriculture holds only true to some extent when agricultural practices are in focus. In order to grasp the livelihood conditions in the Hunza Valley, a non-place-based perspective is essential and irrevocable. Diversifying income structures accompany a high degree of mobility and flexibility by all actors who are under constant challenge from adaptation to local conditions, from Pakistan, and from global economics and politics. Despite a diminishing role of agricultural enterprises for household incomes, land has become the most valuable resource on the local level. The agricultural space is shrinking as habitations and infrastructure assets such as roads and various institutions are expanding and occupying land. These new constructions are often located within the fertile village lands as accessibility and centrality are important in terms of road infrastructure, electricity and water supply. Disputes over land have become related to encroachment, inheritance, libel and property rights conflicts in private and public contexts.

Endnotes of pages 231–350

1 The information was kindly provided by M. Iqbal, Deputy Commissioner's Office, Karimabad, 12 September 2013.
2 The damage was estimated at four crore rupees.
3 See for more details Hermann Kreutzmann (2015a: 470-471).
4 Frederic Drew (1875: 419); Karlheinz Paffen, Wolfgang Pillewizer and Hans-Jochen Schneider (1956: 14); Qudratullah Beg (1962: 151).
5 Qudratullah Beg (1962: 350-351).
6 Qudratullah Beg (1962: 349); Frederic Drew (1975: 416–417).
7 Qudratullah Beg (1962: 349); see James Abbott (1848: 230–232); Frederic Drew (1975: 416–417).
8 Frederic Drew (1875: 419).
9 General Staff India (1928a: 132).
10 John Reid Becher (1859: 223–224); see as well Kenneth Mason (1929).
11 Herbert Todd (1930: 173-174).
12 Kenneth Hewitt (2006: 64–69).
13 Karlheinz Paffen (1956a: 14).
14 Qudratullah Beg (1962: 151).
15 Jenny Visser-Hooft (1926: 79).
16 Kenneth Hewitt (2006: 68); Lasafam Iturrizaga (2011).
17 Gilgit Political Agency Official Diary for the month of June 1918 (IOL/P&S/10/826: 66).

18 During his visit in 1945 Reginald Schomberg perceived Sarat as a 'new cultivation' that had only commenced to become a village after 1935 (RSAA/SC/SCH/2/47/12: 13).
19 Wazir Humayun Beg led the successful construction work of Hunza farmers at Matum Das which was concluded by December 1904; his engineering expertise was admired and especially highlighted in the report of the Political Agent, Gilgit (IOL/P&S/7/172/2263A; IOL/P&S/7/173/283). The cost of construction amounted to 1,600 Rs (IOL/P&S/7/173/351). The distribution of land started in January 1905 (IOL/P&S/7/174/428). During this first phase land was allocated for twenty households in Matum Das (IOL/P&S/7/178/1038).
20 See David Butz and Nancy Cook (2011); Lasafam Iturrizaga (1997; 2005).
21 Gilgit Diary, June 1910 (IOL/P&S/7/241/1118); see Kenneth Hewitt (2014: 254-255, table of events); Hermann Kreutzmann (2012e: 61); Frank Saunders (1983: 107).
22 Lasafam Iturrizaga (2005); Wolfgang Pillewizer (1960: 101).
23 Richard Finsterwalder (1960: 787–788). Wolfgang Pillewizer (1960: 100) gave the date of the outburst as 23 August 1959.
24 Wolfgang Pillewizer (1960: 100).
25 Information by Ghulam-ud-din (2008) who remembered well the history of the 1960s events and recollected all the households that lost land and property; see Hermann Kreutzmann (2012e: 63).
26 See the time series and images in Hermann Kreutzmann (2012e).
27 See the detailed study by Hermann Kreutzmann (2012e).
28 David Lockhart and Robert Woodthorpe (1889: 135–136).
29 See map in Hermann Kreutzmann (2012e: 54).
30 According to Bronislaw Grombchevsky (1888/2015: 120–121) it was abandoned before his visit.
31 Friedrich Ratzel (1909: 282–283); Arved von Schultz (1914: 19–20). More recent repetitions of the same topic are found in Gonzalo Aguirre Beltran (1979); Ronald Skeldon (1985).
32 Klaus Haserodt (1984); Kenneth Hewitt (1988; 1989; 2001; 2005; 2006; 2011; 2014); Walter Immerzeel, Francesca Pellicciotti and Arun Shrestha (2012); Lasafam Iturrizaga (1997; 1999; 2005; 2008a; 2011); Kenneth Mason (1930b); Hans-Jochen Schneider (1962; 1969); Michael Spies (216); Hermann von Wissmann (1960a).
33 Nigel Allan, Gregory Knapp and Christoph Stadel (1988); Stephen Brush (1976); David Guillet (1983); Hermann Kreutzmann (1988; 1989a; 1999); Jon Mathieu (2011); Harald Uhlig (1995).
34 Muhammad Abdullah (1972); Karlheinz Paffen (1956c); Elizabeth Staley (1966); Harald Uhlig (1995); Peter Whiteman (1985).
35 Einar Eberhardt, Bernhard Dickoré and Georg Miehe (2006); Sabine Miehe, Thomas Cramer and Jens Jacobsen (1996); Karlheinz Paffen (1956c).
36 Eckart Ehlers and Hermann Kreutzmann (2000); Hermann Kreutzmann (1989a; 1996b; 2000a).
37 Marc Aurel Stein (1903: 32).
38 Henry Lawrence Haughton (1913: 222–223).
39 Kenneth Mason (1936); see as well Nigel Allan (1986); David Butz and Nancy Cook (2011); Nancy Cook and David Butz (2011); Mahnaz Ispahani (1989); Hasan Karrar and Till Mostowlansky (2018); Hermann Kreutzmann (1991a; 1995c; 1998a; 2004c; 2008c; 2009b, c).
40 Michael Spies (2019: xvii).
41 Sidney Burrard and Henry Hubert Hayden (1907-1908: 98).
42 See David Lorimer (1938); Kenneth Mason (1930a; 1938).
43 See Tandong Yao et al. (2012: 53), Jane Qiu (2013). The 'classical' term 'third pole' goes back to a book on Himalayan mountaineering edited by Günther Oskar Dyhrenfurth under the title 'zum dritten Pol' in 1952 (English translation 1955) and a subsequent monograph titled 'der dritte Pol' in 1961 The recent debate has chosen the term independently of its predecessor.
44 Philips Christiaan Visser and Jenny Visser-Hooft (1938) directed attention towards the extreme setting by estimating the glaciated area in the Karakoram as between 28-43%. See for the latest authoritative analysis of the Karakoram glacial environment Kenneth Hewitt (2014: 6). The estimates about the extent of the Karakoram vary between 51,000 and 60,000 km² (Hermann Kreutzmann 1994: 347).
45 Kenneth Hewitt (2014: 40-41); for a three-dimensional perception of mountains, see Jon Mathieu (2011).
46 The size of the Hunza Basin is estimated at between 13,733 and 14,100 square kilometres; see Kenneth Hewitt (2006) and Adnan Ahmad Tahir et al. (2016).
47 Kenneth Hewitt (2006: 50). Charles Morris (1928) identified the main range as 'Karakoram'. In the map 'Afghānistān, Gilgit Agency and Kashmīr & Jammu' (1930) the main range is identified as 'Karakoram Range' while the massif including Rakaposhi is called 'Kailās Range'. The terms 'Batura-Muztagh' and 'Rakaposhi Range' were used in the map 'Kashmīr & Jammu and Gilgit Agency' (1941) and by Richard Finsterwalder (1960). Rakaposhi Range has been conventionally upheld in the 'Mountaineering map of Pakistan' (1999); see for the appellations 'Hunza-Karakorum' and Saltoro-Karakorum' Karlheinz Paffen, Wolfgang Pillewizer and Hans-Jochen Schneider (1956, map supplement).
48 Walter Lawrence (1909: 15); Mike Searle (2006: 7).
49 Kenneth Hewitt (2014: 40–41).
50 Kenneth Hewitt (2006: 50); the Hispar Glacier covers an area of 622 square kilometres, while the Batura occupies 285 square kilometres. The glaciers of the Karakoram have been well studied; their results can be found in the edited volume on 'Karakoram in transition' (Edward Derbyshire and Monique Fort 2005; Kenneth Hewitt 2006) and Andrew Goudie, David Jones and Denys Brunsden 1984; Kenneth Hewitt (2005; 2014); Lasafam Iturrizaga (1997; 2005; 2011); Umberto Minora, et al. (2016); Michael Spies (2016).
51 Kenneth Hewitt (2006: 53).
52 See for detailed plant species descriptions and their variations Einar Eberhardt, Bernhard Dickoré and Georg Miehe (2006); Bernhard Dickoré (1995: 233–234).

53 A more differentiated assessment is provided by Einar Eberhardt, Bernhard Dickoré and Georg Miehe (2006) who distinguish between sub-humid, semi-arid, sub-arid and eu-arid cross-sections in the Hunza Valley.
54 Data were kindly provided by the Pakistan Meteorological Department; for processed data see Michael Spies (2019: 79).
55 Kenneth Hewitt (2014: 21); see diagrams in Hermann Kreutzmann (1989a: 70; 2015a: 158).
56 Klaus Haserodt (1984: 135); see diagrams in Hermann Kreutzmann (1989a: 72; 2000a: 95).
57 Elizabeth Staley (1966); John Staley (1969).
58 Hermann Kreutzmann (1989a: 99–123).
59 Emily Lorimer (1936g: 264; 1939: 197).
60 Reginald Schomberg in his 'Turkestan Diary' of 1926 (RSAA/SC/SCH/2/47/14: 3).
61 On 22 June 1945 Reginald Schomberg visited Pasu again and for the last time (RSAA/SC/SCH/2/47/12, chapter III: 2).
62 John Biddulph (1880); Frederic Drew (1875); see map in George Abraham Grierson (1900); Karl Jettmar (1989; 2002); Gottlieb Wilhelm Leitner (1877; 1880; 1890; 1891a, d).
63 Quote from the dedication in Jack Ives and Bruno Messerli (1989).
64 Hermann Kreutzmann (1998c: 187).
65 Shah Khan (1985: 1). The quote is taken from the English version of his manuscript that was intended to be part of his autobiography which finally was published in Urdu in 2000.
66 Government of Gilgit-Baltistan, Asian Development Bank and World Bank (2011: 80).
67 Erwin Grötzbach and Christoph Stadel (1997: 30).
68 Erwin Grötzbach and Christoph Stadel (1997: 30).
69 Erwin Grötzbach and Christoph Stadel (1997: 31).
70 John Biddulph (1880: 146-147); Algernon Durand (1899: 162); Edward Frederick Knight (1893: 349); Reginald Schomberg (1935a: 106); see for a summary of the tales Homayun Sidky (1999).
71 Delegations from Greece have been visiting the Kalasha for decades and implemented some rural development projects as a sign of their support for them. More recently the Macedonian government has established a 'diplomatic' exchange between the son of the last *mīr* of Hunza, Ghazanfar Ali Khan, and representatives who have visited Hunza and have invited delegations from Hunza. On 11 July 2008 Ghazanfar Ali Khan and an eight-member delegation visited Skopje (Vasiliki Neofotistos 2011).
72 Peter Bakker and Aymeric Daval-Markussen (2016); see as well Sadaf Firasat et al. (2007) who studied Burusho, Kalash and Pashtuns and could only find slight relations to Europeans by the last-named.
73 Thomas Wilson and Hastings Donnan (2012: 9).
74 George Abraham Grierson (1900, map following page 510).
75 Neil Brenner (1999: 40).
76 Neil Brenner (1999: 41).
77 Hermann Kreutzmann (2015a: 32-36, 303).
78 See for the Pakistan case Tariq Rahman (1996: 210).
79 John Biddulph (1880); Gottlieb Wilhelm Leitner (1877; 1889).
80 Stuart Hill Godfrey (1898).
81 Government of India (1909).
82 Gottlieb Wilhelm Leitner (1877, 1894). The later book contains the 'Map of Dardistan and the Pamirs' by Ernst Georg Ravenstein. See as well John Biddulph (1880: 155–164); Frederic Drew (1875: 393–473).
83 Government of India (1909: 14).
84 Government of India (1909: 13).
85 Government of India (1909: 16).
86 The report was published by the General Staff India in 1928 (IOL/MIL/17/13/73: 15); see as well the lightened version of Dardistan in Pandit Anand Koul (1925: 161) where Dardistan is defined as 'Astor, Bunji, Chilás, Gilgit, Hunza, Nagar, Puniál and Yásin and Chitrál'.
87 IOL/MIL/17/13/73: 20.
88 David Lockhart Robertson Lorimer (1927; 1935-1938; 1939; 1958; 1962).
89 Letter of Emily Overend Lorimer to Gilchrist Thompson, editor of Lovat Dickson's Magazine, dated 2 February 1934 (IOL/Mss Eur/F177/90).
90 First article of four for the BBC published in their magazine (Emily Overend Lorimer 1936d: 100).
91 Reginald Schomberg in 'Karakoram Finale' of 1945 (RSAA/SC/SCH/2/47/12, chapter III: 2).
92 Reginald Schomberg (RSAA/SC/SCH/2/47/12, chapter VII: 35).
93 Reginald Schomberg (RSAA/SC/SCH/2/47/12, chapter VII: 36, 37, 39).
94 Reginald Schomberg (RSAA/SC/SCH/2/47/12, chapter XI: 62).
95 Reginald Schomberg (RSAA/SC/SCH/2/47/12, chapter XI: 62-63).
96 Reginald Schomberg (RSAA/SC/SCH/2/47/12, chapter XI: 64).
97 Reginald Schomberg (RSAA/SC/SCH/2/47/12, chapter XI: 64–65).
98 Reginald Schomberg (RSAA/SC/SCH/2/47/12, chapter XI: 65).
99 Reginald Schomberg (RSAA/SC/SCH/2/47/12, chapter VII: 36).
100 Gerhart Klamert (1954: porter service book) was impressed by the Burusho, but despised the Shin as 'dirty, louse-ridden, and goitrous'.
101 Gottlieb Wilhelm Leitner (1877: 45); see in his tradition Irmtraud Müller-Stellrecht (1979: 22) and explicitly Karl Jettmar (1982a: 255; 2018: 219) who took the opportunity to criticise Graham Clarke's (1977) questioning of the applicability of the term 'Dard'.
102 John Staley (1982); see the map on page 4.
103 Georg Buddruss (1993: 39).
104 Dzoy Edelman and Leila Dodykhudoeva (2009: 773–774).
105 Government of Pakistan (1984a: 5).
106 The Express Tribune, 22 February 2011.

107 Government of Pakistan (1990: 182).
108 Government of Pakistan (1990: 182).
109 See Georg Buddruss (2006).
110 Office for the Population Census of Xinjiang Uygur Autonomous Region (2012: 1–6).
111 Tat'yana Nikolaevna Pakhalina (1960); Hermann Kreutzmann (2015a: 137; 2016b: 25, 2017a).
112 Hermann Kreutzmann (2017b); Lidia Fedorovna Monogarova (1989).
113 The findings of the linguistic survey have been published in Hermann Kreutzmann (1995a; 2005a).
114 Hermann Berger had speculated to find a link between Burushaski and the Basque language of the Pyrenees but gave up this idea as 'a youthful frivolity' later on; Kevin Tuite tried to find connections with Caucasian languages (Karl Jettmar 2018: 307–308).
115 Karl Jettmar (1977a: 429).
116 Hermann Berger (1960) ; Karl Jettmar (1975: 190; 2018: 219).
117 Georg Morgenstierne (1932; 1974).
118 Aleksandr Leonovich Gryunberg and Ivan Mikhailovich Stéblin-Kamensky (1976); Georg Morgenstierne (1938).
119 Peter Bakker and Aymeric Daval-Markussen (2016); Gérard Fussman (1972: 31); Mohammad Ismail Sloan (1981).
120 Richard Strand (2001: 255).
121 Matthias Weinreich (2010).
122 Nazir Ahmad Bülbül (2014); Hermann Kreutzmann (2007).
123 These ascriptions were listed in the 'Military report and gazetteer of the Gilgit Agency and the independent territories of Tangir and Darel' (Government of India 1909: 14).
124 John Biddulph (1880); Matthias Weinreich (2009: 22–23).
125 Government of Pakistan (2000: 26); Hermann Kreutzmann 2012f: 238–239).
126 Marium Karrar and Affan Iqbal (2011: 40).
127 Hermann Kreutzmann 1995a, 2015a). Across the 'line of control' about 10,000 Shina speakers are found in Indian-administered Jammu and Kashmir, the vast majority residing in Gurais in Baramulla District (Kulbushan Warikoo 2007; quoted in Hermann Kreutzmann 2012f: 215).
128 See the map in Hermann Kreutzmann (2015a: 484).
129 For detailed maps and survey data cf. Hermann Kreutzmann 1995a.
130 Hermann Kreutzmann (2017a).
131 Hermann Berger (2008); Matthias Weinreich (2010).
132 Matthias Brenzinger (2007); Dzoy Edelman and Leila Dodykhudoeva (2009); Hakim Elnazarov (2010).
133 Matthias Weinreich (2010).
134 Christopher Moseley (2010).
135 Hakim Elnazarov (2010: 47); see as well Ruth Laila Schmidt and Vijay Kumar Kaul (2008).
136 Hakim Elnazarov (2010: 47).
137 George Abraham Grierson (1908–1928); Clare O'Leary (1992).
138 Quoted after Razwal Kohistani and Ruth Laila Schmidt (2006: 143).
139 Nazir Ahmad Bülbül (2014: 11).
140 Aleksandr Leonovich Gryunberg and Ivan Mikhailovich Stéblin-Kamensky (1976).
141 Georg Buddruss and Almuth Degener (2012).
142 M. Yousuf Shahzad (1990).
143 Razwal Kohistani and Ruth Laila Schmidt (2006: 140).
144 Pamir Times, 2? February 2014.
145 Among the 56,988 persons recorded there were about 31.7 % Maulai, 37.2 % Twelvershii and 31.1 % Sunni (Government of India 1909: 23).
146 These are data for Kuh, Ghizer, Yasin, Ishkoman and Punial as contained in Ghizer District and Gilgit, Hunza and Nagar as the former Gilgit District compared as returned in the 1906 estimates and my own survey in 1990 when we recorded 48.2 % for Ismaili, 31.3 % for Twelvershii and 20.5 % for Sunni adherents (Government of India 1909: 23).
147 Data were collected along with the language survey in the early 1990s. To my knowledge no evidence-based, empirical survey has been conducted since then. Faqir Muhammad Khan (2002) assessed 39 % Twelvershia, 27 % Ismailiya, 18 % Sunni and 16 % Nurbakhshia for the year 1994 in the Northern Areas. Muhammad Feyyaz (2011: 12) estimated respective shares of 40 % Twelvershia, 30 % Sunni, 24 % Ismailiya, and 6 % Nurbakhshia. Manzoom Ali reported in his 'Karakoram map' (2016) 39 % Twelvershia, 27 % Sunni, 18 % Ismailiya and 16 % Nurbakhshia for Gilgit-Baltistan based on the estimates of selected authorities.
148 Ghizer is considerably influenced by developments in neighbouring Chitral where the Ismailiya has a strong representation of nearly half the population in Upper Chitral. A vast majority of more than three quarters of Sunni believers live in Lower Chitral. For tensions in Chitral, see Hermann Kreutzmann (2015a: 271–274; 2015c); on conflict dynamics in Gilgit-Baltistan, see Izhar Ali Hunzai (2013).
149 Sultan Abbas and Aziz Ali Dad (2017: 15). My assessment differs from the estimates, for which no sources are provided by Marium Karrar and Affan Iqbal (2011: 18) who assess for Gilgit City 54 % Sunni, 27 % Ismaili, and 19 % Sunni inhabitants.
150 Marium Karrar and Affan Iqbal (2011: 18); the latest figure was displayed on the website of Gilgit-Baltistan Scouts (http://www.gilgitbaltistanscouts.gov.pk/geodemo.htm; accessed 3 May 2019) and would estimate two thirds of the Gilgit District's population as urban.
151 IOL/MIL/17/13/73: 19.
152 See on conflict dynamics in Gilgit-Baltistan Nosheen Ali (2009); Frédéric Grare (2007); Muhammad Feyyaz (2011); Izhar Ali Hunzai (2013); tables with detailed descriptions of incidents are found in Hermann Kreutzmann (2008a; 2012f).
153 Muhammad Feyyaz (2011: 17).
154 A few exceptions are those settlements in the side valleys such as Naltar, Garmasai (Bar), Hopar, Hispar, Chupursan, Misgar and Shimshal. Some villages such as Altit, Gulmit, Ghulkin and Morkhun have converted former pasture abodes into permanent habitations.

155 Patrick Vans Agnew (1915: 300–302).
156 Census of India 1911 (M. Matin-uz-Zaman Khan 1912: 220).
157 Census of India 1931 (Anant Ram and Hira Nand Raina 1933: 59); Government of Pakistan (1975: 11); Muhammad Masih Pal (1934).
158 Hermann Kreutzmann (2006: 259); Michael Spies (2019: 121-125).
159 Government of Gilgit-Baltistan (2018: 2).
160 See in detail Hermann Kreutzmann (1989a: 48-59).
161 The details of family relationships and connections between Shersabz and Karimabad were provided by Alam Shah, eldest son of Ibadat Shah, our host in Karimabad.
162 In 1981 Dumyal was renamed Mominabad, in 1983 Baltit became Karimabad, and Hindi was relabelled Nasirabad the same year.
163 See John Biddulph (1880: 29–30); Charles Bruce (1892: 768–770); Martin Conway (1894a: 249); Giotto Dainelli (1924); Algernon Durand (1899: 150); Karl Jettmar (1961: 86); Edward Frederik Knight (1893: 463, 475); Friedrich Kussmaul (1965b: 501); Gottlieb Leitner (1891b); David Lorimer (1935–1938, I: xli–xlii); Emily Lorimer (1936a: 175); Irmtraud Müller-Stellrecht (1979: 96–98); Charles du Riche Preller (1924: 338); Peter Snoy (1975: 42–46); Elizabeth Staley (1966: 86); Tahir Ali (1984: 235); Jenny Visser-Hooft (1935: 140).
164 According to the figure in Michael Spies (2019: 103); see for Nagar's early settlement history John Biddulph (1880: 134–135); Jürgen Frembgen (1985b: 35); Gottlieb Wilhelm Leitner (1889: 72–78); Irmtraud Müller-Stellrecht (1973: 2–3); Michael Spies (2019: 95–98).
165 See John Biddulph (1876: 96, 107–108); Reginald Schomberg (1935a: 96–97).
166 Richard Hughes (2005); Richard Hughes and Didier Lefort (1986; 2006).
167 Haji Qudratullah Beg (1962/2006: 43–47); Hermann Kreutzmann (1989a: 51; 2000a: 100–103); David Lorimer (1935–1938, II: 237, 257–259).
168 See Haji Qudratullah Beg (1962: 45-49); John Biddulph (1880: 135); Henry Haughton (1913: 165); Karl Jettmar (2018: 315); M. Nazim Khan (1936: 132–136); David Lorimer (1935–1938, II: 101–179); Irmtraud Müller-Stellrecht (1979: 106–112).
169 Gilgit Agency Political Diary for the month of July 1912 (IOL/P&S/10/826: 232).
170 See Hermann Kreutzmann (1989a: 51); Homayun Sidky (1995; 1997); Irmtraud Müller-Stellrecht (1978: 107–115; 1979: 61).
171 The settlement process and its linkage to irrigation expansion are analysed in Hermann Kreutzmann (1990; 1996b; 2000a: 100–103; 2011d). See for historical evidence about the construction of irrigation networks in Hunza IOL/P&S/7/165/1094; IOL/P&S/7/171/2013A, 2142A; IOL/P&S/7/172/2263A; IOL/P&S/7/173/283, 351; IOL/P&S/7/174/428; IOL/P&S/7/178/1038; IOL/P&S/10/826: 16, 18, 60, 66, 240; IOL/P&S/10/973: 161, 226, 243; IOL/P&S/12/3285; IOR/2/1075/217: 50-54; IOR/2/1083/ 286: 11; M. Nazim Khan (1936: 120); David Lorimer (1935-1938, II: 353); Irmtraud Müller-Stellrecht (1978: 114); PRO/FO65/1507: 29; Thakur Singh (1917: 27).
172 The date is according to Yousaf Khan (2004).
173 Haji Qudratullah Beg (1962/2006: 197–198).
174 William Lockhart and Robert Woodthorpe (1889: 398). For Wakhi migration and settlement across Wakhan's border and in Gojal see Hermann Kreutzmann (2015a: 9, 135; 2017a: 191–197, 233–235).
175 Qudratullah Beg (1962: 442); see Bronislaw Grombchevsky (1888/2015: 116–118) for his meeting with *guśpúr* M. Nafis Khan in Khudabad.
176 Henry Haughton in his diary of 13 and 21 April 1912 (IOL/Mss Eur/D531/5).
177 Bronislaw Grombchevsky (1888/2015: 141).
178 Bronislaw Grombchevsky (1888/2015: 108–110).
179 David Lorimer (1935-1938, III: 357); Irmtraud Müller-Stellrecht (1979: 127).
180 See for the periodical association Jürgen Frembgen (1985b: 37); Michael Spies (2019: 103–104).
181 John Biddulph (1876: 96, 107–108) mentions in his diary that Chalt was held by Hunza from 1866 to December 1875.
182 For an evaluation of the historical developments and its effects on Hunza's exchange relations see Hermann Kreutzmann (1991a; 1993a; 1998a; 2015a, 2017a, b).
183 Contemporary eyewitness accounts of the Hunza-Nagar Campaign have been given by Algernon George Arnold Durand (1899); Edward Frederick Knight (1895); M. Nazim Khan (1936).
184 See for a detailed description of the agricultural and settlement constellation in August 1893 Edmond de Poncins (1897: 206–208).
185 Martin Conway (1920: 140). About the same time Edmond de Poncins (1897: 206) visited the Hunza Valley and made similar remarks by interpreting the war-like behaviour and raiding attitude of Hunzukuč as the result of prevailing food deficiencies: "Peut-être tout cela venait-il surtout de ce que le manque de vivres dans leur pays les obligeait à apporter de l'extérieur ce qu'il fallait; et, comme on n'avait pas de quoi acheter ou échanger, le plus simple était de prendre. Mais si le caractère des habitants est en train de se former, leur pays est toujours le même et leurs mœurs intimes encore presque les mêmes. Au point de vue physique, c'est un des pays les plus durs et difficiles du monde. La vallée est dominée par des montagnes aux escarpements sans pareils, coupées des nombreux glaciers." He perceived the peaceful state of affairs that had emerged after the Hunza-Nagar Campaign as the result of external supplies augmenting the essential demands.
186 Arthur Neve (1913: 169–170).
187 For the importance of the slave trade see Edward Knight (1891: 285); Hermann Kreutzmann (2015a: 206–208, 295–299), Irmtraud Müller-Stellrecht (1981).
188 Henry Haughton related this assessment from his interview with *mīr* Sikandar Khan on 16 April 1912 (IOL/Mss Eur/

D531/5): see William Douglas (1952: 300); Henry Haughton (1913: 216); Mohomed Nazim Khan (1936: 125); Irmtraud Müller-Stellrecht (1979: 150); Reginald Schomberg (1936a: 41). Hugh van Skyhawk (2003: 14) places Hispar's incorporation into Nagar at around 1840. Reginald Schomberg refers to Shimshal in the same manner (RSAA/SC/SCH/2/47: 9); Gerhart Klamert (1954: diary) learnt from the *tham* that Shimshal was destined for the expulsion of trouble-makers who would have the duty to work as his shepherds. Kumara Menon (1947: 43) added one more: "Misgar is a penal settlement."

189 See Frederic Drew (1875: 418); Lasafam Iturrizaga (2008b: 36-39); Charles John Morris (1928: 526); Reginald Schomberg (1935a: 112; 1935c: 225); Marc Aurel Stein (1928, II: 51); Henry Haughton (1913: 235); and the maps in Hermann Kreutzmann (1996a: 281; 2015a: 468).

190 Henry Haughton in his diary of 20 January 1912 (IOL/Mss Eur/D531/5). Nisar Karim (2018: 32) related that the first channel was built in a *rajaáki* effort by the communities of Gojal I under the supervision of *arbāb* Khairullah Baig. Initially some settlers from Gulmit cultivated the land from 1903 onwards, followed by Burusho and Ḍom settlers from Central Hunza who founded Nazimabad.

191 Henry Haughton in his diary of 22 January 1912 (IOL/Mss Eur/D531/5).

192 See Frederic Drew (1875: 418); Lasafam Iturrizaga (2008b: 36–39); Charles John Morris (1928: 526); Reginald Schomberg (1935a: 112; 1935c: 225); Marc Aurel Stein (1928, II: 51); Henry Haughton (1913: 235); and the maps in Hermann Kreutzmann (1996a: 281; 2015a: 468).

193 Henry Haughton in his diary of 23 January 1912 (IOL/Mss Eur/D531/5).

194 Henry Haughton in his diary of 24 January 1912 (IOL/Mss Eur/D531/5).

195 IOL/P&S/12/3288: 247.

196 M. Nazim Khan (1936: 120).

197 Government of India (1909: 84, 88, 120).

198 Gilgit Agency Political Diaries for the months of September 1912, February, April, October, December 1913 (IOL/P&S/10/826: 193, 195, 208, 215, 227).

199 Gilgit Agency Political Diaries for the month of September 1916, (IOL/P&S/10/826: 107); M. Nazim Khan (1936: 120–121); see as well Emily Overend Lorimer (1936a: 178) who regarded the tax as a 'low fixed rent in kind'.

200 Gilgit Agency Political Diaries for the month of September 1916, (IOL/P&S/10/826: 107).

201 Gilgit Agency Political Diary for the month of June 1924 (IOL/P&S/10/973: 161).

202 Gilgit Agency Political Diary for the month of July 1936 (IOL/P&S/12/3285: 260).

203 IOR/2/1075/217: 49.

204 Not all migrants were in search of new cultivation. There was a report in December 1921 that two Wakhi from Gojal escaped the control of the *mīr*, bypassed the *darband*, crossed the northern passes into Sarikol and proceeded to Shughnan where they joined the Bolsheviks (IOL/P&S/10/973: 238).

205 Thakur Singh (1917: 27–28).

206 Thakur Singh (1917: 28). The month of Rabi 1962 is related to the Vikram Samvat Calendar that was used in Kashmir's administration and equals the year 1904 in the Gregorian Calendar.

207 Gilgit Political Agency Official Diary for the month of February 1912 (IOL/P&S/10/826: 245).

208 Peter Snoy (1975: 31).

209 Peter Snoy (1975: 31).

210 See especially Administration reports for the Gilgit Agency for the years 1938, 1939, 1940 (IOL/P&S/12/3288); Hermann Kreutzmann (1989a: 183). The irrigation scheme of Harathingdas (nowadays: Jalalabad) by farmers from Teisot and Bilchar was completed in June 1939 (IOL/P&S/12/3285; IOL/P&S/12/3288: 127).

211 Figures calculated on the basis of Government of Pakistan (1975: 11) and Elizabeth Staley (1966: 69).

212 Henry Haughton in his diary of 14 April 1912 (IOL/Mss Eur/D531/5).

213 Gilgit Agency Political Diary for the month of March 1913 (IOL/P&S/10/826: 212).

214 Gilgit Agency Political Diary for the month of June 1923 (IOL/P&S/10/973: 191).

215 John Biddulph (1876: 123).

216 Bronislaw Grombchevsky (1888/2015: 148).

217 Henry Haughton (1913: 220–221).

218 Algernon Durand (1900: 143–144). The alleged difference in cultivation conditions between Hunza and Nagar has frequently become a hot topic for esoteric interpreters of a Hunza-Nagar divide since the writings of Emily Overend Lorimer (1939); see for example Ralph Bircher (1942); Jerome Irving Rodale (1948). The attempt to prove a differing endowment with agricultural assets by Uwe Schmidt (2006: 118–120) did not yield further insights as too many variables were not properly determined.

219 Letter to Lieutenant-Colonel Frederick Marshman Bailey, sent from Gurais on 3 June 1933 (IOL/Mss Eur/F157/249) who had suggested to recruit more local levies for the Gilgit Scouts.

220 Emily Overend Lorimer (1939: 72, 180, 273–277).

221 Emily Overend Lorimer (1939) and Reginald Schomberg (1935a) inspired popular authors such as Ralph Bircher (1942) and Jerome Irving Rodale (1948) who have created a huge followership of disciples; see as well the well-documented treatise of Jürgen Frembgen (1983a) about the pejorative treatment of Nagarkuč in popular literature.

222 Akhter Hameed Khan (1985: 249).

223 See Michael Spies (2019) and Benno Pilardeaux (1995) for the dissemination of innovations in the agricultural sector.

224 The term 'donkey-tree' (*jakún tom*) refers to non-fruit-bearing trees; see Emily Overend Lorimer (1936g: 264).

225 John Staley (1969).

226 Mohammad Manzar Zarin and Ruth Laila Schmidt (1984).
227 John Staley (1966).
228 Regular lotteries take place in the *wesh* system, which lead to a rotation of lands among *jirga* members who safeguard the maximum equitability of resource allocation; see for societal developments in this region Fredrik Barth (1956b); Zahid Javed Janjua 1998; Karl Jettmar (1961; 1977a, b; 1982b); M. Manzar Zarin & Ruth Laila Schmidt (1984).
229 Udo Schickhoff (2006).
230 Henry Haughton in his diary of 20 April 1912 (IOL/Mss Eur/D531/5).
231 Dendrochronological evidence and ^{14}C measurements of organic matter in building materials support this hypothesis (Richard Hughes and Didier Lefort 2006: 295).
232 For a detailed description of the distributive system see Hermann Kreutzmann (1989a: 80–91; 1990: 15–20; 2000a: 103–110).
233 Ample reference is given to *wazīr* Asadullah Beg in Qudratullah Beg's (1962: 287–345) chronicle which dates the construction of the Barber to the three years after 1852. This history is authored by relatives and functions as family heritage and praise of the achievements of the *wazīrkuč*.
234 Edmond de Poncins (1897: 208) observed the usage of these local tools for agricultural work and channel-building during his visit in August 1893; see Hussain Wali Khan and Izhar Ali Hunzai (2000); Emily Lorimer (1936b: 239; 1936e: 159).
235 See David Lorimer (1979: 126–133).
236 David Lorimer (1935, I: xlii) rated them at the bottom end of Burusho society.
237 See for population data Hermann Kreutzmann (1994: 347).
238 Frank Saunders in The Pakistan Times, 11 November 1984.
239 See Hussain Wali Khan (1994) and Mahrban Ali Sakhi (1997) for the details of the channel project.
240 Mahrban Ali Sakhi (1997: v).
241 Mahrban Ali Sakhi (1997: 13).
242 Mahrban Ali Sakhi (1997: 17–18); see in general John Staley (1982: 23).
243 More than half of the repair cost was contributed from government relief funds, one third by the village organisations (Mahrban Ali Sakhi 1997: 22).
244 See Hermann Kreutzmann (2000a); Michael Spies (2019).
245 Government of Gilgit-Baltistan (2018: 4).
246 Government of Gilgit-Baltistan (2018: 5).
247 Mountain Fruits Private Limited is an offspring of a development-agency-initiated export marketing company that offers three to four times the bazaar price for organic dried apricots, cherries and walnuts at the farm-gate for participating members. In 2013, more than 6,000 farmers were registered with the company in Hunza, Nagar and Ghizer.
248 Hermann Kreutzmann (1989a: 99–123; 2006: 331–339). Bul Mal was mapped in 1985 with first and second crops, in 2001, 2013 and 2015.
249 The potato story was told many times; see Hermann Kreutzmann (1989a: 106–114; 2006: 335–338). For Nagar see Michael Spies (2019).
250 See Stephen Brush (1976); Eckart Ehlers and Hermann Kreutzmann (2000); David Guillet (1983); Hermann Kreutzmann (2012a); Robert Rhoades and Stephen Thompson (1975); Michael Spies (2019); Harald Uhlig (1995).
251 Irmtraud Stellrecht (2007: 55).
252 See the map of the shared pasture settlement for Khudabadkuč in *Shamijerav ilga* for Wakhi and *Burum ter* for Burusho (Hermann Kreutzmann 2000b: 98-99; 2015a: 174).
253 Hermann Kreutzmann (2015a: 184–190).
254 Hermann Kreutzmann (2015a: 478–483).
255 See for the variety of milk products Hermann Kreutzmann (2000b: 99).
256 See for the details of Shimshal Hermann Kreutzmann (2015a: 198–190).
257 William Macandrew Marshall (1913: 1); IOL/P&S/10/826: 214.
258 See the table giving a summary statement of Hunza's revenue generated in the Taghdumbash Pamir in Hermann Kreutzmann (2000b: 116-117; 2004a: 82-85).
259 William Macandrew Marshall (1913: 2-3); IOL/P&S/10/826: 215–216.
260 File on the assessment of potential 'Hunza and Nager subsidies' in the light of revenue generated from agricultural and other taxes (IOR/2/1079/251).
261 Major Richard Alfred Cropper who investigated the situation in 1938 and furnished a 'report on Shimshal reconnaissance' (IOL/P&S/12/3292).
262 IOR/2/1079/251. British India and the Maharaja of Kashmir supported the Hunza ruler in 1896 with subsidies worth 3,000 Rs annually (Stuart Hill Godfrey 1898: 79).
263 IOL/P&S/12/3292: 156; IOR/2/1079/251: 8.
264 Reginald Schomberg (1934a: 211; 1935a: 169). The nomadic connotation probably stems from a military report (General Staff India 1929: 144) and has been afterwards repeated by geographers such as Nigel Allan (1989: 135), David Dichter (1967: 45) and Elizabeth Staley (1966: 322).
265 Quotation from an untitled and undated (app. 1935) file from the Commissioner's office Gilgit about the property rights of the *mīr* of Hunza.
266 David Lorimer (1935–1938, II: 259).
267 Gilgit Agency Political Diary for the month of July 1936 (IOL/P&S/12/3285: 260).
268 See George Cockerill (1922; 1939); David Lorimer (1935–1938), Reginald Schomberg (1935a; 1936a), Jenny Visser-Hooft (1935).
269 Qudratullah Beg (1935); Reginald Schomberg (1936a: 50-59).
270 See David Butz (1996); Hermann Kreutzmann (1986; 1998c; 2000b, c), Reginald Schomberg (1936a: 56, 62).
271 Data according to Karimabad Census 1992 prepared by Karimabad Planning Support Service.
272 Shimshal had about 4,473 goats, 2,547 sheep, 960 yaks, 399 head of cattle and 32 donkeys in 1995. The number of yaks was around 1,000 in 2003, and other livestock numbers were stable as well (information provided by Didar Ali in Shimshal, August 2003).

273 Shah Raees Khan and Syed Ajijur Rahman (2009: 260); Hermann Kreutzmann (2015a: 188).
274 For provision of the data the author is grateful to Einar Eberhardt, Marburg, who conducted the Batura livestock survey in 1998.
275 In Pasu 61 households were recorded in 1985; see Mehjabeen Abidi 1987; Hermann Kreutzmann (1986: 102).
276 For a detailed account of this dispute, see Are Knudsen (1996, 1999); Hermann Kreutzmann (1995b); Zaigham Khan (1996).
277 About four million rupees were generated between 1998 and 2003; information provided by Amjad Bahadur Khan, president KVO, August 2003.
278 See for land disputes Hermann Kreutzmann (1996a: 315–317).
279 The Aga Khan Rural Support Programme (AKRSP) and the FAO/UNDP-sponsored Integrated Rural Development Project (IRDP) advocated such a strategy in the early 1980s. Fencing of irrigated terraces was discussed as one solution, others included the keeping of livestock in stables. Both were not successful with their proposals at the time (Aga Khan Rural Support Programme 1984; Frank Saunders 1983; 1984).
280 In all villages Falai committees were founded and authorised to collect sanctions for trespassing animals. In Haiderabad the fees amounted in 1998 to 5 Rs per sheep, 15 Rs per goat and 25 Rs per head of cattle. The collected fine is handed over to the owner of the respective plot.
281 See Tahir Ali (1983); Anna Schmid (1997).
282 David Lorimer (1979: 90–102) listed the locally available crafts for the mid-1930s.
283 Mehjabeen Abidi-Habib and Anna Lawrence (2007); Are Knudsen (1996; 1999); Hermann Kreutzmann (1995b); George Schaller (1980).
284 For the importance of salt in regional exchange see Gilgit Monthly Political Diaries for August 1901, May-June 1904 (IOL/P&S/7/134/757; IOL/P&S/7/137/757; IOL/P&S/7/167/1319); the 'Military Report and Gazetteer of the Gilgit Agency and the independent territories of Tangir and Darel' by the General Staff India for 1928 (IOL/MIL/17/13/73: 31-32); the 'note on the boundary dispute case Hunza (Gilgit Agency) – Sinkiang in the neighbourhood of the Shimshal Valley and Oprang river' (FO 371/23493: 64–77); Hermann Kreutzmann (1998a: 296; 2015a: 275); Reginald Schomberg (1936a: 78, 84–85); Peter Snoy (1965: 113, 121–122).
285 The amount of Shimshali salt was defined as a 'hundred blocks' (David Lorimer 1935-1938, II: 275) and equalled about one and a half metric tons; see 'Note on the boundary dispute case Hunza (Gilgit Agency) – Sinkiang in the neighbourhood of the Shimshal Valley and Oprang river' by Ian William Galbraith, dated 14 April 1939 (FO 371/23493: 64–77; IOL/P&S/12/3292: 41).
286 Thakur Singh (1917: 129).
287 Thakur Singh (1917: 129); IOL/P&S/12/3291: 17.
288 IOL/P&S/12/3291: 18; IOR/2/1079/253: 62; Arthur Henry McMahon (1898); Reginald Schomberg (1935a: 157).
289 Herbert Todd in the 'Gilgit Agency Political Diary for the month of September 1929' (IOL/P&S/10/973: 36).
290 Thakur Singh (1917: 59).
291 Regularly visited locations in Hunza are the area near the bridge connecting Yal and Nasirabad; another location is close to Dut in the Shimshal Valley; see Hermann Kreutzmann (1991a: 753). See in general George Howard Bretherton (1898: 877); Stuart Hill Godfrey (1898: 97–100); Azam Chaudhary (1997) for the Indus Valley. Government of Gilgit-Baltistan (2011: 48) sources estimate that 200 gold-panning households are involved there.
292 The Chumar Bakhoor mine is operated at an altitude of 4,800 m in the Sumayar Bar by the local community and employs more than 500 persons during summers; see Kuntala Lahiri-Dutt and Hugh Brown (2017). A more comprehensive list of metallic and non-metallic minerals includes gold, copper, antimony, lead, iron, pyrite, sulphur, tungsten, aluminium, palladium, thorium and graphite; in the gemstone sector the list covers emerald, ruby, sapphire, aquamarine, moonstone, amethyst, tourmaline, peridot, topaz, garnet, red spinal, pargasite, diopside, sphene, apatite, azurite, rose quartz, and agate; see Aga Khan Rural Support Programme (2017: 38).
293 Especially the Karakoram Area Development Organisation and the Rupani Foundation have been instrumental in providing skill-based training for gemstone cutting, polishing and marketing.
294 Dawn, 21 June 2017; Government of Pakistan (2017).
295 For details of the Hunza societal layers see Hermann Kreutzmann (1989a: 166–169); Irmtraud Müller-Stellrecht (1978: 116–121; 1979: 116–124; 1983b: 388–415); John Staley (1966: 84–97).
296 For details of the Wakhi taxation system and load-carrying obligations see Hermann Kreutzmann (1996a: 282-289; 2015a: 410–415).
297 See David Butz (2006); Hermann Kreutzmann (1989c). In recent years a shortage of skilled high-altitude porters has led to Nepalese Sherpa again being hired in Baltistan and Diamir for the support of international expeditions.
298 See the photograph in Percy Etherton's (1911: 26) book with the *tham* of Hunza sitting in front of his bodyguards in 1909.
299 Hermann Kreutzmann (2015a: 413–414).
300 David Lorimer (1979: 87–89).
301 Quotes from the personal records of David Lorimer (1979: 88–89) that were written down in 1934 to 1935.
302 Jean Bowie Shor (1955: 279).
303 See as well John Staley (1966: 200).
304 See for a detailed study of Gilgit-Baltistan in general and Gojal in particular Andreas Benz (2014a, b, c; 2016).

Visitors were regularly welcomed with performances of local traditions. In August 1954, sword dancers performed on the polo ground in Baltit. Source: German-Austrian Himalaya-Karakoram expedition 1954. Courtesy of Gerhard Klamert

From factors to actors – Karakoram configurations

The Hunza Valley has acquired a prominence in international relations that is hardly explainable from its position in early periods and in comparison to other regions in the Hindukush and Karakoram. The principalities of Hunza and Nagar had only just grown into a sizable microstate when Patrick Vans Agnew visited the valley in the first half of the 19th century. The Nagar ruler was supposed to have command over six thousand houses, which seems quite an exaggerated figure; its seat of power lay back in the Miatsil Valley with some loose control over remoter villages. Hunza consisted of barely two thousand houses distributed within three villages in the central oasis.[1] The *thamo* had more or less command over Burusho mountain farmers who cultivated the village lands combined with the utilisation of high pastures and hunting-grounds. Some additional income was generated from slave trade, raiding and plundering of caravans. The latter afforded some expressions of loyalty under a unified command and sufficient military strength to allow attacks in remote locations far away from their fortified strongholds that were adequately equipped with defence measures. Early reports by outsiders have emphasised the poor state of domestic affairs and the connected meagre income that 'forced' the people to augment it by tapping other sources. Edward Frederick Knight, the Times correspondent who accompanied British-Kashmiri forces during the 'Hunza-Nagar Campaign' in 1891 summarised the perception of both: "These Hunza-Nagaris, generally known to their neighbours as the Kanjutis, though this name strictly applies to the Hunzas alone, have for centuries been the terror of all the people between Afghanistan and Yarkand. Inhabiting these scarcely accessible defiles, they have been in the habit of making frequent raids across the Hindoo Koosh and earning their livelihood by a well-organised brigandage, the thums, or kings of these two little States deriving the greater portion of their revenue from this source. So great was the dread inspired by these robbers, that large districts have been abandoned by their inhabitants, and land formerly cultivated has lapsed into wilderness, under the perpetual menace of the Kanjut raids. The most profitable hunting-ground of the Kanjutis was the great trade route between Leh and Yarkand over the Karakoram Pass, and many a rich caravan on its way from India to Central Asia has been waylaid and pillaged in the neighbourhood of Shadulah [Shahidulla]. The thums used to maintain their regular agents at Yarkand, who gave them notice of an expected caravan. On one memorable occasion a caravan of fifty laden camels and 500 laden ponies was captured. The Kashmiris and the Chinese found themselves powerless to put a stop to these raids, and the Kanjutis acquired a great prestige, and were considered as quite invincible. The Hunzas, indeed, had never known defeat before Colonel Durand's successful campaign. But this wholesale brigandage, bad as it was, was only a minor offence when put by the side of the systematised slave-dealing in which these scourges of the frontier have been engaged from time immemorial. All prisoners of any commercial value – men, women, and children – captured in these raids were driven across the mountains, to be sold, either directly to the slave-owners in Chinese Turkestan, or to Kirghiz dealers, who served as middlemen in this trade. The forced marches across the snowy ranges that these unfortunate captives were compelled to undertake, thinly clad as they were, and provided with but a minimum of food, caused the death of a considerable proportion; and the abominable cru-

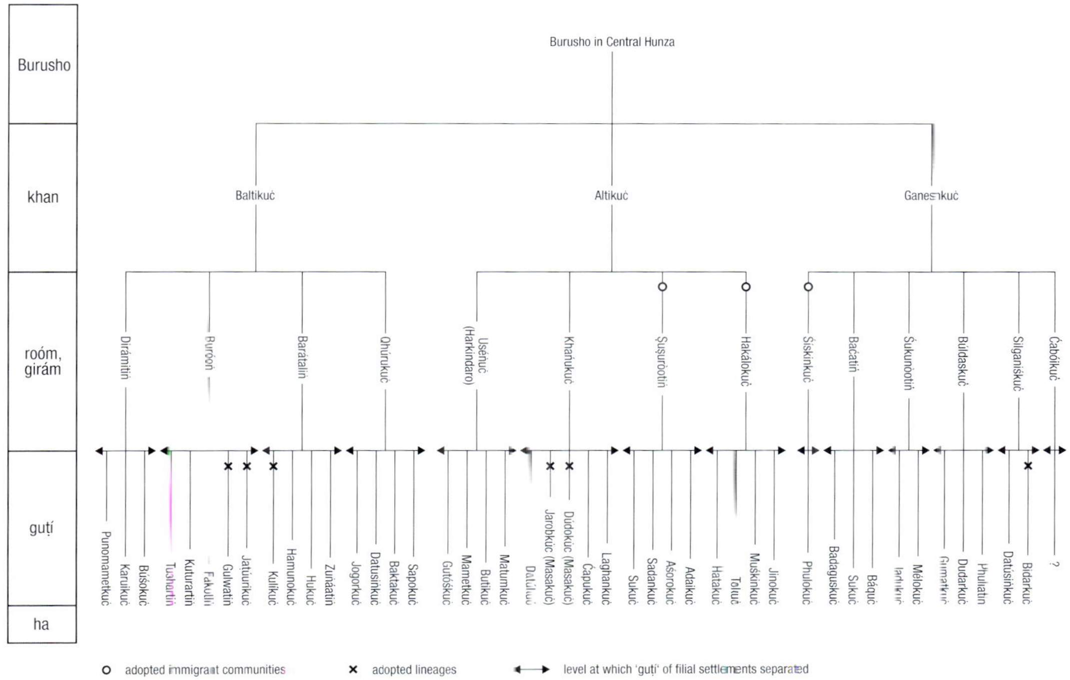

In terms of patrilineal affiliation the Burusho in Central Hunza refer to the three fortified villages (khan) Baltit, Altit and Ganesh. On a sub-level they are divided into lineages (roóm, girám) that are important in terms of water and pasture rights; within Hunza society and their respective villages, their positions are reflected in seating arrangements at public functions or the order of appearance during public dances. These lineages are subdivided into smaller groups that are well known to the villagers and are sometimes visible in settlement patterns. The terminology applied to these patrilineal descent groups refers to the whole group of Burusho as an umbrella. Within this system all terms are related to physical settlements. The khan is the fortified village, the girám is a hamlet outside the khan, while literally the guṭí is a hut in the pasture; the proper house in the village is termed in Burushaski ha and is related to the household; see for a similar terminology in Shin areas Karl Jettmar (1961: 84). Source: Adapted from Hermann Kreutzmann (1989a: 4[illegible]).

The 'Fighting near Gilgit' made it to the front page of 'The Graphic' during the Hunza-Nagar Campaign. Source: The Graphic 19 December 1891, page 717

elty with which the Kanjutis treat their prisoners has been remarked by most travellers on the Pamirs. ... The rulers of these two States were, as might be expected, ignorant and bloodthirsty scoundrels, faithless to their treaty obligations, and incapable of respecting anything but force. They were absolute monarchs, and murdered or sold their subjects into captivity at their own sweet will. The royal families of Hunza and Nagar are descended from two brothers who lived in the fifteenth century, but they trace their ancestry further back, to a divine origin."[2] Although Edward Knight tended to exaggerate the significant threat for frontier security that emanated from Hunza, it seems that he attributed extraordinary fighting skills for executing perfidious assaults to the Kanjutis. The agricultural basis was obviously insufficient to feed the people, at least from a place-based resource-deterministic viewpoint, and had supposedly to be augmented by non-local resource exploitation, in this case slave trade and caravan-raiding.[3] The bravery and courage of the mountain dwellers, their operation of raids with cunning and trickery, and their resilience and valour were admired by outsiders. Reports about the income from such activities seem grossly exaggerated. The rulers were identified as the problematic obstacle for cordial relations; their greediness, ruthlessness, perceived irresponsibility and cruelty provided the pretext for intervention and liberation of the ordinary people from their insinuated yoke. Edward Knight was quite surprised by the bold impertinence and excessive megalomania of the petty rulers: "It is said that it was a point of etiquette in his savage Court, on certain occasions, for a Wazir to ask in the thum's presence, 'Who is the greatest king of the East?' and for another flatterer to reply, 'Surely the Thum of Hunza; unless, perhaps, it be the Khan of China; for these without doubt are the two greatest.' This monarch has a very high opinion of his own importance."[4] A similar statement was related by Francis Younghusband who narrates a story that quotes *tham* Safdar Ali as being 'under the impression that the Empress of India, the Czar of Russia, and the Emperor of China were chiefs of neighbouring tribes' and as assuming 'that 'great kings' like himself and Alexander never left their own country'.[5] Here the rulers pretended to be at eye level with mighty empires and powerful imperial rulers whom they obviously did not fear. Even then, Hunza and Nagar have been quite remote from any prosperity and accountable affluence that would enable them to expand their power much beyond their confines in the Hunza River watershed.

Such narratives about the 'petty' states of 'just enough' have survived to date and kept observers wondering about the impetus and interest in those impoverished microstates.

The dynamism of expansion and broadening the domestic economic base by founding new irrigated oases is one of the early steps in enhancing Hunza's local and sub-regional power by increasing not only its territory but also its population at the same time.

The rulers and their *wazīr* seem to have been the outstanding masterminds, and all related idiographic histories were penned by members of these two elite groups that provided us with their selection of memorable events. Outsiders' reports are mainly corroborated by imperial travellers who have quite different judgements and strong views about the performance of Hunza and Nagar's elites. Their selective views – which entered confidential reports, articles and books for public audiences – are guided by a principal interest in controlling and maintaining their spheres of influence and keeping administration costs low by backing compliant and submissive elites. These limited available written sources can help in identifying certain transformative processes, changes in power structures and their impact on economic performances.

Certain transformative phases can be exposed for microstates in the Hunza Valley on their way from being principalities in 'dual vassalage to both China and Kashmir' to eventually becoming integral districts within Pakistan-administered Gilgit-Baltistan.[6] One workable diachronic approach could be to roughly structure 'secular trends'[7] as pre-colonial, colonial, and postcolonial. If we perceive these three as given, a sequence could be constituted in a first step of formative actions to establish Hunza and Nagar in their spatial extent and shape. The external intervention and establishment of a semi-autonomous status under indirect rule conditions after the 1891 Hunza-Nagar Campaign would represent the intermediate phase, and finally the process of economic and political integration of former principalities into Pakistan's administration leads to the present time. Local elites and colonial officers who performed as main actors in negotiating relationships left written evidence about their observations and interpretations. Less material is accessible from Pakistan's administrators. A different set of sources and observations from academic studies and fieldwork is traceable to shed light on post-colonial times. A growing literature of local narratives and interpretations of the region's past has emerged in recent years.[8] Some spotlights will be introduced to contextualise these transformations.

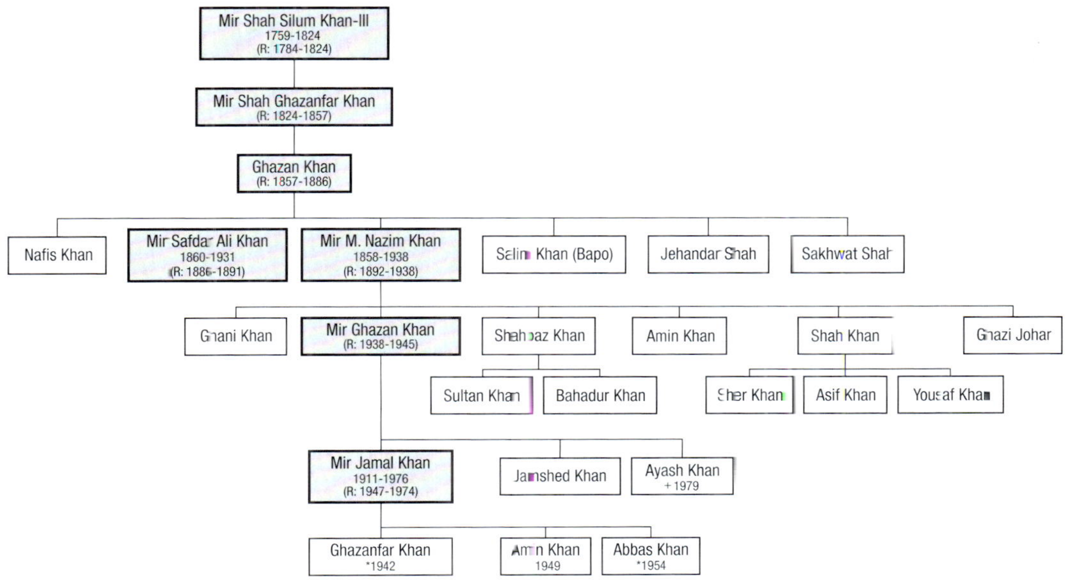

Several versions of Ayáśo genealogical tables exist. The heavenly (ayáśo) dynasty ruled Hunza until 1974 when mīr Jamal Khan was deposed and the Hunza State abolished. The representation here begins with Shah Silum Khan who is identified as the third bearing the same name. Kanjut expanded northwards under his reign and an extensive area was brought under Hunza's control. His son Shah Ghazanfar Khan was notorious for caravan raiding, as was his grandson Ghazan Khan. Their independent and courageous spirit challenged the Kashmiri Maharajas and motivated the British to colonial intervention. The invaders claimed their right to teach the Hunzukuč a lesson as the latter were termed troublemakers and caravan-raiders. The conquest took place in 1891 when Safdar Ali Khan – who had murdered his father in order to usurp power – was deposed and fled to Xinjiang from where he would never return. Under British and Kashmiri domination M. Nazim Khan was installed in the following year. With colonial backing he acquired unchallenged authority and became a strong ruler who substantially expanded Hunza's agricultural space by initiating irrigation schemes and the subsequent establishment of new villages. His 46-year rule shaped Hunza much more than the comparatively short seven years in power of his succeeding son Ghazan Khan. The last Ayáśo tham, mīr Jamal Khan, ruled for 27 years and left a strong imprint on Hunza's reputation beyond its borders. In contrast to neighbouring principalities the ruling elite of Hunza is characterised by a limited number of gušpúr which is probably due to the practice of murdering brothers, fathers and uncles by contenders for prime power. Source: Own compilation based on Haji Qudratullah Beg (1962); M. Nazim Khan (1936), Yousaf Khan (2004); David Lorimer (1979–1980) and and further enquiries

M. Rahber is identifying the graves of mīr Silum Khan (1790–1824) and mīr Ghazan Khan (1938–1945) who passed away in Gulmit. The grave of Silum Khan was removed to a higher location when the rising water from the lake formed after the rock fall of 1858 inundated substantial parts of Gulmit. Later on the grave was restored at its present place. The photograph was taken on 25 March 2019

Opposite
The genealogical table lists the male offspring within five generations of Hunza's Ayáśo ruling class. George Cockerill recorded this table in Baltit during the early years of mīr M. Nazim Khan's rule. Out of twenty-one male descendants, eleven were murdered by their immediate relatives who strove for dominance and acquiring solitary power. Source: George Cockerill (1895: 21)

PRE-COLONIAL EXPANSION FROM CENTRAL HUNZA TO THE NORTH

The pioneering move from the narrowly constituted micro-state of Hunza – mainly located in the central oasis and parts of Shinaki with a *tham* strictly depending on the support of the local elite – with a meagre population of a few thousand inhabitants to an expanding mountain principality is mainly linked to a narrative about *tham* Silum Khan III or Salim Khan. Qudratullah Beg narrates the history of defeating his brother Mirza as *tham* and taking over the rule in Hunza in 1788; other sources place the event around 1790.[9] Qing sources indicate the beginning of the relationship with the arrival of a Hunza delegation sent by *tham* Khisro Khan to Kashgar in 1761. There is no evidence that an annual exchange of gifts began with this event as the next record stems from the year 1792. Various sources seem to indicate that a tribute relationship of Kanjut with Yarkand existed after the Qianlong emperor succeeded in securing his control over the Tarim Basin in the mid-18th century.[10] The extent and importance of these relationships remained rather vague and had initially not been related to a significant northward expansion of Hunza's thrust for dominance. Salim Khan became instrumental for establishing a consolidated principality that enjoyed population growth by broadening its resource base and attracting immigrants from neighbouring valleys. His rule seems to be characterised by fending off Nagar's attacks on Shinaki and by initiating an internal colonisation with the expansion of irrigation systems and the cultivation of barren lands close to the population nuclei in the *khan* settlements of the central oasis. Furthermore, his name is connected with the move into Herbar, later renamed Gojal when Wakhi settlers arrived.[11] The villages of nowadays Lower Gojal from Gulmit to Khaiber were established in this period with settlers from as far away as Chaprot, Ishkoman and Wakhan, while Upper Gojal provided space for settlers from Gharan.[12] Beyond Gojal he and his followers crossed the Karakoram passes bordering with Sarikol and the Chinese realm. The spheres of influence became contact zones especially after the Qing dynasty's expansion into the territories south of Kashgar. In the Hunza tradition *tham* Salim Khan is the pioneering mastermind, while his son Ghazanfar Khan continued to keep up good relations with the Chinese *amban* in Yarkand who conferred the title *khan-e Kanjut* on *tham* Salim Khan.[13] Hunza's presence north of the Karakoram passes was thus established and generated substantial income for the ruler's household. The number of princes and members of the ruling elite was always kept low in Hunza as sons developed a tendency to murder their fathers, uncles and/or brothers on their way to ruling the country. In addition to these killings, Ghazan Khan had murdered his uncle Abdullah Khan, brother of Shah Ghazanfar Khan. Safdar Ali Khan ordered the killing of his two brothers – Jahandar Shah and Sakhawat Shah, eleven and six years old, respectively – in Ghulkin in the spring of 1888 and of his brother Salim Khan when the latter was fourteen years of age.[14] The reasons are not very clear, as not all of them were contenders for the *mīr*-ship. In a genealogical table that was prepared by George Cockerill as an appendix to Edmund Barrow's gazetteer, he listed twenty-one male offspring of *tham* Khisrau Khan down to M. Nazim Khan, the then ruler.[15] Eleven – more than half of them – were victims of fratricide or regicide; Ghazan Khan's brother Rihan Shah escaped to exile in Kashmir, his brother Bakhtawar Shah died of smallpox. Nevertheless, all of the *guśpúr* had to be supported from the meagre resources and revenue Hunza offered. Other elites such as religious elites have never made much impact in Hunza, no *sayid* and *pīr* took residence there; the majority of Ismaili Hunzukuč became followers of *pīr* who resided outside the valley. The elite structure of Hunza contrasted with that of Chitral where *rāja*, *guśpúr* and *akaabīrtiṅ* formed a substantial share of the society and had to be fed from state revenue.[16] A slightly different structure is reported from Nagar where *sayid* or *śaají* form part of the elite society as well as *guśpúr* who have had a significant influence on internal politics and dependency relations.[17]

Ghazanfar Khan, who murdered his father Salim Khan around the year 1825, is remembered as Hunza's consolidator who maintained a certain influence in the Taghdumbash Pamir by securing taxes paid by pastoralists.[18] Even this source of income did not ensure an affluent state of affairs. The diversification of income generation by squeezing pastoralists and traders as well as generating income from slave-trade and raiding caravans remained a life-saving necessity for several generations. Stuart Hill Godfrey assessed its importance: "The most valuable portions of the trophies were brought to Hunza and deposited in the Mir's treasure house. The ordinary merchandise such as 'charas' and wool used to be buried at Baregh Rich and brought over to Hunza at convenient seasons *viá* the Shingshal. The raiders were given shares from the commoner portion of the proceeds in accordance with rank and were exempted from forced labour (begar). The principal raiding season was August. The Mirs of Hunza used, as shewn above, to obtain previous notice of the departure of caravans from friends and merchants in Yarkand and Sarikulis helped the Hunza men in communicating timely notice of passing caravans. Fifteen successful raids were undertaken in Tham Ghazanfur's time."[19]

Ghazan Khan succeeded his father in 1857 and was in no position to tap further sources, but he refined the existing ones. In the colonial records, eighteen raids were attributed to *tham* Ghazan Khan up to 1880 when he supposedly agreed on a moratorium with the Maharaja of Kashmir.[20] He expected some additional wealth when imperial visitors applied for entry to Hunza so that at least the permanent expansion of his territorial control to Chalt and Chaprot would be restored. The prospects were not bright for recovering this valuable source of revenue, and any attempts to get hold of both oases would mainly end with his reign.[21] His neighbours in Nagar and his own *wazīr* were questioning his legitimacy and sternly opposing his grip on the leadership. Already in 1884 the *mīr* of Nagar, *tham* Zafar Zahid Khan, had clandestinely submitted a petition to the Maharaja of Kashmir: "Your Highness you will never get any good out of the Raja of Hunza [Ghazan Khan]. It was he that last year advised Pahlwan to attack Gilgit. Then he himself went to Chaprot until Raja Aman-ul-Mulk caused him to return. It was at his insistence that Pahlwan went to Yasin but by your Highness' good fortune he eventually retreated. The Raja of Hunza makes mischief and with Gilgit for his target secretly forments disturbances. If by your good Highness' good fortune Hunza be taken possession of the way to Badakhshan will be open and this will be to your benefit. If you wish this is the way if not you are powerful. The Gilgit officers did not find my son Kamal land to cultivate in Gilgit, and then the Raja of Hunza deceived him by summoning him to Hunza and telling him that if he went thither he would give him his daughter in marriage and would also give him the Rajaship of Gilgit or Nagar."[22] It seems as if *tham* Ghazan Khan was feared by his rival and that external interference by the Maharaja of Kashmir was urged to contain his expansionist and exploitative doings. Likewise the self-proclaimed *wazīr* of Hunza – it seems that different persons appropriated that title at the same time – wrote to the Maharaja of Kashmir asking for relief and support: "We,

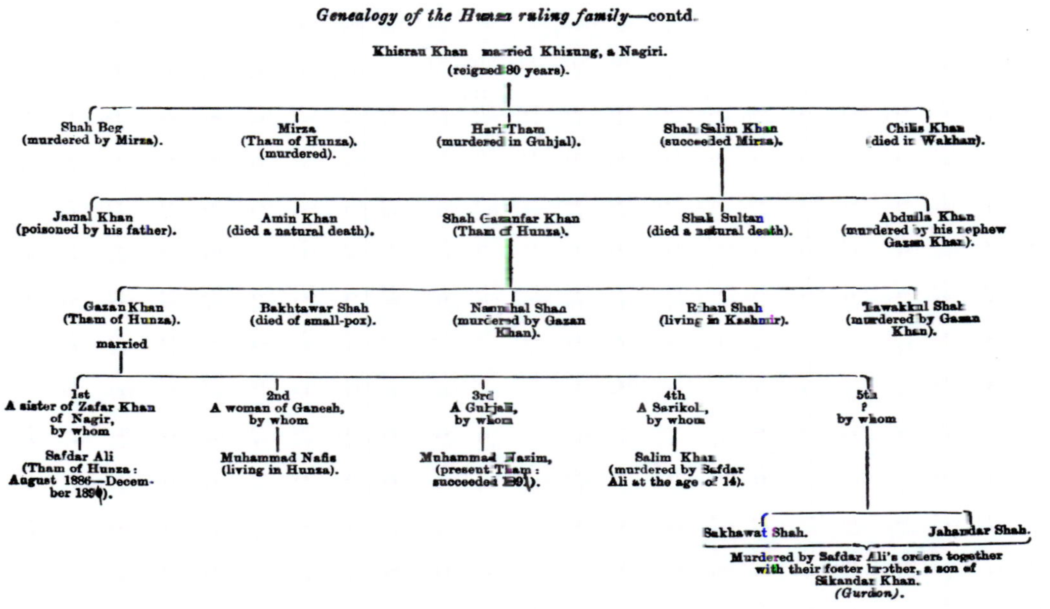

One of the oldest houses in Gulmit is attributed to mīr Silum Khan whose handprints are believed to be engraved in both main pillars (istún) which frame the cooking-place (dildáng) and support the lantern roof (kut). The photograph was taken on 25 March 2019

Portrait of mīr Ghazan Khan who ruled in Hunza from 1857 until 1886 when he was killed by his son Safdar Ali Khan. Source: The picture is a copy of an original painting by Robert Gosset Woodthorpe in April 1886 and photographed by David Lorimer during his stay from 1934 to 1935. Lorimer Collection in the School of Oriental and African Studies, University of London (PP MS 66), Box II.43; reproduced with kind permission

Wazirs and sons of Wazirs of Hunza, beg to submit that we and the other people of Hunza are being ruined, and grieved by the oppressive conduct of Raja Ghazan Khan and trusting you will be pleased to send to Gilgit Raja Rehan Ali Shah for whose coming here we are longing. When once as he arrives we will depose Raja Ghazan Khan and will place Raja Rehan Ali Shah in his stead. A certain Akbar Khan (servant) of Raja Ghazan Khan carried with him 12,000 rupees in cash and the same amount in valuables to Darel for Pahlwan son of Gowhar Aman to assist him in stirring up disturbance and expelling the Maharaja's troops – and he then went to Jummoo on false pretences. We trust that you will arrest him in chains and send Raja Rehan Ali Shah to Gilgit. Please do not deny this act of great kindness otherwise in a short time the Maharaja's troops will come to harm at the hands of Ghazan Khan. We, Wazir Nasiruddin, Karam Shah, Shah Rindar, Wazir Hasna, Wazir Mirza Hassan, Wazir Nadir Kahn, Wazir Mohamad Zamir, Wazir Fazl are dependents of the Sirkar [*sarkār*]. Raja Ghazan Khan during the few years that he has lost hold of the Sirkar's skirt has been behaving badly. All the people look to the Sirkar and have sent Zamir Khan to Your Highness. Kindly send Raja Rehan Ali Shah here having appointed him in place of Raja Ghazan Khan"[23] The *wazīrkuč* had obviously selected a favourite for the succession as *tham* of whom they expected more influence in stately affairs. Rehan Ali Shah was a twenty-five year-old half-brother of Ghazan Khan, who had resided for many years in Jammu since the *tham* exiled him.[24] Their wishes would not be fulfilled by the Maharaja; they had to wait another two years.

Both parties played well into the hands of the British-Kashmiri alliance, but his strength and ability to maintain control of the access to the Pamirian Crossroads qualified the Hunza ruler as a difficult counterpart in any kind of negotiations. Hunza's China-oriented relationships were not amusing for the British and Russian travellers visiting Hunza. In enhancing his demands for gifts, *tham* Ghazan Khan tried to play the China card. The visitors described the *tham* in rather unfavourable terms; William Lockhart remarked during his visit in 1886: "Ghazan Khan is perhaps 60, fat, blackish, but with a rather merry eye. He was dressed in a common *chógha* &c., was supported by servants, and after greeting us was put back on his couch, a raised platform covered with blankets and sheepskins."[25] His appearance was more that of a commoner than of a prince. When it came to discussing politics, *tham* Ghazan Khan was a stern negotiator who repeatedly demanded the return of Chaprot and stubbornly made his position understood: "The old man was, as I have said, very civil, but he ended by saying that he was the subject of the King of China, and could acknowledge no other masters."[26] At this stage Colonel William Lockhart was not aware that *tham* Ghazan Khan would not rule much longer. Four months after they had left Hunza the next regicide happened: "About the same time Safdar Ali, eldest son of Mir Ghazan Khan of Hunza, murdered his father and ruled in his stead. He communicated the circumstance to the Kashmir Government in the following terms: – 'By the will of God and the decree of fate, my late father and I recently fell out. I took the initiative and settled the matter, and have placed myself on the throne of my ancestors. I have now made friends with my mother's brother, Rajah Jáfir Khán of Nagar, and we are of one mind in all things. Nagar is Hunza, and Hunza is Nagar, and we

Colonel Algernon Durand and rāja Akbar Khan discussed possible cooperation measures in Gakuch on 7 May 1891. Source: The Graphic 19 December 1891, page 720

are united in the service of the Maharájah of Kashmír."[27] It seems from his remarks that Safdar Ali Khan is treating the regicide as quite a normal affair: a negligible factor in comparison to the achievement that both rulers have overcome the inherited enmity between Hunza and Nagar and have joined forces to maintain their independence. Both deemed it necessary to fend off British and Kashmiri threats of invasion and control of the Hunza Valley and its northern passages to China and Russia.

The representative of the Maharaja of Kashmir in Gilgit, Bakshi Mulraj, heard about this state of affairs through the *tham* of Nagar and responded to Safdar Ali Khan: "Your letter on the subject of destroying the late Raja Ghazan Khan who had cherished ill will towards you reached me through the medium of Raja Jaffer Khan. The decree of fate inscribed on the forehead of every living being cannot be altered – at all events you are the heir of your forefathers and the expectation is will remain obedient and loyal to the servants of the Sirkar for the sake of your own welfare and peace. You should now send your Motabirs [*moʿtabar ādmī, envoy*] with the usual presents."[28] Bakshi Mulraj treated the affair as a usual change of power according to the local customs and demanded proof of loyalty from the new ruler in the form of gold dust presents for the Maharaja of Kashmir.

Three days later the Kashmiri representative communicated his assessment of the situation to Srinagar: "I have heard from the servants of Hunza chief that 12 days ago [26 August 1886] Safdar Ali Khan and Wazir Dadu son of Doola [Asadullah Beg] agreed together and posted several men armed with muskets outside the fort gate. When in the morning Raja Ghazan Khan, as usual, rode his horse and went from his house towards his garden, 15 muskets were fired (on him) as soon as he came out and when he fell wounded from his horse they beat his head with stones. On Tuesday the 17 Bhadon (31st August) all the country headmen united and appointed Safdar Ali Khan to the Rajaship of Hunza. Safdar Ali Khan is the elder son of Ghazan Khan and nephew (sister's son) of Raja Jaffer and he therefore at first sent his men to Nagar and made friendship and after that sent his Vakil to Gilgit through Raja Jaffer Khan."[29] The news about the circumstances travelled fast: five days after the murder Safdar Ali Khan was acknowledged as the new *tham*, and four days later the representative of the Maharaja of Kashmir demanded from him the 'usual presents' as the symbol for maintaining Hunza's loyalty relationship with Kashmir. Again four days later a telegram was sent by the Maharaja Pratab Singh to Oliver St. John, the influential British Resident in Kashmir.[30] All diplomatic exchanges had taken place within less than a fortnight and raised alarm signals in Srinagar.

The joining of forces between the Hunza and Nagar rulers was perceived as a new phase of hostility towards the British and Kashmir. Communication with Gilgit was at a low point, as Colonel Algernon Durand realised that the

The approach to Baltit Fort follows a narrow, sometimes covered passage through Diramishal Khan. In the foreground are substantial wooden structures normally found only in mosques or habitations of the elite. Less conspicuous are the houses of the formerly fortified village (khan) constructed from clay and stone between the jamáat khana and the fort. Source: German-Austrian Himalaya-Karakoram expedition 1954; Wolfgang Pillewizer (1960, following page 32). Courtesy of Gerhard Klamert

Ali Gauhar, arbāb from Ghulkin, was appointed by mīr Ghazan Khan to accompany the Gilgit Mission on 26 April 1886. Colonel William Lockhart described him in his narrative as 'our guide' and 'chief of Ghulkin': "Alí Gauhar was a fine handsome blue-eyed and brown-bearded Gujálí of about 40, who spoke Persian well. He was always in the highest of spirits, quoting proverbs and making jokes, and, when a difficulty arose with the people of the country, his whip generally solved it – used not brutally, but with a genial jocularity that deprived it of its sting." He accompanied the Gilgit Mission on their way to Wakhan and acted as a valuable expert crossing the Little Pamir. Two years later Ali Gauhar was selected to meet Bronislaw Grombchevsky on his arrival. His impression of him was less favourable, particularly as he was appalled by the state of cleanliness and hygiene of Hunzukuč in general. Coming from Kilik Pass, Ali Gohor was the liaison who guided Bronislaw Grombchevsky to Central Hunza. Source: Bronislaw Grombchevsky (1888/2015: 116); William Stephen Alexander Lockhart and Robert Gosset Woodthorpe (1889: 395). Photograph reproduced with permission © The British Library Board Photo 1040/15

"present ruler, Safdar Ali … is on bad terms with Kashmir owing to the unpolitic treatment he received at the hands of the late Governor of Gilgit. …Probably the best arrangement will be to grant him a small subsidy, and if possible to arrange for his visiting India."[31] Algernon Durand might have misjudged the situation, but sternly refused to accept that *mīr* Safdar Ali Khan claimed that his only allegiance and loyalty relation existed with China.[32] The *tham* sensed that he was in high demand and kept his options open to negotiate with other potential contenders of British-Kashmiri dominance.

The next foreign visitor to the valley after Ghazan Khan's demise arrived from the northern passes. Bronislaw Grombchevsky was met by honourable representatives in Gojal. The first *guśpúr* who welcomed him was *mīrzāda* M. Nafis Khan, a younger brother of the late *mīr* Ghazan Khan from Ghazanfar Khan's fourth wife, and received him at his home in Khudabad: "He is twenty-four years old in the company of twenty-four barefoot men; he is of small stature and fat, not presentable. He is wearing a cotton *toqi* (cap), a thick gown, incredibly dirty pants, and is itching himself all the time. Overall, the Kanjutis are rather unkempt and dirty, lice are everywhere. The *wazīr* [probably the *arbāb*] Ali Gauhar is constantly opening his shirt to remove some vermin; everybody seems to be like that."[33] Bronislaw Grombchevsky was rather surprised about the appalling state of economic and political affairs in Hunza, which had a highly respected reputation of skilful bravery and extreme courage when engaged in raiding and slave-trade. The *arbāb* Ali Gauhar tried to impress him and boasted: "We are a people that have never lost a ruler by the hands of outsiders, but at the same time no ruler passed away naturally. We select our *tham* ourselves and we kill them ourselves. That is because of our greed and poverty. If our *tham* respects one of his followers more than others and gives him a gift, then the others become jealous and think how they can pay him back."[34] The internal power equilibrium seems to have been quite a tricky one based on allegiances and loyalty groups. The remark shows that the *tham* did not display absolute authority; Safdar Ali Khan was rather dependent for better or worse on the mighty lineage groups (*roóm*) and the *wazīrkuč*. Already at this stage, while travelling down the valley towards Baltit, Bronislaw Grombchevsky sensed from his conversations with Ali Gauhar a hidden enmity between Burusho (Kanjuti) and Wakhi. His overall observation was a strict hierarchical setup that placed people at various levels. An extremely low position was occupied by Balti slaves (*čon*, *ġuláam*) who were captured when Hunza forces defeated a Sikh army; most probably the battle had taken place in 1869. Their fate was to be kept in the villages until they were sold on the Central Asian markets. Until then, they would be fed and detained by each household in turn.[35] Bronislaw Grombchevsky experienced poor hygienic conditions during his encounters and observed an appalling state of health; on a daily basis patients were calling on the Russian expedition for treatment of their poverty-induced eye problems, syphilis, rashes and herpes.[36] The expedition was a revelation for the Russian visitors, and *tham* Safdar Ali Khan used their encounter to show his enemies in the south that he was obviously befriended with Russians.[37] Bronislaw Grombchevsky had underestimated *tham* Safdar Ali Khan's courage. Unnoticed by the Russian traveller, a group of Hunzukuč under

Following page
The so-called 10-Verst map R.IX L8 (1903) is based on the surveys by Bronisław Grąbczewski and was included in the map series that covered the Turkestan borderlands in the Hindukush and Karakoram. 'Kanjut' is mainly known along the Hunza River valley while the eastern and western portions remained unsurveyed. Source: Karta desjativerstnaja Turkestanskago Voennago Okruga, Turkestansk. Voenno-Topogr. Otdel, Pod redakciej Polkovn. Rodionova. 1 : 420,000. Tashkent Sheet R[ang]. IX L[ist] 8, 1903. Courtesy of the Pamir Archive Collection

Пер. Сары-курумъ
Пер. Чангъ
Пер. Тугадыръ
Пер. Куромать
Ур. Кызылъ-дукъ
Ур. Девагръ
Пер. Бенкъ
Ур. Янгрыкъ
Пер. Лепъ-лепъ
Пер. Сасыкъ-така
Пер. Сары-бель
Пер. Тышакъ-тазъ
Пер. Илы-су
Пер. Вахджиръ
Пер. Киликъ
Пер. Минъ-тэке
Хребетъ Музъ-тагъ
Ур. Рунхинъ
Пер. Ричъ
Ур. Спандринъ
Пер. Хунджуръ-ябъ
Пер. Упрангъ
Пер. Мурхунъ
Пер. Кунхость
Гильмитъ
Пер. Баскучъ
Нагаръ
Хисперъ
Битермулъ
Дингъ Бранса
Кутче Бранса
Пер. Музъ-тагъ
Ур. Гозъ
Пер. Шимшалъ
Ур. Шимшалъны-аузы
Пер. Урюкъ
Ур. Сары-камышъ
Ур. Сары-тугракъ
Ур. Чунгъ-тукай
Ур. Азгаръ
Пер. Тайпу
Пер. Кукалянъ
Пер. Анъ-казъ
Пер. Чухшунакъ
Пер. Чупъ
Пер. Кара-ташъ
Пер. Агача-яйлякъ
Пер. Сандаль
Ур. Кырчинъ-баши
К А Н Д Ж У Т Ъ

THE FIGHTING NEAR GILGIT—PANORAMA OF CHALT ON THE HUNZA VALLEY

Chalt occupied a prominent position in the approach to Hunza and Nagar from Gilgit. The substantial fort at Chalt, a contested area between Hunza and Nager, and claimed by Kashmir at the same time, dominated the valley. Controlling Chalt offered ample opportunities for backups and supplies from Gilgit on the march up the Hunza Valley. Source: The Graphic 19 December 1891, page 721

the leadership of M. Sakhi had left Pasu for Sarikol where they "... looted a caravan and arrested many people. In the same place, they captured eighteen people of the Yarkend Kyrgyz, nine hundred rams, many horses and all goods. Good subjects of China! All this was done in secret from us. This young sovereign shows extraordinary activity. He tries to enter into relations with powerful neighbours and, at the same time, does not forget the robberies and robberies bequeathed to him by his ancestors. The local Kyrgyz and Sarykol residents fear Kanjut people to death."[38] This information reached him only after they had left Hunza and arrived in Sarikol. Such events proved that *tham* Safdar Ali Khan continued the tradition of caravan-raiding and contributed to the pretext of British-Kashmiri troops for entering Hunza by force after all bilateral negotiations about secure passages through the Hunza Valley had failed.

To analyse the power relationship of Hunza's rulers with their supporters, it is important to note that the expansionist actions of *tham* Salim Khan and Ghazanfar Khan strengthened their position substantially as some wealth could be accumulated by the local elites who needed to strengthen their respective power base. Rulers in Hunza remained quite weak in pre-colonial times as they strongly depended on the loyalty of their *wazīr* and the support of their foster families (*uúšam*) for their acceptance and survival. Such was the situation prior to colonial intervention. Two actors tried to play this dominant role; *tham* Safdar Ali Khan and *wazīr* Dadu thought themselves to be in a strong position against British and Russian demands to enter and trespass on the Hunza Valley. Having been quite successful in earlier times, the combined Hunza and Nagar forces thought they could keep Kashmiri troops at bay. Their partnership enabled them to build a strong position that immediately collapsed when their defence fell into pieces after the deposed *wazīr* Humayun Beg cooperated with the British-Kashmiri forces, thus easing the conquest by treason. Safdar Ali Khan managed to escape across the borders to China after a short and devastating battle; he would never get a chance to return to Hunza. For his predecessors and himself, the Chinese realm had been an arena for looting caravans and bartering goods, but at the same time a loyalty relationship existed which was taken advantage of by the fleeing elites. The invading troops

The Hunza wazīr Humayun Beg is sitting among members of the First Central Asian expedition, photographed by Aurel Stein (1903: 36) who remarked: "He is a tall, well-built man of about fifty years, with an imposing beard, and makes a striking appearance, even in the semi-European costume he has chosen to adopt, evidently as a mark of his progressive ideas." The photograph was taken in Aliabad on 18 June 1900. Source: Stein Collection in the British Library. The latter is reproduced here with kind permission of the International Dunhuang Project: photo 8/1–2

Opposite
Francis Younghusband compiled his route experiences into a map that accompanied his most popular book on the 'heart of a continent' in which the Hunza Valley features prominently. Source: Francis Younghusband (1896, map supplement)

tried to downplay this age-old relationship and mark it as a recent event.

The British Resident in Kashmir Colonel William Francis Prideaux claimed authority on Hunza's history after the conquest in 1891 and narrated the 'official' version of its entering into tribute relationship with China as having taken place much later. His version contrasted with the collective memory of Hunza: "So far as can be ascertained, it does not appear that there was any inter-course between Hunza and China until the insurrection of the seven Khojas in Yarkand in 1847, when Shah Ghazanpur Khan of Hunza rendered assistance to China in overcoming the rebellion. In return for this service a jagir was granted to him in the vicinity of Yarkand, and a brass tablet, inscribed with a record of the friendship of Hunza towards Pekin and its reward, was placed on the gates of the city. A fixed subsidy was paid by China to the Hunza Chief, who in return gave a nominal allegiance. The connection with China was temporarily dissolved when the Atalik Ghazi turned the Chinese out of Yarkand, but the Hunza Chief appears to have retained possession of his jagir, and the former state of things was resumed after the death of Yakub Beg. A yearly tribute appears to have been paid by the Hunza Chief since 1878, consisting of an ounce and a half of gold-dust, in return for which he receives a present of two rolls of satin. This payment has been periodically notified in the 'Pekin Gazette'."[39] William Francis Prideaux tried to downplay the length and strength of Hunza's bond with the representatives of the Chinese Emperor in Kashgar and Yarkand. All efforts were made to initiate a new strategic constellation of southward-oriented dependencies and loyalty relationships.

The Hunza-Nagar Campaign of 1891 put the rule in the Hunza Valley at a crossroads. Both *tham* Safdar Ali Khan and his Nagar counterpart Azhur Khan had fled to China. The Times in London reported in January 1892: "Things are settling down quietly in the Hunza-Nagar country, and Captain Townsend of the Central India Horse, who has been appointed military governor pending future arrangements, will probably have an easy task. The people have accepted the situation with composure. They express gratitude for the leniency shown them, and have sent parties to help in the pursuit of the fugitive chiefs. Those chiefs, Safdar Ali Khan, of Hunza, and Uzr Khan, of Nagar, have not yet been captured. They ... are probably the only real enemies we have in that country."[40] The whole confrontation and conflict was reduced to the misbehaviour of two persons; the Times correspondent felt perfectly all right in Hunza and Nagar, being among friends.

The British had made ample use of Humayun Beg, the former *wazīr* who had provided his colonial masters with strategic expertise and local knowledge.[41] The immediate success of the short military campaign was strongly linked to his collaboration. A new role needed to be found for him as he remained a trusted partner for designing the peaceful future. Humayun Beg had been a close friend and supporter of the Political Agent Algernon Durand since their meeting in Chitral two years earlier.[42] There is evidence that he was considered to have become a trustworthy ally of the colonial forces equipped with administrative power in Hunza.

Bash Gumbaz Ps.
GREAT PAMIR
Gaz Kul
or Victoria L.
13880
LITTLE PAMIR
Andemin Pass
15,500
L. Kara diumer
Ak-su
Chak Mak Kul
13850
Ak Tash
Fort in ruins
Neza-tash Ps.
TAGH DUMBASH PAMIR
Baiyik Pass
Kurghan-i-Ujadbai
Karachukur
Bozai Gumbaz
Daria-i-Wakhan
Wakhjir Pass
Ak Bilis R.
Kilik Ps.
Mintaka Aksai
Ili-su Pass
Kurbu Pass
Mintaka Ps.
Murkush
Misgah
Irshad Ps.
Khora Bhort Ps.
Baroghil Ps.
Gaz Kul
12460
21091
Khaian Aksai
Khunjerab Pass
Darkot Ps.
15000
Darkot
9160
21409
22817
Gircha
KANJUT
HUNZA
Mustagh
Orpang Pass
22590
25870
Shimshal
Shimshal R.
Dalti
Mir Wali
20646
Bar
Gulmit
Darwaza
Afdigar
Yassin
7800
Hunza
Ishkoman R.
Yassin R.
19310
Chaprot
24090
Nagar
NAGAR
Nilt
18340
16767
Gakuch
RAKIPOOSHI
25561
24593
25503
Naltar R.
Hunza R.
Hispar Glacier
Shimshal Pass
Batresgah R.
Cher
Nomal
16379
22448
22278
24486
Nushik La
Chogo Loongma Glacier
Gilgit R.
Doobunni
20168
23914
Kergah R.
16033
Gilgit
Haramosh
24285
Dodargali Ps.
Chonchar Ps.
Hikmul
13850
13950
Vacha

The painting illustrating a 'glacier in the Pamir' captured the sheer ice masses on the rounded peaks towering high above the arid scree slopes interrupted by scanty green patches of Pamirian pastures. Source: Oil on canvas painting by Aleksandr Yakovlev (1931)

Opposite
The portrait of Safdar Ali Khan was taken in Yarkand on 19 April 1930. Source: Image taken by Philips Christiaan Visser or Jenny Visser-Hooft during the Third Dutch Karakoram Expedition and part of the expedition's photo album. Courtesy of the National Archives of the Netherlands, collection 558, No. 2726

EXCURSUS – SAFDAR ALI KHAN IN EXILE, A PLACE OF NO RETURN

Algernon Durand was interested in further investigating Safdar Ali Khan's role during the previous five years. The British Consul-General in Kashgar George Macartney provided him with his findings that were based on the answers provided by the former *tham* and *wazīr* during interrogations by Chinese authorities in Kashgar: "With regard to Ghazan Khan's murder, Safdar Ali denied having had a hand in it at all, but said Wazir Dadu committed the crime. Dadu was then asked … confessed having killed the old Raja; but added that he did so under Safdar Ali's orders. Safdar Ali denied this; but he was also asked why he kept the murderer of his father in his service, to which he replied that he did so out of fear of Dadu. Safdar confessed to having killed his younger brother. Respecting Kuch Muhammad [Kirghiz Beg of Taghdumbash Pamir], Safdar Ali confessed to having kidnapped him from Taghdumbash. He also avowed his complicity in the seizure of Baltis and Kirghiz, and the selling of them as slaves. On the last point, namely, Safdar Ali's relations with the Russians, he claimed having been intriguing with them. Last year ... he sent Kanjutis to Russian territory to purchase some hawks ... but, beyond this, nothing had passed between the Russians and him."[43]

For Safdar Ali Khan an unforeseen odyssey in exile resulted in poverty and substantial financial loss, although the proceeds of a *jagīr,* in Hunza's possession from his grandfather's time – the so-called Rawachi or Robatchi *jagīr* in Ujrik – contributed to his upkeep. He settled near Kuchar where he was closely watched; all his movements were monitored. The Chinese authorities had him arrested as he tried to move to Yarkand and interrogated him about how he judged his civil status. Replying that he was regarding himself as a Chinese citizen, he was then urged to adhere to Chinese laws, remain in Kuchar and not dare to leave.[44] George

Macartney immediately reported any changes in Safdar Ali Khan's family relationships, local connections and the basics of his livelihood. In 1904 the Kashgar files recorded: "Safdar Ali has settled permanently in that town [Kuchar], having taken a wife there by spending two or three Yambus and purchased some land and garden in Kuchar for 35 Yambus. To pay these expenses, Safdar Ali has four or five followers, two of whom are continually travelling backwards and forwards between Kuchar and Yarkand. Safdar Ali receives a monthly subsistence allowance from the Chinese, consisting of 50 chings (66 ½ lbs) of wheat flour, in addition to 2 Taels of silver for each person, i.e., for himself and his followers."[45] George Macartney found out that he possessed some gold and added to his report that 'the gold dust is said to be from Hunza'. He suspected that Safdar Ali Khan had taken the treasure of Hunza along when fleeing to China: "Safdar Ali has, during his stay of two years in Kuchar, sold about 60 taels of gold dust (evidently of the Hunza river) and has ... taken a wife and purchased some land and garden in Kuchar. ... This gold was sent to Safdar Ali by his family in Yarkand through a white bearded Kanjuti ... who is said to be an old and reliable servant. It is said that there are about 1,200 yaks and 5,000 sheep belonging to Safdar Ali, in the charge of his brother-in-law in Wakhan, and for which Ali Mardan Shah is said to have arranged a grazing ground. Safdar Ali and his family in Yarkand generally live on the income derived from these cattle. Safdar Ali spends his money in drink. He has never been heard to speak well of his brother Muhammad Nazim Khan of Hunza; and this should show that no communication passes between them."[46] When Safdar Ali Khan planned to visit his family members in Yarkand in the spring of 1909 he applied for an endorsement to the Acting British Consul at Kashgar; Allen Robert Betham Shuttleworth ordered him to return to Kuchar and to remain there. He treated the affair as a breach of the rules applicable for the ex-ruler's exile. Francis Younghusband, the Resident in Kashmir, supported the consul's action; both British representatives still perceived Safdar Ali Khan as a security threat to the 'Northern Frontier' and the long-lasting 'pacification' of Hunza.[47] In any case Safdar Ali Khan was relatively powerless, being devoid of any substantial pecuniary assets or income. An update on the state of his affairs was recorded in 1911: "... the Amban of Kucha ... gave him a piece of State land, called Kolifangtza. Safdar Ali cultivated this land for one year, and again petitioned the Taotai for more land.

The former tham of Hunza; Safdar Ali Khan, was portrayed by Reginald Schomberg in front of his residence in Kucha in the beginning of January 1928. Owing to the season he is wearing the local-style thick coat and fur hat. Copyright Pitt Rivers Museum, University of Oxford; accession number 1998.292.203

Opposite
After completing his eight-year-long duty as British Consul at Chongqing in 1911, Archibald Rose contributed a paper on 'Inner Asian frontiers of India'; in the appended map he separated the Chinese territories within the Pamirian Crossroads by a yellow line that met the red line of the British sphere of influence. Nepal and Bhutan functioned as buffer states while all parts of the Pamirs, Hindukush and Karakoram Mountains fell into the British realm that contained Afghanistan and Kashmir; even Russian Turkestan was not separated from the red line territory. Source: Archibald Rose (1912, map following page 218). Courtesy of the Pamir Archive Collection

… 100 mow [*mu*; the property equals 6.7 hectares] of State land, called Khanka, was given him. In addition to the above mentioned lands, Safdar Ali has a garden, named Shamal Bagh, in his possession, and he lives there. He has 2-3 Kanjuti followers who work for him."[48] The Chinese authorities allotted some real estate to him, thus guaranteeing him a potentially decent survival in exile. His eldest son, Muzaffar Khan, looked after the family property of the Robatchi *jagīr* as he himself was not permitted to travel to Yarkand.[49] Being himself a brother-in-law of *mīr* Ali Mardan Shah, exiled ruler of Wakhan, he tried to marry off his daughters to respectable families. His eldest daughter Mehtar Nigar married a son of Guda Muhammad, an Ismaili *pīr* of Yarkand; the second daughter Gulmitee married Yakut Shah, son of the former Nagar *tham* Azhur Khan, the third was betrothed to Karim Beg of Sarikol.[50] All the Xinjiang-based matches were meant to root him better in his present socio-political environment. The Chinese administration faced difficulties controlling his drinking habits. The latter broke him and made him the poor fellow seen in Xinjiang's bazaars.

Few people paid him any respect as he rarely received visitors. Reginald Schomberg commiserated with the loser of the Hunza-Nagar Campaign in 1928: "Safdar Ali … was a kindly, jovial old man, but dependently poor and a pathetic figure in his ragged clothes. He had forgotten his native tongue, but he remembered his Persian and neatly capped a quotation from Hafiz. Thirty-five years of exile, penury and hardship had not daunted his courage or spoilt his temper, and whatever his deeds were, he cannot be judged by the standards of Whitehall; and he fared badly at the hands of his friends and his enemies."[51]

Safdar Ali Khan died in March 1931, and the intelligence officer Reginald Schomberg described him in his internal report as a victim of a local power game and external dependency: "In justice to his memory … the events that led to the Hunza-Nagir war were mainly due to his Wazir, and even more to the pressure brought on him by Ju-Sho-we the Governor General of Urumchi after the re-occupation of Sin-Kiang on the fall of the Yaqub Beg, Bedaulat's dynasty. Ju urged Safdar Ali to fight the British, promising money and supplies, of which neither was forthcoming. Safdar Ali kept the letters containing these promises, and in later years when poverty-stricken and desperate, produced them to the Chinese and blamed them for his present plight. The ex-Mir paid bitterly for his action: and it is disagreeable to think that the British treated their old enemy in so ungenerous a fashion, allowing him to end his days in penury. The old man bore no malice, but it was not a pretty sight to see a Mir of Hunza's brother, and the ex-ruler of that state, dressed in rags and often half starved, driving a small donkey in the Kuchar bazar."[52] Later on a dispute came up whether his son Muzaffar-ud-din was entitled to sell the *jagīr* or whether the ruling *tham* in Hunza could claim it. Reginald Schomberg sensed the political implications of the demise of the former *tham* and fate of the refugee's household: "Safdar Ali … [had] some property at Kucha called Shimal Bagh. He gave this to his daughter who now enjoys it, and went himself

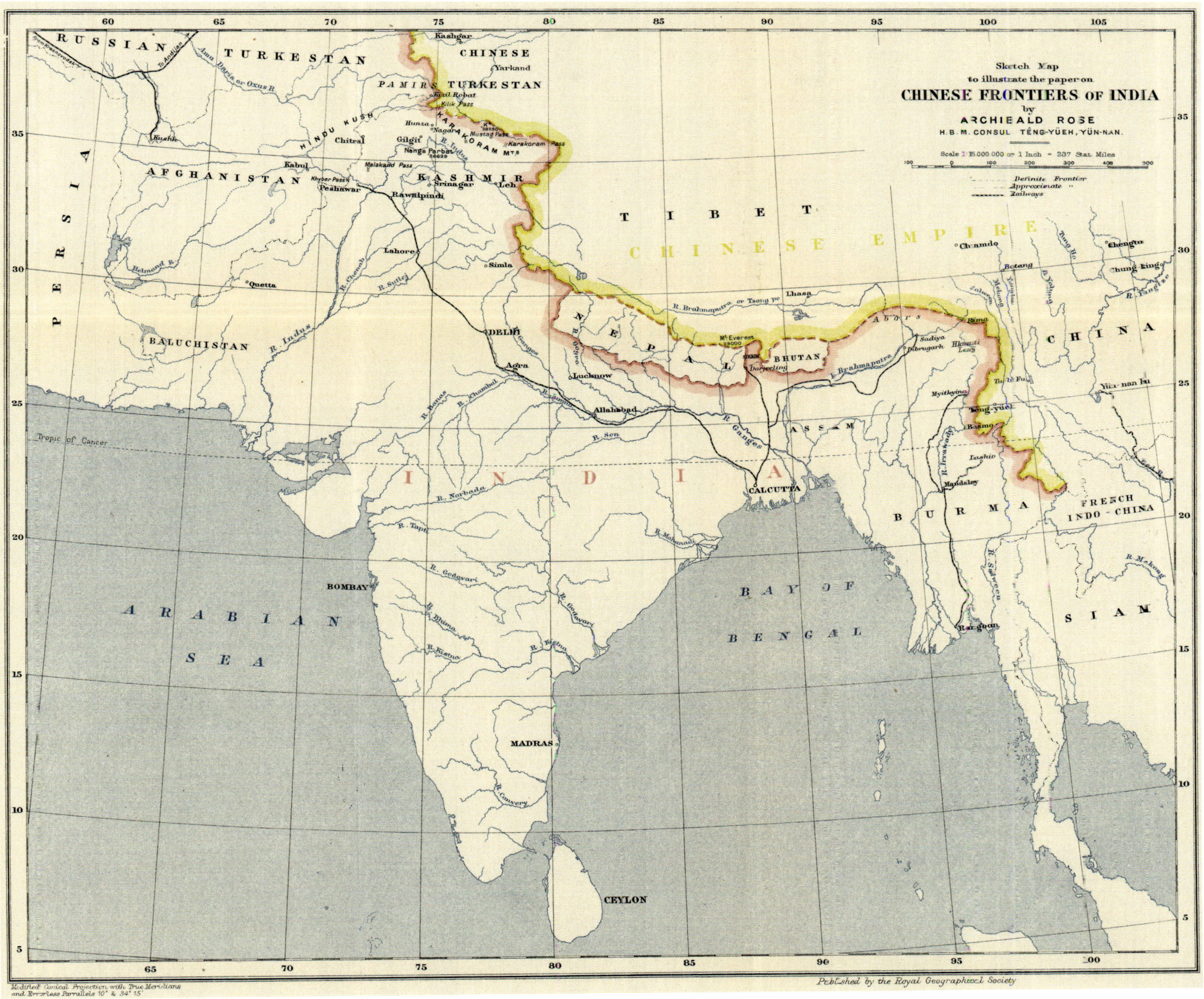
Sketch Map
to illustrate the paper on
CHINESE FRONTIERS OF INDIA
by
ARCHIBALD ROSE
H.B.M. CONSUL TÊNG-YÜEH, YÜN-NAN.
Scale 1:15.000.000 or 1 Inch = 237 Stat. Miles
Definite Frontier
Approximate "
Railways
RUSSIAN TURKESTAN
CHINESE TURKESTAN
PAMIRS
Kashgar
Yarkand
HINDU KUSH
KARAKORAM MTS
Karakoram Pass
AFGHANISTAN
KASHMIR
Kabul
Peshawar
Rawalpindi
Srinagar
Leh
Gilgit
Hunza
Chitral
Lahore
Simla
Quetta
PERSIA
BALUCHISTAN
TIBET
CHINESE EMPIRE
Lhasa
Chamdo
Chengtu
Batang
NEPAL
Mt Everest
BHUTAN
Darjeeling
DELHI
Agra
Lucknow
Allahabad
ASSAM
Sadiya
Dibrugarh
Myitkyina
Têng-yüeh
Bhamo
Lashio
Mandalay
Rangoon
CHINA
Yün-nan fu
BURMA
FRENCH INDO-CHINA
SIAM
INDIA
CALCUTTA
BOMBAY
MADRAS
CEYLON
Tropic of Cancer
ARABIAN SEA
BAY OF BENGAL
R. Indus
R. Ganges
R. Brahmaputra
R. Narbada
R. Godavari
R. Kistna
R. Cauvery
R. Irrawaddy
R. Salween
R. Mekong
Modified Conical Projection with True Meridians and Errorless Parallels 10° & 34° 15'
Published by the Royal Geographical Society

Frederick Marshman Bailey's photograph of Nazim Khan and Ghazan Khan in Karimabad garden was taken on 14 May 1918. During his 'Mission to Tashkent' he was always attracted by groups of soldiers who prominently feature in his photographs, as they do here where Hunza levies from the Gilgit Scouts back the rulers of Hunza. This constellation reflects the political situation: The mīr of Hunza is guarded by Hunzukuč who are remunerated from colonial subsidies, thus safeguarding the appointed ruler's security. Source: Bailey Collection in the British Library. The latter is reproduced here with kind permission of the International Dunhuang Project Neg 1083/44 (14)

to a newly developed district about thirty miles south. … For 35 years Safdar Ali never received any help from his half-brother who succeeded him. Almost in rags, the ex-Mir existed in exile. Now his half-brother proposes to beggar his brother's daughter and her family. There is also a political side. We do not want the Mir of Hunza to have interests so far North. It is bad enough at present with his absurd claims and his mythical rights, and the more he keeps on our side of the frontier, the better."[53] For the reputation of the British Raj his fate seemed tragic, and his treatment inappropriate. Safdar Ali Khan never regained any significant influence and did not detrimentally affect Hunza's diplomatic affairs with British India, China or Russia. However, his half-brother and successor on the throne (*taqt*) tried hard to receive the proceeds from the Robatchi *jagīr* after the demise of Safdar Ali Khan; *tham* M. Nazim Khan corresponded with the British authorities about sending his envoys there to collect the grain taxes even during the troubled times in the second half of the 1930s.[54]

Colonial assistance – strengthening the position of the *tham*

The important geopolitical location of Hunza within the Pamirian Crossroads asked for a reliable and stable power constellation in the Hunza Valley. The new colonial dominators perceived Hunza's relationship to China as a thorn in their side. The full story of Hunza's deep-rooted China connections would become an issue long after the conquest and was a hot topic and much more elaborated on in the colonial correspondence until it ceased in 1937. The British authorities were initially even more concerned with the alleged Russian interference in Hunza's affairs. The 'Great Game' competition and race for dominance had a significant impact on the situation in the Hunza Valley and in neighbouring regions such as Sarikol and Wakhan.[55] The successful Hunza-Nagar Campaign, the fixing of the Durand Line two years later, and the delineation of the Pamir Boundary in 1895 ended a confrontation in the 'Northern Frontier' of the Raj that had quite different results in the various regions involved.[56] The uncertainties connected to *tham* Safdar Ali Khan's attempt to play all three actors – Great Britain, Russia, and China – off against each other had to be paid for dearly by him, resulting in his own downfall and flight into exile. After he had been removed as the major obstacle to implementing a new order, the route through the Hunza Valley could be opened up. Dim and unpleasant pictures of all previous rulers had been painted in finding a pretext for the necessity of conquest. After the British entered the mountain strongholds new loyal actors needed to be identified and appointed.

Contemporary sources differ in narratives as to how the existing power vacuum was to be filled.[57] In the autobiography of *mīr* M. Nazim Khan, who became the first of three rulers appointed and backed by the colonial authorities, a classic justification for his appointment was given by narrating that his father *tham* Ghazan Khan had personally selected him as his successor in his last will, and not the first-born son Safdar Ali Beg. Evidence seems to be at hand that some documents were fabricated to support his claim. Irmtraud Müller-Stellrecht convincingly argued the case and highlighted some contradictions from Nazim Khan's autobiography.[58] The need to forge a document might have been necessary from his point of view. He was an offspring of a morganatic marriage (*arġúndaro*) of his father *tham* Ghazan Khan with Qurbon Bibi from Gulmit, while Safdar Ali Khan had claimed the noble (*qaralí*) origin of Nur Jehan, the daughter of the *rāja* of Gilgit, to become the eligible *tham*. Nazim Khan's earlier acquaintance with the British authorities in Gilgit, his genuine admiration and interest in foreign visitors[59] predestined him for selection.

With Humayun Beg as the influential, loyal and powerful *wazīr* at his side, the expectations of the colonial authorities could be fulfilled to install a functioning pair that would keep administrative costs low. The Hunza-Nagar Campaign obviously appeared and functioned as a pacifying mission; Algernon Durand explained the remarkable result of dominating all externalities as being a major step towards a future roadmap: "My friend Humayun, the ex-Wazir of Hunza, had naturally been with us, and arrangements having been made for the civil government of the country, and a small body of troops left there temporarily, the force was withdrawn, and the headquarters reached Gilgit six weeks after the first move from Chalt. I may mention that no interference in the internal affairs of the states was attempted. Nazim, a half-brother of Safdar Ali Khan, was installed later as Chief of Hunza, with Humayun as his Wazir, and the old King of Nagar [*tham* Zafar Zahid Khan] continued to reign, with his son Sikandur Khan as heir-apparent. Uzr [Azhur] Khan came back, gave himself up, and was deported to Kashmir, where he now lives. Safdar Ali Khan fled to Yarkhand, where he lives on a small property granted to his father by the Chinese, and his truculent Wazir Dadu died in exile."[60] The Hunza-Nagar Campaign had changed all power relationships in the valley in December 1891 and prepared the ground for successful indirect rule. The Acting British Agent in Gilgit; George Robertson, had informed the Viceroy who himself forwarded the encouraging message to the Secretary of State in London on New Year's Eve 1891: "The people, relying on my assurances, are rapidly returning to their homes. Desertions from the fugitives are said to be taking place in large numbers. Certain men who had good cause to fear our resentment have also come back. I have taken possession of the two countries, pending the orders of the Government of India concerning their disposal; and have appointed Lieutenant Townshend, Central India Horse, Military Governor of Hunza. A strong party leaves the day after tomorrow [25 December 1891] for Gulmit, Gircha,

The fifty-year-old ruler of Nagar, mīr Sikandar Khan, is sitting on the left with his wazīr Khuda Aman at his side among notables from Nagar, including rāja Gohor Aman who is wearing the uniform of the Gilgit Scouts. The other person standing behind mīr Sikandar Khan is rāja Khisro Khan from Chalt, his son-in-law, and rāja Baz, nephew of the tham and son of his elder brother M. Khan, at the outside. The photograph was taken by Frederick Marshman Bailey (1946: 16) in Minapin on 13 May 1918. Source: Bailey Collection in the British Library. The latter photograph is reproduced here with kind permission of the International Dunhuang Project Neg 1083/44(12)

The photograph shows George Nathaniel Curzon in viceregal attire during his reign between 1899 and 1905. Source: The half-plate negative was taken after 1899 by Elliott & Fry Studio in London. Courtesy of the National Portrait Gallery

and Misgar, to examine road and impress remote villages with a military display; also with the hope of coming up with runaways. ... People cheerful; full of offers of service; their women move about without fear of molestation."[61] Rarely had a military intervention led to such a complete turn in less than three weeks. It took until 15 December 1892 before George Robertson, Francis Younghusband and George Macartney, three of the most prominent protagonists of the Great Game in the 'Northern Frontier' witnessed the investiture of M. Nazim Khan as *tham* of Hunza. A hope was expressed that was repeated at subsequent similar occasions: "May your rule be a long and happy one and may your people live in peace and prosperity."[62] The desire for internal stability and compliant rule according to the standards of imperial geopolitical interests in the region would be realised in their bilateral future relations.

Hunzukuč and Nagarkuč became loyal supporters of the British Raj immediately after the conquest; under the leadership of Raza Beg, younger brother of *wazīr* Humayun Beg, they willingly and effectively supported British troops in Chitral less than one year after their military defeat.[63] Similarly, the newly appointed local rulers – Nazim Khan and Sikandar Khan – proved to be loyal to the Political Agent in Gilgit for the best part of half a century. Nevertheless, initial mistrust prevailed: "Aliabad has been garrisoned by a company of Kashmir troops, and a Political Officer stationed in the country."[64] Even then, within five years after the successful campaign, its strength was only eighty-five and subsequently further reduced: "This garrison has lately, under the orders of the Government of India, been reduced to a post of 20 men."[65] Bertrand Gurdon was posted as the last Political Officer in Aliabad when Lord Curzon visited Hunza in 1894.[66] The Assistant Political Agent and the Hunza garrison were completely withdrawn from Aliabad by 1897. A local levy post was kept under the command of Raza Beg with two sub-leaders, eight section commanders and one hundred levies from Hunza that were trained for one week per year and otherwise kept without arms.[67] A similar setup was arranged for Nagar. These schemes were mere local employment measures rather than instruments for controlling local residents. Sikandar Khan became the mastermind and powerhouse in Nagar. Although *tham* Zafar Khan remained the nominal ruler until his death in 1904, his son reigned and was eventually formally installed then as *mīr*; the colonial authorities preferred this title from now on, while *tham* remained the honorific address in Burushaski terminology.[68] Both rulers, *tham* Sikandar Khan and Nazim Khan, were initially seen as compliant and malleable puppets who would support the cause of securing the Northern Frontier. While the Hunza Valley was left without garrison, the posts in Bunji, Chilas, Gilgit and Gupis remained under one regiment of Kashmir Imperial Service troops, one Imperial Service Mountain Battery and two companies of Sappers and Miners. The military efforts were quite contained and tightly structured for such a large territory between the Afghan and Chinese boundaries. The trust in the local *tham*, *mīr*, and *rāja* who performed as rulers and governors was substantial; they were closely monitored by monthly reports in the Gilgit Political Agency Diaries. The entries were ranked by importance, and Hunza and Nagar took the lead. Thus, M. Nazim Khan received significant attention and was regarded as the most important and influential personality in the Northern Frontier. At the turn of the century Aurel Stein described him as "a man of about thirty-five, of open and manly bearing, and evidently deserves the reputation for intelligence and firmness which he enjoys. Our conversation, carried on in Persian, turned naturally more to the old conditions of the country than to the reforms about which the Mir is said to be energetic. Road-making, vaccination, and similar Western improvements seem strange as objects of genuine interest in the representative of a family for which intrigue and murder were down to the present generation the main incidents of life. This transformation in its rapidity and evident thoroughness is a striking proof of the results of the pax britannica."[69] It seemed that the firm grip on the *tham* whose investiture was the first in colonial times led to the expected performance of duties and execution of power.

The long rule of *tham* Nazim Khan

Nobody could expect that his rule would last for forty-six years. The investiture of Nazim Khan as *tham* in Hunza coincides with a fundamental change in external and internal relations among stakeholders and actors. The kingship was finely intertwined with quite powerful descent groups in which the *tham* established loyal relationships, a characteristic of precolonial times. The Hunza-Nagar Campaign marks the hiatus; everything would be different afterwards, independent of the person chosen as *tham*. The earlier power struggle was fought between the antipodes of *ayáśo* from whom the *tham* originated and the group of elite seniors (*uyóṅko*) that included the *akaabírtiṅ* and *kurpátiṅ* who were well off and represented their respective peer groups as heads of lineages.[70] The ruling class and the *uyóṅko* had to find an equilibrium of interests while after 1891 two influential players would intervene from outside and modify power relations. The immediate perceivable intervention by the British-Kashmiri troops documented who was in charge of Hunza affairs. All external relations and internalities were monitored from the office of the British Agent in Gilgit who officially became titled 'Political Agent' in November 1896.[71] A kind of 'indirect rule' was installed that allowed a compliant and submissive execution of local power to be contained and controlled. Officially, the Maharaja of Kashmir appointed the *mīr* of Hunza and Nagar 'with the approval and authority of the Governor-General in Council'. Both the *tham* accepted a *sanad* issued by the Maharaja of Kashmir that read: "The chiefship will be hereditary in your family and will descend in the direct line of primogeniture, provided that in each case the succession is approved by the Maharaja of Jammu and Kashmir for the time being and by the Government of India. … Further you are informed that the permanence of the grant conveyed by this Sanad will depend upon the ready fulfilment by you and your successors of all orders given by the Jammu and Kashmir State, with regard to the conduct of relations between the State of Nagir/Hunza and the states and tribes adjoining it, the administration of your territory, the construction of roads through your country, the composition of such troops as you may be permitted to retain, and any other matters in which the Maharaja of Jammu and Kashmir for the time being may be pleased to intervene. Be assured that, so long as your house is loyal to the State of Jammu and Kashmir and to the British Government and faithful to the conditions of this Sanad, you and your successors will enjoy favour and protection".[72] The question of road construction was solved along with a set of rules that needed full compliance with Kashmir's orders. Although it was termed as 'no interference' in domestic affairs, neither legislation nor jurisdiction was fully autonomous and independent. The British authorities reserved their 'right' to guide neighbourly relations, most important in reference to Hunza's connection with China: "No interference is exercised either by the Political Agent or the Gilgit Wazir-i-Wazarat in the internal administration of the two States. In external affairs the Mirs are expected to consult the Political Agent."[73] The sophisticated British reporting system documented in the Political Agent's diaries and annual administration reports ensured that the colonial line departments under the India Office were always aware of what was going on in Hunza and Nagar. A physical proof of the established and reinforced tribute relationship was the fixed exchange of gifts that were to be presented at

Maharaja Pratap Singh ruled in Srinagar from 1885 to 1925. The photograph shows him in full military attire and decorated with medals, it was taken in Jammu shortly before the Hunza-Nagar Campaign started. Source: International Mission Photography Archive

The Lorimers took this portrait of M. Nazim Khan during their stay from 1934 to 1935. Source: David and Emily Lorimer 1934–1935

When the caricature appeared, Sultan M. Shah, the third Aga Khan, had become spokesman of the Muslims in the Indian subcontinent. As President of the League of Nations he enjoyed cosmopolitan personal connections with political leaders in Europe and Asia that allowed him to fulfil an influential role in India's independence struggle and partition. Source: Men of the Day No. 938: The Aga Khan III caricatured by Leslie Ward aka 'Spy' in Vanity Fair 10 November 1904. Courtesy of the National Portrait Gallery

the annual 'Jalsah Darbar' (*jalsa darbār*) in spring in Gilgit and subsequently 'carried down to Srinagar by representatives of the Mirs'. On Hunza was imposed an annual tribute of '16 tolas and 5 mashas of gold, and on Nagir of 17 tolas and 1 masha'; in exchange both principalities received 'annual subsidies of Rs 4,000 each, Rs 2,500 of which is paid by the Government of India and Rs 1,500 by the Kashmir State'.[74] The dual control and supremacy is reflected in the shared payment of subsidies, although the gold dust presents were solely reserved for the Maharaja of Kashmir. The British authorities presented themselves as the paramount power of the 'northern frontier'. The same applied here as in the tribute payment to China. The subsidies were about fifteen times higher in value than the rupee equivalent of the gold dust would have been in exchange.[75] The imperial powers used the tribute relationship mainly as a manifestation of loyalty and of their sphere of influence rather than as a source of income.

Further external interference was based on denominational affiliations. Here Hunza and Nagar significantly differed as nearly the entire population of Nagarkuċ and Ganeshkuċ of Hunza professed Twelvershia faith, while the vast majority of Hunzukuċ were followers of the Ismaili *imām*. Although personal contacts between *mīr* and *imām* had been limited so far, colonial interference and domination would change their relationship. With some time delay a delegation led by M. Raza Beg, brother of *wazīr* Humayun Beg, visited Bombay and the *dar-xāna* of the Aga Khan there. Aga Khan III, Sultan M. Shah, made it clear to the delegation that he expected full obedience from his followers (*murīd*) towards the colonial government and their *mīr*. The highest Ismaili religious authority issued direct decrees (*firmān*) for Hunza as no *pīr* was residing there. The power game that was evolving seemed more straightforward than it was. The Hunza rulers were in no respect natural replacements of absent *pīr* and executed instructions by the Aga Khan. The leaders of the Gilgit Mission had obtained a *firmān* of the Aga Khan that was supposed to offer free passage (*rāhdāri*) through Hunza; *tham* Ghazan Khan had refused to accept external orders and reluctantly let the mission pass through Hunza.[76] Although M. Nazim Khan married daughters of a *sayid* family as his third and fourth wives, it was only in the next generation that his eldest son, the future *tham* Ghazan Khan, married Sabiha, a daughter of *pīr* Shahzada Lais, the most influential *pīr* in Hunza.[77] In connection with *pīr* Shahzada Lais, exiled from Zebak in Badakhshan and residing in Arkari in Chitral, suspicion regularly came up that Ismaili representatives were spying for the British Raj in the Russian Pamirs and Badakhshan. Shahzada Lais was accused of instigating the settlers in the Pamirian Crossroads by suggesting an imminent invasion by British troops and annexation.[78] These events provided early evidence of an Ismaili-British political connection in which the *mīr* of Hunza and the exiled former *mīr* of Wakhan Ali Mardan[79] Shah prominently figured.

All three *mīr* of Hunza who were installed during colonial times were quite ambivalent in their denominational orientation. Marriage relations were never restricted to members of the Ismaili community; several wives and mothers belonged to Twelvershia families. The bond between *mīr* and *imām* became stronger when more secular aspects of community organisation and political orientation were dealt with.

The influence of the Aga Khan was channelled alongside the *pīr* system, which became virulent when subsequent reforms were introduced within the Ismaili Imamat. The growing link between *mīr* and *imām* was merely based on a mutually profitable relationship for both sides in the field of levying tithes in silver and gold. The colonial postal infrastructure was used to transfer the proceeds from collecting the tithe (*daswān*) by money order from Gilgit to the *dar-xāna* in Bombay.[80] Later on the Aga Khan would raise the status of the Hunza *tham* among his people when he made him his Central Asian representative and collector of all tithes from Badakhshan, Chitral and the Gilgit Agency.[81]

The impact of the colonial administration on *mīr* M. Nazim Khan's rule was more complex; their interventions could be selective and change between two antipodes, thus affecting the balance of power between *tham* and *uyóṅko*. Nazim Khan's power game and expansive influence were transformed during this period in respect of his dependency on his *wazīr* and support of the people. A growing liberation from the influence of Hunza's elites and a substantial increase of internal taxation characterised his rule. He significantly raised the state revenue based on growing forced labour services, direct and indirect taxes. Although he claimed the opposite in his autobiography, a biased approach could be observed. While exempting some of his 'state subjects' – especially Burusho and residents of Shinaki – he managed to make Gojal his grain chamber from where the bulk of dues and taxes in grain, butter and livestock could be expropriated. Especially during the first decade of his rule *mīr* M. Nazim Khan stretched the inequalities. An average citizen of Gojal paid ten times higher taxes than anybody in Hunza. Although only every fifth Hunzukuč was living in Gojal, about eighty percent of revenue was generated there, and the *tham* had a vital interest in expanding the settlements there. Nearly all livestock was delivered from the fertile pastures of Gojal and especially the Shimshal Pamir.[82] The value of Hunza's taxes amounted to 1,800 Rs in 1894. Even less was earned from the *tham*'s own lands; they generated 675 Rs, which originated from forced labour (*rajaáki*) contributions by the lower classes. The subsidies paid by the Kashmir Durbar and the Government of India were quite substantial in comparison. They had been doubled by 1896 when the Hunza State received 4,000 Rs distributed between *mīr* Nazim Khan (3,000), his brother M. Nafis Khan (300), and *wazīr* Humayun Beg (600); the remainder was reserved at fifty rupees each for Raza Beg and M. Isa, as the leaders of the Hunza levies.[83] It is important to note that *wazīr* Humayun Beg received a higher subsidy than the brother of Hunza's *tham* and was remunerated with a higher amount than the nominal ruler of Nagar.[84]

In *mīr* M. Nazim Khan's perception of his 'subjects' personal qualities between Burusho and Wakhi were to be distinguished. He himself was born of a Wakhi mother from Bori *qətór* lineage, Qorbon Bibi, and termed an *arġún* in Hunza society. Nevertheless, he developed an unfavourable opinion about Gojali, as he entrusted to Henry Haughton who remarked in 1912: "Suicide is apparently rather common in Gujal, though the men are generally considered to be rather soft and faint-hearted, or, as the Mir of Hunza himself expressed it, the Gujalis are goat-hearted (*buz-dil*) in war but perfect heroes at committing suicide!"[85] His opinion is in tune with a general perception among Burusho who perceived themselves as a class of warriors and caravan looters, while the Gojali (*guíć*) were responsible for providing them with valuable foodstuffs. Being strongly dependent on the support of the *wazīrkuč* and other Burusho elites, *mīr* M. Nazim Khan needed to maintain a working relationship with Humayun Beg. The relations between *tham* Nazim Khan and his *wazīr* were estranged right from the beginning. Few occasions were lost where the *wazīr* would refrain from showing off his supercilious position, backed by his colonial masters. In addition, he enjoyed power con-

The platinum print of a photograph made in the studio of Samuel Bourne and Charles Shepherd was taken in 1903 during the Coronation Durbar in Delhi when the tham of Hunza and Nagar, M. Nazim Khan and Sikandar Khan, represented the two principalities. Source: Bourne & Shepherd's photo biographic album 'Coronation Durbar Delhi 1903'

Humayun Beg was the most influential wazīr from the Tarakuć lineage. He was a personality trusted by the British colonial administration and instrumental in making the Hunza-Nagar Campaign a successful military endeavour for the combined British-Kasmiri forces. Personally he profited by regaining the wazīrship until his death in 1916. Source: Undated photograph provided by Ejazullah Baig, Karimabad

nections embedded in Burusho social structures and his superior experience in political scheming. Between his investiture as *tham* and Humayun Beg's demise in 1916 M. Nazim Khan had experimented with a kind of innovative taxation (*khuṭúkal*) system in 1897 already, but could not succeed owing to the strong influence exerted by *wazīr* Humayun Beg. The new taxation scheme that would generate more revenue from Burusho in Central Hunza was refused by the influential Tarakuć who controlled substantial amounts of land themselves. When *mīr* M. Nazim Khan attended the Delhi Durbar by invitation of the Viceroy in 1911, *wazīr* Humayun Beg used his absence for intriguing and caused an alarmed reporting by the confidant of the *tham* and 'native assistant' Munshi Fida Ali to the colonial authorities that "... party factions broke out headed by Wazir Humayun Beg and his brother, Muhammad Riza Beg, and that recently the situation all but became critical. Matters have since been patched up for the time being and the parties outwardly reconciled, but it is stated that there is much ill-feeling at heart."[86] The quarrels between the Tarakuć brothers continued during the coming months, and it was alleged that *mīr* M. Nazim Khan deliberately caused this trouble to weaken their respective influence.[87] During his travels in the Hunza Valley Henry Haughton sensed the upcoming showdown and noted in his diary: "A storm is brewing in the Hunza tea cup. Put shortly, the Mir having by cunning or bribes got the N. A. [Native Assistant Munshi Fida Ali] entirely on his side, thinks it an excellent opportunity to down Humayun Beg his Wazir of whose influence & connection with the Sahib log he is jealous. This broadly speaking is the 'cause'. The 'occasion' is – a N. A. in tune with himself; a new P. A. [Political Agent] whom he hopes he (the Mir) has taught to sing the same song; & the quarrels of Raza Beg, the Wazir's brother which form a good opportunity for undermining the Wazir's influence in the country."[88] The colonial observers realised that *mīr* M. Nazim Khan was trying to unbalance in his favour the power equilibrium that had been established to 'pacify' the frontier and to reduce administration costs. Henry Haughton put some trust in the *mīrzāda* who might become the successor of Nazim Khan: "I hope Ghūzan Khan will turn out well and shall be interested to hear. He has been spoilt in the past, but has certainly improved since last year. I like the child who I am sure has a lot real good in him."[89] Political relations had developed in such a way that they were understood to be personal bonds at the same time.

The Political Agent Major Archibald Duncan Macpherson reported in summer of 1912: "As the result of the recent intrigues in Hunza the Mir has unquestionably strengthened his position in the country, and now, himself, disposes a great deal of business which formerly the Wazir arrogated to himself. In some respects – provided the Wazir does not lose too much influence, which is unlikely – the change is not an altogether undesirable one, as hitherto the Wazir has probably enjoyed rather too much power, and this has naturally created a constant state of friction and jealousy between him and the Mir. At first the Wazir very strongly resented this weakening of his authority, but there are indications that he is now more or less reconciled to the new position, or, at least he accepts it ..."[90] Different routes were taken to maintain each other's influence; arranged marriages and foster relations played a prominent role of interference. When *wazīr* Humayun Beg accompanied his foster-son

Section of the map 'India (Northern sheet)' prepared by William and Alexander Keith Johnston showing the attribution of Hunza and Nagar to Kashmir that is depicted as a homogenous unit at par with the North-West Frontier Province in 1914. Source: The Handy Royal Atlas of Modern Geography, sheet 28. Edinburgh & London 1914. Courtesy of the Pamir Archive Collection

The photograph shows mīr M. Nazim Khan 'in his characteristic attitude' next to the expedition's leader Kenneth Mason and other members dressed in their fashionable colonial khaki shorts, knee socks and topees. It was taken on May 23, 1913 by Richard William George Hingston, who worked as a surgeon and naturalist with the Indo-Russian Triangulation expedition from April to September 1913. Photograph No. H. 38 is reproduced with permission of the Trinity College Library Dublin (MS 10484/7/38)

mīrzāda Ghazan Khan, the heir apparent, on a mission to Srinagar where they had the honour to attend an audience with the Viceroy Lord Hardinge in October 1912 he obviously managed to strengthen their personal bonds. In his purposeful scheming he convinced Ghazan Khan to let his son be fostered as *uúśam* in the *wazīr* household like himself.[91] Humayun Beg developed a stronger influence over the *mīrzāda* than his father could exert, which substantially contributed to the rift between the two senior rivals. This observation caused *tham* M. Nazim Khan to announce a few months later the complete abolition of the foster system, a move widely accepted except by Humayun Beg.[92] Even the leading Ismaili *pīr* Shahzada Lais – who had a large followership (*murīd*) in Hunza and was the father-in-law of *mīrzāda* Ghazan Khan – was consulted in April 1913 to settle the dispute among the quarrelling parties. The Political Agent Archibald Duncan Macpherson intervened and publicly advised all headmen and the *mīr* to comply and 'to improve the unsatisfactory state of affairs in Hunza'.[93] As an immediate reaction to this destabilising power constellation the Gilgit-based authorities reinstalled a British officer in Hunza to monitor the situation. Lieutenant William Macandrew Marshall was appointed to mediate the conflict and help in setting up two companies of Gilgit Scouts each from Hunza and Nagar.[94]

Visitors during this period did not give the most favourable report about the *tham*. William Hingston on his geodetic mission remarked about his first encounter in May 1913: "Shortly after our arrival at Aliabad we were greeted by the Mir of Hunza who, accompanied by his bodyguard of ruffians, had come from his capital to welcome us. We entertained him to tea. He does not seem an individual in whom one would care to place implicit trust."[95] During public appearances the *tham* was always accompanied and surrounded by his bodyguards, so that he made the impression of a rough gang leader. William Hingston seemed compassionate and looked deeper: "On the whole he appeared a kindly old man and his nice manners and knowledge of English customs and little forms of politeness were surprising in one who lived almost his whole life amongst the primitive and semi-civilised people which occupy his country. ... The Mir seems to be continually surrounded by a band of ruffianly looking blackguards. They followed us into his private rooms, glared in at us through the open door while having tea and followed us in a long line when leaving the palace. We were not sufficiently fortunate to see them in the maroon uniform in which, on State occasions, the Mir dresses them. They looked evil enough when unwashed and in their own filthy clothes and any attempt of decorating them would probably make them look even greater criminals."[96] The gang-like performance might have been due to the conflict-laden juxtaposition of hereditary rule in Hunza and the *tham*'s fear of being assassinated. Overall William Hingston's impression was not that bad, as he recorded on his way out of Central Hunza from Aliabad to Atabad: "While travelling through Hunza territory today we had an opportunity of studying the inhabitants. They appeared a polite and friendly people and crowded out of their little villages to greet us as we passed. The careful cultivation of the crops, the ingenious methods in which the water supply is directed by channels to the fields, the excellent arrangement of superimposing areas under cultivation in tiers one above the other until in some cases they reach high up the steep mountain side all point to the thrift and industry of this interesting people. They appear even a purer type of Aryan extraction than the Kashmiri. Some of the children were almost as fair and of the same cast of countenance as any European. All were well but simply clothed and cleanly in their appearance. They are a fine type of manhood, strong and healthy, but as submissive as children to the authority of their Mir whom they implicitly obey."[97]

Observers might conjecture that a certain grip on the political relations within Hunza was sought because *wazīr* Humayun Beg's health was deteriorating.[98] Both contenders feared that British presence in Hunza would be detrimental to their common and individual interests; therefore, various measures including mediation by *rānī* Urman Pari, wife of *tham* Nazim Khan, were applied.[99] The issue had gained political importance and had occupied much of the reporting about Hunza in the Gilgit Political Diaries. Their armistice was confirmed after much intervention in July 1915 when they publicly took an 'oath in the presence of the leading men not to quarrel again'.[100] Through *wazīr* Humayun Beg's last year of life no further incidents and major confrontations occurred. After contracting pneumonia, which was probably transmitted by his sick son Inayatullah Beg, he died on 25 May 1916 after a short illness. Before his demise he had summoned his family members to his death-bed advising them 'that there was only one door by which they could hope to make a success of life and that was the door of loyalty to the Sirkar [*sarkār*] and he earnestly besought them all to enter thereby.' The Political Agent added in his diary: "In the death of Humayun Beg the Government has lost a loyal and steadfast friend and the Agency a personality of remarkable character that was universally esteemed by all classes."[101] For *tham* M. Nazim Khan the demise of the *wazīr* closed a door of rivalry and opened the door for him to become the authoritarian ruler as he has been remembered since. The members of the *wazīrkuč* took Humayun Beg's advice seriously, and the future generations prominently occupied government posts and high military ranks.

The final two decades – *tham* Nazim Khan's authoritarian rule

The appointed and backed hereditary ruler slowly succeeded in freeing himself from a dependency on the Tarakuč, thus becoming a stronger than ever *tham* after a prolonged time of twenty years into his rule. The British-Kashmiri colonial authorities granted *tham* M. Nazim Khan the required political and substantial financial support, consisting of the funds for the upkeep of his household and for his bodyguards; his standing was manifested during his visit to the Delhi Durbar. The exceptional honour and exchanges with other *rāja* fully convinced him to opt for absolute power. Henceforth, he decided to steer Hunza's affairs without immediately appointing a successor as *wazīr* and the one he selected should be a weaker personality.[102] Shortly after Shukrullah Beg, eldest son of Humayun Beg, was appointed in September 1916, power struggles calmed down. For many years to come no remarks about the dual power relationship appeared in the Gilgit Diaries.[103] After a rare visit by the British Resident in Kashmir, Evelyn Berkeley Howell devoted a special note on the function of hereditary rulers in his report about the Gilgit Agency in 1927.[104] He wholeheartedly praised the performance of *mīr* M. Nazim Khan and his service to the Raj: "I doubt whether even the Government of India fully recognise what a debt of gratitude is due to the Mirs of Hunza and Nagir, especially the former. The Kashmir Darbar certainly do not. The whole Gilgit Agency takes its cue from the Mir of Hunza who is in a

The photograph of tham M. Nazim Khan, his eldest son Ghazan Khan and his then youngest son Amin Khan was taken by the Agency Surgeon Major L. Mackenzie in Baltit in 1924. Source: Lorimer Collection in the School of Oriental and African Studies, University of London (PP MS 66), Box 10; reproduced with kind permission

On 14 May 1918 mīr M. Nazim Khan received the 'Mission to Tashkent' under the leadership of Frederick Marshman Bailey in his typical colonial outfit. Several notables of Hunza attended the reception on the roof of Baltit Fort. Source: Bailey Collection in the British Library. The latter photograph is reproduced here with kind permission of the International Dunhuang Project Neg 1083/44(27)

position to make endless trouble. If he were of an Afghan temperament it is easy to see to what a degree of expensive prominence he might have forced his little state. Instead of seeking to exploit Government's generosity, willing or reluctant, Hunza has an all occasions been uniformly loyal and helpful and has on the whole been well seconded by the Mir of Nagir. So far as one can tell there is general content in both the little kingdoms and no ground on which the seeds of malice from without are likely to spring up. There is however one obvious danger. Both Mirs are now over sixty and senile decay is likely to set in before long. Both have houses full of sons of all ages and both have little Benjamins ... for whom both rulers already show a preference as marked as it is injudicious. Naturally this is bitterly resented by the other sons and before long some one of them may assert himself in the traditional fashion, kill the Mir and anyone else who is in his way and claim the succession for himself. That will raise a very awkward problem, and even if nothing of this sort happens there will in any case be a difficult situation when in the course of nature the demise of either Mir occurs. The Maharaja of Kashmir will probably claim to be consulted with regard to the succession, which will not make things easier, when a quick decision is required. I am of opinion that the Political Agent should be directed to begin to consider the succession question in both Hunza and Nagir, and when he has had time to win the confidence of the Mirs to ascertain what their wishes and intentions are. These can then be considered and, if necessary, steps can be taken to secure the concurrence of the Maharaja of Kashmir. Government will then be free to act when time comes."[105] The British Resident was quite satisfied with the present state of affairs, but he feared that when these great personalities – M. Nazim Khan and Sikandar Khan – stepped down an unstable situation could come up again instigated by fratricide and regicide. Evelyn Berkeley Howell did not mention any problems between *tham* and *wazīr*, but he sensed a far greater danger of instability given by the growing number of princes and members who classified themselves as gentry (*guśpúr*). The experience of Chitral with a huge gentry, unwilling to care and work for themselves, but expecting to be fed from the state budget and supported by the local community, posed a portent and was perceived as a warning sign to think of precautions in the Hunza Valley, which had less affluence and fewer resources in comparison with Chitral. The relatives in person of the elder half-brother M. Nafis Khan, *guśpúr* in Khudabad, and the sons of *tham* M. Nazim Khan were envisaged, but the disturbed relations among the ruling elite in Nagar were much more severe. The enmity between *tham* Sikandar Khan and his nephew *rāja* Khisro Khan of Chalt concerned the colonial authorities in as much as they had avoided trouble in his subsequent succession. They eventually made Mahbub Ali Khan a permanent employee of the Raj, first by his taking a permanent commission in the Gilgit Scouts and second by appointing him governor in Yasin in order to avoid conflict and ease the path for the grandson Shaukat Ali Khan to become *tham* in Nagar.[106] On a somewhat sarcastic note, *mīr* M. Nazim Khan had assessed 'that by the time Mahbub had finished milking his father there will only be dust left for Shaukat Ali'.[107] All of *tham* Sikandar Khan's surviving sons made a career in military service and received some pecuniary income. The Resident had reserved a general comment on the position of *guśpúr*: "Mention of the numerous progeny with which the Mirs of Hunza and Nagir have been blessed, if it is a blessing, serves as a reminder of other problems which the existence of an hereditary aristocracy – known generally in Punial, Yasin and elsewhere as Gushpurs – is likely ere long

Raja Khisro Khan (1873–1938) from Chalt, nephew and son-in-law of mīr Sikandar Khan of Nagar (1866–1940), was an influential relative in Nagar. Allegedly Khisro Khan saved the life of Sikandar Khan by giving him advance warning when tham Azhur Khan tried to eliminate him as well after the previous murder of two other brothers (Henry Haughton 1912). Nevertheless, both seem to have been quarrelling with each other during most of their lifetime. The Political Agents often referred to him; Stuart Hill Godfrey mentioned that he was 'accused of plotting against the life of Raja Sikandar Khan' in 1896; Lieutenant Gough "… considered him as not lacking in intelligence, but he has not much force of character. He has a large following in Nagar, though he was decidedly unpopular in Chalt and Chaprot. Being the eldest son of the eldest son of Tham Zafar Zahid Khan, according to European ideas of the rights of primogeniture, he ought to be the heir to Nagar" (Stuart Hill Godfrey 1898: 84). The latter constellation was obviously the reason for his dissatisfaction with existing power relationships. In the final stages of Raja Khisro Khan's life, George Kirkbride called him 'an obstinate septuagenarian' in July 1937; the following month he reported: "The quarrel between the Mir and Raja Khisro Khan of Chalt … is now in abeyance, if only for the time being. Khisro Khan has begged the Mir's forgiveness and the latter has agreed to restore to the Raja the privileges he enjoyed prior to the quarrel." (IOL/P&S/12/3285: 222, 225). In Chalt he met with Aleksandr Yakovlev; his portrait was drawn on 24 July 1931; he died of pneumonic on 20 January 1938, more than a year before his rival Sikandar Khan passed away on 17 March 1940. Source: Alexandre Evgenievich Iacovleff (1934; plate 25), reproduced with kind permission from Gallica. Bibliothèque nationale de France

to present. No Gushpur will turn his hand to agriculture or any task that savours utility, but expects to be supported by the community to which he belongs in comparative affluence and idleness all days. Now that there has been peace in the land for nearly forty years and the check formerly provided by periodic fratricide on an extended scale has been removed, the number of Gushpurs has increased and will continue to increase far more rapidly than their means of subsistence. Only one in each family can hope to succeed to a position of authority and wealth and the outlook for the remainder is a gloomy one. … I see no hope of their being employed by the Kashmir State. Government cannot support them and unless the military authorities are willing to try the experiment of opening service in the army to them, which I scarcely venture to recommend, since their fighting qualities are untried, I can see no prospect for relief."[108] The same concern was reiterated at later stages by observant political agents. George Van Baerle Gillan entered in the diaries: "Difficulties are, however, already being experienced in finding land to support the Mir's progeny: all previous generations, in both families [Hunza and Nagar], were summarily reduced in numbers by jealous brothers and the novel practice of permitting all to survive until removed by natural causes will undoubtedly entail some trouble to the succeeding Mirs."[109] It sounds as if the representative of a 'civilisation mission' regretted that the practice of fratricide and regicide had ceased. But it shows at the same time that the *tham* had consolidated his reign and exerted his dominance with a high degree of authority over his siblings. M. Nazim Khan had become the sole instance when Hunza or the passage between Gilgit and Kashgar was concerned. Having the only guesthouse in Baltit and small rest-houses along the route, the *tham* was naturally the principal host for every visitor arriving in Hunza. He had built an attractive guest-house on a big natural boulder – Tash Mahal – in his Karimabad garden in 1925, which became the central meeting place for foreign visitors.[110] Later on, he decided to transfer his public receptions and court sessions from Baltit Fort to a new *darbār* hall in this orchard, in which he received and entertained his guests. His monopoly of providing accommodation and full board lacked any notion of hospitality; it was rather a very successful means to absorb all monetary funds that were coming to the valley. All required duties were channelled to the *guśpúr* and other members of the elite. Their expertise in tourism and control of movement that survived for decades after has qualified members of the elite to make a career in external relations and to try to exclude others. The aristocracy provided hospitality and pack animals; in addition, they participated in the polo matches and dancing performances. The sword-dancers were recruited from the ruler's bodyguards. These performances were part of the package deal that foreign visitors enjoyed from the early days of *tham* M. Nazim Khan's rule.[111] All porters were from the subaltern *baldákuyo* and remained unpaid under forced labour (*rajaáki*) obligations. The *Croisière Jaune* was a welcome expedition to arrive in Hunza in August 1931 after many years of having seen few visitors. In his discussion with the expedition leader Georges-Marie Haardt, *mīr* M. Nazim Khan made it clear right from the beginning "… that any money which the Expedition might be intending to offer to his retainers should be handed to him.

The first guesthouse was built by mīr M. Nazim Khan in 1925 in the muḥalla of Karimabad below Suryas Das. He named it 'Tash Mahal' as it was erected on a major boulder. The building still serves as a guesthouse today. The photograph was taken in Karimabad on 22 March 2019

'As for those' – introducing with a vague gesture a group of dignitaries, who bowed low – 'they are the most important functionaries of my State, and I authorise them to receive presents directly from your generous hands.'"[112] His suggestion referred to the old practice that all gifts belonged to the *tham*, and that he would decide whether the *rāja*, *guśpúr*, *wazīr* and *uyóṅko* would be granted the privilege to receive some presents directly. All other gifts had to be delivered in full to the ruler's household. The proceeds from the one hundred and fifty porters that were hired by the expedition on its way to Misgar and to Koshbel via Kilik Pass went directly to the *mīri* purse. Hunza porters were not allowed to pursue the matter further.

The growing number of *guśpúr* went along with their growing demand for services that had to be shouldered in the form of *kār-i-begār* or *rajaáki* by members of the lower classes. The *guśpúr* expected to be treated like princes and to participate in courtly affairs and entertainment. It goes without saying that the most important portrait drawn by Aleksandr Yakovlev was that of *mīr* M. Nazim Khan. According to decreasing rank other members of the elite were portrayed as well; surprisingly, the members of the *wazīrkuċ* do not feature prominently. Shukrullah Beg, successor to his father Humayun Beg, had much less influence as *wazīr* and would soon lose what little he had.

The *ayáśo* dynasty consolidated its absolute authority further by investing in marriage relations with other ruling families. The double marriage of *tham* M. Nazim Khan's grandson, the future *mīr* Jamal Khan, with Shams-un-Nahar, a daughter of the *mīr* of Nagar Sikandar Khan and future *rānī* of Hunza, in July 1934 and at the same time the marital bond between Muzaffar-ud-din Shah, third son of Sikandar Khan, and a daughter of M. Nazim Khan underlines a stronger detachment from previous alliances with the *wazīrkuċ*.[113] With the exception of Ghani Khan, his *arġúndaro* son, who married a daughter of Shukrullah Beg, all other siblings were connected by marriage with the religious elite (*pīr*) or neighbouring ruler's families. Shortly after the double marriage a further important match was arranged; the daughter of *mīrzāda* Subedar Major Ghazan Khan was betrothed to the fourth surviving son of the *mīr* of Nagar, Jamadar Babur Khan; the marriage took place in January 1935.[114] Such a strong bond had never existed before between the ruling dynasties in Hunza and Nagar, and it was seen as a sign of their reinforced strength and common interests. The Political Agent George Kirkbride visited Hunza and Nagar in November 1934 and realised that the rulers seemed to be an assembly of greedy people who tried to extract money from each and everything. Proceeds from trade caravans, visitors' expenses, porterage, customs duties and other taxes contributed mainly to their purse.[115] A major point of discussion was the reorganisation of the levies and scouts from the Hunza Valley. Hitherto six hundred and forty scouts had been trained for two months per year and were thus on the payroll. The new system involved the employment of four hundred scouts permanently and two hundred and forty seasonally for a period of four months, which would significantly increase the income. Major George Kirkbride found grateful partners in both rulers who envisaged a substantial expansion of the job market: "The proposal was bound to find enthusiastic acceptance in Hunza and Nagar whose resources are quite unable to cope with a rapidly increasing population. The only difficulty to be apprehended is the quite erroneous impression that all officers must be sons of the Mirs. Its solution is, however, simplified by the demand being greater than the supply. In regard to the future composition the Mir of Hunza went so far to say that he could easily provide sufficient men for the entire battalion but he admitted that mixed companies of Hunzas and Nagris were for the present a counsel of perfection."[116] The struggle for posts shows how important it was to get these job opportunities to feed a growing number of *guśpúr*. The Hunza ruler tried to take advantage of his favourable geopolitical position, which was regarded as strategically more important than that of remote Nagar. The colonial admin-

Significant amounts of meat needed to be prepared when a double-marriage between members of the ruling families of Hunza and Nagar took place in July 1934. The celebrations were staged in Karimabad, the new summer residence which the visitors from Nagar perceived as small and congested. Source: David and Emily Lorimer 1934–1935

istration, though, was interested in keeping an equilibrium between the two states whenever possible. The sidelining of the *wazīr* continued in both states and enhanced the quest for absolute power. Nagar had a limited representation of *wazīri* influence at the core of decision-making. After the Hunza-Nagar Campaign, Taifoor had succeeded his father Shah Murad as *wazīr* of Nagar. The powerful *tham* Sikandar Khan dismissed Taifoor in 1900 and partly vindicated him by appointing him as *wazīr* of Chalt in 1912; he was officially reinstalled in his post in 1921, which he formally held until his death in June 1936. Taifoor was a frail and weak person late in life and failed to influence Nagar politics that were solely dominated by the *tham*.[117] Hunza's internal politics continued to be confronted with a strong and abiding *wazīri* claim for participation in decision-making, for a share in state revenue and dispensation of justice. The duel reached a showdown in summer 1935 after a Tarakuć, Imamyar Beg, was murdered under suspicious circumstances. A new confrontation was envisaged by George Kirkbride who recorded in the Gilgit Political Diaries: "The relations between the Mir and Wazir Shukrullah Beg are giving cause for some anxiety. ... It has for years been the policy of the Mir to belittle the position of the Wazir. The murder of Jemadar Imamyar Beg, a younger brother of the Wazir ... further embittered the feelings between the families because it was rumoured that the Mir's eldest son and heir, Subadar Major Ghazan Khan, was implicated in the crime. The present position is that the Mir refuses to have the Wazir or his family back in Baltit, the capital of Hunza. Government cannot, however, look on quietly at the ruin of a loyal and honoured family and the Mir will be obliged to give up, as gracefully as may be, his present attitude. Of this he is well

During their journey to Gojal the reconnaissance team stopped in Baltit and was entertained by mīr M. Nazim Khan. The frail and old tham is accompanied by his son M. Ghazan Khan who would succeed him the following year after the death of the reigning tham on 22 July 1938. Sitting on the lawn are his sons Shah Khan and his younger brother Ghazi Johor aka Haritham. They are sons by his third and fourth wives who were sisters and daughters of Syed Jalali Shah, a respected pīr from Chatorkhand. Source: The photograph was taken by Flight Lieutenant Francis Claude Sturgiss on 28 May 1937; File No. 82(2)-X/39, photo 70

aware and it remains to be seen how this very astute ruler will save his face and at the same time effect some sort of reconciliation."[118] Despite having grown old there was still some vigour in the *tham*. George Kirkbride observed: "The Mir is now 70 years old. He is very blind and suffers from diabetes. Obviously he cannot last much longer and even should he live for another ten years his amazing powers are bound to weaken."[119] The colonial administration obviously had clandestinely maintained a certain trust in the *wazīrkuċ*, while the *tham* had severed their previous bond and opted for his sole authority by sidelining them.

The boldness with which *tham* M. Nazim Khan reacted is another proof of his emancipation from Tarakuċ influence and his trust in his acquired absolute powers, which had to be guaranteed by the British Raj and its subsidies for levies, scouts and upkeep. He did not truckle to the Political Agent and dismissed *wazīr* Shukrullah Beg in November 1935, but allowed the family to continue receiving one third of the *wazīri* allowance. With the appointment of Zarparast as *wazīr*, a trusted follower joined the inner power circle who belonged to the Gulwatiṅ lineage within the Burooṅ.[120] The choice of a non-Tarakuċ is ample proof of *tham* M. Nazim Khan's alienation and disappointment with the lineage of trusted *wazīr* who had inherited and occupied their position for such a long time. To select a member of the Gulwatiṅ lineage was especially remarkable as they had the reputation of being later immigrants to Hunza who were subsequently integrated into the Burooṅ *roóm*. This move gave the *tham* even more leverage for sidelining the Tarakuċ. The Political Agent George Kirkbride was not at all satisfied with this outcome, but sensed a conflict-laden constellation in which he would not favour the trusted collaborators from the Tarakuċ, fearing that the whole situation could get out of his hand. Thus, he summarises in his report: "For want of a better I have accepted this proposal. It was made in consultation with the leading men of Hunza excluding Shukrullah Beg's family, and is acceptable to Ghazan Khan which is important. … The future position of Shukrullah Beg is however one that gives cause of anxiety. The loss of a moiety of his income is a severe blow to himself and his family and this coupled with the loss of prestige has aroused the greatest discontent in a family that has always been outstanding in its loyalty to Government. The situation is not easy. … The feud between his family [Nazim Khan's] and that of the ex-Wazir is such that it is impossible to offer the latter any appointment of consequence that would not be construed as weakening the position of the former. For example the Mir would never recommend one of the Wazir's family for a commission in the Scouts. I may be able in course of time to provide minor appointments such as that of primary schoolmaster or jail darogha [*dāroga*, overseer] for the younger members of the ex-Wazir's family but this will only indirectly help Shukrullah Beg himself and his remaining brothers."[121] The elite's overall growing dependence on subsidies and remunerated positions had given the *tham* an additional opportunity of shaping the community of his supporters by inclusion and exclusion. The estranged relationship between *tham* and Tarakuċ was aggravated by the scheming and rift-deepening of *mīrzāda* Ghazan Khan who pursued his own agenda, which was not necessarily in tune with his father's and not at all with the representatives of his Tarakuċ foster family. His actions impeded the attempt of George Kirkbride to maintain a fine balance between the adversaries. The colonial reports mention Ghazan Khan's alcohol problem on several occasions; *mīr* M. Nazim Khan sometimes forcefully tried to combat his son's addiction but success was short-lived and his resumed 'heavily drinking' led to a suicide attempt as well.[122]

All this happened at a time when a number of visitors and guests had coincidentally appeared on the scene as observers in Hunza. The Swiss traveller Ella Maillart had arrived along with the Times reporter Peter Fleming from China. They were met in Hunza by the Political Agent Georg Kirkbride and Lieutenant-Colonel Lionel Edward Lang, Resident in Kashmir, who were on an inspection tour in the Gilgit Agency to figure out what kind of compensation would be needed to make the *tham* sever his relations with China.[123]

Opposite
The image of mīr M. Nazim Khan documents his strong authority and vigour when he was about 65 years old. A series of paintings emerged from the prolonged stay of the remaining members of the 'Croisière Jaune' in Baltit where Aleksandr Yakovlev portrayed the mīr on 24 June 1931. Source: Alexandre Evgenievich Iacovleff (1934; plate 26), reproduced with kind permission from Gallica. Bibliothèque nationale de France

Mir Sir Mohd
Nazim Khan.
K.C.I.E Hunza
Baltit

The son of mīr M. Nazim Khan, Amin Khan, is portrayed here at the age of ten. Tragically he lost his life a year later when he accidentally released a fatal shot from a hunting gun. Source: Pastel drawing on paper by Aleksandr Yakovlev, Baltit 28 July 1931. Reproduction with kind permission of Kunstbibliothek der Staatlichen Museen zu Berlin

Emily Overend and David Lorimer pursued their language and ethnographic studies and had resided in Aliabad for fifteen months until September 1935. The latter were great admirers of *tham* M. Nazim Khan and enjoyed a cordial relationship with him. Their accounts show how M. Nazim Khan diminished the position of Shukrullah Beg as *wazīr* and built up a separate system of advisors and courtiers. Of the 2,300 households in Hunza, 138 were offices preferentially given to Burusho. Every fifteenth household had an office among the Burusho and Shina-speakers, whereas the Wakhi had no *uyóṅko* and only every twentieth house enjoyed an office such as village elder (*arbāb*) or guard (*ćarbú*).[124] First of all, he preferred his inner circle of *traṅphá* – who all originated from Baltit and Haiderabad – whom he made responsible for the non-Burusho *maqsóo* in Shinaki and Gojal. Two *traṅphá* were supervising the Shinaki affairs, while Munshi Hubi from Baltit and Ali Madad from Haiderabad were made responsible for matters related to Gojal I (villages from Gulmit to Pasu), and his close confidant M. Rafi, who was married to a sister of *wazīr* Humayun Beg, was in charge of Gojal II – Khaiber to Chapursan. Imam Yar Beg from the *wazīrkuć* was trusted to deal with Shimshali affairs since 1930. Only recently had *tham* M. Nazim Khan doubled the number of *traṅphá* for Gojal. Places like Shimshal and Misgar were of utmost importance for the border security of Hunza; thus the *tham* entrusted control of the Shimshal Pass to the younger son of the *wazīr* Shukrullah Beg, *ǰamadār* Imam Yar Beg who had received military training; while the Misgar affairs were vested in the hands of the *wazīr* himself and Pfairo. In military affairs and defence questions M. Nazim Khan still relied on the belligerent former elite.

His inner circle kept a close watch on any developments outside the central oasis. On a secondary level, five local *traṅphá* held office in Shinaki supported by four *ćarbú*. M. Nazim Khan had appointed thirty-one *traṅphá* for Central Hunza, thus having a representative of nearly every *roóm* besides a number of elevated *uyóṅko* who totalled forty-sev-

en in Central Hunza compared with nine in Shinaki. These formed the social elite of Hunza, augmented only by two *uyóṅko* from Gojal who were Burusho settlers in Misgar and Reminj.

The fine structure of offices included seventeen *ċarbu* alone for Central Hunza who were guards and instrumental for disseminating public orders by the *tham.*[125] The Gojal system of maintaining loyalty and control as well as stimulating the information flow was quite different; the *arbāb* as village headman was the equivalent to the position of *traṅphá*. In every village there was one of seventeen *arbāb* in Gojal. No Wakhi had ever earned a close and respected position by the *tham* above the village leadership or in the Gilgit Scouts. There had remained a significant mistrust of the Gojal population who had to deliver certain special taxes (*ilbán*) to M. Nazim Khan's siblings, *guśpúr* M. Nafiz Khan in Khudabad and Shahbaz Khan in Gulmit. In reconfiguring his inner and outer circles, M. Nazim Khan had satisfied his immediate relatives, sidelined the *wazirkuċ* and diminished their influence to an extent never experienced before, as well as strengthening the circles of confidants and trusted followers around him. The latter were instrumental for his close watch of any sentiments and roots of ill feelings that needed to be contained. By controlling the movements of all Hunzukuċ at *darband* posts, where only travellers authorised by a written slip of paper (*ċit*) from the *tham* could pass, he made his autocratic rule felt in everyday life and hindered out-migration.

At the same time, other external observers and British spies were sent to the Hunza Valley to have a closer look at the developments there and to assess how they would affect the stability of the 'Northern Frontier' of British India. M. Nazim Khan's position and influence would naturally grow in times of geopolitical challenges, but his actions and deeds would be scrutinised much more closely from now on. Especially when by the mid-1930s the political internal situation worsened in Xinjiang, the 'Northern Frontier' received more attention. Additionally, when the Maharaja of Kashmir was to be convinced to lease the Gilgit Agency to the British Raj, the behaviour of *mīr* M. Nazim Khan was going to be evaluated even more critically. Great admirers and caustic critics were found within the British colonial system who did not hesitate to spread their comments and tried to influence decision-making.

In August 1931 Aleksandr Yakovlev spent some time in Misgar before the remaining members of the Croisière Jaune expedition arrived. He was fascinated by the landscape in the Kilik Valley and the local polo matches which he captured in several studies and paintings. Source: Oil on canvas painting by Aleksandr Yakovlev (1931); see Dessins et peintures d'Asie, exécutés au cours de l'expédition Citroën Centre-Asie. Troisième mission (1934)

Colonel Reginald Charles Francis Schomberg (1880–1958) pursued his military career after he was commissioned in 1901 in the 1st Battalion of the Seaforth Highlanders which he commanded in Mesopotamia and Palestine from 1917 to 1919; after 1922 he returned to India, but retired from army service at the end of 1927. His interest in reconnoitring the Central Asian border areas and the Hindukush, Karakoram, Kun Lun Shan and Himalayas grew stronger during this time. Portraits of him are rarely found; he became more famous for his espionage, commenting on local political practices and travel-writing in the Pamirian Crossroads. After twenty-two years as an observant mind and collector of plant seeds, Reginald Schomberg moved to Rome for theological studies commencing in the late 1940s. Following his ordination to the Catholic priesthood in the Basilica San Giovanni in the Lateran, he served as a chaplain in Hampshire and Oxford from 1953 to 1957. Shortly after returning home to Ross-on-Wye for retirement, he 'had a fall at home and fractured his skull' and died on 1 March 1958. Source: Michael Pollock (2004); Schomberg Collection in the Archives of the British Library (MSS Eur C 371/63; contains public sector information licensed under the Open Government Licence v3.0

The enmity and verdicts of Reginald Schomberg

Lieutenant-Colonel Reginald Schomberg visited the Hunza Valley frequently for two decades up to Pakistan's independence. His distinguished military career began in Mesopotamia with the Seaforth Highlanders in 1901. He retired from the Army in December 1927, two years after he had visited the Himalaya and Karakoram for the first time, and took up the challenge to become an accomplished explorer and daring traveller in the Great Game arena.[126] It was the time when the British Raj sensed a challenge by the 'Bolshevik menace' that could probably affect its 'Achilles heel' as the open flank on the 'Northern Frontier' was perceived in strategic terms. From his Gilgit post, the Political Agent Gordon Loch repeatedly indicated a Russian threat and asked for counter-agency. Subsequently the Resident in Kashmir Evelyn Berkeley Howell took up the topic during his inspection tour in the Gilgit Agency in 1927 and submitted strong indications that this growing challenge should be closely monitored.[127] Taking up the well-known pretext from the Great Game period, he devoted a long chapter to the 'Bolshevik menace'; his readers in the superior line departments realised that 'political' action was required, a euphemism for espionage and intelligence. The 'menace' was explained as the fear of Xinjiang falling into Bolshevik hands, thus losing its neutralised status as envisaged in the Asia Convention.[128] Few persons would be more predestined than this highly decorated retired military officer with explorer's ambition and spying qualities. All his missions and travels that brought Reginald Schomberg to the Hunza Valley and neighbouring regions took place during this eventful period. His extended presence in Xinjiang and the Gilgit Agency during the 1930s made him an authority on regional affairs in the 'Northern Frontier'. Evelyn Howell had demanded measures and packages 'to promote the welfare and contentment of the people and their rulers'[129]; Reginald Schomberg's judgement on the performance of local actors and collaborators was sought and carefully listened to. Despite his disguise as an individual traveller, it became obvious and could hardly be hidden that his prime function was neither that of an explorer nor of a game hunter. As a declared anti-Bolshevik himself, he obviously functioned as an intelligence officer who directly reported to influential clients in the India Office in London. His mentor in Whitehall was John Charles Walton, Joint Secretary in the Political Department of the India Office. Late in life he changed his career path again, gave up espionage, studied theology in Rome and was ordained as a Roman Catholic priest at the age of seventy-one.

Few people have travelled more extensively in the Gilgit Agency and collected a wealth of precise topographical information for map-making in remote areas of difficult access.[130] His travels significantly contrasted with the visits of others. David and Emily Lorimer were based in Hunza's central oasis and very close to *tham* M. Nazim Khan. By contrast, Reginald Schomberg was an occasional and brief visitor in Baltit when returning from long trips to the remote valleys such as Braldu, Muztagh and Shimshal. David Lorimer noted in his diary for 10 October 1934: "Schomberg arrived 2.30." Next day's entry about the restless traveller read: "Schomberg left 9 am."[131] Being polyglot in Urdu, Pashtu, Persian and Turki, he was in a position to interview his interlocutors directly and to compare Hunza's performance and position with that of other principalities and governorships in the Gilgit Agency. He had visited them all and could make out the intentional differences of Sarikol and Hunza's relations with Xinjiang. Reginald Schomberg was a most outspoken visitor, and his remarks on rulers and people were full of unconcealed racism, superficial ethnographic attribution and pejorative distinction of people according to their physical appearance, command of vernaculars, assumed religion and probable origin. Few people have made more sarcastic and derogatory remarks about the aristocracy and mountain dwellers alike; his reports are full of undisguised prejudices and eccentric views. Nevertheless, Reginald Schomberg was a close observer of Hunza affairs and supplied his superiors in London with reports that bypassed the 'proper' channels of Political Agents in Gilgit and Residents in Kashmir on their way to the External Affairs Department in New Delhi and to Whitehall in London. He was critical about each and everybody, thus unveiling the weaknesses and strengths of his counterparts.

The reports of Reginald Schomberg have to be read carefully and with caution. They reveal certain issues that were not addressed by others. From his eyewitness account the transforming role of *tham* in Hunza becomes comprehensible by highlighting his shrinking dependency on the elite and the oppressive increase of taxes and duties. On the basis of repeated conversations, he commented often on *mīr* M. Nazim Khan; initially, he seemed to have admired him, but later on the *tham* was openly belittled and despised. Reginald Schomberg's offensive tone and the pejorative ascriptions might be expressions of his impulsive character, which cut short his career when aiming to become the Consul-General in Kashgar and working for the highest circles in the Intelligence Unit of the Foreign Department.[132] On the other hand, Reginald Schomberg judges people in a 'zeitgeist' manner and does not differ from other contemporaries when classifying people according to questionable categories and parameters derived from physical anthropology and stigmatising ethnography. Often he repeats a behavioural group distinction between Burusho and Wakhi; the latter are judged unfavourably: "He is rather soft and lazy, very good looking, especially when young, with no desire to leave his home, indifferent to serve of any kind, and in many ways he is a human vegetable. On the other hand he is hardy, simple in his habits and method of life, good on rock and cliff, and, as a companion, agree-able in most respects. Perhaps the easiest way of summing up the Wakhi is that he is, in all respects, the complete antithesis of his immediate neighbour and fellow subject, the man of Hunza."[133] Reginald Schomberg should have known better as he had

Reginald Schomberg is approaching a yurt in a Kirghiz settlement near Bulunkul in 1931; one of the rare photographs of him. Source. Schomberg Collection (RSAA/SC/Sch/1/71/4). The image is reproduced with permission of the Royal Society for Asian Affairs in London

The archive's caption reads: "Abdulla, Daulat, Akbar, October 1931 (probably in Murkushi)". Source: Schomberg Collection (RSAA/SC/Sch/1/53/1). The image is reproduced with permission of the Royal Society for Asian Affairs in London

developed a long-term, very close personal relationship with his preferential travel companions and appeared to have full trust in the chosen ones. These 'servants' were his backstoppers during most adventurous journeys, functioned as local experts and furnished information on which his knowledge and writings would be built. From Hunza came Abdulla Beg, a Wakhi *xəlifá* from Shimshal, and Daulat Shah, a Burusho from Central Hunza. Herewith he had two representatives of the communities he liked to compare and despise. The team was completed by Abdulla Rahtar aka Subhana Joo, a Kashmiri who, with the other two, formed his core team for more than two decades. Akbar and Mahbub Ali completed the set of trusted 'servants'. The strongest impression was made by his Burusho companions whom he primarily trusted. They did not refrain from pointing out their Burusho superiority complex and continued to belittle Wakhi and people from Nagar whom they regarded as inferior to themselves. We find numerous allusions to local prejudices in his writings that entered his chapters unreflected and could completely change when other informants appear on the scene.[134] He was the last colonial spy and 'political' traveller to leave a lasting impression on the public opinion and perception about Hunza and the Karakoram until today.[135] After his retirement from the Army he devoted two decades to challenging travelling in extreme mountain borderlands that were of 'political' importance.[136] During his earlier visit in 1933, Reginald Schomberg had some friendly remarks for the *tham*: "Mahomed Nazim Khan, the present Mir, is the outstanding personality of the Gilgit Agency, as his wazir, Humayun Beg, was before him. In 1933 he was about 75 years of age, but still vigorous. He has a cataract in his left eye but time has dealt kindly with him otherwise. All the other rulers take their cue from him and he knows it. 'Why', he said to me once, 'if I threw myself into the river, the Mir of Nagir would do the same.' The Mir is an excellent host and his arrangements are faultless. He speaks Persian well and also Hindustani, besides his native tongues, Wakhi and Burishashki. What is more he is thoroughly conversant with current politics and he has a shrewd, original outlook. He cannot read or write, and anyone who believes that success in life depends on these two accomplishments would reverse his opinion after a stay with the Mir of Hunza. I have never failed to be surprised and diverted by his resource and enterprise."[137] In this quote he made the *tham* nearly ten years older than he was but admired him in the published book as "... a personality seldom met with in the East. He is a thorough Oriental in every respect, and that is to his credit"[138]

The internal report about the same episode sent to his friend Charles Walton sounded completely different and highlighted his deficits: "The Mir is a pleasant old savage but we have adopted a foolish policy of non-interference and this pretty princeling enjoys privileges which Hyderabad and Mysore are denied. The result is that both he and, to a lesser extent Nagir, are liable to become nuisances. There is no limit to their greed and their avarice."[139] Next year's follow-up visit and his expedition to the remote corners beyond Shimshal made him dependent on provisions supplied from the ruler's stocks. As they did not show up in time Reginald Schomberg's mood significantly changed and he embarked on an attack against the hereditary rulers of the Hunza Valley that would initiate a host of controversies: "The two small states of Hunza and Nagir consist of one valley with a population of about 50,000 souls in all. They are ruled over by their chiefs or Mirs, Sir Mohamed Nazim Khan, Mir of Hunza, and Sir Shah Sikandar Khan, of Nagir. Both these chiefs have no claim to their present positions. After the Hunza-Nagir war of 1891, and the occupation of

the valley by the British, these two chiefs were placed on the throne. They were appointed with undesirable and unnecessary haste, and they dispossessed other and rightful claimants. Thus it is that the two Mirs owe their present position solely to the Government of India. Both chiefs are illiterate. The Mir of Hunza is, also, illegitimate. He is about 75 years of age, and is a diabetic. The Mir of Nagir is about ten years his junior. Owing to the accident of his situation on the main route to Kashgar, the Mir of Hunza is better known than his colleague of Nagir. The latter lives a good way off the main road. All British officers speak well of the Mir of Hunza, whether they be officials or visitors who meet the Mir for an hour or two, and only socially, are of no value. The fact is that both these Mirs are greedy, rapacious, and tyrannical rulers, who, thanks to an elaborate system of bribery and espionage in Gilgit, in particular, as well as by arranging that access to the Political Agent by their subjects is well-nigh impossible, have managed to conceal their oppression."[140] Here he challenges the legitimacy of both rulers who had been installed by the British-Kashmiri administration after the Hunza-Nagar Campaign. Furthermore, they are depicted as authoritarian rulers depriving their 'subjects' of necessary resources because of their greed and ruthlessness. No word any more of the 'benevolent ruler'. It follows a long litany of substance that lists all the shortcomings that seem to be related to the constellation caused by the backing of the local *tham* by the colonial authorities who have nothing to fear because of this support: "The chief complaints are as follows: –

(1) Kar-i-begar, forced labour, or the Corvee, precisely such as Lord Cromer had to abolish in Egypt in the middle of the eighties of last century. This has now reached such a state that the population is little better than serfs. Everything is done free at the Mirs Orders. Not merely is the labour free, but the material as well as the workers food are provided by the people. The Chiefs give the order, and the rest is done without any compensation or payment of any kind at all.

(2) The Gilgit Scouts. This is a body of irregular militia, of admittedly no military value. Two companies are raised from Hunza and two from Nagir. Although these men are paid by the Government of India, are embodied for two months yearly under British officers, and are nominally an Imperial commitment, the Mirs have been allowed to usurp complete control. All vacancies are sold by these chiefs at prices varying from ten, twenty, to as much as forty rupees each. The buyer has no security of tenure and often at the end of a year is dismissed. As the cash value of a year's service in this army is about thirty rupees the explanation of such a large payment lies in the desire to escape the now intolerable corvee. The scouts from Hunza are also sent to Kashgar to form a guard on the British consulate there. For this service, the Mir of Hunza exacts forty to sixty rupees. Each candidate has to pay if successful. In 193[illegible], Ghazan Khan, eldest son of the Mir, and Subahdar Major in the scouts, is said to have taken 16,000 rupees with him to India, as the proceeds of these sales. The officers are a preserve of the reigning chiefs. Of the four officers in Nagir company, three are the sons of the Mir, and the fourth vacancy is likely

Shah Sikandar Khan, tham of Nagar, was photographed by Reginald Schomberg in 1933 who remarked that he was "… a good-looking man of sixty-eight, fair-skinned and tall, a great shikari [hunter] and a very pious Mohammedan. Living up a side valley, seeing few visitors, and rather cut off, he lacks the experience and savoir-faire of the Mir of Hunza. His interests are local, he is absorbed in the affairs of his state, and the financial side of administration especially appeals to him." Source: Reginald Schomberg (1935a: 106, photograph on the following page). Courtesy of Jarring Collection at Lund University No. TRV020ap.106

The photograph shows tham M. Nazim Khan with his son Amin Khan probably in 1929. Source: Schomberg Collection (RSAA/SC/Sch/1/49/5). The image is reproduced with permission of the Royal Society for Asian Affairs in London

The photograph, probably taken in 1931, shows the forty-five-year-old mīrzāda Ghazan Khan (sitting) with his nineteen-year-old son Jamal Khan (standing). Source: Schomberg Collection (RSAA/SC/Sch/1/120/5). The image is reproduced with permission of the Royal Society for Asian Affairs in London

to be filled by another son of this prolific chief. The British Commandant is not allowed to enroll a man, to give an order or to make even the most minor appointments. The Mir of each state does all that, for payment. The value of the force, and the scandals and discontent that such a system give rise to, show how little value the force can be.

(4) Marriage cesses. The custom of the country is that on the marriage of the ruler or any of his children each inhabitant is liable to pay three rupees in cash or in kind. That is each household. The present system is to take from Rs. 8 to Rs. 40 from each household, to insist on its payment in cash, which alone is unjust, and levy this toll for the marriage of the grandchildren of the Mir, and even of collaterals. This is a deeply resented and unjustifiable contribution.

(5) No man can go to Gilgit without a permit from his Mir, for which he has to pay in cash or in kind. So a man has sometimes to walk thirty or more miles to the chief, get his permit, and walk back the same distance before starting on his journey. This is a petty and galling abuse of authority.

(6) Tolls. Everyone in Hunza or Nagir has to pay toll on sheep, oxen, or any goods taken to Gilgit for sale. As the roads are kept up by Government it is difficult to justify this imposition. The tax on salt entering Hunza from Gilgit is 20 %. Salt in Gilgit is about six times the cost it is in India. The levying of transit duties through Hunza is extremely severe on trade, and must be controlled by the Political Agent.

(7) The Mir of Hunza has now begun to interfere with the sale of land in Hunza, which has always been one of the rights of the people. The reason given is that the supply of forced labour will be reduced.

(8) Road repairs are paid for by the Government to the Mir, a stupid arrangement, and merely a means for the Divisional Engineer to shirk his duty. The Mir pockets nearly all the money, and merely pays a fraction to the people to save himself if enquiries are made.

(9) Treatment of the members of the ruling families. (a) Raja Nafis Khan, elder half-brother of the Mir of Hunza, and de jure Mir of Hunza, has been reduced to poverty. He is, like the ruler, illegitimate, but of better family. He receives Rs. 300 a year from the Government, but as nearly all his land has been sequestrated by his brother, for no reason, as the people well know, this sum is wholly inadequate. He is not allowed to come to the annual political Durbar. This is solely at the wish of the Mir, who apparently can commit any form of injustice. (b) Raja Khisrau Khan, son the late Mir of Nagir, de jure Mir, was granted the jagir of Chalt and Chaprot as compensation. These two places are not Nagir territory, and have not belonged to that state, but the present Mir of Nagir has been allowed to exercise his authority over these places, to the great detriment of the rightful owner. (c) Shabaz Khan, second son of the Mir of Hunza has not enough to live on. The Mir of Nagir has provided for all his sons, but the numerous, though still young children of the Mir of Hunza should be suitably provided for.

(10) Finally, in brief, the people want justice. As they are always saying, with perfect truth, the Government listens to the rich, and not to the poor. The wretched creatures have no appeal. If they do get to Gilgit, the Indian assistant or the clerks in his office, at once telephone to the Mirs, and the consequence to the petitioners is at once serious and discouraging. Five years ago, there was something in the nature of a revolution in the two villages of Hindi and Mayun,

in Hunza, and the people, who were amply justified, were merely told by the Political Agent that they must obey their Mir. No wonder the people hate the British connection. Before, however, suggesting remedies, it is well to point out that half measures are no good. The people are uneasy under the present oppression. It is unfair to them to procrastinate, or to wait till, say, the death of the Mir of Hunza. Likewise, it is unfair and merely useless to give orders that these manifold abuses should stop. The only effective way of effecting reforms is the presence of a British officer in the Hunza valley, until such time as these reforms are achieved, for the present abuses are largely due to the withdrawal of the political officer who was at one time in Hunza and Nagir. It is not unreasonable that he should be replaced for a time, then when matters have been put right, and a closer control exercised, the Political Agent at Gilgit might spend the hot weather in the valley. The Mirs should be frankly warned that any obstruction will lead to a modification of the present system of rule, and will lead to the appointment of governors, removable at the discretion of the paramount power. This is exactly what the people want, and as the present Mirs are the nominees of Government, they cannot object. The people are not unreasonable. They only ask to be governed as they were before the British came. They ask that their Mirs should rule in future as their ancestors did in the past. The customary laws, especially in Hunza, amply safeguard the rights of the subjects. It is these that should be reinstituted. The present system, whereby in both states, the Mir is the irresponsible arbiter and autocrat, governing solely for his own advantage, and, ignoring and abrogating the ancient customs of the country, must go. We cannot be a party to such abuses."[141]

Reginald Schomberg's roll call is well informed about the situation and addresses a number of salient issues. Motivated by the danger of an open flank in the Hunza Valley, especially with current critical developments in Xinjiang, affecting the trade with Kashgar and Yarkand, he focused on military issues first before evaluating the internal social situation in Hunza and Nagar. A main cause of the greediness of the rulers and the exploitation of the poor peasants is seen in the growing number of *guśpúr* and the termination of fratricide as he remarked ahead of his list: "Until our occupation of the country, the very real fear of insurrection and assassination acted as a wholesome check on the Mirs."[142] The loss of pre-colonial checks and balances fed his venomous argument that was based on the shortcomings of British indirect rule. All British officers were accused of mishandling the situation and enabling the local *tham* to exert their authoritarian domination. The misrule was supposedly the outcome of imperial backstopping that had allowed sidelining the *wazīrkuć* and their neglect of internal power relationships. All valuable economic enterprises, cross-boundary trade, porterage and services contributed to the growing treasure of the *tham* who checked the mail-bag and operated the post-office as well. All commercial activities were charged with special duties, road tolls and tithes. The funds that had been allocated for infrastructure projects and its maintenance were pocketed by the *tham* who commissioned the work as *rajaáki* for which the people supplied the food and labour without remuneration. The non-agricultural source of earning cash income by serving in the Gilgit Scouts had been exploited by the *tham* either by filling the posts with close relatives or by taking briberies from the selected persons. Shortage of labour resulted in the introduction of the *khuṭúkal* system, which enabled *mīr* M. Nazim Khan to receive a guaranteed return from his fields that were leased to individual tenants. The *tham* had refined an exploitative system that provided him with maximum returns from all enterprises and equipped him with the control of passage and movement of his 'subjects' that made use of telephone and telegraph. The imperial power supplied innovative tech-

Farmers from Shimshal village on their way to Pamir-e Tang. They are carrying provisions and warm clothes. The undated photograph was probably taken in autumn of 1931. Source: Schomberg Collection (RSAA/SC/Sch/1/85/6). The image is reproduced with permission of the Royal Society for Asian Affairs in London.

The photograph was taken in 1931 at a darbār held by tham M. Nazim Khan in Aliabad. Source: Schomberg Collection (RSAA/SC/Sch/1/48/11). The image is reproduced with permission of the Royal Society for Asian Affairs in London

nologies and secured his position. Reginald Schomberg had competently and well-informedly listed the grievances of the peasants whom he listened to in many conversations in the villages and discussed with his companions. There is no anti-colonial sentiment in Reginald Schomberg's assessment; even when he demands a return to pre-colonial customary laws and relations, he always has British control and imperial dominance of the situation and region on his mind and demands the presence of a British officer in the valley. His personal superiority complex would not allow the abolishment of indirect rule and the retreat of the British forces, just a year before the lease of the Gilgit Agency was to materialise.

Reginald Schomberg's list of suggestions is even longer and contains a very detailed sixteen-point plan:

"(1) The abolition of all forced labour except so far as was customary before the British occupation. No attention to the plea that conditions are now different. They are not. There is no valid excuse for the continuation of this disgraceful corvee. The ordinary custom was that in each village 6 to 8 men were excused all taxation in return for forced labour. No one else was liable. Likewise the custom of feeding the Mir, his children and retainers irrespective of their numbers, length of stay, etc., should stop. All is laid down by custom, and it is an adequate provision.

(2) The scouts should be under the sole and full control of the Commandant, and should be modelled on the lines of the North West Frontier Province, Militia corps.

(3) Marriage cesses should be limited to Rs. 3 or the equivalent in kind. Only the chief and his children should receive this contribution.

(4) Free ingress and egress from the two states to adjoining places to be enforced. The people should enjoy the same freedom as other inhabitants of the Agency.

(5) No tolls on government road. Freedom of the sale of land, as heretofore.

(7) All roadmaking to be done by the Divisional engineer, who should pay the people himself. He should be an officer of the Royal Engineers as before.

(8) Proper treatment of the Mir's elder brother, proper provision for him, the restoration of his rights in land, and his presence at the annual political durbar insisted on. The restoration to Raja Khisrau Khan of his rights in Chalt and Chaprot, which should be placed under the Political Agent direct. Proper provision for the descendants of the present Mir of Hunza.

(9) A post office to be opened at Aliabad. There was one once, but the Mir of Hunza picketted it, and no letters were posted for six months. This sort of reactionary nonsense should be stopped.

(10) The immediate restoration of the authority of the Wazir in both Hunza and Nagir. In Hunza this is particularly desirable, as the Wazirs have played an important role in the past, and usually to the great benefit of the people. In Hunza, too, owing to the universal unpopularity and rapacity of the eldest son of the Mir, Ghazzan Khan, who is a thoroughly mauvais sujete and is a strong argument for the introduction of governors, this reform is peculiarly needed.

(11) The abolition of the custom of foster parents. This should only be allowed with the political agents sanction, and then on the understanding that no change of parents should be allowed. The practice was forbidden in certain parts of the agency, but the prohibition has become a dead letter.

(12) The overhauling of the taxation system, the review of the abusers of the Mirs lands in Hunza and the limiting of the number of headmen in each Village.

Frederick Marshman Bailey photographed the conditions and support structures during his journey through the Hunza Valley in May 1918. His photographs show the poor equipment of all porters (baldákuyo, borwar) and the difficult road conditions while crossing Batura Glacier and several streams. He reported to his like-minded friend Reginald Schomberg with whom he regularly corresponded (Mss Eur/F157/249). Source: Bailey Collection in the British Library. The photographs are reproduced here with kind permission of the International Dunhuang Project Neg 1083/44 (21, 32, 42, 44)

Several of Frederick Marshman Bailey's photographs capture porters at their resting places during a day's journey. In his report he mentions that they had failed to purchase sufficient rations for the porters: "When still two or three days' journey from the Chinese frontier the coolies started getting restive. They had not enough food. This was, of course, their own fault and I was little inclined to break into the separate rations carried for the small party of sepoys and others. However the situation became serious and at Boihil one Zaheria, the ringleader of the trouble, stripped himself naked and lay down and refused to move as a protest. It was impossible to force hungry men to carry loads in this steep, cold country and in the end I had to give them some of our separate rations, a disquieting business; ..." (Frederick Marshman Bailey 1946: 17). Source: Bailey Collection in the British Library. The photographs are reproduced here with kind permission of the International Dunhuang Project Neg 1083/44 (17, 26, 28, 34)

(13) The monopolisation by the Mir of all spare pasture should be rectified in Hunza and Nagir.

(14) The congestion in Hunza is serious. There is ample land on the left bank of the Gilgit river, with good irrigation, between Bagrot and Chamogah. This is known as Oshikah Das. The Hunza men have always wanted this land, which is now derelict. This new organisation in the Wazarat of Gilgit should remove any obstacles to this reform, but the jurisdiction of the Mir of Hunza must not be tolerated in this area. It would be a dangerous precedent to allow the Mir to claim jurisdiction over emigrants.

(15) The immediate appointment of a junior British officer to ensure that these reforms are carried out, and later the presence in Hunza and Nagir of the Political Agent, or the A.P.A. Chilas, if a British officer.

(16) The abolition of the appointment of Indian Assistant, who is a middlemen between the chiefs and the Political Agent. It is a matter of regret that the Indian Assistant as well as the clerical staff are permanently in Gilgit. The disadvantages of such an arrangement are manifest."[143]

Reginald Schomberg's detailed plea called for restrictive political and fiscal powers for the *tham* to counteract his greed and hunger for wealth accumulation. The demand for a re-positioning of *guśpúr* and *wazīrkuċ* obviously stems from his personal relationship with them and his admiration for their bravery in challenging the sole authority in the states. One of the strongest restrictive measures by the *tham* was their control over the freedom of movement that deprived all Hunzukuċ and Nagarkuċ who would not pay for this favour. It seemed that authoritarian control measures were more severe in the Hunza Valley than anywhere else.

The confrontation between Reginald Schomberg and *mīr* M. Nazim Khan became a grave issue of concern, not as a duel between two stubborn fighters but as strong criticism of imperial rule, the performance of British officers, and their assessment of local conditions. Simultaneously, the former Political Agent David Lorimer was spending a pleasant time as a researcher focusing on language, ethnography and folklore at the court of *mīr* M. Nazim Khan. Emily Overend Lorimer was full of admiration for his rule and performance when she wrote immediately after her return to England: "For forty-three years our trusty friend and faithful ally, Sir Muhammed Nazim Khan, Mir of Hunza has kept this gate of India for us, ruling his people, some 14,000 of them, with firmness and wisdom and understanding kindliness.

Reginald Schomberg titled this photograph 'servants' which was taken in September 1929 during his Xinjiang expedition. He had hired Eli Akhun aka Ali Khan to function as his 'Turki caravan-bashi'. Source: Schomberg Collection (RSAA/SC/Sch/1/49/7). The image is reproduced with permission of the Royal Society for Asian Affairs in London

He is probably the most absolute monarch left on earth, but the smiling faces of his people, the proud dignity of their bearing, the outspoken freedom of their speech in private conversation or in public assembly, give the lie to tales of tyranny and oppression which occasionally titillate the ears of scandal-scavenging travellers."[144] She must have thought of Reginald Schomberg's critique when emphasising their host's benevolent tyranny. David and Emily Lorimer would definitely not share any of the accusations presented by Reginald Schomberg whom she described as 'ever entertaining, informative, and critical' as he had 'passed through on one of his endless wanderings in the wild'.[145] Consequently, it was speculated for a long time what had been the driving force for this enmity; had it been the disappointment about failing support for Reginald Schomberg's challenging expeditions or did he fall prey to the complaints and stories of his local companions? David Lorimer regarded Reginald Schomberg as a personal friend as he mentioned in a letter written nearly two decades later: "The innuendos against the old Mir are, of course, Schomberg. Schomberg is a distinguished and most adventurous and indefatigable traveller, and he presents in his various books a lot of interesting in-

The photograph shows tham M. Nazim Khan sitting on a couch with his son mīrzāda Ghazan Khan in the Karimabad garden. Nazim Khan's son Amin stands on the left, while his younger brother Shah Khan stands on the right close to his elder brother. Jamal Khan, Ghazan Khan's eldest son, is placed behind them. Source: Schomberg Collection (RSAA/SC/Sch/1/48/10). The image is reproduced with permission of the Royal Society for Asian Affairs in London

formation. He is a sturdy partisan and a good friend (I have personally reason to believe), but he is one of those unhappy people who apparently find nothing to like or admire in most of their fellow-beings. There is always a note of querulous disapproval and contempt in his writing. He praises the physical powers of the people of Hunza, but see what he says about their moral qualities. ('Between the Oxus and the Indus', pp. 128-9) which I would not at all endo from my experience."[146] David Lorimer saw his friend as a griper who expressed his dislike for anybody coming close to him. His personal disposition seemed to be read from the wording and accusations in his writings.

David Lorimer offered a further explanation that is more closely related to one of his companions. His 'head man' Daulat Shah was a trusted Hunza companion who had travelled with Reginald Schomberg since 1925 and accompanied him on two journeys to Xinjiang, to the Gilgit Agency, Hunza and Shimshal. When they travelled to Chitral, Daulat Shah had been in personal service with his master for ten years and was the most privileged man in the team. Reginald Schomberg had great trust in him as a person and in his opinions. David Lorimer saw this close relationship as a further point of confrontation: "Some mutual distrust, I think, existed between Mir Muhammad Nazim and Col. Schomberg, but the Mir did not malign Schomberg as S. does him. One cause of the trouble lay probably in Schomberg's pet henchman and factotum, the Hunza man Daulat Shah, whom he had taken about with him everywhere including India. The Mir, I fancy, disliked Daulat as a sophisticated product of travel and as being bumptious under the protective wing of his European master. Daulat on his part probably tried to poison Schomberg's mind against the Mir. This may be purely my imagination. Perhaps I have deduced it from his much reproduced portrait, first appearing in one of Schomberg's books and repeated in Wrench and Rodale, in which he does not look to me at all like a typical fine-drawn Burushin, but more like a self-satisfied, overfed, perhaps, Tyrolese Guide! I do not doubt that he served his master with devoted efficiency. He was probably the medium through which the latter acquired information."[147] Daulat Shah could form his own opinions, as he was not living in Hunza territory any more. Being one of the out-migrants from Hunza, he had settled in the Sunni quarters of Gilgit opposite Konedas and thus was free from the orders of the *tham*. Therefore, this arrangement did not fit in with the usual pattern of hospitality services as they were provided to visitors. Reginald Schomberg's team consisted of individuals who did not belong to the *guśpúr*. The *tham* used always to keep the monopoly of hospitality, tried to be the host of any traveller, and probably was not satisfied with his earnings from this eccentric traveller. During his reign *mīr* M. Nazim Khan built a 'stone bungalow' in Suryas in 1925 to entertain his visitors. The growing number of travellers had motivated him to shift from Baltit Fort to the hamlet of Karimabad and put a guesthouse, *darbār* hall and dancing-floor at the spot where he used to receive his visitors.[148] After the demise of *mīr* M. Nazim Khan his son Ghazan Khan abandoned the location and moved back into Baltit Fort.[149] Only Jamal Khan revitalised Karimabad, where the new palace would be built in the garden next to the Tash Mahal guesthouse.

Preferably *mīr* M. Nazim Khan would usually entrust a *guśpúr*, rarely a member of the *wazīrkuć*, to keep company with his visitors. Reginald Schomberg had selected his own group of trusted fellows against the prevailing rule; thus it was most likely that he was exposed to a different set of information and narratives. Opinions and views like those of

Daulat Shah did not necessarily put the *tham* in the prime position and would fail to reproduce the best and favourable picture of his rule.

From those conversations and his own observations, Reginald Schomberg would definitely not have derived any conclusion coming close to the one David Lorimer arrived at in 1953: "I should say that Hunza under M. Muhd. Nazim came as near to being a true Democracy as any society of human beings can safely go."[150] Both British observers who had spent quite some time in Hunza judged the *tham* of Hunza in their own way. Reginald Schomberg tried to play his trump cards with his Whitehall connections and warned them about the unique status of Hunza being the only principality engaged in cross-border activities. Warning against the threat of Bolshevism, treason and misinformation, he concluded in the final chapter of his report: "The situation in the Gilgit agency is at once complicated and aggravated by the role played by the Mir of Hunza, who, owing to the situation of his State and his claims to trans-frontier rights and privileges has extra-territorial commitments and interests which are at once undesirable and unique. Indeed, the Mir has become the sole arbiter on his part of the frontier. He has no qualification for this role. The writer of this report has observed the intrigues of the Mir, for over nine years and has realised that they are a real potential danger. The Mir maintains an elaborate intelligence system, and passes on to the Political Agent Gilgit such snippets of information, usually false, always incomplete, that his spies give him. The Mir has long had dealings with the Bolshevists, and entered into relations with them before the British Government did so. There is no use giving details about these matters. The fact that he does so, is undoubted. The Sarikolis on the Sarikol Pamir have always given tribute to the Mir. In 1933, when the troubles in Turkestan reached their area, they sent for help to the Mir, in case they were looted. The Mir was foolish enough to promise this to them. When these nomads were pillaged they sent to the Mir, asking him to redeem his pledge. He did not do so, and he could not, so the Sarikolis crossed into the Mir's territory at the Khunjerab Jilga and drove off his flocks. When Shah Mansur was one of the Khotan Amirs, he wrote for help to the Mir, who replied promising it. The murder of the Amir upset the Mir of Hunza who had hoped for grants of land, and other benefits from Amir. His intercourse with Shah Mansur was no hindrance to his remaining on friendly terms with Ma-Shao-

David Lorimer and Reginald Schomberg remained quite impressed by Baltit Fort and the Tarakuč houses below. Nevertheless, as guests of mīr M. Nazim Khan they spent their most comfortable and pleasant days in the Karimabad garden. Source: Lorimer Collection in the School of Oriental and African Studies, University of London (PP MS 66), Box II.30; reproduced with kind permission

Wu, Taotai of Kashgar, who has had many business dealings with the Mir, very much to the Taotais disadvantage. An unpleasant episode in 1934 was the case of a Russian Tartar from Kazan who, landing in India with a Turkish passport, was allowed to travel via Chitral to Sinkiang. In his company were two other Tartars who alleged that they had picked him up at Bombay. This man and his party were robbed of their arms in the Pamirs. It is difficult to say by whom, but one of the arms found its way to the Mir of Hunza. This Tartar who went by the name of Abdul Hamid sent word to the Mir that he would kill any Hunza man he met; whereupon the Mir ordered his flocks to be driven back over the frontier, and let his subjects fend for themselves. This man, Abdul Hamid, after being detained at Misgar, crossed the frontier, and murdered a harmless nomad one march from the border This sort of episode is dangerous, and may lead to a serious incident on the frontier. The Mir's intrigues are futile, and mischievous, and should be stopped entirely. There should be a proper post established and there should

When Aleksandr Yakovlev portrayed M. Rafi in Baltit on 29 July 1931 he considered him to be Hunza's envoy in Kashgar; obviously he was impressed by his looks and posture. M. Rafi was introduced as the 'Ambassador of Hunza to Kashgar', which probably overrated him as Daulat Shah was the leading Hunza representative in Kashgar, and in Tashkurgan Rustam Beg occupied the position at the time when the drawing was made.
M. Rafi's relationship with Sarikol and Raskam became prominent much earlier. He was called in the summer of 1914 and sent as 'havildar' to Raskam where he led the campaign of Hunzukuč to claim and cultivate the mīri lands there. In the winter of 1916 he was commissioned in the rank of sūbedār to safe-guard the mail service between Hunza and Kashgar with a group of Hunza Scouts; his main task was to protect the border passes from Hunza into China at the Paik post (IOL/P&S/10/826: 92, 94, 96, 101, 173–175). M. Rafi, son of Teghun who was married to a sister of wazīr Humayun Beg, was born in 1878 and thus in his fifties when the pencil sketch was made (IOR/V 5116: 20). His image was first published in Alexandre Iakovleff (1934: 14). The finished colour portrait in crayon and pastel was reproduced in Edward Murray and Alexandre Iacovleff (1936: Plate XII) and titled 'A smile would disarrange his neatly brushed moustache!' Jerome Rodale (1948) used M. Rafi's image as title-page vignette for his book on 'healthy Hunzas'.

be no activity allowed by any one beyond the frontier. What is more, no one should be allowed to cross without the leave of the political agent at Gilgit, or in the case of those coming from Kashgar, but from no place nearer the frontier, of the Consul-General there. It will be realised that a guard of local levies will be quite useless for such duties. Gurkhas would be ideal troops. The Soviet has now established itself in the Pamirs and Western Sinkiang and in a very short while conditions there will be the same as now prevail in Outer Mongolia. Some form of supervision and control are needed. On one part of the frontier there is an elaborate system of intelligence, and in a part of the same frontier, within sight of Russia, China and Afghanistan there is no control guard at all. This policy seems inconsistent and hardly prudent. The discontent in Hunza offers a soil ready, even avid, for Bolshevist propaganda. Indeed, it is already there. One man, Kalandar, a Hunza man who fled some years ago to Soviet Russia, has a good post under the Bolshevists. With him went another man who has not been given more than a junior job. The whole situation in the Gilgit Agency calls for some sort of re-organisation, and the situation in Hunza and in Nagir must be dealt with promptly, by a British officer. Even in Gilgit Town itself there is a good deal of Bolshevists propaganda and the appeal is against the Mirs and the Durbar to the free, sympathetic conditions under the red flag. There is nothing in the proposals for administering the new Gilgit Agency that can deal even superficially with any of these problems."[151] Reginald Schomberg made quite clear that he despised the Political Agent's trust in any services rendered by *mīr* M. Nazim Khan. Hunza's part and position in the geopolitical constellation were providing an open flank and a weak spot that could mature into crisis for which the Gilgit Agency's administration was not prepared. Herewith Reginald Schomberg had pulled out all possible stops and used all arrows in his quiver. Provoking such an enmity with *mīr* M. Nazim Khan must really have been the result of a grave personal disappointment. That wound would never heal. The let-down prompted Reginald Schomberg to begrudge the *tham* every honour that was professed. Alarming his contacts in the India Office, he could barely disguise his contempt at the unequal treatment of both *tham*: "I see some mug gave the Mir of Hunza a K.C.S.I. Give the old party a peerage and the Garter if you like, but what you give to him you must give to his foe and rival Nagir. Tweedledum cannot be separated from Tweedledee. There was, of course, no reason whatever to give Hunza anything at all. He is an obscure but interesting minor Chief and is well rewarded with a K.C.I.E. If you did have a K.C.S.I. kicking about loose in the till, give it to someone worth while, say to the Wali of Swat or some big man in the Gulf. Such a waste to no purpose and, as I have said, how sick the Nagir bloke will be."[152] The accomplished explorer and sophisticated spy could not see the compensatory measure in this act; *mīr* M. Nazim Khan was to be compensated with this outstanding order, a higher subsidy and with additional grazing rights in Naltar for his active separation from China by terminating his century-old tribute relationship. In between, it was even contemplated to confer the title 'His Highness' upon him, which would have promoted him to equal standing with the Maharaja of Kashmir. His strongest admirer and supporter was the Political Agent George Kirkbride; nevertheless, the Resident in Kashmir Lionel Edward Lang was sternly opposed to such a move: "I do not consider that the Mir of Hunza has a status sufficiently important to place him on an equality with 11 gun rulers of India. ... our policy towards Hunza should be directed more to the improvement in the condition of the people than to the granting of exalted status to their ruler"[153] George Kirkbride himself had asked for 'intelligence support' two years earlier because of alleged 'infiltration of Soviet agents', but he never considered the widely-travelled 'master spy' Schomberg.[154]

Reginald Schomberg's criticism omitted any respect for the diplomatic aspect and commented solely from a localised perspective. By doing this, he antagonised not only the local rulers, but also provoked counter-action by government servants on various levels of the imperial administration in Gilgit, Srinagar and London. The administration did not care as much for his criticisms as he might have expected. The Gilgit Diaries mention that Lieutenant-Colonel Reginald Schomberg 'has been permitted to tour in the Kuh-Ghizar, Yasin and Ishkoman Districts' in July 1933, and that 'he arrived on May 12th and left for Hunza on May 17th'.[155] At the beginning of July 1934 David and Emily Lorimer arrived in Gilgit and left for Hunza ten days later for their prolonged stay, while Reginald Schomberg 'returned from Hunza to Gilgit on the 14th of October and left for Srinagar on the 16th'.[156] All the observers and protagonists must have met and exchanged their views in Aliabad or Baltit. The newly appointed Political Agent, Major George Kirkbride, went on a two-week inspection tour to the Hunza Valley in November of the same year. He was concerned with the state of health of the two rulers and contemplated the repercussions if succession were to become imminent.[157] A second important point of discussion was the reorganisation of the Gilgit Scouts, which would provide Hunza and Nagar with more jobs and regular income from military services. It lasted until July 1936 when Major George Kirkbride toured the valley again and entered into the diary for that month under the heading 'Political': "Evidence that the Mir is the oppressor of his people is quite impossible to come by. The people, mostly Wakhis from Gulmit to the Frontier, appeared to be contented and, from their physique and general appearance, sufficiently well fed."[158] The remark appears a bit unexpectedly as no context is provided about any tensions between Wakhi 'subjects' in Hunza and their *tham*. No comment was added about similar potential conflicts in Central Hunza or Shinaki, or for Nagar as a whole.

Wearing a woollen Hunza cap (phárčin) mīr M. Nazim Khan is photographed by David Lorimer in his daily attire. In her comment Julie Flowerday described him as in 'undress' wearing an ordinary dress from home-spun wool and a Central Asian robe (PP MS 66). Source: Lorimer Collection in the School of Oriental and African Studies, University of London (PP MS 66), Box II.445; reproduced with kind permission

The administration reports from 1934 to 1937 completely ignored Colonel Reginald Schomberg's visit to the area and did not contain any enquiry about aspects he had mentioned in his notes. The administration reports repeated annually that the rule of the *tham* in Hunza and Nagar was 'satisfactory'; the only references were very indirect notes in 1936: "Though the Mirs are old men their vitality is still remarkable. Sir Muhammad Nazim Khan of Hunza continues to govern his people with measure of success which is the best answer to his critics. His upkeep of the road from Baltit to the Killik and Mintaka Passes is wholly admirable. His economic problem is acute as his cultivable land is barely sufficient for the population while for a few weeks in the winter the people subsist almost entirely on apricots. Of Nagar there is little to report save that relations continue to be strained between the Mir and his cousin in Chalt."[159] The big issue of forced labour was mentioned in passing in the following year: "Both from Hunza and Nagar there have

Probably in 1934 David Lorimer took the photograph of guśpúr Jamal Khan as he noted down on the back of the photo. Source: Lorimer Collection in the School of Oriental and African Studies, University of London (PP MS 66), Box 10; reproduced with kind permission

been some complaints about the old systems of kar-i-begar. Some, at least of these, appear to have been inspired."[160] No mention of anybody who inspired such moves.

The longest-serving *tham* was eager to leave an autobiography of himself. Although *mīr* M. Nazim Khan was not literate he managed to find a scribe in his grandson Jamal Khan who helped him to document his version of his rule. There are ample instances to contradict him – regarding the occurrence of murder during his rule, he admitted only two cases; regarding the levying of taxes, he claimed to have reduced them; regarding the outmigration of Hunzukuć, he insisted on having motivated them instead of exercising strict control – and to find a very selective view on events. But the book, which he freely distributed and sold to visitors, is ample proof of his ability to adjust to the expectation of the Political Agents in Gilgit. The translation into English was furnished by the Indian Assistant Khan Bahadur Muhammad Masih Pal, the most knowledgeable civil servant in the Gilgit Agency office, who supplied David Lorimer with narratives about affairs and demanded demographic data from the Hunza Valley during his thirty years of service in Gilgit until his resignation in 1939.[161] Both Jamal Khan and M. Masih Pal were quite instrumental in shaping the narrative of this 'autobiography' to suit his masters.[162] The English translation of the 'Hunza history' by the chronicler from the *wazīrkuć* Qudratullah Beg was provided by his son Saadullah Beg only in 2006. His edition contains a footnote saying: "He [Nazim Khan] was not a literate/educated man but he got this book written by an educated man from Nagar in Persian."[163] The mimeographed booklet was made available two years before his death and mainly praised his achievements in amelioration projects and settlement expansion, explained his take on the relationships with his half-brother and the *wazīrkuć*, and positioned himself in a favourable light as a benevolent ruler. Definitely, he felt comfortable with his deeds and was pleased at the state of affairs that would be handed over to his son.

When *mīr* M. Nazim Khan finally died, the Political Agent Ian William Galbraith remarked in his annual report about the 'outstanding event' of 1938: "... by his death the Agency loses a ruler who kept his people in strict control by firm and autocratic but kindly rule, and a trusted and loyal friend."[164] The Times published an obituary in which it was stated how Hunza affairs had developed after the installation of *mīr* M. Nazim Khan: "Since that day England has had no more faithful and trusty ally. The policy of permitting Hunza to be governed according to its own customs by its own prince with only the benevolent advice and supervision of the British Political Agent at Gilgit has justified itself by the happiness of the law-abiding unspoilt peasantry of Hunza. The Mir set himself to improve the condition of his people by bringing more land under cultivation through communal schemes of irrigation. He greatly reduced taxation and he abolished the system by which royal lands in every village were cultivated by forced labour. Offices of trust and authority were awarded to merit and efficiency instead of being a prerequisite of a few privileged families."[165] No description could be further away from the ground realities in the valley than that by the anonymous well-wisher who probably might have been Emily Overend Lorimer. Reginald Schomberg contributed an obituary in which he initially regretted that Safdar Ali Khan was deprived of his rightful position as *tham* of Hunza as he was presented as the antithesis of *mīr* M. Nazim Khan. Being of inferior birth, the latter could not live up to his half-brother who supposedly was 'generous and open-handed – which were hardly his successor's virtues – and he was personally popular'.[166] The successes of M. Nazim Khan were attributed to Humayun Beg and the British administration who were determined to guide the 'mere youth' and son of 'doubtful legitimacy'.[167] Reginald Schomberg offers an explanation for M. Nazim Khan's autocratic rule after the demise of the *wazīr* by stating that he converted the external dominance and interference into a state that he became the "... real ruler of his country; and not once, from that time to his death, did he ever allow anyone, not even the new Wazir, Humayun's son, to exercise any control in the State. This was unfortunate. The people did not like it; the Wazir's family were covertly and consistently hostile; and when the Mir grew older, the administration suffered. Feuds and factions were frequent and troublesome, and it is probable that the Mir, astute though he was, often did not know what was happening. As one of the Mir's family told the writer, 'All is now party, party.'"[168] The factional division that can be observed in Hunza until today may originate from this period. Following his criticisms, Reginald Schomberg did not fail to highlight some personal qualities of *mīr* M. Nazim Khan, especially with respect to his fellow-rulers from neighbouring principalities: "He was not the master of a remote and impoverished Highland clan, but the chief personage in his part of North-West India, and as such enjoying an immense prestige. He has gone, and his place can never be filled. His defects are forgotten and he will be long remembered as a sagacious and successful ruler who has carried his people from the calm of the middle ages to the doubtful benefits of the present civilization."[169] His final statement might have been endorsed by the imperial establishment in Gilgit and

The tomb of mīr M. Nazim Khan in Suryas Das in 1954. Courtesy of Gerhard Klamert

Srinagar, while his criticisms did not result in any action. The quotes presented so far have revealed the viewpoints of the *tham* himself and the *wazīrkuč* from a local perspective, and two perspectives from sometimes opposing imperial actors. The content and style of describing local leadership and rule in the official diaries and reports indicate the chief interest of the British administration that aimed at political stability at the Northern Frontier. Autocratic rule by an influential local leader was the ideal model and was backed by subsidies and the colonial apparatus of bureaucrats, bodyguards and soldiers. Protection from major environmental, health and supply crises that otherwise could lead to unrest

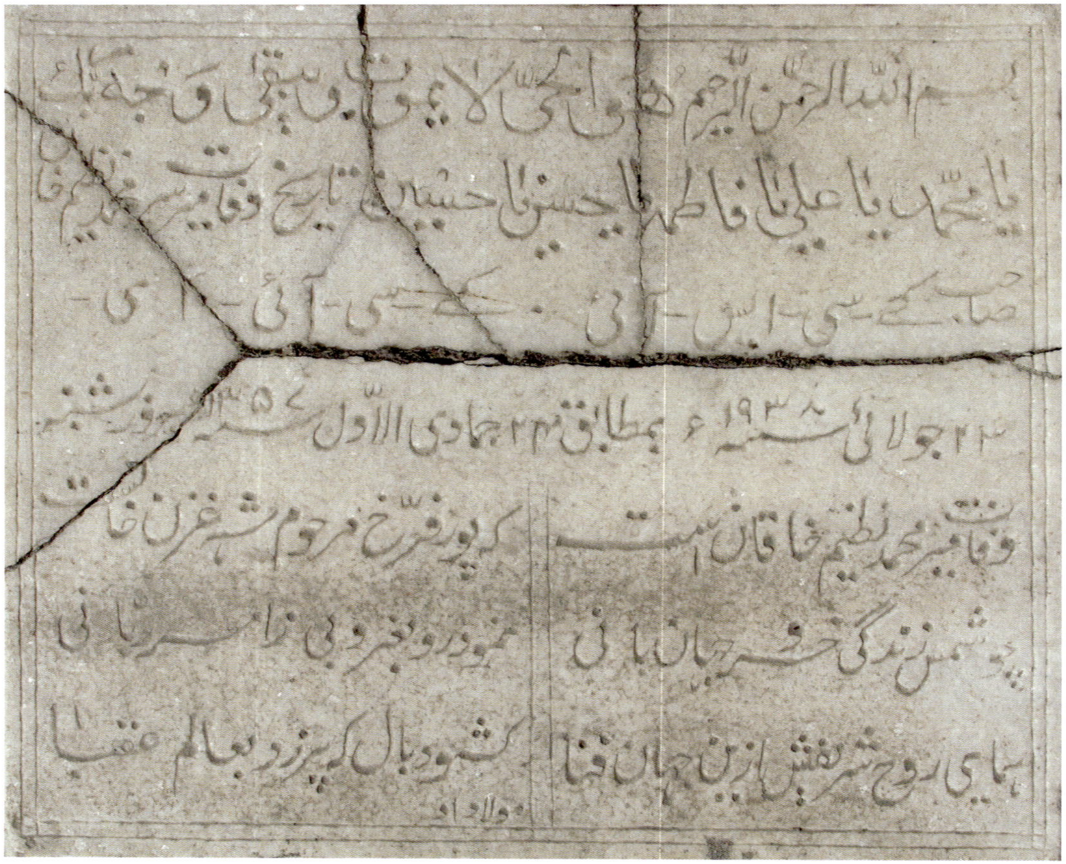

The tombstone of mīr M. Nazim Khan, who died on Saturday, 23 July 1938 (24 Jamadi-Awal 1357 AH), was chiselled by xalifah Mauladad and relates: "He is the son of the fortunate late Ghazan Khan. As the sun of life Khusraw, the guardian of the world, turned his face towards the sunset with the command of God. The Homa [mythical bird] of his noble soul from this perished world opened its wings, flew to the other world" (translated by Chorshanbe Goibnazarov). The photograph was taken in Suryas Das on 21 March 2019

was averted by infrastructure projects that often were interpreted as innovation and modernisation of the local society. These packages contributed to the maintenance of stability that was occasionally challenged by political unrest. It is quite surprising how few conflicts occurred during fifty-five years of British domination. The off-the-record intelligence reports by Reginald Schomberg often pricked the colonial administration and the local leadership at the same time. The Hunza *tham* M. Nazim Khan developed a dynamism during his long rule that was unprecedented. Infrastructure was expanded, new settlements founded by constructing irrigation channels within and outside the Hunza Valley, and he used his degrees of British-backed autocratic freedom to oppress his 'subjects' to such an extent that had never occurred before. He succeeded in utilising his liberties to introduce new and increase existing taxes on various services and commodities, benefited from tolls and dues for external traders, and established a new and favourable tenancy system (*khuṭúkal*). He never failed to promote himself as a benevolent ruler whose leadership was only challenged by the *wazīrkuć*. As Hunza was a comparatively poorly endowed microstate, his wealth accumulation materialised only to a certain degree and never reached luxurious dimensions. The system of 'monarchical absolutism' was refined by *tham* M. Nazim Khan and reached its peak during his reign; people complained to David Lorimer and expressed 'their bitter feelings about the violations of custom and the despotic measures enforced by Nazim Khan with the apparent concurrence of the colonial administration.'[170] The *tham* had refined his agenda and worked by a complex vigilance on information flow and people's mobility, by making people dependent through the allocation of positions in the Gilgit Scouts and other services, and by his draconian court rulings. While *mīr* M. Nazim Khan's rule was characterised by dynamism and power display, his son Ghazan Khan appeared to administer his predecessor's achievements and not to develop them much further.

Successors of Nazim Khan in Hunza and Sikandar Khan in Nagar

The Political Agent and the Resident in Kashmir were concerned about the dubious ability of Ghazan Khan to manage the affairs of Hunza State. It happened that in due course *mīr* Sikandar Khan passed away in Nagar, even before the new *tham* in Hunza could be equipped with his *sanad* and *mīrzāda* Ghazan Khan could be officially approved as the rightful heir. The inauguration ceremony was delayed until 26 August 1940 when the formal investiture of both new *tham* took place in Baltit Fort and three days later in Uyum Nagar. In both cases a strong representation of British officers and local dignitaries witnessed the British Resident in Kashmir, Lieutenant-Colonel Denholm de Montalt Stuart Fraser, handing over the 'khillat' and 'karita' – a ceremonial robe (*xal'at*) and a small bag containing the state subsidy (*xarite*) – on behalf of the Maharaja of Kashmir as the sign of their acknowledged dual authority.[171] In ceremonial matters it seemed to be a Kashmiri affair, in real terms everybody present realised that the ultimate authority was vested in the Political Agent and Resident in Kashmir.

Given this constellation and several incidents casting doubt on *mīrzāda* Ghazan Khan's ability to perform well,

The mīr of Nagar Shaukat Ali Khan, born in 1920, was portrayed in full attire in his palace in Uyum Nagar in 1954. Source: German-Austrian Himalaya-Karakoram expedition 1954. Courtesy of Gerhard Klamert

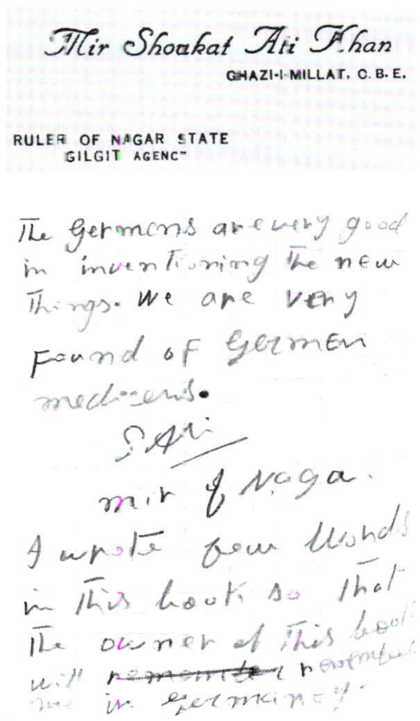

Autograph and business card of mīr Shaukat Ali Khan preserved in Gerhard Klamert's Porter Service Book 1954

the question of indirect rule's practicability had come up again with the colonial masters. The Nagar case was more complex as the colonial administration preferred the young grandson Shaukat Ali Khan to be installed as the new *tham* there, although Mahbub Ali Khan, the second son of *mīr* Sikandar Khan had a legitimate claim against his nephew and stepson Shaukat Ali Khan. A justification for the posting of an 'Assistant Political Agent Hunza-Nagir' was found in 1940, as the Political Agent Gerald Charles Lawrence Crichton pointed out: "With the approval of the Government and His Highness the Maharaja of Kashmir, his grandson, Shaukat Ali Khan succeeded him [Mir Sikandar Khan] as Mir. In view of the youth and inexperience of the new Mir, who is only 20 years of age, it was considered desirable to appoint a British Officer to assist and advise him in the early years of his administration. Captain R[onald] A[llen] McConaghey was accordingly appointed Assistant Political Agent, Hunza-Nagir and took up his duties with headquarters at Nagir, on May 15th. The necessity of the appointment was soon apparent, for the young Mir, clearly influenced by self-seeking private advisers, chiefly his foster-relations, was soon at logger-heads with his step father and eldest uncle, Khan Sahib Mahbub Ali Khan, the most powerful personality in Nagir. ... In general, the internal position in Nagir is unsatisfactory ..."[172] The personal enmity of the two contenders reflected the interests of different factions in Nagar; the necessity to separate the two claims was overcome earlier than expected. Mahbub Ali Khan was promoted and appointed as Governor of Yasin in July 1941, thus leaving Nagar for good. The British officer in Nagar relinquished his post after just one year in station.

The tension between *tham* Shaukat Ali Khan and his step-father was eased by this move and never caused any trouble again.[173] His ability to rule Nagar seems not to have improved much over the following four years as an unfavourable assessment was documented in the annual report: "Shaukat Ali Khan, the present Mir, is young and inexperienced and inclined to over parsimonious. He is an unattractive personality and very distrustful of those around him. His present Wazir, Zafar Khan, his real uncle, feels that the Mir while outwardly showing approval of his work, is inwardly suspicious of him. This attitude on the part of the Mir is unfortunate as it is felt that the Wazir, having the interests of both the Mir and the State at heart, is willing to help but finds himself handicapped by this lack of trust. However, Nagir State has given no trouble during the year and it is hoped that with encouragement and good advice the Mir will improve as time goes on."[174] Shaukat Ali Khan survived all criticisms and vagaries. When his time to rule had come, he had realised that Kashmir was only a minor player and that his power relationship was confined to his British masters. In a rare document authored by him he reflected on colonial interference and summarised the Nagar collective memory on their relationship with the colonial conquerors by stating "... the British requested the ruler of the State to allow them passage through the state, as they wanted to survey the frontiers. The ruler of the state did not agree to this, and the result was that war broke out between Hunza and Nagar on one side, and British on the other side. During

On the occasion of the double marriage – M. Nazim Khan's grandson, the future mīr Jamal Khan, with Shams-un-Nahar, a daughter of the mīr of Nagar Sikandar Khan and future rānī of Hunza, and at the same time the marital bond between Muzaf-far-ud-din Shah, third son of Sikandar Khan, and a daughter of M. Nazim Khan – a major gathering took place in Karimabad garden in July 1934. Jamal Khan's brother Ayash Khan led the dance followed by his half-brother Shah Khan cheered by a big crowd of spectators. Source: Lorimer Collection in the School of Oriental and African Studies, University of London (PP MS 66), Box V. 7; reproduced with kind permission

this war, the people of Nagar were divided into two factions, one in favour of the British, and the other against them. The war ended in the defeat of the Nagar forces and by an agreement the grand father present ruler, Sir, Mir Sikandar Khan, was declared the ruler of the State, a Political Agency was established and the British agreed to give complete independence the ruler in internal administration, while defence and external relations became the concern of the British."[175] This strong bond was the basis of the surviving *thámkuṣ* in his own perception. The hiatus would occur when Pakistan came into existence. The question of accession was unanimously solved on the Nagar side in favour of Pakistan; his rulership survived until he was forced to retire in August 1972. Shaukat Ali Khan was the last *tham* of Nagar and played a secondary role in decision-making about any affairs concerning the Hunza Valley. Nevertheless, his domestic influence in Nagar was much stronger than in Hunza; partly because he survived until 2003.

Ghazan Khan – *tham* without fortune

Hereditary rule and its impact on domestic affairs and outward relations would change after the demise of the two most influential and strong *tham* in Hunza and Nagar. External assessments of Ghazan Khan's personal performance were not favourable even before he succeeded his father. The Political Agent George Kirkbride closely monitored the power constellation and looked down on him in 1935: "Ghazan Khan is a man of about 45. Outwardly he is a fine healthy type but he is not of the same calibre as his father. A few years ago he very nearly had an attack of delirium tremens and though the Mir forced him to mend his ways with an apparent success which is wholly admirable I am doubtful whether he has the moral character to continue in the strait and narrow path once the influence of his father is removed. In other ways Ghazan Khan cannot compare with his father. It is doubtful whether he could continue with any success

The last three tham of Hunza whose investitures took place during colonial times; in the centre M. Nazim Khan (1892–1938); on his left his son Ghazan Khan (1938–1945), on his right the grandson Jamal Khan (1945–1974), photographed by Aurel Stein in Karimabad garden on 13 September 1930. Source: Stein Collection in the British Library. The latter is reproduced here with kind permission of the International Dunhuang Project Photo 392/34 (74)

the policy that the present Mir has conducted for so long in regard to his trans frontier commitments and I believe he would be thankful to be relieved of their responsibility provided he received suitable compensation elsewhere."[176] Such an outside position would probably have helped him to fend off the attacks and intrigues from the *wazīrkuč* with whom he fought a constant battle and it has not been clarified at all who dominated the foul play. The rule of Ghazan Khan was less eventful and did not leave any important visible landmarks in Hunza. Or as Reginald Schomberg would comment: "Ghazan Khan, with the traditional dislike of the Asiatic for inhabiting the dwelling of his predecessor, cared not a doit for this pleasaunce but returned to the dingy, ramshackle castle of Baltit, near which he built himself a small house, which was his sole contribution to the local architecture."[177]

The reputation of *tham* Ghazan Khan was that of a 'kindly old man, I thought, old before his time and a slave to Bacchus' as Kumara Menon, the emissary of the British Indian government to Chiang Kai-shek on his way from Srinagar to Chongqing, reported while resting in Hunza in September 1944.[178] His final years were spent in seclusion among a selected group of followers and friends; he refrained from holding court meetings and retreated to Gulmit where he died of apoplexy. The Hunza *tham* was absent from the annual assembly (*ja'sa*) under the auspices of the Political Agent in Gilgit in November 1944 where all rulers and governors from the 'political districts' normally gathered. The

Gulmit's polo ground forms the centre of the village. The compact settlement (kalha) is visible in the foreground with the house of arbāb Diwane Shah at the open space; on the left is the residence of the mīr of Hunza in which tham Ghazan Khan spent his last six months together with wazīr Inayatullah Beg. He died in this premises on 7 May 1945 and was buried close-by in the fields on the right. At the upper right hand where the road from Central Hunza would enter Gulmit was the residence of Ghazan Khan's half-brother Shah Khan who lived here until his death in 2016. Next to the jamáat khana with the gabled roof was the residence of Ghazan Khan's son Ayash Khan who became the secretary to mīr Jamal Khan. On the opposite side of the polo ground near the dīdārgāh is the residence of rāja Sultan Khan and Bahadur Khan, sons of mīr Ghazan Khan's brother Shahbaz Khan. The photograph was taken on 15 April 1990

incumbent would never return to Baltit, and the Gilgit Diaries recorded: "The Mir of Hunza who proceeded to Gulmit before Jalsa has been unable to return to his capital at Baltit, he is reported to be incapable of undertaking the journey owing to heavy drinking."[179] The final account about *tham* Ghazan Khan is a sad report about his expected demise, strange rumours about the involvement of the Tarakuċ and odd indications about the role of *wazīr* Inayatullah Beg.[180] Major Roger Noel Bacon recorded: "During the year under review unhappy conditions prevailed in the Hunza State. The late Mir Muhammad Ghazan Khan, C.B.E., ruler of the State died at Gulmit on the 7th May 1945. The Mir, encouraged by his Wazir and a few dissolute companions had been drinking hard for some time and alcoholic excess brought about his death. In the autumn of 1944 Mir Muhammad Ghazan Khan left his capital and retired with his Wazir to Gulmit two stages away leaving the State business in the hands of his son Muhammad Jamal Khan, who by his tact and untiring efforts and harassed by intrigue and non-co-operation, saved Hunza from disintegration. On the death of his father, Muhammad Jamal Khan visited the Political Agent [Lieutenant-Colonel Evelyn Hey Cobb] at Gilgit and at the former's request steps were taken to remove the Wazir and his faction from the State. This action proved successful. Muhammad Jamal Khan was installed with full ceremony by the Political Agent in the presence of British Officers and a large number of notables at his capital Baltit, on the 23rd November 1945. The new Mir, intent on following closely in the footsteps of his illustrious grandfather Mir Sir Muhammad Nazim Khan, has now restored the situation in the State to normal. In Gilgit, trouble in Hunza means trouble elsewhere, the Agency taking its cue from the Mir. To Mir Muhammad Jamal Khan is due a debt of gratitude for his good work during the year."[181] Jamal Khan mentioned in his inauguration speech that his grandfather served him as role model and made not a single reference to his father.[182] The power relationship within Hunza had suffered from a renewed confrontation between *tham* and some sections of the *wazīrkuċ*; its detrimental effects after the demise of the non-Tarakuċ *wazīr* Zarparast in 1943 on the peaceful cooperation were realised by the observant imperial administration.[183] The Political Agent Evelyn Hey Cobb ordered that *wazīr* Inayatullah Beg and his two cousins, Tara Beg and Sanaullah Beg, were to be grounded in Gilgit. They lost their share of the *wazīri* allowance of three hundred rupees per annum; at the suggestion of *mīr* Jamal Khan the amount was distributed among five other and loyal Tarakuč.[184]

Anyhow, *mīr* Jamal Khan had greater trust in his brother *guśpúr* Ayash Khan who would function as his chief advisor or as Reginald Schomberg put it 'his Wazir'.[185] The critical observer had a general and final verdict for the aristocracy which he likened to the *guśpúr* whom he termed 'useless'. That statement did not hold true for Ayash Khan who became a respected 'prime minister' of Hunza. Reginald Schomberg had *rāja* Khisrau Khan of Chalt in mind when he met his son Sultan Ashgar. That visit led him to reason again about the function of a *guśpúr* on the occasion of his last stay in Chalt in 1945: "His father was the late Raja Khisrau Khan, a great local celebrity, a fine polo-player and rider, a first-rate judge of a horse, and a very able horse-coper. The poor man suffered all his life from numerous well-founded grievances. He had been consistently unlucky and his misfortunes rankled. It must be unfortunate to be a local royalty and to live in poverty. I found that this talent for grousing had descended not only to his son but to all other members of his family. No account, however trivial, of the Gilgit Agency can omit some mention of the Gushpur, a title or description of those who are related to ruling families, but are not in the line of succession. For example, the late Mir of Nagir had many sons, and all were Gushpurs. The trouble of every Gushpur is the firm conviction that his status entitles him to do no work whatever, but that it is rather the first duty of the country to provide for him and his. The Gushpur ranks therefore as a disaster of full dimensions. He is quite useless, he is devoid of all energy, enterprise and resource, and is, in fact, the worst type of parasite and drone. In the old days before the advent of the British ... there never was a 'Gushpur problem'. There was no workshy upper class. Thus a Gushpur's life was not a happy one, and usually lasted until the dynasty was established, that is until the ruler had heirs when the value of any Gushpur ceased, and an end was then made. In neither of the two world wars did the Gushpur play any part. They steadfastly refused to budge, though well-mannered and well-educated. They were apparently not encouraged by the political authorities to take any share whatever, which was a pity as casualties would have at once solved the ailments of these gentry and have showed they were of some negative value."[186] No potential of any contribution to statecraft and wealth generation was attributed to the aristocracy. The same applied for the *wazīrkuč* who had passed their zenith long ago.

These critical remarks reflect the state of political affairs after the lease of the Gilgit Agency and a quiet 'Northern Frontier'. The previous equilibrium of local powerful actors seemed not to matter anymore. The *guśpúr*, *rāja*, and *wazīrkuč* had lost their previous importance and influence; ideally they would become civil servants and/or promoted to military officers in order to alleviate the burden on the state budget. The *tham* had acquired a standing that had made them quite independent. The imperial masters were quite satisfied with the new team of *tham* and 'Secretary of Hunza State'; Jamal Khan and Ayash Khan represented Hunza before and after partition and managed its entry and incorporation into Pakistan. In the recollections of Ghulam-ud-din

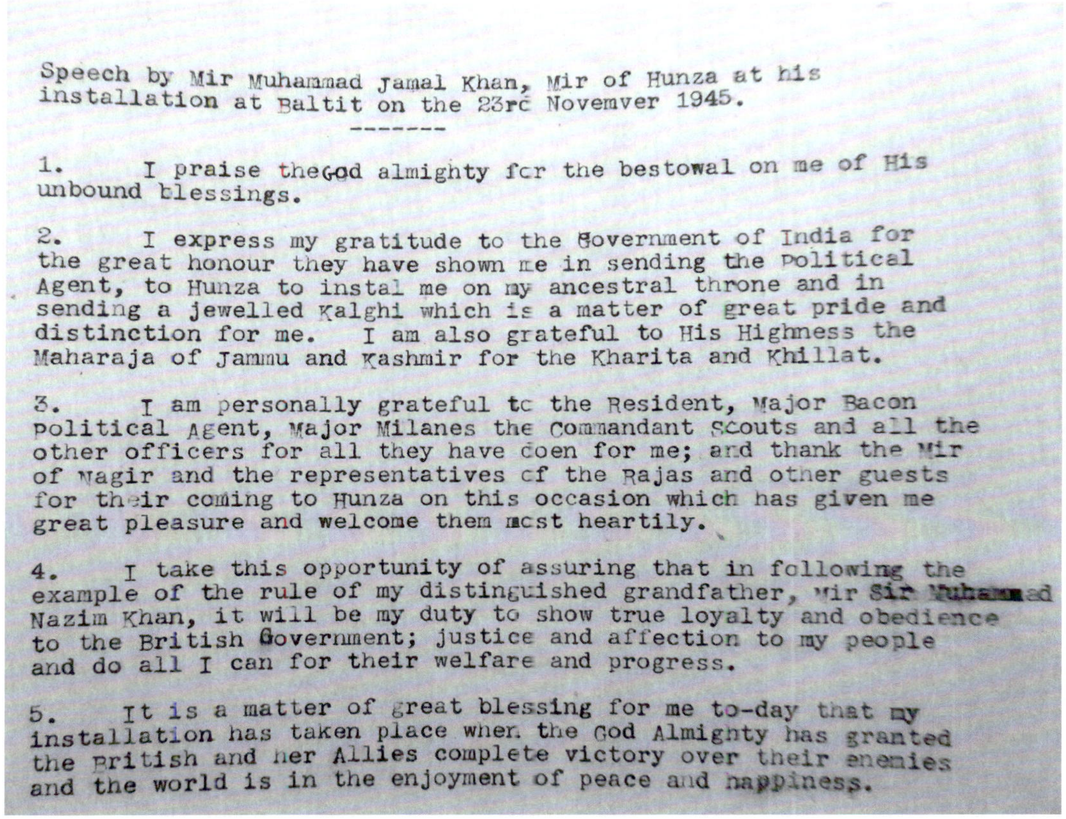
Speech by Mir Muhammad Jamal Khan, Mir of Hunza at his installation at Baltit on the 23rd Novemver 1945.

----

1. I praise the God almighty for the bestowal on me of His unbound blessings.

2. I express my gratitude to the Government of India for the great honour they have shown me in sending the Political Agent, to Hunza to instal me on my ancestral throne and in sending a jewelled Kalghi which is a matter of great pride and distinction for me. I am also grateful to His Highness the Maharaja of Jammu and Kashmir for the Kharita and Khillat.

3. I am personally grateful to the Resident, Major Bacon Political Agent, Major Milanes the Commandant Scouts and all the other officers for all they have doen for me; and thank the Mir of Nagir and the representatives of the Rajas and other guests for their coming to Hunza on this occasion which has given me great pleasure and welcome them most heartily.

4. I take this opportunity of assuring that in following the example of the rule of my distinguished grandfather, Mir Sir Muhammad Nazim Khan, it will be my duty to show true loyalty and obedience to the British Government; justice and affection to my people and do all I can for their welfare and progress.

5. It is a matter of great blessing for me to-day that my installation has taken place when the God Almighty has granted the British and her Allies complete victory over their enemies and the world is in the enjoyment of peace and happiness.

The typescript of tham Jamal Khan's speech on the occasion of his formal installation as mīr of Hunza. Source: India Office file 'Death of Mohammed Ghazan Khan of Hunza and succession of Mohammad Jamal Khan as Mir of Hunza' (IOR/2/1086/308: 92)

After the vacation of Baltit Fort a new dārbar hall was erected in Karimabad palace garden where regular court sessions were held. It was mīr Jamal Khan who followed the vision of his grandfather and developed Karimabad to become the residence of the ruler where all activities and ceremonies would be performed. Petitioners would queue and wait for their turn. Advisors and people close to the tham would attend these meetings and function as experts and witnesses. The same applied for Nagar where the tham received petitioners in his dārbar hall which became a meeting place for idlers. Source: German-Austrian Himalaya-Karakoram expedition 1954. Courtesy of Gerhard Klamert

Opposite
The ruler of Hunza, mīr Jamal Khan, posing in full regal attire in his Karimabad palace garden. Source: German Karakoram expedition 1959. Courtesy of Gerhard Klamert

it was the colonial administration that had asked Colonel Ayash Khan[187] to leave his army career in order to second Jamal Khan.

Jamal Khan – the last *tham*

The new *tham* had been observed for quite some time before he took over the rulership. Kumara Menon had already found him eligible in 1944: "A more forceful and attractive personality than the Mir is his son, Jamal Khan. He has a mind of his own. He unlike the people of Hunza who are all followers of the Agha Khan, has no faith in the Aga Khan's divinity. And, almost alone among his subjects, he observes Roza. ... He has studied only upto the 7th class but speaks English with fluency and vigour. A man to watch, for in his time Hunza will move – let us hope forward."[188]

Never having set his foot into Hunza during the reign of *tham* Ghazan Khan, Reginald Schomberg returned to Hunza for his 'Karakoram Finale' in the year when Jamal Khan followed in the footsteps of his father. The experienced explorer met with his 'personal servant' Mahbub Ali in Srinagar the day after Jamal Khan had succeeded *mīr* Ghazan Khan, who had passed away a day earlier in Gulmit on 7 May 1945. The succession coincided with Nazi Germany's final defeat and surrender in World War II. Reginald Schomberg met with his favoured Hunzukuč in Srinagar the same day and immediately his attention was drawn to Burusho fighting qualities: "They were on leave for their regiments in the Indian Army. They are admirable material, and it has mystified me why in the two great wars no regiment has been raised in Gilgit, or else why no effort has been made to recruit the local men of whom there is ample supply. In neither war did the political officials take the slightest interest of concern in a real war effort which is a poor tribute to their commonsense and sagacity."[189] In Srinagar and Gilgit, the British officers-in-charge had similar thoughts about the German capitulation that was envisaged to mark the beginning of a new and peaceful era in the world and in Hunza. The consequences of the war were felt everywhere. Reginald Schomberg's final journey to Hunza suffered from the effects of the war reflected in the shortage of foodstuffs and other goods. The team was nearly completed when they met with Daulat Shah in Gilgit who had settled close to the Gilgit Fort. All preparations suffered from the high bazaar prices for all re-

Aurel Stein photographed a display of military strength in Karimabad garden on 1 September 1930. The participation of Hunzukuč in British military employ grew after the reorganisation of the Gilgit Scouts in 1935 as more men from Hunza and Nagar were hired. Source: Stein Collection in the British Library. The latter is reproduced here with kind permission of the International Dunhuang Project Photo 392/34 (53)

quired commodities, which surprised the greedy traveller who had returned after a break of nearly a decade.

Reginald Schomberg and his followers arrived in Baltit on 18 June 1945 and spent two days with Jamal Khan: "He is capable and intelligent, and will need all his ability to rule his state as did his grandfather. ... We talked on all matters, lawful and otherwise, and I was very sorry indeed that I could not stay longer ..."[190] It seemed that Reginald Schomberg had made his peace with the Hunza rulers and that he appreciated a stable leadership as much as did the Political Agent in Gilgit. Obviously he was not making his peace with the imperial representatives in Gilgit, Srinagar and Delhi as anticipated. Though praising the support of many officers such as James Glasgow Acheson, Resident in Kashmir, Evelyn Hey Cobb, Political Agent in Gilgit, and Major Reginald Milanes, Commander of the Gilgit Scouts, and acknowledging the *mīr* of Hunza and Nagar, Jamal Khan and Shaukat Ali Khan, in his draft preface to 'Karakoram Finale', he explains the title of the book as a reference to the termination of his relationship with the 'Northern Frontier': "I am afraid that it is a true ending of many years in a delightful country. I had hoped to essay on one more trip in 1946. On my return to India proper in 1945 I was so maladroit as to acquaint H. E. the Viceroy with certain points in the administration of the Gilgit Agency which needed remedying. This report of mine, which I deemed would be privileged, enraged the External Affairs Department of the Government of India, responsible for the conduct of the Gilgit

Agency and in retaliation, I was refused leave to revisit that part of India. It seems incredible that such pettiness could lurk in high places, merely to score off someone who has made an honest complaint."[191] It was really the end of meeting with his acquaintances and friends, criticising each and everybody, and writing reports and letters to imperial officers. 'Karakoram Finale' was not published as a book; disappointed, Reginald Schomberg cancelled his membership in the Royal Geographical Society and the Asiatic Society.[192] Nevertheless, one year after composing the draft of 'Karakoram Finale' in March 1946, Reginald Schomberg planned a new expedition to Badakhshan with his close friend and botanist George Taylor; Daulat Shah and Mubarak Shah were expected to join them. Being opposed to India's and Pakistan's independence, he looked for new destinations and limited his adventurous ambition to being 'solely a floristic one'. Espionage and intelligence gathering had come to an end when the colonial administration left the subcontinent. He wrote to George Taylor: "But as far as India is concerned, it is folly to give these people home rule. What right has any one to sentence a large population to death and misery? It is these cursed sentimentalists and ideologists."[193] After giving up his previous duty, Reginald Schomberg sought the company of friends such as Frederick Bailey and Richard Meinertzhagen, both having been intelligence officers, George Sherriff, former Vice-Consul in Kashgar, and Frank Ludlow.[194] Reginald Schomberg reflected about the fate of his personal records; "What am I to do with all my negatives of Chinese Turkestan, Gilgitia, and so forth? ... Why lose them I've written to the Central Asian, and to the Geographical Societies, but they are not interested – more fools they are."[195] He was quite disappointed at his state of life; two years earlier he had confessed to Frederick Bailey: "I live the existence of a complete troglodyte in this slum. One might be so much worse off, however. Even in this totalitarian state there are some advantages."[196] The longer he was out of business the more he reflected on the character of expeditions and shared his view with Frederick Bailey: "These people who go on expeditions will not take a good cook. Shipton is a great offender in this respect. When you read accounts of these expeditions, one always notices how often the fellows are going sick. I may have a good constitution, but neither I nor any of my men were ever sick – ... I am not a believer in all the masses of stores that are taken – all tins, of course. What you want is a good cook who can

Two levies serving tham Jamal Khan as bodyguards wear the Hunza flag and state emblem on their breast shields. Source: German-Austrian Himalaya-Karakoram expedition 1954. Courtesy of Gerhard Klamert

live on the country; and one, too, that is not forever moaning for the shops in the bazar. Above all, one needs a man who can make bread and live on the country. I always did, and so did you. But no. These tourists to the great peaks – I'm heartily sick of the whole jing-bang – carry masses of tinned chow with them, and then grumble at the cost of transport."[197] With his disposition of living from the country he could never have become a good customer for *mīr* M. Nazim Khan's hospitality services; another proof of their special relationship. A year later he reflected on the meaning of hospitality in his 'Karakoram Finale': "Hospitality is welcome, but it has to be paid for in the East, and always at a high rate. It was expected that I should have to pay rather highly as I could obviously afford to pay if I travelled with a number of coolies. The inference was natural, but the deduction was wrong."[198] Obviously this expectation did not always match with his desires. Reginald Schomberg declares the end of all interest in seeing Hunza and the Gilgit Agency again and confesses to Frederick Bailey: "But I have lost all interest in the place, which I have left a number of years ago. I daresay that you are right in saying that exploration has changed a great deal. But it is the individual that still counts, and the descriptions of many of these journeys only show how ignorant the modern traveller is. That does not prevent him from reading, writing, and being read. But I fear that he reads more about himself, and suffers very much from swelled head. So you see I am not a great admirer of the present explorer. Say what you like, but you can see nothing

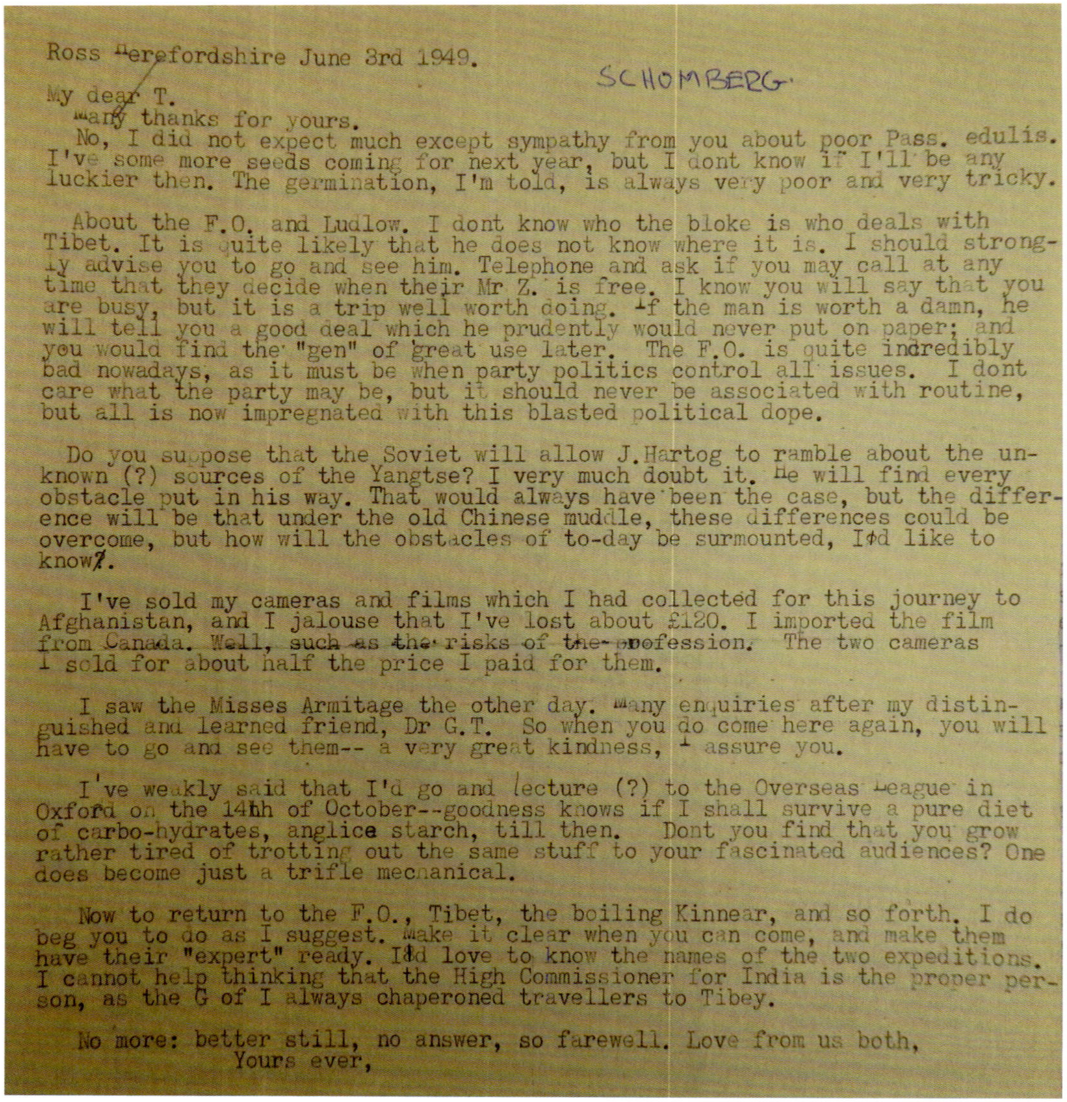

Ross Herefordshire June 3rd 1949.

SCHOMBERG.

My dear T.
Many thanks for yours.
No, I did not expect much except sympathy from you about poor Pass. edulis. I've some more seeds coming for next year, but I dont know if I'll be any luckier then. The germination, I'm told, is always very poor and very tricky.

About the F.O. and Ludlow. I dont know who the bloke is who deals with Tibet. It is quite likely that he does not know where it is. I should strongly advise you to go and see him. Telephone and ask if you may call at any time that they decide when their Mr Z. is free. I know you will say that you are busy, but it is a trip well worth doing. If the man is worth a damn, he will tell you a good deal which he prudently would never put on paper; and you would find the "gen" of great use later. The F.O. is quite incredibly bad nowadays, as it must be when party politics control all issues. I dont care what the party may be, but it should never be associated with routine, but all is now impregnated with this blasted political dope.

Do you suppose that the Soviet will allow J.Hartog to ramble about the unknown (?) sources of the Yangtse? I very much doubt it. He will find every obstacle put in his way. That would always have been the case, but the difference will be that under the old Chinese muddle, these differences could be overcome, but how will the obstacles of to-day be surmounted, I'd like to know.

I've sold my cameras and films which I had collected for this journey to Afghanistan, and I jalouse that I've lost about £120. I imported the film from Canada. Well, such ~~as the~~ risks of ~~the~~ profession. The two cameras I sold for about half the price I paid for them.

I saw the Misses Armitage the other day. Many enquiries after my distinguished and learned friend, Dr G.T. So when you do come here again, you will have to go and see them-- a very great kindness, I assure you.

I've weakly said that I'd go and lecture (?) to the Overseas League in Oxford on the 14th of October--goodness knows if I shall survive a pure diet of carbo-hydrates, anglice starch, till then. Dont you find that you grow rather tired of trotting out the same stuff to your fascinated audiences? One does become just a trifle mechanical.

Now to return to the F.O., Tibet, the boiling Kinnear, and so forth. I do beg you to do as I suggest. Make it clear when you can come, and make them have their "expert" ready. I'd love to know the names of the two expeditions. I cannot help thinking that the High Commissioner for India is the proper person, as the G of I always chaperoned travellers to Tibey.

No more: better still, no answer, so farewell. Love from us both,
Yours ever,

Reginald Schomberg was an acknowledged and prolific letter writer from every destination he travelled to; he regularly kept contact with his friends from the colonial administrative and espionage community such as Frederick Marshman Bailey, Richard Meinertzhagen, George Sherriff and Frank Ludlow. A special bond existed with the Scottish botanist George Taylor who became the Keeper of Botany in the British Museum shortly after this letter was written. British spies seem to have enjoyed a close relationship to ornithology – in the case of John Biddulph and John Scully – or to botany and horticulture. Reginald Schomberg continued to distribute seeds he had collected during his journeys for experiments and planned new expeditions from his home-base in Ross-on-Wye in Herefordshire until the end of the 1940s. He did not mention his move to the Vatican shortly afterwards in order to become ordained as a Catholic priest. Source: Letter from Reginald Schomberg to George Taylor 3 June 1949 (Mss Eur/C371: 54)

from a car or an aeroplane. Daulat, my chief savage, writes from Gilgit now and then. I do not want to go back there, even if I could afford it. I'm no lover of jeeps and the like in a land where only foot-slogging is intended."[199] Here he proves to perceive himself as an 'old explorer' from the imperial school who is pioneering unknown routes: "I crossed into the Shimshal area by Afdigar Pass, never done before by a pale-face."[200] Reginald Schomberg had ended his career as spying explorer and sarcastic critic in order to opt for a new destiny; he moved on to the Collegio Beda in Rome and was ordained as a Catholic priest in 1951.[201] He had realised that no more journeys to Asia would be feasible. His journeys would not be defined as expeditions, rather as 'trips': "I hate the word expedition."[202] In letters to his close friends he mentioned that he is 'suffering from Anno Domini'; thus acknowledging the fact that he might be too old now for travelling to Asia. The winds of global change had reached the Hunza Valley when Jamal Khan faced new challenges in the upcoming phase of decolonisation and the beginning of the Cold War.

The embarrassing interregnum of Ghazan Khan had already been forgotten, and a continuation of the spirit and force of *tham* M. Nazim Khan was seen embedded in the character of Jamal Khan. Several months after Jamal Khan had assumed the hereditary rule, he was furnished with a *sanad* and all honours in Baltit on 23 November 1945. The Political Agent Evelyn Hey Cobb would mention in the draft of his investiture speech: "During the 53 years which have elapsed [since the installation of *mīr* M. Nazim Khan]

The 'new palace' in Karimabad was planned and built by Tham Jamal Khan with financial support from Sultan M. Shah, Aga Khan III, and forced labour (rajaáki) by Hunzukuċ. Since then many modifications and renovations have taken place. The photograph was taken on 22 March 2019

many changes have taken place and the cordial relations between Hunza State and the British Government have been firmly established and strengthened, and the people of Hunza have prospered and become full partners in the British Empire. Your lamented grand-father Mir Sir Mohammad Nazim Khan of illustrious memory was indeed the Father of his people and his strong, loyal and outstanding character had far reaching influence throughout the Agency and even beyond it. On his death in 1938, he was succeeded by your revered father Mir Mohammad Ghazan Khan, who sat on the 'Gaddi' of Hunza for seven years till his unfortunate death on the 8th May 1945, in tragic and lamentable circumstances. I will not dwell at length upon the unsatisfactory state of affairs in Hunza shortly before his death when affairs had drifted into such a state of chaos that he was actually on the point of depriving himself of his Mirship. You have lived through part of the rulership of your illustrious grand-father and also of your revered father and have experienced the changing fortunes of Hunza State. As a young Mir my advice to you today is to follow the advice of the wise men who were selected by your illustrious grandfather and by whose wise counsel Hunza State followed the road of progress."[203] The version that was read on the occasion was a polished one; he had removed some of the implicit criticisms: "… it is impossible to omit a reference to the Great Mir, Sir Muhammad Nazim Khan, whose just and sympathetic rule lasted for nearly half a century . . I can give you no better advice than to follow the footsteps of your illustrious grandfather and so to maintain the influence and prestige of the ruling family of Hunza. … These tasks that

THE MIR OF HUNZA

FAITHFUL ALLY OF ENGLAND

Sir Muhammad Nazim Khan, K.C.I.E., Mir of the State of Hunza, in the Gilgit Agency, has died after ruling his people for nearly half a century as a firm but paternal and benevolent despot. He came to rulership through stormy seas, and thereafter the land had rest. No British officer who was privileged to enjoy his friendship could fail to be struck by his wisdom, geniality, and kindness, his intimate understanding of the life of his subjects, his breadth of vision and his wide culture.

Born in or about 1867, he was the third son of Mir Ghazan Khan, Ruler of Hunza. His youth was promising and his father allowed him, by the time he was seven, to sit with him in council and when cases were being heard in "court." At nine the boy got his first pony and a matchlock, and soon became a first-rate shot and ere long a noted polo player in a land where everyone who can raise a mount plays polo. He was equally good on the mountain side after big game or partridge and in the marshes after wild duck.

His contact with British officers began at the age of 11 when Colonel Biddulph paid his first historic visit to Gilgit and was invited to Hunza by Mir Ghazan Khan. Treachery was feared and the Gilgit authorities tried to dissuade their guest from accepting the invitation.

Ghazan Khan sent the young prince down to Gilgit to offer the invitation in person. He spent an educative month in Gilgit and returned bringing presents for his father from Colonel Biddulph. Two years later he had the pleasure of helping his father to welcome Colonel Biddulph to Hunza. The unsettled state of Hunza-Nagar in those days, when feuds and raiding, slave-trading and war were endemic, made diplomatic missions to neighbouring princes necessary and Muhammad Nazim Khan was frequently employed by his father in such negotiations. At the age of 17 he was sent to Gilgit, accompanied by one notable from each of the Hunza tribes, to act as hostage during Colonel Lockhart's visit to Hunza. He greatly enjoyed this experience and formed a very high opinion of British officers and of the power and good faith of Britain.

When he was about 21 his father Ghazan Khan was murdered by a son Safder Ali, in league with Wazir Thera Beg. Fortunately for himself Nazim Khan was ill at the time. At the sacred shrine of Baba Ghundi he took an oath not to avenge his father's death or to intrigue against the new ruler. This oath he scrupulously observed and by tact soon won the new Mir's favour and was employed in diplomatic missions to China and elsewhere. In 1887 a group of British officers, Durand, Manners-Smith, and Robertson, visited Hunza and were greatly struck by the ability and character of the young prince, which contrasted with the weakness and shiftiness of Safder Ali. The latter, by repeated breaches of his plighted word, provoked the Hunza-Nagar Campaign of 1891-92, and after the capture of Nilt by the combined British and Kashmir forces, fled the country. It was decided to offer the vacant *gadi* to Muhammad Nazim Khan.

Since that day England has had no more faithful and trusty ally. The policy of permitting Hunza to be governed according to its own customs by its own prince with only the benevolent advice and supervision of the British Political Agent at Gilgit has justified itself by the happiness of the law-abiding unspoilt peasantry of Hunza. The Mir set himself to improve the condition of his people by bringing more land under cultivation through communal schemes of irrigation. He greatly reduced taxation and he abolished the system by which royal lands in every village were cultivated by forced labour. Offices of trust and authority were awarded to merit and efficiency instead of being the perquisite of a few privileged families.

The late Mir was wont to remark with pride that he was the first Ruler of Hunza to reign in peace for over 40 years and the only one to celebrate the wedding of a grown-up grandson and heir-apparent. The marriage of Gushpur Jamal Khan in 1934 to a princess of the house of Nagar united the two related rival royal families of Hunza and Nagar, whose ancient feuds are the staple of local history. The two families are descended from twin brothers and came to the country from Badakhshan some 600 years ago. Like so many other Central Asian royal houses they claim a direct, unbroken descent from Alexander the Great. The Mir was keenly interested in the history of his people and preserved and encouraged their traditional sports and festivals. He has fortunately left behind him an autobiography, dictated in Persian, which he entrusted for future publication to Lieutenant-Colonel D. L. R. Lorimer, the first European scholar to make a comprehensive study of the language and the customs of the Burusho of Hunza.

The Mir, who had in succession three royal consorts, is succeeded by his eldest son, Gushpur Muhammad Ghazan Khan, who is the father of three grown-up sons, Jamal Khan, Jamshed Khan, and Ayesho Khan.

The obituary of tham M. Nazim Khan was titled 'The Mir of Hunza' and published in The Times, on 25 July 1938, p. 14. Courtesy of the The Times Digital Archive, http://tinyurl.gale.com/tinyurl/CH35Z2, last access 19 November 2019

lie before you call for patience, firmness and perseverance, qualities that you possess and used in the right way will gain for you the love and respect of your subjects."[204] Without mentioning *tham* Ghazan Khan by name and refraining from mentioning Tarakuč advisors at all, Evelyn Hey Cobb turned to world affairs: "We live in strange and momentous times but at last with the help of God victory over our enemies has been achieved and we look forward eagerly to a lasting peace. In the world war, although only affected remotely, Hunza played her part. She gave what assistance she could to the cause of the Allies both in men and money and it is a happy coincidence that your succession took place at a time when Germany laid down her arms unconditionally in Europe."[205] The Political Agent referred to the attention given in Hunza and Nagar to reporting about the war. In his entry to the Gilgit Diary for May 1940, Gerald Charles Lawrence Crichton had written: "In Hunza and Nagir, the Radio Sets owned, respectively, by the eldest son of the Mir of Hunza and by the Assistant Political Agent, Hunza and Nagir, enable the Mirs, and through them, the populace to follow the progress of the war from day to day. Foreign broadcasts including German, arouse little or no interest and the London and Delhi bulletins are mainly relied upon. In the Political Districts there are no Radio Sets and less interest is taken in the war … In Hunza and Nagir prayers for the success of the Allied arms were held in all mosques on the 26th May."[206] Besides prayers, small sums had been offered as contribution to the imperial war fund. The rulers were mainly concerned with local affairs and the army recruitment as a means of employment; especially after the end of World War II when high commodity prices prevailed in Gilgit bazaar. Evelyn Cobb's remarks put considerable emphasis on Hunza's future within the Gilgit Scouts and its bilateral relations with Nagar under the impression of the just-ended world war. Their expectations were directed towards a renewed strong internal leadership and external compliancy; Jamal Khan's submissiveness would not be questioned during the two final years of the British Raj.

Post-colonial transformations – launching accession to Pakistan

The *mīr* of Hunza and Nagar early on opted for accession to Pakistan; on 3 November 1947 he transmitted a telegram from his winter residence in Gulmit via Gilgit to the Qaid-e Azam M. Ali Jinnah: "I declare with pleasure on behalf of myself and my state accession to Pakistan."[207] The original document typewritten on official state paper titled 'The Coup d'etat' was forwarded to Sardar M. Alam Khan, Pakistan's first Political Agent in Gilgit and copies of the telegram were sent to the Aga Khan Sultan M. Shah and the last British Political Agent in Gilgit, Noel Bacon.[208]

In hindsight, the short-lived 'Independent Republic of Gilgit' that was created two days earlier did not change much in internal rule and external relations. Nevertheless, some political scheming occurred as some signs and symbols were differently interpreted. Pakistan's Political Agent, Sardar M. Alam Khan, had arrived in Gilgit on 16 November 1947 "… bearing letters from the Pakistan Government to the effect that he had been appointed Political Agent of the Gilgit Agency. He was given an enthusiastic welcome by the local population and escorted in triumph to the Agency House. In a few well chosen words he exhorted the people to cast aside their faction feeling and to unite as one in the cause of Islam. … On the 17th November at 9.30 a.m. the Pakistan flag was hoisted on the flag pole at the Agency House whilst a Guard of Honour from the Gilgit Scouts paid full honours and the pipes and drums played the Royal Salute. Large crowds turned out to watch the ceremony and the intense approval of all was very noticeable."[209] Major William Brown's narrative about the transition of power from British rule to Pakistan's takeover suggests a smooth operation: "On the 18th November the Mirs of Hunza and Nagir and the Governors of Punial, Koh Ghazir [Ghizr], Ishkoman and Yasin were introduced to S. Mohd Alam by Major W. A. Brown and successful interviews were completed in the course of which both Mirs signed Instruments of Accession to Pakistan. The Governors expressed their sincerest loyalty to Pakistan on behalf of them selves and their people and the interviews concluded in an atmosphere of cordiality and good will."[210]

This is not the place to discuss and interpret the various narratives about the very important events during the transitional period, which had long-lasting effects on the region.[211]

The once substantial structure of Gilgit Fort has mainly been replaced by the Northern Light Infantry Market with hundreds of bazaar shops. In the centre a single tower survived and is the centrepiece in the lawn of Hayat Shaheed Public School, an educational facility run by the army in the name of Amir Hayat, the first local martyr in the independence struggle of November 1947. The photograph was taken in Gilgit on 13 March 2019

Even during the fighting that immediately set in after independence – the struggle for freedom from the Kashmiri yoke – the active rulers in the Hunza Valley were less engaged than their family members. Subedar-Major M. Babar Khan, uncle of *mīr* Shaukat Ali Khan, was senior to Jamadar Shah Khan, uncle of *mīr* Jamal Khan; both the viceroys and junior commissioned officers of the Gilgit Scouts not only joined the forces, but were instrumental in keeping Hunza's and Nagar's stakes alive. Both have been perceived and revered as heroes who attacked Kashmir proper and kept the Gilgit Agency and Baltistan as part of Pakistan.

The British officers left a functioning administrative and military setup that was described by themselves with pride: "On the 18th November the Political Agent accompanied by Major W. A. Brown and local notables inspected the Agency offices, hospital, Scout Lines, bazaar and other public institutions and the manner in which he was greeted at each left little doubt as to the sentiments of the people. He seemed somewhat astonished that the entire place was intact and that the bazaar was functioning normally with Muslims, Sikhs and Hindus living in communal harmony whilst the various departments carried on the general administration

The renovated monument commemorating the freedom fighters who established the short-lived 'Independent Republic of Gilgit' is located in Chini Bagh, the garden where the first Legislative Assembly of Gilgit-Baltistan was located. The initial monument was inaugurated by Field Marshal Ayub Khan on 27 October 1962. The newly refurbished monument contains black columns with the names of the 'martyrs' and their place of origin. A bronze shield is visible in the background on which the officers from the Gilgit Agency are shown waving a Pakistan flag. On the few historical photographs that are witness of the scene in 1947 the officers are using a different flag, the one of Gilgit and Baltistan which has white stripes and an orange square field next to the moon and star. An image is on display in the Gilgit Municipal Library. The photograph was taken on 13 March 2019

as usual."[212] The formal statements by *mīr* Jamal Khan and *mīr* Shaukat Ali Khan were signed in Gilgit on 19 November 1947; both documents contained the acceptance of the transfer of certain rights to Pakistan: "I accept that in terms of External Affairs, Defences and Communications as specified in the schedule attached hereto the Government of Pakistan has full control and authority, both legislative and executive. Nothing in this instrument affects the continuance of my sovereignty in and over this state or any as provided by, or under this instrument, the exercise of any power, authority and rights now enjoyed by in regard to the internal administration of my State. I hereby declare that I execute this instrument on behalf of this State and my heirs and successors."[213] The Governor-General of Pakistan M. Ali Jinnah endorsed this 'instrument of accession' on 7 December 1947.

Hereditary authority challenged – survival strategies and failures

The power vacuum needed to be filled, but the process took time. In the meantime, *mīr* Jamal Khan ventured to renew his relations with China; his move was noticed by his former British masters and raised an alarm. The reporting about these supposed shifts of loyalty questioned his political intentions. When confronted with his obvious efforts to renew his China relationship after twenty years of abandonment, *mīr* Jamal Khan made clear that he would not leave the already existing anti-communist Cold War alliance against Russia and its influence in Xinjiang.[214] He saw some advantage in resuming his relationship with China and gaining access to his family's property in Xinjiang. In August 1948, the Assistant-Governor Lou from Xinjiang announced his visit to Baltit in order to discuss Hunza's potential accession to China; *mīr* Jamal Khan met the delegation in Murkushi and declared that Hunza had acceded to Pakistan; he kept the Political Agent in Gilgit well-informed about the deliberations and talks before the Chinese delegation returned to Kashgar by mid-August.[215]

During *mīr* Jamal Khan's reign critical changes occurred that substantially affected his standing. Primarily, the colonial backing collapsed after Pakistan's independence. It would be only a matter of time before the central government would withdraw any special support for hereditary rule in principalities. A policy and trend became apparent towards integrating these territories into the nation state. Second, the new regime in down-country Pakistan loosened the grip of the *tham* on the mobility of his 'subjects' who in increasing numbers escaped from the valley to take up outside jobs in urban settings such as the industrial port and new capital city Karachi. A process that had begun with the establishment of irrigated colonies close to Gilgit was amplified by new job opportunities in the Gilgit bazaar economy, by joining the Pakistan army and police, by gaining skills in professional trades and business. Road construction in the context of the Karakoram Highway project created a higher demand in the labour and services market in which many Hunzukuč excelled. The control mechanisms and the exploitation of wage-earners by the *tham* became more complicated; they just simply refused to give a significant share of their wages to the *tham*.[216] More and more migrants managed to escape the stagnant hierarchical social system within the valley. Money-based wealth allowed

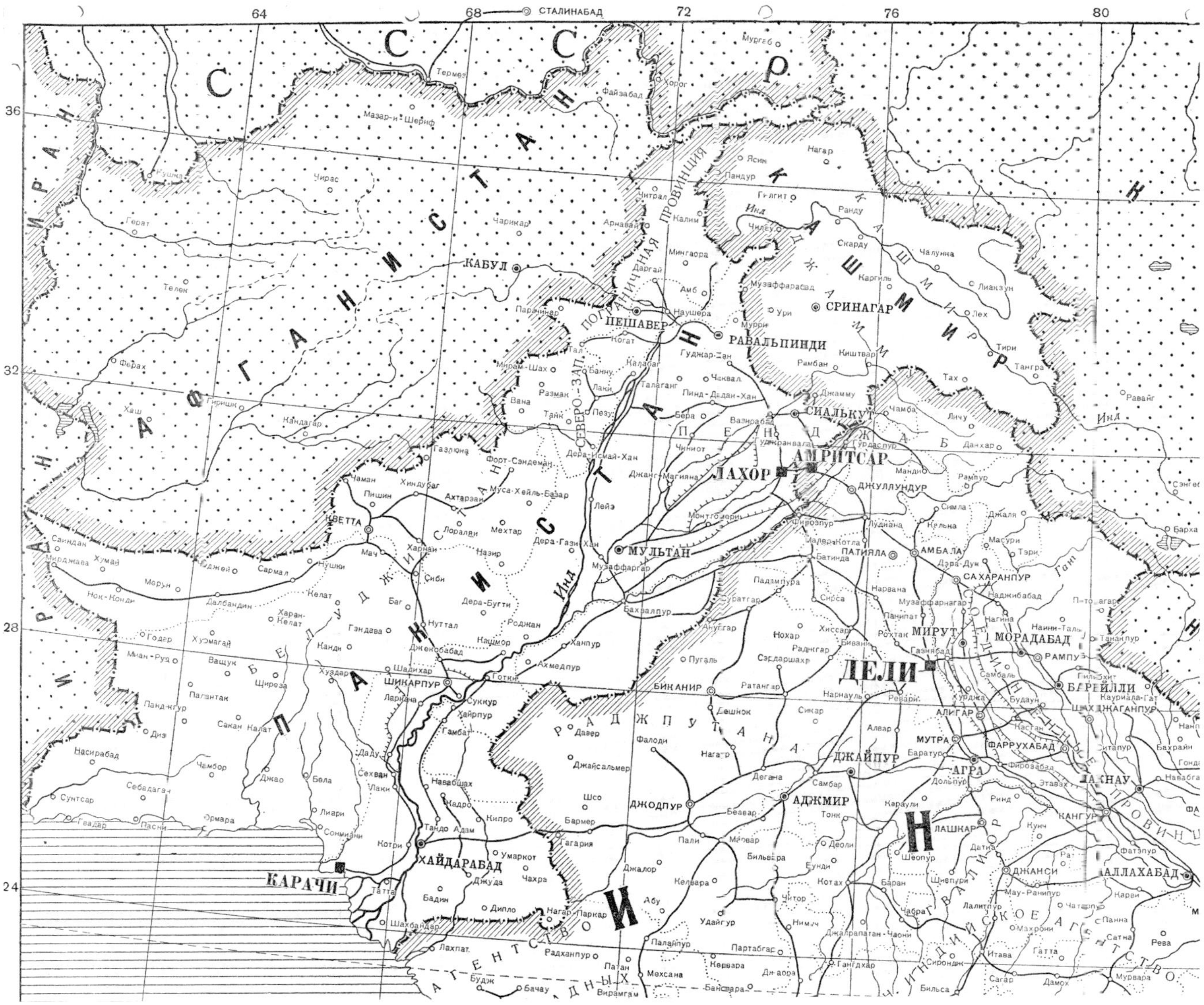

The political position and status of Hunza and Nagar was not always clear to outsiders during the immediate years after Pakistan's independence. In the section of the Soviet map on India and Pakistan an undivided Kashmir is represented including Hunza and Nagar; only the latter is mentioned by toponym. Source: Mikhail Ivanov (1952). Courtesy of the Pamir Archive Collection

people to gain in social standing and political influence that ultimately would challenge hereditary political leadership and established factionalism. This acceleration took place since the mid-1950s, at a time when additional compensatory measures occurred on the scene. A growing number of Hunzukuč opted for joining the Pakistan army which offered a 'pensionable' job opportunity and a legal means to temporarily leave the valley and the grip of the *tham*.

Further financial contributions to Hunza's treasury (*k̲h̲azāna*) were derived from the growing inflow of visitors. Several of them – most prominently John Clark who spent a prolonged time in Hunza in the early 1950s – remark about the heavy demands, greediness and overrated skills of *mīr* Jamal Khan and his Secretary Ayash Khan.[217] Their experience enabled them to supply food items and labour to mountain expeditions – namely the German-Austrian Himalaya-Karakoram Expedition of 1954 and the German Karakoram Expedition of 1959 who spent several months in the Hunza Valley and required substantial amounts of supplies. Building on these international connections, the early establish-

In the garden of their Karimabad Palace, mīr Jamal Khan and rānī Shams-un-Nahar, a daughter of the late mīr of Nagar Sikandar Khan, are portrayed on 13 July 1959. Source: German Karakoram expedition 1959. Courtesy of Gerhard Klamert

ment of fruitful links succeeded in disseminating fancy tales and colourful stories about the 'hidden paradise' that would promote the Hunza Valley and attract further visitors. The process of monopolising the hospitality business for the *tham*, *guśpúr*, *rāja*, and *wazīrkuċ* gained momentum during this period and *mīr* Jamal Khan performed as a perfect host and story-teller about the valley of longevity, healthy food and no crimes where all citizens are satisfied with his fatherly rule.[218] At the end of May 1954 Wolfgang Pillewizer, the scientific leader of the expedition, recorded in his first report about his meeting with *mīr* Jamal Khan and family that he was impressed by the western-style bungalow that featured electric light, water closet and tennis court.[219] Built less than a decade ago, the 'palace' was inhabited by *mīr* Jamal Khan and *rānī* Shams-un-Nahar, their three sons and six daughters who were taught by an American couple. The ruler had just returned from a visit to Paris and accepted an invitation to Munich. The presents in form of a Mauser hunting rifle and a case of Bavarian beer were well received. Nevertheless, *mīr* Jamal Khan related that Hunza is an 'autarkical' state without any imports or exports, a state without beggars in which the *mīri* verdict represents the law.[220] Labour is remunerated in kind and not in cash, although the expedition members soon would realise that the *mīr* Jamal Khan expected payment for his services rendered – provision of porters and food supplies – in Pakistani rupees. The daily charge was the same as levied for the Sherpa who had served the Nanga Parbat expedition the previous year.

As he managed to introduce the dozens of annual visitors only to his family members, entrusted servants and advisors, the vast majority of them got the impression of a land without conflicts, crime and hardship.[221] The romantic picture of the 'healthy Hunzas' became a valuable selling-point and advertising strategy that would carry weight for quite some time.[222] During the first visit of a British High Commissioner to Hunza after Pakistan's independence in 1968, Cyril S. Pickard trusted the explanations of *mīr* Jamal Khan that there 'are no landlords; all land is owned by those who work it; and most disputes are settled by the village community or are referred to the Mir and his Council.'[223] Jamal Khan even succeeded in convincing professional diplomats of his benevolent rule in Hunza, including those who had been sent to evaluate the state of development along the Indus Valley Road and Pakistan's relations with China. Nevertheless, Cyril Pickard sensed that change was overdue and probably inevitable; in his confidential report he stated: "The Mir has a strong sense of his independence and separation from Pakistan. His sons, however, have been educated in Pakistani schools and universities and have been away from the valley for long stretches at a time. His eldest son, who is a charming boy, is much more interested in pop records than the traditional music of the valley. He has taken a degree in politics and political science and is reluctant to take up his permanent residence in the valley. Although the Mir is still comparatively young it is difficult to forecast with any confidence that either he or his successor are likely to retain their hold on their people and defend the separate identity of their state for very long. Nor is it possible to be sure that the Hunzakutz will continue to be satisfied with their traditional way of life as they inevitably come into closer contact with the attractions of the outside world. … It will require

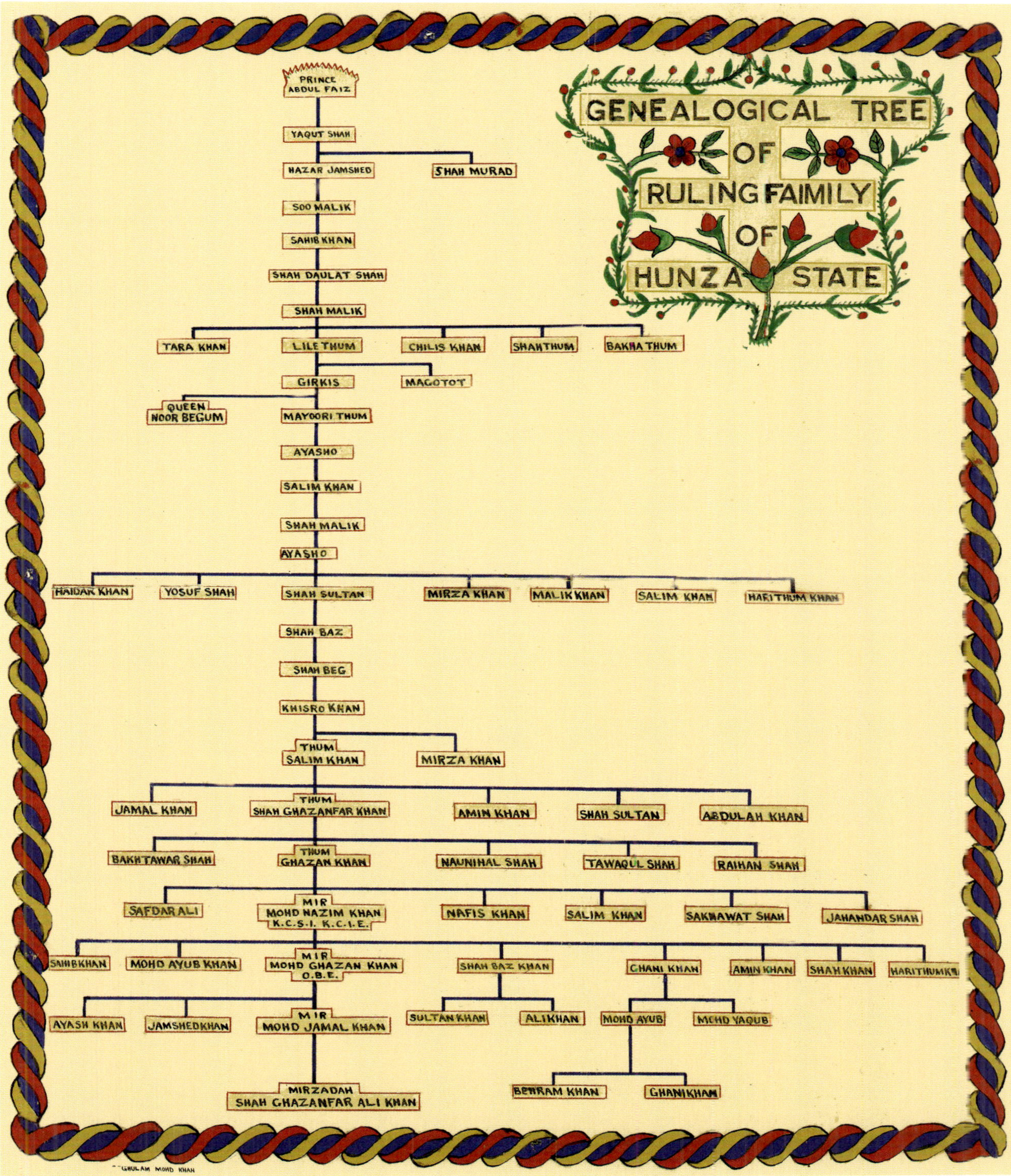

Genealogical table of the Ayášo dynasty drawn by Ghulam M. Khan. Safdar Ali is missing in the central axis of rulers; he was side-lined and stripped off his function and title as tham while all colonial appointments after him were identified as mīr. The Persian title was applied in contrast to the Burushaski denomination as tham. The heir of the last ruler M. Jamal Khan is identified as prince (mīrzāda). Ghazanfar Ali Khan was born in 1942 and his brother Amin Khan in 1949. The table was probably drawn before Pakistan's independence. Source: Original drawing gifted by mīr Jamal Khan to the German-Austrian Karakoram expedition 1954

The shaky bridge connects Hunza with Nagar just below Altit. The bridge was constructed across the Hunza River at a narrow rocky gorge location near Donge Das, with steel cables spanned like a stretched-out ladder. The footway was made of wooden rungs; the traveller would be supported by a rope handrail on either side. The photograph of 1954 shows its partly dilapidated stage. Source: German-Austrian Himalaya-Karakoram expedition 1954. Courtesy of Gerhard Klamert

considerable statesmanship on the Mir's part to continue to retain an independent status for Pakistan."[224] Within less than a decade the Hunza State and its *mīr* would be gone. In a first instance, the separation between the ruling class (*thámkuṣ*) and its territory became obvious and provoked criticism whether an established form of governance could prevail. The colonial structure of indirect rule and semi-autonomy had begun to crumble in independent Pakistan, where princely rule was significantly challenged during the process of nation-building.

The second pillar of external influence grew after the collapse of the colonial system. Aga Khan III, Sultan M. Shah, had slowly increased the responsibility of the *mīr* of Hunza in the management of Central Asian *jamáati* affairs although he was not a strict Ismaili in previously uttered perceptions.[225] The structural replacement of the *pīr* system by the reformed setup attributed the position of chief collector of tithes (*daswān*) to the *mīr* of Hunza. With implementing further structural reforms in the *jamáati* setup – introduction of local and regional councils – in 1951, the Aga Khan assigned a prominent position to *mīr* Jamal Khan. He acted as the 'President of H. R. H. Prince Aga Khan Ismailia Supreme Council for Hunza State, Gilgit Agency, Chitral State and Central Asia' and also functioned as the personal representative of the Ismaili *imām* in Central Asia. His spiritual leader would become instrumental in elevating the social position and financial standing of the *mīr* that he duly acknowledged: "He was the most generous of men and especially to the people of Hunza".[226] Jamal Khan repeatedly mentioned his patronage. Aga Khan III, Sultan M. Shah, invited *mīr* Jamal Khan to Paris in 1950 and introduced him to Egypt, Turkey and other European countries by supporting his travels. After appointing him to the prominent position as his representative for Central Asian affairs, annual invitations for travels to Europe followed.[227] Thus *mīr* Jamal Khan and his wife Shams-un-Nahar became international travellers and gained in reputation.

Female and male education was pioneered through the funds from the Sultan M. Shah's 'Diamond Jubilee' endowment.[228] For Jamal Khan, support from the Aga Khan became a significant part of his income and wealth, especially after the first visit of Aga Khan IV, Shah Karim al-Hussaini, to Hunza culminating with a *darbār* held in Baltit's polo ground on 23 October 1960. The latter became a great sponsor of the last *tham*, and has continued his support for his siblings until today, even after the abolishment of *mīri* rule. Selection as the Aga Khan's representative for Central Asia was deemed a great honour and regarded as an outstanding position among the ruling class in Northern Pakistan.

Because no limousine could reach Hunza mīr Jamal Khan poses with his Hunza State No. 1 car in the yard of Circuit House Lahore. John Clark (1956: 183) mentioned that the Aga Khan had gifted a limousine to mīr Jamal Khan for his down country stays, while he himself purchased a station wagon for his sojourns in Gilgit, paid from his account by the Aga Khan's estate office in Karachi. Source: Photograph taken on 2 September 1954; German-Austrian Karakoram expedition. Courtesy of Gerhard Klamert

The most critical transformations took place in the salient features of Hunza's local economy and resources. The ideal model of rule would put the authority and control over all resources into the hands of the *thámkuṣ* that, seconded by the office-bearers from *wazīrkuć* and the *uyóńko*, would be the chief executive institution at the top of the hierarchy that reached down to each lineage and incorporated every household. This perceived perspective reflects the idea of the *tham*, but his hold on water rights, land titles and pasture access continuously faded after colonial backing was lost. Towards the end of *mīr* Jamal Khan's rule, the claim of villagers towards transferring the property rights of rented lands (*khúṭukal*) and the construction of new channels (*góćil*) grew. The trend of generating more off-farm income outside Hunza led to the exertion of leverage by households on the *thámkuṣ* to promote their claims of perceived equal rights. The growing influence of ideas from down-country Pakistan put severe pressure on the *tham* to maintain his power. Jamal Khan is remembered as a *tham* who became more conciliatory towards the end of his rule, and after his deposition in 1974 is remembered for performing the sym-

Malikamu shikari, the Victoria Monument erected after the British conquest of Hunza in 1891, is a stone heap that towers above the Central Hunza oasis. Its significance was related to its position being the highest monument. Presently it has acquired a different role. Next to it is located a labourers' camp who are constructing and widening the Murku channel system. The reputed Hunzukuċ channel-builders have been replaced by experts such as Waheed from Bunji and Asbar Khan from Tangir. The photograph was taken on 23 March 2019

bolic functions of *bóphao* and *ginάni* within the agricultural calendar in the remaining two years of his life.[229] With the termination of Hunza State, many taxes and the free provision of forced labour were officially abolished, causing a financial challenge for the former ruling house of *ayáśo* that had lost the wealth basis of its nobility. A new business model would be required for future sustenance of the aristocratic household.

Winds of change – abolition of princely states

The crucial date for the final termination of Hunza as a semi-independent state is Zulfiqar Ali Bhutto's decree of 24 September 1974. It marks the end of a process of external interference in domestic affairs that began with the conquest of Hunza and Nagar in 1891. At that time, the colonial forces had occupied Baltit Fort and erected a monument high above at Malikamu Shikari. It became a symbol of the highest authority in Hunza at the highest elevation above Baltit Fort. After the abolition of Hunza State in 1974, the Pakistan administration tried to occupy Baltit Fort as their seat for an Assistant Commissioner (*taḥṣīl-dār*) but failed as *mīr* Jamal Khan refused to allow the civil servant to act from his former seat of power by governing what would become a subdivision within the Gilgit District of the newly formed Northern Areas. During a popular strike, Baltikuċ articulated their outright resistance against any occupation of land for Pakistan's administration and police forces there. They claimed that Baltit Fort was built by Hunzukuċ and that Pakistan's flag should not be hoisted on top of it. The only solution was the shifting of the *taḥṣīl* headquarters and police post to Aliabad where government premises from colonial times were requisitioned.[230]

The colonial setting had supported *mīr* M. Nazim Khan to strengthen his autocratic rule and become more firmly established than any of his predecessors. His efforts to monopolise political and economic power amplified the sidelining of the *wazīrkuċ*, who lost substantial influence after the forced retirement of *wazīr* Shukrullah Beg in 1935. An additional antagonism challenging his authority and dominance emerged from major layers of society and discriminated minority groups. Tahir Ali notes the double marriages between the ruling families of Hunza and Nagar in 1934, from which Jamal Khan's strong nuptial bond with Nagar resulted, as a turning point in the relations of villagers with the *tham*.[231] Until then all residents of the central oasis had been compelled to spend three months in the winter time in the old fortified settlement nuclei (*khan*) and attend certain ceremonies led by the *tham* and would organise the village collective weddings around the winter solstice festival (*thúmuśaliṅ*). It was the time when stores were filled and animals slaughtered. The *tham* would levy marriage taxes and would adopt his role as 'chief-performer-cum-giver of large scale rites and ceremonies'.[232] There was no more danger of mutual attacks since the appointment of the British-backed *tham* in Hunza and Nagar one generation ago. The practice of inhabiting winter quarters in the *khan* nonetheless prevailed; many farmers perceived this as an unnecessary force majeure as they had to shift from their more comfortable houses in the surrounding hamlets (*girám*) back into the old and unpleasant *khan*. Opposition against this practice grew, and it is only one indication of a process of emancipation from *thámkuṣ*. Villages began to act as communities irrespective of existing but fading influences by *uyóṅko* and office-bearers appointed by the *tham*.

Several encounters are on record that illustrate the shifting dominance of inherited power and institutional might on the one hand and the misinterpretation of colonial administrators on the other. The latter had a strong interest in minimising the cost of imperial governance in remote, economically unimportant, but geopolitically sensitive principalities. Indirect rule had proved to be an appropriate and sophisticated strategy of domination aiming at backing the appointed and invested ruler through all vagaries. Incidents of unrest are mainly interpreted as power struggles at the top, communal conflict and ethnicity differences that lead to inevitable temporary destabilisation. The hierarchical setup of invested authority is never challenged, but rather firmly established in order to maintain stability. Consequently, complaints from other social groups or village communities were mainly handled as misunderstandings and acts of disobedience in which the colonial masters pretended to function as unbiased umpires. The growing discontent among farming households with the burden of forced labour (*rajaḱi*) and challenging the inherited modes of expropriation by village leaders can be interpreted as a side-effect of imperial interference. The administrative apparatus offered remunerated wage-labour and salaried jobs in bureaucracy and military service that enabled mountain dwellers from the Hunza Valley to leave the grip of their *tham*, to settle in newly established irrigation colonies outside his jurisdiction, to participate in business in Gilgit bazaar, or to take up services within the political agency. Thus, Hunzukuč and Nagarkuč could gain first experiences with out-migration that contributed to broaden the sources of household incomes and diminished their dependence on their ruler's judgements and orders. Out-migration was an opportunity

Migration patterns diversified since the beginning of the 20th century. Agrarian colonisation had been directed towards the newly established irrigation colonies in the Hunza Valley and near Gilgit Town; some settlers even purchased land and founded new villages as far as in the Ishkoman Valley of Ghizer. About one sixth of Hunza's population had settled outside the state by the 1960s. Seasonal migrant labourers were seeking jobs since the colonial administration offered employment in services in the aftermath of the Hunza-Nagar campaign; a process that accelerated after independence. John Staley (1966: 257) estimated that in 1964 every second household had a migrant; Tahir Ali (1983: 101) calculated that, a decade later, as many Hunzulum Burusho were living outside the state as within. Out-migration from Hunza was amplified and diversified in the first decade after the opening of the Karakoram Highway in 1978.

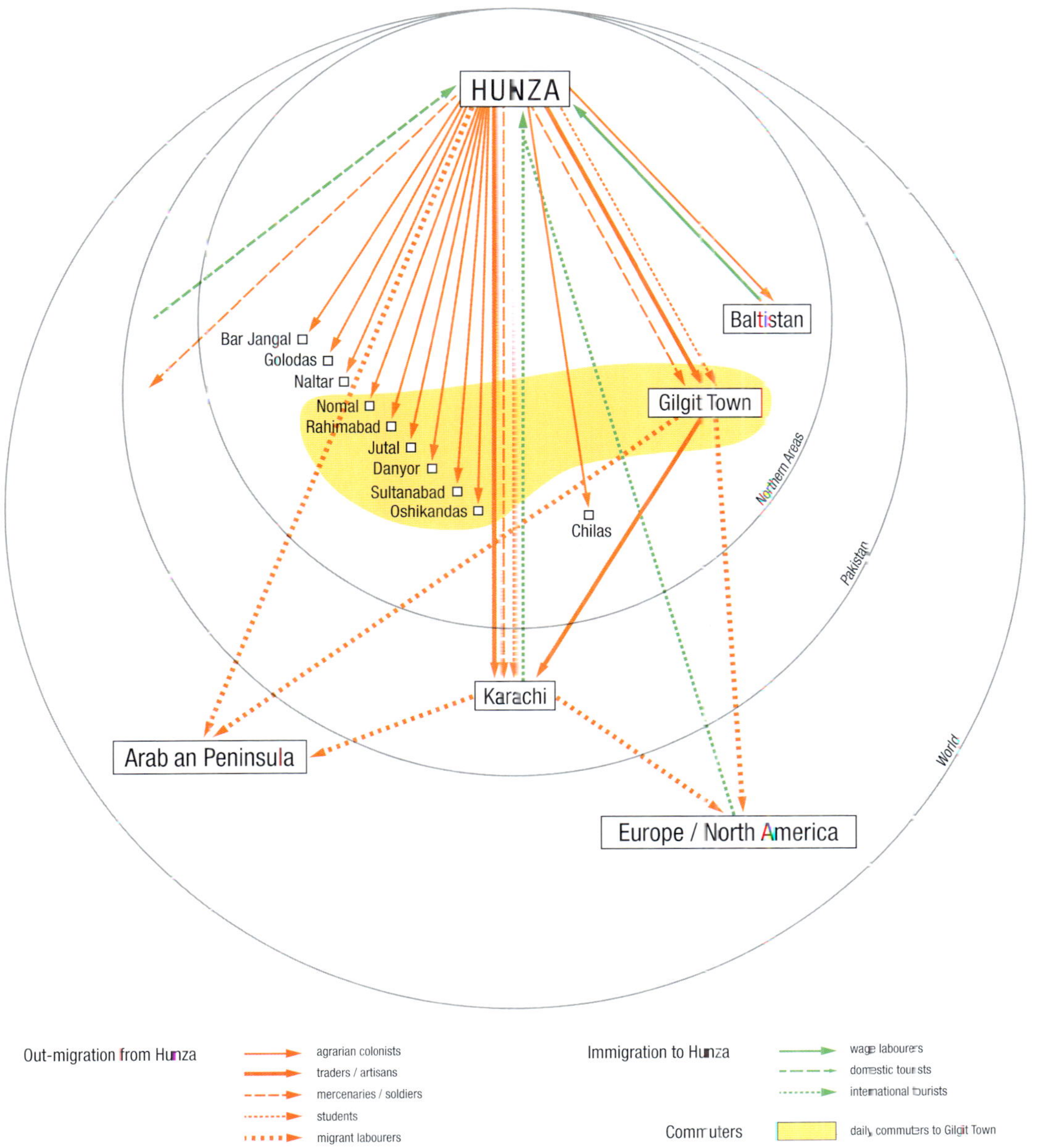

A growing number of labourers and traders sought employment in urban centres of Pakistan; the vast majority went to Gilgit and Karachi. The same applies to student migration; Karachi developed into a prime destination for education and work up to the mid-1980s. A limited number of migrants found opportunities overseas in the Gulf States or in the West. The agricultural labour deficit in Hunza was increasing due to out-migration and partly compensated by seasonal migration of wage-labourers from Baltistan who helped during the sowing and harvesting season. The most valuable contributors in the tourism sector were foreign package deal visitors, mountaineers and trekkers. A sizeable group of commuters from the neighbouring villages of Gilgit sought work in the town. Source: Modified after Hermann Kreutzmann (1989a: 181; 1993a: 31).

Close to Ganesh khan which is hidden by big apricot and walnut trees is the largest building of the village in the extension of the former polo ground. The house of mourning (matám saráa) features prominently in the village which is mainly inhabited by members of the Twelver Shia denomination. Ganesh is their central place in Hunza; other hamlets and villages of this denomination are mainly filial settlements of Ganeshkuč. Tensions in which religious issues of this minority in Hunza played a role have often been linked to Ganesh. The village is located in close proximity to Nagar with its overall Twelver Shia majority. The photograph was taken on 19 October 1984

for ordinary citizens to disentangle their dependency relation with the *tham* who himself was not interested in losing control over a single tax-payer. His financial power was strongly linked to the size of the microstate's population. Two encounters exemplify the handling, growth and scope of increasing discontent and unrest with *thámkuṣ*.

The 'Gilgit diaries' and 'administration reports' recorded instances of communal unrest and growing dissatisfaction among various layers in the Hunzukuč society. In March 1937, some Ganeshkuč rebelled against their *tham*, which the Political Agent Ian Galbraith interpreted as an Ismaili-Twelvershia conflict. From the villages of Ganesh, Silganesh and Gerelt "... some 120 Shiahs ... assembled outside the Agency House with banners flying and supported by some 1,500 co-religionists of Gilgit. ... The leaders were seen and produced three specific grievances. One that they were not allowed to use the principle *Matam Khana* in Ganesh as a general place of prayer; two that Sagi Ali, one of the leaders was not allowed to act as a maulvi and three: that the Mir had confiscated some new flagellatory chains that had been procured for Muharram. It was also stated, though with little hesitation, that the Mir had used blasphemous language about the Shiah sect. The Assistant Political Agent proceeded to Aliabad immediately to enquire into the matter. The Hunza Shiahs were advised to return with him so as to be present to state their case. Fortified by the presence of a British Officer they all returned home though certain apprehension was voiced as to what the Mir would do to them eventually. The Mir naturally denied the whole story and explained that he had asked to see the chains as he suspected they had been manufactured out of stolen bridge material. There is also some reason to believe that the trouble arose with a quarrel between the Shiah lambardar of Ganesh and the other two Shiah villages. The truth is probably that the Mir fears wholesale conversion of his Maulais to the Shiah persuasion and that he has been interfering though not to the extent suggested. There is no reason to hope that an amicable settlement, which was quickly arrived at by the Assistant Political Agent, will prove to be lasting."[233] The entry signifies that there obviously were different layers of unrest in which no 'traditional' office-bearers played a role; neither *wazīr* nor *uyóṅko* were involved, the schoolmaster M. Najat, the Shia *maulawī* Sagi Ali and the colonial *lambardār* represented the interests of different villages. The *tham* sus-

Flagellations and processions during the Shia period of mourning (muḥarram) have frequently caused tension and troublesome encounters until today. State authorities try to ban such public performances; but the religious minority members requested the Karakoram Highway since its opening as their arena on special occasions as well. The photograph was taken on 6 October 1984

pected his 'subjects' of theft; something that would normally not be communicated to outsiders from the alleged peaceful paradise. In addition, he played the religious card and suspected the Shia religious leaders of fostering conversion of his majority Ismaili (*maulai*) subjects. The Assistant Political Agent, Captain George Humphrey Cooke, was sent to Hunza to investigate the case. It appeared that some 'advice' given by the *tham* was taken as his 'order'; thus preventing the Shia Muslims from Silganish and Gerelt from congregating with their fellow believers in the central mosque (*imām-bara*) or mourning place (*matám saráa*) in Ganesh Khan. When both factions were confronted in Aliabad on 22 March 1937, *tham* M. Nazim Khan delivered a long speech in which he accused the Ganeshkuč of having unfavourably compared his *maulai* Ismailis with Hindus. In addition, they had used this occasion to retract their commitment to provide forced labour (*kār-i-begār*) duties for the construction of the new Hindi irrigation channel. Captain Cooke suggested that the *tham* would provide the Ganeshkuč with a *sanad* guaranteeing them freedom of Twelvershia worship if they would refrain from proselytising in Hunza.[234] Captain Cooke provided the writing by stating that the "… matumserais are free for use by all Shiahs. The Shiahs may appoint a maulvi from among them and are free to observe their religious practices." As further measures for reconciliation he gave a verdict on the main instigator: "I consider Saggi Ali is not really a badmash but is extremely bigoted and stupid and really believes that the Mir has been trying to break the Shiahs. I intend to persuade the Mir that Saggi is more fool then knave. If he does not believe this it may go hard for Saggi in the future."[235] The crisis went on, but within six months the Ganeshkuč apologised to the *tham* and paid their respects to him. It seems that both sides tested their strength. During the meeting in Aliabad *tham* M. Nazim Khan tried to behave like the strong ruler he was, but his old age had made him frail and weaker. His son *mīrzāda* Ghazan Khan who would succeed him soon was present at the occasion as well as his grandson Jamal Khan. Both future *tham* would have faced greater difficulties in handling the confrontation.

Such conflicts would occur in different instances and more frequently in future. The Shinaki farmers had complained about the *tham* on several occasions since 1931; the conflict broke out again in the beginning of 1938 November 1939, and January 1941.[236] The villagers from Hindi and Mai-

The Shia mosque in Uyum Nagar is close to the premises of the palace and a regular meeting point. Source: German-Austrian Karakoram expedition 1954; Courtesy of Gerhard Klamert

un repeated their claim that the *tham* had wrongfully taken their barren land (*das*) for making an irrigation channel. The Political Agent immediately sensed foul play and uttered his conjectures: "It appears that they were instigated by Qudratullah Beg a relative of the late Wazir Shukrullah Beg whose family is on bad terms with the Mir. Qudratullah has been forbidden to return to Hunza at present."[237] The Political Agent Ian Galbraith became curious about the roots of the dissent and commissioned a detailed list about taxes paid by different groups in Hunza.[238] The summary of applicable taxes showed a long list of detailed dues and services that had grown towards the end of M. Nazim Khan's rule. In contrast to his autobiographical claim of having reduced most taxation, it records some relief measures for the upper classes (*lambardār, uyóṅko*), while all others are exploited at marriage time, as suppliers of charcoal and firewood, as load-carriers, basket-makers, weavers, millstone-makers, water-mill owners by paying taxes (*bap*) in cash money. At the time of the harvest festival (*gináni*) and at the end of the fasting month (*ᶜid-ul-fiṭr*) clarified butter and milk 'is recovered'. Land taxes (*khuṭúkal*) were counted in grain (*gur, harī*) and lucerne (*śipít*) units, grazing taxes (*ilbán*) were levied in clarified butter and cottage cheese. Every hunter had to deliver part of the prey. In times of land preparation oxen and manual labour were part of the forced labour mechanism as well as for cutting grass and fodder plants. A substantial burden was involved when the *tham*, his relatives (*guśpúr*), messengers (*wakīl*) or office-bearers (*moʿtabar ādmī*) were travelling within Hunza; all of them and their entourage had to be fed (*utáq*) by a number of households (*hakíćan*) that were designated and identifiable for these hospitality services. This assigned duty of providing bread (*ġáne phíṭi*) and meat (*ćhap*) applied to all foreign visitors who were travelling on invitation or under the protection of the *tham*. The pressure on the non-Burusho groups was even higher, and the complaints from the Shinaki and Gojal villages addressed this unequal treatment. Taxes were comparatively higher there than in Central Hunza, but *mīr* M. Nazim Khan had exempted – 'muafed' (*mo'āf*) – heirless people on rented land (*khuṭúkal*) from the special tax, the extra meat tax, and the annual supply of charcoal to upper-class people.

The complaint that was brought to the notice of the Political Agent regarded the ownership of land. After having agreed to pay substantial land taxes, the plaintiffs demanded protection from despotism. The arbitrariness of *mīr* M. Nazim Khan's interventions showed his severe disregard for traditional customs of trust that gave people of various social layers confidence in the reliability of *thámkuṣ*. The pressure was highest on poor people while some upper classes enjoyed exemption from certain duties and taxes. The perception of disappointment and lost confidence prevailed when his son Ghazan Khan took over. When complaints were articulated, the colonial authorities tended to sense only communal, religion-based conflicts rather than social disruptions.

The crisis escalated again in February and March 1941 when Ganeshkuċ threatened to emigrate to Nagar and more than one hundred people tried to reach Gilgit to complain to the Political Agent about ill-treatment. They were not permitted to cross the bridge into Gilgit and had to return empty-handed to Hunza.[239] The incident could have been a sectarian problem between Twelvershia and Ismaili communities, but it appeared to be a question of partisanship and a complaint about the partiality of *mīr* Ghazan Khan in solving disputed cases. The Hunza ruler seemed again to be in no position to handle internal affairs without external backing; all stakeholders tried to use colonial authority for settling their affairs with the *tham*. His system of pitting groups against each other did not benefit his authority.

The preferential positioning of the Tarakuč in particular and the Burusho in general within Hunza's societal layers was perceived as discrimination by Wakhi settlers in Gojal. On and off they had complained about the burden of ever increasing taxes that had grown substantially towards the end of *mīr* M. Nazim Khan's rule and were not relaxed during Ghazan Khan's term on the dais. A rebellion broke out in Gojal in October 1940 when Wakhi demonstrated against *mīri* oppression in front of the Misgar telegraph office where they expected maximum attention by the colonial authorities and threatened to emigrate to Sarikol. Their concern was rooted in this particular case in the local refusal of appointments of village leaders made by *tham* Ghazan Khan and his order to arrest the son of M. Adab Khan, the community leader from Pasu, because of 'impertinence'. The movement was much broader, as Ghulam-ud-din, grandson of Adab Khan, reported. They claimed to belong to the ruling class – at least as *arġúndaro* – an indication of an emerging opposition that had crossed 'traditional' class barriers. With him were M. Aziz and *arbāb* Sami from Hussaini, the latter being the foster-brother of *tham* Ghazan Khan, and Mayum from Pasu. The *arġúndaro* and *guśpúr* from Gulmit and Hussaini did not support the opposition, who based their grievances on severe taxation (*bap*) and forced labour (*rajáaki*) obligations.[240]

The colonial authorities sent the Assistant Political Agent Captain Ronald Allen McConaghey along with a 'platoon of Hunza Scouts' to Gulmit to solve the crisis. The show led to the desired result: "The demonstrators, after some hesitation, came into Gulmit, in a body, on the 29th October and begged unconditional forgiveness to the Mir in the Assistant Political Agent's presence. This was accorded and the demonstrators dispersed to their houses."[241] The colonial diaries regarded the crisis as solved, while the Wakhi community remembers this event as the beginning of the termination of the Hunza State. A weak *tham* in the person of Ghazan Khan could execute authority only with the backing of imperial forces. The occasion to execute further pressure would come with the dawn of independence.

Similar grievances were brought forward by Nagerkuč from Shinber who were suspected to be party in the

The picture shows M Adab in the centre with his son M. Siab on the right and Ibadat Shah, photographed by Aurel Stein for anthropological data collection in Pasu on 16 September 1930. Source: Stein Collection in the British Library. The latter is reproduced here with kind permission of the International Dunhuang Project Photo 392/34 (91)

long-lasting conflict and enmity between *mīr* Sikandar Khan and *rāja* Khisro Khan from Chalt. Ian Galbraith sensed some underlying subtext and realised a growing power imbalance in both principalities: "Although these particular petitions were instigated the old customs of kar-i-begar, etc., in Hunza and Nagar are likely to cause increasing resentment especially in Nagar where the sons of the Mir are so numerous that the pressure of these demands on the people is increasing."[242] The financial (*bap*) and service (*rajaáki*) demands by the ruling elite had grown over time. Available business and job opportunities were increasingly offered with the colonial administration outside the Hunza Valley; the dependency on favours by the *tham* was slightly reduced. The process of exploitation that had earlier been complained about by Reginald Schomberg seemed to have a new turn by evading the vigilance and by refusing to accept land expropriation and finally the payment of exorbitant taxes. Both successors of *tham* M. Nazim Khan could not maintain their predecessor's strong position and had gradually to give up certain claims as criticism and opposition to their style of rule was growing.

The endgame of hereditary rule

The complaints that were listed by Reginald Schomberg in 1934 reappeared on the agenda thirty years later when students from Hunza tried to arrange a meeting with *mīr* Jamal Khan who happened to return to Karachi from one of his several international visits. He and his wife had accepted an invitation to Munich by the organisers of the two German-Austrian Karakoram expeditions. For the Hunzukuċ students in Karachi it seemed an appropriate occasion to put forward their demands and meet with *mīr* Jamal Khan. Two scheduled meetings with Hunza students and Gojal representatives did not take place, as the calling themselves 'his agents' advised him not to confront the students and not to receive a delegation of Wakhi labourers who numbered about forty in Karachi.[243] The first capital of Pakistan became the hub of anti-state activists; the supporters of the Hunza State establishment mainly resided in Hunza, whereas the educational elite and wage-labourers gathered in the port city. Both groups claimed to represent the majority of Hunzukuċ.

Contemporary documents describe a situation that was characterised by a circle of 'advisors' led by *mīr* Jamal Khan's brother Ayash Khan who ostentatiously tried to avoid any change in the existing court system. Since early colonial intervention, the *thámkuṣ* as an institution had been equipped with annual allowances that had financially augmented its upkeep; a smaller contribution was allocated to the *wazīr*. The alienation process between the two elite representatives had critically strengthened the absolute authority and position of the *tham* who could gather a growing number of supporters around himself. The nominal position of the last *wazīr* had been occupied by Sajidullah Beg since 1958; the Tarakuċ had lost the influence they had enjoyed some generations ago.[244] Primarily, *tham* Jamal Khan provided his loyal followers with remunerated positions in the Gilgit Scouts, in the bureaucracy and other services. The preferential positions were occupied by relatives – *rāǰa*, *guśpúr* – but a growing number of traditional office-bearers – *trańphá* and *lambardār* – were incorporated in the inner *mīri* circles that included a few outsiders and supportive members of the Tarakuċ as well. These beneficiaries of the system tried to back the *tham* in refusing any applications for change in matters of forced labour provision, taxation, education and medical support for all. The decision to cancel the meetings in 1963 and the following years contributed to a growing rift between ruler and ruled. The process gained pace, and the oppressed refined their articulation and identified the roots of alienation. Significant suspicion prevailed about the use of development funds from outside that contained contractor fees for labour: payments that would never reach the labourers. The *mīri* responsibility of regularly maintaining the road infrastructure and office buildings was remunerated from the Gilgit Agency's budget, but the *thámkuṣ* had preferred to integrate the work load into its forced labour scheme. The proper functioning of services such as the post office that was controlled and operated by the *tham* and educational institutions that were staffed by his picks was questioned in terms of misuse of funds and nepotism. The heavy taxation in cash and kind had substantially grown over time and augmented the chosen ones in the elite circles. As taxes were levied based on head counts and not on the size of property, the poor suffered and were expropriated more than any other person in Hunza State. The burden of paying horrendous taxes was perceived as feeding the unproductive upper class. The students' demands highlighted the denial of 'basic human rights' and were pointed; they asked for democracy and their right to vote within Hunza as a precondition for further reforms. Hunzukuč residing in Karachi became the spearhead for reiterating their demands. The expressed feelings about the denial of democratic rights totally contradicted the earlier statement of David Lorimer and subsequent narratives of other foreign visitors about a just society in Hunza.[245] The critics of the state addressed the *tham* as Major-General Mohammad Jamal Khan, a title awarded to him by the Government of Pakistan in 1964 and an indication of his fate as he would be retired from his reign as a military officer. The prime complaint of the student leaders was that all monetary funds that reached Hunza were channelled through the *thámkuṣ* and thus created another source of income for the ruler. The Aga Khan had granted support for a student hostel in Gilgit and made available scholarships for higher education. The Pakistan Government had issued funds for educational support; all of those were channelled through the *thámkuṣ* and distributed as favours of *mīr* Jamal Khan to those near to him or people offering something in exchange. A special point of criticism was the building of a German-sponsored hydro-electric power generator in Hassanabad *nālā* in 1967 for which installation labourers were hired. The *tham* functioned as contractor and distributed the workload without paying proper wages. This project added to the observation that any funds that were channelled to Hunza were pocketed by the ruler and rarely redistributed. His expensive lifestyle including international travel and down-country residencies equipped with American limousines were interpreted as a theft from poor people who could hardly feed their household members from the proceeds of agriculture and remittances. Thirteen signatories, ten of whom were students or educated Hunzukuč residing and working in Karachi, handed over their thirteen-point-long 'charter of demands' to the *tham* in the Intercontinental Hotel in Karachi on 24 March 1967.[246] The essential feature of the demands was the creation of an unbiased and transparent institution that would handle all monetary funds and subsidies reaching Hunza. The new board would distribute scholarships on a merit basis, and labourers in public works would be paid nominal wages as fixed per contract. There was grave mistrust of the *tham* and his supporters; the complainants made the assertion that these had taken advantage of their elite position for personal advantages and exploited the general public. Similar reforms to those introduced as 'basic democracies' by President Ayub Khan for Pakistan were demanded for Hunza State. In the following year an expanded delegation met the *tham* at the same hotel in Karachi. When they entered into negotiations their legitimacy as authorised brokers was questioned by the establishment. The twenty-one-person strong delegation contained representatives

The photograph of mīr Jamal Khan and his brother Ayasn Khan who functioned as his advisor and secretary was taken during a journey to Europe in London. Source: The photograph was taken on 6 January 1953

Opposite
The last tham of Hunza, mīr Jamal Khan is sitting while his eldest son, Ghazanfar Ali Khan is standing on his right in a photograph taken in the palace garden in Karimabad on 6 July 1959. Source: German Karakoram expedition 1959. Courtesy of Gerhard Klammert

The opening of the Karakaoram Highway coincided with the abolishment of Hunza and Nagar states. Reforms initiated by Zulfiqar Ali Bhutto contributed to a condition that external food supplies safeguarded sufficient nutrition to a growing population. The generous subsidies have remained in place and have enhanced the acceptance of the Hunza Valley's integration into Pakistan's administration. The photograph was taken in Ghulmet on 10 June 1991

originating from Central Hunza and Gojal and demanded a substantial change in internal affairs. In the summary by Zafar Iqbal who narrated the events, the *tham* should remain the representational head of the state and leader of the government, but 'the system of governance shall be established through an electoral process' and 'should replace the hereditary system of *Trangfa*, *Yarfah* and *lambardar*'. Further demands were directed towards an equitable tax system whose proceeds should be used for the 'welfare of the people rather than to feed the corrupt people around the Mir of Hunza.'[247] In Zafar Iqbal's reading of the events, the *tham* listened sympathetically, but he could not exercise his authority over the circle of his advisors and the elite that surrounded him. The meeting did not yield any negotiated result; the delegation was denied acceptance as representatives of Hunza. Subsequently a demonstration was staged on 4 March 1969 in front of the Intercontinental Hotel in Karachi.[248] This event is remembered as the beginning of the 'Hunza & Nagar Liberation Movement' that did not cease to be active until the abolishment of the Nagar State three years later and the Hunza State in 1974. The movement had gained a momentum that enjoyed growing support among educated outsiders and residents in Hunza; it happened at a time when overall politics in Pakistan favoured a battle against 'feudalism'. Reactions from powerful circles did not wait for long as they observed a development that would curb their resource base. Several active members of the movements were arrested under false accusations and charges that were based on vague evidence and artificial claims. Forty years earlier, Reginald Schomberg had observed a societal split in *mīri* and *wazīri* factions that had created parties based on follower groups.[249] The same happened now as the people in favour of Hunza State formed the pro-*mīr* party that included the majority of hereditary office bearers, *arġúndaro* and *guśpúr*. The opposition was formed by an educated elite and followers of socialist ideologies and land reforms as promoted by Zulfiqar Ali Bhutto. His Pakistan's People's Party became the spearhead of the opposition or the anti-*mīr* party that has remained an institution of self-identification against the establishment until today.[250] The repeated arrest of Ghulam Muhammad from Pasu and his term in jail are relevant features as his peer group was founding the Hunza branch of Pakistan's People's Party in 1972; other party workers such as Karim Khan and Gul Muhammad faced similar sentences.[251] The party had grown to eighty-four members in Gojal alone when Zulfiqar Ali Bhutto commenced his visit to Hunza in September 1974. The hereditary ruler increasingly lost touch with local problems and the means to solve them. The bafflement and perplexity encountered with this new situation is reflected in the strategy to curb the challenge of hereditary rule. A new approach was needed as *mīr* Jamal Khan was obviously not in a position to exercise his authority against influential political activists. Now Ismaili institutions got involved and tried to change pace; *wazīr* Sajidullah Beg who headed the Ismaili Regional Council for Hunza State intervened. Religious authority was employed for solving a secular political crisis by ordering the suspension of any political activities by Ismailis. Other respected members of the Tarakuč such as Haji Qudratullah

Beg were sent to disperse the opposition movement. The result of these interventions was rather counterproductive as the local representatives of the Pakistan's People's Party managed to unify an even stronger opposition under its banner.

The general sentiment that the former elite, centred in the position of *mīr,* used state functions for their own advantage and institutional funds for personal benefits has survived the abolition of Hunza State and is a contemporary feature of dispute when nowadays posts are allocated by political representatives to their chosen clientele and funds are selectively made accessible. A major effect of the integration of Hunza into Pakistan's administrative setup was that educated new elite groups developed parallel to the hereditary institutions. These persons occupied important positions in the 'salariat'[252] or became successful in business. To name just two persons: Qurban Jan from Gulmit was the General Secretary of the 'Hunza & Nagar Liberation Movement' and participated in all events to abolish the Hunza State; as a Gulmitik he began a prestigious career and moved on as the first Hunzukuč to achieve the highest pay-scale and local position in Pakistan's salariat. The president of the movement, Maula Madad from Karimabad, became a successful advocate in Gilgit; many years later, he was nominated by the Aga Khan as President of the Ismaili Regional Council in Gilgit. Several other leading personalities could be named who have occupied influential positions afterwards. The inherited elite of *thámkuṣ* circles gradually lost their grip on successful persons who increasingly challenged the establishment and their leadership. Wealth based on new resources such as proceeds from higher education, professional careers in bureaucracy and success in non-agrarian enterprises contributed to the creation of a new elite that demanded greater participation in decision-making and steering community affairs.

After the abolishment of Hunza State political activities merged with the struggle of activists and representatives of the newly created 'Northern Areas' to reach a provincial status for its territory and to become at par with ordinary citizens of the country in terms of enfranchisement and eligibility for influential posts.[253] The special position of Gilgit-Baltistan and its involvement in the Kashmir conflict has not changed yet. The present civil society continues the struggle for equal rights and participation in decision-making on a quasi-provincial scale.

A scene of the 'Hunza State' movement that is holding up banners with the slogans 'Take the decision back and restore Hunza State' and 'Long live Hunza State and long live the Mir of Hunza'. The undated photograph was taken in late 1974 or early 1975. Courtesy of Zulfiqar Ali Khan

After the Hunza State – elite struggle to maintain influence

Certain elements of previously established influence and power have survived until today despite a disintegrating social hierarchy. The financial leeway and political range of the offspring of Hunza's last *mīr,* Jamal Khan, is of a different order of magnitude. After his forced retirement, the last *tham* fulfilled some ceremonial functions for only two years, and passed away on 18 March 1976. Significant transformations, beginning with the retreat of the colonial administration and Hunza's accession to Pakistan, took place during his reign; he was confronted with the termination of semi-autonomy in the last princely state in the newly independent country and its administrative incorporation into the Northern Areas.

His first-born son Ghazanfar Ali Khan, who was employed in Pakistan's Foreign Service, had to return to Hunza to look after the family affairs although his uncle, his father's 'secretary' Ayash Khan, was still present. Ghazanfar Ali Khan has inherited most of his father's property while two of his sons have pursued careers outside the valley within the government service of Pakistan. The two younger sons, Amin Khan and Abbas Khan, inherited their property mainly as land and houses in Altit and Gulmit.[254] Ghazanfar Ali Khan is a good example of trying to keep as much hereditary power in his hands as possible. The dispute over *khuṭúkal* lands in Akhanabad that lasted well into the 1980s showed the growth of a new elite that based their influence on economic success and prominence in Ismaili institutions. The sophisticated system of revenue generation initiated by his great-grandfather failed to provide a steady income to

The broken tombstone of mīr M. Jamal Khan documents his titles – Hilal-e Pakistan (Crescent of Pakistan), Chief of the Ismaili community of Central Asia (Sardar Jamaat Ismailiya Vasat-i Aasiya) – and the date of his death in the morning on 17 March: 'He took a path with the blessing of the true dawn from this perished world to the settled world'. Recently the family of Ghazanfar Ali Khan placed a new marble stone plate on mīr Jamal Khan's grave, just giving his name and the date of death. The photographs were taken in Suryas Das on 21 March 2015

Eleven years after the abolishment of Hunza State Ghazanfar Ali Khan is attending the National Day celebrations in Karimabad's palace garden which had functioned as the ruler's residence (thámkuṣe diś) until 1974. Two years earlier he had been elected as Hunza's representative for the Northern Areas Council; it was his second term in that capacity. The photograph was taken on 14 August 1985

the son of the deposed and last *mīr* of Hunza. More farmers claimed property rights for the land they had tilled for generations and refused to pay taxes on it. In an interview in 2014, Ghazanfar Ali Khan stated 'there was no tax on anyone. The peasants working on our personal lands would pay one third of the produce called Kutukal' in his father's time.[255] As this source of income was shrinking and endless disputes and court cases claimed their tolls, a shift in economic activities was required. Some investments were made in a 'wagon' transport service between Gilgit and Hunza; former servants in the household of the *tham* were trained as drivers, operated the minibus and charged the passengers. Another source of income was a sawmill where some investment was shared with carpenters; but in both cases the returns were meagre. The private property was partly sold and land ownership had shrunk to about six hectares by the mid-1980s. Greater hopes and aspirations were linked to tourism. The experiences and fame of *tham* Jamal Khan as a hospitality entrepreneur had led to cooperation with tourism operators who channelled a select group of visitors to the hotel operated by the *mīr*, who employed his former bodyguards and other close partners there. After the demise of *tham* Jamal Khan, this business did not thrive as well as expected as other investors started to enter the market. The awkward situation emerged that the former servants of the *tham* who were experienced in all kinds of hospitality services became the main competitors of Ghazanfar Ali Khan's experiments with tourism. As none of these attempts provided a steady cash flow, more lucrative sources needed to be tapped. These were found in political representation and development activities. In both cases infrastructure development and the allocation of funds for contractors (*ṭhekedār*) provided a share for influential and powerful actors.

Ghazanfar Ali Khan succeeded in becoming the longest serving elected political representative of Hunza in the various legislative assemblies that were established after the Northern Areas were formed. The first Advisory Council for Northern Areas (Gilgit Baltistan Agencies) formed in 1971 did not have any representatives of Hunza and Nagar among its fourteen members as both states were still semi-independent.[256] According to Ghazanfar Ali Khan, his uncle Ayash Khan was Hunza's first representative in the assembly elected in 1975 when two more seats were added to the Northern Areas Council for Hunza and Nagar.[257] The opposition members accused the *thamo* representatives including Jamal Khan's uncle Ghazi Johor who was the Additional Deputy Commissioner at that time of widespread rigging to secure the seat for Ayash Khan who later joined the Pakistan People's Party.[258] Ever since, the losers of an election have accused their opponents of rigging.

In the 1979 elections Ghazanfar Ali Khan and Syed Yahya Shah were elected 'un-opposed'. It is important to note that Syed Yahya Shah represented Nagar and not the deposed *tham* Shaukat Ali Khan. As his counterpart, Jamal Khan, had passed away in the same year of the election, it was his son who unanimously won the position to look after Hunza's interests. This was the only occasion of unopposed selection of representatives. Since then fierce fighting for seats and election campaigns has taken place and has been an interesting aspect of power relations within Hunza and Nagar. The 1983 elections returned Shaukat Ali Khan and Ghazanfar Ali Khan as the political representatives; thus the former *thamo* had regained the control of dealing with government institutions and influencing decision-making for infrastructure projects. In the meantime Ghazanfar Ali Khan has been re-elected four times and became the longest standing member in what was called the Gilgit-Baltistan Legislative Assembly since the Governance Order of 2009. Only three times did he lose the elections to candidates from the Pakistan People's Party. In 1994 Nazir Sabir was chosen as Hunza's parliamentarian and gained prominence as an educational advisor to Benazir Bhutto's government. In 1999 and 2009 Wazir Beg from Aliabad, who had already been the opposition's candidate in 1979, won the ticket and promised to break the former elite alliances and hereditary preferences.[259] Although party affiliations are of less importance here because alliances and sides often changed, the principal confrontation is between pro-State and anti-State supporters. The pro-State group has had a stronger leaning to the Pakistan Muslim League, while the opposition has been strongly dominated by members of the Pakistan People's Party. Only in recent years have other parties gained influence; the most prominent new actor on Hunza's political scene is Baba Jan who was able to organise strong support even when he contested the 2015 election from jail and received the second highest number of votes. When Ghazanfar Ali Khan was appointed as Governor of Gilgit-Baltistan in 2016 a by-election for Hunza's seat returned his son Salim Khan who was disqualified after two years, thus leaving Hunza without a member in the Gilgit-Baltistan Assembly since. In

Nagar the leading position of Shaukat Ali Khan was challenged by Qurban Ali who became its representative twice, in 1987 and 1999.[260] Shaukat Ali Khan's grandson, Qasim Ali Khan, contested elections for the Gilgit-Baltistan Assembly in 2014; despite support from *Tehreek-e-Islami* and others he did not succeed in securing a seat In the assembly. The influence of Nagar's ruling dynasty and their related *guśpúr* has survived much longer in local affairs, as is reflected in the election results.[261] Nagar is basically split into a pro-*mīr* and a reformist faction led by *sayid* Yahya Shah who enjoyed majority support in Shinber. In the latest election Islamist parties such as *Tehreek-e-Islami* and the Shia *Majlis Wahdat-e-Muslimeen* gained a substantial share of the vote, the latter secured their first seat. In previous times Nagar was split in two factions just like Hunza, the pro-*mīr* faction had always nominated a candidate from the hereditary ruler's lineage; the opposition was led by *sayid* Yahya Khan. The 2014 election campaign was fought by five parties. The Pakistan Muslim League (Nawaz Sharif's faction) represented Nagar's *mīr* faction, the Pakistan People's Party was again on the side of *sayid* Yahya Khan, both lost significant support to Islamist parties, and Imran Khan's *Pakistan Tehreek-e-Insaf* could harvest a sizeable vote only in Uyum Nagar. Presently, Nagar's two representatives in the Gilgit-Baltistan Assembly are from the *Majlis Wahdat-e-Muslimeen* and the Pakistan People's Party. The change of tides in the centre has affected the political orientation in the Hunza Valley.

Until Ghazanfar Ali Khan resigned as governor in 2018 – merely a result of the 2014 national elections that made Imran Khan's party the dominant factor – he held the most powerful posts in the Gilgit-based assemblies and was influential in Islamabad. These functions have significantly supported his household, especially in times when large hotel-building projects more or less failed and could have ended in bankruptcy, as substantial fixed-term loans were not honoured in time. His major hotel complex in Karimabad's palace garden is ample proof of the diminishing property assets as it had to be transferred. Nevertheless, Ghanzanfar Ali Khan's involvement in shaping Hunza's tourism sector should not be underestimated. From the beginning of the hospitality business a sizeable number of jobs and positions have been allocated to *rāja*, *guśpúr* and *wazīrkuč*. This phenomenon has survived the abolition of *thámkuṣ*. The strong relationship between the ruling elite and tourism operators has shaped participation and led to the exclusion of competitors in the early phases. Tourism entrepreneurship and positions in government-led institutions remained until recently a rather congruent replica of former social hierarchies in Hunza.[262] In general, the filling of posts in the 'salariat' has been strongly dependent on recommendation (*sifārish*) and gift exchange (*rishwat*), resulting in preferential treatment of the former elite, and has only partly been replaced by competition on a merit basis. These structures from the pioneering phases were first broken up when a tourism boom that began after the opening of the Karakoram Highway resulted in a growing demand for accommodation and services. Investments from all layers of society took place, some more successful than others that left unfinished projects. Severe national and international political crises culminating in 9/11 nearly led to the total collapse of international tourism, leaving investment ruins and the inability to repay building loans. The recent renaissance of domestic tourism has opened up the field again; besides local Hunzukuč, external investors from down-country Pakistan are attempting to gain a market share.

Nagar palace is located in Uyum Nagar and has remained the seat of mīr Shaukat Ali Khan until his demise on 23 May 2003. The palace construction is similar to that of Karimabad and its architectural style originates from the same time close to the end of the British Raj. Source: German-Austrian Himalaya-Karakoram expedition on 3 June 1954. Courtesy of Gerhard Klamert

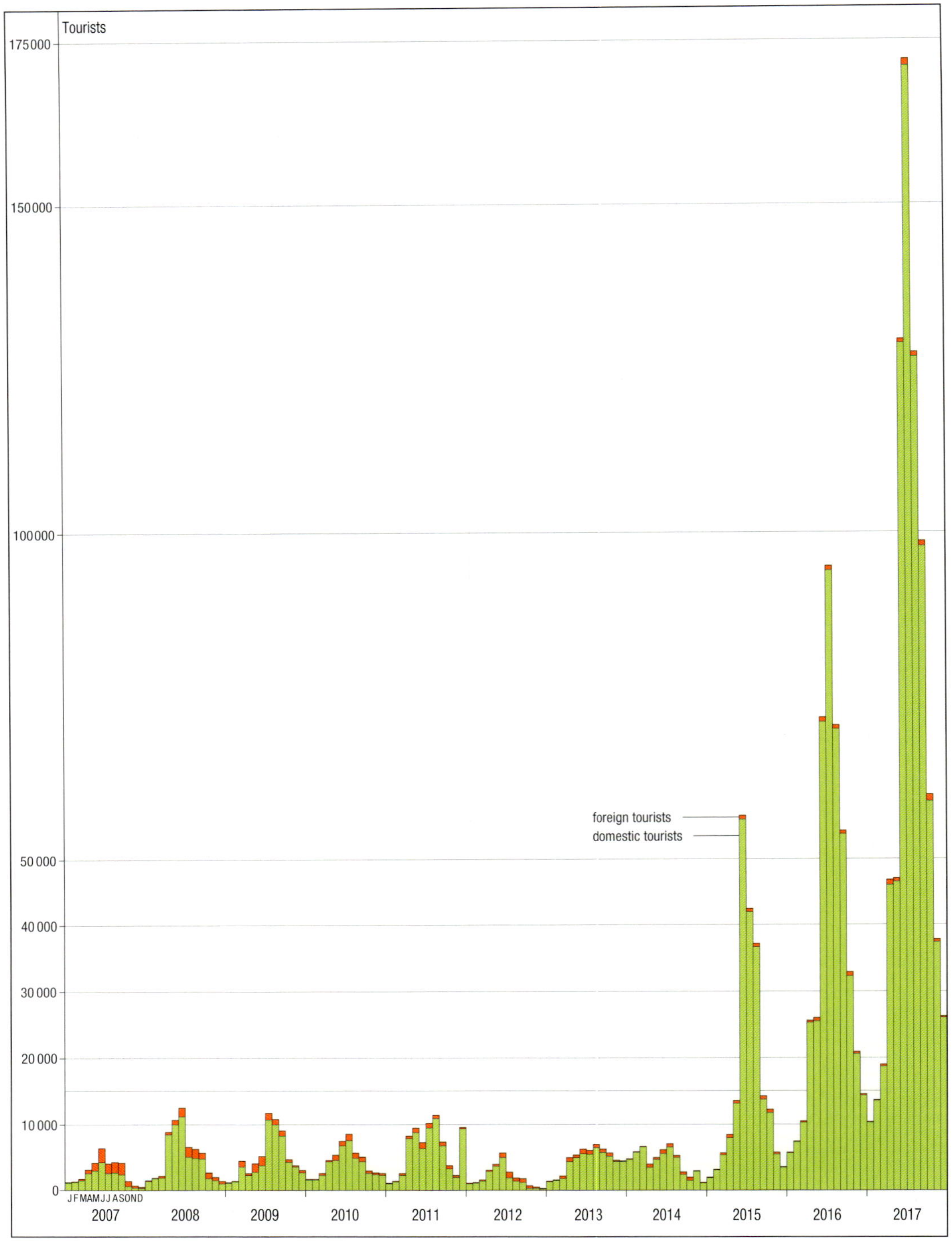

Politics and tourism have been two pillars for sustenance of the heirs of the Hunza State, personified in Ghazanfar Ali Khan. Various Pakistani governments have always extended responsible support to the leading families by providing employment in diplomatic circles and civil service as well as important administrative and military functions. Parallel to such a dependency on changing political tides and newly formed coalitions, a denominational and philanthropic side helped to sustain the former elite. Ghazanfar Ali Khan delayed his decision to leave the Foreign Office service after the demise of his father in 1979 because he was uncertain how to survive in Hunza comfortably. Therefore he travelled to France to meet Shah Karim-al Hussaini, Aga Khan IV, in his *dar-xāna* in Aiglemont. During the audience the Aga Khan assured him that he would take care of his financial situation by confirming 'It is my responsibility to look after you' and suggested that he should return to Baltit and take charge of his father's affairs.[263] The spiritual leader's already strong communal and philanthropic relationship with the predominantly Ismaili Hunzukuċ and the former rulers now entered a new phase of development-related packages. The transition from colonial to postcolonial exchange relations with down-country-based actor is reflected in a multi-tiered change of connectivity and responsibility. Community reforms were implemented after initial visits by Ismaili missionaries and relatives of the Aga Khan in support of the colonial administration. These connections were discontinued after Pakistan's independence. Nevertheless, Hunza has occupied a special position in its relationship with the spiritual and secular leader of the Ismaili *jamáat*. The close bond between Hunza's hereditary rulers and the imamat has been shaped by responses to inner-community dynamics and external challenges. New functions and offices emerged next to top-level agreements and exchanges; thus offering opportunities to a wider group of Hunzukuċ and effecting social transformation in the course of development packages.

Domestic tourism has compensated for the loss of international tourism after its significant decline in the aftermath of 9/11. A tourism boom has occurred since 2015 in Gilgit-Baltistan and rejuvenated this important local source of income-generation. Source: Data based on Government of Gilgit-Baltistan (2018).

At the top left the first tiny guesthouse operated by tham M. Nazim Khan since 1925 is visible; adjacent to it is the 'new' palace built twenty years later. At the bottom below the main dalá channel the Rakakaposhi View hotel is located. It opened in the 1970s and was taken over later by the Serena Group of Hotels, owned and operated by Tourism Promotion Services, an arm of the Aga Khan Fund for Economic Development. The single-storey line of rooms is overshadowed by the substantial structure of the Darbar Hotel which belonged to the heirs of the principality until 2018. The photograph was taken on 18 March 2019

The Aga Khan's connections with Hunza – intensified integration and community reforms

A closer spatial contact and exchange relationship between successive Aga Khans and their followers (*murīd*) in South Asia commenced in the second half of the nineteenth century when new forms of communication were initiated. After his seat (*dar-xāna*) had been established in British India, a closer and stronger link of the *imām* with the *khoja* Ismailis emerged from the contiguous region of Gujarat, Kathiawar, Kutch and Sind which – after some struggles to implement the Aga Khan's authority which was backed by British colonial courts – gained a central position in Ismaili affairs.[264] Once community affairs had been settled and the *dar-xāna* had succeeded in enforcing its secular and spiritual authority backed by colonial power, the Aga Khan attempted to expand his hold on scattered Ismaili communities. It took comparatively long before the *maulai* Ismailis of the Hindukush and Karakoram entered into personal exchange relations with the Aga Khans in Bombay. The forceful and strong leadership of Sultan M. Shah, who had become Aga Khan III in 1885, led to a new phase of *jamáati* relations. He could build on a legal position that had been clarified at a British court, confirming to all Aga Khans the leadership of the Ismaili community with an uncontested authority and right to tithes; in addition, community assets were acknowledged as the personal property of the *imām* of the time (*Hazir Imam*).[265] The position of the Aga Khan had been strengthened and gave him some leverage for implementing cohesion within his 'distinct community' that enjoyed an 'accessible Imam with headquarters in Bombay, and a clearly stated legal status in British India'.[266]

A number of structural reforms were initiated during Sultan M. Shah's seventy-two-year-long imamat. In practices and language use, the reforms followed the traditions of *khoja* Ismailis in Gujarat. From this centre the reforms spread and were implemented in other Ismaili communities, mainly within the British Empire and its sphere of influence.[267] Remoter mountain communities were reached with some delay in the aftermath of the Hunza-Nagar Campaign and partial integration into the British Raj. The *khoja* community structures dominated in the urban areas of the Indian subcontinent and in East Africa. Right from the beginning of their intensified relationship there existed a noticeable wealth gap between the affluent *khoja* as an urban commercial community and their less endowed fellow *maulai* Ismailis. The latter were recognised as faraway *murīd* from the remote rural mountain areas settling in the 'Northern Frontier'. Badakhshan, Central Asia and the *maulai* Ismailis of the Hindukush and Karakoram were accustomed to the decentralised *pīr* system in which Ismaili religious dignitaries residing in various villages had gathered their specific followership. The *murīd* accepted their spiritual guidance in all religious affairs, accepted the performance of rites of passage by *xalīfah* (bur. *qhalipá*, wakh. *xəlifá*) and delivered tithes and taxes, supported their leaders in non-religious activities and engaged in secular agricultural routines for the upkeep of the *pīr* estates. Prior to Hunza's defeat, hardly any authority of the Aga Khan was felt and respected, although there are reports of an enhanced interest by the Aga Khan since the mid-nineteenth century to bring these communities into the fold.[268] The appointment and investiture of M. Nazim Khan as *tham* of Hunza initiated a direct relationship within a few months. During the first visit of a delegation led by M. Reza Beg in March 1892, the Aga Khan expressed his annoyance with the non-obedience of his *maulai murīd*.[269] From then onwards several members of the *guśpúr* and *wazīrkuč* regularly made their way to the *dar-xāna*, paying their respects and delivering Hunza's tithes in the form of gold dust to the Aga Khan. These steps affected and strengthened religious bonds and occurred in quick sequence in an area where religious affiliation had changed not so long ago.

In 1838 *tham* Ghazanfar Ali Khan had made the official declaration that Ismailiya would become Hunza's state religion; consequently, the Badakhshani *maulai* version was

Opposite
Following the visit of the Aga Khan's representative, Captain Abdus Samad Shah, who introduced clerical and liturgical reforms in the Ismaili community in 1922, a second trustworthy representative arrived in Hunza in the following year. Sabz Ali confirmed the message of Sultan M. Shah that jamáat khana should be built all over Hunza (IOL/P&S/10/973: 194). Aurel Stein had photographed the central building in Baltit in September 1930 (Stein Collection in the British Library, Photo 392/34 (80)). Its shape seems not to be modified in the following generation as it was photographed from nearly the same perspective a quarter century later. It still appeared to be a significant landmark as this photograph shows the first and largest jamáat khana of its time in Hunza in 1954; the whole building collapsed and vanished in 1985 as no preservation measures were taken. Courtesy of Gerhard Klamert

Around the time when a regular contact between Sultan M. Shah, the third Aga Khan, and the Ismailis of the Hindukush and Karakoram was established, he had already raised in influence and become a spokesman of the Muslims in the Indian subcontinent. His multi-fold personal connections with political leaders in Europe and Asia allowed him to fulfil an influential role in India's independence struggle and its subsequent partition. Due to his Central Asian followership he was regarded as an authority on regional affairs. The British colonial administration sought his advice in judging Russian and Soviet interests in Central and South Asia. Source: Photograph by Benjamin Stone July 1902. Courtesy of National Portrait Gallery

In the interim between the October Revolution and the end of the First World War a trio of spies under the leadership of Frederick Marshman Bailey was sent along with Major Percy Etherton and Latham Valentine Stewart Blacker from British India on a 'Mission to Tashkent'. One aim of their mission was to contain any expansion of Bolshevism towards the Northern Frontier and to negotiate a buffer. They visited Hunza, inspected the strength of troops and proceeded to Tashkurgan where they inspected the Chinese and Russian garrisons. In Kashgar an important meeting was arranged in which representatives of China, Great Britain and Russia participated. The British Consul-General George Macartney (third from left) is flanked by the Chinese Taoyin and General Ma Titai. Major Percy Thomas Etherton (middle) sits next to the Russian Consul Uspenski; Latham Blacker (left) and Frederick Marshman Bailey (right) augmented the party that took place in Kashgar on 3 July 1918. Source: Bailey Collection in the British Library. The latter photograph is reproduced here with kind permission of the International Dunhuang Project Neg 1083/44 (128).

adopted as all religious proselytising had come from the Pamirs. Although no *pīr* were taking permanent residence in the Hunza Valley the Ismaili among the Hunzukuč became *murīd* of certain *pīr* from neighbouring valleys such as Chitral, Sarikol or Wakhan.[270] There has been some speculation why *pīr* have been residing in Chitral, Ghizer, Wakhan and Sarikol and not in Hunza; a popular explanation was that Hunza was devoid of any additional resources that could have pleased the wealth-oriented *pīr* and their supporters. Their presence in the Hunza Valley would have expanded the upper class that was depending on taxes and services from a rather limited wealth base, thus shrinking the income of the hereditary ruling class. For *pīr*, better living conditions existed in resource-rich regions such as Ghizer and Chitral.

Community integration and reforms

About half a century after the inhabitants of the Hunza Valley had converted and become *maulai*, the Aga Khan attempted to increase his authority and influence with *maulai* Ismailis. The joint venture of communal and colonial expansion based on spiritual and secular authority succeeded in Hunza's case because the British administration provided a fertile environment for collaboration.[271] The joining of forces was intensified after Sultan M. Shah had become a strong voice of the Muslims of British India as co-founder and president of the 'All India Muslim League' in 1906. His position in the diplomatic world was further enhanced as his assistance and cooperation with British colonial authorities was sought, especially along the northwestern periphery of the British Raj. The Anglo-Afghan War of 1919, the Bolshevik October Revolution and its impact on Western Turkestan, and turbulences in Eastern Turkestan affected the remoter corners and buffer zones of the British sphere of influence. In many cases Ismailis inhabited these areas, and the Aga Khan offered his services by spreading news and information, by suggesting proper political behaviour and by taking sides against communism and in favour of British colonial rule. He volunteered to 'send out some carefully chosen emissary and get my people to see that the Southern Star is the guiding light and not the Polar Bear'.[272] This statement of 1921 is part of a regular exchange between Aga Khan III and Edwin Samuel Montagu, the Secretary of State for India in Whitehall. In his outpost in Kashgar, the British Consul-General Percy Thomas Etherton was a stern anti-communist; consequently, he had suggested that the Aga Khan should decree a *firmān* that would bind his followers by spiritual authority to obey and provide 'a check to Bolshevik activity and their designs on India through Pamirs and might assist me materially in my efforts'.[273] The fear of the spread of Bolshevism indirectly shifted the focus of attention and enhanced alertness to the Pamirian Crossroads and the Ismaili communities settling in Chitral and the Gilgit Agency. Hunza's pivotal location in conjunction with its marital, personal and spiritual relationships to nobility and elite groups residing in Afghanistan, Soviet Central Asia and China would push it into a leading role for espionage, intelligence gathering, and spreading orders and news in its neighbourhood. The mission materialised when Captain Abdus Samad Shah, a cousin of Aga Khan III, was appointed as British intelligence officer and was accompanied by Ghaffur Beg and Muhammad Beg from the Central Intelligence Department on an espionage mission to Kashgar. The party left Gilgit on 18 September 1921 and was accompanied by *pīr* Jalali Shah as an indigenous intermediary. After a brief stay in Hunza they proceeded to Sarikol where Abdus Samad Shah "… was enthusiastically received by the Sarikulis. He was the guest there of Pir Nasir Ali Shah, and was presented with a very fine Badakhshi horse by Karim Bai Beg, the leading Beg of Sarikul."[274] After three days,

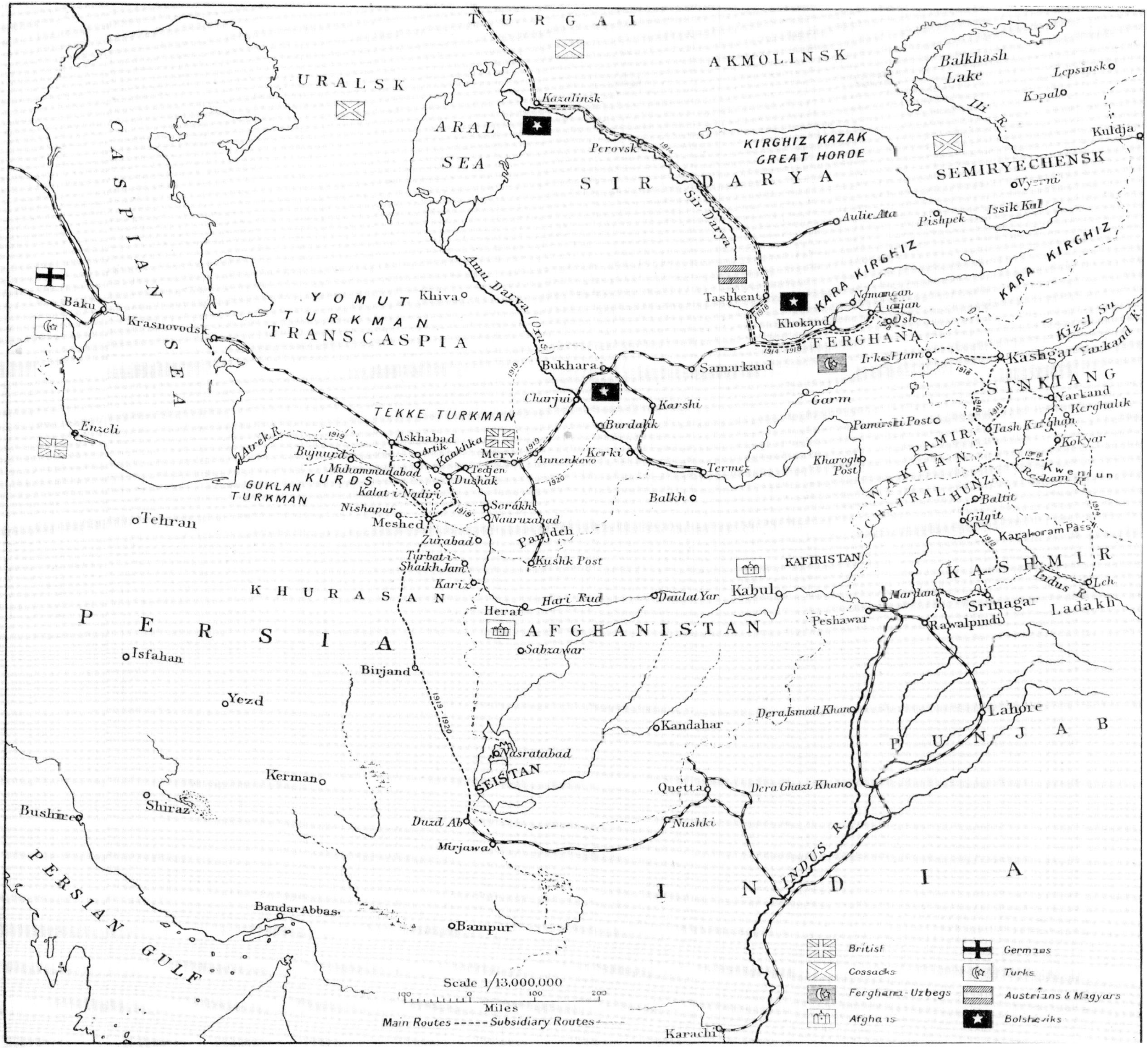

The map shows the perceived geopolitical constellation under which Frederick Bailey, Latham Blacker and Percy Etherton travelled from British India to Fergana via Hunza and Kashgar. Source: Latham Valentine Stewart Blacker (1922). Courtesy of the Pamir Archive Collection

Prior to 1923 wooden mosques were a common element of all villages. Since then new types of community centres and praying places have been built, and various techniques and architectural styles have changed the appearance of mosques, jamáat khana and matám saráa. The photograph was taken on 25 March 2019

they continued their journey and arrived at Kashgar on 22 October 1921. Percy Thomas Etherton wrote in the Kashgar Diary for November 1921 that Abdus Samad Shah 'met all the leading Pirs and followers of the Aga Khan and in conjunction with me is organising an intelligence scheme to cover the Russian Pamirs and all the territory where Maulais are found'.[275] They carried along all kinds of instructions including a *firmān* by Sultan M. Shah for his *murīd*.[276] Abdus Samad Shah would have preferred to set up his station in Mastuj and operate from the centre of *maulai* in Chitral. Mastuj is located in close proximity to the seat in exile of *pīr* Shahzada Lais who had crossed the borders from his ancestral seat in Zebak to Arkari, but the Mehtar of Chitral Shuja-ul-Mulk feared proselytising and unrest in the Yarkhun Valley. Consequently, he banned Captain Abdus Samad Shah from residing there. The tense Sunni-Ismaili relationship in Chitral shifted attention more towards Hunza, where rulers and residents were predominantly Ismailis and had gained a reputation of being supportive of the British Raj. Hunza was the only state with a majority Ismaili population, a factor that would contribute to its unique position within the Ismailiya for many years.

A very strong *pīr – murīd* bond existed between Hunzukuč and the Zebaki *pīr* Shahzada Lais who had gathered a noticeable followership in the valley and was the father-in-law of *mīrzāda* Ghazan Khan.[277] Although related to both ruling families through marriage, by 1909 the Mehtar of Chitral Shuja-ul-Mulk and *tham* M. Nazim Khan had jointly intervened against *pīr* Shahzada Lais' activities fostering insurgency and rifts in the society. By sending another *pīr*, Syed Abdul Hassan, to the Aga Khan in Bombay they expected him to replace *pīr* Shahzada Lais.[278] Until his death in 1916 the influential *pīr* Shahzada Lais had become notorious with the Russian authorities as he was suspected of scheming against their authority in the Pamirs and spying for the British.[279] Abdul Mani, Shahzada Lais' son and successor as *pīr*, visited his *murīd* in Hunza while the intelligence mission stayed in Kashgar. The *pīr* had left when, on his journey back, Captain Abdus Samad Shah arrived in Misgar on 22 January 1922 and in Baltit four days later. The mission had fulfilled its political goals from Kashgar to Srinagar, and a peculiar constellation emerged indicating the winds of change. The overpowering performance of the *pīr* system had critically weakened after the demise of Shahzada Lais, and the secular ruler *tham* M. Nazim Khan received the emissary of Aga Khan III on his way from Kashgar to Srinagar to report in Bombay. During his ten-day stay in Baltit in early February of 1922 they discussed matters of regional politics and ways of strengthening the bond with the *imām*; structural reforms within the Ismailiya were not high on the agenda yet.[280] The previous local and religious office-bearers – *pīr* and *xalīfah* – would be insidiously subordinated under the Aga Khan's umbrella and embedded into a new structure. The upcoming reforms would play into the hands of the British administration. M. Nazim Khan confirmed and endorsed his loyalty towards the Aga Khan and the British Raj. He gathered Hunza's *murīd* on Baltit's polo ground, and Captain Abdus Samad Shah pledged in his public speech that "... he enjoined on all true and devoted followers of the Agha Khan to be always loyal and faithful to His Imperial Majesty, the King Emperor of India, to whom the Agha Khan himself was most sincerely devoted. The assembly expressed their proved devotion to the British Government and promised that whenever called for service they would not be found wanting."[281]

Within less than a year after Abdus Samad Shah's return to Bombay, the restructuring was implemented by the *khoja* Ismaili missionary, *pīr* Sabz Ali Ramzanali who was accompanied by Sayid Munir, an Ismaili preacher from Shughnan in the services of the Aga Khan. Sabz Ali Ramzanali was probably the first *khoja* Ismaili who, as a successful and affluent businessman, had volunteered his services to the Aga Khan; from Hunza he would proceed further to Shughnan, Badakhshan and Kabul.[282] The hierarchical power structure of combined spiritual and secular authority was instrumen-

Roofs of Altit Khan visible from Altit Fort, photographed by Aurel Stein on 14 September 1930. In the background the lantern roof of the first jamáat khana of Altit is visible. By introducing a 'modern' building style with a lantern roof protected by a cupola, its structure differs from the construction of previous houses and mosques. Source: Stein Collection in the British Library. The latter is reproduced here with kind permission of the International Dunhuang Project Photo 392/34 (82)

The front side of Altit's jamáat khana opened to the polo ground. Source: David Lorimer 1934–1935 (PP MS 66: II.47)

tal for further and ongoing reforms within the local community organisation.[283] Thus, *tham* M. Nazim Khan actively supported the visible change from mosque to *jamáat khana* by integrating it into his forced labour scheme (*rajaáki*). The Gilgit Diary reported in May 1923: "In every village throughout Hunza fine 'jamiat khánas' have been built by united efforts of villagers. This is mainly due to the advice of Captain Abdus Samad Shah ... During his stay in Hunza he had explained to the people how the down-country followers of His Highness the Agha Khan stimulated in their community brotherhood, unanimity and morality. It is alleged in well informed quarters that the Maulais in Hunza have to an appreciable extent given up adultery and theft. Not only all the men, but also the women of the villages, for whom a part of the building is reserved, collect at least once a week in their Jamiat Khanas to listen to the sermons of their village 'khalifas'. The Mir is keenly watching the working of these institutions. These may have a beneficial effect on the morality of the community, provided they are maintained on a purely indigenous and religious footing; otherwise they may lead to mischief"[284] Beyond its function as a mosque or prayer room, the *jamáat khana* as a building of 'all-purpose nature' gradually became the centre of village and community activities such as administration, communication, social organisation, with support functions and shelter for travelling members of the *jamáat*, library and study centre.[285] The *jamáat khana*, occasionally addressed as *khoja khana*,[286] became the visible symbol of community reorganisation, received travellers, and contributed to enhancing the personal bonds between the *maulai* and *khoja* communities.

During the 'Golden Jubilee' of Aga Khan III a group of Ismaili missionaries handed over a valuable casket as a personal gift in Bombay in March 1936. The third person from right in the middle row among the seated is Alijah Pir Sabz Ali Ramzanali who had visited Hunza in 1923 to introduce the jamati reforms on behalf of Aga Khan III in the Northern Frontier and Central Asia. Source: The photograph was taken by Golden Art Studio. Courtesy of the Ameer Janmohamed Collection, London

The shifting allegiance from *pīr* to *imām* was followed up by further community reforms; even if their offices were not replaced, *pīr* and *xalīfah* were curtailed in their influence. Based on Gujarati terminology, the positions of *mukhi* and *kamaḍia* not only introduced new liturgical functions for leading the prayers, religious instruction and accounting in the *jamáat khana*, but Gujarati community structures also took over: a transition from inheritable functions to appointment by the *imām*. The Aga Khan docked the entail and reserved his personal right of selecting his office-bearers in a rota system among eligible *murīd*.

A regular delivery of Hunza's tithes and gifts in combination with visits by *guśpúr* and *wazīrkuč* in Bombay commenced. More *pīr* and *sayid* from the Pamirian Crossroads entered the British Raj via Hunza and paid their respects to the *tham*, who received gifts and a share of the proceeds. The customary collection of tithes (*daswān*, *zakāt*) was taken away from the *pīr* and handed over to estate agents: "In every village a Jamat Khana has been built for the preaching and prayers. The representative [Sabz Ali] intimated to the people that in due course he would arrange to send religious teachers to start 'Maktabs' for the benefits of the Jama'at and their children. The new arrangement of collecting the tithes, as far as can be judged, is likely to prove abortive, as the local Pirs who are held in the greatest possible respect are bitterly opposed to it."[287] A few months later, the Gilgit Diary for October 1923 recorded that the missionary 'Sabzali Ramzan Ali … has collected tithes amounting to Rs 6,175 from His Highness' followers in Hunza, Sarikul and Yarkand and has remitted the sum by money order to Bombay'.[288] These reforms caused quite some unrest among the former religious elite that lost influence in favour of a new structure in which the *mīr* of Hunza's status was gradually raised over time.

Sabz Ali Ramzanali's visit had caused some unrest in the Gilgit Agency. The British Resident in Kashmir was asked to submit a report three years later about 'the 'Momin' movement or religious mania in Kuh-Ghizar' where the *pīr* were more influential and saw their inherited position weakened.[289] A letter signed by 'Sabaz Ali Ramzan Ali, representative of His Highness Sir Agha Khan', was distributed in all Ismaili villages. The Gilgit Agency's Munshi M. Masih Pal got hold of a copy and translated 'Ramzan Ali's written instructions': "… by the order of the Maula from the Supreme God, the following order is promulgated for combining together the matters religious and worldly, so that the Jamaat should act upon it. 1. Any thing that has been made Sarkari (i.e. consecrated to Agha Khan or made his property) is the property of the Maula. No one has the right to dispose of such a property. 2. Everything that has been made Sarkari it should be recorded, and at the same time the name of the doner and the name of the Mooki should be written down."[290] This order changed the property rights and established the estate of the Aga Khan in an area where all property was exclusively reserved to the local residents by inherited rules and regulations. The personalisation of real estate commenced in 1923. More worldly matters were addressed in the following instructions: "3. Those who have been appointed for the collection of the tithes should collect it with smoothness from the Jamaat. 4. Any thing that may be in kind should be turned into hard cash. 5. Every year, a month before Nauroz … all the tithes which have been collected should be remitted to the Sacred Darbar (the office of the Agha Khan) by telegraphic money order, and the account and details of the collection duly recorded, submitted by registered post. 6. Any order that may be received from the Durbar-i-Humayun (the Exalted Office) should first be read by those who have been appointed for the Sarkari work. If the order contains any thing for the Jamaat it should be read to the Jamaat, and in case during the reading of the order anyone gives anything it should be credited to the Sarkari Account."[291] Modern technology was applied to convert and transfer tithes to the Aga Khan's personal accounts and it was made clear that all donations and proceeds belonged to him. The Aga Khan worried about protest, unrest or a division of his *jamáat*; so some precautionary rules for arbitration and reconciliation were formulated: "7. If among the Jamaat any quarrel or disagreement should take place those men appointed should with the consultation of the Jamaat settle the quarrel. 8. If, may god forbid, any one among the Jamaat is tyrannized over, all the Jamaat should be on his side and help him and then the result communicated to the Darbar-i-Humayun (Office of the Agha Khan). 9. Since the whole Jamaat is under one Amr (order) it is undesirable that owing to the 'Mookis' a split should occur among the Jamaat in the matter of religion. On the other hand it is incumbent on all that in matters religious or worldly, all should be united in their discharge. 10. Whenever any petition (letter) or Sarkari property is sent it should be sent to the following address: City of Bombay, Post Muzgam, the Office of His Highness Sir Agha Khan, care of Wazir Rahim Basriyah."[292] The 'instructions' do not mention any *pīr* or *xalīfah*. Munshi M. Masih Pal had observed in his report, written three years after the visit, that there was an inbuilt bone of contention right from the beginning: "A perusal of this order will show what a weapon has been given in the hands of the public. The result, as was feared then, is that the Pirs have lost their original influence. They are reduced to the status of a preacher. The Farmans (Orders) now come direct to the councils and being in direct touch with the Agha Khan's office they do not care much of the Pirs or their letters."[293] Whenever the set rules were violated in these three years, immediate action was taken by the *dar-xāna*. The colonial administration feared detrimental repercussions from the expropriation and exclu-

The jamáat khana of Daigiram in Gulmit features the gabled roof of the early reform period. Next to it on the right stands a separate tower with a square structure a so-called farmān khana in which the decrees of the Aga Khan are preserved and venerated. The photograph was taken on 29 April 1990

Just below Baltit Fort was the residence of the wazīrkuć. Their family members played an important role as messengers from Hunza to see the Aga Khan in Bombay. Frederick Marshman Bailey visted Baltit on 14 May 1918 and photographed the house of wazīr Shukrullah Beg. Source: Bailey Collection in the British Library. The latter is reproduced here with kind permission of the International Dunhuang Project Neg 1083/44 (49)

sion of influential *pīr*; in addition, conflicts were brewing in Kuh and Ghizer that affected Chitral with some delay and were probably one trigger for the subsequent sectarian unrest and escape movements there.[294] The remote control of *jamáati* affairs from Bombay posed a critical hiatus and developed a new quality of indirect rule in the Northern Frontier. The decision to attribute a more important role to the *mīr* of Hunza might be connected with the absence of controversial *pīr* in the valley. Munshi M. Masih Pal assessed the situation: "Since the visit of Ramzan Ali Sabz Ali, the Missionary, and the introduction of the system of Councils who have been charged with the duty of collection of tithes, a duty formerly performed by the respective Pirs only, the Pirs have given up visiting their Murids. I have not seen these foreign Pirs in this Agency since the introduction of the new system. Now that their pay has been raised by His Highness the Agha Khan it is just possible that they, to justify their position as Mooki, may visit the Agency and preach among the people. I was informed by Subadar Major Ghazan Khan on his return from India that His Highness the Agha Khan had entrusted the charge of the Maulais in Hunza and Yarkand side to the Mir, his father, and the charge of the Maulais in other Ilakas [districts] of the Agency had been entrusted to Pir Jalali Shah in Kuh-Ghizar and Yasin through the present Governors, but I know the Governors in this matter are powerless. The people owe allegiance to their Pir only."[295] His report confirms that *tham* M. Nazim Khan was early on involved in tithe collection in Hunza and Xinjiang, a function that would be further formalised some years later.

Developments within the British Raj took different paths along an alliance with the Aga Khan, while beyond its borders in Central Asia a stricter control and several objections characterised the local authorities' refusal. In Sarikol and Yarkand the Chinese administration increasingly disapproved of sending religious taxes in a formalised way to Bombay; especially local representatives faced challenges and were arrested after Sabz Ali's visit.[296] The Chinese *amban* made a connection between the tithe collection by the *pīr* and their alleged conspiracy and espionage for the British. The entry into the Kashgar Diary recorded the worries of the Consul General Clarmont Percival Skrine in October-November 1923: "Sarikol – There are indications that the new Amban is anti-British and probably under Bolshevik influence. (i) As reported in Kashgar diary for August, paragraph 13, an agent of the Agha Khan has been collecting tithes from the Maulais in Sarikol. This came to the ears of the Amban, who summoned the Chief Maulai Pir Imam Dad, abused him, spat on his face and charged him not only with being secretly a British subject and sending tribute to the Viceroy but with plotting to hand over Sarikol to the British. Refusing to listen to any explanations, he threw the Pir into prison and has closed his yamen to all Maulais, whom he seemingly advised to seek the aid of their masters, the British."[297] Clarmont Skrine refrained from direct interference and suggested to the maulai of Sarikol to complain to the amban. Two months later, he had collected additional information in the case: "From further reports received by me from Tashkurghan it appears that the Amban not only imprisoned and insulted the Pir, and his followers, but had him thrown down and given 200 strokes on the legs by Chinese soldiers. A recent arrival from Sarikol informs me that he saw the wounds on the legs of the Pir. The latter was released by the Amban only after payment by the Maulais (followers of the Agha Khan) of a large sum of money in kind. This was claimed by the Amban as the equivalent of the tribute alleged by him to have been sent by the Maulais to the British Government."[298] The time had arrived when open tithe collection had to be clandestinely pursued; the intervention by *pīr* Sabz Ali Ramzanali had caused stern reactions by the Chinese authorities who saw their meagre income base dwindling and their influence challenged.

The Bolsheviks banned any transfer from Wakhan and Shughnan; the Soviet officers threatened the *pīr* and *sayid* with 'imprisonment and execution for sending money to the British Government'.[299] They urged their residents and spiritual leaders to spend the money on local charity. By the mid-1920s the formalisation of Ismaili tax collection and monetary transfers made the *tham* of Hunza the acknowledged broker or increasingly a clandestine estate agent for the Aga Khan. Sometimes he had to find businessmen and traders in Gilgit bazaar to convert roubles from Afghan and Soviet Badakhshan into Indian rupees before transmitting the equivalent sum to Bombay.[300] The cross-border tithe collection continued until the early 1930s when the persecution of Ismaili *pīr* led to their arrest or escape from Soviet territory into neighbouring Afghan Badakhshan. Shortly afterwards the hermetic sealing of Soviet boundaries and the closure of the Chinese border by 1935 severely complicated these transactions and communication patterns in general.[301] Nevertheless, tithes still reached the *dar-xāna* in Bombay afterwards. The Aga Khan attempted to formalise this process by enhancing the *mīr* of Hunza's standing. Although the authorities did not particularly esteem Ghazan Khan in his position as *tham* as much as they had appreciated his father, the former received a telegram from Aga Khan III in March 1940 asking the *tham* 'to undertake collection of offerings for him from all places in Central Asia including Badakhshan and Chitral'.[302] His father had conformed with the wishes of the Aga Khan fourteen years earlier, and now the son's portfolio was broadened and expanded. This exceptional position for the *tham* of Hunza was not easily accepted and tolerated by Nasir-ul-Mulk, the Mehtar of Chitral; he successfully intervened and complained to the British authorities, demanding the right to transfer the offerings directly from Chitral to Bombay. These negotiations underline the fact that every collection of tithes and taxes generated some side-income for the collector. From Hunza's perspective it was a welcome perquisite and dearly needed additional income as money flowed from Afghanistan, the Soviet Union and China to its treasury (*khazāna*) and moreover significantly increased its political standing and influence.

During their early travels to Europe in the 1950s mīr Jamal Khan and his rānī Shams-un-Nahar were received by Sultan M. Shah, Aga Khan III, in attractive locations. The undated photograph is contained in the collection of Karimabad palace. Courtesy of Baltit Fort Museum

Ismaili modernisation for Hunza

It would take another twenty years before Aga Khan IV, Shah Karim al-Hussaini, became the first *imām* to visit Hunza and elevate its relationship with the imamat on to another preferential level. Aga Khan III had initiated an educational campaign as a self-appointed 'didactic leader' to mainstream Ismaili citizens by bringing literacy to all of them.[303] The proceeds of his Diamond Jubilee celebrations – during which he twice received donations the equivalent of his personal weight in diamonds, in Bombay in March and in Dar es Salaam in August 1946, by his affluent *murīd* – became the seed money for an endowment financing a network of Diamond Jubilee schools that still exist today. In total more than half a million pounds sterling were collected and earmarked for social projects.[304] In August 1946, Hunza moved a major step forward: the only school in Baltit, founded in 1912 by the colonial administration, was upgraded to a middle school, other villages founded seventeen more educational establishments, and all were sponsored by the Aga Khan's endowment fund.[305] For the running and upkeep of the schools *mīr* Jamal Khan received an annual budget of eight thousand rupees from the fund.[306] Fifteen years earlier only eighty-four male and one female were recorded as literate persons for Hunza; these students originated from *tham*, *guśpúr*, *wazīr* and *uyóṅko* only.[307] Some elite members had been eligible to attend school in Gilgit or Kashmir. By 1938 eighty students were enrolled in Baltit's school, and the population census showed three years later that there was still only one literate female, but 248 male Hunzukuč were recorded.[308] Qudratullah Beg from Baltit and Sultan Ali from Gircha were two prominent members who had enjoyed early exposure to formal education. Aga Khan III appointed Qudratullah Beg to oversee the establishment of schools in Hunza. The management and supervision of the schools rested in the hands of *mīr* Jamal Khan, Sultan Ali and Qudratullah Beg. Although only three teachers had a qualification up to class eight in 1946, the teachers were mainly selected from Hunza and enjoyed a locally generated salary.[309] These valuable jobs were allocated to notables, members of the elite or confidants of the *tham*. Sultan Ali functioned as headmaster of the first Diamond Jubilee Middle School in Baltit from 1946 to 1960; he became one of the most influential and respected educationists of Hunza.[310]

Mawlana Hazir Imam, Karim-al Hussaini, during his first visit to Hunza in October 1960. His local representative mīr M. Jamal Khan is seconded by his brother and secretary Ayash Khan, dressed in military uniform, during the briefing. Source. Photograph taken by Abdul Ismaily, Baltit on 23 October 1960

The Aga Khan's initiative shifted the educational pattern in Hunza from one where limited accessibility was reserved for members of the nobility to a general eligibility for those who could afford to send their children to attend school. With some delay, these would offer female education that aimed to significantly augment public services, which were predominantly focused on boys' schools.[311] Hunza's preferential treatment by the imamat is reflected in its share of resources; out of more than three hundred Diamond Jubilee schools that had been founded worldwide by the 1980s, more than a third were located in northern Pakistan, of which forty-one operated in Hunza alone.[312] The educational initiative was accompanied by support systems in health, house-building, finance and insurance. It was embedded in a reform package that had not only influenced the religious and clergy structure but also noticeably affected and supported living conditions in general. Aga Khan IV had continued and expanded his grandfather's endeavour to improve the livelihoods of the Ismailiya's less affluent communities. His first visit to Hunza was scheduled three years after becoming *imām* and just two years before a major organisational and management reform was implemented within the *jamáati* institutions. The 'Ismaili Constitution for Pakistan' was adopted in 1962, strengthening the hierarchical council system that regulated *jamáati* affairs on federal, regional and local levels.[313] Five years later the Aga Khan Foundation

Opposite
The group photograph in front of the main mosque of Baltit – which had become the school – shows a major gathering of men and pupils in the summer of 1954. Around sixty to eighty pupils were taught here. In the background the gabled roof of Baltit's first jamáat khana is visible, displaying a quite different architectural style. Most people are dressed in a mixture of locally produced clothing with textiles from the Gilgit bazaar that were supplied from down-country Pakistan. Ustad Sultan Ali, first headmaster of the Aga Khan Diamond Jubilee Middle School in Baltit, is the only person in European dress with jacket and tie in the central row standing where teachers and staff of the school are posing. Next to him is Babu Jan; Qudratullah Beg is the second from right in the same row. Source: German-Austrian Himalaya-Karakoram expedition 1954. Courtesy of Gerhard Klammert

Aga Khan IV, Karim-al Hussaini during the first ever meeting of an Ismaili imām with his Hunza jamáat who had arrived from all villages. The dīdār ceremony was celebrated in Baltit in October 1960 where a sizable group of male murīd had gathered. Source: Photograph taken by Abdul Ismaily, Baltit on 23 October 1960

came into being as 'a private, non-denominational institution committed to promoting development and social welfare through philanthropic activities, primarily in the developing countries of Africa and Asia'.[314] This separation of private property from endowment funds was the major precondition for the Aga Khan Foundation's recognition as an eligible recipient of development funds from international donors. The Aga Khan Foundation had to safeguard that the Aga Khan's previous unconditional hold over all sources of income and real estate was split into a private wing of businesses and enterprises worldwide, as well as religious institutions such as the Ismaili Tariqah and Religious Education Board, and into community properties, economic and philanthropic activities, including tourism, community, cultural and development organisations.[315] The advisors and experts employed in these institutions were mainly *khoja* from affluent families working as volunteers. The same affluent community from Sind and Karachi whose leading figures had initially challenged the leadership of the *imām* in the final decades of the 19th century had emerged as the influential inner circle of policymakers, controllers and accountants who have dominated all *jamáati* affairs since.[316] This cooperative and reformist arrangement initiated the emergence of the Aga Khan Development Network as the major non-governmental developmentalist player in northern Pakistan.

These combined efforts have significantly contributed to a surge in educational attainment, the health status and entrepreneurial undertakings in the Hunza Valley. In terms of participation, the process of implementing new rules and regulations conflicted with inherited power structures and traditional influence by elite groups. It took well over a decade to streamline the organisational stage. Again an exceptional rule was reserved for Hunza as decision-making and judicial authority remained vested in *mīr* Jamal Khan; the regional and local level institutions became only fully functional after his demise and helped to fill the power vacuum especially when legal disputes were at stake. The final decade of his rule was characterised by a confrontation between the pioneers in education who had found positions as migrants in Karachi and the old elite in Hunza which profited from being close to *mīr* Jamal Khan and his verdicts.[317] The opposition group was gaining momentum and support year by year and challenged inherited leadership and consultation.

The Aga Khan Development Network offered well-educated and trained people opportunities to influence their chances of prosperity and well-being; besides politics, this was the most effective avenue for social mobility and transformation. The implementation of the council structure might be seen as the turning-point; previously, offices in the Ismaili *jamáat* had been distributed among the *guśpúr* and *wazīr*; with the expansion of new institutions, the demand for more office-bearers and the provision of salaries for professional employees, the personnel basis was broadened, and qualified promising candidates got chances for promotion.

Hunza as one element in the Ismaili global assemblage

Both *imām* who have been instrumental in shaping and transforming Hunza's colonial and postcolonial exchange relations can be seen as proponents of modernising the *maulai* community with a clear embeddedness in British Raj and global networks. During the imamat of Aga Khan III, political influence and commanding power were narrowly linked with respective leaders' territorial control; the international debate among nation states and communities that were struggling to gain self-determination was dominated by the goal of achieving territorial autonomy and independence. During World War I the 'rank and status of First Class Chief with a salute of eleven guns' was conferred on Aga Khan III, who was without state in contrast to the vast majority of his fellow chiefs.[318] In British India the pro and cons were argued whether an Ismaili State would be a desideratum for a personality of such a high esteem and influence in South Asia. During the first two decades of the 20th century the debate gained momentum as Sultan M. Shah proposed certain privileges from the British administration; exemption from taxes in British India was granted to him in 1916, but the endorsement for Great Britain was refused.[319] Dossiers and a new genealogy were prepared in which his rootedness in the most privileged Muslim pedigree were suggested.[320] The only question remained: where could such a territory be found? For that purpose, the unique position of Hunza as the only 'Ismaili State' – a semi-autonomous territory ruled by an Ismaili and having a majority Ismaili population – was put to the fore and emerged as an exceptional asset. After the implementation of the reforms and the appointment of *tham* Jamal Khan, Hunza seemed to be privileged for such an eminent position as all other attempts to allocate a territory to the Aga Khan III or to please his demands had failed. In his memoirs he recollected that a tiny land 'on the lines of Tangier or the Vatican – a scrap of earth of their own which all Ismailis, all over the world, could call theirs in perpetuity, where they could practice all their customs, establish their own laws, and (on the material side) build up their own financial center, with its own banks, investment trusts, insurance schemes and welfare and provident arrangements' would have fulfilled the purpose.[321] Several other options had been suggested to make him a territorial ruler in entities ranging from Afghanistan to Kalat, from Muscat to Sansibar, from Socotra to Syria.[322] With

Representatives of the Women Organisation of Ghulkin have gathered in front of the jamáat khana to receive a delegation from the Aga Khan Rural Support Programme and to discuss the implementation of development packages. The photograph was taken on 21 November 1984

the end of British colonial rule in South Asia, these dreams ceased. With quite some time gap an exalted – some observers say Vatican-like – status was finally granted to Aga Khan IV in Portugal; a new seat was inaugurated on 11 July 2018 in Lisbon as the centre for a multi-cultural and cosmopolitan Ismaili community.[323] The transition from 'cooperative capitalism' to 'corporate globality' might be an appropriate description for how change is promoted and implemented in Gilgit-Baltistan and in Hunza in times of globalisation and neoliberalism.[324] The advantages of community-based co-operatives seem to have translated into communication and network structures that have transgressed national boundaries and spanned continents. The historical and political contexts have changed from colonial to post-colonial; nevertheless, the forces of developmentalism and modernisation persist and prevail.

Between Sultan M. Shah's dream of an independent state and the materialisation of the Lisbon *dar-xāna* in Portugal – lauded as the 'new seat of the Ismaili imamat' in the 'country of opportunity' – there emerged a transformation in valuing certain forms of ex-territoriality.[325] Beyond the spiritual strength of the imamat, the quest for worldly power seemed to be equally important. In the early phase,

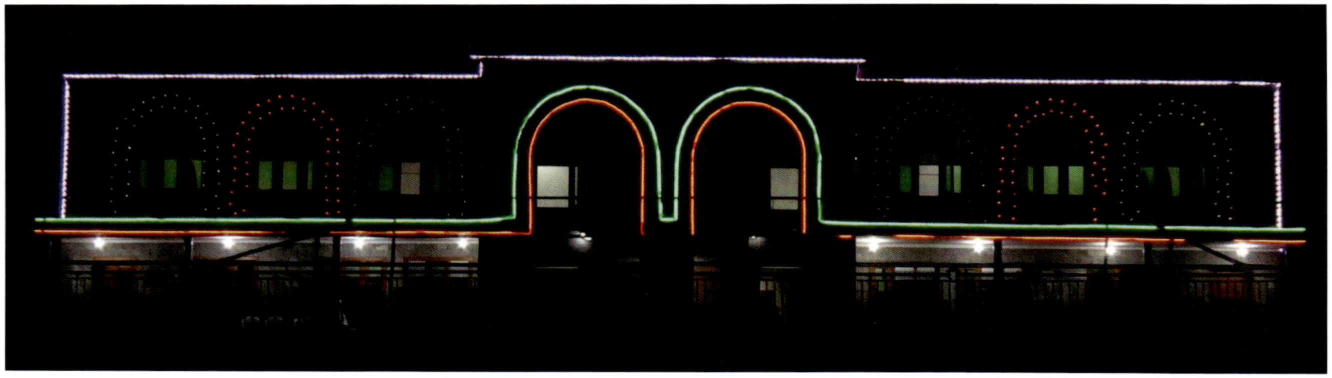

The silhouette of Gulmit jamáat khana is illuminated by night. Besides providing a prayer hall the building contains more than ten offices for Ismaili community organisations and social welfare institutions. The photograph was taken on 30 July 2017

different options were discussed, fully in tune with the 'zeitgeist' and contemporary debates on nations and their states. During colonial rule these territorial limitations dominated the aspirations and were frequently put forward; even if no tangible results emerged, the quest contributed to a more focused perception of Hunza as an Ismaili abode. The colonial partnership – promoted by the joint activities of the British Raj and Aga Khan – has not succeeded in transforming the Hunza microstate into a Central Asian Alamut or a reinvented minority stronghold. The colonial-communal joint venture has critically influenced developments within Hunza and external interventions during the imperial age. The evidence presented here has underlined how the two major forces of modernisation – British colonialism and Ismaili reformism – have shaped an assemblage of various fields. Modernisation as a developmentalist model disguises itself in various designs as the blueprint for change; the powerful concept not only survived the fight for independence but also seems to have remained the major structuring agency and successful tool reinforced by private and state-run development programmes. The position of the Aga Khan Development Network is described by Faisal Devji as a developmentalist model that has a number of repercussions for the community structure and its functioning "... because the bureaucracy that runs the community is resolutely anti-intellectual in character. For institutions that now claim to be global or transnational ones, this bureaucracy is curiously old-fashioned in its very modernity, taking on a vicarious identification with the post-colonial state or the 1960s and '70s, not least in the obsessive attention given the Aga Khan's various meetings with presidents, prime ministers or monarchs and the honours bestowed upon him. Not only is this coverage reminiscent of the official media representation of sundry presidents-for-life, but by bathing the imam in the reflected glory of others, ends up both diminishing and secularising his authority in a potentially damaging way. After all Ismailis have for decades now had a flag, anthem and constitution, and more recently the Aga Khan has managed to secure diplomatic status for his representatives in certain countries, with which he sometimes signs protocols of understanding as if he were a head of state."[326] Faisal Devji hinted at the difference between Catholic 'sacerdotal authority' and a developmentalist approach for wider target groups that is augmented by religious teachings and orders for Ismaili community members: "While it may be argued that the Vatican provides a model for this form or organisation ... the post-colonial model of state followed by the Ismailis is evident in the bureaucracy's almost exclusive focus on social engineering. From the third Aga Khan's colonial model of reform, Ismailis have moved under his successor to that of post-war modernisation theory and development. Although this has resulted in some notable achievements in fields like health or education, it is clear that a statist model stands in complete contradiction with the supposedly transnational community to be 'developed' by such means."[327] Modernisation is here reflected as leading to inertia although modernisation theory claims to overcome stagnation in 'traditional' societies. The effort to provide a role model for modern development that is applicable in all Ismaili settlement regions might have caused a highly abstract modernisation strategy that has neglected local conditions and preferences for the purpose of mainstreaming a diverse community along the lines of the dominant *khoja* group. Their general pursuit of implementing modernisation and especially the agenda of those advisors forming the inner circle of the Aga Khan's decision-making

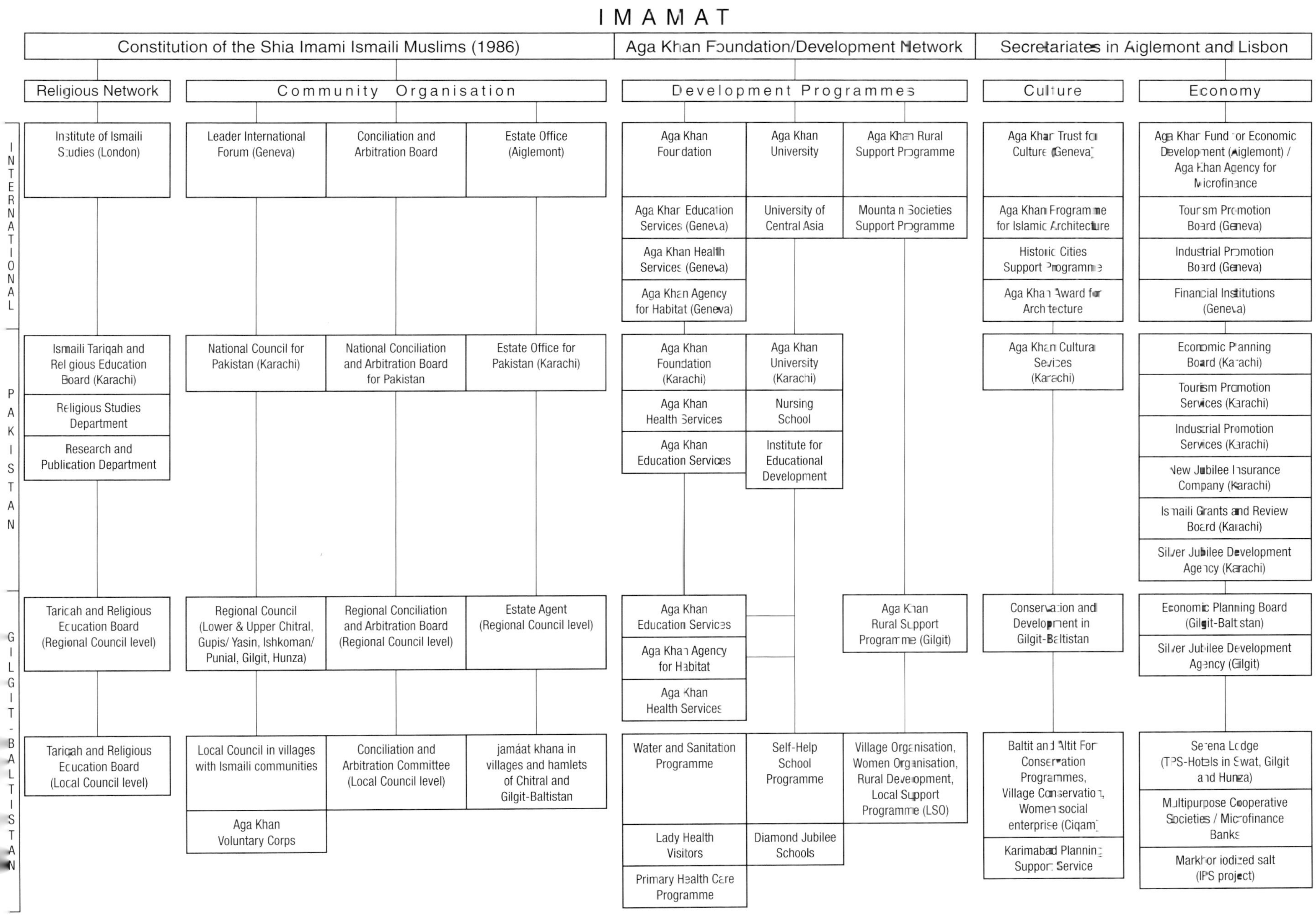

The organisational chart shows the various departments in religious affairs and community organisation, the development activities, cultural and economic enterprises on the international, national and regional levels with a focus on Pakistan and Gilgit-Baltistan. All programmes are managed and steered from European headquarters located in Aiglemont (Paris), Geneva, London and Lisbon. The hierarchical structure streamlines edicts (firmān), initiatives and programmes reaching local councils and jamāat khana on the village level. Since the inception of local councils nearly a century ago, the establishment of the Aga Khan Foundation in 1967, its setup in Pakistan two years later and the adoption of the Ismaili Constitution in 1986, further adjustments and modifications have regularly contributed to its proved effectiveness in philanthropic projects and community development. Source: Modified and updated version based on Hermann Kreutzmann (1996a: 385).

One of the early adopters of the Aga Khan's development message and well performing model village organisations has been Ahmedabad although most of the male workforce had already out-migrated and sought education, employment and service stations outside the Hunza Valley. These efforts were selected by the community as a measure for future land allocation for future generations and as a security for difficult times. Community efforts have contributed to significant land development by constructing a new channel from Gurpi Glacier to feed new land parcels on a steep slope. The Aga Khan Rural Support Programme supplied the initial funding and engineering expertise for the implementation of the project. The photograph was taken on 26 April 1985

body have been inspired by commercial and technocratic considerations rather than by specific aspirations and conditions in certain localities and sensible approaches to not only business-oriented fields such as academia and culture. Perceptions might differ, but these differences matter and are frequently hot topics in discussions.

The impression might prevail that the socio-economic modernisation strategies have been detached from politics. There is ample evidence that the connection between the British colonial administration and Aga Khan III was enforced, instrumentalised and stabilised because of a common anti-communist agenda long before the Cold War in its narrow sense commenced.[328] Both sides expected substantial advantages from their cooperation, be it personal privileges, secured spheres of influence or contributions to a stabilised world order.

The multi-national comprehensive development agenda of Aga Khan IV thrived best when northern Pakistan functioned as a pivot surrounded by a Soviet-occupied Afghanistan, a Soviet-dominated Central Asia, and Xinjiang which was controlled by the Chinese Communist Party during the 1980s. This was the zenith of Western development funds being available and invested in this frontline region. The Aga Khan Development Network functioned as the broker for modernisation programmes that affected nearly all walks of life in the region. Especially in Ismaili settlement regions, it was impossible for the rural population to avoid participation or to decline offers for development packages. The Aga Khan Development Network has functioned as the chief promotor of modernisation; any criticism of the prescribed development path and packages could be interpreted as a personal critique of *Hazir īmām* and his thoughts for his 'spiritual children', the regular form of addressing his *jamáat*.[329] The tremendous transition initiated would effect a break with indigenous healing practices in favour of modern medicine in the health sector. In education, modern pedagogic concepts would promote literacy and augment existing school systems by offering female education; at a later stage the promotion of English-medium instruction followed a popular demand.[330] Even the curriculum implemented by the Tariqah and Religious Education Board was modernised and high-quality instruction material distributed. Engineers trained in house-building were employed to introduce new materials, building techniques and facilities. Preventive measures in hazard-prone settlement regions were promoted on the basis of the latest technological insights. The most comprehensive packages were offered by the rural support programmes that affected agricultural practices and cash crop cultivation, marketing of surplus products and a wide variety of agricultural extension. All these packages were based on village organisations that fulfilled the initial dream of cooperative capitalism when economies of scales encouraged smallholders to invest in their agricultural assets. At a later stage, value chains were expanded and translated into corporate globality. The comprehensive approach of the rural support programmes fostered critical views on the parallel structures that could lead to public-private competition. The Aga Khan Rural Support Programme was building roads and bridges, an undertaking previously reserved for government institutions such as the Public Works Department. Hydro-electrical energy production, water management and afforestation projects, income-generating schemes, marketing support and tourism development have been part of its initiative that provided employment for a growing number of professionals.[331] Experts and advisors are travelling between different areas of activities; quite a number of trained personnel from Gilgit-Baltistan have found employment in projects of the Aga Khan Development Network in Afghanistan, Kyrgyzstan and Tajikistan. Two aspects are connected with the latest developments. First, after the end of the Cold War the centre of activities has shifted away from northern Pakistan towards Afghanistan and Tajikistan; another proof of the interconnectedness of modernisation and politics. Second, during the imamat of Aga Khan IV the notion of diaspora has changed from scattered and remote communities – in former times randomly perceived as 'uncontacted people' – to a multilocal network that is bound and meshed through imamat-established institutions. Presently scattered Ismaili communities that do not have much in common in terms of living standards, economic and socio-cultural practices are managed, positioned and situated under the imamat's umbrella; a constellation that does not necessarily alleviate the pressure on members of small minorities in terms of tolerance and acceptance. Tolerance has come under stress in recent times during which an opaque presence is preferred among Muslim communities who might be minorities in various contexts. A kind of public *taqīya* has emerged, a contradiction in itself, but a practice that has become more common. Public celebrations of religious festivals are avoided or its attendance is limited to community members.

The preservation of built environment has been one of the prime activities of Aga Khan Cultural Services. After Baltit Fort was renovated and inaugurated Altit Fort became a second major effort in Hunza where professionals were trained in cultural preservation, architectural and technical skills as well as tourism services. The photograph was taken on 27 July 2010

Actors and factors in transition

A significant transformation appears to have taken place when considering the Hunza Valley from pre-colonial relations to post-colonial actions and interferences that have affected and triggered internal developments and modified exchange patterns. The emphasis put on the performance of rulers, colonial officers, the Aga Khans and their emissaries in their relationship with local communities, stakeholders and opposition might suggest that elite groups and notables have dominated and shaped external relations and been the main collaborators and/or sole recipients of colonial subsidies and predominant beneficiaries of proceeds from post-colonial investments, offices and jobs. On the one hand, the evidence presented here underlines an authoritative and exclusionary leadership in monopolising wealth-generation during colonial times. Indirect rule profited from its relationship to this small elite group that safeguarded stability and tranquillity within the microstates and guaranteed adequate cooperation and sufficient loyalty. Once a process of broader participation in education and increased out-migration had been triggered, these limitations and the authority of the *thámkuṣ* faded, control weakened and certain restrictions had to be lifted. Since the abolition of Hunza State, this process has been accelerated; multiple groups of migrants and highly qualified professionals from a variety of villages and communities have followed an independent career path that is based on multilocal places of residences and work for men and women from Hunza. The Aga Khan Development Network and governmental scholarships have supported the process of merit-based professionalisation that has opened doors for a growing number of Hunzukuċ. But at the same time certain offices, positions, political representations have continued to be dominated by members of the former elite; *ayáśo*, *guśpúr* and *wazīrkuċ* are still to be found in key positions in the armed forces, bureaucracy, politics, and tourism. Previously dominating monopolies have been broken

Safida Begum has tried to continue the inherited practice of shepherding Shimshal's flocks in its Pamir. The village community has tried to maintain its entrusted resources despite large-scale and improved education, salaried jobs and out-migration. The land resources of Shimshal have remained to be the prime safety measure in risky times and need proper attention. Men and women have cooperated in the villages to find adaptive strategies for optimal use of their heritage. The photograph was taken on 21 July 2017 up. Societal division has been modified by the better economic results of successful entrepreneurs and professionals outside the former elite group. The observation of persisting patterns of domination, dependency and influence is augmented by opportunities provided by development agents, modern technologies and mobility. The split in societal entitlements has been prolonged by the narratives of benevolent rule, a content society of hard-working mountaineers, a caring secular and spiritual leadership. Certain myths have prevailed that have had a stabilising effect for autocratic rule apparently vested in the *tham* and his alleged supernatural abilities that led to a contented community and a valley of respected centenarians.

Endnotes of pages 361–464

1 Patrick Vans Agnew (1915: 291).
2 Edward Frederick Knight (1893: 347–349).
3 The view of robbery as a necessity to augment the meagre locally generated livelihoods is endorsed by Irmtraud Müller-Stellrecht (1979: 84–85), while David Lorimer (1979: 87) refrains from making the same connection.
4 Edward Frederick Knight (1893: 349).
5 Francis Younghusband (1896: 285–286).
6 The quote is retrieved from Arthur Henry McMahon's report of 1898 on the 'territorial claims of the Kanjut tribe' (IOR/2/1079/253: 60).
7 Immanuel Wallerstein (1988).
8 Besides some publications resulting from academic projects and journal articles, local chronicles and treatises on communal affairs need to be mentioned; the most recent contributions are the two volumes by Zafar Iqbal (2018a, b) that document hitherto unpublished letters, reports and statements.
9 Qudratullah Beg (1962: 125); see as well Yousaf Khan (2004); David Lorimer (1979: 54).
10 Hsiao-Ting Lin (2009: 492–493); see as well Wolfgang Holzwarth (2006: 178–179).
11 George Cockerill (1895: 16).
12 Qudratullah Beg (1962: 151–154).
13 Qudratullah Beg (1962: 171).
14 Bronislaw Grombchevsky (1888/2015: 139, 148).
15 George Cockerill (1895: 21); part of the story is narrated by Gottlieb Wilhelm Leitner (1889: 7). Other genealogical tables record further names; see Yousaf Khan (2004); Irmtraud Müller-Stellrecht (1979: 304–308).
16 See for Chitral's social stratification Peter Eggert (1990: 43–59).
17 See for the influence of religious elite groups in Nagar Jürgen Frembgen (1985: 39–42, 87–90).
18 See Hermann Kreutzmann (1999); Irmtraud Stellrecht (1978: 16–17).
19 Stuart Hill Godfrey (1898: 97).
20 Stuart Hill Godfrey (1898: 97).
21 A final major and successful raiding expedition took place in 1888 during the visit of Bronislaw Grombchevsky but Ghazan Khan's son Safdar Ali Khan was the mastermind then.
22 Translation of a petition from Raja Jaffar Khan of Nagar without date [1884] to the Maharaja of Kashmir, via Sir Oliver St. John, Officer on Special Duty in Kashmir to the Secretary Government of India in the Foreign Department, dated Jummoo 5 November 1884 (IOR/2/1079/246).
23 Translation of a petition from the Wazir of Hunza – without date [1884] to the Maharaja of Kashmir, via Sir Oliver St. John, Officer on Special Duty in Kashmir to the Secretary Government of India in the Foreign Department, dated Jummoo 5 November 1884 (IOR/2/1079/246). Messengers often claimed to be *wazīr* although their position was dif-

ferent; see William Lockhart and Robert Woodthorpe (1889: 285, 379).

24 Rehan Ali Shah or Rahim Shah was a son of the third or fourth wife (Qudratullah Beg 1962: 311, 386) of *tham* Ghazanfar Khan; see the genealogical tables prepared by M. Masih Pal and David Lorimer published in Irmtraud Müller-Stellrecht (1979: 305); Yousaf Khan (2004). From the Tarakuč, M. Reza Beg favoured him to become the successor of *tham* Safdar Ali Khan in 1891; see Qudratullah Beg (1962: 5–6).

25 William Lockhart and Robert Woodthorpe (1889: 391).

26 William Lockhart and Robert Woodthorpe (1889: 392).

27 William Lockhart and Robert Woodthorpe (1889: 444). The quote in the text is taken from a letter by Safdar Ali Khan to the representative of Kashmir in Gilgit, Bakshi Mulraj, dated 4 September 1886 (IOR/2/1079/246).

28 Translation of reply to Raja Safdar Ali Khan from Bakshi Mulraj, dated 21st Bhadon 1943 [Vikramsamvat calendar] (4 September 1886) on behalf of the Maharaja of Kashmir (IOR/2/1079/246).

29 Translation of a petition from Bakshi Mulraj dated 24th Badon 1943 (7 September 1886) which was dispatched to Srinagar (IOR/2/1079/246).

30 IOR/2/1079/246.

31 Algernon Durand (1888: 15).

32 Algernon Durand (1899: 138–139, 264).

33 Translated from the diaries of Bronislaw Grombchevsky (1888/2015: 117) that were recently published in the original Russian language.

34 Bronislaw Grombchevsky (1888/2015: 118).

35 Bronislaw Grombchevsky (1888/2015: 122). Stuart Hill Godfrey (1898: 72) mentions the capture of 240 Baltis who were taken as prisoners in a battle that he attributes to a force under *wazīr* Asadullah Beg in 1848. Their fate was not further described.

36 Bronislaw Grombchevsky (1888/2015: 122–123).

37 Bronislaw Grombchevsky (1888/2015: 117).

38 Bronislaw Grombchevsky (1888/2015: 150).

39 Letter from Colonel William Francis Prideaux to the Government of India, dated 12 August 1891; quoted after Stuart Colvin Bayley (1892: 3) in his 'note on the complication with Hunza' (IOL/P&S/18/A82: 3). The quote conforms with the opinion presented by John Biddulph (1880: 28).

40 Report by Edward Knight published in the Times, 11 January 1892.

41 The participation and cooperation of *rājā* Sikandar Khan should not be underestimated in the staging of the assault at Nilt. For the neglect of Nagar actors as collaborators of the British-Kashmiri military force, see Jürgen Frembgen (1985b: 177; 2006).

42 Algernon Durand (1900: 190-191). Humayun Beg fled to Chitral after the murder of *tham* Ghazan Khan in 1886 where he remained in exile for nearly five years; see Qudratullah Beg (1962: 476–478, 517–518); Stuart Hill Godfrey (1898: 82).

43 Letter from G[eorge] Macartney to A[lgernon] Durand, dated Kashgar 6 April 1892 (IOL/P&S/7/66: 894–895).

44 Kashgar Diary, 20 January 1909 (IOL/P&S/7/226/480); Kashgar News-Report, 20 March 1909 (IOL/P&S/7/228/739). Qudratullah Beg (1962: 585–587) narrated his version of the stay of Safdar Ali Khan in Kuchar and of his alleged plot to kill the former *wazīr* Dadu in China. In this narrative Safdar Ali Khan was poisoned by his son Muzaffar Ali Khan who was supposedly instigated by *mīr* M. Nazim Khan. Safdar Ali Khan died in Yarkand in March 1931.

45 Kashgar News-Report, 31 October 1904 (IOL/P&S/7/172/-2264A).

46 Kashgar News-Report, 20 February 1905 (IOL/P&S/7/176/731). His drinking with the local elite was repeatedly an issue; see Kashgar News-Report, 31 January 1905 (IOL/P&S/-7/177/927).

47 Letter from Allen Robert Betham Shuttleworth to Francis Younghusband, dated Kashgar 10 January 1909 in response to Safdar Ali Khan's application four days earlier (NAI/PR_000001415997: 5–6).

48 Kashgar Diary for January 1911 (IOL/P&S/7/247/574).

49 Muzaffar Ali was born in 1884, his younger brother Mohammad Isa in 1888 (IOL/P&S/7/66/689).

50 Letter of George Macartney to Bertrand Evelyn Mellish Gurdon, Political Agent in Gilgit, dated 5 May 1906 (IOL/P&S/7/190/1300).

51 Reginald Schomberg (1933b: 48–49).

52 IOL/P&S/12/2336/116: 29–30). The news about the death of Safdar Ali Khan was reported in the Kashgar Monthly Diary for March 1931 IOL/P&S/12/2332/216); see the published version in Reginald Schomberg's book (1935a: 155–156).

53 Letter of Reginald Schomberg to John Charles Walton, Joint Secretary Political Department India Office, dated 11 October 1933, sent from Bombay (IOL/P&S/12/12/97).

54 Letter by M. Nazim Khan of 14 August 1937 to the Political Agent Gilgit which was forwarded to the Resident in Kashmir and from there to the External Affairs Department in New Delhi (FO/371/22115: 176–182).

55 Hermann Kreutzmann (2015a; 2016a; 2017a, b).

56 Hermann Kreutzmann (2015a: 100–105, 231–245).

57 Qudratullah Beg (1962: 570–571) narrates the option to consider Humayun Beg for the post of 'governor of Hunza'.

58 Irmtraud Müller-Stellrecht (1983a).

59 John Biddulph (1876); Bronislaw Grombchevsky (1888/2015); Francis Younghusband (1896: 288, 345).

60 Algernon Durand (1900: 265); see as well George Cockerill (1895: 18–19), Stuart Hill Godfrey (1898: 73). Reginald Schomberg (1935a: 155) remarked that he was addicted to opium and died of an overdose, while the Tarakuč version is a suicidal use of opium because of mental depression (Qudratullah Beg 1962: 586).

61 Telegram from Viceroy to Secretary of State, 31 December 1891, enclosed in 'correspondence relating to the Operations in Hunza-Nagar. Presented to both Houses of Parliament 1892 (IOR/V/4/6621: 5).

62 Quoted after Evelyn Hey Cobb's draft manuscript prepared

for the 'Political Agent's speech at the Hunza Darbar at the Mir's installation' (IOR/2/1086/308: 73); see as well IOL/P&S/12/3299.

63 Stuart Hill Godfrey (1898: 82).

64 Stuart Hill Godfrey (1898: 73); see as well Francis Younghusband (1896: 347) who mentioned the one 'hundred Kashmir troops under Captain Bradshaw, who now [1893] preserved order in Hunza'.

65 Stuart Hill Godfrey (1898: 73).

66 George Nathaniel Curzon (1926: 194). Francis Younghusband (1896: 345) relates that the post of Political Officer in Hunza was abolished in 1894.

67 Stuart Hill Godfrey (1898: 136); Government of India (1909: 71-74, 84).

68 Charles Umpherston Aitchison (1931: 15).

69 Marc Aurel Stein (1903, I: 40) met with M. Nazim Khan in June 1900.

70 See Tahir Ali (1983: 97–105).

71 George Ogilvie (1929: 6).

72 George Ogilvie (1929: 8).

73 George Ogilvie (1929: 8); see as well IOL/P&S/12/3291: 18.

74 George Ogilvie (1929: 8). A *tolā* can be divided in 12 *māshā* and is the equivalent of about 12 grams. The Hunza and Nagar gold tributes weighed approximately 197 and 205 grams, respectively.

75 The value of one *tolā* of gold dust was given with some variations. For the 1860s, Gottlieb Wilhelm Leitner (1889: 71) distinguished between a *kachchā tolā* of ten rupees value and a *pakkā tolā* of twenty rupees; for a 'good horse' from Nagar he calculated a price of five *kachchā tolā*, equalling fifty rupees, while a Badakhshani or Yarkandi horse would cost 12 *kachchā tolā*. Around 1910 a *tolā* of gold dust – most probably a *pakkā tolā* – was regarded as a monetary equivalent ranging between sixteen to twenty rupees (Government of India 1909: 66; Thakur Singh 1917: 59).

76 Gottlieb Wilhelm Leitner (1890: 119) had suggested obtaining such a letter of safe conduct for the members of the Gilgit Mission, which they obviously did, but it is not mentioned in their own report (William Lockhart and Robert Woodthorpe 1889).

77 Sabiha became the mother of Ghazan Khan's three sons – Jamal Khan, Jamshed Khan, Ayash Khan –, but she already passed away in November 1917 after which the prince (*mīrzāda*) never married again. Her grave is located in Karimabad's Suryas Das cemetery.

78 Especially during 1902 the reports of the Russian commanders of the Pamirian military post in Khorog, Eduard Karlovich Kivekes and Andrei Jevgenievich Snesarev, contained several remarks about the alleged instigation by *pīr* Shahzada Lais (RGVIA, font 1396, opis 2, file 2251).

79 For the circumstances of Ali Mardan Shah's role in border politics see Hermann Kreutzmann (2015a: 2016–207, 222–228, 255–275; 2017a: 193–197).

80 Mohomed Nazim Khan (1936: 131); Hermann Kreutzmann (1989a: 153); David Lorimer (1979: 218–220); Ella Maillart (1937: 287–288); Arthur Neve (1913: 152).

81 Collection by the Mir of Hunza of offerings for his Highness the Aga Khan from places in Central Asia including Badakhshan and Chitral (IOR/2/1086/302); Gilgit Agency Political Diary for the months of May 1923 and August 1928 (IOL/P&S/10/973: 60, 194); see Tahir Ali (1983: 99); Mohomed Nazim Khan (1936: 131); Hermann Kreutzmann (1989a: 154–153); David Lorimer (1979: 218–220).

82 See the report by Bertrand Gurdon (IOR/2/1079: 251: Hunza and Nagir. Grant of subsidies to the Mirs of Hunza and Nagar 1894); IOR/2/1079/252: Hunza and Nagir subsidies (1895); IOL/P&S/12/3291: Northern Frontier: Increase of subsidies in the Gilgit Agency (08. 02. 1926–13. 12. 1928); IOL/P&S/12/3292: Hunza: Situation in Raskam; Darwaza incident (31. 07. 1933–01. 01. 1940): 156; IOR/2/1079/251: 8.

83 Stuart Hill Godfrey (1898: 79). The same amount was paid to Nagar State.

84 Stuart Hill Godfrey (1898: 79–80). In Nagar State *rāja* Zafar Khan were allocated 500 Rs, and *rāja* Sikandar Khan 2,500 Rs.

85 Henry Lawrence Haughton (1913: 232).

86 Gilgit Political Agency Official Diary for the month of February 1912 (IOL/P&S/10/826: 245).

87 Gilgit Political Agency Official Diary for the months of April and June 1912 (IOL/P&S/10/826: 235, 240).

88 Henry Haughton in his diary of 27 April 1912 (IOL/Mss Eur/D531/5).

89 Henry Haughton in his diary of 30 April 1912 (IOL/Mss Eur/D531/5).

90 Gilgit Political Agency Official Diary for the month of August 1912 (IOL/P&S/10/826: 229).

91 Gilgit Political Agency Official Diary for the months of October and December 1912 (IOL/P&S/10/826: 220–222).

92 Gilgit Political Agency Official Diary for the months of December 1913 and January 1914 (IOL/P&S/10/826: 191, 193).

93 Gilgit Political Agency Official Diary for the month of April 1913 (IOL/P&S/10/826: 208).

94 Demi-official letter from Major George Kirkbride, Political Agent Gilgit, to Lieutenant Colonel Lionel Edward Lang, Resident in Kashmir, dated 15 July 1935 (FO 371/19335: 226). William Macandrew Marshall's posting lasted until the outbreak of World War I in 1914; during his stay he prepared the valuable survey about Hunza's revenue from grazing taxes on the Taghdumbash Pamir and in Sarikol (IOL/P&S/10/278).

95 William Hingston (1913: 73).

96 William Hingston (1913: 75).

97 William Hingston (1913: 76).

98 Gilgit Political Agency Official Diary for the month of September 1913 (IOL/P&S/10/826: 197).

99 Gilgit Political Agency Official Diary for the month of July 1914 (IOL/P&S/10/826: 179).

100 Gilgit Political Agency Official Diary for the month of July 1915 (IOL/P&S/10/826: 143).

101 Gilgit Political Agency Official Diary for the month of May 1916 (IOL/P&S/10/826: 115).

102 Mohomed Nazim Khan (1936: 93–98, 129); Gilgit Political Agency Official Diary for the month of June 1916 (IOL/P&S/10/826: 114).
103 Mohomed Nazim Khan (1936: 93–98); Gilgit Political Agency Official Diary for the month of September 1916 (IOL/P&S/10/826: 107).
104 Letter from the Resident in Kashmir, Srinagar, No. F. 85–C./27, dated 14 December 1927: Report of the Resident in Kashmir on his tour in the Gilgit Agency, 1927 (IOL/P&S/12/3288: 260–263).
105 IOL/P&S/12/3288: 262.
106 Shaukat Ali Khan was the son of *tham* Sikandar Khan's first-born son who died early. Mahbub Ali Khan was the second of six sons and the colonial administration appointed him as Governor of Yasin in July 1941 (IOL/P&S/12/3288: 73); see IOR/V 5116: 20–22; the genealogical table in Irmtraud Müller-Stellrecht (1979: 306).
107 Gilgit Agency Political Diary for the month of May 1937 (IOL/P&S/12/3285: 231).
108 IOL/P&S/12/3288: 263.
109 Gilgit Agency Political Diary for the month of February 1932 (IOL/P&S/12/3285: 362).
110 General Staff India (1928a: 164); Barbara Mons (1958: 86); John Tobe (1960: 271); Jenny Visser-Hooft (1926: 33–36); Maynard Owen Williams (1932: 343). The shift from Baltit Fort to Karimabad took nearly twenty-five years. In June 1911 the Gilgit Diaries had recorded: "The Mir of Hunza has completed his new summer-quarters at Karimabad" (IOL/P&S/7/270/1302).
111 See the references to entertainment and hospitality by Clarmont Percival Skrine (1926: 26–27); passing remarks are found in Peter Fleming (1936: 375); Harold Ivan Harding (1922: 30); Ella Maillart (1937: 289); N. N. [Reginald Schomberg] (1938: 593); Jenny Visser-Hooft (1926: 33–37); Maynard Owen Williams (1932: 342–343).
112 George Le Fèvre (1935: 208); Karimabad was described as the summer residence of the *tham*.
113 Gilgit Agency Political Diary for the month of May 1934 (IOL/P&S/12/3285: 315).
114 Gilgit Agency Political Diary for the months of July 1934 and January 1935 (IOL/P&S/12/3285: 301, 311); see for the genealogical connotations IOR/V 5116: 19–22; the genealogical tables prepared by M. Masih Pal and David Lorimer published in Irmtraud Müller-Stellrecht (1979: 306–308).
115 See Eric Teichman (1937: 175) who remarked: "We rested a day in Baltit, enjoying as so many travellers have done, the Mir's unbounded hospitality."
116 Gilgit Agency Political Diary for the month of November 1934 (IOL/P&S/12/3285: 304); see as well for the reorganisation the 'report on the Gilgit Scouts for the period 6th December 1934 to 5th December 1935' (IOL/P&S/12/3288/199).
117 Gilgit Agency Political Diary for the month of September 1936 (IOL/P&S/12/3285: 254). His son Malik Shah succeeded Taifoor as *wazīr* (IOR/V 5116: 21).
118 Gilgit Agency Political Diary for the month of June 1935 (IOL/P&S/12/3285: 292).
119 Demi-official letter from Major G[eorge] Kirkbride, Political Agent Gilgit, to Lt.-Col. L[ionel] E[dward] Lang, Resident in Kashmir No. 166-T, dated 15th July 1935 (IOL/P&S/12/3292: 333).
120 Gilgit Agency Political Diary for the month of November 1935 (IOL/P&S/12/3285: 279); IOR/V 5116: 20. Shukrullah Beg died in Gilgit on 9 April 1939 (Gilgit Agency Political Diary for the month of April 1939; IOL/P&S/12/3285: 175); *wazīr* Zarparast served in this function until his death on 17 February 1943 (IOL/P&S/12/3288: 32) and was followed by the two last *wazīr*, Inayatullah Beg (1943–1958) and Sajidullah Beg (1958–1974); see table in Hermann Kreutzmann (1989a: 222).
121 Letter from the Political Agent, Gilgit, to the Assistant to the Resident in Kashmir, No. 48-S/A-22 of 1936, dated Gilgit, the 30th March 1936 (NAI/PR/4002104: 3–4).
122 Gilgit Agency Political Diary for the months of November 1935, April 1936, and January 1937 (IOL/P&S/12/3285: 240, 268, 279).
123 Demi-official letter by Lieutenant-Colonel Lionel Edward Lang, Resident in Kashmir, dated 30 July 1935 (FO 371/19335: 225).
124 Calculations based on the Population Census of 1931 and David Lorimer's description of offices in Hunza (PP MS 66/01/01/R7 and T4).
125 The information was recorded by David Lorimer in Aliabad on 11 November 1934 (PP MS 66/3).
126 Hugh Leach and Susan Maria Farrington (2003: 141); Michael Pollock (2004).
127 Report of the Resident in Kashmir on his tour in the Gilgit Agency, 1927 (IOL/P&S/12/3288: 261); see the Gilgit Diaries for the period (IOL/P&S/10/973) and the continuation of the debate even after decolonisation and partition as 'The Russian menace to India' in 1947 (IOL/P&S/12/3303).
128 See Hermann Kreutzmann (2015a: 112).
129 IOL/P&S/12/3288: 261.
130 Map of the Shimshal Mustagh Area, drawn by Havildar Sher Hassan and Naik Taqdir Ullah (F. I. C.) 1934 (RSAA/SC/SCH/2/28). A refined version of the map was published along with his book 'Unknown Karakoram' (Reginald Schomberg 1936).
131 David Lorimer's diary in the SOAS archives (PP MS 66/01/01/R7).
132 See Hermann Kreutzmann (1996a: 282–289).
133 The quote is taken from the third chapter of his unpublished reminiscences titled 'Karakoram Finale' (Royal Society for Asian Affairs/SC/SCH/2/47: Archive Box 12: Reginald Schomberg 1945, III: 2).
134 Map of the Shimshal Mustagh Area, drawn by Havildar Sher Hassan and Naik Taqdir Ullah (F. I. C.) 1934 (RSAA/SC/SCH/2/28). A refined version of the map was published along with his book 'Unknown Karakoram' (Reginald Schomberg 1936).

135 See the passing remarks by Gordon Amand (2010: 3); Michael Pollock (2004).
136 He published four books and a number of articles as well as manuscripts of four unpublished books (Hugh Leach and Susan Maria Farrington 2003: 142).
137 Reginald Schomberg (1935a: 114–115).
138 Reginald Schomberg (1935a: 119).
139 Letter of Reginald Schomberg to John Charles Walton, Joint Secretary Political Department India Office, dated 11 October 1933, sent from Bombay (IOL/P&S/12/12/97).
140 The quotes are taken from 'Notes on the Gilgit Agency', a 23-page report written in 1934 and contained in a file titled 'Criticism by Colonel Schomberg of the administration of the Gilgit Agency' (IOL/P&S/12/3293: 52).
141 IOL/P&S/12/3293: 52–57. Reginald Schomberg used capital letters for place names and certain appellation; these have been converted into minuscule. The sequence in his list jumps from point 2 to 4 without mentioning 3.
142 IOL/P&S/12/3293: 53.
143 IOL/P&S/12/3293: 57–59.
144 Emily Overend Lorimer (1936a: 175).
145 Emily Overend Lorimer (1939: 299).
146 Letter of David Lorimer to John R. TeWinkel, Melbourne, Florida, dated 11 November 1953, contained in a file 'Hunza Humbug 1953' subtitled 'The Rise, Decline & Fall of a Myth' (India Office Records and Private Papers Mss Eur D1168/7; David Lorimer 1953: 4).
147 David Lorimer (1953: 4).
148 RSAA/SC/SCH/2/47/12: 60. Tash Mahal is the name of the guesthouse on the major boulder in Karimabad's garden which contains an inscription in English and Persian commemorating the construction time of 1925. Jenny Visser-Hooft (1926: 36) mentioned that is was not yet ready during their visit at the end of May.
149 RSAA/SC/SCH/2/47/12: 12, 60.
150 David Lorimer (1953: 5); a similar attitude towards rule in Hunza was expressed by John Tobe (1960: 292, 361) who compared it with the United Kingdom.
151 IOL/P&S/12/3293: 70–72.
152 Private letter from Lieutenant-Colonel Reginald Schomberg to Eion Pelly Donaldson, India Office, dated Pondicherry, May 13th, 1937 (IOL/P&S/12/3292: 250).
153 Demi-official letter by Lieutenant-Colonel Lionel Edward Lang, Resident in Kashmir, dated 30 July 1935 (FO 371/19335: 225). M. Nazim Khan was envious of the Mehtar of Chitral who had been awarded the title.
154 An agent was posted in Gilgit in June 1935 (Demi-official letter from Major George Kirkbride, Political Agent Gilgit, to Lieutenant Colonel Lionel Edward Lang, Resident in Kashmir, dated 15 July 1935 and report of Lionel Lang on 'Mir of Hunza's rights in the Taghdumbash Pamir and Raskam' 21 October 1935; FO 371/19335: 227–237).
155 Gilgit Agency Political Diaries for the months of July 1933 and May 1934 (IOL/P&S/12/3285: 315, 333).
156 Gilgit Agency Political Diaries for the months of July and October 1934 (IOL/P&S/12/3285: 306, 311).
157 Gilgit Agency Political Diaries for the month of November 1934 (IOL/P&S/12/3285: 305).
158 Gilgit Agency Political Diaries for the month of July 1936 (IOL/P&S/12/3285: 260).
159 Administration Report of the Gilgit Agency for the year 1936 (IOL/P&S/12/3288: 186).
160 Administration Report of the Gilgit Agency for the year 1937 (IOL/P&S/12/3288: 174).
161 Muhammad Masih Pal (1928; 1932; 1934); Administration Report of the Gilgit Agency for the year 1939 (IOL/P&S/12/3288: 143).
162 Mohomed Nazim Khan (1936).
163 Qudratullah Beg (1962/2006: 20).
164 Administration report for the Gilgit Agency for the year 1938 (IOL/P&S/12/3288: 160).
165 The Times, 25 July 1938.
166 N. N. [Reginald Schomberg] (1938: 592).
167 N. N. (1938: 592).
168 N. N. (1938: 593).
169 N. N. (1938: 594).
170 Tahir Ali (1983: 99–100).
171 IOL/P&S/12/3299.
172 Administration report for the Gilgit Agency for the year 1940 (IOL/P&S/12/3288: 125).
173 Administration report for the Gilgit Agency for the year 1941 (IOL/P&S/12/3288: 74).
174 Administration report for the Gilgit Agency for the year 1945 (IOL/P&S/12/3288: 6).
175 'Statement by Mir Shaukat Ali Khan, Mir of Nagar' prepared in 1950 (DAV EXP 2 SG, 133).
176 Demi-official letter from Major G[eorge] Kirkbride, Political Agent Gilgit, to Lt.-Col. L[ionel] E[dward] Lang, Resident in Kashmir No. 166-T, dated 15th July 1935 (IOL/P&S/12/3292: 333).
177 RSAA/SC/SCH/2/47/12: 12.
178 'Report by Mr. K[umara] P[admanbha] S[ivasankara] Menon on first stage of overland journey from Srinagar to Chungking' dated 25th October 1944 (IOL/P&S/12/3299); he stayed for two nights and one day – 5 September 1944 – in Baltit; see as well his travelogue published in 1947.
179 Gilgit Agency Political Diaries for the months of November 1944 and January 1945 (IOL/P&S/12/3285: 33, 35).
180 A summary of the connected events is given in 'Redistribution of Waziri and other allowances in Hunza' (ICR/2/1071/162).
181 Administration report for the Gilgit Agency for the year 1945 (IOL/P&S/12/3288: 6).
182 The file is titled 'Death of Mohammed Ghazan Khan of Hunza and succession of Mohammad Jamal Khan as Mir of Hunza' (IOR/2/1086/308: 92).
183 Zarparast served as *wazir* until his death on 17 February 1943 (IOL/P&S/12/3288: 32); see table in Hermann Kreutzmann (1989a: 222).

184 IOR/2/1071/162: 16.
185 Reginald Schomberg in 'Karakoram Finale' 1945 (RSAA/SC/SCH/2/47/12: 58–59).
186 RSAA/SC/SCH/2/47/12: 9.
187 The 'Administration Report for the year 1945' mentions that 'Subedar Ayash Khan Platoon Commander P1 No. 1 (Hunza) retired from the Corps on 13 Jul 45' (IOL/P&S/12/3288: 12).
188 'Report by Mr. K[umara] P[admanbha] S[ivasankara] Menon on first stage of overland journey from Srinagar to Chungking' dated 25th October 1944 (IOL/P&S/12/3299); see as well Kumara Menon (1947: 29–30).
189 Reginald Schomberg (1945: 3); the unpublished account is found in RSAA/SC/SCH/2/47. Sometimes the date of *tham* Ghazan Khan's demise is given as 7 or 8 May 1945. His death obviously occurred during that night.
190 RSAA/SC/SCH/2/47: 12.
191 The draft is dated 31 March 1946 (RSAA/SC/SCH/2/47).
192 Letter to Frederick Marshman Bailey, dated Ross, Herefordshire 19 December 1946 (IOL/Mss Eur/C371).
193 Letters to George Taylor, dated 14 October, 17 November and 17 December 1947 (IOL/Mss Eur/C371). If Badakhshan would not work – although he claimed to have some connections with the Prime Minister – his alternative suggestions were to travel to British Somaliland or Persia.
194 The latter continued as plant collectors on Himalayan expeditions (IOL/Mss Eur/C371).
195 Letter to Frederick Marshman Bailey, sent from Ross, Herefordshire 2 February 1951 (IOL/Mss Eur/C371).
196 Letter to Frederick Marshman Bailey, sent from Ross, Herefordshire 8 September 1949 (IOL/Mss Eur/C371). Reginald Schomberg disliked the Labour Government and cursed it.
197 Letter to Frederick Marshman Bailey, sent from Ross, Herefordshire on 15 August 1952 (IOL/Mss Eur/C371).
198 Quote from the unpublished manuscript (RSAA/SC/SCH/2/47: 58).
199 Letter to Frederick Marshman Bailey, dated 29 December 1953 (IOL/Mss Eur/C371).
200 Letter to Frederick Marshman Bailey, sent from Bombay on 27 August 1945 (IOL/Mss Eur/C371).
201 RSAA/SC/SCH/2/47; Michael Pollock (2004).
202 Letter to George Taylor, dated 30 October 1948 (IOL/Mss Eur/C371: 47).
203 IOR/2/1086/308: 73.
204 IOR/2/1086/308: 84.
205 IOR/2/1086/308: 84–85.
206 Gilgit Agency Political Diary for the month of May 1940 (IOL/P&S/12/3285: 141). By the end of 1943, the people of the Gilgit Agency had donated 17,026 Rs for the 'War Fund'; there was no involvement of fighting men except for a few serving in the Kashmir State Forces (IOL/P&S/12/3288: 30, 69).
207 Zafar Iqbal (2017: 522) has provided reprints of the original document.
208 Zafar Iqbal (2017: 522).
209 Major William Alexander Brown in his report on 'Precis of events in Gilgit since the arrival of Sardar Mohd. Alam, Political Agent' (IOL/P&S/13/1860: 31).
210 IOL/P&S/13/1860: 32.
211 See Sikandar Khan Baloch (1975; 2004); Yakoob Khan Bangash (2010); William Alexander Brown (1998); Fida M. Hassnain (1977; 1988); Bansi Lal Kak (1977); Hermann Kreutzmann (2008a; 2012f; 2016); Alastair Lamb (1991).
212 IOL/P&S/13/1860: 31.
213 A copy of the original document is attached to the 'Report on the historical perspective of Gilgit-Baltistan', which was submitted by the Convenor Hisamullah Beg and six members to the Government of Gilgit-Baltistan's Chief Secretary on 4 December 2015.
214 In Gordon Walker's telegram to New Delhi on 24 February 1948 (IOL/P&S/13/1860: 2–3) he alleged that Hunza and Nagar might join Russia.
215 See the documents reprinted by Zafar Iqbal (2018a: 729–734).
216 See William Douglas (1952: 307) who related a story of several young men refusing to part with a '*bakshish* to the Mir' from 1951.
217 John Clark (1956); see as well Gerhart Klamert (expedition diary for 7 August 1959).
218 See the popular books of William Douglas (1952); Shahid Hamid (1979: 91); Jewel Henrickson (1960: 16, 54); Barbara Mons (1958: 12, 86-88); Jean Bowie Shor (1955: 280-294); Ian Stephens (1953a); John Tobe (1960: 175); Renée Taylor (1964: 37–40); Irene von Unruh (1955).
219 Wolfgang Pillewizer (1954) in his third expedition report titled 'Expeditionsbesuch beim Mir von Hunza' (DAV EXP 2 SG 237: 2).
220 Wolfgang Pillewizer (DAV EXP 2 SG 237: 3).
221 His visitors' book contained about a dozen entries per year (FCO 21/71: 1).
222 See Allen Banik and Renée Taylor (1960); Ralph Bircher (1942); Jay Hoffman (1968); Jerome Rodale (1948); Hermann Schäfer (1978), Jean Shor and Frank Shor (1953: 490–493).
223 FCO 21/71: 3.
224 The report on 'Pakistan. Visit to Hunza', dated 11 October 1968 was labelled 'Confidential – U. K. eyes only' (FCO 21/71: 4).
225 'Report by Mr. K[umara] P[admanbha] S[ivasankara] Menon on first stage of overland journey from Srinagar to Chungking' dated 25th October 1944. He made his observation of Jamal Khan's Twelvershia leanings an explicit statement in his narrative (IOL/P&S/12/3299); see John Tobe (1960: 270–271).
226 John Tobe (1960: 271).
227 Shams-un-Nahar (2000: 7–9).
228 See Sabine Felmy (1997: 74-76); John Hollister (1953: 412); Hermann Kreutzmann (1989a: 156).
229 Tahir Ali (1984: 234) conjectured that the personal authority of Jamal Khan survived his dismissal and that his judgement ruled Hunza affairs until his death.

230 Opposition and refusal were still fresh in the memories of local leaders during fieldwork in 1984. The former residence of the Lorimers in Aliabad became first a school and then the police station, in addition, there were the office of the former Assistant Political Agent and the hospital constructed during colonial times in the same compound.
231 Tahir Ali (1983: 113–116).
232 Tahir Ali (1983: 114); see IOL/12/3285: 315.
233 Gilgit Agency Political Diary for the month of March 1937 (IOL/P&S/12/3285: 236); see as well the administration report for 1937 (IOL/P&S/12/3288: 174).
234 The notes made by Captain Cooke for the Political Agent in Gilgit are quoted by Zafar Iqbal (2018a: 450–453).
235 Zafar Iqbal (2018a: 452).
236 Gilgit Agency Political Diary for the months of January 1938, November 1939 and January 1941 (IOL/P&S/12/3285: 127, 152, 208).
237 Gilgit Agency Political Diary for the month of January 1938 (IOL/P&S/12/3285: 208).
238 Untitled internal document from the Political Agency Office in Gilgit dated 1 May 1938 collected by the author in 1998; see the remarks by Ian William Galbraith in his 'note on the boundary dispute case Hunza (Gilgit Agency) – Sinkiang in the neighbourhood of the Shimshal Valley and Oprang river', dated 14 April 1939 (FO 371/23493: 75–77; IOL/P&S/12/3292: 41).
239 Gilgit Agency Political Diary for the months of February and March 1941 (IOL/P&S/12/3285: 124–125).
240 Conversation with Ghulam-ud-din in Gulmit on 18 July 1991.
241 Gilgit Agency Political Diary for the month of October 1940 (IOL/P&S/12/3285: 1315).
242 IOL/P&S/12/3285: 208. This episode is not covered in Qudratullah Beg's (1962) history.
243 Zafar Iqbal (2017: 525–526, 551).
244 In an interview in 1984, *wazīr* Sajidullah Beg admitted to the author that his influence had significantly diminished; he could not make his name as a channel-builder as the Dilbar project became a failure under his leadership.
245 David Lorimer (1953: 5); see for contemporary observers during the opposition movement Jewel Hatcher Henrickson (1960: 23); Jay Hoffman (1968: 54); Barbara Mons (1958: 129); John Tobe (1960: 292).
246 The full text with all explanations and demands is reproduced in Zafar Iqbal (2017: 527–531).
247 Zafar Iqbal (2017: 533).
248 Zafar Iqbal (2017: 533).
249 N. N. (1938: 593).
250 In the election manifesto of January 1977 the abolishment of 'princely states' in Hunza and Nagar, their merger with the Gilgit District and the termination of privileges previously enjoyed by the former rulers were highlighted (Pakistan People's Party 1977: 36).
251 Zafar Iqbal (2017: 542-548, 551-558).
252 This term was coined by Hamza Alavi (1989) for persons receiving a salary from government institutions and exercising a certain bureaucratic authority in Pakistan.
253 See Hermann Kreutzmann (2012f; 2015c).
254 Ghazanfar Ali Khan was born on 30 December 1942, his brother Amin Khan seven years later, and the third son of *mīr* Jamal Khan Abbas Khan in 1954.
255 The interview was conducted by Fazal Amin Beg (Ghazanfar Ali Khan 2014: 1).
256 The information about electoral procedures and political representation was provided by the Office of the Election Commission in Gilgit.
257 Ghazanfar Ali Khan (2014: 2).
258 The advocate M. Shafi, who was the opposition candidate, narrated in his autobiography the circumstances of the election process in October 1975 and alleged irregularities: "After complete rigging at the polling station the Additional Deputy Commissioner went to the palace of ex-Mir of Hunza to cast his vote. There he filled the box with fake votes. There were 700 registered voters in Karimabad but the number of casted votes were over 1200" (quoted in Zafar Iqbal 2018a: 564).
259 In 2009 Sheryar Khan, Ghazanfar Ali Khan's second son, was standing against Wazir Beg in the election (Ghazanfar Ali Khan 2014: 4).
260 Shaukat Ali Khan passed away at the age of 86 on 23 May 2003 and had refrained from political activities in his final years.
261 The Express Tribune, 7 August 2016.
262 See Hermann Kreutzmann (1989c; 2005b: 74; 2013d: 11–18).
263 Ghazanfar Ali Khan (2014: 3).
264 Michel Boivin (2013; 2014); John Norman Hollister (1953: 339–370); Shamoon Lokhandwalla (1971: 384).
265 For the position of *imām* see Soumen Mukherjee (2017: 114–145); Raymond Brady Williams (1988: 197).
266 Raymond Brady Williams (1988: 194).
267 See Farhad Daftary (1990: 514).
268 See John Biddulph (1880); Stuart Hill Godfrey (1898: 88); Gottlieb Wilhelm Leitner (1877; III: 41; 1891d: 206); William Lockhart and Robert Woodthorpe (1889).
269 See David Lorimer's notes and records (Irmtraud Müller-Stellrecht 1979: 218–220).
270 For a local version of the advent of the Ismailiya in Gojal see Hermann Kreutzmann (1999); a map showing the respective seat of *pīr* and their escape routes is found in Hermann Kreutzmann (2015a: 271).
271 See Aga Khan (1954: 184).
272 Letter of Aga Khan Sultan M. Shah to Edwin Samuel Montagu, Secretary of State for India, dated 25 March 1921 (IOL/P&S/11/152/P 2732).
273 The Consul-General Percy Thomas Etherton in a letter to Foreign and Political Department in Delhi 14 February 1921 (IOL/P&S/11/152/P 2732); see for his political positioning Daniel Waugh (2007).
274 Gilgit Diary for October 1921 (IOL/P&S/10/973: 244).

275 Kashgar Diaries for October and November 1921 (IOL/P&S/10/976).
276 Letter of Aga Khan Sultan M. Shah to Edwin Samuel Montagu, Secretary of State for India, dated 12 May 1921 (IOL/P&S/11/152/P 2732).
277 Shahzada Lais had personally visited Hunza for the first time in 1905 'to collect alms from his followers' (IOL/P&S/7/185/486).
278 Gilgit Diary for January 1909 (IOL/P&S/7/226/437).
279 Chitral Diaries 21 December 1895 and 11 January 1896 (IOL/P&S/7/84). The Assistant Political Agent in Chitral Bertrand Gurdon certified Shahzada Lais' 'good behaviour' when the Aga Khan contemplated making him his representative for Chitral (IOL/P&S/7/134/783; IOL/P&S/7/179/1198). See for Russian reports about the detrimental activities of *pīr* Shahzada Lais various files from the Russian State Military Historical Archive in Moscow (RGVIA, font 1396, opis 2, file 2251).
280 Gilgit Diaries for January and February 1922 (IOL/P&S/10/973: 232–235).
281 Gilgit Diary for February 1922 (IOL/P&S/10/973: 232).
282 Gilgit Diary for December 1922 and July 1923 (IOL/P&S/10/973: 189, 208). See biographical remarks by Mumtaz Ali Tajddin Sadik Ali (2013) about Sabz Ali who was a director of the Khoja Ismaili Trading Co. in 1918 and was appointed the President of the Ismailia Library of Kharadhar, Karachi the year after. Recently, Nizarali Virani and Shafique Virani (2008) translated and published some episodes from his travelogue.
283 Qudratullah Beg (1962/2006: 256); see as well Stuart Hill Godfrey (1898: 88). For details on the Ismailiya within the Pamirian Crossroads and a special focus on Hunza see Hermann Kreutzmann (1989a: 154–165; 1996a: 282–300; 2015a: 137–139).
284 Gilgit Diary for May 1923 (IOL/P&S/10/973: 194).
285 See Peter Clarke (1976: 487–488). Until the mid-1980s every hamlet had built a *jamáat khana*, one hundred and four were counted in Hunza alone by then. In Gilgit the *jamáat khana* arrived with some delay; the construction for the Gilgit Central Jamatkhana which gave the name to the emerging bazaar in its surrounding was started in 1940 and completed five years later, when *rāja* Jafar Khan of Gilgit opened it on 27 March 1945. The British Political Agent Major Evelyn Cobb and colonial officers were present as well as *rāja* from neighbouring areas except the *mīr* of Hunza (Haji Qudratullah Beg 1978: 49–50).
286 Michel Boivin (2014: 386).
287 Gilgit Diary for July 1923 (IOL/P&S/10/973: 190).
288 Gilgit Diary for October 1923 (IOL/P&S/10/973: 181). Hunza's share in it was about half in 1923, but three years later had increased to 7,000 rupees (Gilgit Diary for January 1926 (IOL/P&S/10/973: 120).
289 The following quotes are from the 'fortnightly report for the first half of August 1926' submitted by the Resident in Kashmir on 14 August 1926 (NAI/PR/5014500: 161).
290 NAI/PR/5014500: 162.
291 NAI/PR/5014500: 162.
292 NAI/PR/5014500: 162.
293 NAI/PR/5014500: 162. The report was written in Camp Gupis on 4 July 1926.
294 NAI/PR/5014500: 163–165; see for the conflict with the Mehtar of Chitral between 1919 and 1925 and related refugee movements Hermann Kreutzmann (1996a: 118–124; 2015a: 271–275).
295 NAI/PR/5014500: 165–166.
296 Gilgit Diaries for December 1923 and January 1924 (IOL/P&S/10/973: 172, 174).
297 Extract from Kashgar Diary for October-November 1923 (IOL/P&S/11/244/823: Kashgar: ill-treatment of followers of the Aga Khan by the Chinese Amban of Sarikol).
298 Letter from C[larmont] P[ercival] Skrine, His Majesty's Consul-General, to the Secretary to the Government of India in the Foreign and Political department, dated December 27, 1923 (IOL/P&S/11/244/823).
299 Gilgit Diaries for July and October 1924 (IOL/P&S/10/973: 153, 159). Subsequently it was permitted again for a short period.
300 Gilgit Diary for January 1928 (IOL/P&S/10/973: 73). In August 1928 the *tham* had received from Soviet Wakhan 40 large silver coins, and 7,852 small ones; 43 large gold coins and 41 small ones. In addition, 472 Indian rupee silver coins were donated (Gilgit Diary for August 1928; IOL/P&S/10/973: 60).
301 See Aga Khan (1954: 183) about the 'closed frontier'; Hermann Kreutzmann (2015a: 271, 364, 445; 2017a: 202-203), with remarks on the *pīr* from Wakhan and their responses to political pressure. However, there are reports as well about *pīr* that were instrumentalised by Soviet authorities to spy on British 'subjects'; see Marc van Grondelle (2009: 47-52).
302 Telegram of Political Agent Gilgit to Resident in Kashmir 16 March 1940, contained in 'Collection by the Mir of Hunza of offerings for his Highness the Aga Khan from places in Central Asia including Badakhshan and Chitral' (IOR/2/1086/302). The source confirms his appointment a decade earlier than other authors suspected; see Wolfgang Holzwarth (1986: 33); Irmtraud Müller-Stellrecht (1979: 219).
303 Hanna Papanek (1962: 158–170, 311–329). It was the Aga Khan's declared goal to make all his *murīd* literate by the 1960s.
304 See Aga Khan (1954: 207) for his intentions to invest the donated funds. Similar weighing ceremonies were staged during the golden jubilee in Bombay in January, in Nairobi in March 1936 and the platinum jubilee in Karachi in February 1954.
305 See Aafiyat Nazar (2009: 36–40); Zafar Iqbal (2018a: 525); Hermann Kreutzmann (1989a: 156–165). All the central places in Shinaki, Central Hunza and Gojal would receive primary schools while Baltit's primary was raised to middle school level. The number of Diamond jubilee schools would grow to sixty by 1973; its number diminished due to the merging of

existing schools to 41 during fieldwork in 1985; the majority of the 4,500 students were female.

306 The amount was an equivalent of 2,300 US dollars (John Clark (1960: 22, 39).

307 Aafiyat Nazar (2009: 24); Anant Ram and Hira Nand Raina (1933: 312–313); in Nagar only forty-four male and two female persons were listed as being literate.

308 Aafiyat Nazar (2009: 86); Reginald George Wreford (1943); in Nagar only hundred thirty-seven male and two female persons were recorded as being literate. Gilgit's first primary school opened in 1892; see John Clark (1956: 24, 39); Hermann Kreutzmann (1989a: 162); Emily Overend Lorimer (1939: 217); Arthur Neve (1913: 153); Thakur Singh (1917: 16); John Staley (1966: 200–202).

309 Aafiyat Nazar (2009: 39–42).

310 Four teachers were employed in Baltit during the visit of Jewel Hatcher Henrickson (1960: 80-81) and Barbara Mons (1958: 117); sixty to eighty students were taught in Baltit. Sultan Ali aka *ustād* Samarqand continued his services from 1960 to 1972 in the Government High School at Baltit, from 1972 to his retirement in 1980 as the headmaster in Gulmit; for another ten years he volunteered in Khudabad and Sost. Being one of the most competent English speakers in the 1950s he was selected by *mīr* Jamal Khan to frequently accompany foreign guests in Central Hunza; see Jewel Hatcher Henrickson (1960: 54); Barbara Mons (1958: 88) and John Tobe (1960: 237), thus becoming one of the best known Hunzukuč to travel writers and journalists.

311 The *firmān* about the urgency of improving female education was issued in 1976; three years earlier the first 'girls' school' of Gojal had opened in Gulmit.

312 Ismaili Horizon 11 July 1983.

313 See the schematic diagram in Hermann Kreutzmann (1989a: 158). A 'Regional Education Board' was introduced in the final years of the Hunza State in 1973. The 'Constitution of the Shia Imami Muslims' was adjusted for all communities in 1986.

314 Aga Khan Foundation (1983: 1).

315 Hermann Kreutzmann (1996a: 385); see as well Karim H. Karim (2014: 149-154).

316 See Michel Boivin (2013; 2014) for insights into the challenging relationship between Aga Khan III and the *khoja* community in Sind and Karachi.

317 See Zafar Iqbal's (2018a: 525-540) interpretation of governance issues and documentation of the growing demand for the abolition of Hunza State out of this movement.

318 Memorandum by the Aga Khan for the Viceroy Rufus Daniel Isaacs, submitted on 4 August 1925 (IOL/P&S/10/588); see as well Jonah Steinberg (2011: 54–55).

319 The refusal of tax exemption in Great Britain led to the constellation that the imamat took its seat in the outskirts of Paris and that the Aga Khan Foundation was registered in Geneva. France and Switzerland had granted him tax-free status (John Norman Hollister 1953: 370; IOL/P&S/10/588/1).

320 See the Aga Khan's genealogical table by Sulaiman Ghulam Husain Haji published on 21 August 1905 in which certain non-Muslim traits were eliminated (Michel Boivin 2014: 393).

321 Aga Khan (1954: 305).

322 Rana Bashir Ahmad Frank (1934); a wide range of options was discussed and documented in a file on 'Aga Khan and Aly Khan' (IOL/P&S/13/64).

323 See Faisal Devji (2014: 57) and the announcement about the new seat of the imamat (https://the.ismaili/imamat; last access 23 May 2019).

324 Jonah Steinberg (2011: 54–58) coined both terms for the imamat of Aga Khan IV.

325 Both terms were used by Aga Khan IV in his speech in the Portuguese Parliament the day before the inauguration that took place three years after signing the agreement in 2015 (https://www.akdn.org/speech/sua-alteza-o-aga-khan/address-his-highness-aga-khan-parliament-portugal; last access 20 May 2019).

326 Faisal Devji (2014: 57).

327 Faisal Devji (2014: 57).

328 The novel by Georgij Tushkan (1944) identified Aga Khan III as being the 'hunter of the Pamirs'; an early and popular exemplary version of a conspiracy theory centred around the Ismaili leader.

329 Abdus Salam Picklay (1940: 108); the term is used in official addresses until today.

330 Andreas Benz (2013; 2014a); Sabine Felmy (2006); Hermann Kreutzmann (1996a).

331 Shoaib Sultan Khan (2009); Hermann Kreutzmann (1989a; 1996a; 2005b); Jonah Steinberg (2011: 60–68); Geoffrey Wood, Abdul Malik and Sumaira Sagheer (2006); World Bank (2002; 2011).

A plethora of Hunza myths revisited

Hunzukuċ and Nagarkuċ would easily have qualified for inclusion in the large group of 'people without history' if a criterion such as the lack of a written language had been applied in pre-colonial and colonial times.[1] European conquerors and colonialists used this term for declaring otherness in old and new worlds in terms of ethnicity, language, script and certain administrative and cultural traits. Modernisation was supposed to be the vehicle for change and the engine of transformation to overcome the denial of one's own history. Membership of the modernity club was strongly linked to the existence and practice of written languages and to the existence and performance of state institutions. Prior to the conquest of the Hunza Valley in 1891 these supposedly lacking attributions were a substantial part of a spectrum of outsider's ascriptions such as backward, barbarous, insignificant, marginal and remote which partly survived until and even after independence. The evidence presented in the previous chapters supports my argument of an early involvement and participation of these microstates and their inhabitants in all kinds of power struggles, armed encounters, robberies and trade. Expertise in plundering and slave-trading, participation in marriage relationships with neighbouring potentates, regular diplomatic and economic exchange relations, the circulation of letters and treaties written in the Persian language have provided ample evidence about a wide spectrum of outward contacts. All this was enabled by highly mobile communities dispatching messengers, raiders and traders. Regular gift exchanges with Chinese representatives in Kashgar and Yarkand and the Kashmir Court in Srinagar were combined with other types of business and relations between peoples. Episodic population movements across borders into the Hunza Valley resulted in the acceptance of persecuted refugees, their integration and resettlement; flows in the reverse direction provided options of being accepted as mercenaries, migrants or people seeking shelter.[2] Living between ordering and bordering in the Karakoram, the mountain dwellers from stigmatised remote valleys have been competitors who were involved in local supremacy struggles, sometimes affected by external intervention, performing and presenting themselves as players in historical power games for control and dominance. It would be too simple to stigmatise people as being without history on the basis of a one-dimensional explanation of the lack of acknowledged scripts for all local vernaculars, namely Burushaski, Domaaki, Shina, and Wakhi. The Hunza and Nagar court scribes used Persian for cross-border exchanges, and local chroniclers had collected material in Chitral, Gilgit, and Hunza since the 19th century which contributed to the historiographies subsequently compiled by their descendants.[3] All areas concerned were part of contested arenas between neighbouring actors and imperial powers.

In various instances local rulers tried to appear or pretend to be on par with their neighbours and potential invaders. As *tham,* Safdar Ali Khan had been notorious for claiming equal standing with the Chinese emperor, the Russian Tsar and Queen Victoria as Empress of India; the two imperial commentators and prominent observers from the Hunza-Nagar Campaign – Edward Knight and Francis Younghusband – were quite outraged about Safdar Ali's boldness and impertinence in regarding the rulers of mighty and huge empires as 'chiefs of neighbouring tribes'.[4] It had been standard practice to enhance one's own importance and strength by such comparisons. During the same conversation and encounter Safdar Ali Khan presented his pedigree's descent and alleged link with Alexander the Great. History

The section taken from the map 'Eastern Turkestan and its neighbours' by Demetrius Boulger (1878) represents a true 'white spot' in the Pamirian Crossroads between the Russian Empire, Bokhara, Afghanistan, Cashmere and the Chinese Empire; mountain ranges constitute and feature as prominent frontiers. The white-coloured interface appears as a dividing space that keeps neighbours at bay. Neither the Hunza Valley nor any relationship between Gilgit and Kashgar is apparent. Courtesy of the Staatsbibliothek Munich

Opposite
Clouds are surrounding Baltit Fort after its renovation and reopening as a museum. Above towers the colonial monument of Malikamu Shikari, commemorating the British conquest of Hunza in 1891. The house of Hunza's last wazir, Sajidullah Beg, is located below the fort in a prominent location where once the oldest polo ground was found. The photograph was taken on 29 July 2010

The oasis of Pasu is nestled between the Pasu Glcier and the Hunza River which is undercutting the bank and has significantly reduced the village lands of Pasu. In the backdrop the Pasu Group is towering above the scenery with its dominant peak Tupopdan (6,100 m). The photograph was taken on 26 July 1989

was conceived and created by referring to ancient genealogical roots, thus affirming not only the existence of a state, but assuring the rule of the heavenly (*ayáśo*) dynasty since time immemorial.[5]

As long as Hunza and Nagar were supposed to be enemies of the British Raj their rulers' extraordinary arrogance and unparalleled presumption were taken as a pretext to teach them a lesson. After the conquest more racist overtones stressed their belonging to the 'people without history' and modernity. The outspoken quibbler Reginald Schomberg constantly classified and regularly divided the communities of the Hindukush and Karakoram; he preferred some and did not refrain from diminishing others. He still noted at the time of the British lease of the Gilgit Agency in 1935: "The Nagiris are thus to all intents a people without a history, and when the stirring events that enlivened life in other states close by are considered, it is remarkable that Nagir should have lived so placidly. But its seclusion as well as its unwarlike people insured its insignificance."[6] The imperial power had matured beyond being the controller of indirect rule and ventured to function as the instructor on the path to modernity. Nevertheless, their local counterparts from the Hunza

Valley had learnt their lesson and attempted to uplift themselves and become equal partners. In this context the creation and promotion of certain myths supported their case. Historical rooting in a European tradition of conquest was offered by constructing and promoting the Alexander myth. Mountain exceptionalism was embedded in their affiliation with an isolated language such as Burushaski. This extraordinary and salient position was loaded with another myth; the one about longevity and health responded to Western projections in their search for an ideal Shangri-La type society in a remote mountain location. Defining an exceptionally endowed community came at the price of belittling and denigrating others. A third dimension in myth construction related to Hunza's perception as an Ismaili state. Statehood was an important measure for one's own position and for communicating on equal terms in times when nation states were the role model for development and modernity. The Hunza nobility played all these cards with differing persistence and varying success. Sometimes it appeared as if they were staging a play in a puppet theatre.

Irrigated terraces form the basis for survival in the Hunza Valley. The oldest known systems are barely seven hundred years old; the process of expanding village lands by constructing new terraces in higher elevations continues. The photograph was taken in Duiker on 4 April 1985

Gratitude to Alexander the Great

On 11 July 2008, the Governor of the Northern Areas, Ghazanfar Ali Khan, arrived with an eight-member delegation in Skopje, the capital of the Republic of Macedonia. The international press headlined 'the latest campaign for Alexander' and saw the Hunza delegation as a party in the dispute between Greece and Macedonia about the republic's official name and Alexander's heritage.[7] After landing in Macedonia the visitors were confronted with 'about a dozen men who were dressed in uniforms, helmets, shields, and spears as soldiers of Alexander the Great.'[8] Here the delegation was treated as royal and blessed by the orthodox archbishop. Some spectators welcomed them home; ten scholarships to Hunza students were offered during an audience with the Macedonian Prime Minister.[9] Such a royal treatment was a long-time bygone for Ghazanfar Ali Khan and he enjoyed the special attention given to Hunza as the long-lost relative in the East. Suddenly Hunza was back on the geo-political stage, assumed and occupied an unexpected role on the world map. International coverage was referring to the dispute both countries were fighting on Pakistan's soil. Greece tried to form an alliance stimulated by the provision of cultural support and development aid for Chitral's Kalasha valleys, while Macedonia played the Hunza card. In the meantime the conflict has cooled off since Greece successfully demanded the renaming of its neighbour into Northern Macedonia (*Republika Severna Makedonija*), which was endorsed by parliament. Presently the political dimension of the Alexander topos has lost in attraction. During a return visit by an official delegation from Skopje support was offered for internally displaced persons in the aftermath of the Atabad landslide. Nevertheless, no Hunzukuč volunteered to be resettled in Macedonia.

Genetic science has given another blow to the pedigree and root question; meanwhile it has been proved that neither Hunzukuč nor Kalasha are genetically related to Macedonians.[10] Nevertheless, the Alexander tale and the alleged genetic imprint of his soldiers is one of the oldest myths connected to Hunza, and it was already brought up in earlier times when travellers enquired about historical roots and ethnic traits: their supposed affiliation was printed on 19th century maps and described in folklore, gazetteers and travelogues.[11] It developed into a sales pitch for giving Hunza the advantage as an attractive destination for a visit to Pakistan. The last aspect has survived until today and is a welcome asset for folk traditions, materialising in tourism-related activities.

One of the famous faces from Hunza is Juno whose image has been reproduced by several journalists and photographers. Juno – estimated by his neighbours to be a septuagenarian – is sitting here with his son Haji Jan and grandson on a ćhar-pā'ī in his house's guestroom in Karimabad. He always carried a handwritten paper (ćiṭ) with him stating: 'I am 105 years old, if you want to take my photograph please give me five rupees.' Juno pleased many visitors with his attitude and fixed price, keeping this slip of paper unchanged for more than a decade. Based on his anatomic knowledge about the human body he functioned as a healer in his village. The photograph was taken on 20 October 1985

Mythical longevity—'Hunza humbug'

The colonial dream of discovering Shangri La-like places in the peripheries of newly conquered territories focused on the mountain belt from the Pamirs to the Himalayas and the Tibetan Plateau. Reaching the 'lost horizon'[12] was expected to discover idealised communities in which the fallacies and shortcomings of 'modern' societies would not exist. James Hilton's novel became the role model for a genre in literature and film-making where an unidentified Shangri La was discovered by stranded westerners who would be confronted with a community of contented, happy and healthy people. Long before the novel was published during the interwar years the search for healthy spots in isolated and remote valleys had begun. In the context of the Hunza Valley the work and publications of Robert McCarrison had inspired advocates of a healthy and long life who searched for evidence that an isolated and secluded life far away from the distractions of the industrialised world with a diet based on home-grown fresh food could prevent certain 'modern' illnesses.[13] Earlier on, the pioneer visitor from the north Bronislaw Grombchevsky was not at all impressed by the cleanliness and health status of Burusho and Wakhi in September 1888. He described his counterparts as constantly itching and full of fleas and lice. Cleanliness was not part of his observations among the Hunzukuć; all of them supposedly were using *nos* (*nasawáar*) a chewing tobacco which detrimentally affected the teeth and caused bad breath.[14] The Gilgit Mission had observed quite normal patterns of disease, Hunza being no exception in any respect. Its surgeon George Michel Giles hinted especially at the prevalence of goitre which became a special topic for Robert McCarrison's research while posted in Gilgit.[15] Fifty years later Emily Overend Lorimer was convinced that she lived in 'the cleanest and the tidiest spot on earth'.[16] What had happened in the meantime? To be honest, not much had changed except some outsiders' perceptions. There is no evidence of a significant change in daily behavioural patterns regarding hygiene and sanitation.

Subsequent narratives, reports and stories linked to Hunza contributed to the myth creation that this valley was peopled by an exceptional number of vigorous and healthy centenarians. Little evidence was presented from an area that had never participated in a population census based on birth certificates. John Clark, who lived in Baltit in the early 1950s, operated a clinic and treated several thousand patients for a wide spectrum of diseases. In his writings he questioned the superficial and unfounded attributions made.[17] Neither age nor dates of birth were of any relevance in Hunza nor in neighbouring valleys. Outside visitors created the myth of longevity and instrumentalised it for their publications. James Hilton's novel was published just a year before the Lorimers started their long stay in Hunza from 1934 to 1935. Emily Overend Lorimer was extremely satisfied with her chance and fate to be in Hunza: "But to get down to more personal matters: there is no money and there are no shops and there are no dogs (for the people have no spare food for parasites), no thieves, no prostitutes and no crime. In the forty-three years that the present Chief—or the Mir, as he is called—our trusty ally, Sir Muhammad Nazim Khan, has been on the throne, there have been two murders. So considering that there is no police and no law courts and no prisons. You will agree that it is a remarkable little country, and strangely different from ours."[18] Despite probably knowing better, she used the attractive topic of a happy and just community for her articles and books designed for a British audience: "The peasants in Hunza are the jolliest, happiest lot of people I have ever met ..."[19] Her sympathising description of the benevolent rule of *mīr* M. Nazim Khan as heading a society where everybody had a place in the social setup, where dietary patterns were based on farmers' own produce, and where physical challenges met with a demanding and perfect mountain environment contributed to the likening of Hunza to Shangri La.

The 'zeitgeist' accepted her descriptions and made her publications successful contributions from which subsequent books would take their inspiration. Emily Overend Lorimer's romanticising and marketing skills did not at all affect her husband's writing as he had a different curiosity and interest; he had to struggle with his personal health issues during their stay. Immediately after their arrival in Aliabad, David Lorimer 'went down with a serious but undiagnosed illness' for nine weeks.[20] For months he was sick, kept to his bedroom and could not walk around the Hunza oasis. Nevertheless he continued his language studies, never referred to the longevity myth and was more sceptical about his wife's romanticising descriptions and especially what other people interpreted and read into it. The reception of her writing focused on health and supported the longevity myth, but Emily Overend Lorimer enhanced the bias by introducing an uplifting differentiation that attributed to Hunza exceptional and unique features while degrading neighbouring Nagar. She confessed her sympathy for Hunza women and society; she found the antipode across the river in Nagar: "I've seen a little of woman's life in Muhammadan countries ... I should not be all afraid to be reborn as a Hunza peasant woman. In fact, I should rather like it! I never saw a lot of women—old or young—who looked so happy and contented, and I never lived in any place where the women were more independent or more respected. Though the Hunza people have been Muhammadans for three centuries or so, they have kept many of their ancient customs. They don't veil their women or keep them in the background. The women and girls move about just as freely as any English countrywoman, work in the fields alongside their men folk, and pass the time of day with a stranger without feeling self-conscious or embarrassed. Just across the river, in Nagir, though they talk the same language as our Hunza friends, you never see a woman's face. If you happen to ride past, the women fly off, crouch down and cover their heads."[21] Her biased views on Hunza and Nagar were endorsed in health terms: "Since early times the Hunza people have been divided into several clans and it has always been the custom for a man to marry a woman from a different clan from his own. This has prevented any close inbreeding, and I fancy this partly accounts for the very high standard of health. We did not see a single person suffering from goitre, we did not see a single imbecile-looking person in all the time we lived amongst them—though again there are lots of both in Nagir."[22] Emily

Emily Overend Lorimer could meet in Hunza with men and women alike; she contributed a number of images that showed everyday life. This photograph is titled 'Sanghi Khan's house. Group of women' at Dorkhan. Source: Lorimer Collection in the School of Oriental and African Studies, University of London (PP MS 66, Box IV.16); reproduced with kind permission

Overend Lorimer's comparison and differentiation of people according to environmental considerations was not the first of its kind; Henry Haughton and Arthur Neve dwelled on geographically determined differences and hinted at the suggested advantages for Hunzukuč living on slopes with southern aspect and an alleged disadvantage for Nagar being exposed to the north.[23] Shortly after the first mercenaries were recruited from Hunza and Nagar, military gazetteers adopted landscape determinism to characterise people in 1909: "Although of the same race and origin, the people of Hunza and Nagir vary considerably from one another both in physique and physiognomy. The Hunzas are of a cheerful open disposition, generally powerfully built and of medium stature. Their complexion is rather fair and individuals may be found who would pass for Europeans. Among them red and sandy hair is often seen, and the women are said to be beautiful. The men are splendid mountaineers and never seem to tire. One man on foot will bring in an urgent dâk from Hunza to Gilgit in less than 24 hours the distance being over 60 miles. The people of Gilgit would invariably ride at least part of the way, and could not possibly cover the distance on foot in that time. The Nagir men do not possess the fine physique of those in Hunza, and are generally of darker complexion This is probably due to inter-marriages with the Gilgitis who are physically a more feeble race than those of the pure Burish stock."[24] Emily Overend Lorimer's contemporary visitor Reginald Schomberg used the usual strong language filled with prejudices, abusive and pejorative

statements: "The people [Nagarkuċ], stupid, dirty, and but indifferent imitations of monkeys, cannot mar the noble prospect."[25] In his opinion Hunza people had developed into what they had become because of the constant nutritional and physical challenges denied to the Nagarkuċ who were endowed with a more fertile agro-ecological environment: "Nevertheless the people of Nagir are poor husbandmen, believing rather in the kindness of Providence than in hard work, and their lovely fertile country owes but little to its owners."[26] For him Hunzukuċ were quite different, not a surprising fact as all of his companions were from Hunza, none from Nagar. When returning for his last visit in 1945, Reginald Schomberg reflected on the superiority of the Hunza people for which he would not have necessarily found the same words earlier on when he was cross with the then ruling *tham*: "We were glad to be in Hunza, and amongst perhaps the pleasantest and least spoilt people of Asia. I have described this state and its people elsewhere but it is never possible to visit the country without being struck by the intelligence manifest on all sides. The cultivation of the land, the siting of the villages, the construction of the 'kuls' or irrigation channels, and the general set-up of the whole state, show the industry, alertness and energy of the people. There is a happy absence of the casual, offhand and sloppy. The people are neither literate, sophisticated nor particularly polished, but they have some[thing] better than these superficial and unremunerative qualities. They are enterprising, hard-working and ingenious, and their practical, progressive spirit has hitherto defeated the clumsy efforts to spread useless instruction and teach them foolish lessons. In the end they may succumb to the paltry nonsense of modern civilization, and that will be the end of these people."[27] During this last visit in the summer of 1945 he became nostalgic and had forgotten his troubles with *mīr* M. Nazim Khan that had developed into enmity. Not depending on anybody's official support, he was quite pleased with *mīr* Jamal Khan who seemed to better fit the post-war circumstances. The achievements of the Burusho were praised in a romanticising manner: "We were now in the real centre of Hunza, with the white walls of the fort-palace of Baltit, the capital, rising like a lighthouse above the green sea of billowing barley and wheat, and the startlingly fresh emerald of the new-born apricot leaves. It is a wonderful expanse of carefully, well and patiently tended arable and orchard land. Tier upon tier, with their retaining walls so hidden as to present an incline of unbroken vegetation, are the fields, the living granary of the Hunza folk, spreading for miles with the surly brown river roaring unseen below, the boundary on the East, as are the barren walls of the valley on the West. A skilful and elaborate system of water courses irrigates this fertile area, the wealth, the home, the very life of Hunza."[28] Reginald Schomberg had not lost his biased and prejudiced view on local affairs in his latter days as the most widely travelled visitor, but Hunza remained to him the most likeable place.

Especially since the 1940s the development of the Hunza longevity myth gained pace. Emily Overend Lorimer's 'Language hunting in the Karakoram' was the blueprint from which writers from the health-food niche such as Ralph Bircher, Jerome Irving Rodale, and Guy Theodore Wrench derived their inspiration and copied what they paraphrased in their own books.[29] Although all three never visited Hunza they used her material for a narrative about Hunza's health secrets that influenced and inspired their followers in adopting and sustaining the myth. The construction of an idealised mountain kingdom where a benevolent rule and an industrious, healthy population demonstrated to the rest of the world that a different life form was possible served the demand in the war-stricken Western world and fell on sympathetic ears. If Shangri La were to exist in the Karakoram Hunza would fulfil the purpose of role model and would be attractive for elite travellers to visit.[30] In response to these expectations on the receivers' end the *mīr* of Hunza discovered a growth potential in the economic niche of hospitality services and prepared his facilities accordingly. His monopoly in providing all kinds of required services for the Western visitors included his control of interpretations about the region's history and society; he commanded the sovereignty in explaining health and wealth in Hunza. The visitors were served with the expected set of tales about longevity and his benevolent rule in the hidden mountain abode of 'just enough'.[31] Film teams, travel writers and journalists have continued their coverage and carried this trend for more than half a century well into the 1990s.

Although an ardent admirer of *mīr* M. Nazim Khan, David Lorimer maintained a more critical and distant view when it came to longevity. He survived his wife for many years and as an authority on Burushaski and Hunza was contacted by a number of potential visitors who planned a trip to the Karakoram. David Lorimer kept a separate file marked as 'Hunza humbug' and subtitled 'The rise, decline & fall of a myth' in which he collected these letter exchanges.[32] The number of visitors significantly increased during the 1950s when Pakistan seemed to be an attractive destination as a newly independent country to which Hunza was affiliated. Some visitors had difficulties to narrow down the meaning of the term Hunzukuċ as they were interested in those who were eligible to be termed exceptional. David Lorimer explained to an American letter writer: "The term Hunzukuts ought literally to cover all the inhabitants of Hunza, including the Wakhi Iranian people in the North, and the Shèn people in the S. W. Now the people with whom all our enthusiasts are concerned are the Burushaski-speaking people living mostly in the S. E. part of the Hunza River valley."[33] Other potential travellers contacted David Lorimer for all kinds of practical and esoteric questions. In 1953 he finally realised the detrimental effects of Ralph Bircher's book that had obviously reached a wider readership than his wife's account, although the Swiss author had paraphrased most of her work as he had no personal knowledge of Hunza at all and never tried to change this. In response to an enquiry by a Swedish artist and traveller David Lorimer stated: "I strongly depreciate the Myth-Building about Hunza which has been going on in recent years in Switzerland, the U.S.A., and Sweden. It appears to me in most cases to be based on quite inade-

Two women visiting Afiato's house are spinning fine wool. A basket with wool was regularly carried along on all sojourns in the village. Source: Emily Lorimer 1934–1935 (PP MS 66. IV. 21)

Opposite
Mosques and their verandas were a common meeting place and playground for children, like here in Altit. Source: Lorimer Collection in the School of Oriental and African Studies, University of London (PP MS 66. Box III. 6); reproduced with kind permission

Emily Overend Lorimer captured everyday scenes during her daily sojourns in the central oasis of Hunza and she knew all the people photographed in person. Daulat is weaving homespun wool for the preparation of coarse cloth (paṭṭū); Juma and another boy are carrying home bundles of fodder; Afiato and his sons are preparing a threshing floor (darċ); Rahman Beg is leading the winnowing on his family's threshing floor. Source: Lorimer Collection in the School of Oriental and African Studies, University of London (PP MS 66, Box III. 31 & 39, IV.5 & 22); reproduced with kind permission

quate data by people who have some theories of their own to push, whether relating to Agriculture, Health, or Dietetics, etc., and who have never even seen the country or the people."[34]

In a letter to Hans de Meiss Teuffen he disclosed his reservations and stated his feelings about these approaches: "I am sorry to shatter your 'Castle in Spain' built on the insecure foundations of Dr. Ralph Bircher's enthusiasms. I fear he suffers from a number of hallucinations about me, which I must dispel. 1. I am now old and tired-to-death. 2. My main objects have always been linguistic, and I do not claim or believe myself to be an authority on life in Hunza. … 3. Naturally I made some enquiries and recorded some notes about the life and habits of the people, but my enquiries were neither comprehensive, nor systematic, nor directed scientifically. I know little about nutrition, dietetics and cooking, whether in Hunza, or in this country. …

7. Of course I took photos, but I do not particularly want to see them published outside their proper context in my own work. Requests for photos by persons who are exploiting the journalistic value of Hunza are not unknown! … 9. My wife, who died in 1949, on the other hand published a popular account of our stay in Hunza in 1934-35. She saw more of the people, and especially of the women, than I did. The book, which is out of print, is entitled 'Language hunting in the Karakoram', By E. O. Lorimer …

It was this book that drew Bircher's attention to us and from which he derived a considerable part of the contents of his book. … 10. In recent years Hunza seems to have become the Promised Land, or Paradise of the protagonists of many 'Causes' – Composters, Healthy Food-mongers, Dietetic Enthusiasts, Soil Theorists and Vegetarians, and generally the admirers of the Simple Life all turn to it as Muslims turn to Mecca. It bids fair to become a Land of Romantic Myth. All this publicity will, of course, play its part in hastening the decay and disappearance of the old ways of life in Hunza which are so admired! …"[35] For David Lorimer it was quite obvious that both books had played into the hands of esoteric protagonists, some of whom planned to visit Hunza. A potential companion of Hans de Meiss Teuffen was Yvonne Gräfin Kanitz who received a response to her enquiry about road conditions and the comfort of travel a week later: "I might also add that the dissemination of Dr. Bircher's misconceptions about me in recent years, reinforced by his personal exhortations, has landed me in a great deal of trouble and seems likely to make my latter days intolerable. Yours is the seventh application I have received in the last two months relating to Hunza. A considerable part of his book consists of paraphrases of passages from a book written by my Wife, and he took illustrations from the same source."[36]

These responses are follow-up replicas after David Lorimer and his publisher Philip Unwin had discovered that Ralph Bircher had deliberately and unauthorisedly used Emily Overend Lorimer's narratives and photographs for his own successful books which had been translated into several languages except English by then.[37] Ralph Bircher was so bold and demanding that he requested the publisher should be provided with the original images as his copies taken from the published book did not comply with the required quality standards. In 1952 an exchange of letters began that reveals the demands and requests of Ralph Bircher and his enthusiasm about profit-making. When the fifth German edition of his highly successful book was envisaged, ten years after its first release, Ralph Bircher contacted David Lorimer: "Now for this 5th edition the publisher and I have a desire which I strongly hope you will see your way to comply with. All the pictures in the book are reproduced after book reproductions and not after photographs, that means screen after screen, and though relatively well made they are poor technically. If you would be so kind to lend me the photographs and perhaps some new ones with them, we would be very obliged. The publisher is, of course, prepared to pay the usual price for them. … My book on the Hunzas intended never to be one of those autistic and uncritical adaptation of earnest work to the public taste as there are many, but a respectful discussion of an immensely interesting phenomenon. I hope recognized this my intention and felt it a fair picture as much as that can be drawn by one who was not there. When writing I made it plain that I remained open to any criticism at any moment, and that I wanted to eliminate any prejudiced idea though maintaining our viewpoint which seemed particularly fruitful in this case. If this book had some success this is of course partly due to the charming traits taken out of your wife's book and therefore I indeed owe you your part of my royalties … What I however want to tell you is that, as I see ever more, a considerable part of the medical profession in Germany and Switzerland indeed takes this my book very seriously and call it, at medical congresses and in books, a serious and

David Lorimer was photographed on 27 March 1924 when he was serving as Political Agent in Gilgit. Source: Lorimer Collection in the School of Oriental and African Studies, University of London (PP MS 66, Box 10); reproduced with kind permission

Bibi Anjir is posing with her grandchild in front of her husband Nadiro's new house in Aliabad. Source: David Lorimer 1934–1935 (PP MS 66: IV. 10)

worthwhile contribution to the health problem, though, of course, they agree with me that it is only an outlining of problems while the facts would have to be the target of an expert team of research workers."[38] David Lorimer reluctantly responded to this letter and pretended not to know exactly where her photographs were located: "... that I do not know where Mrs. Lorimer's original photos of the illustrations are, and I do not know whether I would be able to trace them without undue labour. ..."[39] The next day he gave vent to his anger in a letter to Philip Unwin: "Our troublesome friend Dr. Ralph Bircher of the Bircher-Benner Establishment in Zürich ... is again on the warpath, or is it the peacepath? There doesn't seem to be much difference nowadays. ... Personally I am not much interested in his book. I have not read it, but have only glanced at the French version of it—which still remains uncut. ... Royalties too, unless amounting to something appreciable, do not interest me except in principle. So far they have been of the 'intangible' variety. ... I do not propose to saddle myself with any responsibility, direct or indirect, for the accuracy of his book. If he has misrepresented us in any way, the damage has been done and there is no worrying about it at the thirteenth hour."[40] The publisher Philip Unwin followed up the royalties case with Ralph Bircher about the 'infringement of copyright' and could finally report: "... we have recently received payment from that Swiss publisher, Ralph Bircher, who made use of considerable extracts from Language Hunting in the Karakoram for a book of his own".[41] David Lorimer continued to be involved in further controversies especially when scientific expeditions reached Hunza. The German-Austrian Himalaya Karakoram Expedition of 1954 seemed to have become a challenge for Ralph Bircher as they brought their own physician Paul Bernett along. In addition, the qualified medical doctor Irene von Unruh who had come to Hunza as an ardent admirer of Maximilian Bircher-Benner, the mastermind and pioneer of the Swiss nutrition movement who had passed away three years before his son Ralph Bircher published his Hunza book in 1942. Irene von Unruh travelled to Hunza and enjoyed some support from expedition members as she arrived in Hunza during their stay. Having arrived as a romantic admirer, she left as a critical observer, acknowledging certain aspects of the Hunza diet as a more compulsory adjustment to environmental conditions and agricultural potential. During her visit she explored several villages in the valley, observed a mismatch between dietary patterns and hygienic conditions and compared here experiences with the existing literature which made her a stern critic of 'occultist' writers presenting false evidence that was based on imagination without ever visiting the Hunza Valley. Nevertheless, she fell into Emily Overend Lorimer's trap when distinguishing the 'less handsome and strong' Nagarkuċ from the 'friendly and beautiful' Hunzukuċ.[42] Irene von Unruh claimed in her book to be the first 'biological' medical doctor to visit Hunza and challenged the empirical and narrative foundations of longevity story-tellers and might have threatened their success in sales. The events reached dramatic dimensions when Ralph Bircher tried to make her a *persona non grata* in Hunza by humiliatingly blackmailing her. Irene von Unruh had observed a mismatch between the romantic praise of healthy and strong people leading an exemplary life and the high demand of Hunzukuċ seeking her help for all kinds of medical treat-

Opposite
Four pictures taken during a séance in which the popular shaman (biṭán) Ibrahim had fallen into a trance after consuming goat's blood and the fumes of juniper branches. The special tunes of the Ḍom musicians are very important for the biṭán to make his connection to the fairies (parí). The kettle drum (ḍaámal) and the cylindrical drum (ḍaḍáṅ) are accompanying the double reed instrument (suranái) which is playing the tune. Rahimo had realised at the age of fifteen that he was privileged to be a biṭán (Homayun Sidky 1994: 75). The scene was photographed on 8 July 1959 and commented by Gerhard Klamert (1959: 55–56) in his diary: "Accompanied by the sound of drums juniper twigs are lighted, the biṭán inhales the fumes and immediately reaches a state of trance ... In the meantime a goat was slaughtered; while continuing his dancing and running in a circle the biṭán takes the goat's head and sucks the remaining blood out of it. He approaches the musicians, hits the drums with his blood-stained head, forcefully the goat's head is snatched away from him, his hollow eyes search the horizon for any sign of fairies who might sit on trees and inform him about their prophecies. He pretends to fly like a bird, swirls in circles, jumps and lashes out, returns to the musicians to carefully listen to the flautists, suddenly stops and begins to sing verses in vernacular Shina of which he would be unable to do when awake. The performance is repeated two or three times during which his dancing deteriorates, finally switching off, flashing and fainting. This moment has to be cared for by his assistants. The prophetic singing commences with a special tune attributed to the fairies. One gathers the impression that a majority of the people believes in the biṭán and would not call his performance humbug." (Gerhard Klamert 1959: 55–56, translated from German)

The tug-of-war game is part of festivities in Hunza and Nagar. The display in July 1954 was in honour of the guests of the German-Austrian Karakoram expedition who had attempted several peaks in Nagar and were received by mīr Shaukat Ali Khan at his residence in Uyum Nagar. Courtesy of Gerhard Klamert

ment. After her return to Germany she innocently wrote to Ralph Bircher informing him about the state of health affairs in Hunza based on her own observations. Ralph Bircher saw his goldmine of substantial book sales fading and wrote to *mīr* Jamal Khan for support of his case and of all the others promoting the longevity myth.[43] He quotes from Irene von Unruh's letter by stating that she writes "… in a deep depression about the hygienic and health condition of your people. She found them 'dirty and sick' and 'going to perish in filth', ridden by chronic eye disease in adults and children, by skin and mucus membrane infection, stomach and intestinal diseases and increasing incidence of tuberculosis. This seem to be appalling news. Everywhere people followed her, she writes, asking for treatment. Contrary again to what we knew before she found a large consumption of meat, sugar, tea and salt, and even of alcohol and tobacco. Dr. v. Unruh says she met the German-Austrian expedition and learned they also were dismayed about these things."[44] In his last sentence he referred to Dr. Paul Bernett who was the medical doctor of the expedition and who had made similar observations. The scientific expedition leader, Wolfgang Pillewizer, endorsed the findings as did subsequent visitors such as Naohiko Harada and Atsuko Miyoshi, who stated that the 'predominant diseases are goiter, conjunctivitis, rheumatism, and asthma'. Many observers suspected that new diseases such as 'heart disease, malignant diseases and appendicitis' could have been new additions to the long list of illnesses for which no treatment was available in Hunza.[45] In Ralph Bircher's worldview this could not be possible as he had promoted the Hunzukuċ as the last survivors of a specific healthy species. If anything had changed it must be attributed to external contacts: diseases brought in by outsiders and modernisation in form of road access would have to be blamed for it. Ralph Bircher tried to find an accomplice and ally in *mīr* Jamal Khan, as both derived significant profits from maintaining the longevity myth: "All this looks indeed so strange and in part, at least, incredible that I feel urged to ask your opinion."[46] Referring to the contrasting literature that promoted the myth "… about the buoyant health, stamina, food tidyness and happyness of your people, but even to brush new highlights into the picture. They wrote that you were fully aware of the danger involved in the road connection and were careful to prevent any bad influence. Could it be possible that in these two years since the road reaches Hunza, things have taken such a turn?"[47] It really seems that Ralph Bircher believed in societal and health changes that could be brought about in such a short span; he never doubted the romanticising reports about Hunza as Shangri La, he really needed it to support his mission: "Of course, nobody hereover among the many friends and admirers of you and your people would not like the Hunzas free of want and getting all things they like, but they have become a symbol of buoyant health for the world and it was hoped that one day science which was neglecting the study of health for one century, would find in Hunza the last stronghold of real health, for its research into these thrilling problems. How this hope seems to become obsolete. I hope these news, however to be greatly exaggerated. But even then, the news that are going to be published by Dr. v. Unruh and Dr. Bernett may cause deep confusion and dejection among the friends of your people. So I would like to know whatever can be said to disprove them."[48] A serious situation had occurred when the *mīr* of Hunza was asked to provide evidence about his healthy people when medical doctors that had visited the previous year had been constantly busy in treating people wherever they came. A month later Paul Bernett felt obliged to write an apologetic letter to *mīr* Jamal Khan explaining the conflict and confusion by distancing himself and the expedition members from Irene von Unruh's observations and writings. He referred to her as 'a woman who does not know exeuropean conditions and could not succeed in getting the real view on the Hunzas during the short time she spent there'.[49] Paul Bernett tried in his letter to placate Jamal Khan, refused any responsibility for the misunderstandings and tried to convince him that the Bircher-von Unruh dispute would not affect Hunza's reputation in the world. In addition he approached him a few months later and invited the *tham* and the *rānī* of Hunza to visit Munich.[50] It had been the explicit aim of the German-Austrian expedition to use their enterprise as the starting operation for a long-term mountaineering and scientific endeavour that should not be affected by a discussion about health and longevity.

Members of the academic expedition group were convinced that scientific approaches were necessary for a comprehensive understanding of the geographical conditions of the Hunza Valley. They were quite aware of the obvious facts regarding health and hygiene in a fascinating mountain world that did not need to be a paradise at the same time. Two worldviews contrasted here. The romanticising, selective and partly blind commentators told a story about nutrition and health located in an area where neither an elaborated indigenous tradition of healing was developed nor sufficient facilities of colonial and post-colonial medical infrastructure existed. In terms of nutrition it was quite cynical how non-affected and non-participating Western writers interpreted the periods of starvation that regularly occurred in spring as fasting by choice and not by need. These writers selfishly cooperated with Hunza's hospitality service providers in marketing their products and offers.[51]

The second group of development experts and researchers well realised deficits in agricultural production, seasonal food consumption and quality; they related health issues to deficiencies and shortages in adequate supply. Hunza was one of the least well-endowed principalities in the Hindu-

The wazīr Sarwar Khan is portrayed as a member of Nagar's upper class during a visit of members of the German-Austrian Karakoram expedition in July 1954. His portrait prominently featured in Wolfgang Pillewizer's book (1960, following page 96) book. Courtesy of Gerhard Klamert

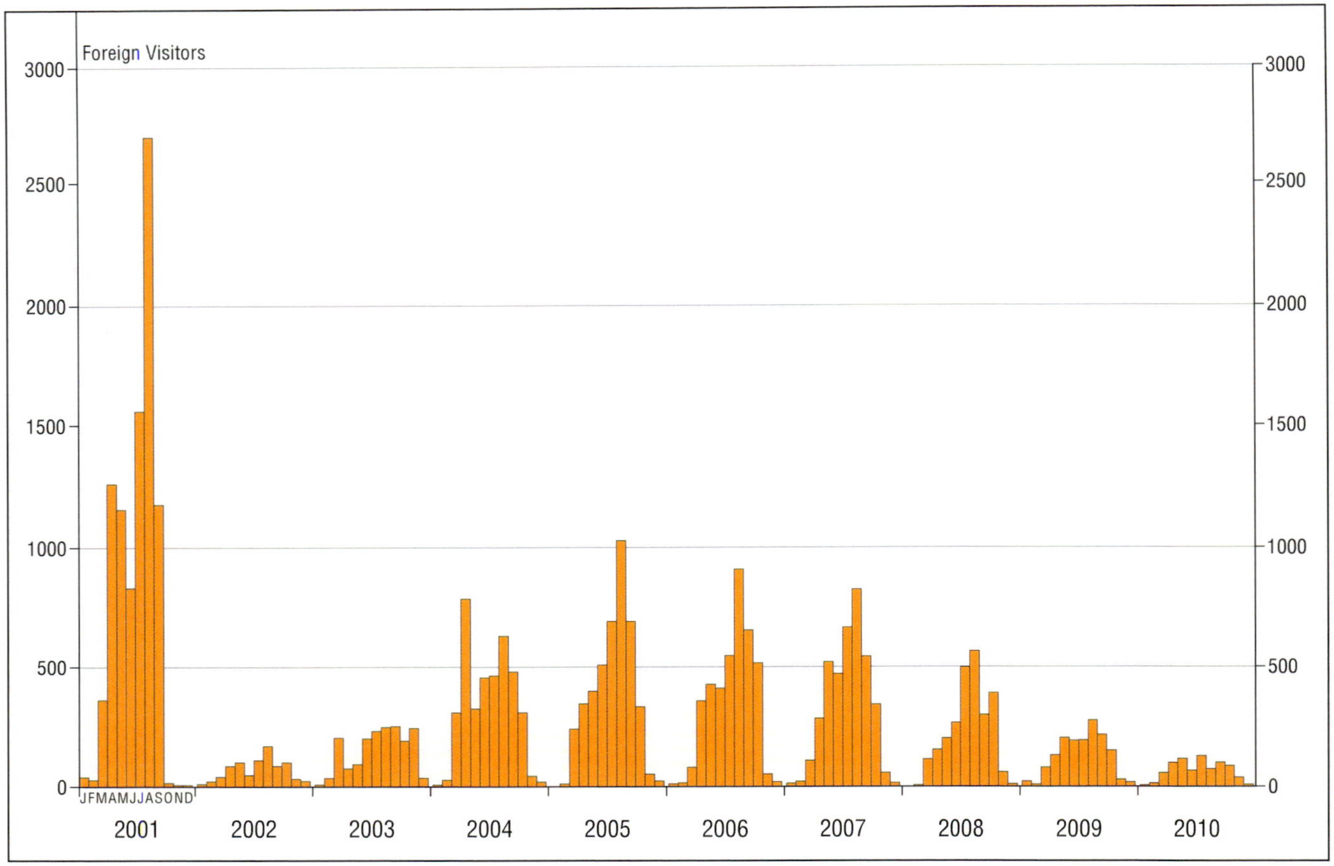

The monthly figures for foreign guests who visited Baltit Fort have shown a significant decline since 9/11. International tourism collapsed for nearly one and a half decades and has not fully recovered from the loss involved. A further decline is connected with the Atabad landslide in January 2010 and the temporary interruption of road traffic along the Karakoram Highway for five years. Source: Data provided by Baltit Fort office; modified and updated from Hermann Kreutzmann (2013d: 15)

kush and Karakoram when it came to agricultural potential. Significant numbers of Hunzukuċ had begun out-migrating long before *mīr* Jamal Khan came to power. Finding herself between a rock and a hard place, Irene von Unruh desperately wrote to David Lorimer who probably agreed with her estimate that 'Bircher is a very good businessman' who is convinced 'that Hunza people got spoiled, ill and dirty since crowds of reporters and expeditions overflow the country'.[52] Subsequent travel writers continued to promote the longevity myth. The joint-venture successfully functioned when John Tobe quoted *mīr* Jamal Khan as saying: "I think my people are the happiest and healthiest in the world, and that is more than any other nation can claim."[53] Evidence did not increase because of repeating the same mantra, but the marketing strategy stimulated a steady flow of visitors from Western countries who well into the 1990s contributed to the main cash flow derived from tourism in Hunza.[54] Presently the longevity myth plays only a secondary role as domestic tourism has significantly overtaken and sidelined international visitors' flows in terms of revenue generated. The longevity myth was closely associated with an orientalist view and related to a romanticist perception that nothing should be altered in this society to preserve Hunza as Shangri-La, to allow the *tham* to continue his apparently benevolent rule, and to restrict outside contacts to visitors who had become pilgrims to the mountain abode.

In recent years Shargri-La aspirations have moved away from Central Hunza towards the remoter pasture areas such as Shuwert in the Shimshal Pamir. The international visitors tend to be members of trekking groups and mountaineering teams rather than of culture tours. The photograph was taken on 20 July 2017

The Baltit lambardār Mirza Hassan in front of a wooden mosque below Baltit Fort, photographed by Aurel Stein on 13 September 1930. Source: Stein Collection in the British Library. The latter is reproduced here with kind permission of the International Dunhuang Project Photo 392/34 (79)

From longevity to developmentalism myth

Colonialism and developmentalism both promote forceful interventions into existing structures and are based on external forces causing fundamental changes in existing systems. All elements of society would be significantly affected and reconfigured to such an extent that new constellations would be likely to emerge. The Hindukush and Karakoram valleys are no exception to these infringing rules. Intruders' activities to develop the agricultural potential, to broaden the nutrition base and to utilise so far untapped sources go back to early colonial times. Irrigation channels for expansion of village lands and land development for grain and grass production to supply the garrisons, road development and control stations, a network of bazaars and post offices were all part of these 'modernisation strategies' that involved early formal education and health services on a low level. After the lease of the Gilgit Agency in 1935 that was stimulated by strategic and defence considerations, plans were drawn up and especially designed when irrigation projects, afforestation schemes, weaving schools and vocational training, supply of agricultural implements and schemes for financing agricultural inputs were fostered to broaden the economic base. The catalogue of packages reads like later comprehensive approaches for integrated rural development. Colonial authorities were directed and enabled to improve the 'pedigree stock' and encourage local farmers to produce more wool in order to meet the demands of the Gilgit Bazaar. At the same time the first-ever cattle show was held in Gilgit and merino rams were introduced to improve the breeding stock.[55] This event might mark the beginning and also a principal change of development activities in the Hindukush and Karakoram. As Hunza had always been a vulnerable principality in producing sufficient nutrition from grain crops and Nagar had just met its basic requirements a central aim of these colonial plans focused on increasing yields from mountain farming and coincided with a phase when the second pillar of combined mountain agriculture – seasonal livestock migration to high-lying pastures – was at its peak in Hunza and Nagar. It is no surprise that Mirza Hassan and Qudratullah Beg prepared an inventory of Hunza pastures (*ter*) for David Lorimer at that point in time when animal husbandry was still a significant subsistence strategy for generating sustenance in the Karakoram and had become a target for development packages. The year 1935 has remained an important benchmark as livestock-keeping in Hunza has declined since.[56] Up to then the *tham* of Hunza derived a substantial share of his revenue from livestock herds within and grazing rights outside his territory in Sarikol.

At the same time the impact of colonial rule on local administration and livelihoods had detrimentally affected a previously encountered, somewhat less unequal power balance in the relationship between rulers, local elite groups and mountain farmers-cum-pastoralists. Colonial authorities trusted in the leadership of local *tham*, and contributed their share to institutionalise and stabilise indirect rule. This came at a rather cheap cost; of the more than half a million rupees-worth budget for the Gilgit Agency in 1935 to 1936, hardly one percent was reserved for the subsidies of the Hunza and Nagar *tham* each; that share had not changed in principle since the early occupation.[57] Henceforth local rulers could rely on their privileged relationship with the Political Agent in Gilgit under the Resident in Kashmir who provided them with a spectrum of fringe benefits such as bodyguards, the monopoly of sanctioned contracts for in-

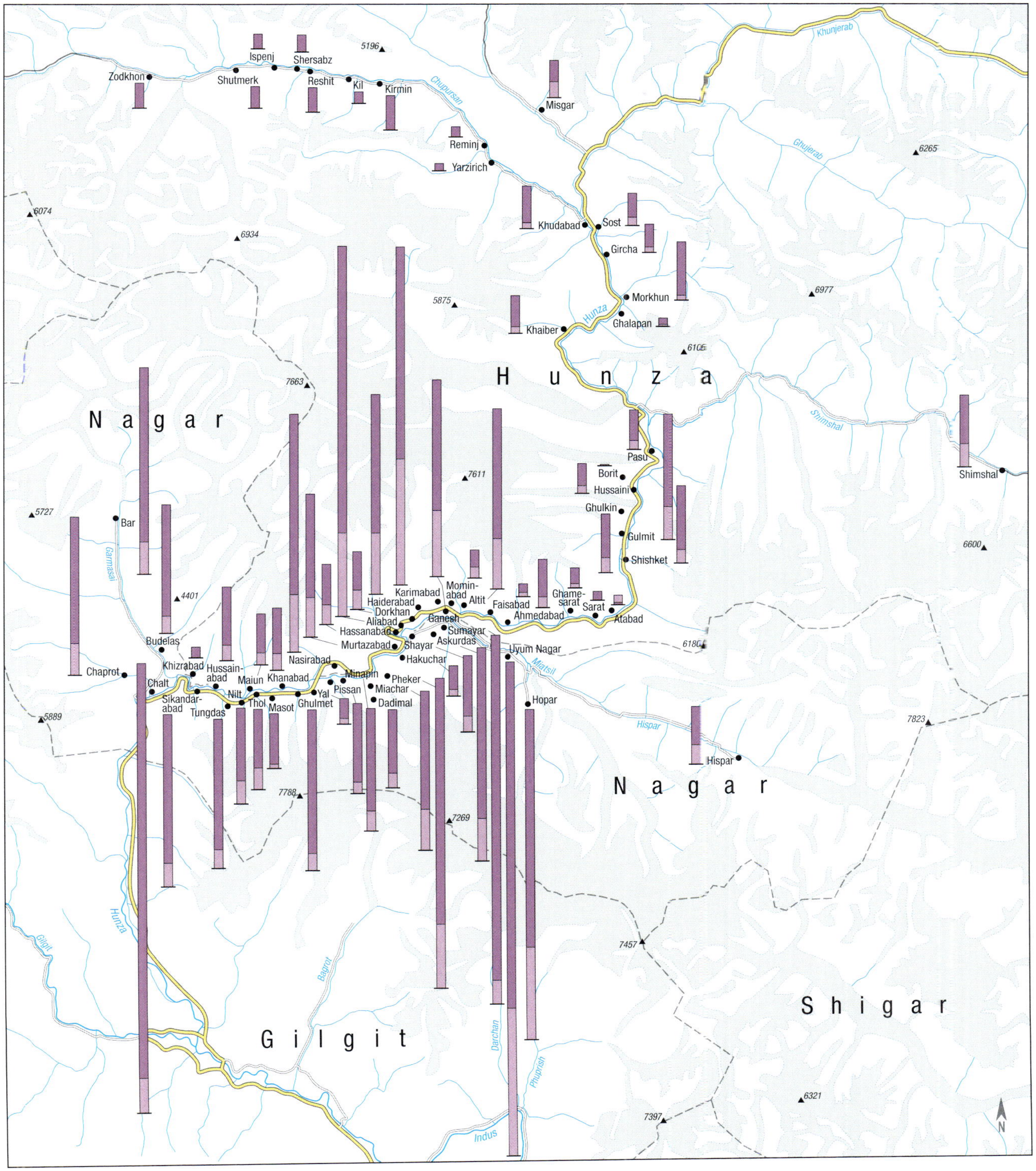

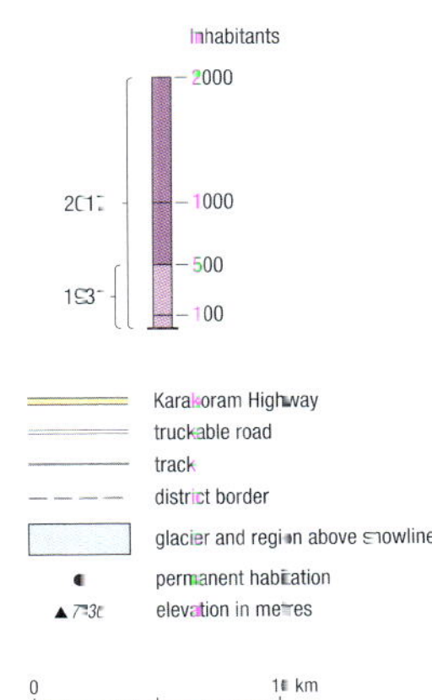

All villages of the Hunza Valley had a higher population in 2017 than in 1931 with two exceptions: Atabad and Ainabad which were more or less abandoned after the 2010 Atabad landslide. Existing settlements have been expanded and new ones founded despite increased out-migration during the 20th century. Comparing Hunza with Nagar it seems obvious that out-migration and resettlement has slowed down village growth in Hunza while all settlements in Nagar show a significant population increase that sometimes seems exaggerated. The historical census data are based on Muhammad Masih Pal (1934). Design by Hermann Kreutzmann

After Pakistan's independence John Clark had come to Hunza to expose the valley to 'development' by exploring its mineral resources, introducing new seed varieties, providing health services and fostering skills development. Accepting his offer, mīr Jamal Khan allowed him to operate a professional centre under his vigilant eyes in Hunza. He let him use the prominent, but empty Baltit Fort until the project was suddenly terminated in autumn 1951. The photograph of the again abandoned Baltit Fort was taken in July 1954. Courtesy of Gerhard Klamert

frastructure projects, privileged involvement in attractive support schemes, in addition to monetary subsidies; however, more important was their position as broker and their function as sole contractor (*ṭhekedār*) for these development schemes. The default setting in semi-autonomous principalities consisted of local rulers who operated as prime imperial auxiliaries. These fillings of posts and key positions survived the struggle for independence and formed the basis for future engagement and involvement in subsequent development packages that outsiders introduced in the mountain valleys.

The embodiment of the colonial officer bringing welfare to the remote mountain dwellers was John Clark, an American geologist who had worked for General Stillwell's road-making endeavour in the north-eastern Indian borderlands during the final years of World War II. A year after Pakistan's independence he suddenly appeared on the scene in Hunza with the suggestion to improve education and livelihood conditions for the poor mountain dwellers.[58] His American-funded projects were aimed at developing human resources in the mountain valley. John Clark, who spent nearly twenty months in Hunza, acted as a colonial traveller and post-colonial moderniser at the same time.[59] His impetus of visiting Hunza was driven by American anti-communism in order to prevent the border states from changing sides and joining their competitor's ideological fold in a perceived hostile sphere of influence. The rationale for the Hunza-Nagar Campaign more than half a century early seems to have been the modernised and updated blueprint for his agenda. His agency and mission conformed well with the tradition conceptually pioneered by Rudyard Kipling: that development was part of the 'white man's burden' by bringing civilisation and modernisation to remote corners of the empire after having conquered and integrated them into the administrative colonial system.[60] As a non-romantic follower of modernisation theory John Clark despised the Shangri-La and longevity myths; he truly believed in the positivist agency of modernisation and effects of technology transfer. John Clark and the Associate Justice of the Supreme Court of the United States, William Orville Douglas, who had visited Hunza in 1951 specified their developmentalist approach to 'save Asia from Communism' in a meeting after their return to Washington later in the year.[61] The comprehensive programme included grassroots development on the village level as well as foreign experts who would live in the village for a prolonged time; training institutions in formal education, health and agricultural extension would be based there on the scene.[62] John Clark summarised five aspects that he believed were the 'fundamentals of Western philosophy': objective consideration of the situation, dissatisfaction with the achieved results, creative confidence in one's own ability, value of the individual, and a sense of responsibility.[63] On the basis of these principles he envisaged the potential for development in Hunza, but failed to implement them further as his criticism of hereditary rule and the suspicion of his pursuit of espionage terminated his project in the autumn of 1951. Local presence and professional expertise in salient fields would be the precondition and remedy that both modernisers were convinced were the only way forward. Their agenda was quite similar to later packages and programmes that were implemented in the Northern Areas from the 1970s onwards by national and international development agencies that depended on a limited number of expatriates and included local professionals.[64]

In Hunza the external process of implementing development packages with a local and village focus started late and can be linked to a concurrent constellation. Pakistan's independence had failed to have an immediate and significant impact in the far-flung mountains. Coincidentally the Bhutto reforms triggered a societal transformation in line with the opening of the Karakoram Highway. The fading-out of the longevity myth is strongly linked with the abolishment of Hunza State, the demise of the last *mīr*, Jamal Khan, and enhanced people's participation in decision-making. Over long periods of time two political factions have been opposing each other in Hunza; both are linked to the autonomous state and anti-state movements that resulted in the termination of *thamo* rule in 1974, a major success for the followers of Zulfiqar Ali Bhutto. A new narrative was required to accompany the transformative phase. A perfect constellation occurred when the Karakoram Highway was opened and offered additional opportunities for further infrastructure development such as building link roads connecting side-valleys and villages to the traffic artery. In terms of supply and marketing, new avenues were opened, business enterprises, land and resource development were initiated, all development efforts supported the expansion of educational and health facilities. The enhanced connectivity and infrastructure facilitated the regular arrival of consultants, experts and civil servants who depended on motor transport and other amenities. Civil society activists profited from the coincidental abolishment of hereditary rule and the advent of comprehensive efforts in infrastructural improvements; their voices were heard in local decision-making processes and negotiations with higher authorities and development agents.

The geopolitical constellation in the aftermath of the Red Army invasion and presence in Afghanistan, the collapse and termination of the Cultural Revolution leading to Deng Xiaopeng's reforms (apostrophised as 'four modernisations' (*si hua*) in China), the pro-Soviet leaning and state-led economy in India had all catapulted Northern Pakistan into a frontline position of the Cold War where the World Bank and western donors attempted to show presence and could achieve a lot with limited investment scope. Pakistan's Northern Areas were located on the line drawn between the Cold War geopolitical pivots of Asian politics that Owen Lattimore had identified in Urumchi and Milan Hauner in Kabul.[65] The Hindukush and Karakoram had matured to become the arena for the performance of international donors that showcased here a community-driven effort for cooperative and equitable development. Projects by the Food and Agriculture Organisation of the United Nations, the Aga Khan Rural Support Programme and the Community Basic Services Programme were inspired by

Colonial agencies introduced new seed varieties for grain and vegetables especially after 1935. Pakistan's Agricultural Department has continued these efforts. A growing number of seeds are purchased in the bazaars. The Pashtun vendor is offering local varieties and Chinese imports. The photograph was taken on 29 March 2019

grass-roots level cooperative efforts that were rooted in previous attempts by local government, but that only became successful with the establishment and widespread coverage of village-based organisations (*tanzīm*) as their nucleus of agency and representation. The resulting development myth focused on isolated, long-neglected, equally poor mountain communities with extremely small land-holdings and meagre livestock-keeping in a subsistence-oriented strategy of combined mountain agriculture with 'traditional' limitations. Hunza and Nagar represented ideal areas with ruling elites that did not enjoy significantly larger landholdings than the average farming household. While *mīr* Jamal Khan claimed landholdings of one hundred and twenty hectares to be in his property in Hunza, his son's ownership was estimated at six hectares less than two decades later.[66] This was the effect of dividing the inherited property from *mīr* Jamal Khan among three brothers and Ghazanfar Ali Khan's final defeat in a long struggle about property rights connected to tenancy occupation of *mīri* lands (*khuṭúkal*) in several villages in Hunza. The villagers had succeeded in claiming their entitlement over farmland that they had tilled since *tham* M. Nazim Khan's invention of this taxation scheme in 1926. When development agencies started their operations in the 1980s Ghazanfar Ali Khan still controlled seven times more property than the average household's landholding in Hunza. But the property-based social stratification is not an expression of a very significant wealth difference in terms of agricultural assets in comparison with other parts of Pakistan. The success of the *tanzīm* model has often been related to the prevalent, less significant differences in land property that

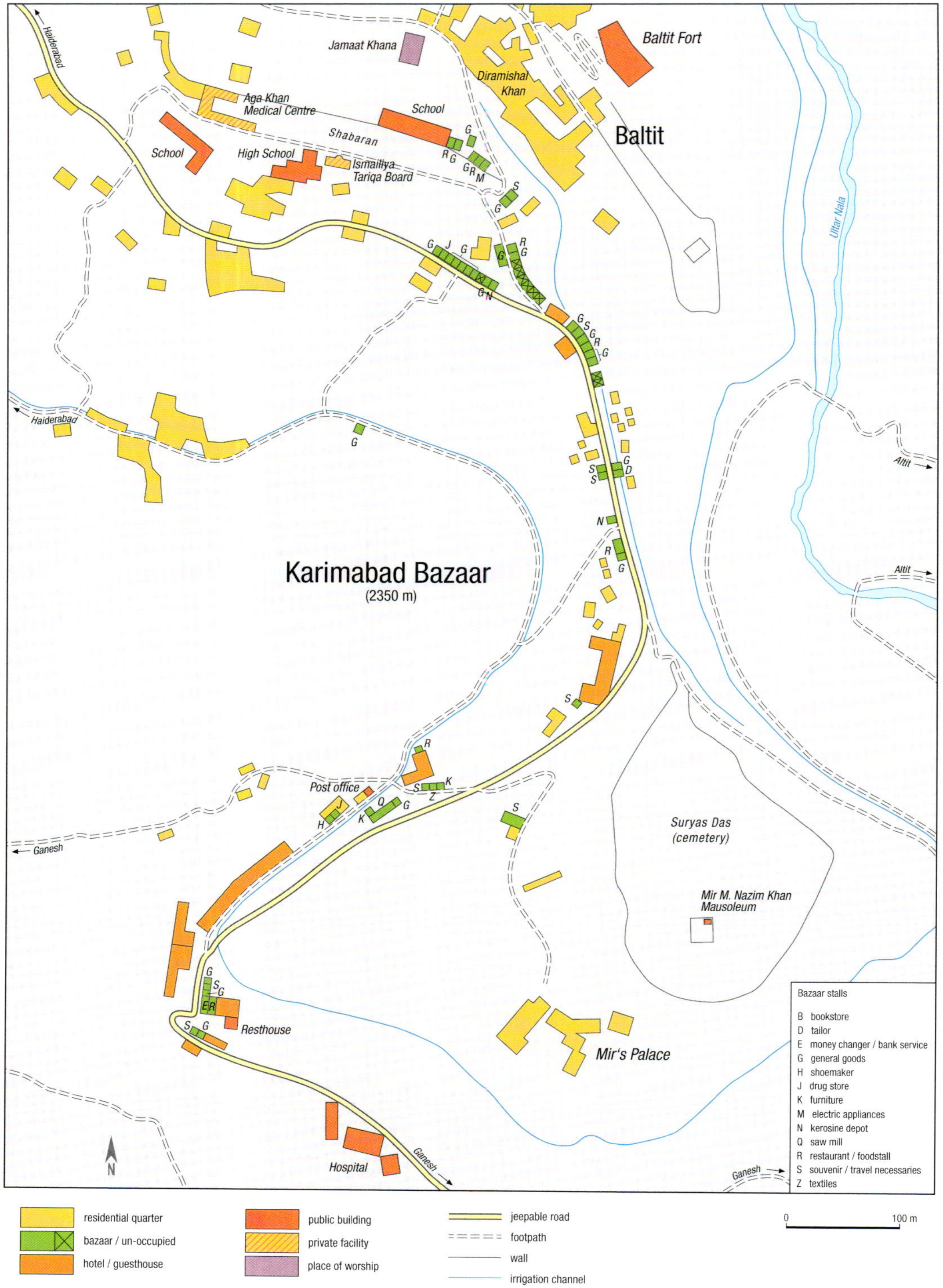

made it easier to organise village communities of mutual interests.[67] Besides educational and health packages the main impetus was laid on agricultural development and resource utilisation. The majority of physical infrastructure projects included irrigation channels and related water works, link roads and land development from which basically all households could profit. Newly developed resources were equally distributed, either by mutual agreements or by lottery. Their contribution to reducing vulnerability was highlighted in all assessments and evaluations that reported a significant uplift in rural living conditions.[68] Savings accrued to such huge monetary assets that economies of scale became feasible in applying for loans. Finally the saving schemes led to village-banking arrangements and contemporary micro-finance banks. A narrative highlighting small land-holdings and paltry animal flocks emerged that suggested local community efforts were able to change and lead poverty-stricken mountain societies into a better future. The tales of 'first harvests' in 'valleys of transition' excluded or at least neglected the contributions made by off-farm income based on migrants' remittances, profits from business and other enterprises, incomes from public and military services that were instrumental for the upkeep of the vast majority of households and from which saving groups derived their assets.[69] The residents of Northern

In 1984 (left), a link road from Aliabad had reached Karimabad and connected the upper villages and hamlets with the Karakoram Highway in Ganesh. Along the link road between the main channel (dalá) and the fortified settlement (Baltit khan), a small bazaar had been established close to the seat of the last mīr of Hunza reflecting the growing economic activities after the opening of the Karakoram Highway. Public buildings were located around the polo ground and along the narrow road.

Pakistan belonged to communities significantly more stratified in social terms than the development myth alleged.

The Northern Areas of Pakistan developed into a role model for the application of a classical modernisation paradigm projected on a remote mountain society. International donors lined up for contributions during the 1980s. The hope and trust in such a development model suggested success by forming village organisations whose members cooperated for a mutual goal of enhanced welfare by mobilising and utilising locally available resources. Somehow external relationships and the households' participation in networks of economic exchange were neglected and faded out in public awareness. The business-school model of implementing packages of general design that are not modified for local conditions and do not take into account specific social constellations supports an externally driven approach to modernisation by reinforcing prevailing structures rooted in heredity rule and derived from 'traditional' power constellations.[70] The classical modernisation approach—formal education, health improvement, out-migration, remittances—to overcome 'traditional' behavioural patterns in connection with infrastructure development and input-dominated agricultural practices fulfils the aspirations of international donors and has made local participants supportive actors in a play of modernisation theory in practice.

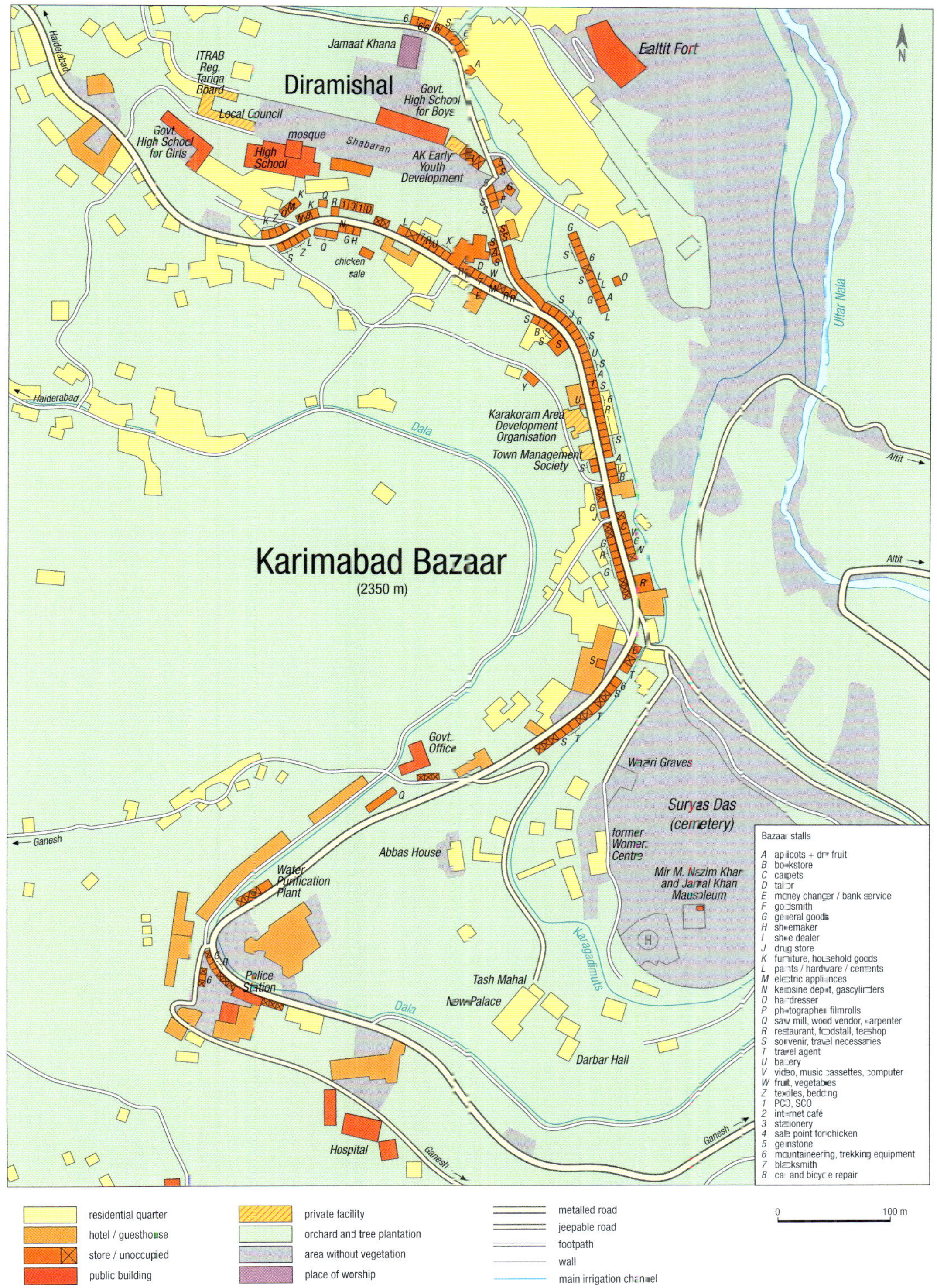

By 2013 (right), Karimabad had developed into the complex structure of a congested road-side bazaar where commercial enterprises dominated. Owing to lack of available building space, the administration and communal institutions relocated most of their offices to Aliabad. Karimabad has become the tourism centre at the foot of the restored Baltit Fort, the most prominent cultural landmark. The Karakoram Area Development Organisation had established a training and promotion centre next to the Town Management Society; on popular demand and initiative both institutions were founded to guarantee people's participation in local decision-making and entrepreneurship. The major hotel had been taken over by the China Road Building Company that coordinated and operated their Karakoram Highway Improvement Project from here. Source: Mapped by Sabine Felmy and Hermann Kreutzmann in September 1984 and September 2013; modified representations based on Hermann Kreutzmann (1989, supplement; 2013d: 16).

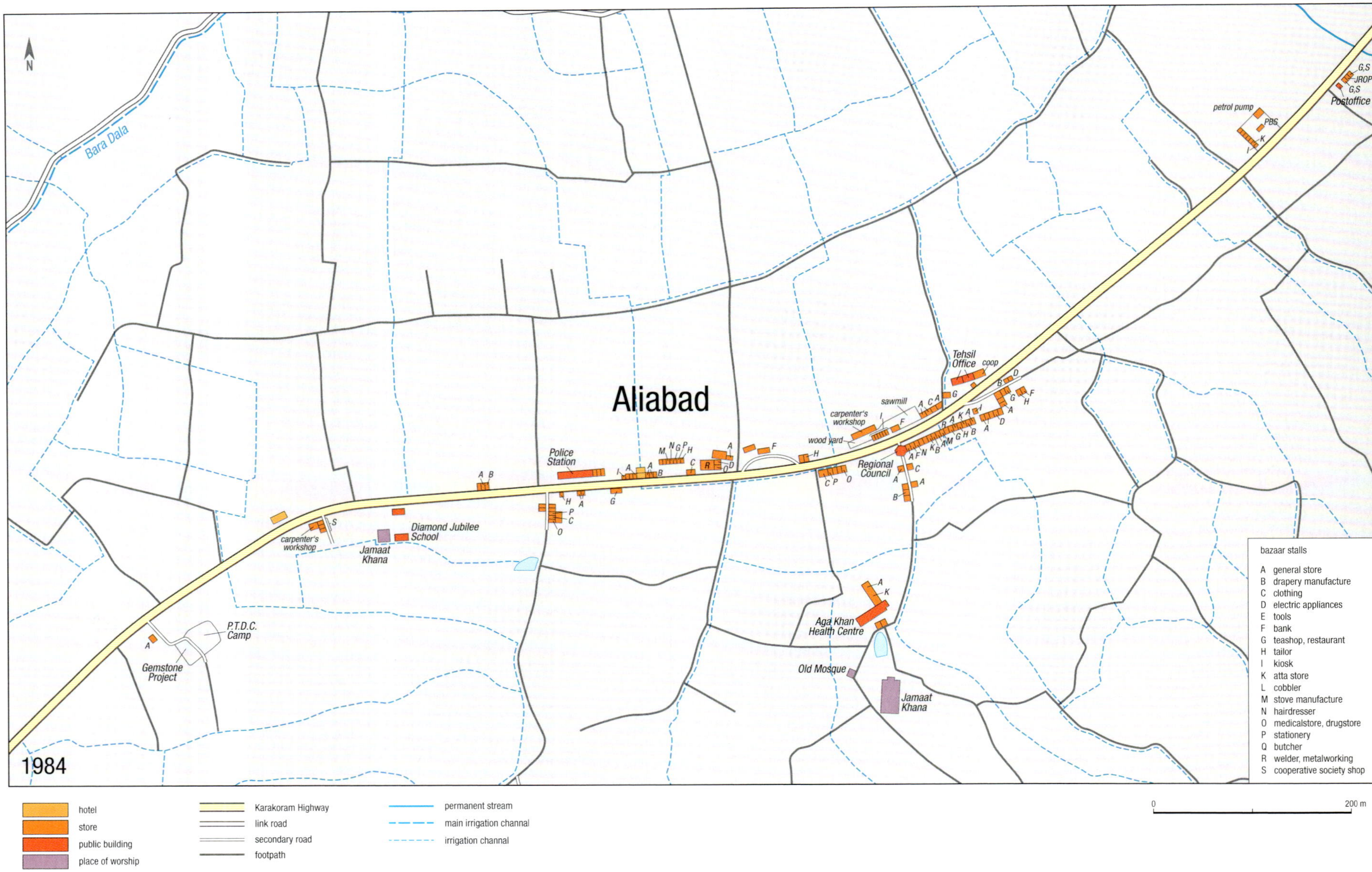

Aliabad Bazaar was an assembly of approximately one hundred shops scattered along the Karakoram Highway. Less than a decade after its opening, Aliabad was in a nascent state of becoming the administrative headquarters for Hunza and the commercial centre of Hunza and Nagar. The process of shifting governmental and Ismaili institutions from Karimabad to Aliabad had already begun when the Regional Council took office there in a prominent location and as one of the few two-storey buildings. Source: Mapping by Sabine Felmy and Hermann Kreutzmann October–November 1984

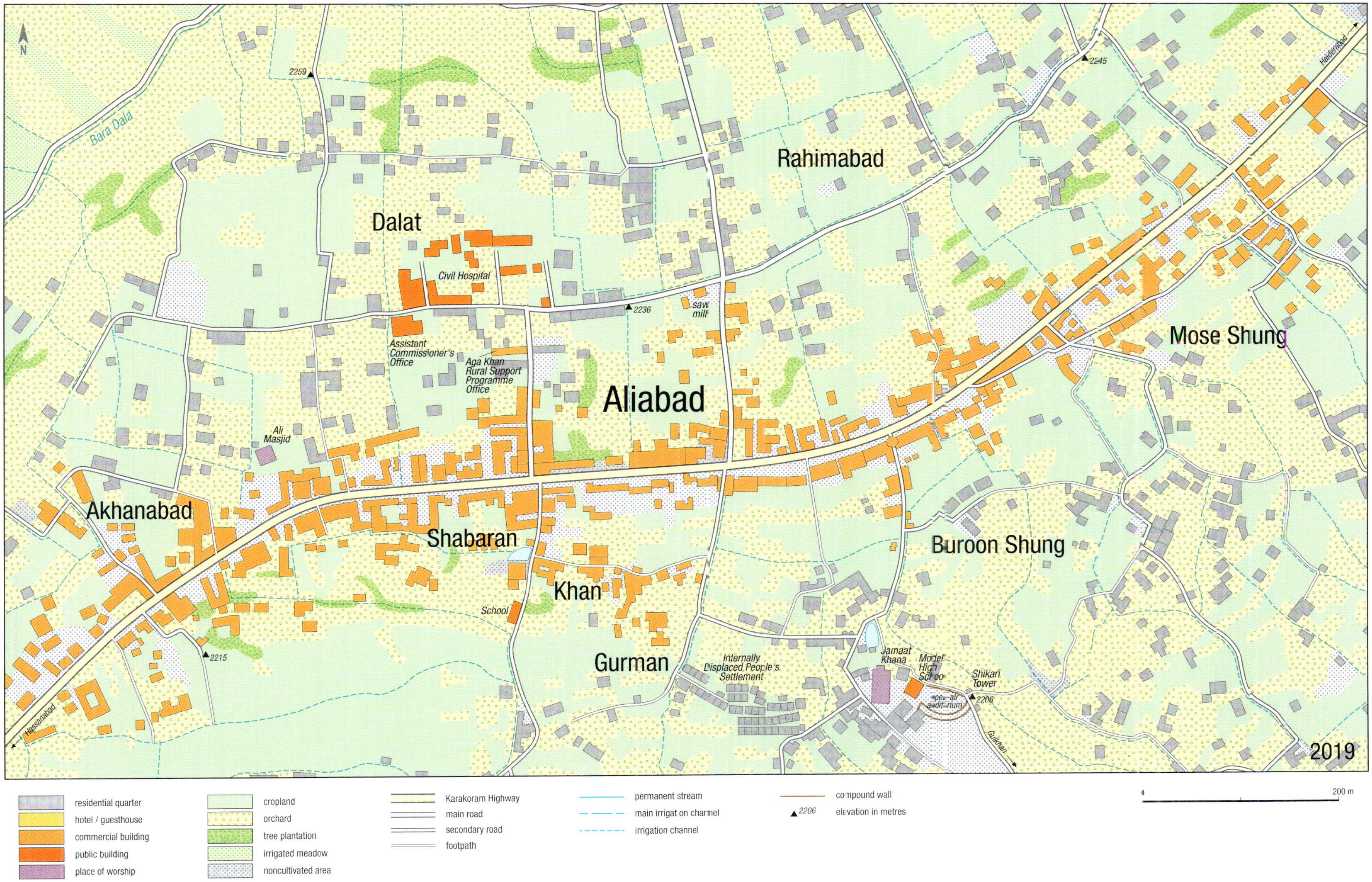

By 2019 Aliabad had emerged as the dominant administrative, commercial and educational hub of the Hunza Valley. Many functions that were previously concentrated in Gilgit Town are now accessible and available in Aliabad. The district headquarters is located here; banks, colleges, hospitals and other important services are available. A wide selection and variety of commodities and goods of regular need and episodic purchase are in stock; agricultural inputs such as seeds and fertilisers, mechanical tools and car repair services are offered as well as a wide variety of imported goods from China. From its bus station, direct buses leave daily for down-country Pakistan and towards the Chinese border. The Karakoram Highway's roadside is completely blocked by office buildings, restaurants, hotels and shops. Commercial markets and multistorey buildings have emerged and spread out to the side lanes. Aliabad provided space for some internally displaced persons in the aftermath of the Atabad landslide disaster; some victims have settled here permanently. Design by Hermann Kreutzmann

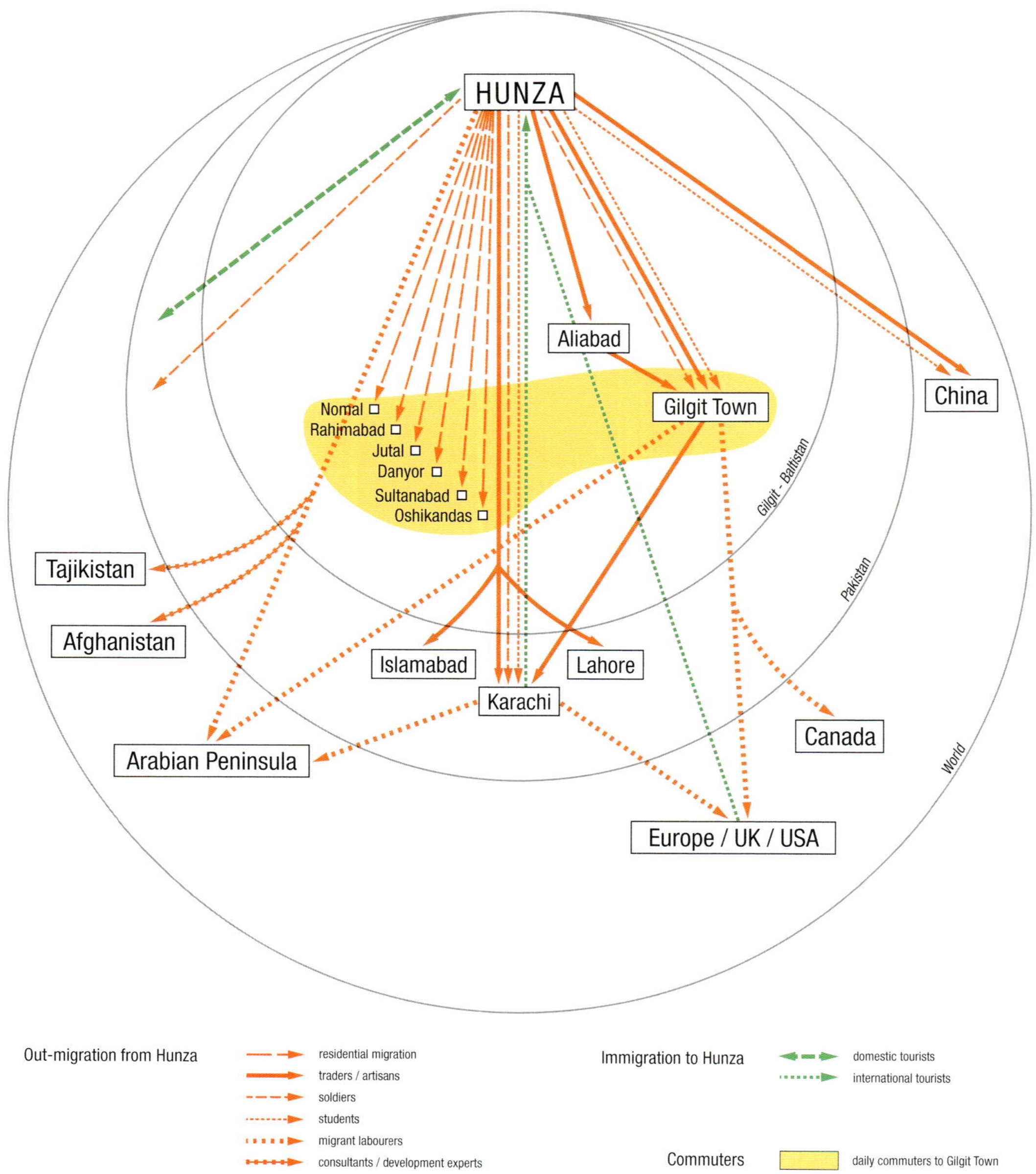

Out-migration patterns have been modified in recent years. Agrarian colonisation has lost in importance, but the villages near Gilgit Town play an important role in circulation for office work, services and entrepreneurs. More urban destinations are covered within Pakistan for education and work. Overseas migration has become more complex; China has become a destination for students and businesspersons. Afghanistan and Tajikistan employ quite a number of development experts from Hunza. Others continue to seek jobs in the Arabian Gulf States and the West; more students search for scholarship opportunities world-wide. International tourism plays a minor role nowadays in comparison to domestic tourism that brings hundred of thousands of visitors to the Hunza Valley, mainly during summers. The provision of border passes for Gilgit-Baltistan's citizens allows easier travel to the People's Republic of China for business and education. Design by Hermann Kreutzmann

The structure of village organisations resembles cooperative ideas and principles of community efforts albeit they have come into being as a top-down strategy based on the consent of ruling elites and external donor organisations. They have undoubtedly benefited quite some households in each and every village, but at the same time interrupted a process of community development in the aftermath of abolishing hereditary rule. International donors have contributed to the re-appearance of hereditary elite members in the boards of their organisations, respecting their roles by assigning privileges to them such as helicopter flights, vehicle support and preferential selection for exposure visits. Priority access to resources and non-meritocracy-based overrepresentation of their family members in well-remunerated positions have been another outcome of these external interventions. The Pakistani state would only offer careers as military officers or influential positions in the bureaucracy that allowed certain fringe benefits and side-incomes. The parallel world of non-governmental organisations and international donors would offer significantly higher monetary salaries and substantial consultancy fees for short-term schemes in a contested mountain region in the final years of the Cold War. Its termination has subsequently led to donor fatigue and shifting interests of development actors. The beginning of their activities was explained as utilising a 'power vacuum' that had emerged after the abolishment of hereditary rule and prior to the functioning of elected local bodies.[71] After three and half decades of western donor-driven development, the baton seems to have been handed over to the mighty neighbour in the north. The bilateral agreement about the China-Pakistan Economic Corridor has put the Hunza Valley into a bottleneck position that has provided a key position and attracted other players to enter the development stage.

Gilgit Town has developed into the major city of Gilgit-Baltistan. From a mere 6,000 inhabitants prior to the construction of the Karakoram Highway it has grown into an administrative and commercial centre of more than two lākh population within half a century. Roads are lined by offices and shops; places of worship have grown into substantial buildings over time. Governmental and non-governmental offices are concentrated here; with its airport and traffic infrastructure Gilgit Town is the hub for all development, tourism and trade activities between down-country Pakistan and Gilgit-Baltistan and for cross-boundary exchange with China. The photograph was taken on 15 March 2019

The map of Taghdumbash Pamir shows Highway No. 314 from Tashkurgan to Khunjerab Pass. The Wakhi village of Dafdar is the last major settlement before the border with Pakistan. On the Chinese side the road is coloured red; after crossing into the Hunza Valley the Karakoram Highway is depicted in green. The map represents mainly the road connection and the topography. The Hunza Valley appears as a comparatively less populated place than Sarikol; only a few villages are identified and given names. Source: Xinjiang Atlas Xeritler Toplimi (2005: 206)

The Chinese occupation of Gilgit-Baltistan myth

It starts in the 1970s when Indian reports of Chinese road-building in the Hunza Valley suggested not only a breach of international law in a disputed area, but claimed that Chinese engineers were disguised army officers intending to occupy Hunza and to integrate the principality into China.[72] Such a suggestion was initially made by *mīr* Jamal Khan who feared that the Chinese would 'eventually take over Hunza'.[73] The Daily Telegraph reported in 1970 that the 'town of Gilgit in Pakistani-controlled Kashmir is reported to have been turned into a military base and storehouse for Chinese military hardware' and added: "The Chinese-built road system is for trade according to Pakistan, but the Indians say the roads can also become invasion routes."[74] The secretiveness of Chinese road-builders, who forced the Pakistan authorities to avoid any press coverage and visits of international observers during the construction phase of the Karakoram Highway, had raised significant suspicions although after its opening and the retreat of Chinese engineers and road-builders traffic and economic exchanges proceeded with normalcy. Until the end of the Cold War the Karakoram Highway was discussed in strategic terms challenging India to build similar roads in Jammu and Kashmir. From a tourism point of view the Karakoram Highway was seen as a 'silk route' or a 'highway of dreams' that from 1986 onwards allowed the passage across high mountain passes connecting Central and South Asia by a unique and comfortable road.

This narrative changed in the aftermath of the Cold War when the alignment of the Karakoram Highway was improved and when Chinese road-building companies transferred their equipment into northern Pakistan. When the respected journalist and political analyst Selig Harrison published an article in the New York Times in 2010 that was titled 'China's discreet hold on Pakistan's northern borderlands', he suggested an 'influx of an estimated 7,000 to 11,000 soldiers of the People's Liberation Army'. Although the Cold War had ended two decades earlier he advised: "Precisely because the Gilgit-Baltistan region is so important to China, the United States, India and Pakistan should work together to make sure that it is not overwhelmed, like Tibet, by the Chinese behemoth."[75] The article made a big stir and was immediately reprinted in the Indian press and defence journals and distributed through websites. Its mes-

The federally-administrated Gilgit-Baltistan presently features ten districts: Astor, Diamir, Ghanche, Ghizer, Gilgit, Hunza, Kharmang, Nagar, Shigar, and Skardu. Design by Hermann Kreutzmann

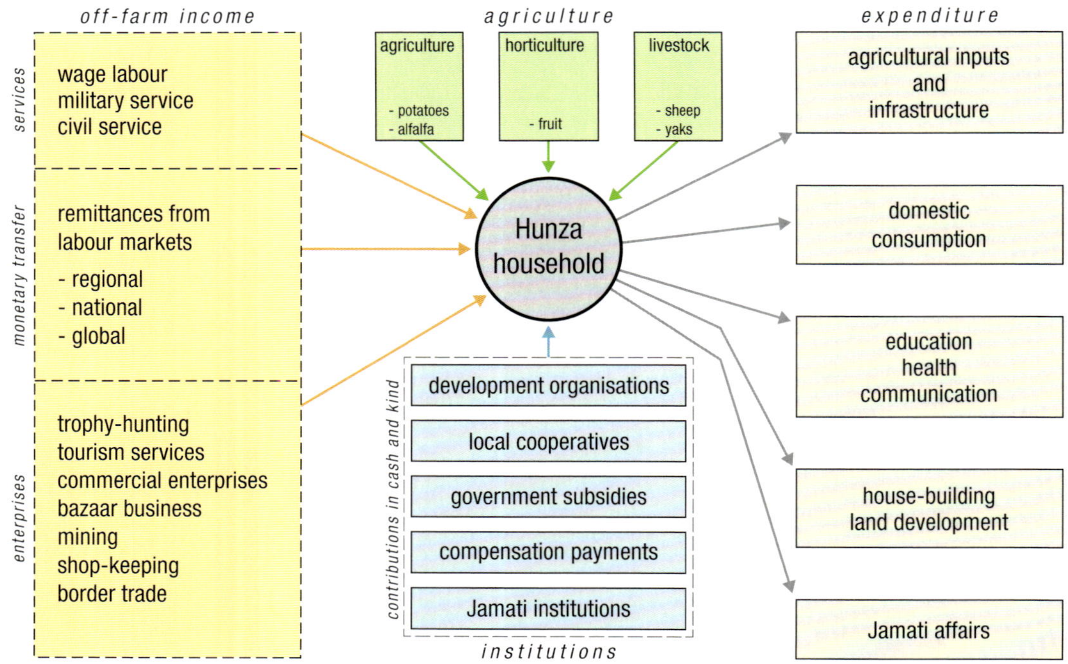

Diversification of income sources has expanded in favour of off-farm income and institutional support in recent years. As higher monetary incomes are generated, the expenses invested for the upkeep of the households, for agricultural inputs enabling cash crop production and especially for health and education have significantly grown. The monetary transfer of non-local household members remains a valuable source augmenting local employment opportunities. Design by Hermann Kreutzmann

sage was gladly taken up by all kinds of anti-Pakistan movements such as the Balawaristan movement and self-declared advocates of the people of Gilgit-Baltistan.[76] Somehow naïve narratives were produced of Chinese soldiers disguised as engineers and road-builders; the Karakoram Highway was identified as a landing ground for aircraft even in places where bends and steep slopes would never allow an aircraft to touch down. The Chinese occupation myth was gladly adopted by protagonists who advocated the independence of Gilgit-Baltistan from Pakistan, who opted for a united Kashmir, or who are financed by dubious sponsors.[77] Their activities were aimed at changing regional politics within Gilgit-Baltistan by self-appointed representatives residing in foreign countries.

The Chinese occupation myth has been rejuvenated in recent years when the China-Pakistan Economic Corridor initiated a number of joint ventures, not only in business and economy, but also in strategy and public security. China Aid has become a welcome development agent in the Hunza Valley, especially after the Atabad Lake disaster when the cut-off villages of Gojal were supplied with valuable commodities for the upkeep of households. When the Pakistan highway patrol was equipped with China Aid-sponsored vehicles labelled as 'CPEC patrol GB police', a discussion commenced again whether salient features of a sovereign state had been outsourced to the mighty neighbour. Pakistan has definitely become more dependent on Chinese loans, investments, technology transfer and development efforts; nevertheless, the number of international and national protagonists has risen who are interested in developing and investing in Gilgit-Baltistan and especially in the Hunza Valley.

The new China Gate marks the border between Gilgit-Baltistan and Xinjiang. Regular traffic plies from Sost to Tashkurgan. The photograph was taken on 10 July 2017

Epilogue

Hunza Valley's external relations have been shaped by exchanges with immediate neighbours and imperial stakeholders. Far-reaching forces of colonial interventions even affected external and internal relations when the principalities in the valley isolated themselves, denying entry to potential outside attackers, unwelcome intruders and undesirable traders. Thus their respective rulers attempted to maintain a high degree of autonomy and independence. Mutual relationships were based on the exchange of gifts and marriage partners in order to strengthen loyalties and diplomatic bonds. Commodity exchange and provision of transport services, political negotiations and military encounters had a limited relevance for internal affairs, but managed to keep external interference at bay. A high degree of formal dependent relationships was established between China and Hunza, based on the exchange of gold dust and gifts. Similar forms of loyalty relations were testified with Kashmir. During British colonial times Kashmir and the British Raj allocated subsidies for the principalities and issued legal documents (*sanad*) to acknowledge rulers appointed by them.

Between China and Hunza a long-lasting relationship has existed for more than 250 years. It has been transformed from a modest gift exchange to a costly and hopefully lucrative partnership within the framework of the China-Pakistan Economic Corridor of today. Roads and exchange corridors have gained in importance since the 20th century. Three major undertakings exemplify the transformation of their bilateral relationship. Within five years after the forced termination of their mutual gift exchange and in the final years of World War II, the Hunza Valley became part of a grand and secret British-Chinese plan to build a road across the Karakoram in less than six months. The desperate state of mutual ideological and strategic interests resulted in unrealistic plans to support Chiang Kai-shek's troops in the fight against communist forces by means of an overland route from British India. This road was part of several attempted British projects to keep Japanese attacks at bay

Cherries have become one of the most valuable cash crops from Gilgit-Baltistan. Initially they were packed and shipped to down-country markets alone; in recent years the local demand of domestic tourists has motivated shopkeepers to offer cherries in popular cardboard boxes in Hunza. The photograph was taken in Karimabad on 7 July 2017

and to stop communist advances. The year 1942 was the beginning of precise road-planning that materialised only during the second phase when the Karakoram Highway was conceived. The strategic role of this modern artery challenged neighbouring countries to invest in mountain roads in order to manifest their territorial claims; especially the Kashmir conflict arena was shaped by these endeavours. Pakistan succeeded as one of the first countries in High Asia to link the vast majority of mountain villages to motorable tracks. As a welcome side-effect the Karakoram Highway became an instrument for regional development. This road is symbolic and visible evidence for Gilgit and Baltistan's strong link with down-country Pakistan, a growing commodity exchange in both directions and a trigger for enhanced mobility. The Karakoram Highway has shaped the position of the Hunza Valley within Pakistan and in its exchange relations with China and Central Asia. Although hardly one percent of bilateral trade flows through this artery, it has had a tremendous effect on Hunza and Nagar. Since the official launch of the China-Pakistan Economic Corridor in 2015, a third phase of bilateral cooperation and dependency has commenced that seems to alter all kinds of relationships. Hectic activities that resemble a gold-rush-like movement into the Hunza Valley have brought significantly multiplied numbers of domestic tourists and have attracted down-country investors to enter the land property market. They expect high returns for stakes in commodity trade, tourism, energy production and claims in mineral extraction. Certain tasks for the development of infrastructure have been outsourced to Chinese companies and entrepreneurs; administrative and security measures have been tightened by Pakistan civil and military authorities. The bottleneck function of the Hunza Valley as the sole terrestrial thoroughfare to link the markets of down-country Pakistan with China has increased interest in the region.

Civil society activists from Gilgit-Baltistan have perceived a growing danger of expropriation and ask whether the China-Pakistan Corridor is a game changer.[78] Living in a liminal state of affairs has created an atmosphere of insecurity and inequality when the granting and provision of constitutional and provincial rights are at stake and when enfranchisement and proper political representation are denied by the central government. It seems to be a retrograde step when existing rights are even curtailed by a new governance order. Proper compensation for the land that was allocated for the Karakoram Highway realignment project has not been fully paid and transferred to the former proprietors.[79] International development agencies have begun to display a 'donor fatigue' when it comes to packages for Gilgit-Baltistan. In this state of affairs Chinese initiatives such as the provision of aid and technical support in the aftermath of natural hazards events have been welcomed; potential opportunities resulting from the China-Pakistan Economic Corridor's regional impact are linked to rays of hope for increasing future welfare and are loaded with scepticism about domestic political affairs and the role local residents might play when it comes to decision-making and participation. When the 'horizons of the China-Pakistan Economic Corridor'[80] were debated, the participants advocated community ownership, payment for ecological services, respect for local labour and skills improvement, promotion of peace and sustainability, preservation of local identities and cultural heritage; these aspects were selected as the dominant fields of attention and given high priority.

The aspects of infrastructure assets and potentials for economic uplift are stressed in the public debate about this new project of China-Pakistan interaction. Local conditions and participation are grossly neglected as the Karakoram area in between has been termed a 'zone of border logistical channel' in the 'long-term plan for China-Pakistan Economic Corridor (2017-2030)'.[81] Its function was restricted to 'resource exploration and ecological conservation'.[82] It could have not been stated more clearly that the areas in between function only as a thoroughfare for container trucks, offer natural beauty for tourism development and can be exploit-

ed for their mineral wealth by external investors. The result might be an additional step to reduce Gilgit-Baltistan's participation in national affairs and economic enterprises. The planned developments project a smaller share of local involvement. The region is envisaged only when natural disasters and earthquakes cause the interruption of the thoroughfare or trigger destructive floods. For local residents an awkward situation has occurred. For decades now the importance of their inherited agricultural resources has been dwindling and major efforts have been initiated to boost formal education and non-agrarian professionalisation. Both efforts have resulted in significant growth of out-migration and a societal setup that is based on multi- or trans-local households. Various shifts in agricultural practices leading to less time-consuming procedures or the involvement of hired labour, to cash-crop fruit and potato production and marketing, and to the shrinking importance of irrigation and pastoralism in a system of combined mountain agriculture have affected the inherited resource base and security system. The present generation is facing multiple challenges in decision-making at a time when potent external investors are trying to get hold of their remaining land property in attractive locations. Connectivity with the main commodity and employment markets has exposed the participants from the Hunza Valley to a variety of newly emerged constraints and repercussions.

The clearly defined hierarchical community configuration of the past that was reinforced by colonial backstopping for hereditary rule and dominating elites has broken up on the one hand, but is felt in nearly all relationships to date. Some representatives of the 'old system' still claim to be privileged in our time. Their bonds and connections with influential parts of Pakistan's bureaucracy and military establishment provide them with some leverage in economic enterprises and political affairs. Their influence is dwindling owing to successful entrepreneurs, civil servants and well-educated

Supporters of Baba Jan are campaigning on the Karakoram Highway for his election. Baba Jan was in jail in Punial when he contested a seat in the Gilgit-Baltistan Assembly. The photograph was taken in Nasirabad on 25 May 2015

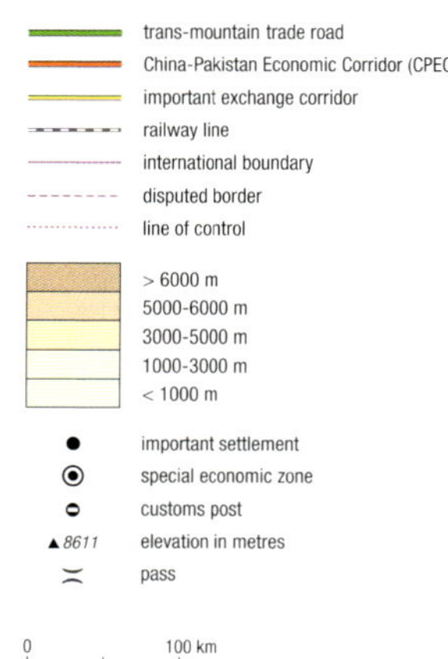

professionals who often have made their fortunes outside the valley. Still it is surprising how strongly felt their influence is when it comes to political representation and occupying influential offices. Pakistan's governmental institutions and non-governmental development organisations have significantly contributed to retarding and obstructing a process that would bring different layers of society at par. Efforts to reduce socio-political inequality were slowed down by supporting overprivileged groups in a colonial manner in post-colonial times. The case of putting people such as Baba Jan in jail for crimes that they could not have committed is an indicator for unequal treatment of political activists, strong influence by traditional elites and the perseverance of withholding democratic rights from the citizens of a disputed area. The liminal state of affairs provides ample leverage for such distortions and interventions.

For a long time maintenance of the Hunza myth was monopolised by representatives of the inherited elites who controlled the hospitality business and thus the relations with outside visitors. Modern forms of travel and communication have destroyed such a monopoly. Interestingly, certain aspects of the previous myth creation have survived and are still utilised to emboss the perception of the Hunza Valley. The attempt to look at present developments and structures from a historical perspective has opened some avenues for linkages that have shown persistence and resilience that are considered in other contexts to be a general feature of mountain dwellers.

Endnotes of pages 475–508

1 See for the wider debate the seminal contributions of Ranajit Guha (2002: 8–9); Walter Mignolo (2011); Eric Wolf (1982). David Spurr (1993) hinted at the negation of the region's own history by emphasising contested spaces and regions as battlegrounds between outsiders only and from the authoritative position of the privileged visitor.
2 See for more evidence the preceding volumes on 'Pamirian Crossroads' and the 'Wakhan Quadrangle' (Hermann Kreutzmann 2015a; 2017a).
3 The chronologies of Haji Qudratullah Beg (1962) and Mirza Mohammed Ghufran (1962/1974) were published in the 1960s, but the material collections began much earlier by their forefathers. Wolfgang Holzwarth (1998; 2006) has contributed valuable insights into sources of pre-colonial and colonial origin; see as well Ahmed Hasan Dani (1987; 1989); Mohomed Nazim Khan (1936); Irmtraud Stellrecht (2006).
4 Edward Frederick Knight (1893: 349); Francis Younghusband (1896: 285–286).
5 Karl Jettmar (1975: 288; 2019: 315) was convinced that the connotation of *ayáśo* was related to an old substratum that under the influence of Ismailism received a reinterpretation as a trick to hide an illicit progeny involving a prince of Shughnan as reported by John Biddulph (1880: 27–28).
6 Reginald Schomberg (1935a: 159).
7 International Herald Tribune, 1 October 2008; New York Times, 1 October 2008.
8 Vasiliki Neofotistos (2011: 292).
9 Vasiliki Neofotistos (2011: 292).
10 Sadaf Firasat et al. (2007).
11 See for example William Martin Conway (1920: 141); Algernon Durand (1900: 162): Gottlieb Wilhelm Leitner (1877); Kumara Menon (1947: 40); Reginald Schomberg (1935a: 106, 124, 145); as a cartographic specimen the map 'Central Asia; comprising Bokhara, Cabool, Persia, the River Indus, & Countries eastward of it' printed in 1838. Travel writers such as Daud Beg (1964: 20); Henry Haughton (1913: 97–102); Ella Maillart (1937: 301); Barbara Mons (1958: 90–92); researchers such as Tahir Ali (1982: 139), Karl Jettmar (2018: 317) and Homayun Sidky (1994: 88) refer to the Alexander connection as well.
12 James Hilton (1933); four years later Frank Capra adapted the Shangri La theme for a movie.
13 Guy Theodore Wrench (1938: 26).
14 Bronislaw Grombchevsky (1888/2015: 150).
15 See the chapter on 'prevalent diseases' by George Michel Giles in William Stephen Alexander Lockhart and Robert Gosset Woodthorpe (1889: 185–213).
16 Emily Overend Lorimer (1939: 146).
17 John Clark (1956; 1963).
18 Emily Overend Lorimer (1936e: 160). See the four articles published by Emily Overend Lorimer in the popular BBC magazine 'The Listener' from July to August 1936 titled: Peasant life in Hunza; Civilisation without shops; An unspoiled people; Thrifty housekeeping in the Karakoram.
19 Emily Overend Lorimer (1936d: 98).
20 David and Emily Lorimer (1936) in their 'Report on expedition to Gilgit and Hunza, India 1934-1935' (RAS/LOR/1/2/2: 2–3).
21 Emily Overend Lorimer (1936f: 208)
22 Emily Overend Lorimer (1936f: 208).
23 Henry Haughton (1913: 222-223) and Arthur Neve (1913: 157–158).
24 The 'Military report and gazetteer of the Gilgit Agency and the independent territories of Tangir and Darel' (Government of India 1909: 15).
25 Reginald Schomberg (1935a: 102); see the remarks by Jürgen Frembgen (1983a 147–148).
26 Reginald Schomberg (1935a: 103).
27 RSAA/SC/SCH/2/47: 11.
28 RSAA/SC/SCH/2/47: 11. During his final visit in 1945 Reginald Schomberg (RSAA/SC/SCH/2/47: 39) disparaged the people of Baltistan who now were assigned the position of being the antithesis of Burusho and Wakhi, although the Burusho still appeared to be the most eminent and superior people of all.
29 Ralph Bircher (1942); Emily Overend Lorimer (1939); Jerome Irving Rodale (1948); Guy Theodore Wrench (1938); see the reflections of Shiuhhuah Serena Chou (2013) about 'agricultural travels and the rise of organic farming discourse' then and now.
30 Daud Beg (1964); John Clark (1956); William Douglas (1952); Jewel Henrickson (1960); Barbara Mons (1958); Allah Bakhsh Rajput (1962); Jean and Frank Shor (1953); Eric Shipton (1951: 14–15); Jean Bowie Shor (1955); Ian Stephens (1953a); John Tobe (1960); Irene von Unruh (1955).
31 Shahid Hamid (1979); see Allen Banik and Renée Taylor (1960); Jay Milton Hoffman (1968); Hermann Schäfer (1978); Renée Taylor (1964).
32 David Lockhart Robertson Lorimer (1953), letters assembled in the file 'Hunza Humbug 1953' (Mss Eur D1168/7).
33 Letter to John R. TeWinkel, Melbourne, Florida, dated 11 November 1953 contained in David Lockhart Robertson Lorimer (1953), letters assembled in the file 'Hunza Humbug 1953' (Mss Eur D1168/7: 9).
34 Letter to Mr. Mart-Henrik Wagner, Stockholm, dated 20 January 1953 (Mss Eur D1168/7).
35 Letter to Hans de Meiss Teuffen, dated 6 March 1953 in response to a letter from 20 February 1953 (Mss Eur D1168/7).
36 Letter to Yvonne Graefin Kanitz [from Freiburg planning to accompany Hans de Meiss-Teuffen on his journey to Hunza], dated 12 March 1953 (Mss Eur D1168/7).
37 The French edition was published a year after the German in Switzerland; Dutch, Italian and Portuguese editions also exist.

Opposite
The China-Pakistan Economic Corridor connects Pakistan with the Central Asian communication and trade networks. In addition to its main line across the Karakoram Mountains the western continuation of the Grand Trunk Road across the Khyber Pass to Kabul and Central Asia could become an important route as well, while India would initially be a potential participant via the Pakistan road network. Given the India-China stalemate in the Aksai Chin question and their border disputes, it seems rather unlikely that a revival of the old Leh route as a motor highway across the Karakoram Pass (5,575 m) would be feasible in the near future. Design by Hermann Kreutzmann

38 Letter of Ralph Bircher, dated 27 December 1952 (Mss Eur D1168/7).
39 Reply by David Lorimer to Ralph Bircher, dated Welwyn Garden City 15 January 1953 (Mss Eur D1168/7).
40 Letter to Philip Unwin, dated 16 January 1953 (Mss Eur D1168/7).
41 Letters by Philip Unwin, dated 22 and 31 December 1953 (Mss Eur D1168/7).
42 Irene von Unruh (1955: 74): "… doch erscheinen mir die Nagarbewohner auch rassisch nicht so ansehnlich und kräftig und nicht so freundlich wie ihre besonders schönen Nachbarn."
43 Ralph Bircher explicitly refers in his letter to the article of Jean and Frank Shor (1953) and claims Robert McCarrison and David Lorimer as his witnesses. The latter would not be, but his wife would qualify. A copy of the letter from Ralph Bircher to Mir M. Jamal Khan, Erlenbach-Zürich 8 September 1954 is kept in the archives of the German Alpine Museum (DAV EXP 2 SG 100).
44 DAV EXP 2 SG 100.
45 Wolfgang Pillewizer (1960: 80-84); Naohiko Harada and Atsuko Miyoshi (1963: 14); see as well Gottfried Neureuther (1961: 1643).
46 DAV EXP 2 SG 100.
47 DAV EXP 2 SG 100.
48 DAV EXP 2 SG 100.
49 Letter from Paul Bernett to Mir Jamal Khan, dated 8 October 1954 (DAV EXP 2 SG 100).
50 Letters from Paul Bernett to Mir Jamal Khan, dated 5 February 1954 (DAV EXP 2 SG 100) and 3 April 1955 (DAV EXP 2 SG 99).
51 For example Jay Milton Hoffman (1968); Hermann Schäfer (1978); Renée Taylor (1964).
52 Letters by Irene von Unruh to David Lorimer, dated 25 June 1956 (Mss Eur D1168/7).
53 John Tobe (1960: 368).
54 Hermann Schäfer's (1978) book about a 'people without disease' might be considered to represent a sizeable group of like-minded publications; see for critical remarks Jürgen Frembgen (1983a, b; 1989); Hermann Kreutzmann (1989c).
55 The comprehensive programme is summarised in 'Letter No. 13-T of 1936, dated Camp Chilas, the 21st February 1936 by Major George Kirkbride to the Assistant to the Resident Kashmir' (IOL/P&S/12/3294); Administration Report for 1936 (IOL/P&S/12/3288).
56 Mirza Hassan and Qudratullah Beg in Lorimer-Personal Records (SOAS: PP MS 66/01/01/R3); see other contemporary emphasis on animal husbandry by Reginald Schomberg (1936a: 50–59); Eric Shipton (1938); see Hermann Kreutzmann (2000b; 2004a).
57 The estimated budget expenditure is detailed in the file 'Future of Gilgit Agency: position of Kashmir Darbar, ceremonial Nagars, withdrawal of Kashmir troops from Gilgit' (IOL/P&S/12/3287: 52–68).
58 John Clark (1956: 4).
59 John Clark (1956; 1963).
60 See for a detailed discussion of Rudyard Kipling and his impact Hermann Kreutzmann (2015a: 52; 2017a: 39–40).
61 William O. Douglas (1952: 315).
62 William O. Douglas (1952: 314–318); John Clark (1956: 265–270).
63 John Clark (1956: 267–268).
64 During his final days in Hunza John Clark (1956: 232) expected a visit by the Ford Foundation Exploratory Committee in search for appropriate development projects. The Ford Foundation remained a long-term actor in Pakistan's development packages; its representative became the head of the Aga Khan Foundation's development packages later on.
65 Milan Hauner (1989); Owen Lattimore (1950); see Hermann Kreutzmann (2013a: 161–162).
66 Frank Saunders (1983a: 71–110); Renee Taylor (1964: 46); see Hermann Kreutzmann (1989a: 177–179).
67 Akhter Hameed Khan (1985: 248–259). Geoffrey Wood characterised the social differences in Gilgit-Baltistan as 'class differences (stratification) rather than exploitative class relations' (Geoffrey Wood, Abdul Malik and Sumaira Sagheer 2006: 45).
68 Arif Hasan (2009); Mahmood Hasan Khan (2009); Shoaib Sultan Khan (2009); Geoffrey Wood, Abdul Malik and Sumaira Sagheer (2006); Frank Saunders (1983; 1984); World Bank (2002, 2011).
69 The promotion of development achievements used such phrases; see Geoffrey Wood, Abdul Malik and Sumaira Sagheer (2006).
70 See Faisal Devji (2014: 57) for the general design of the Aga Khan Development Network and the lack of spirituality within Ismaili religious institutions.
71 Akhter Hameed Khan (1985: 248–252); see as well Geoffrey Wood, Abdul Malik and Sumaira Sagheer (2006: 42–44).
72 Times of India 26 June 1970; FCO 37/694.
73 Minutes by Ian Sutherland, Head of South Asia Department in the Foreign Commonwealth Office, dated 11 May 1970 (FCO 37/694).
74 Daily Telegraph, 21 April 1970; FCO 37/694.
75 New York Times, 26 August 2010.
76 See the extensive discussion of these issues in Hermann Kreutzmann (2015c).
77 See for example Abdul Hamid Khan (2010); Senge Sering (2012); similar anti-Pakistan activities are promoted by the Institute of Gilgit-Baltistan Studies in Washington or the European Foundation for South Asian Studies in Amsterdam pretending to be neutral think tanks for policy advice and representing the people of Gilgit-Baltistan; see Hermann Kreutzmann (2015c).
78 High Asia Herald, 13 September 2017; Pamir Times, 24 June 2019.
79 Pamir Times, 24 June 2019.
80 Aga Khan Rural Support Programme (2017).
81 Dawn, 21 June 2017.
82 Dawn, 21 June 2017.

Glossary of important terms and toponyms

Abbreviations used in the glossary: Arabic (A); Burushaski (B for *burúśaski*); Chinese (C for termini in *hanyu pinyin*); Gujarati (G); Hindi (H); Kirghiz (K); Khowar (Kh); Persian (P); Russian (R); Shina (S for *ṣinaá*); Tajik (T); Turk languages, such as Uigur and Uzbek (Tu); Urdu (U); Wakhi (W for *x̌ikwor*)

abi	P	irrigated
abodí	W	prosperity, from Persian *abadi*
abrīšam	P	silk
afsar	W	forced labour demanded by a ruler
ag üy	K	white yurt
aġél	B	livestock pen, corral for keeping sheep and goats during the night
akaabírtiṅ	B	member of the elite
akoi	K	Kirghiz yurt, yurt community
ālim, ulemā	A, U	religious scholar
alsī	H	linseed (*Linum usitatissimum*)
ālu	B, H, U	potato (*Solanum tuberosum*)
amban	C	a Manchu word for a high official, used during the Qing Dynasty for a Chinese resident in Altishar
ambór	W	grain store
amir	P, W	ruler, king
ānā	U, H	one sixteenth of a rupee
aqiq	W	ruby
aqsaqal	Tu, P	literally: white beard; elected and accepted representative of a rural community, representative of an ethnic group or social sub-group
arbāb	A, P	village elder, representative of a village community
arġún	B	morganatic wife, child of a morganatic marriage
arġún bépay	B	crossbred yak (*zoi, zomo*)
arġúndaro	B	being a child of a morganatic marriage; offspring of a marriage between partners of unequal social rank
arsan	P	millet (*Panicum miliaceum, Setaria italica*)
asāmī	U	cultivator, brief in land assessment reports for tenant with right of occupancy (*asāmī-i-mustaqil*)
aṣtïr	W	Bactrian camel (*Camelus bactrianus*)
at	K	horse
āṭā	U	bread
avlād	A, W	lineage
ayáśulum	B	heavenly, coming from heaven (*ayáśo*)
ayıl	K	camp, sometimes written as *aul*
aylóq, aylaq	W, P	summer pasture and the related settlement
báare sis	B	member of Burusho society
bādām	P	almond
bād-e Gojal	W, P	the (strong) wind of Gojal, (B. *tiṣ*)
badi pon	S	big road, Karakoram Highway
badli	P	exchange; here: the exchange of hostages, from P. *badal*
bahār	P	spring
bālā	P	upper (part of a settlement)
baldá	B	load
baldákuyo	B	load-carrier
baldí	B	veranda, open terrace with some protective walls
bām-e dunyā	P	roof of the world
bang	W	hashish (*Cannabis indica*)
bap	B	tax, tribute, due
bar	B	tributary side valley, gorge
bar	W	door
baráai	B	fairy (*parí*, in Shina *barāi*)
barani	U	rain-fed agriculture
báṛi pon	S	often spoken as *badi pon*, big road
barj	U	Himalayan birch (*Betula utilis*)
barú	B	buckwheat (*Fagopyrum esculentum* or *F. tataricum*); sweet (*císki*) or bitter (*ġaqáy*) varieties
basíkiṣ	B	orchard
baṣ	B	bridge
baṣ-talénum maqsoó	B	the people on the other side of the bridge; the meaning is Gojal seen from Central Hunza; it is the section beyond the Gulmit bridge
bap	B	tax
bay	K	head of a community
bay	B	panicled millet (*Panicum miliaceum*)
baÿ	W	garden, orchard
be-qasab	W	cloth
be-paderan	P	fatherless
bépay	B	yak (*Bos grunniens*)
bérico	B	blacksmith
bïc	W	adult female goat
biík	B	willow (*Salix*)
bilás	B	female demon (W. *bəlhas, bəlas;* Sariqoli *bïlas*)
bilís	B	to weed crop fields for thinning and fodder-making
bïng car	W	winnowing of harvested grains
biṭán	B	shaman (Shina *ḍayál*)
bixunemon	W	refugee (derived from T., P. *be-xonumon* = without shelter)
bóphao	B	sowing festival in spring related to barley

borī	U	sack, a measure of three *maund*, regular sack for potatoes
borwar	P	load-carrier
bozganj	P, T	shell of pistachio nut
brənj	W	rice
buġór	B	female court dancer
bukák	B	broad beans (*Vicia faba*)
bulak	K	source
buna, bunegi	W	new settlement, process of establishing a new settlement
burui	W	buckwheat (*Fagopyrum tataricum*)
burúm	B	white
burúui	B	first watering of field after the seeds have germinated
buxāri	B, P, W	fire place, stove (U. *buk̲h̲ārī*)
bəqlá	W	broad beans (*Vicia faba*)
bərənǰ	W	rice
ċak	B	sluice, opening from irrigation channel to the field
ċhíl	B	water
ċhíle yáṭkuin	B	water inspector, guard
ċhilgálas	B	controller and water warden in the irrigation system
ċon	B	slave
čal	W	desert, muddy area after rain
čaras	H, U	hashish (*Cannabis indica*)
časma	P	source
čat	W	bovide (excluding yak)
činaei pon	S	road to/from China, Karakoram Highway
čikar	W	willow (*Salix*)
čikíṣ̌, čkíš	B, W	a store which can be sealed or locked
čikor, čakor	W, H	partridge (*Tetragallus himalayensis*)
čít	W	printed cotton cloth
čowan, čwan	W	apricot (*Prunus armeniaca*)
čpan	W	long robe
čujan	W	steel (P. *fulād*), cast iron
čuṭ	W	partition
čəkmən	W	soft cloth, long gown
čəy̌	W	lamb
ćarbú	B	watchman (*ćaukī-dār*), harbinger
ćauk	U	square, crossing, open place (*chowk*)
ćaukī-dār	U	guard, watchman, sentinel
ćha	B	millet (*Setaria italica*)
ćhap	B	meat
ćhar-pā'ī	H, P, U	bedstead
ćhaṣ	B	sea buckthorn (*Hippophae rhamnoides*)
ćhíkas	B	garnet
ćiṭ	H, U	small piece of paper containing a message
ćūlhā	U	fireplace, hearth, here: household
ç̣harwélu	Kh	official, taxation officer in Chitral
dadui	C	unit, production brigade within a people's commune
dāl	U	pulses
dalá	B	main irrigation channel in Hunza
darband	P	lock, gate, closure, narrow gap, control post
darbār	P, U	court, royal audience
darċ	B	threshing floor
dárċe tiṣ	B	winnowing
dāroga	U	overseer, inspector
darwāza	P	door, gate
dar-xāna	P, U	palace, seat of the Ismaili Imam
daryā	P	river
daswān	H, U	tithe
dašt	P	plain, steppe
dīdār	P	appearance, public religious blessing
dīdārgāh	P	area reserved for meetings with the spiritual leader to receive *dīdār*
dildúng	W	platform dedicated to food preparation
dildúnge ben	W	special seat close to the cooking platform
díltar	B	buttermilk
díltare taríṅ	B	inflated skin for churning
dirán, ḍir	B	secondary irrigation canal
diyór	W	homestead, permanent settlement
diyorčí	W	villager, settler in homestead
do-faṣlī	P, U	two harvests in one year
dotor	W	string instrument (P. *dotar*)
drukṣ̌	W	oxen
dut	W	rope bridge
dəmá	W	wind, storm
dəmbá	T	fat-tailed sheep
ḍaámal	B	kettle drum, membranophone
ḍaḍáṅ	B	cylindrical drum, membranophone
ḍāk	H, U	post station for conveyance of letter bags, relay station for changing horses, porters or runners
ḍáṅalaṭhás	B	female ghost, demon, giant
δart	W	manure, dung
δīm	W	kaolin, china clay
δiy	W	yoghurt, soured milk
δiydor	W	woman responsible for milk-processing
ek-faṣlī	P, U	single harvest
faiz	P	blessing
faráaji	B, P	treasurer, tax collector (U. *farrāsh*)
firmān	P	order, edict, decree (U. *farmān*)
firz	W	birch (*Betula pamirica*, *Betula utilis*), birch bark
gáam	S	village
gal	B	juniper (*Juniperus*); bridge made from plaited juniper twigs
galgí	B	fin, here a short ridge in irrigated field

gan	B	path
gan-giánum	B	legal deposit, tribute
garaḍí	B	a gondola fixed to a steel rope to cross a river (U. *garārī*); small wooden gondolas were common for the transport of individuals and light goods
gark	B	broad peas (*Pisum sativum*)
gʰī	H, U	butter fat
gináni	B	harvesting festival related to barley
girám	B	cattle-shed outside the fortified *khan*; at the same time the meaning of lineage
girán	B	woven basket from willow twigs, container for carrying harvest products from the field and pastures to the homestead
góċil	B	irrigation channel
gongshe	C	people's commune
gorá	W	cradle
gorezgah	P	refuge, hide-out
gaíċ	B	Burusho term for a member of the Wakhi group
gaíski	B	Burusho term for the Wakhi language
gaǰáali	B	Burusho term for a member of the Wakhi group
gāl	U	irrigation channel
gar	B	wheat (*Triticum aestivum*)
gaśpúr	B	prince, member of the ruling elite
gaṭí	B	literally: hut; in this case: lineage
gáuço	S	irrigation channel
gərwá	W	cradle
ġamú	B	glacier
ġáne phíṭi	B	tax paid in bread
ġaláam	B, U	slave, servant
yazal	W	ode, poem
yazg	W	green vegetable
ylom, yəlom	W	slave
y̌anj	W	open store
y̌ef	W	fireplace in the house
y̌el	W	livestock pen, corral for animals
y̌av	W	literally: underground pit for the storage of grain, in this case: store above ground, constructed with stones and mud
y̌az	W	firewood
y̌azá	W	capsule of poppy (*Papaver somniferum*)
y̌ədím, y̌dím	W	wheat (*Triticum aestivum*)
y̌īw	W	cow
ha	B	house, number of houses (*hakíċan*)
hākim	P	local representative, judge
hále ġaṣ	B	trail of the fox; here the place name Halegush between Atabad and Goshben
har	B	side valley, gorge

harí	B	barley (*Hordeum vulgare*)
hárki	B	ploughing
hárkiṣ	B	cropland
havildār	H, U	corporal, sergeant
heṭiṅ	B, S	area in village lands for free grazing
hír-bilas	B	one-eyed giant
Hunzae róɔṭ	B	Hunza Road
huyéltarċ	B	shepherd
huyes	B	sheep and goats
ḥāǰ	A, P, U	pilgrimage to Mecca
iǰín	W	felt mat produced from sheep wool
ikta	W	shirt for men
il	B	sluice between channel and field
ilbán	B	animal taxes (derived from Mongolian language)
imām	A, P	religious leader
imām-barga	A, P	Shia mosque, sometimes *imām-bāṛā*
ishan, īshān	R, P	religious Ismaili leaders and clergy such as *pīr* and *xəlifá* (from P. *īshān*)
istín	W	pillar, support column
íċit	B	village land
íl	B	secondary irrigation canal
írkiṣ	B	secondary irrigation channel
ʿid, ʿīd	A, P, U	Muslim festival, end of the fasting month (*ʿīd-ul-fiṭr*)
jáalo	B, S	raft (in colonial reports sometimes *jhula*)
jaalwáan	B	rafts man
jalsa	P, U	assembly, gathering
jamáat	B, W	community, family, spouse, community meeting, group
jamáat khana	B, W	Ismaili community centre (from P. *jamāt-xāna*)
jarpá	B	poplar (*Populus alba, Populus pamirica*)
jaylö	K	summer pasture
jılga	K	valley
jirga	P, U	assembly, council
jum'arāti	U	Friday evening, migrants returning for the weekends to their village
jut	K	storm
juú	B	apricot (*Prunus armeniaca*)
jakún tom	B	tree without fruit; 'donkey-tree'
ǰagīr	P, U	tax-free land, fiefdom
ǰagīrdar	P, U	beneficiary of a *ǰagīr*
ǰamadār	P	official, representative
ǰul, ǰhul	P, H, K	saddle, horse-blanket
ǰut	S	meadow, pasture
ǰərab	W	Pamirian sock
ǰəráv, ǰərab	W	river, valley
ǰəngál	W, H, U	forest, tree plantation
kachchā	U	crude, unrefined; here clay-built
kaḍáko	B	headman; superior representative of four *roóm*
kāfila	U	caravan (from P. *qāfila*)

kafir	U	infidel, non-believer (from P. *kofr*, A. *kufr*)
kāh-pūli	P	pasture taxes (from P. *kāh* = grass, *pūl* = money)
kalān	P	big, large
kambal	H, U	woollen blanket
kəmpál	W	woollen blanket
kanāl	U	spatial unit equalling 505 square metres
kandir	K	hashish (*Cannabis indica*)
kaprā	U	clothes
karagáḍimuć	B	wooden pillar to support water
karawal	K	military post (*karaul*)
karoṛ	U	ten million or one hundred *lākh*, E. *crore*
karun	C	Manchu expression for guard-post (C. *ka-lun*)
kār-i-begār	P	forced labour, corvée
khaám	P	coarse material (from P. *xām* = raw), cotton cloth
khái	B	fortified village, mainly used in Nagar and Yasin
khan, khanánć	B	fortified village/s
khan	P, U	honorific title, leader
k̲h̲azāna	P, U	treasury
khiraǰ	P	pasture taxes (from P. *xarāǰ*= tax, tribute)
khoǰa	G	denomination for Ismailis who converted from the Hindu Lohana caste to the Ismaili faith, originally from Gujarat
khuṭúkal	B	*mīri* land that was let to farmers who deliver a fixed amount in kind to the *tham*
kibitka	K, R	yurt, felt tent, tent community, household, nowadays also in use for the Central Asian mud house
kirpas	K, P, H	cotton cloth (from P. *kirpās*, H. *karpās*)
kïk	W	source, fountain
kïnǰ	W	house entrance
kla	W	small livestock (sheep and goats)
klae ǰel	W	night pen for sheep and goats
köchmön	K	nomad, pastoralist
köl	K	lake
kolkhoz	R	*колхоз*, *kollektivnoe khozyaistvo* (= collective economy)
kotal	P	pass
krəst	W	pelisse, leather gown from sheepskin, posteen (P. *pustīn*)
kroš	W	chickling pea, grass pea (*Lathyrus sativus*)
ktic, kətíč	W, T	simple hut, shelter in the high pasture
kuč	W, T	seasonal migration to and from summer pastures
kum	W	palate
kurpátiṅ	B	elder, experienced person
kut	B, W	lantern roof
kəriyár, kəryár	W	mutual neighbourhood support in times of planting, harvesting etc., remuneration in food which is consumed in community
kərk xun	W	chicken house
kərpá	W	quilt, cotton blanket
kətíč	W	hut
lākh	H, U	one hundred thousand, E. *lakh*
lambardār	B, U	village representative; the superior *a'lā-lambardār* is equivalent to *kaḍáko*
lojǰwár	W	lazurite (*lapis lazuli*)
lup žarž	W	'big' foster relationship
mahalá	B	hamlet, village
maidān	U	open field, plain, polo ground
makái	B, H	maize (*Zea mays*)
makān	A, U	house
mákućim gan	B	middle path
mal	K	livestock
mál-i-sarkár	P	tithe paid by Ismailis to the Aga Khan
maltáṣ	B	butter fat
mamú ćhil	B	glacier meltwater, glacier milk
man	H, U	a measure of 40 *sər*, a *maund* (37.32 kilogrammes)
maqsòo	B	section of society and/or population
mara	W	store
marúć	B	gold panner (S. *marúuć*)
māshā	H, U	gold equivalent of about one gram; twelve *māshā* equal one *tolā*
mašak	W	mung bean, green gram (*Vigna radiata*)
maská	B, P, W	fresh butter
matám saráa	B, U	Twelvershia mosque, house of mourning; sometimes *matám sarái*
maulai	G	the opposite of *khoja*, term used to designate Central Asian Ismailis
maulawī	A	preacher, learned man, here: Shia religious leader
mauza	P	village
mazar	P	shrine, mausoleum
mehmon xona	P, W	guest-room
mehtar	P, U	prince, high-ranking personality, another meaning is sweeper; in this case: from P. *mihtar*, title of the ruler of Chitral
merik, mrik	W	cream
mimón	W	guest
minzu	C	nationality, minority, ethnic group, comprehensive term for all non-Han-Chinese groups in the People's Republic of China
mīnār	H, P	tower, pillar

mīr	B, P	local ruler
mīri	P	ruling class, elite
mīrzāda	P	prince, heir
mīskal	P	gold weight, measure
mīr	W	apple (*Malus sylvestris)*
moʿāf	P, U	exempted, privileged (P . *moʿāfijat* privilege; colonial U. *muafed* exempted)
mon, móiṅ	B	shrub (*Artemisia fragrans*) used as fodder and fuel
mos	B	mudflow
moʿtabar ādmī	A, P, H	respected person, in colonial reports *motabir* or *mehtabar*
mozdur	P	hired labour
mozduri	P	wage labour
muḥalla	A, U	hamlet, part of a village
muḥarram	A, U	first month of the Islamic calendar
mujaheddin	A, P	struggler, fighter for a cause; from Arabic *mujāhidīn*
mukhi	G	leader of the prayers in *jamáat khana*
mullā	K, P, T	Sunni clergyman
mum	W	grandmother
munshī	A, H, U	secretary, translator, scribe
murīd	A, U	follower of a *pīr*
muzdūr	U	load-carrier, wage labourer
muzdūri	U	wage labour
məǰək, məžək	W	variety of peas (*mazhak*)
naʿib	A, P	state representative, administrator
Nai Tareekh-e-Chitral	P	new history of Chitral
nā-farmān	U	disobedient
naik	U	corporal
nālā	B, U	side valley, rivulet
namdā, namad	H, P	felt, large woollen blanket, E. *numdah*
nau rōz, nouruz	P	new year, spring
nasawáar	B	chewing tobacco (U. *naswār*)
nazr, názar	B, U	gift, present, offering
nikard	W	space designated for the consumption of food
noṣpətí	W	pear; *Pyrus communis*
očṡo	B	guest, visitor
očṡo basá	B	guesthouse, inn, shelter, resthouse
ovring	W	suspended mountain path, fixed in mountain cliff, difficult passage, sometimes termed *aviring, avaring*
pāīn, pāyān	T, P	down (for the description of the lower part of a settlement)
paipakh	K	socks made from felt
Pak-Cheen dosti	U	Pak-China friendship
pakkā	U	genuine, perfect, finished
pakkā pul	U	suspension bridge
palās, palós	B, W, K	woven mats from yak and/or goats' hair, used as carpet for bedding

pamir yingdi	C	Pamirian flute made of eagle bone
pamér, paméer	W	extended pasture area, Pamir
paṛā'o	U	stage, halting-place, encampment
pardah	P, U	screen, veil, cover of the face
parí	B	fairy
parīi, parri	B	suspended mountain path, scaffold
paṭṭū	U	coarse woollen cloth
paṭwārī	U	recorder of land property
pāu	P, W	weight, about 450 grams
pay	W	yoghurt
perien	W	difficult passage, scaffolded mountain path (*ovring*), sometimes referred to as *parīi* or *rafik*
phárčin	B	local cap made from homespun sheep wool
phári	B	water reservoir
phíṭi	B	bread loaf, prepared in a cast-iron baking tin
phurúṅ	B	cultivation bed in the field
pīr	P, U	religious dignitary
pustīn	P	posteen, pelisse
pundit	H, U	learned man, from Sanskrit *paṇḍita*
pus	W	adult female sheep
pəxtá	W	mulberry (*Morus nigra, Morus alba)*
pəy	W	wooden shovel
pïx̌t	W	stable, shed
qal'a	P	fortified settlement, sometimes *qil'a*
Qaśgare gan	B	the road to Kashgar, a name used in Hunza for Karakoram Highway
qaralí	B	child of noble descent
qhaám	B	cotton cloth
qhalipá	B	Ismaili lay priest, leader of rituals
qhazína, qhazaná	B, P, T	treasure-house, stock
qir	W, T	mountain ridge
qišlóq	W	homestead, permanent settlement
qurút	P, W	dehydrated buttermilk
qətór	W, T	lineage of Wakhi
rāhdāri	P, H	transit fee, escort, in this case: passport, travel document
ra'īyat	P	commoner, peasant, tenant
rāǰa	H, S, U	local ruler
rajaáki	B	forced labour, communal work for the ruler
rānī	H, U	wife of *tham*
raž	W	platform for sitting, working and sleeping
renmin gongshe	C	people's commune
renmin ribao	C	people's daily
rezg	W	saltish
ricn	W	roof hole, light hole, smoke hole
rin	W	cupboard
rishwat	U	bribe

rizabárg	T, W	poplar (*Populus pamirica*)
robat	K	pass
roóm	B	social group, lineage
roṭi	U	bread
roxsat	P	permit, permission to leave
rukhṣat	U	leave, permission, dismissal
rúy̌un	W	butter fat (P. *rōyan)*
ruṅ	B, S	high pasture
safidor	T, W	poplar (*Populus pamirica*)
sal	T,W	raft of inflated animal skins
sāl	P	year
sālgirah	P	birthday, festive day to commemorate Aga Khan IV's birthday
sanad	A, U	contract, document, legal instrument, record
sandóq	P, U, W	storage box
sangar	P	observing post, fortified breastwork made of stones
sanglowi	W	otter (*Lutrinea*)
sanjāb	P, W	ermine (*Mustela erminea*), grey squirrel
sar	B	channel head
sarhad	P	boundary, border province
sarkār, sərkór	P, U, W	rule, dominant and powerful superior instance, colonial administration of the East India Company (*sarkār-kampanī*), imperial power, the British-Indian government
sarkārī, sərkorí	P, U, W	dues for the dominant power, the colonial administration and/ or the Aga Khan
sarké baldí	B	partly covered room or veranda with a good view
sarúu	B	to carry a load
sayid	W, P, T	lineage group (from A. *saiyid)* of the Wakhi upper class; religiously legitimated ruling class
sepoy	U	soldier, mercenary
shīʿa	P	Shia
shikārī	U	hunting
sifārish	P, U	recommendation, connection
sipāhī	U	soldier
skid	W	cap
skord	W	bridge
soyo	W	butter churner (B. *sayó* = big hollow container made from a tree trunk, container for grain and flour)
sor	W	winnowed and cleaned cereals, prepared for storage
sowar	U	cavalry
sū	K	water, river
subahdār, sūbedār	U	mid-level junior commissioned officer, sometimes anglicised to *subedar*
sunnī	A, P	orthodox Muslim, Sunni

sur	W	turn system, rotating shepherding duty
suranái	B	double reed, musical instrument
swolək	W	trousers
sər	P, T, U	weight measure of 7.06 kg in Afghanistan and 0.933 kg in Pakistan
səróy	W	house, camp for the night, caravanserai, place offered to guests to stay overnight
syir	W	orphan
śaají	B, U	religiously legitimated ruling class (*sayid*)
śaníe khúṭkus	B	vegetable garden, kitchen garden
śardáne baṣ	B	suspension bridge, steel bridge
śarmá	B	coarse rug made from goat or yak hair
śask	B	willow (*Salix*); twigs are used for the making of baskets and trays
śikáar	B	hunt
śikáarí	B	tower, isolated or in fortified village
śipít	B	alfalfa, lucerne (*Medicago sativa*)
śuqá	B	big woollen gown with extended arms
šabáran	B	polo ground, Hunza-Burushaski *śabáran*, Nagar-Burushaski *śawáran*
šaftolú	W	peach; *Prunus persica*
šarā-e rešum	W	silk road, derived from P. *xāh-rāh-e abrīšam*
šardáṅ	S	suspension bridge
šawaáran	S	polo ground
širčóy	W	salt tea
špod	W	rhubarb
špïn	W	shepherd in high pasture
ṣádariśo	B	servitor, member of Burusho society
ṣaṅ	B	breach, opening in a wall, border between villages
ṣiqá	B	spellbound, bewitched
ṣiqámaṭiṅ	B	a local festival, in former times celebrated during apricot blossom
ṣïṣk	W	leather boot
ṣïw	W	black
ṣuṅ	B	steep path, route for secondary irrigation channel providing the connection between main and field channels
ṣ̌ax	W	broad peas (*Pisum sativum*)
ṣ̌óy̌un	W	functionary, announcing that spring has come
ṣ̌tïr, aṣ̌tïr	W	Bactrian camel (*Camelus bactrianus*)
ṣ̌əlmək	W	resin of apricot tree
ṣ̌əy̌d	W	first harvest, first sheaf of ears (*shaggat*) done by a man called *ṣ̌óy̌un* (*shágóni)*
taḥṣīl	A, U	administrative unit, here: subunit of a subdivision
taḥṣīl-dār	A, U	revenue collector, head of an administrative unit

takht	P, U	throne, seat of power
tal	B	birch (*Betula*)
talx čāi/čóy	P, W	green or bitter tea
talpaq	P	cap
talwār	H, U	sword
tanga	A, P, T	coin, one *tanga* equals 1.25 Rs or 20 *ānā; tanga siyāh* (black coin)
tanúr	P, W	oven, fireplace
tanzīm	A, U	organisation
taqīya	A, P, U	devoutness, piety; in Ismaili terms: secret religious practice
taqsīm	A, U, H	partition; e.g. partition deed (*taqsīm nāmah*)
taqt	B	throne
tarbúz	T, W	water melon; *Citrullus lanatus*
tār-garh	H, P, U	telegraph office
taríṅ	B	inflated skin for churning or for using on a raft
tepa	P	small hill
ter	B	pasture
tertsi	B, W	immigrant (from Wakhan)
teryók	P, W	opium, processed from poppy (*Papaver somniferum*)
teryoki	P, W	consumer of opium
tham, thamo	B	title of the local rulers in Hunza and Nagar
thámkuṣ	B	principality, position of ruler
thámkuṣe diś	B	capital, residence of *tham*
thuáaṅ thanánċ	B	new (*thos*) settlements
thúmuśaliṅ	B	festival at winter solstice
tík	W	coarse cotton cloth
tiskan	W	low shrub with a wooden root (*Krascheninnikovia ceratoides*)
tolā	P, U	measurement of weight for gold and silver; a common standard is the equivalent of about twelve grams, regularly used for *čaras* portions
tōp	P	cannon
toq	B	meadow
traṅ	B	partition
traṅphá	B, S	village elder
truk təl	W	bitter apricot kernel oil
tsungling	C	onion mountain
tuy̌	W	male goat
tul	B	hole, underground passage in irrigation channel
tumúro	B, W	herbal tea
tuxm	W	sowing of seeds
tuxmwór	W	seed-sowing ceremony
təbəsk	W	forage plant, *Kochia prostata* or *Artemisia fragrans*
təl	W	apricot kernel oil
təlpək	W	cap made from animal skins and cloth
tər xundaq, xunən daq	W	to offer a meal, traditional duty to feed the ruling family while visiting Gojal as part of the local tax system
tərək	W	poplar (*Populus alba*)
tüd	W	mulberry (*Morus alba, Morus nigra,* P. *tut*)
ṭhekedār	H, U	contractor
ṭis	B	pit in a field for the storage of potatoes
ṭor	W	walnut (*Juglans regia*)
ṭóri	B	plug, wooden peg for closing an outlet from an irrigation channel
üy	K	yurt, house
ulamā	A	religious leaders
ustād	U	expert, master craftsman, teacher, driver
utáq	B	food supply during travelling
uúśam	B	foster relationship, wet nurse
uyóṅko	B	member of upper class
ʿušr	A	tenth, tithe
wakīl	A, P, U	ambassador, counsellor, representative
wandan	W	night pen for yak
war	W	male sheep
waxī	T	term for Wakhi person
waxīwor	T	Wakhi language
waxonī	T	term for Wakhi person
wazārat	A, P, U	subdivision, district
wazīr	A, P, B, U	prime minister, advisor of the ruler
wazīr-i-wazārat	A, P, U	senior administrative officer, representing the Maharaja of Kashmir in Gilgit (= Gilgit *wazārat*)
wazīrkuċ	B	the close relatives of the *wazīr* forming a descent group
winək	W	sable (*Marmotta ondata*)
wod	W	irrigation channel
wuč	W	upper, high above
wujərk	W	alfalfa, lucerne (*Medicago sativa*)
wundr	W	field parcels composed of *racn* and *rag* respectively
wurk	W	lamb
wušk	W	calf
wiṣ̌	W	fodder grass
wiṣ̌dún	W	storage for fodder grass
xāh-rāh-e abrīśam	P	silk road, literally: royal road of silk
xalʿat	P	robe of honour
xalīfah	U	Ismaili lay priest, leader of rituals
xana	P	house
xarite	P	purse, map; here the small bag containing the subsidy in colonial times
xibu da kaifa	C	great development of the West
xirgó	W	yurt, household, felt tent (from P. *xār-gāh*)
xitóy, xitái	W, T	(Han)-Chinese
xujá	W	Wakhi lineage (from P. *xwāja*), Ismaili elite of Wakhan

xun	W	house
xunedor	W	master or mistress of the house
xur	W	donkey
xur vdək	W	donkey trail
xurd	P	small
xuš xun	W	guest-room
xədórg	W	water mill for grain
xədoyí	W	ritual meal
xəlifá	B, W	Ismaili lay priest, leader of rituals
xïč, x̌əč, x̌eč	W	food
x̌ik	W	self-designation by Wakhi
x̌ikwor	W	Wakhi language
x̌ik zik	W	self-designation of Wakhi language
yabá	B	male yak (*Bos grunniens*), often left alone on high pastures
yabú	B	good mountain horse, pack animal (pony in U. *yābū*)
yáiṅ	B	water-mill
yambū	C	silver ingot equalling five *yambucha*
yarpá	B	supervisor of the herds of the local ruler
yarz	W	juniper (*Juniperus*)
yaš	W	horse
yaš yirk	W	barley for horses
yáṭċil	B	second watering of fields
yáṭkuin	B	inspector, controller
yāyla, yāylow	K	summer grazing ground
yaz	W	glacier
yaze yupk	W	glacier meltwater
ying	C	five hundred, battalion
yirí	W	store, place for household utensils
yirk	W	barley (*Hordeum vulgare*)
yirz	W	millet (*Panicum miliaceum*, *Setaria italica*)
yorč	W	platform
yoš	W	shed for newly-born livestock
yukṣ̌	W	ibex (*Capra ibex*)
yumǰ	W	flour
yupk	W	water
yurġol	B	female court dancer
zak	W	green meadow
zakāt	A, P, U	charity in tune with the Holy Quran
zamindār	H, U	cultivator, farmer
zax̌	W	sea buckthorn (*Hippophae rhamnoides*)
zhonghua renmin gonghe guo	C	People's Republic of China
ziarat	P	religious shrine
ziča rāž	W	elevated store in the house
žarž	W	milk
ʒay	W	sea buckthorn (*Hippophae rhamnoides*)
ʒuγ̌	W	yak (*Bos grunniens*)
ʒəqláy žarž	W	'small' foster relationship (related to taxation)

PRONUNCIATION

Burushaski, Gujarati, Hindi, Shina

ć	*ch* as in 'cheap'
ċ	*ts* as in 'lets'
ç, ṣ	*ch* as in 'cheap' and *sh* as in 'shush', but the tongue is placed behind the alveolar ridge, and may be even curled back to touch the palate
ś, š	*sh* as in 'shush'
s	*s* as in 'set'
z	*z* as in 'zebra'
ṅ	*ng* as in 'long'
ṭ, ḍ	*t* and *d* but the tongue is curled back
ġ	like the French *r* as in 'grasseyé'
q	*q* as in 'queen', but guttural
j̣	*y* as in 'yarn', but the tongue is placed behind the alveolar ridge, and may be even curled back to touch the palate
j	*j* as in 'James'
ỵ	pronounced as a combination of *y* and *r* in English pronunciation
y	*y* as in 'yarn'

Wakhi

ə	reduced central vowel, similar to *u* in English 'but', but different from *a*
c	*ts* as in 'lets'
č	*ch* as in 'cheap'
ḍ	retroflex *d*, with retracted tongue
δ	voiced as *th* as in English 'that'
γ	*ġ* like French *r* in 'grasseyé', North German 'Tage'
γ̌	close to *ch* in German 'ich', but voiced
ǰ	*j* as in 'James'
ï	between *i* and *u*
š	similar to English *sh*, but more fronted, palatal
ṣ̌	similar to English *sh*, but retroflex, retracted
ṣ	similar to English *sh*
ṭ	like *t* in English, but more retracted, retroflex
x	like German 'ach', Scottish 'loch'
x̌	the most frequent sound like 'ich' in German, the voiced counterpart is *γ̌*
z	as in English 'zebra', voiced *s* as in German 'Rose'
ž	*zh* as in 'azure'
ʒ	pronounced like *dz* in 'adze'

Kirghiz, Uigur and Uzbek

č	*ch* as in 'cheap'
ı	*i* as in 'children'
x	*ch* as in Scottish 'loch'
ö	*eu* as in French 'peu'
š	*sh* as in French 'chaud'

Urdu, Persian, Tajik

ā, ī	long vowels	*ṛ*	retroflex *r*
x, <u>kh</u>	*kh* as in Scottish 'loch'	*š*	*sh* as in 'shush'
h	h as in 'hat'	*c*	glottal stop, not pronounced in Urdu
j	*j* as in 'James'		

Sources: compiled on the basis of Abdus Salam (1988); Fazal Amin Beg word lists (2013); Hermann Berger (1983, 1998); Georg Buddruss word lists (1974, 1984, 1985, 1987, 1992, 1993, 2006, 2012, 2018); Georg Buddruss and Almuth Degener (2012); Rémy Dor (1975); Rémy Dor and Clas Naumann (1978); Aleksandr Leonovich Gryunberg and Ivan Mikhailovich Stéblin-Kamensky (1976, 1988); Sven Hedin (1899, I: 156, 191); Heinrich Junker and Bozorg Alavi (1986); Georg Morgenstierne (1932: 67–74; 1938); M. Nazif Shahrani (1979); Ivan Mikhailovich Stéblin-Kamensky (1999); Henry Yule and Arthur Burnell (1903) and author's collections.

TOPONYMS AND MEANINGS

Abaasíin	Indus
-abad	suffix for place name, settlement (U, P)
Altit Rat	plain of Altit (B), former cultivated land of *mīr* of Hunza
Barashal	hamlet of the Barátaliṅ, filial settlement
Bul Mal	field of springs (B), cultivated land in Karimabad
Búlulo	spring field (B), pasture area above the Buróoṅ villages in Baltit
Burondo Bar	ring valley (B)
Buroonshal	hamlet of the Buróoṅ, filial settlement
Chaman Gul	meadow flower (P), filial settlement of Gulmit
Chumar Khan	iron fort (*ćhumár khan*), hamlet of Haiderabad
Cikar	willows (*čikar* = *Salix turanica*)
Chikas	garnets (*ćhíkas*)
Dal Haram	upper grain field (*dal harúm*)
Darband	gate, closed passage (P), control post
Darwaza	gate, door (P), control post
Diramishal Khan	settlement of Dirámitiṅ
Diráćhil	water of the Dirámitiṅ (B)
Dorkhan	fort of the ram (B)
Dut	rope bridge (W)
Fatma hel	Fatima, daughter of the Holy Prophet, high pasture in Batura
Gara Bares	? and vein (B), filial settlement in Karimabad
Gojal	Little Gojal as opposed to Big Gojal which is Wakhan
Gókomuċe kóor	the cave of the young donkey
Haldeikish	rock of the billy goat, 'sacred rock' of Hunza
Halegush	place with trails of foxes (B ḱále ġaṣ)
Hanuman Mun	rock of Hanuman (B)
Herbar	the upper valley, older term (B) for Gojal
Imamabad	*imām*, in honour of the Aga Khan, new village in Gojal
Kampire Diyor	village of the old woman (W), hamlet of Shutmerk
Karagadimuċ	irrigation channel (*thámo ṭór*) leading across a gate with wooden support structure (*karagáḍimuċ*) providing water to the (new) palace of the *mīr* of Hunza
Kairabad	place of good people (U)
Khudabad	place of God (P, U), Burusho village in Upper Gojal
Khunjerab	red valley (W) in popular etymology
Kukhel	high pasture with a spring (W)
Langar	literally kitchen, resting place and shrine

Malikamu Shikari	royal tower, in honour of Queen Victoria (B), above Karimabad
Mamu Har	milk valley (B)
Momhil	pasture of the grandmother (W)
Murku	pebble (B); name of an irrigation channel crossing a cliff of pebbles
Pratabsinghpura	settlement founded during the reign of Pratab Singh, Dogra-Maharaja of Kashmir
Shabaran	polo ground
-shal	suffix for place names
Shersabz	green lion, Timur Leng's birth place, a village in Chupursan Valley
Shinaki	the settlement region in lower Hunza where predominantly Shina-speakers reside
Shinber	the settlement region of lower Nagar, probably a contraction of Shin and Bar, the side valley joining the Hunza River near Chalt
Shuijerab	black valley (W), high pasture at the foot of Shimshal Pass
Shuwart	black mill stone (W); high pasture close to Shimshal Pass
Silganish	hamlet of the Silgániškuċ
Suryas Das	the cemetery above the hamlet of the old hamlet of Suryas, later renamed Karimabad
Tarkichan shun	path to the cattle-sheds (B) outside of Baltit Khan
Ulobassi	inner orchard (B), filial settlement in Karimabad
Ultar	inner pasture (B), glacier above Karimabad
Ultar Shan	access to Ultar, covered passage connecting Altit and Baltit underneath Karagadimuċ
Virjerab	long valley (W))
Yashpirt	opening, horse pasture in Batura (W))
Ziarat	shrine (B, P, W), resting place

Source: based on Hermann Berger (1998), Georg Buddruss (word lists), Bronislaw Grombchevsky (1888/2015), Hermann Kreutzmann (1989a: 219) and local enquiries

TOPONYMS AND PREVIOUS TRANSLITERATIONS AND USAGES

present name	explanation	former name	explanation
Ainabad		Goshben	the village below the pasture of Gaush
Agakhanabad	Aga Khan IV	Akhanabad	
Aliabad	village of Hazrat Ali	Burum Mos	white mudflow (B)
Ahmedabad		Hamatabad	
Ahmedabad		Muhammadabad	
Aminabad	Prince Amyn, brother of Aga Khan IV (renamed in 1982)	Herra	
Aminabad		Nomal	
Anwarabad		Golodas	
Astana	shrine (renamed in 1975)	Shikakienċ	place where Syed Shah Wali performed
Bar Valley		Garmasai	
Beriċo Ċok	pasture of the Ḍom	Altikuċe Sat	high pasture above Altit
Buldas	desert with spring	Ganesh Giram	filial hamlet (*muḥalla*)
Burum Bun	white rock	Altit Giram	filial hamlet
Dala	big irrigation channel	Khul [*gūl*] Samarkand	Samarkand channel
Faisabad		Koshi	filial village of Altit
Gojal	more precisely 'Little Gojal' in contrast to 'Big Gojal' for Wakhan	Herbar	Upper Hunza (B)
Haiderabad		Bimal Das	
Hunza		Haháyul	
Hunza		Khajuna, Khajuni	
Hunza		Kanjut, Hunzije	
Hussainabad	named after Hussain, son of Ali	Spiko, Seko	
Hussaini		Sussani	
Imamabad	hamlet of the *imām*	Tonich	hamlet in Altit Giram
Imamyarabad	friend of the *imām*	Shukshal	

present name	explanation	former name	explanation
Imanabad		Mangushi	
Jangi Dur	checkpoint and observation post beyond the Shimshal Pass on Braldu River	Darwaza, Darband	door, checkpoint, control post
Karimabad	hamlet in Baltit	Suryas	summer residence of the *tham*
Karimabad	named after Karim-al-Hussaini, Aga Khan IV (renamed in 1983)	Baltit	
Khairabad		Reminj	
Khanabad	place of a *khan*	Mose (More) Das	filial village of Hindi
M(a)iatsil		Hispar Valley	
Mominabad	place of the believers, Dumyal (renamed in 1981)	Berishal	village of artisans
Mominabad		Domel, Dumyal	
Muhammadabad		Jutal	
Muhammadabad		Diding Das	barren land
Murtazabad		Naray Das	
Nasirabad	Nazir Khusraw (renamed in 1983)	Hindi	
Nazimabad	founded by *mīr* Nazim Khan	Shishket	
Nazimabad		Sost Bala	
Oshikandas	windy desert (S)	Chamogah	
Princeabad	Prince Aga Khan IV	Danyor (hamlet)	place of the mill-stone
Rahimabad	Prince Rahim, eldest son of Aga Khan IV	Matum Das	black desert
Reminj		Khairabad	village in Chupursan
Rezgine Ben	salty place	Shimshal (hamlet)	
Roshanabad	place of light	Dal Haram	
Sadruddinabad	Prince Sadruddin, uncle of Aga Khan IV	Nomal (hamlet)	
Salmanabad		Gamesarat	
Sarat	lake		
Seko/Spiko		Hussainabad	
Shaboikushal	hamlet of Shaboikuċ	Ganesh Giram	
Sherabad	village of lions	Muchuċil	water from Muchu Har
Shuknoshal	hamlet of Shuknotiṅ	Ganesh Giram	
Sogural	fief	Ganesh Giram	
Sultanabad	Sultan M. Shah, Aga Khan III	Altit Giram	
Sultanabad		Gujur Das	seasonal settlement of Gujur shepherds close to Gilgit in Hunza Valley
Sultanabad		Kutukal	leased land, owned by the *tham*
Talmushi	name of a fairy, valley and irrigation channel above Altit		

Source: compiled and expanded from Hermann Kreutzmann (1989a: 220–221; 2015a: 494–495) and based on own enquiries during fieldwork

Bibliography

Archival collections

Archive of the German Alpine Club – Archiv des Deutschen Alpenvereins

Deutsch-Österreichische Himalaya-Karakorum-Expedition (DÖHKE) 1954 (German-Austrian Himalaya-Karakoram Expedition 1954); Deutsche Karakorum-Expedition (DKE) 1959 (German Karakoram Expedition 1959)

DAV DOK 1 SG 142/1 – Expeditionstagebuch Martin Schließler

DAV EXP 2 SG 69 – 1954 DÖHKE und 1959 Karakorum

DAV EXP 2 SG 78 – Presseberichte

DAV EXP 2 SG 99 – DHS – DÖHKE 1954, Diverses

DAV EXP 2 SG 100 – DHS – DÖHKE 1954, Organisation/Schriftverkehr

DAV EXP 2 SG 104 – DÖHKE 1954

DAV EXP 2 SG 105 – Schriftverkehr

DAV EXP 2 SG 106 – Presseberichte

DAV EXP 2 SG 109 – Presse-Korrespondenz

DAV EXP 2 SG 132 – Deutsche Karakorum Expedition 1959

DAV EXP 2 SG 133 – Besuch des Mir von Hunza in München

DAV EXP 2 SG 237 – Presse

DAV EXP 2 SG 244 – Rückkehr

DAV EXP 2 SG, 293 – Karakorum-Karten

DAV NAS 2 SG 132 – Berichterstattung über Deutsch-Österreichische Karakorum Expedition 1954 (DÖHKE)

DAV NAS 9 SG 23 – DÖHKE 1954

Barrow, Edmund George. 1885. *Extracts of the diary of Captain E. G. Barrow, Chitral-Kafiristan Mission 1885/6*. Gilgit: unpublished diary, 110 pp. India Office Records and Private Papers Mss-Eur/E420

Barrow, Edmund George. 1886. *Chitral-Hunza diary*. Simla: unpublished diary, 83 pp. India Office Records and Private Papers MssEur/E420

Barrow, Edmund George. 1887a. *Gazetteer of the Eastern Hindú Kush. Confidential. Part I. The Tajik provinces of Wakhán, Ishkásham, and Zébák*. Simla: Government Central Branch Press

Barrow, Edmund George. 1887b. *Gazetteer of the Eastern Hindú Kush, Part II. Dárdistán*. Simla: Government Central Branch Press

Barrow, Edmund George. 1888. *Gazetteer of the Eastern Hindu Kush in four parts: Part I. Wakhán, Ishkásham, and Zébák. Part II. Dárdistán. Part III. Káfiristán. Part IV. Routes in the Hindú Kush; compiled under the orders of the Quartermaster-General in India*. Simla: Government Central Press

Barrow, Edmund George. 1889. *Confidential. Gilgit and the Eastern Hindu Kush*. Calcutta, 6 pp. Appendix to India Office Records and Private Papers MssEur/E420

Barrow, Edmund George 1890: *Gazetteer of Kashmír and Ladák together with routes in the territories of the Maharája of Jamú and Kashmír*. Calcutta: The Superintendent of Government Printing, 871 pp. IOL/P&S/20/B 20

Bayley, Stuart Colvin. 1892. *Note on the complication with Hunza*. Simla, 12 pp. IOL/P&S/18/A82

Biddulph, John. 1876. *Personal Diaries: 'Trip to the Chitral Passes' 25 April to 16 November 1876: 1–150. Diaries and miscellaneous papers 1866–1903*. Hereford: Herefordshire Record Office (A 87/6), 150 pp

Biddulph, John. 1878. *Report by Captain John Biddulph of a journey to Gilgit, Hunza, and the Karoombur valley*. In: Précis of correspondence, &c., relating to affairs in Central Asia, Biluchistan, Persia, &c. 1875–77. With appendices. IOL/P&S/20/253; Appendix 13: 235–243

Biddulph, John. 1881. *Memorandum by Major Biddulph on the present condition of affairs in Gilgit. March 1881. Appendix to Arthur Naylor Wollaston: Memorandum on Chitral, including the Frontier States of Gilgit and Yassin*. IOL/P&S/18/A 18/2: 23–32

Cockerill, George Kynaston. 1895. *Appendix to Barrow's gazetteer of Eastern Hindu Kush*. Simla: Government Central Printing Office

Cockerill, George Kynaston. 1896. *A further appendix to Barrow's gazetteer of Eastern Hindu Kush*. Simla: Government Central Printing Office

Ghazanfar Ali Khan. 2014. *Biographical interview of Mir Ghazanfar Ali Khan of Hunza*. Recorded by Fazal Amin Beg. Karimabad, 5 pp

Ghulam-ud-din. 2011. *Recollections of Ghulam-ud-din*. Recorded and compiled by Fazal Amin Beg. Gilgit, 11 pp.

Godfrey, Stuart Hill. 1898. R*eport on the Gilgit Agency and Wazarat and the countries of Chilas, Hunza-Nagar, and Yasin, including Ashkuman, Ghizr, and Koh. 1896–97. (Confidential)*. Calcutta: Superintendent of Government Printing, 152 pp

Government of India. 1909. *Military report and gazetteer of the Gilgit Agency and the independent territories of Tangir and Darel*. Simla: Government Monotype Press, 208 pp

Government of Pakistan. 2000. *1998 District Census Report of Gilgit*. Islamabad: Population Census Organization, Statistics Division

Haughton, Henry Lawrence. 1912. *Diaries of Henry Lawrence Haughton. Volume 5: 1 January 1912 to 3 June 1912*. India Office Records and Private Papers Mss Eur/D531/5

Hentig, Werner Otto von. 1912–1917. *Das nie geschaute Land (AT): Asiatische Reisen 1912–1917*. Unpublished manuscript; edited by Hans Wolfram von Hentig in 1994

Hisamullah Beg, Usman Ali, Imtiaz ul Haq, M. Dil Pazeer, Sher Baz Ali Bercha, Qasim Naseem, Israruddin Israr. 2015. 'Report on the historical perspective of Gilgit-Baltistan' submitted to the Secretariat Office of the Chief Secretary, Government of Gilgit-Baltistan on 4 December 2015, 61 pp

Howell, Evelyn Berkeley. 1927. *Report of the Resident in Kashmir on his tour in the Gilgit Agency, 1927. Northern Frontier: Administration Report for the Gilgit Agency for the year 1927*. Srinagar, 25 November 1927 (IOL/P&S/3288: 260–263), 8 pp

India Office Library & Records: Military Department Library: North-West Frontier, Baluchistan

IOL/MIL/17/13/58: *General Staff India 1942: Routes in Chitral, Gilgit and Kohistan.* Simla: Government of India Press, fourth edition, 485 pp

IOL/MIL/17/13/73: *General Staff India 1928: Military Report and Gazetteer of the Gilgit Agency and the independent territories of Tangir and Darel.* Simla: Government of India Press, second edition, 232 pp

India Office Library & Records: Military Department Library: Afghanistan, Central Asia, Tibet

IOL/MIL/17/14/72: Intelligence Branch 1871–1875: Central Asia, Calcutta: Quarter-Master General's Department, nine volumes. Part 1: Charles Metcalfe MacGregor: *A contribution towards the better knowledge of the topography, ethnography, statistics & history of the North-West Frontier of British India, three volumes.*

IOL/MIL/17/14/74: Intelligence Branch, Division of the Chief of Staff 1907: *Military Report on Kashgaria.* Simla: Government of India Press, 199 pp

IOL/MIL/17/14/75: General Staff India 1929: *Military Report on Sinkiang (Chinese Turkestan).* Calcutta: Government of India Press, 308 pp

IOL/MIL/17/14/76: General Staff India 1926: *Routes in Sinkiang.* Simla: Government of India Press, 352 pp

India Office Library & Records: Files relating to Indian States extracted from the Political and Secret Letters from India 1881–1911

IOL/P&S/7/57/19–20: *Reestablishment of a British Agency at Gilgit, 1889*

IOL/P&S/7/57/33: *Report on the present military position in Gilgit* by Captain A[lgernon] Durand, Simla 1888, 50 pp

IOL/P&S/7/59: *Gilgit Agency Report 1889* by Captain A[lgernon] Durand, 29 December 1889 Gilgit, 9 pp

IOL/P&S/7/63/1363: *Letter* from A[lgernon] G[eorge] Durand to Colonel W[illiam] F[rancis] Prideaux, Resident in Kashmir, Gilgit 30 July 1891

IOL/P&S/7/66/689: *Letter* from Jan Muhammed to Colonel Durand, Tashkurghan 20 March 1892

IOL/P&S/7/66/699: *Translation of a letter from the Taotai of Kashgar to Muhammed Nazim [Khan]*

IOL/P&S/7/66/701: *Letter* from J[ohn] Manners-Smith to Resident in Kashmir, Gilgit 4 April 1892

IOL/P&S/7/66/702: *Note* by F[rancis] E[dward] Younghusband, 20 April 1892

IOL/P&S/7/66/894–895: *Letter* from G[eorge] Macartney to A[lgernon] Durand, Kashgar 6 April 1892

IOL/P&S/7/66/906: *Letter* from Lieutenant Colonel A[lgernon] G[eorge] Durand, British Agent at Gilgit to the Resident in Kashmir, Gilgit 13 May 1892

IOL/P&S/7/66: *Translation of a letter from Jan Muhammed Khan to the British Agent at Gilgit, 6 May 1892*

IOL/P&S/7/67/1131: *Letter* from Jan Muhammed to Dr. Robertson, Officiating British Agent, Gilgit 26 June 1892

IOL/P&S/7/67/621–663: *Hunza's claims in Taghdumbash, Safdar Ali's role etc.; letters found in Baltit Fort in 1891 and their translations*

IOL/P&S/7/67/1142: *Letter* from G[eorge] Macartney to H[enry] [Mortimer] Durand, the Secretary to the Government of India. Foreign Department, Tashkurghan, Sarikul, 20 July 1892

IOL/P&S/7/68/209: *Letter* from Jan Muhammed to British Agency, 12 August 1892

IOL/P&S/7/69/934: *Letter* from G[eorge] Macartney to British Agent, Gilgit 1 September 1892

IOL/P&S/7/70/1753: *Letter* from British Agent, to Resident in Kashmir, Gilgit 1 September 1892

IOL/P&S/7/72/26: *Letter* from G[eorge] Macartney to Resident in Kashmir, Kashgar 29 July 1893

IOL/P&S/7/72/29: *Letter* from Jan Muhammed to British Agent, (Translation), Gilgit 1 August 1893

IOL/P&S/7/72: *Emancipation of British Indian subjects in slavery in Chinese Turkestan, 18 October 1893*

IOL/P&S/7/73/1174–1175: *Emancipation of British Indian subjects in slavery in Chinese Turkestan, 21 March 1892*

IOL/P&S/7/73/1199–1200: *Letter* from G[eorge] Macartney to Resident in Kashmir, Kashgar 11 September 1893

IOL/P&S/7/75: *Emancipation of British Indian subjects in slavery in Chinese Turkestan, 21 August 1894*

IOL/P&S/7/75: *Gilgit Diary*, 21 July 1894

IOL/P&S/7/75: *Memorandum of information received during the month of June 1894, regarding affairs beyond the North-West frontier of India*

IOL/P&S/7/76: *Gilgit Diary*, 18 August 1894

IOL/P&S/7/77: *Gilgit Diary*, 20 October 1894

IOL/P&S/7/77: *Gilgit Diary*, 3 November 1894

IOL/P&S/7/78: *Emancipation of slaves in Chinese Turkestan, 27 March 1895*

IOL/P&S/7/78/1395: *Memorandum of proposed concessions to be obtained from the Chinese Government regarding trade between India and Kashgaria*, G[eorge] Macartney to S[tuart] H[ill] Godfrey, British Joint Commissioner Ladakh, Kashgar 18 July 1894

IOL/P&S/7/78/1549: *Gilgit Agency Diary for the week ending the 5th January 1895*

IOL/P&S/7/80: *Emancipation of slaves in Chinese Turkestan, 25 June 1895*

IOL/P&S/7/81: *Gilgit Diary*, 9 June 1895

IOL/P&S/7/81: *Gilgit Diary*, 22 June 1895

IOL/P&S/7/82: *Gilgit Diary*, 17 August 1895

IOL/P&S/7/84: *Gilgit Diary*, 7 December 1895

IOL/P&S/7/84: *Gilgit Diary*, 14 December 1895

IOL/P&S/7/84: *Chitral Diary*, 21 December 1895

IOL/P&S/7/84: *Chitral Diary*, 11 January 1896

IOL/P&S/7/85: *Gilgit Agency Diary*, 8 February 1896

IOL/P&S/7/85: *Memorandum of Information received during the month of February 1896, regarding affairs beyond the North-West frontier of India*

IOL/P&S/7/87: *Gilgit Diary*, 9 May 1896

IOL/P&S/7/87: *Gilgit Diary*, 13 June 1896

IOL/P&S/7/87: *Gilgit Diary*, 11 July 1896
IOL/P&S/7/88: *Gilgit Diary*, 8 August 1896
IOL/P&S/7/88: *Gilgit Diary*, 15 August 1896
IOL/P&S/7/88: *Gilgit Diary*, 22 August 1896
IOL/P&S/7/89: *Statistics of Trade of India with Chinese and Russian Turkestan and Tibet for the year ending 31st of March 1896.* Letter from G[odfrey] C[henevix] Trench, Ass. to Res. Kashmir for Leh to Res. Kashmir, Leh 12 August 1896
IOL/P&S/7/89: *Gilgit Diary*, 31 October 1896
IOL/P&S/7/90/169: *Pol. Diary Gilgit*, 5 December 1896
IOL/P&S/7/90/293: *Pol. Diary Gilgit*, 8 February 1897
IOL/P&S/7/90/314: *Pol. Diary Gilgit*, 23 January 1897
IOL/P&S/7/92/539: *Gilgit Diary*, 17 April 1897
IOL/P&S/7/92: 539: *Gilgit Diary*, 24 April 1897
IOL/P&S/7/94/807: *Gilgit Diary*, 3 July 1897
IOL/P&S/7/94: 807: *Gilgit Diary*, 10 July 1897
IOL/P&S/7/95/937: *Gilgit Diary*, 7 August 1897
IOL/P&S/7/95/952: *Gilgit Diary*, 14 August 1897
IOL/P&S/7/98/1244: *Gilgit Diary*, 16 November 1897
IOL/P&S/7/98: 1256: *Liberation of slaves of Indian origin in Chinese Turkestan*, 23 December 1897
IOL/P&S/7/99/179: *Gilgit Diary*, 30 November 1897
IOL/P&S/7/99: 179: *Gilgit Diary*, 8 December 1897
IOL/P&S/7/100/298: *Gilgit Diary*, 22 December 1897
IOL/P&S/7/100: *Gilgit Diary*, 8 January 1898
IOL/P&S/7/100: *Gilgit Diary*, 16 January 1898
IOL/P&S/7/103/541: *Gilgit Diary*, 31 March 1898
IOL/P&S/7/103/597: *Gilgit Diary*, 16 April 1898
IOL/P&S/7/104/674: *Gilgit Diary*, 16 May 1898
IOL/P&S/7/107/1009: *Gilgit Diary*, 16 September .1898
IOL/P&S/7/109/1076: *Gilgit Diary*, 8 October 1898
IOL/P&S/7/110/1155: *Gilgit Diary*, 31 October 1898 & 6 November 1898
IOL/P&S/7/113/439: *Gilgit Diary*, 14 March 1899
IOL/P&S/7/115/740: *Gilgit Diary*, 10 June 1899
IOL/P&S/7/115/803: *Gilgit Diary*, 7 July 1899
IOL/P&S/7/115/846: *Gilgit Diary*, 22 July 1899
IOL/P&S/7/116/877: *Gilgit Diary*, 5 August 1899
IOL/P&S/7/116/894: *Gilgit Diary*, 12 August 1899
IOL/P&S/7/117/1004: *Gilgit Diary*, 30 September 1899
IOL/P&S/7/117/1048: *Gilgit Diary*, 6 October 1899
IOL/P&S/7/118/1080: *Gilgit Diary*, 21 October 1899
IOL/P&S/7/118/1085: *Ladakh Trade Report* 1899
IOL/P&S/7/118/1117: *Gilgit Diary*, 4 November 1899
IOL/P&S/7/118/1117: *Gilgit Diary*, 11 November 1899
IOL/P&S/7/119/264: *Gilgit Diary*, 6 January 1900
IOL/P&S/7/122/535: *Gilgit Diary*, 7 April 1900
IOL/P&S/7/123/605: *Gilgit Diary*, 28 April 1900
IOL/P&S/7/123/646: *Gilgit Diary*, 12 May 1900
IOL/P&S/7/123/698: *Gilgit Diary*, 26 May 1900
IOL/P&S/7/124/784: *Gilgit Diary*, 9 June 1900
IOL/P&S/7/124/812: *Gilgit Diary*, 30 June 1900
IOL/P&S/7/125/881: *Gilgit Diary*, 7 July 1900
IOL/P&S/7/126/967: *Gilgit Diary*, 28 July 1900
IOL/P&S/7/128/1202: *Gilgit Diary*, 28 September 1900
IOL/P&S/7/128/1202: *Gilgit Diary*, 29 September 1900
IOL/P&S/7/128/1221: *Gilgit Diary*, 6 October 1900
IOL/P&S/7/128/1313: *Gilgit Diary*, 10 November 1900
IOL/P&S/7/129/166: *Gilgit Diary*, 8 December 1900
IOL/P&S/7/129/282: *Gilgit Diary*, 12 January 1901
IOL/P&S/7/130: *Russian soldiers working at Tashkurghan*, 13 March 1901
IOL/P&S/7/131/395: *Gilgit Diary*, 2 February 1901
IOL/P&S/7/131/432: *Gilgit Diary*, 16 February 1901
IOL/P&S/7/132/499: *Gilgit Diary*, 9 March 1901
IOL/P&S/7/133/593: *Gilgit Diary*, 6 April 1901
IOL/P&S/7/133/690: *Gilgit Diary*, 27 April 1901
IOL/P&S/7/134/757: *Gilgit Diary*, 4, 11 & 18 May 1901
IOL/P&S/7/134/783: *Chitral Diary*, 18 May 1901
IOL/P&S/7/134/784: *Gilgit Diary*, 25 May 1901
IOL/P&S/7/135/903: *Gilgit Diary*, 22 June 1901
IOL/P&S/7/135/935: *Gilgit Diary*, 2 July 1901
IOL/P&S/7/136/1009: *Gilgit Diary*, 13 July 1901
IOL/P&S/7/137/757: *Gilgit Diary*, August 1901
IOL/P&S/7/138/1233: *Gilgit Diary*, 7 September 1901
IOL/P&S/7/138/1323: *Gilgit Diary*, 28 September 1901
IOL/P&S/7/140/1467a: *Gilgit Diary*, 19 October 1901
IOL/P&S/7/140/1494a: *Gilgit Diary*, 26 October 1901
IOL/P&S/7/141/266: *Gilgit Diary*, 28 December 1901
IOL/P&S/7/142/425: *Statement of savings, Gilgit Agency, 1902*
IOL/P&S/7/143/468: *Chitral Diary*, 18 February 1902
IOL/P&S/7/143/504: *Gilgit Diary*, 15 February 1902
IOL/P&S/7/144/602: *Gilgit Diary*, 1 March 1902
IOL/P&S/7/144/604: *Gilgit Diary*, 11 March 1902
IOL/P&S/7/144/727: *Gilgit Diary*, 12 April 1902
IOL/P&S/7/145/856: *Gilgit Diary*, 17 May 1902
IOL/P&S/7/145/886: *Gilgit Diary*, 24 May 1902
IOL/P&S/7/146/1051: *Gilgit Diary*, 28 June 1902
IOL/P&S/7/147/1117: *Gilgit Diary*, 5 July 1902
IOL/P&S/7/151/201: *Gilgit Diary*, 20 December 1902
IOL/P&S/7/151/337: *Gilgit Diary*, 24 January 1903
IOL/P&S/7/152/377: *Gilgit Diary*, 7 February 1903
IOL/P&S/7/153/576: *Gilgit Diary*, 14 March 1903
IOL/P&S/7/154/789: *Gilgit Diary*, 2 May 1903
IOL/P&S/7/154/875: *Gilgit Diary*, 16 May 1903
IOL/P&S/7/155/973: *Gilgit Diary*, 13 June 1903
IOL/P&S/7/155/1040–1041: *Gilgit Diary*, 20 June 1903
IOL/P&S/7/156/1078: *Gilgit Diary*, 27 June 1903
IOL/P&S/7/157/1277: *Gilgit Diary*, 1 August 1903
IOL/P&S/7/157/1399: *Gilgit Diary*, 5 September 1903
IOL/P&S/7/158/1459: *Gilgit Diary*, 19 September 1903
IOL/P&S/7/159/1667: *Gilgit Diary*, 31 October 1903
IOL/P&S/7/160/1728: *Gilgit Diary*, 21 November 1903
IOL/P&S/7/160/1773: *Gilgit Diary*, 5 December 1903
IOL/P&S/7/161/334: *Gilgit Diary*, 26 December 1903
IOL/P&S/7/162/597: *Gilgit Diary*, 6 February 1904
IOL/P&S/7/162/634: *Gilgit Diary*, 13 February 1904
IOL/P&S/7/165/1094: *Gilgit Diary*, 7 May 1904

IOL/P&S/7/167/1319: *Gilgit Diary*, 18 June 1904
IOL/P&S/7/168/1525: *Gilgit Diary*, 23 July 1904
IOL/P&S/7/168/1612: *Gilgit Diary*, 6 August 1904
IOL/P&S/7/169/1659: *Gilgit Diary*, 30 July 1904
IOL/P&S/7/169/1659: *Gilgit Diary*, 13 August 1904
IOL/P&S/7/170/1845: *Gilgit Diary*, 15 September 1904
IOL/P&S/7/171/2113A: *Gilgit Diary*, 15 October 1904
IOL/P&S/7/171/2142A: *Gilgit Diary*, 22 October 1904
IOL/P&S/7/172/2263A: *Gilgit Diary*, 12 November 1904
IOL/P&S/7/172/2264A: *Kashgar News-Report*, 31 October 1904
IOL/P&S/7/173/283: *Gilgit Diary*, 10 December 1904
IOL/P&S/7/173/351: *Gilgit Diary*, 10 December 1904
IOL/P&S/7/173/371: *Gilgit Diary*, 23 April 1904
IOL/P&S/7/173/371: *Foreign Department. Secret. Frontier. Proposed composite arrangement for settlement of outstanding questions with Chinese in Kashgaria*, Fort William, dated 26 January 1905
IOL/P&S/7/173/371: *Foreign Department. Secret. Frontier. Enclosures of letter to His Majesry's Secretary of State for India, No. 20*, Fort William, dated 26 January 1905
IOL/P&S/7/174/428: *Gilgit Diary*, 7 January 1905
IOL/P&S/7/174/477: *Ladakh/Kashgar trade reports 1903–1904*, 1905
IOL/P&S/7/175/568: *Gilgit Diary*, 28 January 1905
IOL/P&S/7/176/731: *Kashgar News-Report*, 20 February 1905
IOL/P&S/7/177/849: *Gilgit Diary*, 8 April 1905
IOL/P&S/7/177/927: *Kashgar News-Report*, 31 January 1905
IOL/P&S/7/177/1038: *Gilgit Diary*, 6 May 1905
IOL/P&S/7/178/1008: *Kashgar Diary*, 20 April 1905
IOL/P&S/7/178/1144: *Kashgar Diary*, 20 May 1905
IOL/P&S/7/179/1198: *Chitral Diary*, 22 June 1905
IOL/P&S/7/179/1199: *Gilgit Diary*, 17 June 1905
IOL/P&S/7/179/1229: *Gilgit Diary*, 24 June 1905
IOL/P&S/7/179/1270: *Gilgit Diary*, 1 July 1905
IOL/P&S/7/180/1338: *Gilgit Diary*, 15 July 1905
IOL/P&S/7/180/1397: *Gilgit Diary*, 29 July 1905
IOL/P&S/7/180/1426: *Gilgit Diary*, 5 August 1905
IOL/P&S/7/180/1428: *Translation of telegram received by Mir of Hunza from Taotai of Kashgar and Amban of Sarikol*, 11 August 1905
IOL/P&S/7/181/1641: *Gilgit Diary*, 23 September 1905
IOL/P&S/7/181/1668: *Gilgit Diary*, 30 September 1905
IOL/P&S/7/182/1786: *Gilgit Diary*, 4 November 1905
IOL/P&S/7/183/1917: *Gilgit Diary*, 2 December 1905
IOL/P&S/7/184/338: *Gilgit Diary*, 16 December 1905
IOL/P&S/7/185/413: *Gilgit Diary*, 6 January 1906
IOL/P&S/7/185/413: *Gilgit Diary*, 30 December 1905
IOL/P&S/7/185/453: *Gilgit Diary*, 13 January 1906
IOL/P&S/7/186/696: *Gilgit Diary*, 10 February 1906
IOL/P&S/7/188/1024: *Gilgit Diary*, 8 May 1906
IOL/P&S/7/190/1300: *Letter* of George Macartney to B[ertrand] E[velyn] M[ellish] Gurdon in Gilgit, 5 May 1906
IOL/P&S/7/191/1497: *Gilgit Diary*, 25 July 1906
IOL/P&S/7/192/1618: *Gilgit Diary*, 15 August 1906
IOL/P&S/7/193/1654: *Gilgit Diary*, 22 August 1906
IOL/P&S/7/193/1654: *Gilgit Diary*, 29 August 1906
IOL/P&S/7/194/1869: *Kashgar Diary*, 20 September 1906
IOL/P&S/7/199/570: *Letter from Younghusband, Resident in Kashmir, proposing troop reductions in Gilgit and introduction of local levies*, 1907
IOL/P&S/7/201/940: *Report on the conditions of the existing trade between the Punjab and Tibet and of the trade routes traversed* by Mr. H[ubert] Calvert 6 November 1906, map plus 36 pp
IOL/P&S/7/205/1506: *Gilgit Diary*, 27 July 1907
IOL/P&S/7/205/1507: *Gilgit Diary*, 3 August 1907
IOL/P&S/7/206/1760: *Gilgit Diary*, 8 September 1907
IOL/P&S/7/209/2153: *Gilgit Diary*, 13 November 1907
IOL/P&S/7/209/2193: *Gilgit Diary*, 23 November 1907
IOL/P&S/7/210/333: *Gilgit Diary*, 28 December 1907
IOL/P&S/7/210/440: *Gilgit Diary*, 11 January 1908
IOL/P&S/7/216/1135: *Gilgit Diary*, 25 April 1908
IOL/P&S/7/218/1469: *Gilgit Diary*, 20 June 1908
IOL/P&S/7/218/1469: *Gilgit Diary*, 27 June 1908
IOL/P&S/7/219/1558: *Gilgit Diary*, 4 July 1908
IOL/P&S/7/226/437: *Gilgit Diary*, January 1909
IOL/P&S/7/226/480: *Kashgar Diary*, 20 January 1909
IOL/P&S/7/227/592: *Gilgit Diary*, February 1909
IOL/P&S/7/228/702: *Gilgit Diary*, March 1909
IOL/P&S/7/228/738: *Kashgar News-Report*, 20 March 1909
IOL/P&S/7/228/842: *Gilgit Diary*, April 1909
IOL/P&S/7/229/980: *Gilgit Diary*, May 1909
IOL/P&S/7/230/1275: *Gilgit Monthly Diary*, July 1909
IOL/P&S/7/231/1399: *Kashgar News-Report*, 10 August 1909
IOL/P&S/7/233/1556: *Gilgit Diary*, September 1909
IOL/P&S/7/235/239: *Gilgit Diary*, December 1909
IOL/P&S/7/240/824: *Gilgit Diary*, April 1910
IOL/P&S/7/241/1118: *Gilgit Diary*, June 1910
IOL/P&S/7/242/1281: *Gilgit Diary*, July 1910
IOL/P&S/7/243/1477: Captain [Douglas Giffard] Oliver: *Report on a journey from Gilgit to Leh via Kashgar and Yarkand*, 1910
IOL/P&S/7/243/1599: *Gilgit Diary*, September 1910
IOL/P&S/7/246/815: *Gilgit Diary*, March 1911
IOL/P&S/7/247/574: *Kashgar Diary*, January 1911
IOL/P&S/7/250/1302: *Gilgit Diary*, June 1911
IOL/P&S/7/250/1654: *Gilgit Diary*, 1911

India Office Library & Records: Departmental Papers: Political and Secret Subject Files 1902–1931
IOL/P&S/10/55: *Chinese Turkestan (1904–1916)*
IOL/P&S/10/241: *China Revolution; the situation at Kashgar* (1912)
IOL/P&S/10/278: Pt. 2, *Northern Frontier: Taghdumbash Pamir, Hunza rights (1904–1916)*
IOL/P&S/10/278: *Letter* of P. A. Gilgit to Resident in Kashmir, Gilgit, 2 October 1909
IOL/P&S/10/278: *Note on the history of Hunza rights on the Taghdumbash Pamir* by William Macandrew Marshall, Gilgit, 1913
IOL/P&S/10/278: *Letter* from the hon'ble Mr. S[tuart] M[itford]

Fraser, C.S.I., C.I.E., Resident in Kashmir, to A[lfred] H[amilton] Grant, Officiating Foreign Secretary to the Government of India, Foreign and Political Department, Delhi, 24 February 1914
IOL/P&S/10/278: *Letter* by P[ercy] M[olesworth] Sykes, C[onsul] [General] Kashgar, to H[enry] V[enn] Cobb, Resident in Kashmir, Kashgar, 10 August 1915
IOL/P&S/10/279: *Northern Frontier: Hunza-China frontier; Kashmir-China frontier; Kanjut, Raskam, Taghdumbash (1904–1911)*
IOL/P&S/10/279: Subject File 2426/4–6, 1912 (Part 2), *Letter* from Captain B[ertrand] E[velyn] M[ellish] Gordon, P. A. in Gilgit to E[lliot] G[raham] Colvin, Resident in Kashmir, No. 6C, 19 May 1904
IOL/P&S/10/279: *Letter* to John Morley, Secretary of the State for India, 27 September 1906
IOL/P&S/10/330: Chinese Turkestan. Part 1: *Affairs at Kashgar (1913)*; Part 2: *Chinese Turkestan: registration of British subjects (1910–1915)*
IOL/P&S/10/453: *Kashgar Consulate General officiating appointments of Lt Col [Roger Lloyd] Kennion, Lt Col P[ercy] M[olesworth] Sykes; Sir G[eorge] Macartney, Maj P[ercy] T[homas] Etherton, Mr C[larmont] P[ercival] Skrine, Lt Col R[obert] A[dolphus] Lyall (1914–1925)*
IOL/P&S/10/588/1: *H. H. the Aga Khan. Status etc.* (1915–1925).
IOL/P&S/10/588/2: *The Aga Khan – the Khoja community* (1912–1929)
IOL/P&S/10/825: *Kashgar Monthly Diaries* (1912–1920)
IOL/P&S/10/826: *Gilgit Agency Monthly Diaries* (1912–1920)
IOL/P&S/10/924: *Northern Frontier: Hunza affairs; situation in Raskam (1920–1932)*
IOL/P&S/10/973: *Gilgit Agency Diaries* (1921–1930)
IOL/P&S/10/976: *Kashgar Diaries* (1921–1930)
IOL/P&S/10/1018: *Chinese Turkestan: Travellers* (1921–1929)
IOL/P&S/10/1152: *Soviet menace to Northern Frontier* (1925–1934)
IOL/P&S/10/1153: *Northern Frontier: Supply of arms to Sinkiang* (1928–1932)
IOL/P&S/10/1192: *Revision of Aitchison's Treaties: Bhutan, Kashgar, Nepal, Siam and Tibet sections* 1926–1933
IOL/P&S/10/1217: *Chinese Turkestan: Travellers* (1927–1931)
IOL/P&S/10/1218: *Chinese Turkestan: Travellers* (1930–1933)

India Office Library & Records: Political and Secret Annual Files 1912–1930
IOL/P&S/11/8: *Slave trade expenditure* (1890–1912)
IOL/P&S/11/85: *Chinese Turkestan: alleged German intrigues with Muhammadans* (1914)
IOL/P&S/11/142: *Memorandum by India Office on 'The Future of Russian Central Asia'* (1918)
IOL/P&S/11/152/P 2732: *Northern Frontier: tour of Captain Samad Shah among the local followers of the Aga Khan* (20 May 1919 – 26 April 1922)
IOL/P&S/11/157/P 5983: *Telegram* from the Aga Khan (29 September 1919 – 30 September 1919)
IOL/P&S/11/184/P 511: *Afghanistan: Afghan refugees in India* (1919–1928)
IOL/P&S/11/243: *Chinese Turkestan: opium traffic* (1923–1928)
IOL/P&S/11/244: *Kashgar: ill-treatment of followers of the Aga Khan by the Chinese Amban of Sarikol* (1924)

India Office Library & Records: Afghanistan Consulate 1927–1947
IOR/12/50/394 I: *Trade of Turkestan* (from 1927)
IOR/12/51/394 II: *Trade of Chinese Turkestan and Sinkiang* (1934–1938)
IOR/12/52/394 III: *Trade of Chinese Turkestan and Sinkiang* (1937–1942)
India Office Library & Records: Departmental Papers: Political & Secret Internal Files & Collections 1931–1947
IOL/P&S/12/97: *Northern Frontier, Property of late Safdar Ali, Ex-Mir of Hunza* (1933–1934)
IOL/P&S/12/1321: *Memorandum 'A survey of Anglo-Afghan Relations' 1747–1919* (Part 1), 1919–1947 (Part 2), *The Indo-Afghan Frontier* (Part 3) (1948)
IOL/P&S/12/1619: *Russia: treaties between H. M. G. and Russia regarding the northern frontier of Afghanistan, since 1869* (1922–1945)
IOL/P&S/12/1619: *Convention between the United Kingdom and Russia relating to Persia, Afghanistan and Tibet*, signed at St. Petersburg, 31 August 1907. London
IOL/P&S/12/1632: *Opium: Export of Afghan opium through India* (1924–1928)
IOL/P&S/12/1633: *Opium: Export of Afghan opium through India* (1940–1947)
IOL/P&S/12/1672: *Northern Frontier: Soviet activities on the northern frontier of Afghanistan's Oxus river; Soviet-Afghan frontier agreement* (1934–1948)
IOL/P&S/12/1673: *Northern Frontier: Soviet activities on the northern frontier of Afghanistan; Soviet-Afghan frontier agreement (Oxus river)* (1948)
IOL/P&S/12/1753: *Trade. Stoppage of Trade between Chitral and Badakhshan* (1935–1939)
IOL/P&S/12/1993: *Aircraft & aviation. Flight of R. A. F. aeroplanes to Gilgit; flights of foreign aircraft over Gilgit and Chitral* (1932–1938)

India Office Library & Records: Departmental Papers: Political & Secret Internal Files & Collections 1931–1947 – Collection 10: Central Asia
IOL/P&S/12/2276: *Railways, roads and air routes in Central Asia (including the Soviet Union)* (15 December 1925 – 2 September 1946)
IOL/P&S/12/2277: *Soviet Policy in Central Asia: organisation of Central Asia intelligence system; propaganda, etc., in regard to the USSR* (22 September 1939 – 4 November 1942)

India Office Library & Records: Departmental Papers: Political & Secret Internal Files & Collections 1931–1947 – Collection 12: Chinese Turkestan

IOL/P&S/12/2331: *Internal situation* (September 1930 – November 1933)
IOL/P&S/12/2332: *Chinese Turkestan. Kashgar Monthly Diaries* (January 1931 – December 1938)
IOL/P&S/12/2333: *Danger of Soviet penetration on Northern Frontier: question of appointment of a British representative at Urumchi* (2 October 1925 – 27 July 1937)
IOL/P&S/12/2335: *Kashgar Consulate* (25 June 1930 – 14 October 1940)
IOL/P&S/12/2336: *Tours: Colonel Schomberg's reports; Political report of 1930, report by Mr Gillett in 1937* (2 January 1930 – 11 February 1941)
IOL/P&S/12/2337: *Grant of honorary decorations to friendly local and Chinese officials at Kashgar* (23 September 1929 – 27 July 1931)
IOL/P&S/12/2342: *Kashgar Annual Reports* 1922–1931 (14 January 1924 – 2 June 1945)
IOL/P&S/12/2346: *Trade: development of trade with India* (1 September 1927 – 19 September 1939)
IOL/P&S/12/2348: *Kashgar mail* (20 May 1927 – 29 April 1948)
IOL/P&S/12/2354: File copy of: *Letter* from his Britannic Majesty's Consul-General, Kashgar, No. 8-M./37, dated Kashgar, 01 July 1937, *Trade reports 1928–1939: statement of trade 1943/44* (25 June 1929 – 12 July 1944)
IOL/P&S/12/2357: *Chinese Turkestan, Sinkiang, Internal situation and affairs 1937–1938 (1937 Rebellion)* (16 November 1936 – 4 November 1938)
IOL/P&S/12/2358: *Chinese Turkestan, Sinkiang, Internal situation 1938–1940* (7 November 1938 – 28 December 1940)
IOL/P&S/12/2358/14: *Letter* from H[arry] H[all] Johnson, Consul General Kashgar, to Government of India, External Affairs Department Simla, Kashgar 15 August 1940
IOL/P&S/12/2359: *Chinese Turkestan, Sinkiang, Internal situation* (17 August 1939 – 28 June 1946)
IOL/P&S/12/2360: *Chinese Turkestan, Sinkiang, Internal situation*, 1946–1948
IOL/P&S/12/2361: *Chinese Turkestan, Sinkiang, Internal situation (Kashgar Reports)* (22 June 1948 – 3 January 1949)
IOL/P&S/12/2364: *Sinkiang Internal situation (mail reports)* 15 November 1933 – 12 November 1936)
IOL/P&S/12/2378: *Chinese Turkestan: Export of opium and charas from Sinkiang to India* (15 January 1936 – 19 August 1944)
IOL/P&S/12/2382: *Chinese Turkestan, Communications: improvements of roads between India and Chinese Turkestan* (22 June 1935 – 2 August 1938)
IOL/P&S/12/2383: *Internal situation: Kashgar weekly letter October 1937 to April 1943* (24 January 1938 – 17 June 1942)
IOL/P&S/12/2384: *Chinese Turkestan. Internal Situation: Kashgar weekly letter April 1942 – July 1946* (8 June 1942 – 24 November 1946)
IOL/P&S/12/2403: *Trade with India. Provincial regulations; Khotan silk; development of trade* (4 October 1942 – 4 October 1946)
IOL/P&S/12/2407: *Report on journeys from India to Kashgar: Mr K[umara] P[admanbha] S[ivasankara] Menon and Mr Etherington-Smith; opening of Pakistan Consulate-General at Kashgar* (17 January 1945 – 17 July 1946)

India Office Library & Records: Departmental Papers: Political & Secret Internal Files & Collections 1931–1947 – Collection 24: Northern Frontier

IOL/P&S/12/3285: *Gilgit Agency: Political diaries January 1931 to May 1947* (30 May 1931 – 26 July 1947)
IOL/P&S/12/3286: *Chitral and Kashmir: Mastuj Agreement of 1914, Kashmir suzerainty over Chitral* (16 February 1932 – 6 April 1933 and 18 April 1945 – 26 April 1945)
IOL/P&S/12/3287: *Future of Gilgit Agency: position of Kashmir Darbar, ceremonial Nagars, withdrawal of Kashmir troops from Gilgit* (15 June 1931 – 11 December 1941)
IOL/P&S/12/3288: *Northern Frontier: Gilgit Administration Reports 1927 to 1945* (7 August 1928 – 29 October 1945)
IOL/P&S/12/3291: *Northern Frontier: Increase of subsidies in the Gilgit Agency* (8 February 1926 – 13 December 1928)
IOL/P&S/12/3292: *Hunza: Situation in Raskam; Darwaza incident* (31 July 1933 – 1 January 1940)
IOL/P&S/12/3292: *Report on Shimshal Reconnaissance* by Major R[ichard] [Alfred] Cropper 1938
IOL/P&S/12/3292: *Note on the boundary dispute case Hunza (Gilgit Agency) – Sinkiang in the neighbourhood of the Shimshal Valley and Oprang river* by Political Agent Ian William Galbraith 17 April 1939, 14 pages
IOL/P&S/12/3293: *Criticism by Colonel Schomberg of the administration of the Gilgit Agency* (14 May 1934 – 6 May 1946)
IOL/P&S/12/3294: *Economic development of the reconstituted Gilgit Agency: grant by Government of India* (31 August 1936 – 16 September 1936)
IOL/P&S/12/3297: *Gilgit Agency: reorganization of Gilgit Corps of Scouts* (20 May 1935 – 27 March 1940)
IOL/P&S/12/3299: *Hunza Affairs: new Mir of Hunza, 1938* (2 August 1938 – 6 June 1945)
IOL/P&S/12/3301: *Northern Frontier. Political Agent at Gilgit* (14 June 1939 – 5 January 1940)
IOL/P&S/12/3302: *Northern Frontier. Kazaks* (8 June 1942 – 21 December 1945)
IOL/P&S/12/3303: *Northern Frontier. Gilgit Affairs 1948: Stalin Road; frontier beyond Gilgit* (21 January 1948 – 2 March 1948)

India Office Library & Records: Departmental Papers: Political & Secret Internal Files & Collections 1931–1947 – Collection 37: Travellers

IOL/P&S/12/4264: *Chinese Turkestan: Lt Col R[eginald] C[harles] F[rancis] Schomberg* (19 May 1932 – 26 March 1933 and 19 November 1946 – 14 April 1947)
IOL/P&S/12/4323: *Supply of maps of Karakorum Range* (2 October 1936)

India Office Library & Records: Departmental Papers: Political & Secret Internal Files & Collections 1931–1947 – Collection 46: Routes
IOL/P&S/12/4609: *Road routes via the Karakorum Mountains to Sinkiang; supplies to Russia and China* (24 November 1941 – 4 December 1946)
IOL/P&S/12/4610: *Air routes from India to Sinkiang via the Karakorum Mountains* (22 May 1942 – 26 February 1943)
IOL/P&S/12/4616: *Map: Central Asia; supply routes to China* (c. January 1943)
IOL/P&S/12/4618: *Supplies to China and Russia: allied Supplies Executive Sub-Committee on supplies to China* (10 July 1942 – 18 September 1945)
IOL/P&S/12/4621: *Routes through Afghanistan to Russia. Supplies to Russia* (14 July 1941 – 30 March 1944)

India Office Library & Records: Departmental Papers: Political Internal/Indian States Files and Collections
IOL/P&S/13/63: Aga Khan and Aly Khan (1929–1940)
IOL/P&S/13/64: Aga Khan and Aly Khan (1940–1947)

India Office Library & Records: Political and Secret Files and Collections – Collection 48: Indian Independence
IOL/P&S/13/1860: 5–6: *The Times: The high Karakoram. Ancient states and modern communications* (22 June 1949)
IOL/P&S/13/1860: 31–40: *Precis of events in Gilgit since the arrival of Sardar Mohd. Alam, Political Agent*. Report by Major William Alexander Brown, 10 pp
IOL/P&S/13/1860: 53–59: *The Gilgit Agency.* Memorandum prepared by Lieutenant John Benedict Eden, 7 pp (1947)
IOL/P&S/13/1860: 64–71: *Gilgit.* Memorandum on the Gilgit Agency forwarded by Lieutenant Horace Algernon Fraser Rumbold, 8 pp (10 February 1948)
IOL/P&S/13/1860: 114–123: *Precis on recent events in the Gilgit Agency.* Report by Major William Alexander Brown, 10 pp (December 1947)
IOL/P&S/13/1936: *Pinnell Files V. Memoranda on Gilgit and Kashmir, the Indian States and transfer of power, with annexures* (October 1947 – January 1948)

India Office Political and Secret Department: Political and Secret Department Memoranda c1840–1947
IOL/P&S/18/A 83: *Note on the complication with Hunza* by Stewart C[olvin] Bayley, 12 pp (25 January 1892)
IOL/P&S/18/A 88: *Reorganization of the Gilgit Agency* by S[tewart] C[olvin] Bayley, 15 pp (17 November 1892)
IOL/P&S/18/A 92: *Gilgit Agency, Chilas and Chitral* by Stewart C[olvin] Bayley, 9 pp (11 July 1893)
IOL/P&S/18/A 95: *The northern frontier of India. Roads and passes. Measures for defence of frontier.* Francis Edward Younghusband February 1895
IOL/P&S/18/A 96: *Chitral and the Gilgit Agency* by Sir A[lfred] C[omyn] Lyall, 7 pp (11 March 1895)
IOL/P&S/18/A 98: *Chitral and Gilgit* by D[onald] M[artin] Stewart, 3 pp (26 March 1895)
IOL/P&S/18/A 160: *Frontier between Hunza and the Chinese Dominions* by P[ercy] H[arald] Dumbell, 10 pp (19 July 1904)
IOL/P&S/18/A 170: *Frontier between Hunza and the Chinese Dominions* by J[ohn] E[velyn] Shuckburgh, 8 pp (18 November 1911)
IOL/P&S/18/A 172: *Suggestions for a convention between Great Britain and Russia, regarding the Chinese Province of New Dominion* by G[eorge] Macartney, 5 pp (23 August 1915)
IOL/P&S/18/C14: *Mission to Yarkand*, T[homas] D[ouglas] Forsyth, 1876, 11 pp
IOL/P&S/18/C15: *Kashgar Mission [Yarkand]. Secret Letter from India No. 50*, dated 16 October 1876, A[dolphus] W[arburton] Moore, 15 November 1876, 15 pp
IOL/P&S/18/C 104: *Memorandum on the present position in Central Asia* by Sir H[enry] M[ortimer] Durand, 31 pp (21 May 1887)
IOL/P&S/18/C142: *Revision of Anglo-Russian Convention of 1907* [1915], 20 pp
IOL/P&S/18/C 202: *The political situation in Russian and Chinese Central Asia* by Lieutenant Colonel P[ercy] T[homas] Etherton, 5 pp (20 October 1920)

India Office Library & Records: Political and Secret Department Library. Foreign Office Prints
IOL/P&S/20/B 20: Barrow, Edmund G. 1890: *Gazetteer of Kashmír and Ladák together with routes in the territories of the Maharája of Jamú and Kashmír*. Calcutta: The Superintendent of Government Printing, 871 pp
IOL/P&S/20/B 275/2: Government of India 1925. *Jammu and Kashmir State. List of ruling princes, chiefs and leading personages.* Calcutta: Government of India Central Publication Branch, 31 pp [with Addendum and Corrigendum inserted on 28 October 1926]
IOL/P&S/20/B 309: General Staff India 1941: *Military report and gazetteer of Indus Kohistan, Black Mountain and adjacent territories.* 1940. Simla: Government of India Press, 263 pp

India Office Records – Crown Representative Indian States Residencies Records
IOR/1/29/1695 (new: IOR/1/1/3065): *Proposals made by the Kashmir Govt., regarding the appointment of the Political Agent, Gilgit and holding of Darbars at Gilgit, Srinagar or Jammu (1937–1938)*
IOR/1/29/2322 (new: IOR/1/1/3688): *Status of Hunza and Nagir and Political Districts of Chilas, Koh, Ghizar, Ishkoman and Yasin vis-a-vis the Kashmir state*

India Office Library & Records: Crown Representative's Records – Indian State Residencies: Kashmir Residency Files
IOR/2/1064/42: *Abolition of slavery in Chitral and Chilas* (1897)
IOR/2/1064/44: *Emancipation of slaves in Chinese Turkestan* (1897)
IOR/2/1064/45: *Indo-Chinese Turkestan Trade (1897).* Report by S[tuart] H[ill] Godfrey, Political Agent Gilgit, to A[delbert] C[ecil] Talbot, Resident Kashmir, Gilgit 17 December 1896

IOR/2/1067/89: *Discussion of various questions between the resident H. B. Majesty's Consul General, Kashgar, (Col. P[ercy] T[homas] Etherton) on certain matters which affect the Gilgit Agency & India and notes recorded thereon* (1922)

IOR/2/1067/93: (1) *Political control of the independent territory lying between the North-West Frontier Province and the Gilgit Agency.* (2) *Political and administrative status of the Gilgit Agency and the Gilgit Wazarat.* (3) *Boundary between the Gilgit Agency and Chitral* (1927–1933)

IOR/2/1069/129: (1) *Jurisdiction of the Joint Commissioner appointed under the Treaty of 1870 between the British Government and His Highness the Maharaja of Jammu and Kashmir over the Central Asian Trade Route.* (2) *Rules regarding check of goods in transit in the Treaty High Road (Central Asian Trade)* (1939)

IOR/2/1070/137: *Area of Jammu and Kashmir State – Census of 1941* (1942)

IOR/2/1071/162: *Redistribution of Waziri and other allowances in Hunza.* File No. 821 (3)/45

India Office Library & Records: Crown Representative's Records – Indian State Residencies: Kashmir Residency Files – Confidential Files

IOR/2/1075/217: *Proceedings of the Russians and Chinese with respect to the Raskam valley and the Taghdumbash. Negotiations for the lease of land in Raskam to the Mir of Hunza. Possibility of providing land for the surplus population of Hunza in the Gilgit Agency* (1900)

IOR/2/1075/217: 40: *Letter* from Political Agent Gilgit to Resident in Kashmir, Bunji, 6 June 1900

IOR/2/1076/219: *The political situation in Kashgar* (1913)

IOR/2/1076/226: *Position of Afghan subjects in Chinese Turkestan and Kashgaria* (1920)

IOR/2/1076/230: *Raskam valley dispute* (1925)

IOR/2/1076/231: *Kashgar Consulate Guard* (1936)

IOR/2/1076/232: *Development of trade between India and Chinese Turkistan* (1938)

IOR/2/1076/233: *Indo-Sinkiang mail and Kashgar Consulate supplies* (1947)

IOR/2/1078/241: *Chitral-Gilgit Defence Scheme. Revision of the Chitral-Gilgit Defence Scheme. Proposed withdrawal of the Kashmir State troops from the Gilgit Agency and embodiment of Gilgit Corps of Scouts. Proposed transfer by the Kashmir Government to the Government of India of the entire control of the civil and military administration including the Gilgit Wazarat, i.e. the unsettled districts plus that portion of the Wazarat which lies north of the Indus. Estimate by the Political Agent, Gilgit, of revenue and expenditure connected with the proposed transfer of control* (1931–1934)

IOR/2/1078/242: *Chitral-Gilgit Defence Scheme. Withdrawal of the Kashmir troops from the Gilgit Agency. Decision of the Government of India to take over the civil and military administration of the portion of the Gilgit Wazarat lying on the north of the Indus river. Estimates of revenue and expenditure in connection with the proposed transfer of the administration* (1931–1934)

IOR/2/1079/244: *Activities of His Highness the Mehtar of Chitral. Repatriation of Chitrali refugees in the Gilgit Agency. Maulai situation in Chitral* (1925)

India Office Library & Records: Crown Representative's Records – Indian States Residencies – Gilgit, Chilas, Hunza and Nagir Files (Confidential)

IOR/2/1079/245: *Captain Durand's Mission to Hunza and Nagir* (1886)

IOR/2/1079/246: *News that Safdar Ali Khan murdered by a gun shot his father Raja Ghazan Khan of Hunza and with the advice of the Wazirs and officials succeeded the chiefship of Hunza* (1886)

IOR/2/1079/247: *Increase in the permanent staff of the Gilgit Agency. Political separation of Chitral from Gilgit and the future status of the British Agent, Gilgit* (1889)

IOR/2/1079/248: *Correspondence* with the British Agent, Gilgit (1890)

IOR/2/1079/251: *Hunza and Nager subsidies. Proposed renewal of the subsidies paid by the British Government and by the Kashmir State to the Hunza and Nagir States prior to 1891. Sanction of the Government of India, with effect from the 1st April, 1895, to the grant of an annual allowance of Rs.2,000 each to the States of Hunza and Nagir* (1894)

IOR/2/1079/252: *Hunza and Nagir: Grant of subsidies to the Mirs of Hunza and Nagir* (1895)

IOR/2/1079/253: *Political relations of Hunza with China; payment of the annual present of gold by the Mir of Hunza to China. Claims of the Mir of Hunza in the Tagdumbash Pamir and the Raskam Valley; question of the settlement with China of the boundaries of Kashmir, Hunza and Afghanistan, and the establishment of a post at Darwaza by the Hunza men* (1880–1898)

IOR/2/1079/254: *Proceedings of the Russians and Chinese with respect to the Raskam valley and the Taghdumbash Pamir. Negotiations for the transfer of land in Raskam to the Mir of Hunza* (1899)

IOR/2/1080/256: *Hunza claims to the Raskam valley. Proposal to sever Hunza's connections with China and compensate Hunza. Proposals to provide for the redundant population of Hunza* (1904)

IOR/2/1080/257: *Allotment and distribution of land among the people of Hunza – Hunza's relation with China – Hunza-China Boundary. Mr. Macartney's position at Kashgar* (1905)

IOR/2/1081/263: *Hunza's rights to grazing fees in the Taghdumbash Pamir* (1913)

IOR/2/1082/265: *Hunza rights in Raskam* (1914)

IOR/2/1082/266: *Mir of Hunza's rights in Raskam* (1915)

IOR/2/1082/267: *Hunza's right to grazing fees in the Taghdumbash Pamir* (1915)

IOR/2/1082/276: *Mir of Hunza's right to grazing fees in the Taghdumbash Pamir* (1917)

IOR/2/1083/280: *Reports of Lieutenant-Colonel C[hetwynd] H[enry] Haswell, R. E., on the forts, roads etc., in the Gilgit Agency.* 1921

IOR/2/1083/284: *Increase of subsidies of certain Chiefs and Governors of the Gilgit Agency.* 1926

IOR/2/1083/286: *Future of the Gilgit Agency. Memorandum on the future of the Gilgit Agency* prepared by Lieut.-Colonel G[eorge] D. Ogilvie, CIE, Resident in Kashmir (1929)

IOR/2/1085/292: *Mir of Hunza's rights in the Taghdumbash Pamir and the Raskam valley* (1935)

IOR/2/1085/295: *Question of the issue of regular passports (with photographs and visas) for carriers of Kashgar Consulate General official mail from Gilgit. Establishment of regular postal service between India and Sinkiang. Conditions of service of mail runners on the Misgar-Tashkurghan section* (1936–1939)

IOR/2/1085/296: *Compensation to the Mir of Hunza for the loss of his rights in Chinese Turkestan (5 April 1937). Grant of a jagir to the Mir of Hunza in the Gilgit Sub-division and increase in his subsidy* (1937)

IOR/2/1085/298: *Air navigation in the Gilgit Agency. Report on the Hunza valley ground reconnaissance. Proposed construction of a landing ground at Pasu in preference to Sost in the Hunza Valley* (1937)

IOR/2/1086/299: *Accession of Indian States to Federation. Position of Hunza, Nagir, Yasin, Koh Ghizar, Ishkuman, Chilas, etc., in the event of Kashmir acceding to federation. (View that these States and districts are not an integral part of Kashmir)* (1938)

IOR/2/1086/302: *Collection by the Mir of Hunza of offerings for his Highness the Aga Khan from places in Central Asia including Badakhshan and Chitral* (1940)

IOR/2/1086/303*: Status of Hunza, Nagir, Chilas Koh Ghizar, Iskhoman and Yasin vis-à-vis the Kashmir State. Decision that these territories are not part of the Kashmir State* (1941)

IOR/2/1086/304: *Status of Hunza, Nagir, Chilas Kuh-Ghizar, Iskhoman and Yasin vis-à-vis the Kashmir State. Attestation Kashmir State subject certificates by the Political Agent, Gilgit* (1941)

IOR/2/1086/306: *Proposed reopening of the Leh-Karakorum Trade Route to Sinkiang. Visit of the Chinese Survey Party to Gilgit* (1942)

IOR/2/1086/308: *Death of Mohammed Ghazan Khan of Hunza and succession of Mohammad Jamal Khan as Mir of Hunza* (1945)

IOR/2/1086/311: *Matters relating to the measures found necessary to strengthen the position of Gilgit, Hunza and Nagar* (1891)

IOR/2/1086/312: *Hunza rights in the Raskam valley* (1923)

India Office Records – Official Publications: British Parliamentary Papers 1802–1955

IOR/V/4/6621: House of Commons. Parliamentary Papers 1892: *East India (Hunza expedition). Correspondence relating to the operations in Hunza-Nagar*. London: Eyre and Spottiswoode, 12 pp

IOR/V/15/52: Rai Bahadur Pandit Bhag Ram 1893: *The Kashmir State (Vol. XXVIII): Report on the census and Imperial and supplementary tables*. Lahore: Mufid-i-Am Press, 200 pp

IOR/V/15/84: Ghulam Ahmed Khan 1902: *Census of India 1901 (Vol. XXIII). Kashmir, Part I: Report*, 109 pp. *Part II: Tables*, 325 pp. Lahore: Civil and Military Gazette Press,

IOR/V/15/103: Md. Matin-Uz-Zaman Khan 1912: *Census of India (Vol. XX). Kashmir, Part I: Report*, 256 pp. *Part II: Tables*, 223 pp. Lucknow: Newul Kishore Press

IOR/V/15/119: Chaudri Khushi Mohammed 1923: *Census of India (Vol. XXII). Kashmir, Part I: Report*, 209 pp. *Part II: Tables*, 235 pp. *Final census report of census administration in Jammu and Kashmir State, 1921*, 29 pp. Lahore: Mufid-i-Am Press

IOR/V/27/60/160: Mason, Kenneth 1922: *Routes in the Western Himālaya, Kashmīr, &c. Volume I: Pūnch, Kashmīr & Ladākh*. Dehra Doon: Government of India Press, 223 pp

IOR/V/27/400/3–4: De Schlagintweit, Hermann, Adolphe and Robert 1861–1866: *Results of a scientific mission to India and High Asia*. Vol 2: *General hypsometry of India, the Himalaya, and Western Tibet with sections across the chains of the Karakorum and Kuenlun*. Vol. 3: *Route-book of the western parts of the Himalaya, Tibet and Central Asia; and geographical glossary from the languages of India and Tibet*. Leipzig: Brockhaus, London: Trubner

IOR/V 5116: *Revised List of Ruling Prices, Chiefs and leading Personages of the Jammu and Kashmir State and the Gilgit Agency*. Published by Manager of Publications, Government of India Press, New Delhi 1939

Kirkbride, George. 1936. *Report on visit of Air Marshal Sir Edgar Ludlow-Hewitt … Air Officer Commanding the Royal Air Force in India to Gilgit, October 21st to 24th 1936* (IOL/P&S/12/3285: 252–253), 4 pp

Klamert, Gerhart. 1954. *Diary of the German-Austrian Himalaya-Karakorum Expedition 1954*, diary and porter service book (3 volumes)

Klamert, Gerhart. 1959. *Diary of the German Karakorum Expedition 1959*, 100 pp

Lorimer, David Lockhart Robertson. 1906–1948. *Papers of Lieutenant-Colonel David Lockhart Robertson Lorimer*. School of Oriental and African Studies, University of London, 62 boxes and 2 albums; PP MS 66 (previously MS 181247)

Lorimer, David Lockhart Robertson. 1921–1957. *Papers of Lt.-Col. David Lockhart Robertson Lorimer (1876–1962), Indian Army 1896, Indian Political Service 1903–27, Political Agent, Gilgit 1920–24, including notes on Shina grammar and vocabulary and correspondence on the lifestyle of the Hunzas*. India Office Records and Private Papers Mss Eur D1168

Lorimer, David Lockhart Robertson. 1953. *File 'Hunza Humbug 1953' subtitled 'The Rise, Decline & Fall of a Myth'*. India Office Records and Private Papers Mss Eur D1168/7

McMahon, Arthur Henry. 1898. *Report on the subject of the claims of the Kanjut tribe (i.e., the people of Hunza) to territory beyond the Hindu Kush, i.e., to the Tagdumbash Pamir and the Raskam Valley* (IOR/2/1079/253)

Mir Shaukat Ali Khan. 1950. *Statement by Mir Shaukat Ali Khan, Mir of Nagar*, 5 pp (DAV EXP 2 SG, 133/0)

Muhammad Riza Beg. 1920. *Tarikh-i Hunza*. Manuscript in Bodleian Library, Oxford. Pers d. 62

National Archives (Public Record Office), Kew – Foreign & Commonwealth Office

FCO 21/71: *China – Political Affairs (External): Bilateral Sino-Pakistan relations*. January 1967 to October 1968, 101 pp

FCO 37/314: *Briefs for Secretary of State's visit to India*, Foreign Office Files for India, Pakistan and Afghanistan 1968
FCO 37/694: *Chinese interests in Kashmir. Road routes between Pakistan and China*. Foreign Office Files for India, Pakistan and Afghanistan 1970
FCO 37/1453: *Political situation in Kashmir 1974. Confidential report on visit to Jammu and Kashmir, between 6–12 October 1974* by Robin Christopher, 19 pp
FCO 37/1498: *Internal situation in Pakistan 1974*
FCO 37/1502: *Political situation in Pakistan Kashmir 1974*
FCO 37/1514: *Dispute between India and Pakistan over Kashmir 1974*
FCO 37/1653: *Internal situation in Pakistan 1975*
FCO 37/1664: *Dispute between India and Pakistan over Kashmir 1975*
FCO 37/2035: *Internal political situation in Pakistan 1977*
FCO 37/2108: *Internal political situation in Pakistan 1978*

National Archives (Public Record Office), Kew – Foreign Office
FO 371/6625: *Employment of an Indian officer as intelligence officer at Kashgar*, 43 pp
FO 371/19335: *Claim of Chinese administrator in Kashgar over sovereignty over Hunza 1935, Frontier between India and Chinese Turkestan, Hunza affairs*, 38 pp
FO/371/22115: China F 3092: 175–182: *Situation in Sinkiang: Mir of Hunza's Robatchi Estate at Yarkand, Kashmir*, 8 pp
FO 371/31640: *Supply routes to China between Sinkiang and India 1942*, 5 pp
FO 371/101199: *Chinese claims on Pakistan territory*, 11 pp
FO 371/105173: *Establishment of autonomous regions in the border area of China, statistics of national minorities 1953*, 203 pp
FO 371/133389: Far Eastern Department FC 1081: *Extract from a report on a visit to Gilgit Agency by Brigadier B[rian] Kingzett, Military adviser to the H. C. Karachi. Activity on the Chinese border*. 30 June 1958, 3 pp
FO 371/170676: *Sino-Pakistan relations 1963*, 176 pp
FO 371/23493: *Chinese claims to Darwaza and to suzerainty over Hunza*. 14 August 1939, 40 pp

National Archives of India – Abilekh Patal. Digitized Public Records External Affairs
NAI/PR/1415997. Foreign Department April 1909 No. 139: *Visit of Sardar Ali Khan, Ex-Mir of Hunza, to Yarkand, when he has been returned to Kuchar as it is undesirable that he should be anywhere near the Hunza frontier*, 8 pp
NAI/PR/4001671. File No. 583-X: *Government of India; Foreign & Political Department. External (Secret). 1927. Note on Chinese Turkistan by Colonel Schomberg*, 16 pp
NAI/PR/4001675. File No. 37-X (Secret) of 1928: *Government of India. Foreign and Political Department. 1928. India's concern with Sinkiang*, 20 pp
NAI/PR/4001990. File No. 32-X: *Administration of Nager, on death of Sir Sikandar Khan, Mir of Nager*, 118 pp
NAI/PR/4001993. File No. 112-P: *Status of Hunza and Nagir and Political Districts of Chilas, Koh Ghizar, Ishkoman and Yasin vis-à-vis the Kashmir. Question of (a) an appeal by the Kashmir Govt. to the Sect. of State or reference to a Court of Arbitration and (b) inclusion of area and population of these areas (under the suzerainty of Kashmir) in those of Kashmir State and their treatment in the Political map of India*, 39 pp
NAI/PR/4001996. File No. 286-C.A: *Proposals for deposition of Mir of Hunza on grounds of Kashimir misrule in favour of his and capable heir Mirzada Jamal Khan*, 76 pp
NAI/PR/4002097. File No. 123-X (Secret): *Note by G. S. Branch on the political and strategically situation in Hunza State*, 21 pp
NAI/PR/4002104. File No. 58-X/37 (Secret): *Replacement of Shukrullah Beg by Zarparast as Wazir of Hunza. Proposed re-distribution in three equal shares of that part of the Hunza subsidy which comprises the Waziro allowance, between Shukrullah Beg, Zarparast and Jumal Khan*, 11 pp
NAI/PR/4002107. File No. 82(2)-X/39 (Secret): *Proposed establishment of further R.A.F. landing grounds in the Gilgit Agency in general and at Pasu and Sost in particular (1937)*, 189 pp
NAI/PR/5014500. File No. 112-IV (Confidential): *Fortnightly report for the first half of August 1926. From the Resident in Kashmir, dated 14 August 1926*, 6 pp

НАРГО – Научный Архив Русского Географического Общества – Nauchnyi Arkhiv Russkogo Geograficheskogo Obshhestva – Scientific Archive of the Russian Geographical Society St. Petersburg
Photographs of Bronislaw Grabczewski [Grombchevsky]: NARGO, font 45, opis 1, file 4, sheets 17–78

N. N. 1950. *Kashmir in maps*. [probably printed in Karachi, prepared for the United Nations], 12 pp
N. N. 1952. *Kashmir. Basic information in maps*. [probably printed in Karachi, prepared for the United Nations], 14 pp
Nazim Khan, Mohomed. 1936. *The autobiography of Sir Mohomed Nazim Khan, K.C.I.E. Mir of Hunza*. Translated from Persian into Urdu by M. Jamal Khan, translated from Urdu into English by Khan Bahadur Muhammad Maseh Pal and completed by B. K. Baltit: mimeographed, 145 pp
Ogilvie, George D. 1929. *Future of the Gilgit Agency. Memorandum on the future of the Gilgit Agency* (reprinted in IOL/P&S/12/3287; IOR/2/1083/286). Srinagar, 27 pp
Pakistan People's Party 1977. *Pakistan People's Party Manifesto January 1977*. Islamabad, 66 pp
Pal, Muhammad Masih. 1928. *Letters* from Muhammad Masih Pal to David Lorimer dated 5 and 19 January, 14 February, 20 April, 8 July 1928. Lorimer Personal Records in SOAS (PP MS 66/01/01, previously MS 181247)
Pal, Muhammad Masih. 1932. *Letter* from Muhammad Masih Pal to David Lorimer dated 4 July 1932. Lorimer Personal Records in SOAS (PP MS 66/01/01, previously MS 181247)
Pal, Muhammad Masih. 1934. *Letters* from Muhammad Masih Pal to David Lorimer dated 24 and 28 July 1934. Lorimer Personal Records in SOAS (PP MS 66/01/01, previously MS 181247)

Public Record Office (now National Archives)
PRO Russia. Proceedings in Central Asia 1873–1898
PRO/FO 65/1202/8: *No. 32 C., dated Srinagar, 29th September 1883*. From Lieutenant Colonel Sir O[liver] St. John, K. C. S. I., R. E., on Special Duty in Kashmir, to C[harles] Grant, Esq., C.S.I., Secretary to the Government of India, Foreign Department
PRO/FO 65/1202/8: *Translation of a letter from the Joint-Governors of Gilgit, to the Jummoo Darbar*, dated 31st Bhadon 1940 (14th September 1883)
PRO/FO 65/1202/10: *No. 33 C., dated Srinagar, 1st October 1883*. From Lieutenant Colonel Sir O[liver] St. John, K. C. S. I., R. E., on Special Duty in Kashmir, to C[harles] Grant, Esq., C.S.I., Secretary to the Government of India, Foreign Department
PRO/FO 65/1202/10: *Translation of a letter from the Joint-Governors of Gilgit, to the Jummoo Darbar*, dated 2nd Asoj 1940 (16th September 1883)
PRO/FO 65/1202/15: *Translation of a vernacular letter from the Joint-Governor of Gilgit, to the Jummoo Darbar*, dated 4th Asoj 1940 (18th September 1883)
PRO/FO 65/1202/15: *Translation of a letter from Raja Ghazan Khan of Hunza to the Joint-Governors of Gilgit*
PRO/FO 65/1202/15: *Translation of a letter from the Joint-Governors of Gilgit, dated 15th Asoj 1940* (29th September 1883)

Public Record Office: China. Confidential Prints. Trade with Chinese Turkestan 1888–1898
PRO/FO 405/44–45
PRO/FO 405/53–54
PRO/FO 405/67, 83
PRO/FO 418/51–52: *Russia and Central Asia*, 1912–1913

Qudratullah Beg, Haji. 1935. *Hunza ters*. Lorimer Personal Records in SOAS (PP MS 66/01/01/R3).
Research Department, Foreign Office. 1947. *The Russian menace to India*, 1 December 1947 (IOL/P&S/12/3303), 15 pp

Royal Asiatic Society Archives London
RAS/LOR/1: Lorimer-Gilgit Papers
RAS/LOR/1/1/1–5: Lorimer letters
RAS/LOR/1/2/1: Lorimer; David Lockhart Robertson. 1934. *The Burushaski Language. Introduction*. Aliabad, 1 September 1934, 41 pp
RAS/LOR/1/2/2: Lorimer; David Lockhart Robertson and Emily Overend Lorimer.1936. Report on expedition to Gilgit and Hunza, India 1934–1935. Welwyn Garden City, 7 January 1936, 24 pp
RAS/LOR/2: Lorimer-Carter Papers
RAS/LOR/2/1/1: Lorimer; David Lockhart Robertson. 1946: Note on the study of the languages of the Gilgit Agency, N.W. India. Welwyn Garden City, 21 January 1946, 13 pp
RAS/LOR/2/1/2–5: Letters

Royal Society for Asian Affairs London
RSAA/SC/SCH/2/47: Archive Box 12: Schomberg, Reginald Charles Francis. 1945. *Karakoram finale*. London, 90 pp
RSAA/SC/SCH/2/47: Archive Box 14: Schomberg, Reginald Charles Francis. 1926. *A Turkestan Diary*. Part 1 ten pages covering the journey of 1926; in total 162 pp
RSAA/SC/SCH/2/47: Archive Box 14: Schomberg, Reginald Charles Francis. 1931. *A Turkestan Diary*. Part 2; 91 pp

Russian State Military Historical Archive Moscow – Российский Государственний Военно-Исторический Архив (РГВИА)
RGVIA, font 1396, opis 2, file 2168, sheets 527–530: Report of Head of Pamir troops, Yagello, to Military Governor Fergana
RGVIA, font 1396, opis 2, file 2251, sheets 6–7: Memorandum about the Pamirs by Captain of the General Staff Andrei Jevgenievich Snesarev. Observations from two journeys 1899 and 1900
RGVIA, font 1396, opis 2, file 2251, sheet 19: Report No. 77 from Eduard Karlovich Kivekes to the General Staff Turkestan
RGVIA, font 1396, opis 2, file 2251, sheet 24: Report No. 79 from Eduard Karlovich Kivekes to the General Staff Turkestan 25 February 1902
RGVIA, font 1396, opis 2, file 2251, sheet 44: Report No. 102 from Eduard Karlovich Kivekes to the General Staff Turkestan 4 April 1902
RGVIA, font 1396, opis 2, file 2251, sheet 47: Report No. 105 from Eduard Karlovich Kivekes to the General Staff Turkestan 17 April 1902
RGVIA, font 1396, opis 2, file 2251, sheet 51: Report No. 108 from Eduard Karlovich Kivekes to the General Staff Turkestan 23 April 1902
RGVIA, font 1396, opis 2, file 2251, sheet 55: Report No. 121 from Eduard Karlovich Kivekes to the General Staff Turkestan 15 May 1902
RGVIA, font 1396, opis 2, file 2251, sheet 110: Report No. 163 from Eduard Karlovich Kivekes to the General Staff Turkestan 6 July 1902
RGVIA, font 1396, opis 2, file 2251, sheets 113–114: Report No. 1946 from Eduard Karlovich Kivekes to the Military Governor Margilan 6 July 1902
RGVIA, font 1396, opis 2, file 2251, sheet 115: Report No. 163 from Eduard Karlovich Kivekes to the General Staff Turkestan 6 July 1902
RGVIA, font 1396, opis 2, file 2251, sheets 130–134: Report No. 14 from Andrei Jevgenievich Snesarev to the General Staff Turkestan 5 August 1902
RGVIA, font 1396, opis 2, file 2251, sheet 137: Report No. 16 from Andrei Jevgenievich Snesarev to the General Staff Turkestan 5 August 1902
RGVIA, font 1396, opis 2, file 2251, sheets 141–144: Report No. 20 from Andrei Jevgenievich Snesarev to the General Staff Turkestan 17 August 1902
RGVIA, font 1396, opis 2, file 2251, sheet 149: Letter of the Chief Commander in the General Staff Turkestan Vsevolod Viktor-

ovich Sacharov to Captain Andrei Jevgenievich Snesarev to the General Staff Turkestan 13 September 1902

RGVIA, font 1396, opis 2, file 2251, sheets 190–191: Report No. 23 from Andrei Jevgenievich Snesarev to the General Staff Turkestan 26August 1902

RGVIA, font 1396, opis 2, file 2251, sheets 227–235: Report about the activities of the Pamir troops between 27 October 1901 and 31 August 1902 by Eduard Karlovich Kivekes to the General Staff Turkestan 15 October 1902

Schomberg, Reginald Charles Francis. 1927–1928. *Chinese Turkestan road reports*. Notebook transferred from the Cabinet Office to the Director of Military Intelligence in the War Office in 1951, from the Ministry of Defence to the Royal Geographical Society in London in 1959 (LMS S2), 173 pp

Schomberg, Reginald Charles Francis. 1933–1958. *Correspondence with Lieutenant-Colonel Frederick Marshman Bailey*. London. India Office Records and Private Papers Mss Eur/F157/249

Schomberg, Reginald Charles Francis. 1934. *Notes on the Gilgit Agency*. London, 23 pp

Schomberg, Reginald Charles Francis. 1947–1950: *Letters to George Taylor*, 58 pp. India Office Records and Private Papers Mss Eur/C371

School of Oriental and African Studies Archive, London

Lorimer Collection PP MS 66: Personal records of David Lorimer and Emily Overend Lorimer

PP MS 66/01/01/R2: *Mapping data of Aliabad* by David Lorimer

PP MS 66/01/01/R3: *Notes on various subjects: principally 'Ters'* by Mirza Hassan and Qudratullah Beg

PP MS 66/01/01/R7: *Diary, Hunza 1934–35* by David Lorimer

PP MS 66/01/01/R8: *Diary and note books 1934–35* by Emily Lorimer

PP MS 66/01/01/T1: *Lecture notes on Hunza* by David Lorimer

PP MS 66/01/01/T4: *Gilgit census 1931, list of villages* etc. by M. Masih Pal

PP MS 66/02/02/01: *65 sleeves of assorted photographs. Hunza, Aliabad*

PP MS 66/02/02/02: *Hunza Samples 1935* [2 photographs and rough notes].

PP MS 66/02/02/03: *Loose assorted photographs, including – Hunza, Aliabad, Burzil, Raka*

PP MS 66/02/02/04: *Photograph index book labelled 'Aliabad'*

PP MS 66/02/05/I: *Photographs from Kashmir, Bandipur to Gilgit, Gilgit to Aliabad in Hunza*

PP MS 66/02/05/II: *Photographs from Aliabad, Baltit Fort and Altit*

PP MS 66/02/05/III: *Photographs from Hunza: Altit, irrigation, agriculture, fruit, wine*

PP MS 66/02/05: *Glass lantern slides* by David and Emily Lorimer

Shah Khan. 1985. *The liberation of Gilgit and Baltistan*, mimeographed unpublished typescript. Gilgit, 87 pp

Shams-un-Nahar. 2000. *A thrilling voice from inside the palace: Alluring recollections of the last Rani of Hunza State*. Recorded by Fazal Amin Beg. Karimabad, 19 pp

United States State Department: *Tourist restrictions on travel to Hunza*. State Department Cable 1974–258157, Islamabad

Wu, K. Chaucer. 1944. *Memorandum on the proposed motor road from Northwest India to Sinkiang via Gilgit*, prepared by the Chinese Ministry of Communications, 8 June 1944 (IOL/P&S/12/4609: 6–14), 9 pp

Yousaf Khan. 2004. *Genealogy of the Ayesho (Ruling family of Hunza)*. Islamabad, 1 pp

Published sources

Aafiyat Nazar. 2009. *Evolution of education in Hunza, the Northern Areas of Pakistan. Views of early teachers*. Karachi: Aga Khan University, 98 pp

Aase, Tor. 1999. The theological construction of conflict: Gilgit, Northern Pakistan. In *Muslim diversity. Local Islam in global contexts*, ed. Leif Manger, 58–79. Richmond: Nordic Institute of Asian Studies

Aasim Akhtar. 1997. *Indus Kohistan. The distant steppe*. Islamabad: Alliance Française d'Islamabad, 169 pp

Abbott, James. 1848. Inundation of the Indus, taken from the lips of an eye-witness, A. D. 1842. *Journal of the Asiatic Society of Bengal* 17 (1): 230–232

Abdullah, Muhammad. 1972. *Report on prospects of agricultural development in Northern Areas (Gilgit and Baltistan)*. Islamabad: Pakistan Agricultural Research Council, 20 pp

Abdus Salam, ed. 1988. *Ferozsons Urdu-English dictionary. A comprehensive dictionary of current vocabulary*. Lahore, Karachi, Rawalpindi: Ferozsons, 831 pp

Abidi, S. Mehjabeen. 1987. *Pastures and Livestock Development in Gojal*. Gilgit: Aga Khan Rural Support Programme: AKRSP Rural Science Research Programme Reports 5

Abidi-Habib, Mehjabeen and Anna Lawrence. 2007. Revolt and remember. How the Shimshal Nature Trust develops and sustains social-ecological resilience in Northern Pakistan. *Ecology and Society* 12 (2): 35–54

Aga Khan Rural Support Programme. 2017. *Horizons of CPEC in Gilgit Baltistan. A prospective study*. Gilgit: AKRSP, 55 pp

Ahmed, Shama S. 1975. *Religious leadership and social change in the Ismaili Khoja community (unpublished Master Thesis)*. Birmingham: University of Birmingham, 164 pp

Aitchison, Charles Umpherston. 1909. *A collection of treaties, engagements and sanads relating to India and neighbouring countries, Vol. XI: The treaties &c, relating to the North-West Frontier Province, Baluchistan, Jammu and Kashmir, Eastern Turkistan and Afghanistan*. Calcutta: Superintendent Government Printing, 452 pp

Aitchison, Charles Umpherston. 1929. *A collection of treaties, engagements and sanads relating to India and neighbouring countries, Vol. XIV: Eastern Turkestan, Tibet, Nepal, Sikkim, Bhutan and Siam*. Calcutta: Government of India Central Publication Branch, 220 pp

Aitchison, Charles Umpherston. 1931. *A collection of treaties, engagements and sanads relating to India and neighbouring countries, Vol. XII: The treaties, &c., relating to Jammu and Kashmir, Sikkim, Assam and Burma*. Calcutta: Government of India Central Publication Branch, 283 pp

Alder, Garry John. 1963. *British India's northern frontier 1865–95. A study in imperial policy*. London: Longmans, 392 pp

Alder, Garry John. 1974. Introduction. In: Algernon Durand: *The making of a frontier*. Graz: III-XXXVI

Alder, Garry John. 1980. Standing alone: William Moorcroft plays the Great Game, 1808–1825. *The International History Review* 2 (2): 175–215

Ali, Ameneh Azam. 1988. Beyond the next mountain. *The Herald (January 1988)*: 70–76

Ali, Nosheen Kassam. 2009. *States of struggle. Politics, religion, and ecology in the making of the Northern Areas, Pakistan*. Cornell: Dissertation Cornell University, 252 pp

Ali, Nosheen Kassam. 2010. Books vs bombs? Humanitarian development and the narrative of terror in Northern Pakistan. *Third World Quarterly* 31 (4): 541–559

Allan, Nigel J. R. 1986. Accessibility and altitudinal zonation models of mountains. *Mountain Research and Development* 6 (3): 185–194

Allan, Nigel J. R. 1989. Kashgar to Islamabad: the impact of the Karakorum Highway on mountain society and habitat. *Scottish Geographical Magazine* 105 (3): 130–141

Allan, Nigel J. R. 1990. Household food supply in Hunza Valley, Pakistan. *Geographical Review* 80 (4): 399–415

Allan, Nigel J. R., Gregory W. Knapp and Christoph Stadel, eds. 1988. *Human impact on mountains*. Totowa, New Jersey: Rowman & Littlefield, 308 pp

Almgren, Bertil. 1962. Geographical aspects of the Silk Road especially in Persia and East Turkestan. *Bulletin of the Museum of Far Eastern Antiquities* 34: 93–108

Amand, Gordon. 2010. Reginald Schomberg. Public enemy number one. *Herefordiensis. The official journal of Herefordshire Family History Society* 11 (1): 3–5

Appadurai, Arjun. 1996. *Modernity at large. Cultural dimensions of globalization*. Minneapolis: University of Minnesota Press, 248 pp

Ardener, Edwin. 2012. 'Remote areas'. Some theoretical considerations. *HAU. Journal of Ethnographic Theory* 2 (1): 519–533

Asani, Ali S. 1987. The Khojas of Indo-Pakistan: The quest for an Islamic identity. *Journal of the Institute of Muslim Minority Affairs* 8 (1): 31–41

Audouin-Dubreuil, Ariane. 2002. *La Croisière Jaune – Sur la route de la soie*. Grenoble: Éditions Glénat, 192 pp

Aziz, Siddiqui. 1994. Too little, too late. *Newsline* 6 (5): 49–50

Baig, Ikram M. 1992. *An introduction and annual report of N. A's Traders Cooperative Ltd. Gilgit*. Gilgit: NATCL, 5 pp

Bailey, Frederick Marshman. 1946. *Mission to Tashkent*. London: Jonathan Cape, 312 pp

Bailey, Thomas Grahame. 1924. *Grammar of the Shina (Sinā) language; consisting of a full grammar, with texts and vocabularies of the main or Gilgiti dialect and briefer grammars (with vocabularies and texts) of the Kohistani, Guresi, and Drasi dialects*. London: Royal Asiatic Society, 285 pp

Bakker, Peter and Aymeric Daval-Markussen. 2016. Linguistic and genetic roots of the Kalasha. In *In the footsteps of Halfdan Siiger: Danish research in Central Asia*, eds. Ulrik Høj Johnsen, Armin W. Geertz, Svend Castenfeldt and Peter Birkelund Andersen, 93–114. Aarhus: Moesgaard Museum

Baloch, Sikandar Khan. 1975. *Liberation of Northern Areas (Gilgit and Baltistan)*. Rawalpindi: Army Education Press, 130 pp

Baloch, Sikandar Khan. 2004. *In the wonderland of Asia: Gilgit & Baltistan*. Lahore: Sang-e-Meel Publications, 352 pp

Band, George. 1955. *Road to Rakaposhi*. London: Hodder & Stoughton, 192 pp

Bandini-König, Ditte. 2003. *Die Felsbildstation Thalpan (Materialien zur Archäologie der Nordgebiete Pakistans 6–9)*. Mainz: Philipp von Zabern, 4 vols.

Bangash, Yakoob Khan. 2010. Three forgotten accessions: Gilgit, Hunza and Nagar. *The Journal of Imperial and Commonwealth History* 38 (1): 117–143

Banik, Allen E. and Renée Taylor. 1960. *Hunza land. The fabulous health and youth wonderland of the world*. Long Beach: Whitehorn, 239 pp

Banks, Michael Edward Borg and E. James E. Mills. 1960. The ascent of Rakaposhi. *The Geographical Journal* 126 (1): 1–10

Barrow, Edmund G. 1885. *Extracts of the diary of Captain E. G. Barrow, Chitral-Kafiristan Mission 1885/6*. Gilgit: unpublished diary, 110 pp

Barrow, Edmund G. 1886. *Chitral-Hunza diary*. Simla: Unpublished diary, 83 pp

Barrow, Edmund George. 1887a. *Gazetteer of the Eastern Hindú Kush. Confidential. Part I. The Tajik provinces of Wakhán, Ishkásham, and Zébák*. Simla: Government Central Branch Press, 32 pp

Barrow, Edmund George. 1887b. *Gazetteer of the Eastern Hindú Kush. Confidential. Part II. Dárdistán*. Simla: Government Central Branch Press, 232 pp

Barrow, Edmund George. 1888. *Gazetteer of the Eastern Hindu Kush in four parts: Part I. Wakhán, Ishkásham, and Zébák. Part II. Dárdistán. Part III. Káfiristán. Part IV. Routes in the Hindú Kush; compiled under the orders of the Quartermaster-General in India*. Simla: Government Central Press, 424 pp

Barrow, Edmund George. 1890. *Gazetteer of Kashmír and Ladák together with routes in the territories of the Maharája of Jamú and Kashmír*. Calcutta: The Superintendent of Government Printing, 871 pp

Barth, Fredrik. 1956a. Ecologic relationships of ethnic groups in Swat, North Pakistan. *American Anthropologist* 58 (6): 1079–1089

Barth, Fredrik. 1956b. Indus and Swat Kohistan–an ethnographic survey. *Studies honouring the centennial of Universitetets Etnografiske Museum, Oslo 1857–1957*. Oslo: Universitetets Etnografiske Museum: 5–98

Barth, Fredrik, ed. 1969. *Ethnic groups and boundaries. The social organization of culture difference*. Boston: Little, Brown and Co., 153 pp

Bashir, Elena. 2009. Wakhi. In *The Iranian Languages*, ed. Gernot Windfuhr, 825–862. London: Routledge

Bashkanov, Mikhail Kazbekovich [Басханов Михаил Казбекович], Alexander Kolesnikov and Maria Matveeva. 2015. *Дервиш Гиндукуша Путевые дневники центральноазиатских экспедиций генерала Б.Л. Громбчевского [Dervish of the Hindukush. Diaries of the Central Asian expeditions of the General Grombchevsky [Bronislaw Ludwig Grabçzewski]*. St. Petersburg: Nestor History Publishing House, 375 pp

Bashkanov, Mikhail Kazbekovich [Басханов Михаил Казбекович], Alexander Kolesnikov, Maria Matveeva and Alexander Ivanovich Glukhov. 2017. *Памир, Хунза и Кашгария в экспедиционных фотографиях генерала Б.Л. Громбчевского [Pamir, Hunza and Kashgaria in the expedition photographs of General B. L. Grombchevsky]*. Moscow: Pelican, 188 pp

Baskhanov, Mikhail. 2018. Bronislav Liudvigovich Grombchevskii (1855–1926). In *The quest for forbidden lands. Nikolai Przhevalskii and his followers on Inner Asian tracks*, eds. Alexandre Andreyev, Mikhail Baskhanov and Tatiana Yusupova, 255–310. Leiden: Brill

Bayley, Stuart Colvin. 1891. *Note by Sir Stewart Bayley on the Pamir question and the north-east frontiers of Afghanistan*. Simla: Government of India, 30 pp

Bayley, Stuart Colvin. 1892. *Note on the complication with Hunza*. Simla: Government of India, 12 pp

Becher, John Reid. 1859. Letter addressed to R. H. Davies, Esquire, Secretary to the Government of the Punjab and its Dependencies. *Journal of the Asiatic Society of Bengal* 18 (1): 218–229

Beg, Fazal Amin. 2004. *Kinship and marriage in the Chinese Wakhi society, Xinjiang*. Peshawar Area Study Centre (unpublished M. Phil. thesis), 178 pp

Beg, Ghulam Amin. 2011. Legacy, journey and story of a community leader: Rai Ghulam ud din Khan (1944–2011). *Simurgh; http://simurgh-simurgh.blogspot.de/2011_12_01_archive.html*

Beg, Ghulam Amin and Inam-ur-Rahim. 2011. Production and marketing of livestock products in the Hindu Kush-Karakoram-Himalaya. In *Regional workshop in Lhasa 2010 – Pastoralism and rangeland management on the Tibetan Plateau in the context of climate and global change*, eds. Hermann Kreutzmann, Yang Yong and Jürgen Richter, 177–199. Bonn: GIZ

Beg, Rehmat Karim. 2004. *Chitral. A study in statecraft (1320–1969)*. Islamabad: IUCN, 114 pp

Beltran, Gonzalo Aguirre. 1979. *Regions of refuge*. Washington: The Society for Applied Anthropology, 172 pp

Benz, Andreas. 2013. Education and development in the Karakorum. Educational expansion and its impact in Gilgit-Baltistan, Pakistan. *Erdkunde* 67 (2): 123–136

Benz, Andreas. 2014a. *Education for development in Northern Pakistan. Opportunities and constraints for rural households*. Karachi: Oxford University Press, 452 pp

Benz, Andreas. 2014b. Multilocality as an asset. Translocal development and change among the Wakhi of Gojal, Pakistan. In *Tracing connections. Explorations of spaces and places in Asian contexts*, eds. Andreas Benz and Henryk Alff, 111–138. Berlin: Wissenschaftlicher Verlag Berlin

Benz, Andreas. 2014c. Mobility, multilocality and translocal development. Changing livelihoods in the Karakoram. *Geographica Helvetica* 69 (1): 1–12

Berger, Hermann. 1960. Bericht über sprachliche und volkskundliche Forschungen im Hunzatal. *Anthropos* 55: 657–664

Berger, Hermann. 1968. Zwei Erzählungen aus dem Hunza-Tal. *Zeitschrift für Kulturaustausch* 18 (2–3): 224–225

Berger, Hermann. 1974. *Das Yasin-Burushaski (Werchikwar): Grammatik, Texte, Wörterbuch*. Wiesbaden: Harrassowitz, 228 pp

Berger, Hermann. 1983. Etymologische Bemerkungen zu einigen auf Geister und Geisterglaube bezügliche Wörter im Burushaski. In *Ethnologie und Geschichte. Festschrift für Karl Jettmar*, ed. P. Snoy, 29–33. Wiesbaden: Steiner (= Beiträge zur Südasienforschung 86)

Berger, Hermann. 1985. A Survey of Burushaski Studies. *Journal of Central Asia* 8 (1): 33–37

Berger, Hermann. 1992. Das Burushaski. Schicksale einer zentralasiatischen Restsprache. *Sitzungsberichte der Heidelberger Akademie der Wissenschaften, Philosophisch-historische Klasse* 1: 5–25

Berger, Hermann. 1994. Kombinatorischer Lautwandel im Burushaski. *Studien zur Indologie und Iranistik (= Festschrift für Georg Buddruss)* 19: 1–19

Berger, Hermann. 1998. *Die Burushaski-Sprache von Hunza und Nager. Teil I: Grammatik; Teil II: Texte mit Übersetzungen; Teil III: Wörterbuch*. Wiesbaden: Harrassowitz, 646 pp

Bhushan, Shashi. 1978. *Karakoram road. Dominationistic conspiracy of China in Kashmir*. New Delhi: Congress Socialist Forum, 16 pp

Bhutto, Zulfikar Ali. 1972. *Speeches and statements: July 1, 1972 - September 30, 1972*. Karachi: Produced by the Dept. of Films and Publications, Govt. of Pakistan, 240 pp

Bianca, Stefano. 2006. Reconciling culture and development in the Hunza Valley. Concepts and programmes of the Aga Khan Trust for Culture. In *Karakoram in transition. Culture, development, and ecology in Hunza Valley*, ed. Hermann Kreutzmann, 287–291. Karachi: Oxford University Press

Biddulph, John. 1874a. *The Chitral passes*. London: India Office Library & Records (= IOL/P&S/7/4, Part 2), 4 pp

Biddulph, John. 1874b. *Report on the army, resources, and defences of Kashghar. Military*. London: India Office Library & Records (= IOL/P&S/7/4, Part 2), 5 pp

Biddulph, John. 1876. *Biddulph diary 1876*. Hereford: Hereford Record Office (A 87/6), 150 pp

Biddulph, John. 1880. *Tribes of the Hindoo Koosh*. Calcutta: Office of the Superintendent of Government Printing, 164 pp

Biddulph, John. 1881. On the birds of Gilgit. *Ibis. A Quarterly Journal of Ornithology* 5 (17): 35–102

Biddulph, John. 1882. On the birds of Gilgit. *Ibis. A Quarterly Journal of Ornithology* 6 (22): 266–290

Biddulph, John. 1899. *The nineteenth and their times*. London: John Murray, 330 pp

Bircher, Ralph. 1942. *Hunsa. Das Volk, das keine Krankheit kennt.* Bern: Hans Huber, 135 pp

Birdwood, Christopher. 1952. The Asian frontiers of Kashmir. *Journal of The Royal Central Asian Society* 39: 241–245

Black, Charles E. D. 1892. M. Henri Dauvergne's explorations in the Pamirs. *The Scottish Geographical Magazine* 8: 362–367

Blacker, Latham Valentine Stewart. 1917. From India to Russia. *The Geographical Journal* 50 (6): 393–418

Blacker, Latham Valentine Stewart. 1922. *On secret patrol in High Asia.* London: John Murray, 302 pp

Blacker, Latham Valentine Stewart. 1933. La croisière jaune. *The Geographical Journal* 81 (1): 53–58

Blackwood, William. 1895. Little wars on the Indian frontier. *Blackwood's Edinburgh Magazine* 157 (956): 930–938

Boivin, Michel. 2003. *La rénovation du shî'isme ismaélien en Inde et au Pakistan – d'après les écrits et les discours de Sultan Muhammad Shâh Aga Khan (1902–1954).* London: Routledge Curzon Press, 475 pp

Boivin, Michel. 2013. *L'âghâ khân et les Khojah. Islam chiite et dynamiques sociales dans le sous-continent indien (1843–1954).* Paris: Khartala, 351 pp

Boivin, Michel. 2014. The Isma'ili – Isna 'Ashari divide among the Khojas. Exploring forgotten judicial data from Karachi. *Journal of the Royal Asiatic Society* 24 (3): 381–396

Bonvalot, Gabriel. 1888. *Du Caucase aux Indes à travers le Pamir.* Paris: Librairie Plon, 458 pp

Bonvalot, Gabriel. 1889. *Through the heart of Asia. Over the Pamïr to India, Vol. 2.* London: Chapman and Hall, 255 pp

Bouzas, Antía Mato. 2012. Mixed legacies in contested borderlands: Skardu and the Kashmir dispute. *Geopolitics* 17 (4): 867–886

Bouzas, Antía Mato. 2017. Territorialisation, ambivalence, and representational spaces in Gilgit-Baltistan. *Transcultural Studies* 1/2017: 197–223

Bouzas, Antía Mato. 2019. *Kashmir as a borderland. The politics of space and belonging across the Line of Control.* Amsterdam: Amsterdam University Press, 206 pp

Brenner, Neil. 1999. Beyond state-centrism? Space, territoriality, and geographical scale in globalization studies. *Theory and Society* 28: 39–78

Brenzinger, Matthias. 2007. *Language diversity endangered.* Berlin: Mouton de Gruyter, 454 pp

Bréon, Emmanuel. 2004. *Alexandre Iacovleff: Itinérances.* Paris: Somogy Editions d'Art, 244 pp

Brescius, Moritz von. 2018. *German science in the age of empire. Enterprise, opportunity and the Schlagintweit brothers.* Cambridge: Cambridge University Press, 414 pp

Bretherton, George Howard. 1898. Life in Gilgit. *The Contemporary Review* 74 (December 1898): 872–881

Bridges, Fitzstephen Henry. 1930. An unpublished report on Hunza glaciers. *The Geographical Journal* 75 (2): 166–172

Brophy, David. 2016. *Uyghur nation. Reform and revolution on the Russia-China frontier.* Cambridge: Harvard University Press, 347 pp

Brower, Barbara and Barbara Rose Johnston, eds. 2007. *Disappearing peoples? Indigenous groups and ethnic minorities in South and Central Asia.* Walnut Creek: Left Coast Press, 275 pp

Brown, William Alexander. 1998. *The Gilgit rebellion 1947.* St. Boswells: Ibex, 297 pp

Bruce, Charles Granville. 1892. Extracts from two letters to his father, June/July 1892. On the Karakoram expedition. *Proceedings of the Royal Geographical Society, N. S.* 14: 768–770

Bruce, Charles Granville. 1910. *Twenty years in the Himalaya.* London: Edward Arnold, 335 pp

Brush, Stephen B. 1976. Introduction: Symposium on Cultural Adaptation of Mountain Ecosystems. *Human Ecology* 4 (2): 125–134

Buddruss, Georg. 1974. Neuiranische Wortstudien. Zur Wakhi-Sprache in Hunza. *Münchener Studien zur Sprachwissenschaft* 32: 9–40

Buddruss, Georg. 1977. Nochmals zur Stellung der Nuristan-Sprachen des afghanischen Hindukusch. *Münchener Studien zur Sprachwissenschaft* 36: 19–38

Buddruss, Georg. 1984. Domaakí-Nachträge zum Atlas der Dard-Sprachen. *Münchener Studien zur Sprachwissenschaft* 43: 9–24

Buddruss, Georg. 1985. Linguistic research in Gilgit and Hunza – Some results and perspectives. *Journal of Central Asia* 8 (1): 27–32

Buddruss, Georg. 1986. Wakhi-Sprichwörter aus Hunza. In *Studia Grammatica Iranica – Festschrift für Helmut Humbach*, eds. Rüdiger Schmitt and Prods Oktor Skjærvø, 27–44. München: Röll (= Beiheft zur Zeitschrift für vergleichende Sprachwissenschaft)

Buddruss, Georg. 1993. Deutsche sprachwissenschaftliche Forschung in den Nordgebieten Pakistans. In *Neuere deutsche Beiträge zu Geschichte und Kultur Pakistans/ Contemporary German Contributions to the history and culture of Pakistan* eds. Stephanie Zingel-Avé Lallemant and Wolfgang-Peter Zingel, 22–37. Bonn: Schriftenreihe des Deutsch-Pakistanischen-Forums, 10

Buddruss, Georg. 2006. Linguistic diversity in the Hunza Valley. In *Karakoram in transition. Culture, development, and ecology in the Hunza Valley*, ed. Hermann Kreutzmann, 236–245. Karachi, Oxford: Oxford University Press

Buddruss, Georg and Almuth Degener. 2012. *The meeting place. Radio features in the Shina language of Gilgit by Mohammad Amin Zia.* Wiesbaden: Harrassowitz, 308 pp

Buddruss, Georg and Sigrun Wiehler-Schneider. 1978. Wakhi-Lieder aus Hunza. In *Jahrbuch für musikalische Volks- und Völkerkunde*, 89–110. Köln

Bülbül, Nazir Ahmad. 2014. *Biyoze Bülbül.* Islamabad: Pamir Media Group, 132 pp

Burrard, Sidney Gerald and Henry Hubert Hayden. 1907–1908. *A sketch of the geography and geology of the Himalaya mountains and Tibet.* Calcutta: Superintendent of Government Printing, 308 pp

Butz, David. 1989. *Pastures and pastoralism in Shimshal.* Gilgit: Aga Khan Rural Support Programme

Butz, David. 1996. Sustaining indigenous communities: symbolic and instrumental dimensions of pastoral resource use in Shimshal, Northern Pakistan. *The Canadian Geographer* 40 (1): 36–53

Butz, David. 2006. Tourism and portering. Labour relations in Shimshal, Gojal Hunza. In *Karakoram in transition. Culture, development, and ecology in Hunza Valley*, ed. Hermann Kreutzmann, 394–403. Karachi: Oxford University Press
Butz, David and Nancy Cook. 2011. Accessibility interrupted: the Shimshal road, Gilgit-Baltistan, Pakistan. *The Canadian Geographer* 55 (3): 354–364
Butz, David and Nancy Cook. 2016. Political ecology of human-environment change in Gojal, Gilgit-Baltistan, Pakistan. In *Mapping transition in the Pamirs. Changing human-environmental landscapes*, eds. Hermann Kreutzmann and Teiji Watanabe, 197–214. Cham: Springer
Cacopardo, Alberto M. and Augusto S. Cacopardo. 2001. *Gates of Peristan. History, religion and society in the Hindu Kush*. Rome: Istituto Italiano per l'Africa e l'Oriente, 327 pp
Carey, Arthur Douglas. 1887. A journey round Chinese Turkistan and along the northern frontier of Tibet. *Proceedings of the Royal Geographical Society* 9 (12): 731–752
Caroe, Olaf Kirkpatrick. 1960. The geography and ethnics of India's northern frontiers. *The Geographical Journal* 126 (3): 298–308
Census of Pakistan 1953. *1951 Village list: North West Frontier – Chitral State.* Karachi: Government of Pakistan Press, 43 pp.
Chambers, Robert. 1983. *Rural development: Putting the last first*. London: Longman, 246 pp
Chambers, Robert. 2006. *Poverty unperceived. Traps, biases and agenda*. Brighton: Institute of Development Studies, 46 pp
Chambers, Robert and Gordon Conway. 1991. *Sustainable rural livelihoods: practical concepts for the 21st century*. Brighton: IDS (Discussion Paper 296), 29 pp
Chapman, John A. et al. 1972. Endemic goitre in the Gilgit Agency, West Pakistan. *Philosophical Transactions of the Royal Society of London. Series B, Biological Sciences* 263 (856): 459–490
Charles, Christian. 1985. *La vallee de Hunza, Karakorum – Pakistan – Milieu naturel, aménagement traditionel et mutations récentes dans une vallée aride du Nord-Ouest de L'Ensemble Himalayen.* Grenoble: Université de Grenoble, 430 pp
Chaudhary, M. Azam. 1997. Maruts. Gold-washers of the Indus. In *Perspectives on history and change in the Karakorum, Hindukush, and Himalaya*, eds. Irmtraud Stellrecht and Matthias Winiger, 431–462. Cologne: Köppe
Chaudhuri, Kirti Narayan. 1981. The world-system east of longitude 20°. The European role in Asia, 1500–1750. *Review* 5 (2): 219–245
Chaudhuri, Kirti Narayan. 1990. *Asia before Europe. Economy and civilisation of the Indian Ocean from the rise of Islam to 1750*. Cambridge: Cambridge University Press, 477 pp
Chohan, Amar Singh. 1998. *The Gilgit Agency 1877–1935*. New Delhi (undated, based on doctoral thesis at University of Jammu 1980): Atlantic Publishers and Distributors, 241 pp
Chou, Shiuhhuah Serena. 2013. Agricultural travels and the rise of organic farming discourse. *Comparative Literature Studies* 50 (1): 108–119
Choudhury, Deba Prosad. 1996. *Trade and politics in the Himalaya-Karakoram borderlands*. London, Hyderabad: Sangam Books, Orient Longman, 130 pp
Christian, David. 2000. Silk roads or steppe roads? The silk roads in world history. *Journal of World History* 11 (1): 1–26
Citroën, André. 1935. Preface. In *An eastern odyssey. The third expedition of Haardt and Audouin-Dubreuil*, ed. Georges Le Fèvre, 9–17. London: Victor Gollancz
Clark, John. 1956. *Hunza – Lost kingdom of the Himalayas*. New York: Funk and Wagnalls Company, 271 pp
Clark, John. 1963. Hunza in the Himalayas. Stories Shangri-La undergoes scrutiny. *Natural History* 72: 38–45
Clark, L. P. 1960. Progress in the Gilgit Agency. *Eastern World* 14: 21–22
Clarke, Graham E. 1977. Who were the Dards? A review of the ethnographic literature of the North-Western Himalaya. *Kailash* 5 (4): 323–356
Clarke, Peter B. 1976. The Ismailis: A study of community. *The British Journal of Sociology* 27 (4): 484–494
Clarke, Peter B. 1978. The Ismaili sect in London: religious institutions and social change. *Religion* 8 (1): 68–84
Cobb, Evelyn Hey. 1951. The frontier states of Dir, Swat and Chitral. *Journal of the Royal Central Asian Society* 38 (2–3): 170–176
Cobbold, Ralph Patterson. 1900. *Innermost Asia: travel and sport in the Pamirs*. London: W. Heinemann, 354 pp
Cockerill, George Kynaston. 1895. *Appendix to Barrow's gazetteer of Eastern Hindu Kush*. Simla: Government Central Printing Office, 75 pp
Cockerill, George Kynaston. 1896. *A further appendix to Barrow's gazetteer of Eastern Hindu Kush*. Simla: Government Central Printing Office, 230 pp
Cockerill, George Kynaston. 1922. Byways in Hunza and Nagar. *The Geographical Journal* 60 (2): 97–112
Cockerill, George Kynaston. 1939. Pioneer exploration in Hunza and Chitral. *The Himalayan Journal* 11: 15–41
Community Basic Services. 1984. *Basic services for children and women in Northern Areas through community participation. Progress report: January 1982 – March 1984*. Gilgit: Community Basic Services (CBS) Programme, 17 pp
Community Basic Services. 1985. *Basic services for children and women in Northern Areas through community participation. Annual progress report 1984*. Gilgit: Community Basic Services (CBS) Programme, 12 pp
Conway, William Martin. 1894a. *Climbing and exploration in the Karakoram-Himalayas*. London: T. F. Unwin, 709 pp
Conway, William Martin. 1894b. The development of mountain exploration. *The New Review* 10 (61): 736–745
Conway, William Martin. 1897. Formation of a lake in Nagar. *The Geographical Journal* 10 (1): 107
Conway, William Martin. 1920. *Mountain memories. A pilgrimage of romance*. New York: Funk and Wagnalls, 282 pp
Cook, Nancy and David Butz. 2011. Narratives of accessibility and social change in Shimshal, Northern Pakistan. *Mountain Research and Development* 31 (1): 27–34

Cook, Nancy and David Butz. 2013. The Atta Abad landslide and everyday mobility in Gojal, Northern Pakistan. *Mountain Research and Development* 33 (4): 372–380

Cook, Nancy and David Butz. 2015. The dialectical constitution of mobility and immobility: recovering from the Attabad Landslide disaster, Gojal, Gilgit-Baltistan, Pakistan. *Contemporary South Asia* 23 (4): 388–408

Curzon, George Nathaniel 1896. The Pamirs and the source of the Oxus. *The Geographical Journal* 8: 15–54, 97–119, 239–264

Curzon, George Nathaniel. 1926. *Leaves from a viceroy's notebook and other papers*. London: Macmillan, 414 pp

Curzon, George Nathaniel. 2012. *On the Indian frontier*. Karachi: Oxford University Press, 283 pp

Daftary, Farhad 1990. *The Isma'ilis: their history and doctrines*. Cambridge: Cambridge University Press, 804 pp

Dainelli, Giotto, ed. 1924. *Le condizioni delle genti*. Spedizione Italiana de Filippi nell'Himàlaja, Caracorùm e Turchestàn Cinese (1913–1914). Serie II: Resultati Geologici e Geografici. Vol. 8. Bologna: Nicola Zanichelli, 418 pp

Dainelli, Giotto, ed. 1934a. *La esplorazione della regione fra l'Himàlaja occidentale e il Caracorùm*. Relazioni Scientifiche della Spedizione Italiana de Filippi nell'Himàlaia, Caracorùm e Turchestàn Cinese (1913–1914). Serie II: Resultati geologici e geografici. Vol. 1. Bologna: Nicola Zanichelli, 428 pp

Dainelli, Giotto, ed. 1934b. *Studi sul glaciale. Tavole*. Relazioni Scientifiche della Spedizione Italiana de Filippi nell'Himàlaia, Caracorùm e Turchestàn Cinese (1913–1914). Serie II: Resultati geologici e geografici. Vol. 3. Bologna: Nicola Zanichelli, 177 pp

Dani, Ahmad Hasan 1981. *Karakorum Highway unfolds the romance of the past*. Islamabad: Barqsons, 16 pp

Dani, Ahmad Hasan. 1983. *Human records on Karakorum Highway*. Islamabad: Centre for the Study of Civilizations of Central Asia, 95 pp

Dani, Ahmad Hasan. 1985. The sacred rock of Hunza. *Journal of Central Asia* 8 (2): 5–124

Dani, Ahmad Hasan, ed. 1987. *Shah Rais Khan's history of Gilgit*. Islamabad, Gilgit: Centre for the Study of the Civilizations of Central Asia, 49 & 514 pp

Dani, Ahmad Hasan. 1989. *History of Northern Areas of Pakistan*. Islamabad: National Institute of Historical and Cultural Research (= Historical Studies (Pakistan) Series 5), 532 pp

Daud Beg. 1964. Hunza and Nagar Valleys. *Pakistan Quarterly* 12 (3): 18–22

Dauvergne, Henri. 1892. Exploration dans l'Asie centrale. *Bulletin de la Société de Géographie* 7e series XIII: 5–40

Davies, Cutbert Collin. 1932. *The problem of the North-West Frontier 1890–1908, with a survey of policy since 1849*. Cambridge: Cambridge University Press, 220 pp

Davies, Lewis Merson. 1940. Note on three Himalayan Rivers. *Geological Magazine* 77: 410–412

Davies, Robert Henry. 1862. *Report on the trade and resources of the countries on the north-western boundary of British India*. Lahore: Government Press, 538 pp

Davies, Robert Henry. 1867. *Report on trade routes of the North-West Frontier*pp

Degener, Almuth. 2008. *Shina-Texte aus Gilgit (Nord-Pakistan)*. Wiesbaden: Harrassowitz, 333 pp

Delaney, Keith B. and Stephen G. Evans. 2017. The evolution (2010–2015) and engineering mitigation of a rockslide-dammed lake (Hunza River, Pakistan); characterisation by analytical remote sensing. *Engineering Geology* 220: 65–75

Denker, Debra. 1983. The last migration of the Kirghiz of Afghanistan? *Central Asian Survey* 2–3: 89–98

Derbyshire, Edward and Monique Fort. 2006. Geomorphology and mountain hazards in the Hunza Valley. In *Karakoram in transition. Culture, development, and ecology in Hunza Valley*, ed. Hermann Kreutzmann, 73–95. Karachi: Oxford University Press

Derbyshire, Edward, Monique Fort and Lewis A. Owen. 2001. Geomorphological hazards along the Karakoram Highway: Khunjerab Pass to the Gilgit River, northernmost Pakistan. *Erdkunde* 55 (1): 49–71

Derbyshire, Edward and Keith J. Miller. 1981. Highway beneath the Ghulkin. *The Geographical Magazine* 53 (11): 626–635

Devji, Faisal. 2014. The idea of Ismailism. *Critical Muslim* 10 (April-June): 51–62

Dichter, David. 1967. *The North-West Frontier of West Pakistan. A study in regional geography*. Oxford: Clarendon Press, 231 pp

Dickoré, Bernhard. 1991. Zonation of flora and vegetation of the northern declivity of the Karakoram/Kunlun mountains (SW Xinjiang China). *GeoJournal* 25 (2/3): 265–284

Dickoré, Wolf Bernhard. 1995. *Systematische Revision und chorologische Analysen der Monocotyledoneae des Karakorum (Zentralasien, West-Tibet). Flora Karakorumensis. I. Angiospermae, Monocotyledoneae*. Linz: Botanische Arbeitsgemeinschaft am Oberösterreichischen Landesmuseum (= Stapfia 39), 427 pp

Dor, Rémy. 1975. *Contribution à l'étude des Kirghiz du Pamir Afghan*. Paris: Publications Orientalistes de France, 341 pp

Dor, Rémy and Clas Michael Naumann. 1978. *Die Kirghisen des afghanischen Pamir*. Graz: Akademische Druck- und Verlagsanstalt, 124 pp

Douglas, William Orville. 1952. *Beyond the high Himalayas*. Garden City: Doubleday, 352 pp

Drew, Frederic. 1875. *The Jummoo and Kashmir territories. A geographical account*. London: Edward Stanford, 568 pp

Driver, Felix. 1998. Scientific exploration and the construction of geographical knowledge: Hints to travellers. *Finisterra* 33 (65): 21–30

Driver, Felix. 2001. *Geography militant. Cultures of exploration and empire*. Oxford: Blackwell, 258 pp

Driver, Felix. 2015. Intermediaries and the archive of exploration. In *Indigenous intermediaries. New perspectives on exploration archives*, eds. Shino Konishi, Maria Nugent and Tiffany Shellam, 11–29. Acton: Australian National University Press

Du Riche Preller, Charles. 1924. The racial and economic conditions of Transhimalaya (Upper Indus Basin: Ladàk and Baltistan). *Scottish Geographical Magazine* 40 (6): 334–344

Duckworth, W. Laurence H. 1894. Description of two skulls brought by Mr. Conway from Nagyr. In *Climbing and exploration in the Karakoram-Himalayas containing scientific reports*, ed. William Martin Conway, 95–108. London: T. Fisher Unwin

Durand, Algernon George Arnold. 1888. *Report on the present military position in Gilgit*. Simla (= IOL/P&S/7/57/33). Government Central Press, 50 pp

Durand, Algernon George Arnold. 1889. *Gilgit Agency report 1889* Gilgit (= IOL/P&S/7/59).

Durand, Algernon George Arnold. 1894a. The eastern Hindu Kush. In *Climbing and exploration in the Karakoram-Himalayas containing scientific reports*, ed. William Martin Conway, 1–18. London: T. Fisher Unwin

Durand, Algernon George Arnold. 1894b. The eastern Hindu Kush. *Contemporary Review* 66: 685–693

Durand, Algernon George Arnold. 1899. *The making of a frontier. Five years' experiences and adventures in Gilgit, Hunza, Nagar, Chitral, and the eastern Hindu-Kush*. London: John Murray, 298 pp

Dyhrenfurth, Günther Oskar. 1952. *Zum dritten Pol*. Graz: Stiasny, 285 pp

Dyhrenfurth, Günther Oskar. 1955. *To the third pole. The history of the High Himalaya*. London: Werner Laurie, 233 pp

Dyhrenfurth, Günther Oskar. 1961. *Der dritte Pol. Die Achttausender und ihre Trabanten*. Frankfurt/Main: Büchergilde Gutenberg, 262 pp

Eberhardt, Einar, Bernhard Dickoré and Georg Miehe. 2006. Vegetation of Hunza Valley. Diversity, altitudinal distribution and human impact. In *Karakoram in transition. Culture, development, and ecology in Hunza Valley*, ed. Hermann Kreutzmann, 109–122. Karachi: Oxford University Press

Eckenstein, Oskar. 1896. *The Karakorams and Kashmir. An account of a journey*. London: T. Fisher Unwin, 253 pp

Edelman, Dzoy I. and Leila R. Dodykhudoeva. 2009. The Pamir languages. In *The Iranian languages*, ed. Gernot Windfuhr, 773–786. London: Routledge

Edney, Matthew H. 2009. The irony of imperial mapping. In *The imperial map: cartography and the mastery of empire*, ed. James R. Akerman, 11–45. Chicago: University of Chicago Press

Edwards, John. 1960. The Batura Muztagh expedition, 1959. *Alpine Journal* 65 (300): 43–52

Edwards, John. 1961. Die Batura-Mustagh-Expedition 1959. *Berge der Welt. Das Buch der Forscher und Bergsteiger* 13: 106–122

Eggert, Peter. 1990. *Die frühere Sozialordnung Moolkhos und Turkhos*. Stuttgart: Franz Steiner (= Beiträge zur Südasienforschung 134), 271 pp

Ehlers, Eckart and Hermann Kreutzmann, eds. 2000. *High mountain pastoralism in Northern Pakistan*. Stuttgart: Franz Steiner (= Erdkundliches Wissen 132), 209 pp

Elnazarov, Hakim. 2010. Western and Central Asia. In *Atlas of world's languages in danger*, ed. Christopher Moseley, 43–47. Paris: Unesco Publishing

Elphinstone, Mountstuart. [3]1839. *An account of the Kingdom of Caubul and its dependencies, in Persia, Tartary, and India: Comprising a view of the Afghaun nation and a history of the Dooraunee monarchy; 2 Vols.* London: Richard Bentley, 422 & 465 pp

Embree, Ainslee. 1977. Pakistan's imperial legacy. In *Pakistan's western borderlands. The transformation of a political order*, ed. Ainslee Embree, 24–40. New Delhi: Vikas Publishing House

Engelbrecht, Thies Hinrich. 1914. *Die Feldfrüchte Indiens in ihrer geographischen Verbreitung*. Hamburg: L. Friederichsen, 271 pp

Etherton, Percy Thomas. 1911. *Across the roof of the world. A record of sport and travel through Kashmir, Gilgit, Hunza, the Pamirs, Chinese Turkistan, Mongolia and Siberia*. London: Constable and Co., 437 pp

Etherton, Percy Thomas. 1925. *In the heart of Asia*. London: Constable and Co., 305 pp

Everest-Phillips, Max. 1991. British consuls in Kashgar. *Asian Affairs* 22 (1): 20–34

Felmy, Sabine. 1986. *Märchen und Sagen aus Hunza*. Köln: Eugen Diederichs, 143 pp

Felmy, Sabine. 1993. Division of labour and women's work in a mountain society. Hunza Valley in Pakistan. In *Women and work in South Asia. Regional patterns and perspectives*, eds. Saraswati Raju and Deipica Bagchi, 196–208. London, New York: Routledge

Felmy, Sabine. 1997. *The voice of the nightingale. A personal account of the Wakhi culture in Hunza*. Karachi, Oxford: Oxford University Press, 109 pp

Felmy, Sabine. 2006. Transfer of education to the mountains. In *Karakoram in transition. Culture, development, and ecology in the Hunza Valley*, ed. Hermann Kreutzmann, 370–381. Karachi, Oxford: Oxford University Press

Feyyaz, Muhammad. 2011. *Sectarian conflict in Gilgit-Baltistan*. Islamabad: Pakistan Institute of Legislative Development and Transparency, 20 pp

Feyyaz, Muhammad. 2011. *Sectarian conflict in Gilgit-Baltistan*. Islamabad: Fildat, 20 pp

Filippi, Filippo de. 1912. *Karakoram and Western Himalaya. An account of the expedition of H. R. H. Prince Luigi Amadeo of Savoy, Duke of Abruzzi*. London: Harrison and Sons, 469 pp

Filippi, Filippo de. 1932. *The Italian expedition to the Himalaya, Karakoram, and Eastern Turkestan (1913–14)*. London: Edward Arnold & Co., 528 pp

Finsterwalder, Richard. 1960. German glaciological and geological expeditions to the Batura Mustagh and Rakaposhi Range. *Journal of Glaciology* 3 (28): 787–788

Fleming, Peter. 1936. *News from Tartary: A journey from Peking to Kashmir*. London: Jonathan Cape, 384 pp

Flowerday, Julie. 2005. Framing change: Hunza (Pakistan) in treble vision – 1930s and 1990s. *Visual Communication* 4 (3): 296–303

Flowerday, Julie. 2006. Changes over time. Hunza in treble vision: 1930s and 1990s. In *Karakoram in transition. Culture, development, and ecology in Hunza Valley*, ed. Hermann Kreutzmann, 273–284. Karachi: Oxford University Press

Forbes, Andrew D. W. 1986. *Warlords and Muslims in Chinese Central Asia. A political history of republican Sinkiang 1911–1949*. Cambridge: Cambridge University Press, 376 pp

Frank, Andre Gunder. 1978. *Dependent accumulation and underdevelopment*. London: Macmillan, 226 pp

Frank, Andre Gunder. 1980. *Crisis in the world economy*. London: Heinemann, 366 pp

Frank, Andre Gunder. 1992. *The centrality of Central Asia (= Comparative Asian Studies 8)*. Amsterdam: VU University Press, 57 pp

Frank, Andre Gunder. 1998. *ReOrient: Global economy in the Asian age*. Berkeley: University of California Press, 352 pp

Frank, Rana Bashir Ahmad. 1934. *Why not a territory for the Aga Khan in India? The history of the dynasty*. Lahore

Frembgen, Jürgen 1983a. Die Nagerkuc̆ im Licht der populären Reiseliteratur. Ein Beitrag zur Vorurteilsforschung in Nordpakistan. In *Ethnologie und Geschichte. Festschrift für Karl Jettmar*, ed. Peter Snoy, 147–161. Wiesbaden: Franz Steiner (= Beiträge zur Südasienforschung 86)

Frembgen, Jürgen. 1983b. Tourismus in Hunza: Beziehungen zwischen Gästen und Gastgebern. *Sociologus* 33 (2): 174–185

Frembgen, Jürgen. 1985. *Zentrale Gewalt in Nager (Karakorum). Politische Organisationsformen, Ideologische Begründungen des Königtums und Veränderungen in der Moderne*. Wiesbaden: Franz Steiner (= Beiträge zur Südasienforschung 103), 240 pp

Frembgen, Jürgen. 1986a. Ethnographical field-research on the history and culture of Nager: Some preliminary remarks on the process of settlement. *Central Asiatic Journal* 30 (1/2): 22–34

Frembgen, Jürgen. 1986b. Aspekte der Oralität und Literalität. Ihre Implikationen für das Geschichtsbewußtsein der muslimischen Nagerkuts in Nordpakistan. *Anthropos* 81 (4/6): 567–581

Frembgen, Jürgen. 2006. The battle of Nilt in 1891. Oral tradition as sources of history in Nager. In *Karakoram in transition. Culture, development, and ecology in Hunza Valley*, ed. Hermann Kreutzmann, 217–224. Karachi: Oxford University Press

Frembgen, Jürgen Wasim. 1985. Handel zwischen Nager und Hunza. *Archiv für Völkerkunde* 39: 199–213

Frembgen, Jürgen Wasim. 1989. Hunza und Shangri-la. Ein Bergvolk in der Tourismuswerbung. *Münchener Beiträge zur Völkerkunde* 2: 51–68

Frembgen, Jürgen Wasim. 1996. Local dignitaries as historians: Guardians of traditional culture from Gilgit, Hunza, and Nagar (Northern Pakistan). In *Proceedings of the Second International Hindukush Cultural Conference*, eds. Elena Bashir and Israr-uddin, Karachi: Oxford University Press

French, Patrick. 1994. *Younghusband. The last great imperial adventurer*. London: Harper Collins, 440 pp

Friese, Kai. 2000. The Aryan handshake. *Transition* 83: 4–35

Frischauer, Willi. 1970. *The Aga Khans*. London, Sydney, Toronto: The Bodley Head, 286 pp

Fussman, Gérard. 1972. *Atlas linguistique des parlers Dardes et Kafirs: Commentaire*. Paris: École Française Êxtreme Orient, 451 pp

Fussman, Gérard. 1978. Inscriptions de Gilgit. *Bulletin de l'Ecole Française d'Extrême Orient* 65: 1–64

Fyzee, Asaf Ali Aghar. 1971. The religion of the Ismailis. In *India and contemporary Islam*, ed. Shamoon Tayyib Lokhandwalla, 71–87. Simla: Indian Institute of Advanced Study

Gannon, Jack Rose Compton. 1932. A frontier tour (being extracts from a diary written during a journey made with H. E. Lord Rawlinson, C.-in-C. in India, through Dir, Chitral, and the Gilgit Agency, in 1923). *The Himalayan Journal* 4: 75–90

Gardezi, Hassan Nawaz. 1991. *Understanding Pakistan. The colonial factor in societal development*. Lahore: Maktaba, 162 pp

Garlick, Jeremy. 2018. Deconstructing the China–Pakistan Economic Corridor. Pipe dreams versus geopolitical realities. *Journal of Contemporary China* 27 (112): 519–533

Garver, John W. 2006. Development of China's overland transportation links with Central, South-west and South Asia. *The China Quarterly* 185 (March 2006): 1–22

Ge, Yonggang, Fenghuan Su, Xiaoqing Chen and Jianqiang Zhang. 2015. Destructions on the Karakoram Highway (KKH) from Sost to Khunjerab induced by geo-hazards and prevention. *Applied Mechanics and Materials* 744–746: 1234–1243

General Staff India 1928a. *Military report and gazetteer of the Gilgit Agency and the independent Territories of Tangir and Darel. Second edition*. Simla: Government of India Press, 232 pp

General Staff India 1928b. *Military report and gazetteer on Chitral*. Calcutta: Government of India Press, 328 pp

Ghiasy, Richard and Jiayi Zhou. 2017. *The Silk Road economic belt. Considering security implications and EU–China cooperation prospects*. Solna: Stockholm International Peace Research Institute, 56 pp

Ghosh, Suchandra. 2000. The route of horse trade early in India (up to c. 500 AD). *Proceedings of the Indian History Congress* 61: 126–131

Ghufran, Mirza Mohammed. 1962/1974. *Tareekh-e-Chitral [New history of Chitral, edited by Ghulam Murtaza]*. Peshawar [Urdu 1962; English translation 1974]: Public Arts Press, 288 pp

Ghulam Ali. 2017. *China-Pakistan relations. A historical analysis*. Karachi: Oxford University Press, 263 pp

Gibb, Hamilton Alexander Rosskeen and Johannes Hendrik Kramers. 1974. Khodja. In *Shorter Encyclopaedia of Islam*, ed. Johannes Hendrik Kramers, 256–257. Leiden: E. J. Brill

Gillett, Michael C. 1937. *Report on Vice-Consul Gillett's tour to Keria via Yarkand, Karghalik and Khotan 1st February to 5th April 1937*. Kashgar: India Office Library & Records (= IOL/P&S/12/2336), 13 pp

Gillett, Michael C. 1965. Ancient routes through the Pamirs. *Central Asian Review* 13 (1): 44–54

Gilley, Bruce. 2017. The case for colonialism. *Third World Quarterly*: 17 pp

Goodwin, Peter. 1950. Baltistan: Northern gateway of Kashmir. *The Geographical Magazine* 23: 350–360

Goodwin, Peter. 1951. Gilgit Bazaar. *The Geographical Magazine* 23 (9): 400, appendix

Goudie, Andrew. 1981. Fearful landscape of the Karakoram. *The Geographical Magazine* 53 (5): 306–312

Goudie, Andrew, David K. C. Jones and Denys Brunsden. 1984. Recent fluctuations in some glaciers of the western Karakoram Mountains, Hunza, Pakistan. In *The International Karadoram*

Project; Vol. 2, ed. Keith J. Miller, 411–455. Cambridge: Cambridge University Press

Goudie, Andrew S., D. Brunsden, D. N. Collins, E. Derbyshire, R. I. Ferguson, Z. Hashmet, D. K. C. Jones, F. A. Perrott, et al. 1984. The geomorphology of the Hunza Valley, Karakoram mountains, Pakistan. In *The International Karakoram Project*, ed. Keith J. Miller, 359–410. Cambridge, Vol. 2

Government of Azad Kashmir. 1952. *Census of Azad Kashmir, 1951. Azad Kashmir, Gilgit & Baltistan. Report & tables.* Murree

Government of Gilgit-Baltistan. 2018. *Gilgit-Baltistan at a glance.* Gilgit: Planning & Development Department, 16 pp

Government of Gilgit-Baltistan, Asian Development Bank and World Bank. 2011. *Gilgit-Baltistan Economic report. Broadening the transformation.* Islamabad: Asian Development Bank, 202 pp

Government of India. 1925. *Jammu and Kashmir State. List of Ruling Princes, Chiefs and Leading Personages.* Calcutta: Government of India Central Publication Branch (= IOL/P&S/20/B 275/2), 31 pp

Government of India. 1939. *Revised list of ruling princes, chiefs and leading personages of the Jammu and Kashmir State and the Gilgit Agency.* New Delhi: Manager of Publications, 27 pp

Government of Pakistan. 1975. *Population Census of Northern Area 1972. District Census Report Gilgit.* Islamabad: Census Organization. Ministry of Interior, States and Frontier Regions, 232 pp

Government of Pakistan. 1984. *1981 District Census Report of Gilgit.* Islamabad: Population Census Organization. Statistics Division, 59 pp

Government of Pakistan. 1988. *Government-sponsored corporations 1986–87.* Islamabad: Finance Division, 380 pp

Government of Pakistan. 1990. *The constitution of the Islamic Republic of Pakistan.* Islamabad: Ministry of Law, Justice and Parliamentary Affairs, Law and Justice Division, 271 pp.

Government of Pakistan. 2000. *1998 District Census Report of Gilgit.* Islamabad: Population Census Organization, 223 pp

Government of Pakistan. 2009. *An order to provide greater political empowerment and better governance to the people of Gilgit-Baltistan.* Islamabad: Ministry of Kashmir Affairs and Northern Areas, 61 pp

Government of Pakistan. 2017. *Long term plan for China-Pakistan Economic Corridor (2017–2030.* Islamabad: Ministry of Planning, Development and Reform, 33 pp

Government of Pakistan. 2018. *An order to provide for political empowerment and good governance in Gilgit-Baltistan.* Islamabad: Ministry of Kashmir Affairs and Gilgit-Baltistan, 62 pp

Government of Pakistan. 2019. *Pakistan Economic Survey 2018–19.* Islamabad: Finance Division, 185 pp

Grare, Frédéric. 2007. The evolution of sectarian conflicts in Pakistan and the ever-changing face of Islamic violence. *South Asia. Journal of South Asian Studies* 30 (1): 127–143

Grierson, George Abraham. 1900. On the languages spoken beyond the North-Western frontier of India. *Journal of the Royal Asiatic Society of Great Britain and Ireland*, July 1900: 501–510

Grombchevsky [Grabçzewski], Bronislaw L. 1888/2015. Дневник экспедиции в Канджут и Раскем в 1888. In *Дервиш Гиндукуша. Путевые дневники центральноазиатских экспедиций генерала Б. Л. Громчевского*, eds. Басханов М. К., Колесников А. А. and М. Ф. Матвеева, 77–185. Sankt Peterburg: Nestor-Istoriya

Grondelle, Marc van. 2009. *The Ismailis in the colonial era. Modernity, empire and Islam.* London: Hurst, 139 pp

Grötzbach, Erwin. 1984. Mobility of labour in high mountains and the socio-economic integration of peripheral areas. *Mountain Research and Development* 4 (3): 229–235

Grötzbach, Erwin and Christoph Stadel. 1997. Mountain peoples and cultures. In *Mountains of the world*, eds. Bruno Messerli and Jack D. Ives, 17–38. New York: Parthenon

Gryunberg, Aleksandr Leonovich and Ivan Mikhailovich Stéblin-Kamensky. 1976. *Ваханский Язык. Тексты, словарь, грамматический очерк (Языки Восточного Гиндукуша) [Vakhan language. Texts, dictionary, grammar (Languages of the Eastern Hindukush)].* Moskva: Главная Редакция Восточной Литературы [Glavnaya Redakziya Vostochnoi Literatury], 669 pp

Gryunberg, Aleksandr Leonovich and Ivan Mikhailovich Stéblin-Kamensky. 1988. *La langue Wakhi [French translation of the Russian edition by Dominique Indjoudjian, 2 Vols].* Paris: De la Maison des Sciences de l'Homme, 290 & 376 pp

Guha, Ranajit. 2002. *History at the limit of world-history.* New York: Columbia University Press, 116 pp

Guillet, David. 1983. Toward a cultural ecology of mountains: The Central Andes and the Himalayas compared. *Current Anthropology* 24 (5): 561–574

Gurdon, Bertrand Evelyn Mellish. 1933. Chitral memories. Events leading up to the siege. *Himalayan Journal* 5: 1–27

Gurdon, Bertrand Evelyn Mellish. 1934. Chitral memories. The siege of Chitral. *Himalayan Journal* 6: 1–30

Gurdon, Bertrand Evelyn Mellish. 1936. Early explorers of Kafiristan. *Himalayan Journal* 8: 25–43

Gyr, Hans. 1949. Karakorum-Expedition 1947. *Berge der Welt* 4: 3–95

Haines, Chad. 2004. Colonial routes: Reorienting the northern frontier of British India. *Ethnohistory* 51 (3): 535–565

Haines, Chad. 2012. *Nation, territory, and globalization in Pakistan. Traversing the margins.* London: Routledge, 176 pp

Haji, Sulaiman Ghulam Husain. 1905. *Genealogical table of H. H. the Hon'ble Sir Aga Sultan Muhammad Shah Aga Khan, G.C.I.E., K.C.I.E. &c. &c.* Karachi: Mercantile Steam Press

Hamid, S. Shahid. 1979. *Karakuram Hunza. The land of just enough.* Karachi: Ma'aref, 175 pp

Hamza Alavi. 1971. The crisis of nationalities and the state in Pakistan. *Journal of Contemporary South Asia* 1 (3): 42–66

Hamza Alavi. 1989. Politics of ethnicity in India and Pakistan. In *Sociology of 'developing societies': South Asia*, eds. Hamza Alavi and John Harriss, 222–246. London: Macmillan

Hamza Alavi. 1991. Nationhood and the nationalities in Pakistan. In *Economy and culture in Pakistan: Migrants and cities in a Muslim society*, eds. Hastings Donnan and Pnina Werbner, 163–187. London: Macmillan

Harada, Naohiko and Atsuko Miyoshi. 1963. Is the 'healthy Hunza' true? In *Personality and health in the Hunza Valley*, ed. Kinji Imanishi, 1–14. Kyoto: Nippon Printing

Harding, Harold Ivan. 1922. *Diary of a journey from Srinagar to Kashgar via Gilgit*. Kashgar: Swedish Mission Press, 80 pp

Harris, Leslie. 1971. The frontier route from Peshawar to Chitral: Political and strategic aspects of the 'forward policy' 1889–1896. *Contributions to Asian Studies* 1: 81–108

Hasan, Arif. 2009. *The unplanned revolution. Observations on the processes of socio-economic change in Pakistan*. Karachi: Oxford University Press, 307 pp

Haserodt, Klaus. 1984. Abflußverhalten der Flüsse mit Bezügen zur Sonnenscheindauer und zum Niederschlag zwischen Hindukusch (Chitral) und Hunza-Karakorum (Gilgit, Nordpakistan). *Mitteilungen der Geographischen Gesellschaft in München* 69: 129–161

Hassnain, Fida M. 1978. *Gilgit – The Northern Gate of India*. New Delhi: Sterling Publishers, 184 pp

Hassnain, Fida M. 1988. *Freedom struggle in Kashmir*. New Delhi: Rima Publishing House, 168 pp

Haughton, Henry Lawrence. 1913. *Sport and folklore in the Himalaya*. London: Edward Arnold, 332 pp

Hauner, Milan. 1989. Central Asian geopolitics in the last hundred years: A critical survey from Gorchakov to Gorbachev. *Central Asian Survey* 8: 1–19

Hauptmann, Harald. 2007. Pre-Islamic Heritage in the Northern Areas of Pakistan. In *Karakoram. Hidden treasures in the Northern Areas of Pakistan*, ed. Stefano Bianca, 21–40. Turin: Umberto Allemandi

Hayden, Henry Hubert. 1907. Notes on certain glaciers in North-West Kashmir. *Records of the Geological Survey of India* 35: 127–137

Headrick, Daniel R. 1981. *The tools of empire. Technology and European imperialism in the nineteenth century*. New Delhi: Oxford University Press, 221 pp

Heckler, Karl. 1954. Als Geodät im Karakorum. Aus dem Tagebuch von Karl Heckler. *Jahrbuch des Österreichischen Alpenvereins* 79: 15–25

Hedin, Sven Anders. 1899. *Durch Asiens Wüsten. Drei Jahre auf neuen Wegen in Pamir, Lop-Nor, Tibet und China; 2 Vols.* Leipzig: F. A. Brockhaus, 236 & 275 pp

Hedin, Sven Anders and Albert Herrmann. 1922. The Ts'ung-ling mountains. In *Southern Tibet. Discoveries in former times compared with my own researches in 1906–1908*, ed. Sven Anders Hedin, 3–88. Stockholm: Lithographic Institute of the General Staff of the Swedish Army

Heichel, Wolfgang. 2010. *Chronik der Erschließung des Karakorum Teil II – Central Karakorum I*. Neubiberg: Institut für Geodäsie der Universität der Bundeswehr München, 460 pp

Henderson, George and Allan Octavius Hume. 1872. *Lahore to Yarkand. Incidents of the route and natural history of the countries traversed by the expedition of 1870, under T. D. Forsyth*. London: L. Reeve, 370 pp

Henrickson, Jewel Hatcher. 1960. *Holiday in Hunza*. Washington DC: Review and Herald Publishing Association, 128 pp

Hewitt, Kenneth. 1982. Natural dams and outburst floods of the Karakoram Himalaya. *International Hydrological Association Proceedings (Hydrological aspects of alpine and high mountain areas)* 138: 259–269

Hewitt, Kenneth. 1988. The study of mountain lands and peoples: A critical overview. In *Human impact on mountains*, eds. Nigel J. R. Allan, Gregory W. Knapp and Christoph Stadel, 6–23. Totowa N. J.: Rowman & Littlefield

Hewitt, Kenneth. 1989. The altitudinal organisation of Karakoram geomorphic processes and depositional environments. *Zeitschrift für Geomorphologie N. F. Supplementband* 76: 9–32

Hewitt, Kenneth. 1992. Mountain hazards. *GeoJournal* 27 (1): 47–60

Hewitt, Kenneth. 1998. Catastrophic landslides and their effects on the Upper Indus streams, Karakoram Himalaya, northern Pakistan. *Geomorphology* 26: 47–80

Hewitt, Kenneth. 2001. Catastrophic rockslides and the geomorphology of the Hunza and Gilgit River valleys, Karakoram Himalaya. *Erdkunde* 55 (1): 72–93

Hewitt, Kenneth. 2005. The Karakoram anomaly? Glacier expansion and the 'elevation effect'. *Mountain Research and Development* 25 (4): 332–340

Hewitt, Kenneth. 2006. Glaciers of the Hunza Basin and related features. In *Karakoram in transition. Culture, development, and ecology in Hunza Valley*, ed. Hermann Kreutzmann, 49–72. Karachi: Oxford University Press

Hewitt, Kenneth. 2008. Rediscovering colonised landscape: The first Europeans at the Mustagh Pass, Karakoram Himalaya, Inner Asia. In *The exploitation of the landscape of Central and Inner Asia*, eds. Michael Gervers, Uradyn Erden Bulag and Gillian Long, 41–67. Toronto: Asian Institute, University of Toronto (= Toronto Studies in Central and Inner Asia 9)

Hewitt, Kenneth. 2010. Gifts and perils of landslides. Catastrophic rockslides and related landscape developments are an integral part of human settlement along upper Indus streams. *American Scientist* 98: 410–419

Hewitt, Kenneth. 2011. Glacier change, concentration, and elevation effect in the Karakoram Himalaya, Upper Indus Basin. *Mountain Research and Development* 31 (3): 188–200

Hewitt, Kenneth. 2014. *Glaciers of the Karakoram Himalaya. Glacial environments, processes, hazards and resources*. Dordrecht: Springer, 363 pp

Hewitt, Kenneth and Jingshi Liu. 2010. Ice-dammed lakes and outburts floods, Karakoram Himalaya. Historical perspectives and emerging threats. *Physical Geography* 31 (6): 528–551

Heidelberger Akademie der Wissenschaften. Forschungsstelle 'Felsbilder und Inschriften am Karakorum Highway'. 1994–2013. *Materialien zur Archäologie der Nordgebiete Pakistans = Materials for the archaeology of the northern regions of Pakistan, 11 vols.* Mainz: Philipp von Zabern (https://digi.hadw-bw.de/view/manp)

Hilton, James. 1933. *Lost Horizon*. London: Macmillan, 281 pp

Hingston, Richard William George. 1913. *Private journal of Captain R. W. G. Hingston, Indian Medical Service, while surgeon and naturalist with the Indo-Russian Triangulation Expedition. April 11th, 1913 to September 16th, 1913*. Dublin: Unpublished manuscript. Trinity College Library Dublin, 185 pp

Hinüber, Oskar von. 1989. 'Brāhmī inscriptions on the history and culture of the Upper Indus Valley' and 'Buddhistische Inschriften aus dem Tal des Oberen Indus'. In *Antiquities of Northern Pakistan 1: Reports and studies*, ed. Karl Jettmar, 41–60. Mainz: Philipp von Zabern

Hoffman, Jay Milton. 1968. *Hunza. Fifteen secrets of the healthiest and oldest living people*. Escondido: Professional Press Publishing Association, 236 pp

Holdich, Thomas Hungerford. 1899. The use of practical geography illustrated by recent frontier operations. *The Geographical Journal* 13 (5): 465–479

Holdich, Thomas Hungerford. 1900. An orographic map of Afghanistan and Baluchistan. *The Geographical Journal* 16 (5): 527–530, 596

Hollister, John Norman. 1953. *The Shi'a of India*. London: Luzac, 440 pp

Holzwarth, Wolfgang. 1994. *Die Ismailiten in Nordpakistan. Zur Entwicklung einer religiösen Minderheit im Kontext neuer Außenbeziehungen*. Berlin: Verlag das Arabische Buch, 136 pp

Holzwarth, Wolfgang. 1998. Change in pre-colonial times: An evaluation of sources on the Karakorum and Eastern Hindukush regions (from 1500 to 1800). In *Karakorum-Hindukush-Himalaya: Dynamics of change*, ed. Irmtraud Stellrecht, 297–335. Cologne: Köppe (= Culture Area Karakorum Scientific Studies 4/II)

Holzwarth, Wolfgang. 2006. Sources of Gilgit, Hunza and Nager history (1500–1800) and comments on the oral roots of local historiography. In *Karakoram in transition. Culture, development, and ecology in Hunza Valley*, ed. Hermann Kreutzmann, 171–190. Karachi: Oxford University Press

Hong, Caylee. 2013. Law and liminality in Gilgit-Baltistan. Managing natural resources in constitutional limbo. *Canadian Journal of Poverty Law* 2 (1): 71–103

Hudson, Brian. 1977. The new geography and the new imperialism: 1870–1918. *Antipode* 9 (2): 12–19

Hughes, Richard. 2005. The restoration of Baltit Fort. In *Karakoram. Hidden treasures in the Northern Areas of Pakistan*, ed. Stefano Bianca, 185–214. Turin: Umberto Allemandi

Hughes, Richard and Didier Lefort. 1986. At the roof of the world. The Baltit Fort. *Mimar* 20: 10–19

Hughes, Richard and Didier Lefort. 2006. The conservation of Baltit Fort 1981–1996. In *Karakoram in transition. Culture, development, and ecology in Hunza Valley*, ed. Hermann Kreutzmann, 293–304. Karachi: Oxford University Press

Humboldt, Wilhelm von. 1838. *Über die Verschiedenheit des menschlichen Sprachbaus und ihren Einfluß auf die geistige Entwicklung des Menschengeschlechts*. Berlin: Druckerei der Königlichen Akademie der Wissenschaften, 511 pp

Hunzai, Izhar Ali. 2013. *Conflict dynamics in Gilgit-Baltistan*. Washington DC: United States Institute of Peace, 16 pp

Husain, Tariq. 1987. *Innovation adoption and choice of technique in a mountain farming system. A study of the diffusion of new wheat varieties in the Gilgit District of Pakistan*. Chicago: Department of Economics, 156 pp

Hussain Wali Khan and Izhar Ali Hunzai. 2000. Bridging institutional gaps in irrigation management. The 'post-ibex-horn' innovations in Northern Pakistan. In *Sharing water. Irrigation and water management in the Hindukush-Karakoram-Himalaya*, ed. Hermann Kreutzmann, 132–145. Oxford, London Karachi: Oxford University Press

Hyder, Nasir. 1961. Gilgit in Winter. *Pakistan Quarterly* 10 (3): 18–23

Iacovleff, Alexandre. 2004. *Itinérances*. Boulogne-Billancourt: Musée des années 30, 239 pp

Iacovleff, Alexandre, Caroline Haardt de La Baume and Jean-François Ruchaud. 2012. *Les croisières Citroën*. Lyon: Fage éditions, 108 pp

Iacovleff, Alexandre Evgenievich [Yakovlev, Aleksandr]. 1934. *Dessins et peintures d'Asie, exécutés au cours de l'expédition Citroën Centre-Asie. Troisième mission, G.-M. Haardt, L. Audouin-Dubreuil*. Paris: Meynial, 50 plates & 17 pp

International Fund for Agricultural Development. 2015. *Economic transformation initiative Gilgit-Baltistan. Programme design report: Main report and appendices*. Islamabad: IFAD, 208 pp

Iqbal, Muhammad Jawed, Fida Hussain Shah, Anwar-ul Haq Chaudhry and Muhammad Naseem Baig. 2014. Impacts of Attabad Lake (Pakistan) and its future outlook. *European Scientific Journal* 10 (8): 107–120

Ispahani, Mahnaz Z. 1989. *Roads and rivals. The politics of access in the borderlands of Asia*. London: Tauris, 286 pp

Iturrizaga, Lasafam. 1997. The valley of Shimshal – a geographical portrait of a remote high mountain settlement and its pastures with reference to environmental habitat conditions in the North-West Karakorum (Pakistan). *GeoJournal* 42 (2–3): 303–328

Iturrizaga, Lasafam. 1999. Typical debris accumulation forms and formations in High Asia: A glacial-history-based concept of the origin of postglacial debris accumulation landscapes in subtropical high mountains with selected examples from the Hindu Kush, the Karakoram and the Himalayas. *GeoJournal* 47 (1/2): 277–339

Iturrizaga, Lasafam. 2005. New observations on glacier lake outbursts in the Shimshal valley. *Journal of Asian Earth Sciences* 25: 545–555

Iturrizaga, Lasafam. 2008a. Paraglacial landform assemblages in the Hindukush and Karakoram Mountains. *Geomorphology* 95: 27–47

Iturrizaga, Lasafam. 2008b. *Die Talschaft Chapursan (Karakorum). Zur Frage der geomorphologischen Genese von Schutthügellandschaften im Zusammenhang mit historischen Siedlungslandzerstörungen*. Aachen: Shaker, 111 pp

Iturrizaga, Lasafam. 2011. Hummocky debris landforms in the Chapursan Valley (Karakoram range, Pakistan): A glacio-geomorphological investigation. *Geomorphology* 169–170: 1–16

Ivanow, Mikhail Sergeevich. 1952. *Очерк истории Ирана*. Moskva: Gospolitizdat 467 pp

Ivanow, Wladimir. 1938. Ismailiya. *Enzyklopaedie des Islam, Ergänzungsband*: 101–106

Ives, Jack D. and Bruno Messerli. 1989. *The Himalayan dilemma. Reconciling development and conservation*. London, New York: Routledge, 295 pp

Jacobs, Justin M. 2016. *Xinjiang and the modern Chinese state*. Seattle: University of Washington Press, 297 pp

Janjua, Zahid Javed. 1998. Tradition and change in Darel and Tangir valleys. In *Karakorum-Hindukush-Himalaya. Dynamics of change. Part I*, ed. Irmtraud Stellrecht, 415–427. Köln: Rüdiger Köppe

Jettmar, Karl. 1958. Völkerkundliche Forschungen im Haramoshgebiet (Gilgit Agency). *Zeitschrift für Ethnologie* 83: 252–256

Jettmar, Karl. 1960. Soziale und wirtschaftliche Dynamik bei asiatischen Gebirgsbauern (Nordwestpakistan). *Sociologus* 10 (2): 120–138

Jettmar, Karl. 1961. Ethnological research in Dardistan 1958. Preliminary report. *Proceedings of the American Philosophical Society* 105 (1): 79–97

Jettmar, Karl. 1973. Weinbereitung und Weinrituale im Hindukusch. In *Festschrift zum 65. Geburtstag von Helmut Petri*, ed. Kurt Tauchmann, 191–205. Köln, Wien: Böhlau

Jettmar, Karl. 1975. *Die Religionen des Hindukusch*. Stuttgart: W. Kohlhammer, 525 pp

Jettmar, Karl. 1977a. Bolor – a contribution to the political and ethnic geography of North Pakistan. *Zentralasiatische Studien* 11: 411–448

Jettmar, Karl. 1977b. Sozialer Wandel am Karakorum-Highway. *Indo-Asia* 19: 48–55

Jettmar, Karl. 1978. Brücken und Flöße im Karakorum. Aus dem Material der Heidelberger Expeditionen (1964, 1968, 1971, 1975). *Heidelberger Jahrbücher* 22: 59–70

Jettmar, Karl. 1980. *Bolor and Dardistan*. Islamabad: National Institute of Folk Heritage, 100 pp

Jettmar, Karl. 1982a. Kafiren, Nuristani, Darden: Zur Klärung des Begriffsystems. *Anthropos* 77: 254–263

Jettmar, Karl. 1982b. Indus Kohistan. An ethnographic and linguistic overview. *Kohistan Development Board News* 2: 6–13

Jettmar, Karl. 1983. Indus-Kohistan. Entwurf einer historischen Ethnographie. *Anthropos* 78: 501–518

Jettmar, Karl. 1989. Northern Areas – an ethnographic sketch. In *History of Northern Areas of Pakistan*, ed. Ahmad Hasan Dani, 59–88. Islamabad: National Institute of Historical and Cultural Research (= Historical Studies (Pakistan) Series 5)

Jettmar, Karl. 1993. Voraussetzungen, Verlauf und Erfolg menschlicher Anpassung im nordwestlichen Himalaya mit Karakorum. In *(= Erdkundliches Wissen 112)*, ed. Ulrich Schweinfurth, 31–47. Stuttgart: F. Steiner

Jettmar, Karl. 2002. *Beyond the gorges of the Indus. Archaeology before excavation*. Karachi: Oxford University Press, 248 pp

Jettmar, Karl and Volker Thewalt. 1985. *Zwischen Gandhara und den Seidenstraßen. Felsbilder am Karakorum Highway. Entdeckungen deutsch-pakistanischer Expeditionen 1979–1984*. Mainz: Zabern, 36 pp

Johnson, Basil Leonard Clyde. 1979. *Pakistan*. London: Heinemann, 214 pp

Joldan, Eliezer 1985. Leh as entrepot of Central Asian trade. In *Harvest festival of Buddhist Dards of Ladakh and other essays*, ed. Eliezer Joldan, 58–84. Srinagar: Kapoor Brothers

Jones, Max. 2005. Measuring the world: exploration, empire and the reform of the Royal Geographical Society, c. 1874–93. In *The organisation of knowledge in Victorian Britain*, ed. Martin Daunton, 313–336. Oxford: Oxford University Press

Junker, Heinrich and Bozorg Alavi. 1965. *Persisch-Deutsches Wörterbuch*. Leipzig: Verlag Enzyklopädie, 864 pp

Kak, Bansi Lal. 1977. *The fall of Gilgit: the untold story of Indo-Pak affairs from Jinnah to Bhutto, 1947 to July 1977*. New Delhi: Light & Life Publishers, 122 pp

Kamal, Nazir A. 1979. Karakoram Highway: A nation-building effort. *Strategic Studies* 2 (3): 18–31

Kappeler, Robert. 1950. Im Banne des Rakaposhi. Originalbericht von der englisch-schweizerischen Karakorum-Expedition 1947. *Prisma. Illustrierte Monatsschrift für Natur, Forschung und Technik* 5 (2): 69–77

Karim, Karim H. 2014. Aga Khan Development Network. Shia Ismaili Islam. In *Global religious movements across borders. Sacred service*, eds. Stephen M. Cherry and Helen Rose Ebaugh, 143–160. Farnham: Ashgate Press

Karrar, Hasan H. and Till Mostowlansky. 2018. Assembling marginality in northern Pakistan. *Political Geography* 63: 65–74

Kazak, Fuad. 1937. *Ostturkistan zwischen den Großmächten. Ein Beitrag zur Wirtschaftskunde Ostturkistans*. Königsberg: Ost-Europa-Verlag (= Osteuropäische Forschungen N. F. 23), 160 pp

Kendall, Nicholas and Jack Silverman. 1983. *First harvest: valleys in transition [videogram]*. London: TVE International

Kennedy, Dane. 2007. British exploration in the nineteenth century: A historiographical survey. *History Compass* 5 (6): 1879–1900

Kennion, Roger Lloyd. 1910. *Sport and life in the further Himalaya*. Edinburgh: William Blackwood and Sons, 330 pp

Khalid, Muhammad Mumtaz. 2006. *History of Karakoram Highway, Volume I: The Indus Valley Road*. Rawalpindi: Hamza Pervez Printers, 340 pp

Khalid, Muhammad Mumtaz. 2009. *History of Karakoram Highway, Volume II: The Karakoram Highway*. Rawalpindi: Hamza Pervez Printers, 400 pp

Khan, Aga. 1954. *The Memoirs of Aga Khan. World enough and time*. New York: Simon and Schuster, 366 pp

Khan, Ahmad Ali. 1890a. Hunza. In *Gazetteer of Kashmír and Ladák: together with routes in the territories of the Maharája of Jamú and Kashmír*, ed. Quartermaster General in India, 367–371. Calcutta: Superintendent of Government Printing

Khan, Ahmad Ali. 1890b. Nagar. In *Gazetteer of Kashmír and Ladák: together with routes in the territories of the Maharája of Jamú and Kashmír*, ed. Quartermaster General in India, 614–617. Calcutta: Superintendent of Government Printing

Khan, Akhter Hameed. 1985. *Rural development in Pakistan*. Lahore: Vanguard Books, 360 pp

Khan, Abdul Hamid. 2010. *'Untold story' of Balawaristan (Pakistan and China occupied Gilgit Baltistan). The epicentre of a new Great Game*. Brussels: Balawaristan National Front, 114 pp

Khan, Faqir Muhammad. 2002. *The story of Gilgit-Baltistan and Chitral. A short history of two millenniums AD 7 – 1999*. Gilgit: Eejaz, 256 pp

Khan, Ghulam Ahmed, ed. 1902. *Census of India 1901; Vol. 23: Kashmir; Part I: Report; Part II: Tables.* Lahore: Civil and Military Gazette Press, 325 pp

Khan, Masood A. 2006. Conceptual planning in the Karakoram. The built heritage and dynamics of institution building. In *Karakoram in transition. Culture, development, and ecology in Hunza Valley*, ed. Hermann Kreutzmann, 305–319. Karachi: Oxford University Press

Khan, Mahmood Hasan. 2009. *Participatory rural development in Pakistan. Experience from rural support programmes.* Karachi: Oxford University Press, 534 pp

Khan, Mahmood Hasan and Shoaib Sultan Khan. 1992. *Rural change in the Third World. Pakistan and the Aga Khan Rural Support Program.* New York: Greenwood Press, 179 pp

Khan, M. Matin-uz-Zaman, ed. 1912. *Census of India 1911; Vol. 20: Kashmir. Part I: Report.* Lucknow: Newul Kishore Press, 256 pp

Khan, M. Shah. 1999. *Gilgit Scouts [Urdu].* Gilgit: self-published, 219 pp

Khan, Shah Raees. 2012. *Linking conservation with sustainable mountain livelihoods. A case study of Northern Pakistan.* Winnipeg: University of Manitoba dissertation, 338 pp

Khan, Shah Raees and Syed Ajijur Rahman. 2009. Integrating yak herding a resource for community livelihood in protected area management: A case study of Northern Pakistan. *Global Journal of Environmental Research* 3 (3): 258–263

Khan, Shoaib Sultan. 2009. *The Aga Khan Rural Support Programme. A journey through grassroots development.* Karachi: Oxford University Press, 502 pp

Kitamura, Siro. 1964. *Plants of West Pakistan and Afghanistan.* Kyoto: Kyoto University (= Results of the Kyoto University Scientific Expedition to the Karakoram and Hindukush 1955), 283 pp

Knight, Edward Frederick. 1891. Slavery amongst the Hunza-Nagyrs. *The Anti-Slavery Reporter* 11 (6): 285

Knight, Edward Frederick. 1893. *Where three empires meet. A narrative of recent travel in Kashmir, Western Tibet, Gilgit, and adjoining countries.* London: Longmans, Green & Co., 528 pp

Knight, Edward Frederick. 1923. *Reminiscences. The wanderings of a yachtsman and war correspondent.* London: Hutchinson & Co., 320 pp

Knudsen, Are. 1996. State intervention and community protest. Nature conservation in Hunza, Northern Pakistan. In *Asian perceptions of nature. A critical approach*, eds. Ole Bruun and Arne Kalland, 103–125. London: Routledge

Knudsen, Are. 1999. Conservation and controversy in the Karakoram. Khunjerab National Park, Pakistan. *Journal of Political Ecology* 6: 1–30

Koul, Pandit Anand. 1925. *Geography of the Jammu and Kashmir State.* Calcutta: Thacker, Spink & Co., 164 pp

Krader, Lawrence. 1959. Recent trends in Soviet anthropology. *Biennial Review of Anthropology* 1: 155–184

Krader, Lawrence. 1963. *Peoples of Central Asia.* Bloomington: Indiana University (= Uralic and Altaic Series 26), 319 pp

Kreutzmann, Hermann. 1986. A note on yak-keeping in Hunza (Northern Areas of Pakistan). *Production Pastorale et Société* 19: 99–106

Kreutzmann, Hermann. 1987. Die Talschaft Hunza (Northern Areas of Pakistan): Wandel der Austauschbeziehungen unter Einfluß des Karakoram Highway. *Die Erde* 118: 37–53

Kreutzmann, Hermann. 1988. Oases of the Karakorum: Evolution of irrigation and social organization in Hunza, North Pakistan. In *Human impact on mountains*, eds. Nigel J. R. Allan, Gregory W. Knapp and Christoph Stadel, 243–254. Totowa N. J.: Rowman and Littlefield

Kreutzmann, Hermann. 1989a. *Hunza – Ländliche Entwicklung im Karakorum.* Berlin: Reimer, 272 pp

Kreutzmann, Hermann. 1989b. Der Karakoram Highway. Eine Straße der Freundschaft? In *Pakistan – Destabilisierung durch Kontinuität? Eine Dokumentation des Südasienbüro.* Wuppertal, 40–45

Kreutzmann, Hermann. 1989c. Entwicklung und Bedeutung des Fremdenverkehrs in Hunza (Northern Areas of Pakistan). In *Beiträge zur regionalen Geographie 2 (= Hochgebirgsräume Nordpakistans im Hindukusch, Karakorum und Westhimalaya)*, ed. Klaus Haserodt, 19–31.

Kreutzmann, Hermann. 1990. Oasenbewässerung im Karakorum. Autochthone Techniken und exogene Überprägung in der Hochgebirgslandwirtschaft Nordpakistans. *Erdkunde* 44 (1): 10–23

Kreutzmann, Hermann. 1991a. The Karakoram Highway: The impact of road construction on mountain societies. *Modern Asian Studies* 25: 711–736

Kreutzmann, Hermann. 1991b. The southern frontier of the Bactrian camel in High Asia. *Camel Newsletter* 8: 8–10

Kreutzmann, Hermann. 1993a. Challenge and Response in the Karakoram. Socio-economic transformation in Hunza, Northern Areas, Pakistan. *Mountain Research and Development* 12 (1): 19–39

Kreutzmann, Hermann. 1993b. Sozio-ökonomische Transformation und Haushaltsreproduktion in Hunza (Karakorum). In *Neuere Forschungen im Himalaya*, ed. Ulrich Schweinfurth, 75–98. Stuttgart: Steiner

Kreutzmann, Hermann. 1994. Habitat conditions and settlement processes in the Hindukush-Karakoram. *Petermanns Geographische Mitteilungen* 138 (6): 337–356

Kreutzmann, Hermann. 1995a. Sprachenvielfalt und regionale Differenzierung von Glaubensgemeinschaften im Hindukusch-Karakorum. Die Rolle von Minderheiten im Konfliktfeld Nordpakistans. *Erdkunde* 49 (2): 106–121

Kreutzmann, Hermann. 1995b. Globalization, spatial integration and their impact on sustainable development in Northern Pakistan. *Mountain Research and Development* 15 (3): 213–227

Kreutzmann, Hermann. 1995c. Communication and cash crop production in the Karakoram: Exchange relations under transformation. In *Pak-German workshop: Problems of comparative high mountain research with regard to the Karakorum, Tübingen October 1992 (= Occasional Papers 2)*, ed. Irmtraud Stellrecht, Tübingen, 100–117

Kreutzmann, Hermann. 1996a. *Ethnizität im Entwicklungsprozeß. Die Wakhi in Hochasien*. Berlin: Reimer, 488 pp

Kreutzmann, Hermann. 1996b. Siedlungsprozesse und territoriale Aneignung im zentralen Hunza-Tal. Kulturgeographische Anmerkungen zur Publikation der Karte Hunza-Karakorum 1:100 000. *Erdkunde* 50 (3): 173–189

Kreutzmann, Hermann. 1996c. Wasser als Entwicklungsfaktor in semiariden montanen Siedlungsräumen. Systemansatz und Entwicklungspotential. *Zeitschrift für Wirtschaftsgeographie* 40 (3): 129–143

Kreutzmann, Hermann. 1997. Vom Great Game zum Clash of Civilizations? Wahrnehmung und Wirkung von Imperialpolitik und Grenzziehungen in Zentralasien. *Petermanns Geographische Mitteilungen* 141 (3): 163–186

Kreutzmann, Hermann. 1998a. The Chitral Triangle: Rise and decline of trans-montane Central Asian trade, 1895–1935. *Asien-Afrika-Lateinamerika* 26 (3): 289–327

Kreutzmann, Hermann. 1998b. Trans-montane exchange patterns prior to the Karakoram Highway. In *Karakorum-Hindukush-Himalaya: Dynamics of change*, ed. Irmtraud Stellrecht, 21–43. Cologne: Köppe (= Culture Area Karakorum Scientific Studies 4/II)

Kreutzmann, Hermann. 1998c. From water towers of mankind to livelihood strategies of mountain dwellers. Approaches and perspectives for high mountain research. *Erdkunde* 52 (3): 185–200

Kreutzmann, Hermann. 1999. The origin of our belief: How the Ismailiya came to Ghojal in Hunza. *European Bulletin of Himalayan Research* 15–16: 39–40

Kreutzmann, Hermann, ed. 2000a. *Sharing water. Irrigation and water management in the Hindukush-Karakoram-Himalaya*. Oxford, London, Karachi: Oxford University Press, 283 pp

Kreutzmann, Hermann. 2000b. Livestock economy in Hunza. Societal transformation and pastoral practices. In *High mountain pastoralism in Northern Pakistan*, eds. Eckart Ehlers and Hermann Kreutzmann, 89–120. Stuttgart: Franz Steiner

Kreutzmann, Hermann. 2000c. Animal husbandry in High Asia. Yak-keeping at the upper pastoral limits. In *Environmental changes in High Asia. Proceedings of an International Symposium at the University of Marburg, Faculty of Geography 29th of May to 1st of June 1997 under the auspices of the Unesco*, eds. Georg Miehe and Yili Zhang, 361–375. Marburg/Lahn

Kreutzmann, Hermann. 2001. Societal developments in high mountain regions of South Asia exemplified by the case of geolinguistic diversity. *Revue de Géographie Alpine* 89 (2): 148–151

Kreutzmann, Hermann. 2003a. Ethnic minorities and marginality in the Pamirian knot. Survival of Wakhi and Kirghiz in a harsh environment and global contexts. *The Geographical Journal* 169 (3): 215–235

Kreutzmann, Hermann. 2003b. Yak-keeping in Western High Asia: Tajikistan, Afghanistan, Southern Xinjiang, Pakistan. In *The yak. Second edition*, eds. Gerald Wiener, Han Jianlin and Long Ruijun, 323–336. Bangkok: FAO

Kreutzmann, Hermann. 2003c. Yak-keeping in the Pamirs: strategies under changing frame conditions. In *Natur und Landnutzung im Pamir*, ed. Siegmar-W. Breckle, 54–63. Bielefeld: Bielefeld University Press

Kreutzmann, Hermann. 2004a. Pastoral practices and their transformation in the North-Western Karakoram. *Nomadic Peoples* 8 (2): 54–88

Kreutzmann, Hermann. 2004b. Ellsworth Huntington and his perspective on Central Asia. Great Game experiences and their influence on development thought. *GeoJournal* 59: 27–31

Kreutzmann, Hermann. 2004c. Accessibility for High Asia. Comparative perspectives on Northern Pakistan's traffic infrastructure and linkages with neighbours in the Hindukush-Karakoram-Himalaya. *Journal of Mountain Science* 1 (3): 193–210

Kreutzmann, Hermann. 2005a. Linguistic diversity in space and time: a survey in the Eastern Hindukush and Karakoram. *Himalayan Linguistics 4 (= http://www.uwm.edu/ Dept/CIE/HimalayanLinguistics/Journal_2005/Kreutzmann_HLJ4.pdf)*

Kreutzmann, Hermann. 2005b. The Karakoram landscape and the recent history of the Northern Areas. In *Karakoram. Hidden treasures in the Northern Areas of Pakistan*, ed. Stefano Bianca, 41–76. Turin: Umberto Allemandi

Kreutzmann, Hermann, ed. 2006. *Karakoram in transition. Culture, development and ecology in the Hunza Valley*. Oxford, New York, Karachi: Oxford University Press, 419 pp

Kreutzmann, Hermann. 2007. The Wakhi and Kirghiz in the Pamirian Knot. In *Disappearing peoples? Indigenous groups and ethnic minorities in South and Central Asia*, eds. Barbara Brower and Barbara Rose Johnston, 169–186. Walnut Creek: Leftcoast Press

Kreutzmann, Hermann. 2008a. Kashmir and the Northern Areas of Pakistan: Boundary-making along contested frontiers. *Erdkunde* 62 (3): 201–219

Kreutzmann, Hermann. 2008b. Boundary-making and geopolitical diversity in the Pamirian Knot. In *Diversity in mountain systems*, eds. Jörg Löffler and Jörg Stadelbauer, 155–175. St. Augustin: Asgard-Verlag (= Colloquium Geographicum 31)

Kreutzmann, Hermann. 2008c. Trade links in the Eastern Hindu Kush. The Chitral route. In *Proceedings of the Third International Hindu Kush Cultural Conference*, ed. Israr-ud-din, 305–316. Karachi: Oxford University Press

Kreutzmann, Hermann. 2008d. *Geographic comparison of pasture land use patterns in the Afghan and Tajik Pamirs*. GTZ-CCD Conference on Desertification, Khorog, GTZ

Kreutzmann, Hermann. 2009a. Transformations of high mountain pastoral strategies in the Pamirian Knot. *Nomadic Peoples* 13 (2): 102–123

Kreutzmann, Hermann. 2009b. Geopolitical perspectives on cross-border exchange relations. In *Experiences with and prospects for regional exchange and cooperation in mountain areas*, eds. Hermann Kreutzmann, Ghulam Amin Beg and Jürgen Richter, 20–57. Bonn: InWEnt

Kreutzmann, Hermann. 2009c. The Karakoram Highway as a prime exchange corridor between Pakistan and China. In *Integrated tourism concepts to contribute to sustainable development in mountain regions*, eds. Hermann Kreutzmann, Ghulam Amin Beg, Lu Zhaohui and Jürgen Richter, 13–36. Bonn: InWEnt

Kreutzmann, Hermann. 2009d. Weidewirtschaftliche Transformationen in zentralasiatischen Hochgebirgswüsten. *Nova Acta Leopoldina* 108 (373): 79–109

Kreutzmann, Hermann. 2010. An inevitable disaster. *Himal Southasian* 23 (6): 24–25

Kreutzmann, Hermann. 2011a. Pastoralism in the Pamirs. Regional contexts, political boundaries and market integration in Central Asia. In *Economic spaces of pastoral production and commodity systems: Markets and livelihoods*, eds. Jörg Gertel and Richard Le Heron, 175–193. Aldershot: Ashgate

Kreutzmann, Hermann. 2011b. Pastoralism in Central Asian mountain regions. In *Regional workshop in Khorog and Kashgar – Pastoralism and rangeland management in mountain areas in the context of climate and global change*, eds. Hermann Kreutzmann, Kishwar Abdulalishoev, Lu Zhaohui and Jürgen Richter, 38–63. Bonn: GIZ

Kreutzmann, Hermann. 2011c. Pastoral practices on the move – Recent transformations in mountain pastoralism on the Tibetan Plateau. In *Regional workshop in Lhasa 2010 – Pastoralism and rangeland management on the Tibetan Plateau in the context of climate and global change*, eds. Hermann Kreutzmann, Yang Yong and Jürgen Richter, 200–224. Bonn: GIZ

Kreutzmann, Hermann. 2011d. Scarcity within opulence: Water management in the Karakoram Mountains revisited. *Journal of Mountain Science* 8 (4): 525–534

Kreutzmann, Hermann, ed. 2012a. *Pastoral practices in High Asia. Agency of 'development' effected by modernisation, resettlement and transformation*. Dordrecht: Springer, 350 pp

Kreutzmann, Hermann. 2012b. Pastoral practices in transition – Animal husbandry in High Asian contexts. In *Pastoral practices in High Asia. Agency of 'development' effected by modernisation, resettlement and transformation*, ed. Hermann Kreutzmann, 1–29. Dordrecht: Springer

Kreutzmann, Hermann. 2012c. Kirghiz in Little Kara Köl – the forces of modernisation in Southern Xinjiang. In *Pastoral practices in High Asia. Agency of 'development' effected by modernisation, resettlement and transformation*, ed. Hermann Kreutzmann, 109–125. Dordrecht: Springer

Kreutzmann, Hermann. 2012d. Pastoralism – a way forward or back? In *Pastoral practices in High Asia. Agency of 'development' effected by modernisation, resettlement and transformation*, ed. Hermann Kreutzmann, 323–336. Dordrecht: Springer

Kreutzmann, Hermann. 2012e. After the flood – Mobility as a survival strategy in high mountain oases. The case of Pasu in Gojal, Hunza Valley, Karakoram. *Die Erde* 193 (1–2): 49–73

Kreutzmann, Hermann. 2012f. Democracy for Gilgit-Baltistan? Kashmir's legacy and administrative challenges on route to equal rights within Pakistan. In *Understanding Pakistan*, eds. Hermann Kreutzmann and Talat Mahmood, 211–245. Berlin: Wissenschaftsverlag Berlin

Kreutzmann, Hermann. 2013a. Boundary-making as a strategy for risk reduction in conflict-prone spaces. In *The spatial dimension of risk. How geography shapes the emergence of riskscapes*, ed. Detlef Müller-Mahn, 154–171. Milton Park: Routledge

Kreutzmann, Hermann. 2013b. The significance of geopolitical issues for internal development and intervention in mountainous areas of Crossroads Asia. *Crossroads Asia Working Paper Series* 7: 1–31

Kreutzmann, Hermann. 2013c. The tragedy of responsibility in High Asia. Modernizing traditional pastoral practices and preserving modernist worldviews. *Pastoralism: Research, Policy and Practice* 3 (7): 11 pp.

Kreutzmann, Hermann. 2013d. Preservation of Gilgit-Baltistan's cultural heritage as a key to development. In *Preservation of built environment and its impact on community development in Gilgit-Baltistan*, ed. Hermann Kreutzmann, 1–40. Berlin: Centre for Development Studies

Kreutzmann, Hermann. 2014. Wakhi in transition – From Wakhan to multilocal distribution. In *Biyoze Bыlbыl*, ed. Nazir Ahmad Bыlbыl, 5–10. Islamabad: Pamir Media Group

Kreutzmann, Hermann. 2015a. *Pamirian crossroads. Kirghiz and Wakhi of High Asia*. Wiesbaden: Harrassowitz, 559 pp

Kreutzmann, Hermann. 2015b. Boundaries and space in Gilgit Baltistan. *Contemporary South Asia* 23 (3): 276–291

Kreutzmann, Hermann. 2015c. Die Brüder Schlagintweit im Great Game. Prolog und Epilog zum tragischen Ende von Adolph Schlagintweit in Kashgar. In *Über den Himalaya – Die Expedition der Brüder Schlagintweit nach Indien und Zentralasien 1854 bis 1858*, eds. Moritz von Brescius, Friederike Kaiser and Stephanie Kleidt, 97–112. Köln, Weimar, Wien: Böhlau-Verlag

Kreutzmann, Hermann. 2016. Kashmir. In *The Blackwell encyclopedia of race, ethnicity and nationalism*, eds. John Stone, Rutledge M. Dennis, Polly Rizova and Anthony D. Smith, Chichester: Wiley-Blackwell

Kreutzmann, Hermann. 2017a. *Wakhan quadrangle. Exploration and espionage during and after the Great Game*. Wiesbaden: Harrassowitz, 282 pp

Kreutzmann, Hermann. 2017b. Historical geography of the Pamirs. In *Oxford Research Encyclopedia of Asian History*, ed. David Ludden, New York: Oxford University Press (online edition)

Kreutzmann, Hermann. 2017c. Iranische Gemeinschaften in der Pamirregion. In *Handbuch der Iranistik 2*, ed. Ludwig Paul, 225–229. Wiesbaden: Reichert

Kreutzmann, Hermann, Ghulam Amin Beg and Jürgen Richter, eds. 2009. *Experiences with and prospects for regional exchange and cooperation in mountain areas*. Bonn: InWEnt, 324 pp

Kreutzmann, Hermann, Ghulam Amin Beg, Lu Zhaohui and Jürgen Richter. 2009. *Integrated tourism concepts to contribute to sustainable development in mountain regions*. Bonn: InWEnt, 184 pp

Kreutzmann, Hermann and Sabine Felmy. 2006. Wakhan Woluswali – Winds of change in Badakhshan. In *Indo-Iranian linguistics and the typology of linguistic situations. Prof. Alexander L. Gruenberg memorial volume*, ed. Mikhail N. Bogolyubov, 165–176. St. Petersburg: Nauka

Kreutzmann, Hermann and Stefan Schütte. 2010. Persistence and change of Afghan-German economic cooperation. The case of the New Baghlan Sugar Company. *Erdkunde* 64 (1): 1–16

Kreutzmann, Hermann and Stefan Schütte. 2012. Human security, vulnerability and development in Afghanistan. *Orient. German Journal for Politics, Economics and Culture of the Middle East* 53 (2): 69–76

Kreutzmann, Hermann and Teiji Watanabe, eds. 2016. *Mapping transition in the Pamirs. Changing human-environmental landscapes*. Cham: Springer, 274 pp

Kussmaul, Friedrich and Peter Snoy. 1972. *Bergvölker im Hindukusch. Eine Dokumentation des Instituts für Auslandsbeziehungen und des Linden-Museums*. Stuttgart: Institut für Auslandsbeziehungen, 104 pp

Lamb, Alastair. 1964a. *The China-India border. The origins of the disputed boundaries*. London: Oxford University Press (= Chatham House Essays 2), 192 pp

Lamb, Alastair. 1964b. The Sino-Pakistani Boundary Agreement of 2 March 1963. *Australian Outlook* 18: 299–312

Lamb, Alastair. 1966a. *Crisis in Kashmir, 1947–1966*. London: Routledge, Kegan Paul, 163 pp

Lamb, Alastair. 1966b. *The Kashmir problem: A historical survey*. New York: Praeger, 163 pp

Lamb, Alastair. 1967. Pakistani Kashmir since the 1965 war. *Journal of The Royal Central Asian Society* 54 (2): 151–155

Lamb, Alastair. 1991. *Kashmir. A disputed legacy 1846–1990*. London: Oxford University Press, 368 pp

Lansdell, Henry. 1887. *Through Central Asia. With a map and appendix on the diplomacy and delimitation of the Russo-Afghan frontier*. London: Sampson Low, Marston, Searle, and Rivington, 668 pp

Lattimore, Owen. 1950. *Pivot of Asia. Sinkiang and the Inner Asian frontiers of China and Russia*. New York: Little Brown, 288 pp

Lawrence, Walter R., ed. 1908. *Imperial Gazetteer of India. Provincial Series, Vol. 13, Kashmir and Jammu*. Oxford (reprint: Lahore 1983): Clarendon Press, 140 pp

Le Fèvre, Georges. 1933. *Expédition Citroën Centre-Asie'. 3me Mission G.-M. Haardt, Audouin-Dubreuil. Le but atteint de Beyrouth à Pékin le retour*. Paris: L'édition Artistique, 20 pp

Le Fèvre, Georges. 1935. *An eastern odyssey. The third expedition of Haardt and Audouin-Dubreuil*. London: Victor Gollancz, 358 pp

Leitner, Gottlieb Wilhelm. 1877. *The languages and races of Dardistan*. Lahore: Government Central Book Depôt, 109 pp

Leitner, Gottlieb Wilhelm. 1889. *The Hunza and Nagyr Handbook, being an introduction to a knowledge of the language, race and countries of Hunza, Nagyr, and a part of Yassin, in two parts*. Calcutta: Superintendent of Government Printing, 247 pp

Leitner, Gottlieb Wilhelm. 1890. On 'the sciences of language and of ethnography', with general reference to the language and customs of the people of Hunza. *Journal of the Transactions of the Victoria Institute* 23: 109–122

Leitner, Gottlieb Wilhelm. 1891a. The races and languages of the Hindukush. *The Imperial and Asiatic Quarterly Review, N. S.* 2: 139–156

Leitner, Gottlieb Wilhelm. 1891b. Rough accounts of itineraries through the Hindukush and to Central Asia. *The Imperial and Asiatic Quarterly Review, N. S.* 2: 243–248

Leitner, Gottlieb Wilhelm. 1891c. Hunza, Nagyr, and the Pamir regions. *The Imperial and Asiatic Quarterly Review, N. S.* 3: 52–76

Leitner, Gottlieb Wilhelm. 1891d. On the ethnographical basis of language, with special reference to the customs and language of Hunza. *The Journal of the Anthropological Institute of Great Britain and Ireland* 20: 204–210

Leitner, Gottlieb Wilhelm. 1893a. Dardistan. *The Imperial and Asiatic Quarterly Review, N. S.* 6: 202–207

Leitner, Gottlieb Wilhelm. 1893b. Dardistan in 1893 and the treaty with Kashmir. *The Imperial and Asiatic Quarterly Review, N. S.* 6: 422–425

Leitner, Gottlieb Wilhelm. 1894. *Dardistan in 1866, 1886 and 1893*. Woking: Oriental University Institute, 254 pp

Lin, Hsiao-Ting. 2009. The tributary system in China's historical imagination: China and Hunza, c. 1760–1960. *Journal of the Royal Asiatic Society, Series 3*, 19 (4): 489–507

Llewellyn-Jones, Rosie. 2018. Letters from the RSAA Archive. Sir Aurel Stein, Colonel Reginald Schomberg and a secret mission to Central Asia. *Asian Affairs* 49 (3): 492–506

Lockhart, William Stephen Alexander and Robert Gosset Woodthorpe. 1889. *The Gilgit Mission 1885–86*. London: Eyre and Spottiswode, 452 pp

Lokhandwalla, Shamoon Tayyib. 1971. Islamic law and Isma'ili communities (Khojas and Bohras). In *India and contemporary Islam*, ed. Shamoon Tayyib Lokhandwalla, 379–397. Simla: Indian Institute of Advanced Study

Long, Francis William. 1948. A recent visit to Hunza-Nagir. *Journal of the Royal Central Asian Society* 35: 161–167

Lorimer, David Lockhart Robertson. 1927. A Burushaski text from Hunza. *Bulletin of the School of Oriental Studies* 4 (3): 505–531

Lorimer, David Lockhart Robertson. 1929. The supernatural in the popular belief of the Gilgit region. *Journal of the Royal Asiatic Society of Great Britain and Ireland* 3: 507–536

Lorimer, David Lockhart Robertson. 1931. An oral version of the Kesar saga from Hunza. *Folklore* 42 (2): 105–140

Lorimer, David Lockhart Robertson. 1935–1938. *The Burushaski language*. Oslo: Instituttet for Sammenlignende Kulturforskning (= Serie B: Skrifter XXIX-1–3), 1398 pp

Lorimer, David Lockhart Robertson. 1936. Nugae Burushaskicae. *Bulletin of the School of Oriental and African Studies* 8 (2–3): 627–636

Lorimer, David Lockhart Robertson. 1937. Burushaski and its alien neighbours. Problems in linguistic contagion. *Transactions of the Philological Society* 35: 63–98

Lorimer, David Lockhart Robertson. 1938. A note on various Hunza and Shimshali names. *The Himalayan Journal* 10: 121–125

Lorimer, David Lockhart Robertson. 1939. *The Dumaki language. Outlines of the speech of the Doma, or Bericho, of Hunza*. Nijmegen: Dekker & van de Vegt (= Comit, International Permanent des Linguistes. Publications de la Commission d'Enquête linguistique 4), 244 pp

Lorimer, David Lockhart Robertson. 1958. *The Wakhi language, 2 Vols*. London: School of Oriental and African Studies, 163 &163 pp

Lorimer, David Lockhart Robertson. 1962. *Werchikwar-English vocabulary*. Oslo: Norwegian University Press (= Instituttet for Sammenlignende Kulturforskning, Serie B: Skrifter LI), 391 pp

Lorimer, David Lockhart Robertson. 1979–1980. Lorimer Personal records (= Collection of notes, letters and material from 1923–1924 and 1934–1935, archived as personal records in SOAS PP MS 66; mostly published, edited and commented upon by Irmtraud Müller-Stellrecht 1979, 1980).

Lorimer, Emily Overend. 1936a. The road to Hunza. *The Geographical Magazine* 3 (3): 161–178

Lorimer, Emily Overend. 1936b. Peasant life in Hunza. *The Geographical Magazine* 3 (4): 235–254

Lorimer, Emily Overend. 1936c. Note on the wheel-barrow of Hunza. *Antiquity* 10 (40): 464–465

Lorimer, Emily Overend. 1936d. Peasant life in Hunza. *The Listener* 15 July 1936: 98–101

Lorimer, Emily Overend. 1936e. Civilisation without shops. *The Listener* 22 July 1936: 159–161

Lorimer, Emily Overend. 1936f. An unspoiled people. *The Listener* 29 July 1936: 208–210

Lorimer, Emily Overend. 1936g. Thrifty housekeeping in the Karakoram. *The Listener* 5 August 1936: 261–269

Lorimer, Emily Overend. 1937. Bleak Hunza laughs at hardship. *Travel* 69 (3): 16–20, 45

Lorimer, Emily Overend. 1938. The Burusho of Hunza. *Antiquity* 12 (45): 5–15

Lorimer, Emily Overend. 1939. *Language hunting in the Karakoram*. London: G. Allen & Unwin, 306 pp

Luknitsky, Pavel. 1954. *Soviet Tajikistan*. Moscow: Foreign Languages Publishing House, 254 pp

Mahmud, Ershad. 2008. The Gilgit-Baltistan reforms package 2007: Background, phases and analysis. *Policy Perspectives* 5 (1): 23–40

Maillart, Ella. 1937. *Forbidden journey. From Peking to Kashmir*. London: W. Heinemann, 306 pp

Malik, Abdul, Ali Effendi and Muhammad Darjat. 2006. Community infrastructure. In *Valleys in transition. Twenty years of AKRSP's experience in Northern Pakistan*, eds. Geoff Wood, Abdul Malik and Sumaira Sagheer, 196–225. Karachi: Oxford University Press

Malik, Abdul and Mujtaba Piracha. 2006. Economic transition in Hunza and Nager valleys. In *Karakoram in transition. Culture, development, and ecology in Hunza Valley*, ed. Hermann Kreutzmann, 359–369. Karachi: Oxford University Press

Malik, Shabnam Bahar. 2012. The anthropology of headdress. Use of caps and ceremonial headdress by women of Gilgit-Baltistan region as a symbol of their unique cultural heritage. *International Journal of Humanities and Social Science* 2 (1): 143–164

Malleson, Wilfred. 1907. *Frontier and overseas expeditions from India. Selections from Govt. records. Vol. 1: Tribes north of the Kabul River*. Simla: Government Monotype Press, 591 pp

Manzar Zarin, Mohammad and Ruth Laila Schmidt. 1984. *Discussions with Hariq: Land tenure and transhumance in Indus Kohistan*. Islamabad, Berkeley: Lok Virsa, Center for South and Southeast Asia Studies, University of California (= Berkeley working papers on South and Southeast Asia), 70 pp

Marsden, Magnus. 2016. *Trading worlds. Afghan merchants across modern frontiers*. London: Hurst Publishers, 432 pp

Marsden, Magnus and Benjamin D. Hopkins. 2013. *Fragments of the Afghan frontier*. Karachi: Oxford University Press, 301 pp

Marsh, Hippisley Cunliffe. 1876. Description of a trip to the Gilgit Valley, a dependancy of the Mahárájá of Kashmir. *Journal of the Asiatic Society of Bengal* 45: 119–138

Marshall, William Macandrew. 1913. *Note on the history of Hunza rights on the Taghdumbash Pamir*. Gilgit: unpublished (= IOL/P&S/10/278)

Mason, Kenneth. 1929. Indus floods and Shyok glaciers. *The Himalayan Journal* 1: 10–29

Mason, Kenneth 1930a. The proposed nomenclature of the Karakoram Himalaya. *The Geographical Journal* 75 (1): 38–44

Mason, Kenneth. 1930b. The glaciers of the Karakoram and neighborhood. *Records of the Geological Survey of India* 63 (2): 214–278

Mason, Kenneth. 1935. The study of threatening glaciers. *The Geographical Journal* 85 (1): 24–41

Mason, Kenneth. 1936. The Himalaya as a barrier to modern communication. *The Geographical Journal* 87 (1): 1–16

Mason, Kenneth. 1938. Karakoram nomenclature. *The Geographical Journal* 91: 123–152

Mason, Kenneth. 1955. *Abode of snow. A history of Himalayan exploration and mountaineering*. London: Rupert Hart-Davis, 372 pp

Mathieu, Jon. 2011. *The third dimension. A comparative history of mountains in the modern era*. Cambridge: The White Horse Press, 194 pp

Maxwell, Neville. 1972. *India's China war*. Harmondsworth: Penguin, 546 pp

McCormick, Arthur David. 1895. *An artist in the Himalayas*. London: T. Fisher Unwin, 306 pp

McGibbon, Keith J. 1982. *Karakoram expedition 1982. A scientific expedition to the Hunza Valley Northern Pakistan (74° E, 36° N)*. Newcastle: Newcastle University Exploration Society, 88 pp

Medley, Major Ernest James. 1901. India to England via Central Asia and Siberia. *Scottish Geographical Magazine* 17 (6): 281–192

Menon, Kumara Padmanabha Sivasankara 1947. *Delhi – Chungking. A travel diary*. Bombay: Oxford University Press, 257 pp

Michaud, Sabrina and Roland Michaud. 1975. Trek to lofty Hunza and beyond. *National Geographic Magazine* 148 (5): 644–668

Mielke, Katja and Anna-Katharina Hornidge. 2017. *Area studies at the crossroads. Knowledge production after the mobility turn*. New York: Palgrave Macmillan, 363 pp

Mignolo, Walter D. 2011. *The darker side of Western modernity*. Durham: Duke University Press, 408 pp

Miller, Keith J., ed. 1984. *The International Karakoram Project, 2 Vols*. Cambridge: Cambridge University Press, 412 & 635 pp

Millward, James A. 2000. Historical perspectives on contemporary Xinjiang. *Inner Asia* 2 (2): 121–135

Millward, James A. 2007. *Eurasian crossroads. A history of Xinjiang*. New York: Columbia University Press, 440 pp

Minora, Umberto et al. 2016. Glacier area stability in the Central Karakoram National Park (Pakistan) in 2001–2010: The 'Karakoram Anomaly' in the spotlight. *Progress in Physical Geography* 40 (5): 629–660

Mohammed, Chaudri Khushi, ed. 1923. *Census of India 1921; Vol. 22: Kashmir. Part I: Report*. Lahore: Mufid-i-Am Press, 209 pp

Mohammed, Wazir and Ashraf Khan. 1969. Heavenly Hunza. *Pakistan Quarterly* 16 (1–4): 81–86

Mommsen, Wolfgang J. 1981. Das Britische Empire. Strukturanalyse eines imperialistischen Herrschaftsverbandes. *Historische Zeitschrift* 233: 317–361

Monogarova, Lidia Fedorovna. 1989. Памирцы – народности или субэтносы Таджиков? [The Pamiri – Nationalities or ethnic subgroups of the Tajikis?]. *Советская Этнография [Sovetskaya Etnografiya]* 5: 28–35

Mons, Barbara Hilda. 1958. *High road to Hunza*. London: The Travel Book Club, 157 pp

Montgomerie, Thomas George. 1871. *Report on the Trans-Himalayan explorations in connection with the Great Trigonometrical Survey of India, during 1870*. Dehra Doon: Office of the Superintendent Great Trigonometrical Survey of India, 11 pp

Moore, Raymond A. 1979. *Nation building and the Pakistan Army 1947–1969*. Lahore: Aziz Publishers, 384 pp

Morgan, Edward Delmar. 1886. Recent geography of Central Asia from Russian sources. *Supplementary Papers of the Royal Geographical Society* 1: 203–267

Morgenstierne, Georg. 1932. *Report on a linguistic mission to North-Western India*. Oslo: Instituttet for Sammenlignende Kulturforskning (= Serie C: III-1), 74 pp

Morgenstierne, Georg. 1938. *Indo-Iranian frontier languages Vol. II: Iranian Pamir languages*. Oslo: Instituttet for Sammenlignende Kulturforskning (= Serie B: Skrifter XXXV), 564 pp

Morgenstierne, Georg. 1974. Languages of Nuristan and surrounding regions. In *Cultures of the Hindukush*, ed. Karl Jettmar, 1–10. Wiesbaden: Franz Steiner (= Beiträge zur Südasienforschung 1)

Morris, Charles John. 1928. Some valleys and glaciers in Hunza. *The Geographical Journal* 71: 513–537

Mott, Peter G. 1950. Karakoram survey 1939: A new map. *The Geographical Journal* 116: 89–95

Mueezuddin Hakal. 2019. Inheritance rights and tribal governance of innermost Hunza (Gilgit-Baltistan). *Naveiñ Reet: Nordic Journal of Law and Social Research* 8: 107–122

Muhammad, Peer. 2014. From intra-sectarianism to fragile peace. The Gilgit-Baltistan model. *Conflict and Peace Studies* 6 (2): 88–103

Mukherjee, Soumen. 2016. *Ismailism and Islam in modern South Asia: Community and identity in the age of religious internationals*. Cambridge: Cambridge University Press, 203 pp

Mukherjee, Soumen. 2017. *Ismailism and Islam in modern South Asia. Community and identity in the age of religious internationals*. Cambridge: Cambridge University Press, 203 pp

Müller-Stellrecht, Irmtraud. 1973. *Feste in Dardistan. Darstellung und kulturgeschichtliche Analyse*. Wiesbaden: Franz Steiner, 354 pp

Müller-Stellrecht, Irmtraud. 1978. *Hunza und China (1761–1891)*. Wiesbaden: Franz Steiner (= Beiträge zur Südasienforschung 44), 139 pp

Müller-Stellrecht, Irmtraud. 1979–1980. *Materialien zur Ethnographie Dardistans (Pakistan). Aus den nachgelassenen Aufzeichnungen von D. L. R. Lorimer. Teil 1: Hunza. Graz 1979. Teil 2/3: Gilgit, Chitral und Yasin*. Graz: Akademische Druck- und Verlagsanstalt, 645 pp

Müller-Stellrecht, Irmtraud. 1981. Menschenhandel und Machtpolitik im westlichen Himalaya. *Zentralasiatische Studien* 15: 391–472

Müller-Stellrecht, Irmtraud. 1983. Der Thronantritt König Muhammad Nazim Khans von Hunza (Nordpakistan): Ein Beispiel 'indirekter Herrschaft' im kolonialen Indien. In *Ethnologie und Geschichte. Festschrift für Karl Jettmar*, ed. Peter Snoy, 423–437. Wiesbaden: Franz Steiner (= Beiträge zur Südasienforschung 86)

Mumtaz Ali Tajddin Sadik Ali. 2013. *101 Ismaili Heroes*. Karachi: Islamic Book Publisher, 440 pp

Murray, Edward Stevenson and Alexandre Iacovleff. 1936. With the nomads of Central Asia. A summer's sojourn in the Tekes Valley, plateau paradise of Mongol and Turkic tribes. *National Geographic Magazine* January 1936: 1–57 plus 15 plates

Musharraf, Pervez. 2006. *In the line of fire*. London: Simon & Schuster, 352 pp

Muzaffar-ud-din. 2009. Cross-border value chains and competitive advantages for mountain areas. In *Experiences with and prospects for regional exchange and cooperation in mountain areas*, eds. Hermann Kreutzmann, Ghulam Amin Beg and Jürgen Richter, 212–218. Bonn: InWent

Myrdal, Jan. 1981. *Die Seidenstraße*. Wiesbaden: F. A. Brockhaus, 325 pp

N. N. 1879. Major Biddulph's tour in Chitral and Yassin. *Proceedings of the Royal Geographical Society N. S.* 1: 794–796

N. N. 1880. The roof of the world. *Blackwood's Edinburgh Magazine* 128: 462–477

N. N. 1882. Biddulph's routes near Gilgit. *Proceedings of the Royal Geographical Society N. S.* 4: 548–549

N. N. [Edward Frederik Knight]. 1891. Slavery amongst the Hunza-Nagyrs. *The Anti-Slavery Reporter and Aborigines' Friend* 11 (6): 285

N. N. 1908. Hunza-Nager. In *The Imperial Gazetteer of India. Provincial Series, Vol. 13: Kashmir and Jammu*, ed. Walter Roper Lawrence, 225–226. Calcutta: Superintendent of Government Printing

N. N. 1922. Obituary John Biddulph. *Ibis. A quarterly journal of ornithology* 4 (2): 348–349

N. N. 1978. Friendship notes. Karakoram Highway. *Peking Review* 27 (July 7, 1978): 42–43

N. N., [Reginald Schomberg]. 1938. The late Mir of Hunza. *Journal of the Royal Central Asian Society* 25 (4): 592–594

Namus, Muhammad Shuja. 1961. *Gilgit aur Shina zaban (Urdu)*. Bahawalpur: Urdu Academy, 499 pp

Nanji, Azim. 1974. Modernization and change in the Nizari Ismaili community in East Africa. A perspective. *Journal of Religion in Africa* 6 (2): 123–139

Nanji, Azim. 1978. *The Nizari Isma'ili tradition in the Indo-Pakistan subcontinent*. Delmar: Caravan Books, 215 pp

Nasr-ul-Mulk, Shahzada. 1935. The Ismailis or Maulais of the Hindu Kush. *Journal of the Royal Central Asian Society* 22: 641–645

Nazaroff, Pavel Stepanovich. 1935. *Moved on! From Kashgar to Kashmir*. London: G. Allen & Unwin, 317 pp

Nazim Khan, Mohomed. 1936. *The autobiography of Sir Mohomed Nazim Khan, K.C.I.E., Mir of Hunza*. Karimabad: mimeographed, 145 pp

Nazim Khan, Muhammad. 2014. *An autobiography of the years 1867 to 1938. Originally recorded in Persian. Translated by Khan Bahadur Muhammad Maseh Pal. New edition of Nazim Khan 1936 published by Zahoor Hussain Khan*. Gilgit, 144 pp

Nazir Sabir. 2006. Story of our transformation. In *Karakoram in transition. Culture, development, and ecology in Hunza Valley*, ed. Hermann Kreutzmann, 404–409. Karachi: Oxford University Press

Neelis, Jason. 2006. Hunza-Haldeikish revisited. Epigraphical evidence for trans-regional history. In *Karakoram in transition. Culture, development, and ecology in Hunza Valley*, ed. Hermann Kreutzmann, 159–170. Karachi: Oxford University Press

Neelis, Jason. 2011a. *Early Buddhist transmission and trade networks. Mobility and exchange within and beyond the northwestern borderlands of South Asia*. Leiden: Brill, 371 pp

Neelis, Jason. 2011b. Capillary routes of the Upper Indus. In *Early Buddhist transmission and trade networks. Mobility and exchange within and beyond the Northwestern borderlands of South Asia*, ed. Jason Neelis, 257–288. Leiden: Brill

Neelis, Jason. 2012. Overland shortcuts for the transmission of Buddhism. In *Highways, byways, and road systems in the pre-modern world*, eds. Susan E. Alcock, John Bodel and Richard J. A. Talbert, 12–32. New York: John Wiley & Sons

Neofotistos, Vasiliki. 2011. Going home to Pakistan. Identity and its discontents in Southeastern Europe. *Identities. Global Studies in Culture and Power* 18 (4): 291–316

Neureuther, Gottfried. 1961. Als Arzt im Karakorum. *Münchener Medizinische Wochenschrift* 103: 1639–1644

Neve, Arthur. 1913. *Thirty years in Kashmir*. London: E. Arnold, 316 pp

Neve, Arthur. 1979. *The legacy of Kashmir, Ladakh & Skardu*. Lahore: Civil and Military Gazette, 192 pp

Nevill, Hugh Lewis. 1912. *Campaigns on the North-West Frontier*. London: John Murray, 413 pp

Newby, Laura J. 2013. Bondage on Qing China's northwestern frontier. *Modern Asian Studies* 47 (3): 968–994

Nichols, Robert. 2013. *The Frontier Crimes Regulation. A history in documents*. Karachi: Oxford University Press, 253 pp

Nisar Karim. 2018. *Gojal between the Pamirs & the Karakorams with contrast of tribes*. Rawalpindi: Sigma Press, 334 pp

Office for the Population Census of Xinjiang Uygur Autonomous Region. 2012. *Tabulation on the 2010 population census of Xinjiang Uygur Autonomous Region (Book I)*. Beijing: China Statistics Press, 1936 pp.

Paffen, Karlheinz. 1956a. Geomorphologische Beobachtungen im Hunza-Karakorum. *Erdkunde* 10: 12–15

Paffen, Karlheinz. 1956b. Zum Klima des NW-Karakorum. *Erdkunde* 10: 22–23

Paffen, Karlheinz. 1956c. Die Pflanzenwelt des Hunza-Landes und die Vegetationsstufen und -zonen des NW-Himalaya und -Karakorum im Profil. *Erdkunde* 10: 23–30

Paffen, Karlheinz. 1956d. Die Kulturlandschaft des Hunza-Tales. *Erdkunde* 10: 30–32

Paffen, Karlheinz, Wolfgang Pillewizer and Hans-Jochen Schneider. 1956. Forschungen im Hunza-Karakorum. Vorläufiger Bericht über die wissenschaftlichen Arbeiten der Deutsch-Österreichischen Himalaya-Karakorum-Expedition 1954. *Erdkunde* 10 (1): 1–33

Pakhalina, Tat'jana Nikolaevna. 1960. *The relationship between Sarikoli dialect and other dialects of Shughni-Roshani group*. Moscow: Oriental Literature Publishing House, 23 pp

Papanek, Hanna. 1962. *Leadership and social change in the Khoja Ismaili Community*. Cambridge Mass.: Radcliffe College Dissertation, 363 pp

Parker, Edward Harper. 1897. A few Chinese observations about Chitral, Hunza, &c. *The China Review* 22: 787–789

Parveen, Sitara et al. 2015. Irrigation in Upper Hunza. Evolution of socio-hydrological interactions in the Karakoram, Northern Pakistan. *Erdkunde* 69 (1): 69–85

Picklay, Abdus Salam. 1940. *History of the Ismailis*. Bombay: Popular Printing Press, 175 pp

Pilardeaux, Benno. 1995. *Innovation und Entwicklung in Nordpakistan*. Saarbrücken: Verlag für Entwicklungspolitik, 308 pp

Pilardeaux, Benno. 1998. Surviving as a mountain peasant. Innovation, development and the dynamics of global change in a high mountainous region (Punial, Northern Pakistan). In *Transformation of social and economic relationships in Northern Pakistan*, ed. Irmtraud Stellrecht, 355–426. Köln: Rüdiger Köppe

Pillewizer, Wolfgang. 1957. Bild und Bau des NW-Karakorum. *Photographie und Forschung* 7: 185–192

Pillewizer, Wolfgang. 1960. *Zwischen Wüste und Gletschereis. Deutsche Forscher im Karakorum*. Gotha: Hermann Haack, 211 pp

Pillewizer, Wolfgang. 1986. *Zwischen Alpen, Arktis und Karakorum. Fünf Jahrzehnte kartographische Arbeit und glaziologische Forschung*. Berlin: Dietrich Reimer, 129 pp

Pollock, Michael J. 2004. Schomberg, Reginald Charles Francis (1880–1958), army officer and explorer. *Oxford Dictionary of National Biography*. Oxford: Oxford University Press (http://dx.doi.org/10.1093/ref:odnb/61923)

Popiel-Machnicki, Wawrzyniec, Pleskaczyński and Konstancja Pleskaczyńska. 2017. *Podróże nieodkryte – dziennik ekspedycji Bronisława Grabczewskiego 1889–1890 jako świadectwo historii i elemen. dziedzictwa kulturowego*. Poznan: Wydawnictwo Naukowe UAM, 595 pp

Prescott, John Robert Victor. 1975. *Map of mainland Asia by treaty*. Melbourne: Melbourne University Press, 518 pp

Prescott, John Robert Victor, Harold John Collier and Dorothy Francis Prescott. 1977. *Frontiers of Asia and Southeast Asia*. Melbourne: Melbourne University Press, 106 pp

Prescott, John Robert Victor and Gillian D. Triggs. 2008. *International frontiers and boundaries. Law, politics and geography*. Leiden: Martinus Nijhoff Publishers, 504 pp

Puri, Luv. 2009. Pakistan's Northern Areas: Time for a reality check. *Economic and Political Weekly* 44 (39): 13–15

Qadeer, Mohammed. 1977. *An evaluation of the Integrated Rural Development Programme*. Islamabad: The Pakistan Institute of Development Economics (= Monographs in the Economics of Development No. 19), 95 pp

Qamar, A. Sayeed Khan. 1975. *The lure of the Karakorams*. Rawalpindi: American Book Company, 111 pp

Qamar, A. Sayeed Khan. 1980. Polo in the North. *Pakistan Pictorial* 9: 38–40

Qiu, Jane. 2008. China. The third pole. *Nature* 454: 393–396

Qudratullah Beg, Haji. 1962 (2006). *History of ancient era Hunza State (Part-I) (English Edition)*. Baltit: S. T. Printers (English translation by Saadullah Beg Rawalpindi 2006), 603 pp

Qudratullah Beg, Haji. 1978. *Tarikh-e-tamir-e-Sentral Jama'at Khana-ye Gilgit [The History of the Central Jamatkhana in Gilgit]*. Rawalpindi, 85 pp

Qureshi, Muhammad Ateeq, Chaolu Yi, Xiangke Xu and Yingkui Li. 2017. The evolution (2010–2015) and engineering mitigation of a rockslide-dammed lake (Hunza River, Pakistan); characterisation by analytical remote sensing. *Quaternary International* 30: 1–12

Rajput, Allah Bakhsh. 1962. Hunza and Gilgit. The mountain paradise. *Pakistan Quarterly* 11 (2): 44–53, 66

Ram, Anant and Hira Nand Raina. 1933. *Census of India 1931; Vol. 24: Jammu & Kashmir State. Part I: Report*. Jammu: Ranbir Government Press, 337 pp

Ram, Rai Bahadur Pandit Bhag, ed. 1893. *Census of India 1891; Vol. 28: The Kashmir State.* Lahore: Mufid-i-Am Press, 200 pp

Raman, Anita D. 2004. Of rivers and human rights: The Northern Areas, Pakistan's forgotten colony in Jammu and Kashmir. *International Journal on Minority and Group Rights* 11 (1/2): 187–228

Rasul, Gholam and Abid Hussain. 2015. Sustainable food security in the mountains of Pakistan. Towards a policy framework. *Ecology of Food and Nutrition* 54: 625–643

Razvi, Mujtaba. 1971. *The frontiers of Pakistan. A study of frontier problems in Pakistan's foreign policy*. Karachi, Dacca: National Publishing House, 339 pp

Rebitsch, Matthias and Wolfgang Pillewizer. 1955. Deutsch-Österreichische Karakorum-Expedition 1954. *Berge der Welt. Das Buch der Forscher und Bergsteiger* 10: 19–38

Reinhold, Beate. 2006. *Neue Entwicklungen in der Wakhi-Sprache von Gojal (Nordpakistan). Bildung, Migration und Mehrsprachigkeit*. Wiesbaden: Harrassowitz (= Iranica 10), 333 pp

Repp, Gertraud. 1963. Waldökologische Studien im westlichen Himalaya. In *Mitteilungen der Floristisch-Soziologischen Arbeitsgemeinschaft N.F. 10:* 207–222.

Rhoades, Robert E. and Stephen I. Thompson. 1975. Adaptive strategies in alpine environments: Beyond ecological particularism. *American Ethnologist* 2 (3): 535–551

Ritter, Carl. 1832. *Die Erdkunde von Asien. Dritter Theil. Zweites Buch. Asien. Band 2: Der Norden und Nord-Osten von Hoch-Asien*. Berlin: Reimer, 1143 pp

Rodale, Jerome Irving. 1948. *The healthy Hunzas*. Emmaus: Rodale Press, 257 pp

Ruchaud, Jean-François and Caroline Haardt de la Baume. 2012. *Alexandre Iacovleff et les croisières Citroën* Paris: Fage Editions, 108 pp

Sadaf Firasat et al. 2007. Y-chromosomal evidence for a limited Greek contribution to the Pathan population of Pakistan. *European Journal of Human Genetics* 15 (1): 121–126

Saidula, Amier. 2011. The Nizari Ismailis in China in modern times. In *A modern history of the Ismailis. Continuity and change in a Muslim community*, ed. Farhad Daftary, 77–92. London: I. B. Tauris

Sajid Ali, Sascha Schneiderwind and Klaus Reicherter. 2017. Structural and climatic control of mass movements along the Karakoram Highway. In *Advancing Culture of Living with landslides*, eds. Matjaž Mikoš, Binod Tiwari, Yueping Yin and Kyoji Sassa, 509–516. Cham: Springer

Sakhi, Mahrban Ali. 1997. *Socio-economic impact of Aliabad irrigation channel*. Gilgit: Aga Khan Rural Support Programme, 53 pp

Sarkar, Rinki. 2010. Rural accessibility and development: Sustainability concerns in an ecologically fragile mountain belt. *Economic and Political Weekly* 45 (21): 63–71

Saunders, Frank. 1983. *Karakoram Villages. An agrarian study of 22 villages in the Hunza, Ishkoman & Yasin valleys of Gilgit District*. Gilgit: Integrated Rural Development Project of the Food and Agriculture Organisation of the United Nations, 221 pp

Saunders, Frank. 1984. *Integrated Rural Development. Pakistan (Northern Areas). Project findings and recommendations*. Rome: United Nations Development Programme, 35 pp

Schäfer, Hermann. 1978. *Hunza. Ein Volk ohne Krankheit*. Köln: Diederichs, 271 pp

Schaller, George B. 1980. *Stones of silence. Journeys in the Himalaya*. London: André Deutsch, 292 pp

Schendel, Willem van. 2002. Geographies of knowing, geographies of ignorance: jumping scale in Southeast Asia. *Environment and Planning D: Society and Space* 20: 647–668

Schendel, Willem van. 2014. Making the most of 'sensitive borders'. In *Borderland lives in Northern South Asia*, ed. David N. Gellner, 266–271. New Delhi: Orient Blackswan

Schickhoff, Udo. 2006. The forests of Hunza Valley. Scarce resources under threat. In *Karakoram in transition. Culture, development, and ecology in Hunza Valley*, ed. Hermann Kreutzmann, 123–144. Karachi: Oxford University Press

Schmid, Anna. 1997. *Die Dom zwischen sozialer Ohnmacht und kultureller Macht. Interethnische Beziehungen in Nordpakistan.* Stuttgart: Steiner (= Beiträge zur Südasienforschung 179), pp

Schmid, Anna. 2006. The landscapes of Mominabad. Meanings and identities. In *Karakoram in transition. Culture, development, and ecology in Hunza Valley*, ed. Hermann Kreutzmann, 320–328. Karachi: Oxford University Press

Schmidt, Ruth Laila and Vijay Kumar Kaul. 2008. A comparative analysis of Shina and Kashmiri vocabularies. *Acta Orientalia* 69: 231–302

Schmidt, Uwe. 2006. *Modellierung des kurzwelligen solaren Strahlungshaushalts im Hochgebirge auf der Basis von digitalen Geländemodellen und Satellitendaten am Beispiel des Hunza-Karakorum/Nordpakistan*. Bonn: Asgard, 133 pp

Schneider, Hans-Jochen. 1956. Bild und Bau des NW-Karakorum. *Photographie und Forschung* 7: 80–91

Schneider, Hans-Jochen. 1959. Zur diluvialen Geschichte des NW-Karakorum. *Mitteilungen der Geographischen Gesellschaft München* 44: 201–216

Schneider, Hans-Jochen. 1961. Die Deutsche Karakorum-Expedition 1959. *Berge der Welt. Das Buch der Forscher und Bergsteiger* 13: 86–105

Schneider, Hans-Jochen. 1962. Die Gletschertypen. Versuch im Sinne einer einheitlichen Terminologie. *Geographisches Taschenbuch* 1962/63: 276–283

Schneider, Hans-Jochen. 1969. Minapin – Gletscher und Menschen im NW-Karakorum. *Die Erde* 100: 266–286

Schneider, Hans-Jochen and Hans Baumert. 1968. Die Expeditionskarte Minapin/NW-Karakorum 1 : 50 000. *Zeitschrift für Vermessungswesen* 93: 446–457

Schomberg, Reginald Charles Francis. 1933a. The Kuruk Tagh or dry mountains of Chinese Turkestan. *Scottish Geographical Magazine* 49: 154–161

Schomberg, Reginald Charles Francis. 1933b. *Peaks and plains of Central Asia*. London: Martin Hopkinson, 288 pp

Schomberg, Reginald Charles Francis. 1935a. *Between the Oxus and the Indus*. London: M. Hopkinson, 274 pp

Schomberg, Reginald Charles Francis. 1935b. Passes – Northern Yasin. *Alpine Journal* 47: 316–322

Schomberg, Reginald Charles Francis. 1936a. *Unknown Karakoram*. London: Martin Hopkinson, 244 pp

Schomberg, Reginald Charles Francis. 1936b. Lutkuh and Hunza. *The Alpine Journal* 48 (252): 124–133

Schomberg, Reginald Charles Francis. 1936c. Derdi and Chapursan valleys: mountains of N. W. Chitral. *Alpine Journal* 48: 295–310

Schomberg, Reginald Charles Francis. 1938. *Kafirs and glaciers. Travels in Chitral*. London: M. Hopkinson, 287 pp

Schomberg, Reginald Charles Francis. 1946. The Afdigar Pass in Hunza. *The Himalayan Journal* 13 (8): 78–86

Schomberg, Reginald Charles Francis. 1947. North Karakoram. A journey in the Muztagh-Shaksgam area. *The Geographical Journal* 109: 94–98

Schulin, Ernst. 1958. Die weltgeschichtliche Erfassung des Orients bei Hegel und Ranke. *Veröffentlichungen des Max-Planck-Instituts für Geschichte* 2: 52–53

Schultz, Arved von. 1912. Bericht über den bisherigen Verlauf meiner Pamirexpedition 1911/12. *Petermanns Geographische Mitteilungen* 58: 190–193, 261–265

Schultz, Arved von. 1914. *Die Pamirtadschik*. Giessen: Alfred Töpelmann, 97 pp

Schultz, Arved von. 1921. *Kaschgar (Chinesisch-Turkestan), Stadt und Landschaft*. Hamburg: Meißner, 51 pp

Scott, James C. 2009. *The art of not being governed. An anarchist history of upland Southeast Asia*. New Haven, London: Yale University Press, 442 pp

Searle, Mike P. 1991. *Geology and tectonics of the Karakoram Mountains*. Chichester: Wiley, 358 pp

Sering, Senge. 2012. *Expansion of the Karakoram corridor. Implications and prospects*. New Delhi: Lancer Books, 64 pp

Shafqat Hussain. 2006. Small players in the great game: Marginality and representation on the northern frontiers of nineteenth-century colonial India. *South Asia: Journal of South Asian Studies* 29 (2): 235–253

Shafqat Hussain. 2016. *Remoteness and modernity. Transformation and continuity in Northern Pakistan*. New Haven: Yale University Press, 262 pp

Shahrani, M. Nazif. 1979. *The Kirghiz and Wakhi of Afghanistan. Adaptation to closed frontiers*. Seattle and London: University of Washington Press, 264 pp

Shaw, Robert B. 1871. *Visits to High Tartary, Yârkand and Kâshghar (formerly Chinese Tartary) and return journey over the Karakoram Pass*. London: John Murray, 486 pp

Shazad, Mohammad Yousuf. 1990. Anjuman-E-Taraqqi-Khowar. A profile. *Terich Mir. Annual Journal of Chitral Students Peshawar University*: 18–22

Sherullah, Beg. 1976. *An introduction to Northern Areas of Pakistan*. Gilgit: Planning and Development Department, 16 pp

Shipton, Eric. 1938. *Blank on the map*. London: Hodder and Stoughton, 299 pp

Shipton, Eric. 1951. *Mountains of Tartary*. London: Hodder & Stoughton, 224 pp

Shipton, Eric. 1969. *That untravelled world – an autobiography*. London: Hodder & Stoughton, 286 pp

Shor, Jean and Frank Shor. 1953. At world's end in Hunza. *National Geographic Magazine* 104: 485–518

Shor, Jean Bowie. 1955. *After you, Marco Polo*. New York, Toronto, London: McGraw-Hill, 294 pp

Sial, Safdar. 2014. The China-Pakistan Economic Corridor. An assessment of potential threats and constraints. *Conflict and Peace Studies* 6 (2): 11–40

Siddiqa, Ayesha. 2007. *Military Inc. Inside Pakistan's military economy*. London: Pluto Press, 292 pp

Sidky, Homayun. 1994. Shamans and mountain spirits in Hunza. *Asian Folklore Studies* 53 (1): 67–96

Sidky, Homayun. 1995. Irrigation and state formation in Hunza. *Central Asiatic Journal* 39 (2): 246–269

Sidky, Homayun. 1997. Irrigation and the rise of the state in Hunza: A case for the hydraulic hypothesis. *Modern Asian Studies* 31 (4): 995–1017

Sidky, Homayun. 1999. Alexander the Great, the Graeco-Bactrians, and Hunza. Greek descents in Central Asia. *Central Asiatic Journal* 43 (2): 232–248

Singh, Gurnam. 1981. The Karakoram Highway and its strategic implications for India. *The Indian Journal of Political Science* 42 (1): 18–26

Singh, Thakur. 1917. *Assessment report of the Gilgit Tahsil*. Lahore: Khosla Brothers, 178 pp

Skrine, Clarmont Percival. 1925. The roads to Kashgar. *Journal of the Royal Central Asian Society* 12: 226–250

Skrine, Clarmont Percival. 1926. *Chinese Central Asia*. London: Methuen, 306 pp

Skrine, Clarmont Percival and Pamela Nightingale. 1973. *Macartney at Kashgar. New light on British, Chinese and Russian activities in Sinkiang (1890–1918)*. London: Methuen, 282 pp

Skyhawk, Hugh van. 2003. *Burushaski-Texte aus Hispar. Materialien zum Verständnis einer archaischen Bergkultur in Nordpakistan*. Wiesbaden: Harrassowitz, 506 pp

Small, Andrew. 2015. *The China-Pakistan axis. Asia's new geopolitics*. New York: Oxford University Press, 319 pp

Snoy, Peter. 1965. Nuristan und Mungan. *Tribus* 14: 101–148

Snoy, Peter. 1975. *Bagrot. Eine dardische Talschaft im Karakorum*. Graz: Akademische Druck- und Verlagsanstalt, 245 pp

Snoy, Peter. 1983. Thronfolge in den Fürstentümern der Grossen Scharungszone. In *Ethnologie und Geschichte. Festschrift für Karl Jettmar (= Beiträge zur Südasienforschung 86)*, ed. Peter Snoy, 573–581. Wiesbaden: Steiner

Sökefeld, Martin. 2005. From colonialism to postcolonial colonialism. Changing modes of domination in the Northern Areas of Pakistan. *The Journal of Asian Studies* 64 (4): 939–973

Spies, Michael. 2016. Glacier thinning and adaptation assemblages in Nagar, Northern Pakistan. *Erdkunde* 70 (2): 125–140

Spies, Michael. 2018. Changing food systems and their resilience in the Karakoram Mountains of Northern Pakistan: A case study of Nagar. *Mountain Research and Development* 38 (4): 299–309

Spies, Michael. 2019. *Northern Pakistan. High mountain farming and changing socionatures*. Lahore: Vanguard, 394 pp

Spurr, David. 1993. *The rhetoric of empire. Colonial intercourse in journalism, travel writing, and imperial administration*. Durham: Duke University Press, 224 pp

Staley, Elizabeth. 1966. *Arid mountain agriculture in Northern West Pakistan. A geographical study*. Lahore: Dissertation University of the Panjab, 417 pp

Staley, John. 1966. *Economy and society in Dardistan: Traditional systems and the impact of change*. Lahore: University of the Punjab (unpublished dissertation), 285 pp

Staley, John. 1969. Economy and society in the high mountains of Northern Pakistan. *Modern Asian Studies* 3: 225–243

Staley, John. 1982. *Words for my brother. Travels between the Hindu Kush and the Himalayas*. Karachi, New York: Oxford University Press, 287 pp

Stéblin-Kamensky, Ivan Mikhailovich. 1999. *Etymological dictionary of the Wakhi language*. St. Petersburg: St. Petersburg State University, 478 pp

Stein, Marc Aurel. 1903. *Sand-Buried Ruins of Khotan: personal narrative of a journey of archaeological and geographical exploration in Chinese Turkestan*. London: Hurst and Blackett, 485 pp

Stein, Marc Aurel. 1916a. A third journey of exploration in Central Asia, 1913–1916. *The Geographical Journal* 48 (2): 97–130

Stein, Marc Aurel. 1916b. A third journey of exploration in Central Asia, 1913–1916. *The Geographical Journal* 48 (3): 193–225, 228

Stein, Marc Aurel. 1928. *Innermost Asia. Detailed report of explorations in Central Asia, Kan-Su, and Eastern Iran; 4 Vols.* Oxford: Clarendon Press, 1160 & 1137 & 1159 & 1147 pp

Steinberg, Jonah. 2011. *Isma'ili modern. Globalization and identity in a Muslim community*. Chapel Hill: University of North Carolina Press, 234 pp

Stellrecht, Irmtraud, ed. 1998. *Karakorum – Hindukush – Himalaya: dynamics of change*. Culture area Karakorum scientific studies 4, Vols I & II. Cologne: Köppe, 608 & 678 pp

Stellrecht, Irmtraud. 2006. Passage to Hunza. Route nets and political process in a mountain state. In *Karakoram in transition. Culture, development, and ecology in Hunza Valley*, ed. Hermann Kreutzmann, 191–216. Karachi: Oxford University Press

Stellrecht, Irmtraud. 2007. 'Gemischte' Grenzgängerinnen – Schamaninnen in Hunza/Nordpakistan. *Kulturwissenschaften und Frauenstudien. Aktuelle Arbeiten und Vorträge an der Fakultät für Kulturwissenschaften der Universität Tübingen* 2: 49–112

Stephens, Ian. 1953a. *Horned moon*. London: Chatto and Windus, 288 pp

Stephens, Ian. 1953b. Politics and apricots (Hunza and Nagir). *The Geographical Magazine* 25: 489–495

Steveni, William Barnes. 1891. Colonel Grambcheffsky's Pamir explorations and the Indian Government. *The Imperial and Asiatic Quarterly Review and Oriental and Colonial Record N. S.* 2: 255–260

Steveni, William Barnes. 1892. Colonel Grambcheffsky's expeditions in Central Asia, and recent events on the Pamirs. *The Imperial and Asiatic Quarterly Review and Oriental and Colonial Record N. S.* 3: 17–52

Stöber, Georg. 2007. Religious identities provoked: The Gilgit 'textbook controversy' and its conflictual context. *Internationale Schulbuchforschung* 29 (4): 389–411

Strand, Richard F. 2001. The tongues of Peristân. In *Gates of Peristan. History, religion and society in the Hindu Kush*, eds. Alberto M. Cacopardo and Augusto S. Cacopardo, 251–257. Rome: Istituto Italiano per l'Africa e l'Oriente

Sultan Abbas and Aziz Ali Dad. 2017. *Negotiating change: Recognizing the role of customary laws for sustainable livelihoods and development in Gilgit-Baltistan*. Islamabad: Heinrich Böll Stiftung, 19 pp

Syed, Anwar Hussain. 1974. *China & Pakistan. Diplomacy of an entente cordiale*. London: Oxford University Press, 259 pp

Sykes, Ella Constance. 1916. Seven months in High Asia. *Journal of the Central Asian Society* 3 (1): 16–24

Sykes, Ella Constance and Percy Molesworth Sykes. 1920. *Through deserts and oases of Central Asia*. London: Macmillan, 340 pp

Tahir, Adnan Ahmad, Jan Franklin Adamowski, Pierre Chevallier, Ayaz Ul Haq and Silvia Terzago. 2016. Comparative assessment of spatiotemporal snow cover changes and hydrological behavior of the Gilgit, Astore and Hunza River basins (Hindukush–Karakoram–Himalaya region, Pakistan). *Meteorology and Atmospheric Physics* 128: 793–811

Tahir Ali. 1983. *The Burusho of Hunza: Social structure and household viability in a mountain desert kingdom*. Rochester: Dissertation University of Rochester, 383 pp

Tahir Ali. 1984. Ceremonial and social structure among the Burusho of Hunza. In *Asian highland societies in anthropological perspective*, ed. Christoph von Fürer-Haimendorf, 231–249. New Delhi: Sterling Publishers

Tariq Ali. 1983. *Can Pakistan survive? The death of a state*. Harmondsworth: Penguin Books, 237 pp

Tariq Rahman. 1996. *Language and politics in Pakistan*. Karachi: Oxford University Press, 320 pp

Taylor, Renee. 1964. *Hunza health secrets for long life and happiness*. New York: Prentice-Hall, 173 pp

Taylor, Renee and Mulford J. Nobbs. 1962. *Hunza, the Himalayan Shandri-La*. El Monte: Whitegorn Publishing, 56 pp

Teichman, Eric. 1937. *Journey in Turkistan*. London: Hodder and Stoughton, 221 pp

Tharoor, Shashi. 2016. *An era of darkness. The British Empire in India*. New Delhi: Aleph Book Company, 333 pp

Tharoor, Shashi. 2017. *Inglorious empire. What the British did to India*. London: Hurst, 295 pp

Thauren, Johannes. 1931. *Die Missionen in Kansu, Ost-Turkestan und Honan: Apostolisches Vikariat Lanchowfu und Mission Sinkiang*. Steyl: Missionsdruckerei, 55 pp

Thiel, Erich. 1934. *Verkehrsgeographie von Russisch-Asien*. Königsberg: Ost-Europa-Verlag, 324 pp

Tobe, John H. 1960. *Hunza. Adventures in a land of paradise*. Emmaus: Rodale Books, 646 pp

Todd, Herbert. 1930. Correspondence: Gilgit and Hunza River floods. *The Himalayan Journal* 2: 173–175

Todd, Herbert. 1931. Tours in the Gilgit Agency. *The Himalayan Journal* 3: 110–115

Todd, Herbert. 1951. The Sinkiang-Hunza frontier. *Journal of the Royal Central Asian Society* 38: 73–81

Trench, Charles Chevenix. 1986. *The Frontier Scouts*. Oxford: Oxford University Press, 298 pp

Trevelyan, Walter Raleigh Fetherstonhaugh. 1931. Polo in Gilgit. *Blackwood's Magazine* 229: 561–565

Tushkan, Georgij. 1944. *The hunter of the Pamirs. A novel of adventure in Soviet Central Asia*. London: Hutchinson, 327 pp

Uhlig, Harald. 1995. Persistence and change in high mountain agricultural systems. *Mountain Research and Development* 15 (3): 199–212

Unruh, Irene von. 1955. *Traumland Hunza. Erlebnisbericht von einer Asienreise*. Mannheim: Verlagsgenossenschaft der Waerland-Bewegung, 143 pp

Vans Agnew, Patrick Alexander. 1915. Diaries of Mr. P. A. Vans Agnew, Assistant to the Agent, Governor-General, North-West Frontier, on deputation to Gilgit, June to September 1847. In *Political diaries of Lieutenant Reynell G. Taylor, Mr. P. Sandys Melvill, Pandit Kunahya Lal, Mr. P. A. Vans Agnew, Lieutenant J. Nicholson, Mr. L. Bowring and Mr. A. H. Cocks, 1847–1849*, ed. A. Raynor, 275–302. Lahore: Pioneer Press

Vertzberger, Yaacov. 1983. The political economy of Sino-Pakistani relations: Trade and aid 1963–82. *Asian Survey* 23 (5): 637–652

Vertzberger, Yaacov. 1985. *China's southwestern strategy: Encirclement and counterencirclement*. London: Praeger, 239 pp

Visser, Philips Christiaan. 1926. Explorations in the Karakoram. *The Geographical Journal* 68: 457–473

Visser, Philips Christiaan. 1935. *Durch Asiens Hochgebirge. Himalaya, Karakorum, Aghil und K'un-lun*. Frauenfeld: Huber & Co., 256 pp

Visser, Philips Christiaan and Jenny Visser-Hooft. 1935. *Wissenschaftliche Ergebnisse der Niederländischen Expeditionen in den Karakorum und die angrenzenden Gebiete in den Jahren 1922, 1925 und 1929/30. Vol. 1: Geographie, Ethnographie, Zoologie*. Leipzig: F. A. Brockhaus, 499 pp

Visser, Philips Christiaan and Jenny Visser-Hooft. 1938. *Wissenschaftliche Ergebnisse der Niederländischen Expeditionen in den Karakorum und die angrenzenden Gebiete in den Jahren 1922, 1925 und 1929/30. Vol. 2: Glaziologie*. Leiden: Brill, 216 pp

Visser-Hooft, Jenny. 1926. *Among the Kara-Korum glaciers in 1925*. London: Edward Arnold, 303 pp

Visser-Hooft, Jenny. 1935. Ethnographie. In *Wissenschaftliche Ergebnisse der Niederländischen Expeditionen in den Karakorum und die angrenzenden Gebiete in den Jahren 1922, 1925 und 1929/30, Vol. 1*, eds. Philips Christiaan Visser and Jenny Visser-Hooft, 121–155. Leipzig: Brockhaus

Walker, James Thomas. 1873. Notes on the maps of Central Asia and Turkestan which have been compiled and published in the Office of the Great Trigonometrical Survey of India, under the superintendence of Colonel J. T. Walker, R. E. Туркестанский Сборник. Собрание Сочинений о Туркестанском Крае вообще и сопредельныхъ съ нимъ странъ Средней Азии [Turkestan anthology. A collection about the Turkestan Region in general and the adjoining Middle Asian countries] 101: 137–143

Walker, James Thomas. 1892. Notes on M. Dauvergne's travels in Chinese Turkestan. *Proceedings of the Royal Geographical Society and Monthly Record of Geography N.S.* 14 (11): 779–785

Wallerstein, Immanuel. 1974. *The Modern World-System I: Capitalist agriculture and the origins of the European world-economy in the sixteenth century*. New York: Academic Publishers, 412 pp

Wallerstein, Immanuel. 1988. Inventions of TimeSpace realities: Towards an understanding of our historical systems. *Geography* 73 (4): 289–297

Wang, Wenying, Maohuan Huang and Jianming Chen. 1984. A surging advance of Balt Bare glacier, Karakoram mountains. In *The International Karakoram Project; Vol. 1*, ed. Keith J. Miller, 76–83. Cambridge: Cambridge University Press

War Office. 1902. *Secret. Notes on the Central Asian question and the cost and frontier defences of India*. London: Harrison and Sons, 326 pp

Watanabe, Teiji, Yasuhiro Ochiai, Kazuo Mizushima, Shuji Iwata and Hermann Kreutzmann. 2011. Current status of tourism and roles of proposed local guide association in Pasu, Northern Areas of Pakistan. *Geographical Studies [Hokkaido Geographical Society]* 86: 41–54

Watson, Geoff. 2009. 'Failed states' on the 'perilous frontier'. Historical bases of state formation in Afghanistan and Central Asia.

In *China, Xinjiang and Central Asia. History, transition and crossborder interaction into the 21st century*, eds. Colin Mackerras and Michael Clarke, 75–93. London: Routledge

Waugh, Daniel C. 2007. *Etherton at Kashgar. Rhetoric and reality in the history of the 'Great Game'*. Seattle: Bactrian Press, 72 pp

Weber, Siegfried. 2007. *Die persische Verwaltung Kaschmirs (1842–1892)*. Wien: Verlag der Österreichischen Akademie der Wissenschaften, 1104 pp.

Weinreich, Matthias. 2009. *'We are here to stay'. Pashtun migrants in the Northern Areas of Pakistan*. Berlin: Klaus Schwarz Verlag, 120 pp

Weinreich, Matthias. 2010. Language shift in Northern Pakistan: The case of Domaakí and Pashto. *Iran & the Caucasus* 14 (1): 43–56

Whiteman, Peter T. S. 1985. *Mountain oases. A technical report of agricultural studies (1982–1984) in Gilgit District, Northern Areas, Pakistan* Gilgit (FAO/UNDP - Pak80/009)

Wilbanks, Thomas J. 1972. Accessibility and technological change in Northern India. *Annals of the American Association of Geographers* 62 (3): 427–436

Williams, Laurence Frederic Rushbrook. 1964. Hunza and Nagar. Yesterday and today. *Journal of the Royal Central Asian Society* 51: 228–235

Williams, Maynard Owen. 1932. First over the roof of the world by motor. The Trans-Asiatic Expedition sets new records for wheeled transport in scaling passes of the Himalayas. *National Geographic Magazine* March 1932: 321–363

Williams, Raymond Brady. 1988. *Religions of immigrants from India and Pakistan. New threads in the American tapestry*. Cambridge: Cambridge University Press, 326 pp

Wilson, Thomas M. and Hastings Donnan. 2012. Borders and border studies. In *A companion to border studies*, eds. Thomas M. Wilson and Hastings Donnan, 1–25. Oxford: Blackwell Publishing

Wissmann, Hermann von. 1960. *Die heutige Vergletscherung und Schneegrenze in Hochasien*. Wiesbaden: Franz Steiner (= Abhandlungen der Akademie der Wissenschaften und Literatur. Mathematisch-Naturwissenschaftliche Klasse 14), 331 pp

Wolf, Eric R. 1982. *Europe and the people without history*. Berkeley, Los Angeles, London: University of California Press, 503 pp

Wolf, Eric R. 1994. Perilous ideas. Race, culture, people. *Current Anthropology* 35 (1): 1–12

Wood, Geoffrey, Abdul Malik and Sumaira Sagheer, eds. 2006. *Valleys in transition. Twenty years of AKRSP's experience in Northern Pakistan*. Karachi: Oxford University Press, 522 pp

Woodman, Dorothy. 1969. *Himalayan frontiers: a political review of British, Chinese, Indian and Russian rivalries*. London: Barry and Rockliff, The Cresset Press, 423 pp

Workman, William Hunter and Fanny Bullock Workman. 1910. *The call of snowy Hispar. A narrative of exploration and mountaineering on the Northern Frontier of India*. London: Constable and Company, 297 pp

World Bank. 2002. *The next ascent. An evaluation of the Aga Khan Rural Support Program, Pakistan*. Washington: World Bank Operations Evaluation Department, 139 pp

World Bank. 2011. *Gilgit-Baltistan economic report. Broadening the transformation*. Islamabad: Government of Gilgit-Baltistan, 202 pp

Wreford, Reginald George. 1943. *Census of India 1941; Vol. 22: Jammu and Kashmir*. Jammu: Ranbir Government Press, 388 pp

Wrench, Guy Theodore. 1938. *Wheel of health*. London: C. W. Daniel Company, 105 pp

Wriggins, Sally Hovey. 1980. Karakoram Highway: 500 miles of men against glaciers. *Orientations* 11 (4): 20–25

Yao, Tandong et al. 2012. Third pole environment (TPE). *Environmental Development* 3: 52–64

Yibulaying, Dilimaolati 2009. Developing eco-tourism in border areas and promoting socio-economic development in mountain areas. The case of Taxkorgan Tajik Autonomous County. In *Integrated tourism concepts to contribute to sustainable development in mountain regions*, eds. Hermann Kreutzmann, Ghulam Amin Beg, Lu Zhaohui and Jürgen Richter, 135–139. Bonn: InWEnt

Younghusband, Francis Edward. 1892. Journeys in the Pamirs and adjacent countries. *Proceedings of the Royal Geographical Society* 14: 205–234

Younghusband, Francis Edward. 1895a. *The northern frontier of India. Roads and passes. Measures for defence of frontier. India Office Library & Records: Political and Secret Department Memoranda (= IOL/P&S/18/A95)*, 12 pp

Younghusband, Francis Edward. 1895b. Chitral, Hunza and the Hindukush. *The Geographical Journal* 5: 409–426

Younghusband, Francis Edward. 1896. *The heart of a continent*. London: John Murray, 409 pp

Younghusband, Francis Edward. 1926. The problem of the Shaksgam Valley. *The Geographical Journal* 68 (3): 225–230

Younghusband, George John. 1898. *Indian frontier warfare*. London: Kegan, Paul, Trench, Trübner, 254 pp

Yule, Henry and Arthur Coke Burnell. 1903. *Hobson-Jobson. A glossary of colloquial Anglo-Indian words and phrases, and of kindred terms, etymological, historical, geographical and discursive*. London: John Murray, 1021 pp

Zafar Iqbal. 2018a. *Diverse and adventurous history of Hunza*. Gilgit: North Books, 769 pp

Zafar Iqbal. 2018b. *Diverse & tribal domains Gojal Hunza*. Gilgit: North Books, 516 pp

Zaidi, S. Akbar. 2015. *Issues in Pakistan's economy. A political economy perspective*. Karachi: Oxford University Press, 836 pp

Zaigham Khan. 1996. Saviours of the lost park. *The Herald* September 1996: 142–143

Zhang, Xiangsong. 1984. Recent variations of some glaciers in the Karakoram mountains. In *The International Karakoram Project; Vol. 1*, ed. Miller. Keith J., 39–50. Cambridge: Cambridge University Press

Zutshi, Chitralekha. 2004. *Languages of belonging: Islam, regional identity, and the making of Kashmir*. London: Hurst, 359 pp

Map bibliography

A map of the Kingdom of Caubul and some of the neighbouring countries, altered from a map constructed in the year 1809. By Lieut. John Macartney, 5th Regt. Bengal Native Cavalry. A New Edition: Re-drawn and engraved with considerable improvements. Published in: Mountstuart Elphinstone. *An account of the Kingdom of Caubul*. London ^{3}1838

Afghānistān, China, Gilgit Agency, U. S. S. R. Little Pamir, No. J-43 O. Scale one inch to four miles. Published by the War Office in 1942; Geographical Section General Staff No. 3919; originally published by the Survey of India 1926 and revised to 1941; heliographed by O. S. O. 1942

Afghānistān, Gilgit Agency and Kashmīr & Jammu. Baltit (Hunza), No. 42 L [J-43–U]. Scale one inch to four miles. Published by the Survey of India and heliozincographed at the Survey Offices, Calcutta 1930 [revised edition from the first edition 1923]

An orographic map of Afghanistan and Baluchistan. Published by Thomas Hungerford Holdich. *The Geographical Journal* 16 (5): 596

Asia. *The London atlas of universal geography, exhibiting the physical & political divisions of the various countries of the World, constructed from original materials.* Published by John Arrowsmith 1842. London, Plate 27.

Atlas of Pakistan. Published by the Survey of Pakistan; edited by Anis Ali Syed. Rawalpindi 1985, 129 pp

Atlas of the Northern Frontier of India. Published by the Ministry of External Affairs, Government of India. New Delhi 1960, 39 maps

Bagrot-Hunza-Nagyr and the Hispar Glacier. Scale 1 : 126,720. Published in: Walter Martin Conway. *Climbing and Exploration in the Karakoram-Himalayas.* London 1894

Carta ipsometrica del Bacino del Cashmìr. Scale 1 : 300,000. Published in: Giotto Dainelli. *Relazioni Scientifiche della Spedizione Italiana de Filippi nell'Himàlaia, Caracorùm e Turchestàn Cinese (1913–1914). Serie II. V.* Bologna 1924

Central Asia. Scale 207 miles to an inch. A sketch map classified 'most secret' and enclosed in: *Central Asia; supply routes to China, c. January 1943* (IOL/P&S/12/4616)

Central Asia. Afghanistan and her relation to British and Russian territories. Published for the Library of Congress by G. W. & C. B. Colton & Co. New York in 1885

Central Asia & the Indian Frontier. Plate 109–110 of the *Harmsworth universal atlas and gazetteer.* 1895 edition, published for the London Geographical Institute by George Philip & Son

Central Asia; comprising Bokhara, Cabool, Persia, the River Indus, & Countries eastward of it. Constructed from numerous authentic documents, but principally from the original M. S. surveys of Lieut. Alex. Burnes, F. R. S. to whom this map is most respectfully dedicated, by his obliged Servant, J. Arrowsmith, June 1834. *The London atlas of universal geography, exhibiting the physical & political divisions of the various countries of the World, constructed from original materials.* Published by John Arrowsmith 1838. London, Plate 29.

China, Gilgit Agency and Kashmir & Jammu. Scale one inch to four miles. No. 42–P. Surveyed in 1930. Published by the Survey of India and heliozincographed at the Survey Offices, Calcutta 1930

China-Pakistan boundary. Published by United States Central Intelligence Agency 1968. Retrieved from the Library of Congress, https://www.loc.gov/item/2007627235/

China's southern frontier. Secret map, January 1958 (Briefs for Secretary of State's visit to India, (Folder 2); Appendix 9; FCO 37/314)

Colonel Grambchevsky's expeditions in Central Asia, and the recent events on the Pamirs [К сообщению Б. Л. Громческаго о путешествии его в 1889–1890 (Памир)]. Published by W. Barnes Steveni 1892. *The Asiatic Quarterly Review.* N. S., January-April 1892

Eastern Turkestan and its neighbours. Published in: Demetrius Charles Boulger. *The life of Yakoob Beg. Athalik Ghazi, and Badaulet, Ameer of Kashgar.* London 1878

Gilgit Agency, Kashmir & Jammu, N. W. Frontier Province. Scale one inch to four miles. No. 43–I (I-43 C). Surveyed in 1912, 1930, 1931; first edition 1923, second 1934, reprint 1944. Published by the Survey of India and heliozincographed at the Survey Offices, Calcutta 1934

Gilgit Agency and Jammu & Kashmir. Scale 1 : 1,000,000. Survey of Pakistan, Rawalpindi. Editions 1980, 1985, 1992 and 1995

Gilgit and surrounding country. One inch equals four miles, 1881. India Office Records – Political & Secret Department: Map Collection X/9496

Hispar Glacier (1925). India Office Records – Political & Secret Department: Map Collection Y/104/42P

Hunza-Karakorum 1: 100,000. Topographische Ergebnisse der Deutsch-Österreichischen Himalaya-Karakorum-Expedition 1954 und der Deutschen Karakorum-Expedition 1959. Published by Deutscher Alpenverein (German Alpine Club), Munich 1995 (being a supplement to the journal *Erdkunde* 50 (3) 1996

Index map to routes in Chitral, Gilgit and Kohistan. Scale eight miles to an inch. Simla: General Staff India, July 1942

India (Northern sheet). Scale 94 miles to an inch. Published by William and Alexander Keith Johnston in the *Handy Royal Atlas of Modern Geography*, sheet 28. Edinburgh & London 1914

India. Showing railways open and under construction on 31st March 1945. Scale 67 miles to an inch. Published by the Survey of India, Dehra Dun 1945

India and Pakistan [Индия и Пакистан]. Published by Mikhail Ivanov 1952

India and Pakistan 1 : 250,000. Series U 502: Sheet NI-43–2 Gilgit. Edition 1963; sheet NJ-43–14. Baltit. Edition 1963 sheet NJ-43–15. Shimshal. Edition 1959

International World Map Series. Scale 1 : 1,000,000 Series 1301: Sheets NI-42 (Kabul), NI-43 (Srinagar), NJ-42 (Stalinabad), NJ-43 (Su-fu). Different editions (1940–1971)

Karakoram. Orographical sketch map. Scale 1 : 250,000 (two sheets). Swiss Foundation for Alpine Research. Edition 1990

Karakoram map series. Gilgit Baltistan. Scale 1 : 500,000. Manzoom Ali. Gilgit, edition 2016

Karte des Pamir von Arved v. Schultz. Auf Grund der russischen Karte in 1 : 420,000 unter Benutzung der neuesten Quellen und nach eigenen Aufnahmen entworfen. Scale 1 : 750,000. Published by Arved von Schultz. *Petermanns Geographische Mitteilungen* 1912, II, Tafel 28

Karte Inner-Asien's zu Carl Ritters Erdkunde Buch 3. Published by Carl Zimmermann. *Karte Inner-Asien's zu C. Ritter's Erdkunde Buch 3: Übergang von Ost- nach West-Asien*. Berlin 1841

Karte von Afghanistan mit den angrenzenden russischen und englischen Gebieten. Bearbeitet von Gustav Freytag. Published in: Hermann Roskoschny. *Afghanistan und seine Nachbarländer*. Leipzig 1885

Kashmīr. Southern Asia Series. Scale 1 : 2 million. Engraved, heliozincographed and printed at the Survey of India Offices, Calcutta 1929

Kashmīr & Jammu. Kashmīr North (Bārāmūla) and Kashmīr South (Anantnāg) Districts (No. 43–J-16). Scale one inch to a mile. Surveyed in 1911, heliozincographed at the Survey of India Offices, Calcutta 1912

Kashmīr & Jammu and Gilgit Agency. Scale one inch to 15,783 miles. Survey of India. Published under the direction of Brigadier C[linton] G[resham] Lewis, Surveyor General of India. Printed at the Survey of India Offices, Dehra Dun. First edition 1940; printed 1941

Map illustrating Sir John Ardagh's proposal for settlement of Northern Frontier from the Pamirs to Tibet. No date, no place. India Office Records – Political & Secret Department: Map Collection W/LPS/21/J45

Map illustrating the languages spoken beyond the North-Western Frontier of British India. Published by George Abraham Grierson. *Journal of the Royal Asiatic Society of Great Britain and Ireland*, (July 1900), map following page 510

Map of Afghanistan, and the adjacent countries. Scale 1 : 2,217,125. Published by authority of the Hon[oura]ble Court of directors of the East India Company. London 1844. Drawn and engraved by J[ohn] & C[harles] Walker

Map of Central and Western Asia 1873. Published by Edward Stanford, London 1873. Scale 110 miles to an inch.

Map of Central Asia. Published by John Arrowsmith. *Journal of the Royal Geographical Society* 45, 1875: 420

Map of Chitrál, Hunza, and parts of Wakhán and Káfiristán surveyed by Colonel R[obert] G[osset] Woodthorpe, R. E., assisted by Subsurveyor Bápu Jádu. Scale 12 miles to an inch. Dehra Dun: Survey of India 1888. India Office Records – Political & Secret Department: Map Collection W/LPS/21/J9

Map of Chitrál, Hunza, and parts of Wakhán and Káfiristán surveyed by Colonel R[obert] G[osset] Woodthorpe, R. E., assisted by Subsurveyor Bápu Jádu. Scale 12 miles to an inch. Dehra Dun 1888 (Public Record Office FO 925/2661)

Map of Dardistan and the Pamirs. Scale 1 : 4,000,000. Published by Ernst Georg Ravenstein. Published in: Gottlieb Wilhelm Leitner. *Dardistan in 1866, 1886 and 1893*. Woking 1894 (reprint: New Delhi 1978, Karachi 1985)

Map of Hunza, Nagar, &c. by Colonel R[obert] G[osset] Woodthorpe. Scale 12 miles to an inch. Intelligence Division No. 875. Reproduced & Lithographed at I. D. W. O., December 1891. India Office Records – Political & Secret Department: Map Collection W/LPS/21/J16

Map of Kashmir with its passes; Ladakh & Little Tibet, the mountain course of the Indus; and the Alpine Panjab generally. Taken principally from the MSS surveys of G. T. Vigne, Esqr. FGS. Compiled by order of the Court of Directors of the East India Company. Engraved and published by John Walker 16 May 1842

Map of Muztāgh-Ata and Lake Little Karakul. Prepared by Lieutenant F. B. Tillard, R. E., from Photo-Theodolite Survey of M[arc] A[urel] Stein, Ph. D. Surrounding district from plane-table survey Sub-surveyor S.-R. 1900. Drawn by Karim Bakhsh. Scale one inch to four miles. Calcutta 1902

Map of the boundary between China's Sinkiang and the contiguous areas the defence of which is under the actual control of Pākistān. Scale 1 : 500,000. Published under the direction of G. H. Khan, Surveyor General of Pākistān at the Survey of Pākistān Offices, Rawalpindi. First edition 1966

Map of the country between the Oxus and the Indus. Scale 1 : 731,520. Published in: Reginald C. F. Schomberg. *Between the Oxus and the Indus*. London 1935

Map of the Kingdom of Kashmir. Scale 1 : 2,027,520. Published in: Walter Martin Conway. *Climbing and Exploration in the Karakoram-Himalayas*. London 1894

Map of the North Western Frontier of India showing the Pamir region and part of Afghanistan. Scale: 1 : 2,011,680. Published in: Francis E. Younghusband. *The Heart of a Continent*. London 1896

Map of the Pamir Steppe and neighbouring districts to illustrate the letters of Colonel Gordon & members of the Kashgar Mission. Extracts of letters from members of Mr. Forsyth's mission to Kashgar relating to the geographical results of the mission. *Proceedings of the Royal Geographical Society and Monthly Record of Geography* 18 (4) 1873–1874

Map of the Shimshal Mustagh Area, drawn by Havildar Sher Hassan and Naik Taqdir Ullah (F. I. C.) 1934. Scale one inch equals four miles. Royal Society for Asian Affairs, catalogue number SC/SCH/2/28

Map of the Shingshāl Mustāgh Area. Scale 1: 253,440. Published in: Reginald C. F. Schomberg. *Unknown Karakoram*. London 1936 (in pocket)

Map of valleys and glaciers in Hunza. Scale 1 : 400,000. *The Geographical Journal* 71 (1928)

Map to illustrate the explorations of Captain F. E. Younghusband. King's Dragoon Guards on the Northern Frontier of Kashmir. Published in Francis Edward Younghusband. *A mission to the Northern Frontier of Kashmir in 1889*. London 1890

Map showing portions of Chinese Turkestan and Kansu to illustrate the explorations of Aurel Stein and his assistants Lal Singh and Ram Singh, 1906–1908. *The Geographical Journal* 37 (3), 1911, map following page 280

Map showing position of glaciers in Hunza, Nagir and Bagrot. Scale one inch to four miles. *Records of the Geological Survey of India* 35, 1907, plate 39

Minapin (Rakaposhi Range) NW-Karakorum 1 : 50,000. Berlin 1967

Mountaineering map of Pākistān. Scale 1 : 1,000,000. Survey of Pakistan, Rawalpindi. First edition 1999

North-West India & The Pamirs. Plate 115–116 of the *Harmsworth universal atlas and gazetteer*. 1908 edition, published for the London Geographical Institute by George Philip & Son

N.-W. F. Province & Afghanistan 1 : 250,000. Sheet No. 38M Chitral. Edition 1978

N. W. Frontier Province. Dīr, Swāt & Chitrāl Agency (No. 43–B-13). Scale one inch to a mile. Second edition 1944, published and printed at the Survey of Pakistan Offices Murree 1951

NW-Karakorum (Batura-Muztagh). Scale 1 : 100,000. Prepared by Ing. Hans Baumert and Prof. Dr. Hans-Jochen Schneider, not dated, contour line plan

Ordnance map (Soviet Union) [Генеральный штаб (Советский Союз)] Afghanistan. Scale 1 : 200,000. No. 3640_7400_GEO

Pakistan 1 : 500,000. Survey of Pakistan. Rawalpindi. Sheet NI-43/NW. Gilgit. Edition 1986; sheet NJ-43/SE. Khunjerab Pass. Edition 1986; sheet NJ-43/SW. Hunza. Edition 1983

Parts of Wakhan and Chitral. Northern Trans-Frontier. Sheet 2, south-west. Confidential. Scale 1 : 253,440. Published under the direction of Colonel H[enry] R[avenshaw] Thuillier, Surveyor General of India, Survey of India Offices. November 1893. Photozincographed at the Trigonometrical Branch, Survey of India Offices, Dehra Dun (Public Record Office FO/925/2025)

Provisional sketch map showing portions of Chinese Turkestan, Kansu and the Oxus regions to illustrate the explorations of Sir Aurel Stein. Indian Archaeological Survey, and his assistants R[ai] B[ahadur] Lal Singh, Muhammad Yakub Khan and Afrazgul Khan. Survey of India Department 1913–15. Scale 1 : 7,500,000. *The Geographical Journal* 48 (1916); map following page 225

Railways in Asia. Sketch map published in *Journal of the United Service Institution of India* 78 (1889)

Raskam & Mustagh Rivers. Mustagh Karakoram 1934. Drawn by Naik Taqdir Ullah (F. I. C.). Scale one half inch equals one mile. Royal Society for Asian Affairs, catalogue number SC/SCH/2/29

Rough sketch map of the Pámírs and Wakhán to accompany a report by Lieut. Colonel T[homas] E[dward] Gordon. Lithographed in the Quarter Master General's Office from an original prepared & drawn by Lt. Col. Gordon & Capt. Biddulph. Simla, September 1874, India Office Records – Political & Secret Department: Map Collection W/LPS/21/J4

Russland (Turkestan) J-43–SW/4. Scale 1 : 200,000. Map prepared for the General Staff of the German Army, Department for War Maps and Surveys (II) 1941. First special edition September 1941 (classified). The map is an enlarged and translated version of the Russian map J-43–B. Scale 1 : 500,000

Skeleton route map of the Pamirs and adjacent countries, showing routes traversed by British and other European explorers, 1868–1892. Scale 16 miles to an inch. Photozincographed from an original supplied by the Intelligence Branch, Quarter Master General's Department, at the Office of the Trigonometrical Branch, Survey of India, Dehra Dun, August 1892. India Office Records – Political & Secret Department: Map Collection W/LPS/21/J19

Sketch map of motor route taken from Peking to Kashgar by Sir Eric Teichman's mission (west sheet), 10 February 1936 (FO 676/241)

Sketch map of countries surrounding Gilgit. *Journal of the Asiatic Society of Bengal* 45, 1876: 119, Plate IV

Sketch map of the countries north of Iskardo and Ladak. Insert map to 'Map of Kashmir with its passes … Taken principally from the MSS surveys of G. T. Vigne, Esqr. FGS. Compiled by order of the Court of Directors of the East India Company by John Walker'. Engraved and published by John Walker 16 May 1842

Sketch map of the territory between Cabul and the Mustagh Pass. Scale 1 : 1,018,760. Published in: John Biddulph. *Tribes of the Hindoo Koosh*. Calcutta 1880

Sketch map to illustrate the geology of the districts traversed by the Chitral Mission by Surgeon Geo[rge] M. Giles Produced by Dangerfield Lith, London. Published in: William Stephen Alexander Lockhart and Robert Gosset Woodthorpe. *The Gilgit Mission 1885–86*. London 1889, map 3

Sketch map to illustrate the paper on Chinese frontiers of India. Scale 1 : 15 million. *The Geographical Journal* 39 (3), 1912, map following page 218

Surveys in Swāt, Chitrāl, Gilgit and neighbouring territories, 'A' Company, 1925 to 1931. Scale 16 miles to an inch. Published by the Survey of India. Dehra Dun 1934

Tash-Kurghan (1926). India Office Records – Political & Secret Department: Map Collection Y/104/42O

The map of Batura Glacier. Scale 1 : 60,000. Institute of Glaciology, Cryopedology and Desert Research. Academia Sinica, Lanchow 1978

The Pamirs. Tourist map of Gorno-Badakhshan with adjacent areas of Afghanistan, China, Kyrgyzstan and Pakistan. Scale 1 : 500,000. Markus Hauser, The Pamir Archive. Third edition 2016

The territories of the Maharaja of Jummoo and Kashmir. Five maps on faith, language, race, snow and politics to illustrate Mr. Drew's work 'Jummoo and 'Kashmir Territories'. Scale 32 miles to an inch. Published in: Frederic Drew. *The Jummoo and Kashmir territories. A geographical account*. London 1875, Appendix

Xinjiang Atlas Xeritiler Toplimi [Atlas of Xinjiang Uygur Autonomous Region, in Uigur]. Published by Sinopress Beijing 2005, 307 pp

Xinjiang Weiwuer Zizhiqu di-tu: Xinjiang Uighur Aptonom Rayonining heritisi [Map of the Uigur Autonomous Region Xinjiang, toponyms in Uigur script]. Scale 1 : 1,300,000. First edition 1985 (six sheets)

Zhonge jiaojie quantu [The full map of Chinese-Russian border], published by Xia Hongjun in the 16th year of the Guangxu Empire (in 1890), based on a Russian map from 1884. Scale 1 : 1,729,729 in ten sheets. Staatsbibliothek Berlin (Kart D 2285)

Zhongua Halk Jumhuriyiti. *Xinjiang Uygur Avtonom Rayoni Heritiliri Toplimi*. [Atlas of Xinjiang Autonomous Uigur Region, toponyms in Romanised Uigur; Urumchi] 1966

Newspaper article collections

Birmingham Post 18 January 1933. *Air pictures of the Himalaya*. Birmingham

Daily News 12 December 1891. *Fighting in Cashmere. British officers wounded*. London

Daily News 14 April 1892. *The Hunza-Nagar* rising through Reuter's Agency. London

Daily Telegraph 21 April 1970. *Kashmir road brings arms from China*. London

Dawn 4 March 1963. *A victory for peace*. Karachi

Dawn 22 October 1967. *Overland trade between Gilgit and Sinkiang. Pakistan, China sign accord to open ancient road*. Karachi

Dawn 29 July 1968. *W. Pakistan, China 'silk route' to be reopened*. Karachi

Dawn 2 November 1968. *Historic 'silk route' to be opened next summer*. Karachi

Dawn 14 November 1968. *Trade thru' legendary 'Silk Route' from June*. Karachi

Dawn 8 May 1970. *Silk route trade. Chinese team due on 16th for talks*. Karachi

Dawn 12 May 1970. *Silk route trade*. Karachi

Dawn 18 May 1970. *Sinkiang-Gilgit border trade*. Karachi

Dawn 10 October 1970. *Chinese trade team leaves by Silk Route*. Karachi

Dawn 27 August 1982. *Khunjerab Pass opens on 27th: more trade with China*. Karachi

Dawn 28 August 1982. *Khunjerab Pass opened to traffic*. Karachi

Dawn 9 February 1994. *Reasons for not finalising N. Areas' status*. Karachi

Dawn 11 September 2000. *Pindi-Gilgit distance to be reduced by 150 km*. Karachi

Dawn 25 September 2000. *Karakorum Highway's role in Pak-Central Asian trade* by Moonis Ahmar. Karachi

Dawn 24 October 2002. *New power sharing formula. N. Areas, Kashmir division delinked*. Karachi

Dawn 21 September 2007. *Proposed names for the Northern Areas* by Fazal Amin Beg

Dawn 3 November 2007. *Northern Areas. An illusion of autonomy* by Aziz Ali Dad

Dawn 28 October 2008. *N[orthern]A[rea']s chief loses Sost dry port battle* by Zulfiqar Ali Khan

Dawn 21 June 2017. *CPEC masterplan revealed* by Khurram Husain. Karachi

Dawn 25 May 2018. *GB reforms leave people lesser citizens: HRCP*. Karachi

Dawn 27 May 2018. *Many injured during protest against new Gilgit-Baltistan law*. Karachi

Dawn 17 May 2018. *Storm in Gilgit-Baltistan* by Ibn Abdur Rehman. Karachi

Dawn, 28 May 2018. *Govt. ready to accommodate more Gilgit-Baltistan demands: Abbasi*. Karachi

Dawn Overseas Weekly 24–30 June 1978. *Opening of Karakoram Highway. Sino-Pakistan co-operation crystallised* by A. T. Chaudhri. Karachi

Dawn Overseas Weekly 21–27 June 1980. *Highways – the national lifeline of Pakistan* by M. Amiruddin. Karachi

Dawn Overseas Weekly 12–18 June 1981. *Border trade accord with China*. Karachi

Dawn Overseas Weekly 5–11 February 1982. *Tourists can now travel on Karakoram Highway*. Karachi

Dawn Overseas Weekly 3–9 September 1982. *Khunjerab Pass opens to civilian traffic. Another milestone in Pakistan-China cooperation* by Mohsin Ali. Karachi

Dawn Overseas Weekly 5 January 1989. *KKH. Highway of hazards* by B. A. Jeddy

Economic and Political Weekly 1 July 1978. *Karakoram Highway* by Mohan Ram

Far Eastern Economic Review 7 August 1969. *An unholy alliance* by Werner Adam

Far Eastern Economic Review 14 October 1977. *Letter from Gilgit* by Salamat Ali. Hong Kong

High Asia Herald 13 September 2017. *From the Great Game to the game-changer* by Aziz Ali Dad

Hindustan Times 2 September 2010. *China has Major Presence in Pakistan-occupied Kashmir: Gilgit Activist*

International Herald Tribune 1 October 2008. *The latest campaign for Alexander. In name dispute, Greece and Macedonia vie for his legacy* by Matthew Brunwasser

Le Monde 6–7 October 1974. *Lettre de Hunza. La fin d'un 'Paradis terrestre'*. Paris

New China News Agency 18 June 1978. *Karakoram Highway ceremony*. Beijing

New China News Agency 18 June 1978. *Keng-Piao remarks*. Beijing

New York Times 15 December 1891. *The Hunza-Niger revolt. Suspicion that Russia is at the bottom of the trouble*. New York

New York Times 12 September 1940. *Soviet opens mountain road*. New York

New York Times Magazine 2 December 1979. *Opening the high road to China* by Seymour Topping. New York

New York Times 1 October 2008. *Macedonia dispute has an Asian flavor* by Matthew Brunwasser. New York

New York Times 26 August 2010. *China's discreet hold on Pakistan's northern borderlands* by Selig Harrison

Pamir Times 6 January 2008. *Passu glacier burst caused flood*. Islamabad

Pamir Times 22 May 2008. *KKH flooded near Ghulkin Gojal*. Islamabad

Pamir Times 17 September 2008. *CBRC team inspects bridges on the KKH*. Islamabad

Pamir Times 22 October 2008. *The significance of October 23* by Zulfiqar Ali Khan. Islamabad

Pamir Times 28 February 2009. *Massive avalanches blocks KKH near Gulmit*. Islamabad

Pamir Times 19 March 2009. *Landslide damages bridge on KKH between Khyber and Ghalapan*. Islamabad
Pamir Times 4 January 2010. *Massive landslide destroys most of Attabad – Hunza, KKH blocked*. Islamabad
Pamir Times 9 October 2010. *Landslides at Hussainabad continue*. Islamabad
Pamir Times 11 September 2011. *Gilgit – Hunza section of KKH blocked due to landslides*. Islamabad
Pamir Times 26 July 2012. *KKH realignment: Chinese survey team reaches Gojal*. Islamabad
Pamir Times 28 September 2012. *NHA plans to complete KKH expansion by 2013, land owners await compensation*. Islamabad
Pamir Times 28 September 2012. *Pervez Ashraf reaching Gilgit today, will inaugurate KKH realignment work*. Islamabad
Pamir Times 29 September 2012. *Mayun landslide: 20 families evacuated from a village of Hunza Valley*. Islamabad
Pamir Times 26 May 2013. *Heavy rain across Gilgit-Baltistan, traffic on KKH suspended*. Islamabad
Pamir Times 16 July 2013. *Hunza. Landslide kills two people, including Chinese national, near Attabad spillway*. Islamabad
Pamir Times 31 August 2013. *Erosion threatens Passu settlement, KKH*. Islamabad
Pamir Times 11 February 2014. *Miachar landslide: Barjees Tahir wants precautionary measures taken*. Islamabad
Pamir Times 7 May 2014. *Gojal: GLOF wreaks havoc in Hussaini village, water scarce, KKH blocked*. Islamabad
Pamir Times 4 July 2014. *Sost: Protesters block KKH, demand opening of road at Yazyupk*. Islamabad
Pamir Times 3 October 2014. *Landslide: 25 households shifted to safe areas in Ghulmet village of Nagar Valley*. Islamabad
Pamir Times 21 October 2014. *Hunza: KKH remains blocked since yesterday due to landslide*. Islamabad
Pamir Times 27 June 2015. *KKH re-alignment: 94% work on the project completed so far, remaining to be completed by Sep. 25 this year* by Asghar Khan and Zulfiqar Ali Khan
Pamir Times 14 September 2015. *PM inaugurates KKH realignment project in Hunza*. Islamabad
Pamir Times 5 April 2016. *Pak-China friendship tunnel complex blocked by landslide, floods*. Islamabad
Pamir Times 17 April 2016. *Rahimabad: 25 families forced out of their houses after continuous rockfall, landslide*. Islamabad
Pamir Times 19 May 2018. *Landslide blocks the road to Misgar Valley, Gojal*. Islamabad
Pamir Times 20 November 2018. *Status of Provisional Province for GB* by Iftikhar Ali. Islamabad
Pamir Times 24 June & 3 July 2019. *Gilgit-Baltistan at the crossroad of CPEC* by Zulfiqar Ali Khan
Peking Review 7 July 1978. *Friendship notes. Karakoram Highway*. Beijing: 42–43
People's Daily 18 June 1978. *New China News Agency reporters laud highway* by Li Yun-fei and Yuan Chang-fa. Beijing
Renmin Ribao – China Daily 19 June 1978. *Opening ceremony of Karakoram Highway at Thakot: General Zia-ul-Haq and Chinese Vice-Minister Geng Biao made speeches; Major General Ali hosted the ceremony and gave a speech* [in Chinese]. Beijing
Renmin Ribao – China Daily 19 June 1978. *Karakoram Highway* [in Chinese]. Beijing
Renmin Ribao – China Daily 30 August 1982. *Khunjerab Pass officially opened* [in Chinese]. Beijing
Shanghai Daily 8 May 2015. *China's highest customs starts operation in Xinjiang*
Tehran Times 26 August 1982. *India protests Pakistan-China protocol*. Teheran
The Economist 17 May 1986. *Roof with a view*. London
The Express Tribune 22 February 2011. *Mother tongue: Shina group endorses language bill* by Shabbir Mir. Islamabad
The Express Tribune 2 October 2015. *G-B govt to purchase wheat from local farmers* by Shabbir Mir. Islamabad
The Express Tribune 7 August 2016. *The ruling Mir of G-B's Nagar district* by Shabbir Mir. Islamabad
The Friday Times 11–17 June 2010. *A man-made disaster* by Aziz Ali Dad. Lahore
The Frontier Post 3 December 1989. *Govt. studying representation of Northern Areas in N. A.* Peshawar
The Frontier Post 5 December 1989. *Govt. to give constitutional status to Northern Areas*. Peshawar
The Graphic 19 December 1891. *The little war in Kashmir*. London
The Graphic 14 May 1892. *The fighting at Nilt*. London
The Illustrated London News 27 September 1890. *The Hunza people of the Pamir*. London
The Illustrated London News 9 January 1892. *Gilgit and the Hunza frontier*. London
The Illustrated London News 20 August 1892. *The trouble in Gilgit: On march*. London
The Illustrated London News 26 August 1893. *The Hunza and Nagyr*. London
The Illustrated London News 19 October 1935. *The crafts and husbandry of the Burusho. A mountain people in the 'apex' of India* by Emily Overend Lorimer. London, p. 642
The Illustrated London News 19 October 1935. *Domestic and outdoor life in Hunza. A peasant community whose language is unique* by Emily Overend Lorimer. London, p. 643
The Illustrated London News 19 October 1935. *Speakers of a language that is, perhaps, related to no other. The Burusho of Hunza – A mountain people of Northern Kashmir* by Emily Overend Lorimer. London, p. 644
The Muslim 11 August 1985. *Exchange of mail with China from September*. Islamabad
The Muslim 1 May 1986. *Opening of the Khunjerab Pass. The modern Silk Route* by M. Hanif Raza. Islamabad
The Muslim 1 May 1986. *Opening of the Khunjerab Pass. Khunjerab opening a major breakthrough* by M. F. Khan. Islamabad
The Muslim 28 August 1993. *KKH reopened for traffic. Normalcy returning to Gilgit*. Islamabad
The Muslim 5 October 1993. *New administrative set-up for Northern Areas* by Muhammad Saleh Zaafir. Islamabad

The Muslim 18 May 1996. *Prime Minister's reforms package for Northern Areas* by Mehmood Khan. Islamabad
The Muslim 18 May 1996. *N[orthern]A[rea']s people await a fair deal* by Akbar Hussain Akbar. Islamabad
The Muslim 19 May 1996. *KKH upgradation vital to defence* by Sher Mohammad Khan. Islamabad
The Nation 15 December 1992. *The long-neglected Northern Areas* by Aziz-ud-din. Lahore
The News 5 May 1994. *Package for N[orthern] Areas to ensure people's involvement in political set up, says Afzal.* Islamabad
The News 2 August 2007. *Giving the Northern Areas their due rights* by Afzal A. Shigri
The News 11 September 2007. *Renaming the Northern Areas* by Aziz Ali Dad
The News 5 May 2018. *Economic geology of CPEC* by Aziz Ali Dad
The News on Sunday 17 June 2018. *A disorderly order* by Aziz Ali Dad
The News on Sunday 10 November 2019 *Unimagining Gilgit-Baltistan* by Aziz Ali Dad
The Pakistan Times 12 December 1974. *Isolation of large parts of country to be broken. Rs. 100 crore plan for roads in Northern Areas, Baluchistan.* Islamabad
The Pakistan Times 24 September 1983. *The little victory of the Karakorams* by Frank Saunders. Islamabad
The Pakistan Times 11 November 1984. *Channelbuilders of Hunza* by Frank Saunders. Islamabad
The Pakistan Times Overseas Weekly 27 July 1986. *Federal Govt. deeply interested in Northern Areas.* Islamabad
The Pakistan Times Overseas Weekly 1 November 1987. *Kashgar to Khunjerab Pass road to be completed by Oct. '88* by APP correspondent. Islamabad
The Statesman 16 January 1948. *The October Revolution in Gilgit. Maharaja repudiated: British officers avert bloodshed.* Calcutta
The Times 2 November 1891. *Indian Affairs.* London
The Times 16 December 1891. *The Hunza-Nagar expedition.* London
The Times 19 December 1891. *The Hunza-Nagar expedition.* London
The Times 21 December 1891. *The Hunza-Nagar expedition.* London
The Times 25 December 1891. *The Hunza-Nagar expedition.* London
The Times 29 December 1891. *The Hunza-Nagar expedition.* London
The Times 31 December 1891. *The Hunza-Nagar expedition.* London
The Times 1 January 1892. *Indian frontier tribes* by correspondent. London
The Times 11 January 1892. *India.* London
The Times 19 March 1892. *The recent Hunza-Nagar expedition.* London
The Times 24 March 1892. *The operations in Hunza-Nagar* from Parliamentary Paper. London
The Times 13 July 1892. *The Hunza-Nagar expedition* from the London Gazette, War Office 12 July 1892. London
The Times 28 December 1932. *Over Mount Rakaposhi. Notable R.A.F. flight in the Himalayas.* London
The Times 9 January 1935. *The Gilgit Agency. New administrative arrangement.* London
The Times 29 March 1935. *The Gilgit Jalsa. Last of the old regime.* London
The Times 10 December 1935. *Interlude at Kashgar The last of Chinese territory* by Peter Fleming. London
The Times 11 December 1935. *The end of the journey. To civilization in Kashmir* by Peter Fleming. London
The Times 25 July 1938. *The Mir of Hunza. Faithful ally of England.* London
The Times 2 January 1948. *Revolution in Gilgit. Communal strife beneath the Hindu Kush. Bloodshed averted by British officers* from our Peshawar correspondent. London
The Times 22 June 1949. *High Karakoram. Ancient states and modern communications* by our special correspondent in Gilgit. London
The Times 31 December 1954: *Peking map of China. Apparent claim to Indian soil.* London
The Times 14 August 1963. *The intolerable beauty of Hunza-Nagar* by Ian Stephens. London
The Times 2 December 1963. *China pledges aid to Pakistan.* London
The Times 13 February 1965. *Caravan route to China opening* by correspondent. London
The Times 31 December 1974. *4,700 dead in Pakistan earthquake* by Hasan Akhtar. London
The Times 17 June 1978. *Sino-Pakistan route illegal, Delhi claims* by correspondent. London
The Times 20 November 1978 *Travels on the great road of China* by Paul Routledge. London
The Times 17 September 1982. *Khunjerab Pass opens. Finishing the road on roof of the world* by Trevor Fishlock. London
The Welwyn Times 6 February 1936. *The Lorimers triumph of research. A language that is only spoken.* Welwyn Garden City
The Welwyn Times 27 October 1938. *Some points of Hitler's philosophy. Views on Britain, religion, law and justice. 'Promoting the victory of the strong'* by Emily Overend Lorimer, Welwyn Garden City
Times of India 8 April 1933. *Gilgit Frontier. Reorganisation of defence.* New Delhi
Times of India 8 March1963. *Pak cession of area to China. Kashmir cease-fire pact violated.* New Delhi
Times of India 27 June 1970. *Protest to Pindi, Peking. Laying road in Kashmir.* New Delhi

Index

A

B